Moonshine Beyond the Monster

The Bridge Connecting Algebra, Modular Forms and Physics

Moonshine forms a way of explaining the mysterious connection between the monster finite group and modular functions from classical number theory. The theory has evolved to describe the relationship between finite groups, modular forms and vertex operator algebras. *Moonshine Beyond the Monster*, the first book of its kind, describes the general theory of Moonshine and its underlying concepts, emphasising the interconnections between modern mathematics and mathematical physics.

Written in a clear and pedagogical style, this book is ideal for graduate students and researchers working in areas such as conformal field theory, string theory, algebra, number theory, geometry and functional analysis. Containing almost two hundred exercises, it is also a suitable textbook for graduate courses on Moonshine and as supplementary reading for courses on conformal field theory and string theory. This title, first published in 2006, has been reissued as an Open Access publication on Cambridge Core.

TERRY GANNON is currently Professor of Mathematics at the University of Alberta. Over the years he has had extended research visits to the Erwin-Schrodinger Institute in Vienna, IHES Bures, Max-Planck Institute in Bonn, St. John's College, Cambridge, the Feza Gürsey Institute in Istanbul, Universität Hamburg and the University of Wales. He became an assistant professor at York University in 1996 and then moved to the University of Alberta in 1998. His research interests cover a wide range of mathematics, in particular the interactions of algebra, number theory and mathematical physics, with a focus on conformal field theory.

CAMBRIDGE MONOGRAPHS ON
MATHEMATICAL PHYSICS

General editors: P. V. Landshoff, D. R. Nelson, S. Weinberg

[†]Issued as a paperback

Moonshine Beyond the Monster

The Bridge Connecting Algebra, Modular Forms and Physics

TERRY GANNON
University of Alberta

CAMBRIDGE
UNIVERSITY PRESS

CAMBRIDGE
UNIVERSITY PRESS

Shaftesbury Road, Cambridge CB2 8EA, United Kingdom

One Liberty Plaza, 20th Floor, New York, NY 10006, USA

477 Williamstown Road, Port Melbourne, VIC 3207, Australia

314–321, 3rd Floor, Plot 3, Splendor Forum, Jasola District Centre, New Delhi – 110025, India

103 Penang Road, #05-06/07, Visioncrest Commercial, Singapore 238467

Cambridge University Press is part of Cambridge University Press & Assessment,
a department of the University of Cambridge.

We share the University's mission to contribute to society through the pursuit of
education, learning and research at the highest international levels of excellence.

www.cambridge.org
Information on this title: www.cambridge.org/9781009401586

DOI: 10.1017/9781009401548

First published 2006
Reissued as OA 2023

A catalogue record for this publication is available from the British Library.

ISBN 978-1-009-40158-6 Hardback
ISBN 978-1-009-40155-5 Paperback

To the children in my life

L'homme [. . .] passe à travers des forêts de symboles
Qui l'observent avec des regards familiers.
Comme de longs échos qui de loin se confondent
Dans une ténébreuse et profonde unité,
Vaste comme la nuit et comme la clarté

BAUDELAIRE, *Les Fleurs du Mal*

Contents

Acknowledgements

The original suggestion to write this book was made to me by Abe Schenitzer, and Akbar Rhemtulla talked me into it, so to them I send my thanks (and my wife her curses). This book is a grandchild of 16 lectures I gave at the Feza Gursey Institute in Istanbul, and which were published in [**233**]. I warmly thank them, and in particular Teo Turgut, for their hospitality during my month-long stay. The book is a child of a course I taught at the University of Alberta. The book itself was partly written at St. John's College (Cambridge), the Universities of Cardiff, Swansea and Hamburg, IHES (Bures), and MPIM (Bonn), and for me they were all great environments to work. My research has been supported in part by NSERC, EPSRC and Humboldt. Innumerable people, scattered over the last dozen years, helped me understand various parts of the following story; in particular I should mention J. Conway, C. Cummins, D. Evans, M. Gaberdiel, A. Ivanov, J. Lepowsky, J. McKay, A. Ocneanu, C. Schweigert, M. Walton and J.-B. Zuber. Most of all I thank my loving wife Geneviève who shared with me all of the pain, but none of the pleasure.

0

Introduction: glimpses of the theory beneath Monstrous Moonshine

When you are collecting mushrooms, you only see the mushroom itself. But if you are a mycologist, you know that the real mushroom is in the earth. There's an enormous thing down there, and you just see the fruit, the body that you eat. In mathematics, the upper part of the mushroom corresponds to theorems that you see, but you don't see the things that are below, that is: *problems, conjectures, mistakes, ideas*, etc.

V. I. Arnold [17]

What my experience of mathematical work has taught me again and again, is that the proof always springs from the insight, and not the other way around – and that the insight itself has its source, first and foremost, in a delicate and obstinate feeling of the relevant entities and concepts and their mutual relations. The guiding thread is the inner coherence of the image which gradually emerges from the mist, as well as its consonance with what is known or foreshadowed from other sources – and it guides all the more surely as the 'exigence' of coherence is stronger and more delicate.

A. Grothendieck.[1]

Interesting events (e.g. wars) always happen whenever different realisations of the same thing confront one another. When clarity and precision are added to the mix, we call this mathematics. In particular, the most exciting and significant moments in mathematics occur when we discover that seemingly unrelated phenomena are shadows cast by the same beast. This book studies one who has been recently awakened.

In 1978, John McKay made an intriguing observation: $196\,884 \approx 196\,883$. *Monstrous Moonshine* is the collection of questions (and a few answers) that it directly inspired. No one back then could have guessed the riches to which it would lead. But in actual fact, Moonshine (albeit non-Monstrous) really began long ago.

0.1 Modular functions

Up to topological equivalence (homeomorphism), every compact surface is uniquely specified by its genus: a sphere is genus 0, a torus genus 1, etc. However, a (real) surface can be made into a complex curve by giving it more structure. For a sphere, up to

---

[1] Translated in *Geometric Galois Actions* 1, edited by L. Schneps *et al.* (Cambridge, Cambridge University Press, 1997) page 285.

complex-analytic equivalence there is only one way to do this, namely the Riemann sphere $\mathbb{C} \cup \{\infty\}$. Surfaces of genus > 0 can be given complex structure in a continuum of different ways.

Any such complex curve Σ is complex-analytically equivalent to one of the form $\Gamma \backslash \overline{\mathbb{H}}$. The *upper half-plane*

$$\mathbb{H} := \{\tau \in \mathbb{C} \mid \text{Im} \, \tau > 0\} \tag{0.1.1}$$

is a model for hyperbolic geometry. Its geometry-preserving maps form the group $\text{SL}_2(\mathbb{R})$ of 2×2 real matrices with determinant 1, which act on $\mathbb{H}$ by the familiar

$$\begin{pmatrix} a & b \\ c & d \end{pmatrix} \cdot \tau = \frac{a\tau + b}{c\tau + d}. \tag{0.1.2}$$

Γ is a discrete subgroup of $\text{SL}_2(\mathbb{R})$. By $\overline{\mathbb{H}}$ here we mean $\mathbb{H}$ with countably many points from its boundary $\mathbb{R} \cup \{i\infty\}$ added – these extra boundary points, which depend on Γ, are needed for $\Gamma \backslash \overline{\mathbb{H}}$ to be compact. The construction of the space $\Gamma \backslash \overline{\mathbb{H}}$ of Γ-orbits in $\overline{\mathbb{H}}$ is completely analogous to that of the circle $\mathbb{R}/\mathbb{Z}$ or torus $\mathbb{R}^2/\mathbb{Z}^2$. See Section 2.1.1 below.

The most important example is $\Gamma = \text{SL}_2(\mathbb{Z})$, because the moduli space of possible complex structures on a torus can be naturally identified with $\text{SL}_2(\mathbb{Z}) \backslash \mathbb{H}$. For that Γ, as well as all other Γ we consider in this book, we have

$$\overline{\mathbb{H}} = \mathbb{H} \cup \mathbb{Q} \cup \{i\infty\}. \tag{0.1.3}$$

These additional boundary points $\mathbb{Q} \cup \{i\infty\}$ are called cusps.

Both geometry and physics teach us to study a geometric shape through the functions (fields) that live on it. The functions f living on $\Sigma = \Gamma \backslash \overline{\mathbb{H}}$ are simply functions $f : \overline{\mathbb{H}} \to \mathbb{C}$ that are periodic with respect to Γ: that is,

$$f(A.\tau) = f(\tau), \qquad \forall \tau \in \overline{\mathbb{H}}, \; A \in \Gamma. \tag{0.1.4}$$

They should also preserve the complex-analytic structure of Σ. Ideally this would mean that f should be holomorphic but this is too restrictive, so instead we require meromorphicity (i.e. we permit isolated poles).

Definition 0.1 *A modular function f for some Γ is a meromorphic function $f : \overline{\mathbb{H}} \to \mathbb{C}$, obeying the symmetry (0.1.4).*

It is clear then why modular functions must be important: they are the functions living on complex curves. In fact, modular functions and their various generalisations hold a central position in both classical and modern number theory.

We can construct some modular functions for $\Gamma = \text{SL}_2(\mathbb{Z})$ as follows. Define the *(classical) Eisenstein series* by

$$G_k(\tau) := \sum_{\substack{m,n \in \mathbb{Z} \\ (m,n) \neq (0,0)}} (m\tau + n)^{-k}. \tag{0.1.5}$$

For odd k it identically vanishes. For even $k > 2$ it converges absolutely, and so defines a function holomorphic throughout $\mathbb{H}$. It is easy to see from (0.1.5) that

$$G_k\left(\frac{a\tau + b}{c\tau + d}\right) = (c\tau + d)^k\, G_k(\tau), \qquad \forall \begin{pmatrix} a & b \\ c & d \end{pmatrix} \in \mathrm{SL}_2(\mathbb{Z}) \qquad (0.1.6)$$

and all τ. This transformation law (0.1.6) means that G_k isn't quite a modular function (it's called a *modular form*). However, various homogeneous rational functions of these G_k *will* be modular functions for $\mathrm{SL}_2(\mathbb{Z})$ – for example, $G_8(\tau)/G_4(\tau)^2$ (which turns out to be constant) and $G_4(\tau)^3/G_6(\tau)^2$ (which doesn't). All modular functions of $\mathrm{SL}_2(\mathbb{Z})$ turn out to arise in this way.

Can we characterise all modular functions, for $\Gamma = \mathrm{SL}_2(\mathbb{Z})$ say? We know that any modular function is a meromorphic function on the compact surface $\Sigma = \mathrm{SL}_2(\mathbb{Z}) \backslash \overline{\mathbb{H}}$. As we explain in Section 2.2.4, Σ is in fact a sphere. It may seem that we've worked very hard merely to recover the complex plane $\mathbb{C} \cong \mathrm{SL}_2(\mathbb{Z}) \backslash \mathbb{H}$ and its familiar compactification the Riemann sphere $\mathbb{P}^1(\mathbb{C}) = \mathbb{C} \cup \{\infty\} \cong \mathrm{SL}_2(\mathbb{Z}) \backslash \overline{\mathbb{H}}$, but that's *exactly* the point!

Although there are large numbers of meromorphic functions on the complex plane $\mathbb{C}$, the only ones that are also meromorphic at ∞ – the only functions meromorphic on the Riemann sphere $\mathbb{P}^1(\mathbb{C})$ – are the rational functions $\frac{\text{polynomial in } z}{\text{polynomial in } z}$ (the others have essential singularities there). So if J is a change-of-coordinates (or *uniformising*) function identifying our surface Σ with the Riemann sphere, then J (lifted to a function on the covering space $\overline{\mathbb{H}}$) will be a modular function for $\mathrm{SL}_2(\mathbb{Z})$, and any modular function $f(\tau)$ will be a rational function in $J(\tau)$:

$$f(\tau) = \frac{\text{polynomial in } J(\tau)}{\text{polynomial in } J(\tau)}. \qquad (0.1.7)$$

Conversely, any rational function (0.1.7) in J is modular. Thus J generates modular functions for $\mathrm{SL}_2(\mathbb{Z})$, in a way analogous to (but stronger and simpler than) how the exponential $e(x) = e^{2\pi i x}$ generates the period-1 smooth functions f on $\mathbb{R}$: we can always expand such an f in the pointwise-convergent Fourier series $f(x) = \sum_{n=-\infty}^{\infty} a_n\, e(x)^n$.

There is a standard historical choice j for this uniformisation J, namely

$$
\begin{aligned}
j(\tau) &:= 1728\, \frac{20\, G_4(\tau)^3}{20\, G_4(\tau)^3 - 49\, G_6(\tau)^2} \\
&= q^{-1} + 744 + 196\,884\, q + 21\,493\,760\, q^2 + 864\,299\,970\, q^3 + \cdots \qquad (0.1.8)
\end{aligned}
$$

where $q = \exp[2\pi i \tau]$. In fact, this choice (0.1.8) is canonical, apart from the arbitrary constant 744. This function j is called the absolute invariant or *Hauptmodul* for $\mathrm{SL}_2(\mathbb{Z})$, or simply the *j-function*.

0.2 The McKay equations

In any case, one of the best-studied functions of classical number theory is the j-function. However, its most remarkable property was discovered only recently: McKay's approximations $196\,884 \approx 196\,883$, $21\,493\,760 \approx 21\,296\,876$ and $864\,299\,970 \approx$ $842\,609\,326$. In fact,

$$196\,884 = 196\,883 + 1, \tag{0.2.1a}$$

$$21\,493\,760 = 21\,296\,876 + 196\,883 + 1, \tag{0.2.1b}$$

$$864\,299\,970 = 842\,609\,326 + 21\,296\,876 + 2 \cdot 196\,883 + 2 \cdot 1. \tag{0.2.1c}$$

The numbers on the left sides of (0.2.1) are the first few coefficients of the j-function. The numbers on the right are the dimensions of the smallest irreducible representations of Fischer–Griess's *Monster finite simple group* $\mathbb{M}$.

A *representation* of a group G is the assignment of a matrix $R(g)$ to each element g of G in such a way that the matrix product respects the group product, that is $R(g)\,R(h) = R(gh)$. The dimension of a representation is the size n of its $n \times n$ matrices $R(g)$.

The *finite simple groups* are to finite groups what the primes are to integers – they are their elementary building blocks (Section 1.1.2). They have been classified (see [22] for recent remarks on the status of this proof). The resulting list consists of 18 infinite families (e.g. the cyclic groups $\mathbb{Z}_p := \mathbb{Z}/p\mathbb{Z}$ of prime order), together with 26 exceptional groups. The Monster $\mathbb{M}$ is the largest and richest of these exceptionals, with order

$$\|\mathbb{M}\| = 2^{46} \cdot 3^{20} \cdot 5^9 \cdot 7^6 \cdot 11^2 \cdot 13^3 \cdot 17 \cdot 19 \cdot 23 \cdot 29 \cdot 31 \cdot 41 \cdot 47 \cdot 59 \cdot 71 \approx 8 \times 10^{53}. \tag{0.2.2}$$

Group theorists would like to believe that the classification of finite simple groups is one of the high points in the history of mathematics. But isn't it possible instead that their enormous effort has merely culminated in a list of interest only to a handful of experts? Years from now, could the Monster – the signature item of this list – become a lost bone in a dusty drawer of a forgotten museum, remarkable only for its colossal irrelevance?

With numbers so large, it seemed doubtful to McKay that the numerology (0.2.1) was merely coincidental. Nevertheless, it was difficult to imagine any deep conceptual relation between the Monster and the j-function: mathematically, they live in different worlds.

In November 1978 he mailed the 'McKay equation' (0.2.1a) to John Thompson. At first Thompson likened this exercise to reading tea leaves, but after checking the next few coefficients he changed his mind. He then added a vital piece to the puzzle.

0.3 Twisted #0: the Thompson trick

A nonnegative integer begs interpretation as the dimension of some vector space. Essentially, that was what McKay proposed. Let $\rho_0, \rho_1, \ldots$ be the irreducible representations of $\mathbb{M}$, ordered by dimension. Then the equations (0.2.1) are really hinting that there is an infinite-dimensional graded representation

$$V = V_{-1} \oplus V_1 \oplus V_2 \oplus V_3 \oplus \cdots \tag{0.3.1}$$

of $\mathbb{M}$, where $V_{-1} = \rho_0$, $V_1 = \rho_1 \oplus \rho_0$, $V_2 = \rho_2 \oplus \rho_1 \oplus \rho_0$, $V_3 = \rho_3 \oplus \rho_2 \oplus \rho_1 \oplus \rho_1 \oplus \rho_0 \oplus \rho_0$, etc., and the j-function is essentially its graded dimension:

$$j(\tau) - 744 = \dim(V_{-1})\,q^{-1} + \sum_{i=1}^{\infty} \dim(V_i)\,q^i. \tag{0.3.2}$$

Thompson [**525**] suggested that we twist this, that is more generally we consider what we now call the *McKay–Thompson series*

$$T_g(\tau) = \mathrm{ch}_{V_{-1}}(g)\,q^{-1} + \sum_{i=1}^{\infty} \mathrm{ch}_{V_i}(g)\,q^i \tag{0.3.3}$$

for each element $g \in \mathbb{M}$. The character 'ch$_\rho$' of a representation ρ is given by 'trace': $\mathrm{ch}_\rho(g) = \mathrm{tr}(\rho(g))$. Up to equivalence (i.e. choice of basis), a representation ρ can be recovered from its character ch$_\rho$. The character, however, is much simpler. For example, the smallest nontrivial representation of the Monster $\mathbb{M}$ is given by almost 10^{54} complex matrices, each of size $196\,883 \times 196\,883$, while the corresponding character is completely specified by 194 integers (194 being the number of 'conjugacy classes' in $\mathbb{M}$).

For any representation ρ, the character value ch$_\rho(id.)$ equals the dimension of ρ, and so $T_{id.}(\tau) = j(\tau) - 744$ and we recover (0.2.1) as special cases. But there are many other possible choices of $g \in \mathbb{M}$, although conjugate elements g, hgh^{-1} have identical character values and hence have identical McKay–Thompson series $T_g = T_{hgh^{-1}}$. In fact, there are precisely 171 *distinct* functions T_g. Thompson didn't guess what these functions T_g would be, but he suggested that they too might be interesting.

0.4 Monstrous Moonshine

John Conway and Simon Norton [**111**] did precisely what Thompson asked. Conway called it 'one of the most exciting moments in my life' [**107**] when he opened Jacobi's foundational (but 150-year-old!) book on elliptic and modular functions and found that the first few terms of each McKay–Thompson series T_g coincided with the first few terms of certain special functions, namely the Hauptmoduls of various genus-0 groups Γ. Monstrous Moonshine – which conjectured that the McKay–Thompson series *were* those Hauptmoduls – was officially born.

We should explain those terms. When the surface $\Gamma \backslash \overline{\mathbb{H}}$ is a sphere, we call the group Γ *genus 0*, and the (appropriately normalised) change-of-coordinates function from $\Gamma \backslash \overline{\mathbb{H}}$ to the Riemann sphere $\mathbb{C} \cup \{\infty\}$ the *Hauptmodul* for Γ. All modular functions for a genus-0 group Γ are rational functions of this Hauptmodul. (On the other hand, when Γ has positive genus, two generators are needed, and there's no canonical choice for them.)

The word 'moonshine' here is English slang for 'insubstantial or unreal', 'idle talk or speculation',[2] 'an illusive shadow'.[3] It was chosen by Conway to convey as well the

---

[2] Ernest Rutherford (1937): 'The energy produced by the breaking down of the atom is a very poor kind of thing. Anyone who expects a source of power from the transformation of these atoms is talking moonshine.' (quoted in *The Wordsworth Book of Humorous Quotations*, Wordsworth Editions, 1998).

[3] *Dictionary of Archaic Words*, J. O. Halliwell, London, Bracken Books, 1987. It also defines moonshine as 'a dish composed partly of eggs', but that probably has less to do with Conway's choice of word.

impression that things here are dimly lit, and that Conway and Norton were 'distilling information illegally' from the Monster character table.

In hindsight, the first incarnation of Monstrous Moonshine goes back to Andrew Ogg in 1975. He was in France discussing his result that the primes p for which the group $\Gamma_0(p)+$ has genus 0, are

$$p \in \{2, 3, 5, 7, 11, 13, 17, 19, 23, 29, 31, 41, 47, 59, 71\}.$$

(The group $\Gamma_0(p)+$ is defined in (7.1.5).) He also attended there a lecture by Jacques Tits, who was describing a newly conjectured simple group. When Tits wrote down the order (0.2.2) of that group, Ogg noticed its prime factors coincided with his list of primes. Presumably as a joke, he offered a bottle of Jack Daniels whisky to the first person to explain the coincidence (he still hasn't paid up). We now know that each of Ogg's groups $\Gamma_0(p)+$ is the genus-0 modular group for the function T_g, for some element $g \in \mathbb{M}$ of order p. Although we now realise why the Monster's primes must be a subset of Ogg's, probably there is no deep reason why Ogg's list couldn't have been longer.

The appeal of Monstrous Moonshine lies in its mysteriousness: it unexpectedly associates various special modular functions with the Monster, even though modular functions and elements of $\mathbb{M}$ are conceptually incommensurable. Now, 'understanding' something means to embed it naturally into a broader context. Why is the sky blue? Because of the way light scatters in gases. Why does light scatter in gases the way it does? Because of Maxwell's equations. In order to understand Monstrous Moonshine, to resolve the mystery, we should search for similar phenomena, and fit them all into the same story.

0.5 The Moonshine of E_8 and the Leech

McKay had also remarked in 1978 that similar numerology to (0.2.1) holds if $\mathbb{M}$ and $j(\tau)$ are replaced with the Lie group $E_8(\mathbb{C})$ and

$$j(\tau)^{\frac{1}{3}} = q^{-\frac{1}{3}} (1 + 248\, q + 4124\, q^2 + 34\,752\, q^3 + \cdots). \qquad (0.5.1)$$

In particular, $4124 = 3875 + 248 + 1$ and $34\,752 = 30\,380 + 3875 + 2 \cdot 248 + 1$, where 248, 3875 and 30 380 are all dimensions of irreducible representations of $E_8(\mathbb{C})$. A *Lie group* is a manifold with compatible group structure; the groups of E_8 type play the same role in Lie theory that the Monster does for finite groups. Incidentally, $j^{\frac{1}{3}}$ is the Hauptmodul of the genus-0 group $\Gamma(3)$ (see (2.2.4a)).

A more elementary observation concerns the Leech lattice. A lattice is a discrete periodic set L in $\mathbb{R}^n$, and the Leech lattice Λ is a particularly special one in 24 dimensions. 196 560, the number of vectors in the Leech lattice with length-squared 4, is also close to 196 884: in fact,

$$196\,884 = 196\,560 + 324 \cdot 1, \qquad\qquad (0.5.2a)$$
$$21\,493\,760 = 16\,773\,120 + 24 \cdot 196\,560 + 3200 \cdot 1, \qquad\qquad (0.5.2b)$$
$$864\,299\,970 = 398\,034\,000 + 24 \cdot 16\,773\,120 + 324 \cdot 196\,560 + 25\,650 \cdot 1, \quad (0.5.2c)$$

where $16\,773\,120$ and $398\,034\,000$ are the numbers of length-squared 6- and 8-vectors in the Leech. This may not seem as convincing as (0.2.1), but the same equations hold for any of the 24-dimensional even self-dual lattices, apart from an extra term on the right sides corresponding to length-squared 2 vectors (there are none of these in the Leech).

What conceptually does the Monster, E_8 and the Leech lattice have to do with the j-function? Is there a common theory explaining this numerology? The answer is yes!

It isn't difficult to relate E_8 to the j-function. In the late 1960s, Victor Kac [325] and Robert Moody [430] independently (and for entirely different reasons) defined a new class of infinite-dimensional Lie algebras. A *Lie algebra* is a vector space with a bilinear vector-valued product that is both anti-commutative and anti-associative (Section 1.4.1). The familiar vector-product $u \times v$ in three dimensions defines a Lie algebra, called $\mathfrak{sl}_2$, and in fact this algebra generates all Kac–Moody algebras. Within a decade it was realised that the graded dimensions of representations of the *affine* Kac–Moody algebras are (vector-valued) modular functions for $SL_2(\mathbb{Z})$ (Theorem 3.2.3).

Shortly after McKay's E_8 observation, Kac [326] and James Lepowsky [373] independently remarked that the unique level-1 highest-weight representation $L(\omega_0)$ of the affine Kac–Moody algebra $E_8{}^{(1)}$ has graded dimension $j(q)^{\frac{1}{3}}$. Since each homogeneous piece of any representation $L(\lambda)$ of the affine Kac–Moody algebra $X_\ell{}^{(1)}$ must carry a representation of the associated finite-dimensional Lie group $X_\ell(\mathbb{C})$, and the graded dimensions (multiplied by an appropriate power of q) of an affine algebra are modular functions for some $\Gamma \subseteq SL_2(\mathbb{Z})$, this explained McKay's E_8 observation. His Monster observations took longer to clarify because so much of the mathematics needed was still to be developed.

Euler played with a function $t(x) := 1 + 2x + 2x^4 + 2x^9 + 2x^{16} + \cdots$, because it counts the ways a given number can be written as a sum of squares of integers. In his study of elliptic integrals, Jacobi (and Gauss before him) noticed that if we change variables by $x = e^{\pi i \tau}$, then the resulting function $\theta_3(\tau) := 1 + 2e^{\pi i \tau} + 2e^{4\pi i \tau} + \cdots$ behaves nicely with respect to certain transformations of τ – we say today that Jacobi's theta function θ_3 is a modular form of weight $\frac{1}{2}$ for a certain index-3 subgroup of $SL_2(\mathbb{Z})$. More generally, something similar holds when we replace $\mathbb{Z}$ with any other lattice L: the theta series

$$\Theta_L(\tau) := \sum_{n \in L} e^{\pi i\, n \cdot n \tau}$$

is also a modular form, provided all length-squares $n \cdot n$ are rational. In particular, we obtain quite quickly that the theta series of the Leech lattice, divided by Ramanujan's modular form $\Delta(\tau)$, will equal $J(\tau) + 24$.

For both E_8 and the Leech, the j-function arises from a uniqueness property ($L(\omega_0)$ is the only 'level-1' $E_8{}^{(1)}$-module; the Leech lattice Λ is self-dual), together with the empirical observation that $SL_2(\mathbb{Z})$ has few modular forms of small level. In these examples, the appearance of the j-function isn't as significant as that of modularity.

Monster, lattices, affine algebras, ...   Hauptmoduls, theta functions, ...

Fig. 0.1 Moonshine in its broader sense.

0.6 Moonshine beyond the Monster

We've known for many years that lattices (quadratic forms) and Kac–Moody algebras are related to modular forms and functions. But these observations, albeit now familiar, are also a little mysterious, we should confess. For instance, compare the non-obvious fact that $\theta_3(-1/\tau) = \sqrt{\frac{\tau}{i}}\theta_3(\tau)$ with the trivial observation (0.1.6) that $G_k(-1/\tau) = \tau^k G_k(\tau)$ for the Eisenstein series G_k. The modularity of G_k is a special case of the elementary observation that $\mathrm{SL}_n(\mathbb{Z})$ parametrises the change-of-bases of n-dimensional lattices. The modularity of θ_3, on the other hand, begs a *conceptual* explanation (indeed, see the quote by Weil at the beginning of Section 2.4.2), even though its *logical* explanation (i.e. proof) is a quick calculation from, for example, the Poisson summation formula (Section 2.2.3). Moonshine really began with Jacobi and Gauss.

> *Moonshine should be regarded as a certain collection of related examples where algebraic structures have been associated with automorphic functions or forms.*

Grappling with that thought is the theme of our book. Chapters 1 to 6 could be (rather narrowly) regarded as supplying a context for Monstrous Moonshine, on which we focus in Chapter 7. From this larger perspective, illustrated in Figure 0.1, what is special about this single instance called *Monstrous* Moonshine is that the several associated modular functions are all of a special class (namely Hauptmoduls).

The first major step in the proof of Monstrous Moonshine was accomplished in the mid-1980s with the construction by Frenkel–Lepowsky–Meurman [**200**] of the Moonshine module $V^\natural$ and its interpretation by Richard Borcherds [**68**] as a *vertex operator algebra*. A vertex operator algebra (VOA) is an infinite-dimensional vector space with infinitely many heavily constrained vector-valued bilinear products (Chapter 5). It is a natural, though extremely intricate, extension of the notion of a Lie algebra. Any algebra $\mathcal{A}$ can be interpreted as an assignment of a linear map $\mathcal{A} \otimes \cdots \otimes \mathcal{A} \to \mathcal{A}$ to each binary tree; from this perspective a VOA $\mathcal{V}$ associates a linear map $\mathcal{V} \otimes \cdots \otimes \mathcal{V} \to \mathcal{V}$ with each 'inflated' binary tree, that is each sphere with discs removed.

In 1992 Borcherds [**72**] completed the proof of the original Monstrous Moonshine conjectures[4] by showing that the graded characters T_g of $V^\natural$ are indeed the Hauptmoduls identified by Conway and Norton, and hence that $V^\natural$ is indeed the desired representation

---

[4] As we see in Chapter 7, most Moonshine conjectures involving the Monster are still open.

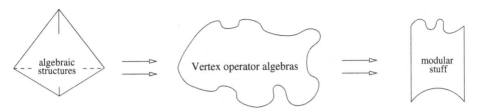

Fig. 0.2 The algebraic meaning of Moonshine.

V of $\mathbb{M}$ conjectured by McKay and Thompson. The explanation of Moonshine suggested by this picture is given in Figure 0.2. The algebraic structure typically arises as the symmetry group of the associated VOA – for example, that of $V^\natural$ is the Monster $\mathbb{M}$. By Zhu's Theorem (Theorem 5.3.8), the modular forms/functions appear as graded dimensions of the (possibly twisted) modules of the VOA. In particular, the answer this framework provides for what $\mathbb{M}$, E_8 and the Leech have to do with j is that they each correspond to a VOA with a single simple module; their relation to j is then an immediate corollary to the much more general Zhu's Theorem.

It must be emphasised that Figure 0.2 is primarily meant to address Moonshine in the broader sense of Figure 0.1, so certain special features of, for example, Monstrous Moonshine (in particular that all the T_g are Hauptmoduls) are more subtle and have to be treated by special arguments. These are quite fascinating by themselves, and are discussed in Chapter 7. Even so, Figure 0.2 provides a major clue:

> *If you're trying to understand a seemingly mysterious occurrence of the Monster, try replacing the word 'Monster' with its synonym 'the automorphism group of the vertex operator algebra $V^\natural$'.*

This places the Monster into a much richer algebraic context, with numerous connections with other areas of mathematics.

0.7 Physics and Moonshine

Moonshine is profoundly connected with physics (namely conformal field theory and string theory). String theory proposes that the elementary particles (electrons, photons, quarks, etc.) are vibrational modes on a string of length about 10^{-33} cm. These strings can interact only by splitting apart or joining together – as they evolve through time, these (classical) strings will trace out a surface called the *world-sheet*. Quantum field theory tells us that the quantum quantities of interest (amplitudes) can be perturbatively computed as weighted averages taken over spaces of these world-sheets. Conformally equivalent world-sheets should be identified, so we are led to interpret amplitudes as certain integrals over moduli spaces of surfaces. This approach to string theory leads to a conformally invariant quantum field theory on two-dimensional space-time, called *conformal field theory (CFT)*. The various modular forms and functions arising in Moonshine appear as integrands in some of these genus-1 ('1-loop') amplitudes: hence their modularity is manifest.

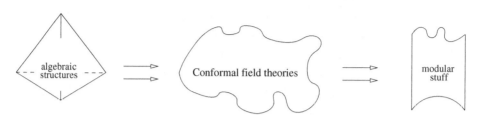

Fig. 0.3 The stringy picture of Moonshine.

Many aspects of Moonshine make complete sense within CFT, something which helps make the words of Freeman Dyson ring prophetic:

> I have a sneaking hope, a hope unsupported by any facts or any evidence, that sometime in the twenty-first century physicists will stumble upon the Monster group, built in some unsuspected way into the structure of the universe [**167**].

All that said, here we are, sometime in the twenty-first century, and alas the Monster still plays at best a peripheral role in physics. And some aspects of Moonshine (e.g. the Hauptmodul property) remain obscure in CFT. In any case, although this is primarily a mathematics book, we often sit in chairs warmed by physicists. In particular, CFT (or what is essentially the same thing, perturbative string theory[5]) is, at least in part, a machine for producing modular functions. Here, Figure 0.2 becomes Figure 0.3. More precisely, the algebraic structure is an underlying symmetry of the CFT, and its graded dimensions are the various modular functions. VOAs can be regarded as an algebraic abstraction of CFT, since they arise quite naturally by applying the Wightman axioms of quantum field theory to CFT. The lattice theta functions come from bosonic strings living on the torus $\mathbb{R}^n/L$. The affine Kac–Moody characters arise when the strings live on a Lie group. And the Monster is the symmetry of a string theory for a $\mathbb{Z}_2$-orbifold of free bosons compactified on the Leech lattice torus $\mathbb{R}^{24}/\Lambda$.

Physics reduces Moonshine to a duality between two different pictures of quantum field theory: the Hamiltonian one, which concretely gives us from representation theory the graded vector spaces, and another, due to Feynman, which manifestly gives us modularity. In particular, physics tells us that this modularity is a topological effect, and the group $SL_2(\mathbb{Z})$ directly arises in its familiar role as the modular group of the torus.

Historically speaking, Figure 0.3 preceded and profoundly affected Figure 0.2. One reason the stringy picture is exciting is that the CFT machine in Figure 0.3 outputs much more than modular functions – it creates automorphic functions and forms for the various mapping class groups of surfaces with punctures. And all this is still poorly explored. We can thus expect more from Moonshine than Figure 0.2 alone suggests. On the other hand, once again Figure 0.3 can directly explain only the broader aspects of Moonshine.

---

[5] Curiously, although nonperturbative string theory should be *physically* more profound, it is the perturbative calculations that are most relevant to the *mathematics* of Moonshine.

0.8 Braided #0: the meaning of Moonshine

In spite of the work of Borcherds and others, the special features of Monstrous Moonshine still beg questions. The full conceptual relationship between the Monster and Hauptmoduls (like j) arguably remains 'dimly lit', although much progress has been realised. This is a subject where it is much easier to speculate than to prove, and we are still awash in unresolved conjectures. But most important, we need a second independent proof of Monstrous Moonshine. In order to clarify the still murky significance of the Monster in Moonshine, we need to understand to what extent Monstrous Moonshine determines the Monster. More generally, we need to go beneath the algebraic explanation of Moonshine in order to find its more fundamental meaning, which is probably topological. Explaining something (Moonshine in this case) with something more complicated (CFT or VOAs here) cannot be the end of the story. Surely it is instead a beginning.

To Poincaré 125 years ago, modularity arose through the monodromy of differential equations. Remarkably, today CFT provides a similar explanation, although the relevant partial differential equations are much more complicated. The monodromy group here is the braid group $\mathcal{B}_3$, and the modular group $SL_2(\mathbb{Z})$ arises as a homomorphic image.

Today we are taught to lift modular forms for $SL_2(\mathbb{Z})$ to the space $L^2(SL_2(\mathbb{Z})\backslash SL_2(\mathbb{R}))$, which carries a representation of the Lie group $SL_2(\mathbb{R})$. However, $SL_2(\mathbb{R})$ is not simply connected; its universal cover $\widetilde{SL_2(\mathbb{R})}$ is a central extension by $\mathbb{Z}$, and the corresponding central extension of $SL_2(\mathbb{Z})$ – the fundamental group of $SL_2(\mathbb{Z})\backslash SL_2(\mathbb{R})$ – is the braid group $\mathcal{B}_3$. By all rights, these central extensions should be more fundamental. Indeed, modular forms of fractional weight, such as the Dedekind eta, certainly see $\mathcal{B}_3$ more directly than they do $SL_2(\mathbb{Z})$ (Section 2.4.3). Similar comments hold for other Γ – for example, the congruence subgroup $\Gamma(2)$ lifts to the pure braid group $\mathcal{P}_3$.

The best approach we know for relating the Monster and the Hauptmodul property is Norton's action of $\mathcal{B}_3$ on $G \times G$. This associates a genus-0 property with '6-transposition groups', which in turn points to a special role for $\mathbb{M}$, as the Monster is expected to be essentially the largest such group (Section 7.3.3). Incidentally, the number '6' arises here because the principal congruence subgroup $\Gamma(N)$ is genus 0 iff $N < 6$.

For these reasons and others we explore on the following pages, we expect a new proof for Moonshine to involve the braid group $\mathcal{B}_3$. The modular groups $SL_2(\mathbb{Z})$ and $PSL_2(\mathbb{Z})$ arise only indirectly as quotients. We also identify other promising places to look for alternate arguments for Moonshine – for example, the partial differential equations of CFT are built from the heat kernel, which has a long historical association with modularity.

0.9 The book

Borcherds' paper [72] and the resulting Fields medal close the opening chapter of the story of Moonshine. Now, 25 years after its formulation in [111], we are in a period of consolidation and synthesis, flames fanned I hope by this book.

Most of us might liken much of our research to climbing a steep hill against a stiff breeze: every so often we stumble and roll to the bottom, but with persistence we eventually reach the summit and plant our flag amongst the others already there. And before our bruises fade and bones mend, we're off to the next hill. But perhaps research in its purest form is more like chasing squirrels. As soon as you spot one and leap towards it, it darts away, zigging and zagging, always just out of reach. If you're a little lucky, you might stick with it long enough to see it climb a tree. You'll never catch the damned squirrel, but chasing it will lead you to a tree. In mathematics, the trees are called theorems. The squirrels are those nagging little mysteries we write at the top of many sheets of paper. We never know where our question will take us, but if we stick with it, it'll lead us to a theorem. That I think is what research ideally is like. There is no higher example of this than Moonshine.

This book addresses the theory of the blob of Figure 0.1. We explore some of its versatility in Chapter 6, where we glimpse Moonshine orthogonal to the Monster. Like moonlight itself, Monstrous Moonshine is an indirect phenomenon. Just as in the theory of moonlight one must introduce the sun, so in the theory of Moonshine one must go well beyond the Monster. Much as a book discussing moonlight may include paragraphs on sunsets or comet tails, so do we discuss fusion rings, Galois actions and knot invariants. The following chapters use Moonshine (Monstrous and otherwise) as a happy excuse to take a rather winding little tour through modern mathematics and physics. If we offer more questions and suggestions than theorems and answers, at least that is in Moonshine's spirit.

This is not a textbook. The thought bobbing above my head like a balloon while writing was that the brain is driven by the *qualitative* – at the deepest level those are the only truths we seek and can absorb. I'm trying to share with the reader my understanding (such as it is) of several remarkable topics that fit loosely together under the motley banner *Moonshine*. I hope it fills a gap in the literature, by focusing more on the ideas and less on the technical minutiae, important though they are. But even if not, it was a pleasure to write, and I think that comes across on every page.

This book is philosophic and speculative, because Moonshine is. It is written for both physicists and mathematicians, because both subjects have contributed to the theory. Partly for this reason, this book differs from other mathematics books in the lack of formal arguments, and differs from other physics books in the lack of long formulae. Without doubt this will froth many mouths. Because the potential readership for this story is unusually diverse, I have tried to assume minimal formal background. Hence when you come to shockingly trivial passages or abrasively uninteresting tangents, please realise they weren't written for you.

In modern mathematics there is a strong tendency towards formulations of concepts that minimise the number and significance of arbitrary choices. This crispness tends to emphasise the naturality of the construction or definition, at the expense sometimes of accessibility. Our mathematics is more conceptual today – more beautiful perhaps – but the cost of less explicitness is the compartmentalism that curses our discipline. We have cut ourselves off not only from each other, but also from our past. In this book I've tried to balance this asceticism with accessibility. Some things have surely been lost, but some perhaps have been gained.

The book endures some glaring and painful omissions, due mostly to fear of spousal reprisals were I to miss yet another deadline. I hope for a second edition. In it I would include a gentle introduction to geometric Langlands. I'd correct the total disregard here for all things supersymmetric – after all, most of the geometric impact of string theory involves supersymmetry. The mathematical treatment of CFT in Chapter 4 is sparser than I'd like. Section 5.4 was originally planned to include brief reviews of the chiral algebras of Beilinson–Drinfel'd [48] and the coordinate-free approach to VOAs developed in [197]. Cohomological issues arise in every chapter, where they are nonetheless quietly ignored. The lip-service paid to subfactors does no justice to their beautiful role in the theory.

I will probably be embarrassed five years from now as to what today I feel is *important*. But at worst I'll be surprised five years from now at what today I find *interesting*. The topics were selected based on my present interests. Other authors (and even me five years from now) would make different choices, but for that I won't apologise.

So let the chase begin...

1

Classical algebra

In this chapter we sketch the basic material – primarily algebra – needed in later chapters. As mentioned in the Introduction, the aspiration of this book isn't to 'Textbookhood'. There are plenty of good textbooks on the material of this chapter (e.g. [162]). What is harder to find are books that describe the ideas beneath and the context behind the various definitions, theorems and proofs. This book, and this chapter, aspire to that. What we lose in depth and detail, we hope to gain in breadth and conceptual content. The range of readers in mind is diverse, from mathematicians expert in other areas to physicists, and the chosen topics, examples and explanations try to reflect this range.

Finite groups (Section 1.1) and lattices (Section 1.2.1) appear as elementary examples throughout the book. Lie algebras (Section 1.4), more than their nonlinear partners the Lie groups, are fundamental to us, especially through their representations (Section 1.5). Functional analysis (Section 1.3), category theory (Section 1.6) and algebraic number theory (Section 1.7) play only secondary roles. Section 1.2 provides some background geometry, but for proper treatments consult [113], [104], [527], [59], [478].

Note the remarkable unity of algebra. Algebraists look at mathematics and science and see structure; they study *form* rather than *content*. The foundations of a new theory are laid by running through a fixed list of questions; only later, as the personality quirks of the new structure become clearer, does the theory become more individual. For instance, among the first questions asked are: What does 'finite' mean here? and What plays the role of a prime number? Mathematics (like any subject) evolves by asking questions, and though a good original question thunders like lightning at night, it is as rare as genius itself. See the beautiful book [504] for more of algebra presented in this style.

1.1 Discrete groups and their representations

The notion of a group originated essentially in the nineteenth century with Galois, who also introduced normal subgroups and their quotients G/N, all in the context of what we now call *Galois theory* (Section 1.7.2). According to Poincaré, when all of mathematics is stripped of its contents and reduced to pure form, the result is group theory.[1] Groups are the devices that act, which explains their fundamental role in mathematics. In physics like much of Moonshine, groups arise through their representations. Standard references for representation theory are [308], [219]; gentle introductions to various aspects of group theory are [162], [421] (the latter is especially appropriate for physicists).

---

[1] See page 499 of J.-P. Serre, *Notices Amer. Math. Soc.* (May 2004).

1.1.1 Basic definitions

A *group* is a set G with an associative product gg' and an identity e, such that each element $g \in G$ has an inverse g^{-1}. The number of elements $\|G\|$ of a group is called its *order*, and is commonly denoted $|G|$.

If we're interested in groups, then we're interested in *comparing* groups, that is we're interested in functions $\varphi : G \to H$ that respect group structure. What this means is φ takes products in G to products in H, the identity e_G in G to the identity e_H in H, and the inverse in G to the inverse in H (the last two conditions are redundant). Such φ are called *group homomorphisms*.

Two groups G, H are considered equivalent or *isomorphic*, written $G \cong H$, if as far as the essential group properties are concerned (think 'form' and not 'content'), the two groups are indistinguishable. That is, there is a group homomorphism $\varphi : G \to H$ that is a bijection (so φ^{-1} exists) and $\varphi^{-1} : H \to G$ is itself a group homomorphism (this last condition is redundant). An *automorphism* (or symmetry) of G is an isomorphism $G \to G$; the set $\text{Aut}\, G$ of all automorphisms of G forms a group.

For example, consider the cyclic group $\mathbb{Z}_n = \{[0], [1], \ldots, [n-1]\}$ consisting of the integers taken mod n, with group operation addition. Write $U_1(\mathbb{C})$ for the group of complex numbers with modulus 1, with group operation multiplication. Then $\varphi([a]) = e^{2\pi i a/n}$ defines a homomorphism between $\mathbb{Z}_n$ and $U_1(\mathbb{C})$. The group of positive real numbers under multiplication is isomorphic to the group of real numbers under addition, the isomorphism being given by logarithm – as far as their group structure is concerned, they are identical. $\text{Aut}\,\mathbb{Z} \cong \mathbb{Z}_2$, corresponding to multiplying the integers by ± 1, while $\text{Aut}\,\mathbb{Z}_n$ is the multiplicative group $\mathbb{Z}_n^\times$, consisting of all numbers $1 \le \ell \le n$ coprime to n (i.e. $\gcd(\ell, n) = 1$), with the operation being multiplication mod n.

Field is an algebraic abstraction of the concept of number: in one we can add, subtract, multiply and divide, and all the usual properties like commutativity and distributivity are obeyed. Fields were also invented by Galois. $\mathbb{C}$, $\mathbb{R}$ and $\mathbb{Q}$ are fields, while $\mathbb{Z}$ is not (you can't always divide an integer by, for example, 3 and remain in $\mathbb{Z}$). The integers mod n, i.e. $\mathbb{Z}_n$, are a field iff n is prime (e.g. in $\mathbb{Z}_4$, it is not possible to divide by the element [2] even though [2] $\ne$ [0] there). $\mathbb{C}$ and $\mathbb{R}$ are examples of fields of characteristic 0 – this means that 0 is the only integer k with the property that $kx = 0$ for all x in the field. We say $\mathbb{Z}_p$ has characteristic p since multiplying by the integer p has the same effect as multiplying by 0. There is a finite field with q elements iff q is a power of a prime, in which case the field is unique and is called $\mathbb{F}_q$. Strange fields have important applications in, for example, coding theory and, ironically, in number theory itself – see Sections 1.7.1 and 2.4.1.

The *index* of a subgroup H in G is the number of 'cosets' gH; for finite groups it equals $\|G\|/\|H\|$. A *normal* subgroup N of a group is one obeying $gNg^{-1} = N$ for all $g \in G$. Its importance arises because the set G/H of cosets gH has a natural group structure precisely when H is normal. If H is a normal subgroup of G we write $H \lhd G$; if H is merely a subgroup of G we write $H < G$. The kernel $\ker(\varphi) = \varphi^{-1}(e_H)$ of a homomorphism $\varphi : G \to H$ is always normal in G, and $\text{Im}\,\varphi \cong G/\ker \varphi$.

By the *free group* $\mathcal{F}_n$ *with generators* $\{x_1, \ldots, x_n\}$ we mean the set of all possible words in the 'alphabet' $x_1, x_1^{-1}, \ldots, x_n, x_n^{-1}$, with group operation given by concatenation. The identity e is the empty word. The only identities obeyed here are the trivial ones coming from $x_i x_i^{-1} = x_i^{-1} x_i = e$. For example, $\mathcal{F}_1 \cong \mathbb{Z}$. The group $\mathcal{F}_2$ is already maximally complicated, in that all the other $\mathcal{F}_n$ arise as subgroups.

We call a group G *finitely generated* if there are finitely many elements $g_1, \ldots, g_n \in G$ such that $G = \langle g_1, \ldots, g_n \rangle$, that is any $g \in G$ can be written as some finite word in the alphabet $g_1^{\pm 1}, \ldots, g_n^{\pm 1}$. For example, any finite group is finitely generated, while the additive group $\mathbb{R}$ is not. Any finitely generated group G is the homomorphic image $\varphi(\mathcal{F}_n)$ of some free group $\mathcal{F}_n$, i.e. $G \cong \mathcal{F}_n / \ker(\varphi)$ (why?). This leads to the idea of *presentation*: $G \cong \langle X \mid \mathcal{R} \rangle$ where X is a set of generators of G and $\mathcal{R}$ is a set of relations, that is words that equal the identity e in G. Enough words must be chosen so that $\ker \varphi$ equals the smallest normal subgroup of $\mathcal{F}_n$ containing all of $\mathcal{R}$. For example, here is a presentation for the dihedral group $\mathcal{D}_n$ (the symmetries of the regular n-sided polygon):

$$\mathcal{D}_n = \langle a, b \mid a^n = b^2 = abab = e \rangle. \tag{1.1.1}$$

For two interesting presentations of the trivial group $G = \{e\}$, see [**416**]. To define a homomorphism $\varphi : G \to H$ it is enough to give the value $\varphi(g_i)$ of each generator of G, and verify that φ sends all relations of G to identities in H.

We say G equals the (internal) *direct product* $N \times H$ of subgroups if every element $g \in G$ can be written uniquely as a product nh, for every $n \in N, h \in H$, and where N, H are both normal subgroups of G and $N \cap H = \{e\}$. Equivalently, the (external) direct product $N \times H$ of two groups is defined to be all ordered pairs (n, h), with operations given by $(n, h)(n', h') = (nn', hh')$; G will be the internal direct product of its subgroups N, H iff it is isomorphic to their external direct product. Of course, $N \cong G/H$ and $H \cong G/N$. Direct product is also called 'homogeneous extension' in the physics literature.

More generally, G is an (internal) *semi-direct product* $N \rtimes H$ of subgroups if all conditions of the internal direct product are satisfied, except that H need not be normal in G (but as before, $N \lhd G$). Equivalently, the (external) semi-direct product $N \rtimes_\theta H$ of two groups is defined to be all ordered pairs (n, h) with operation given by

$$(n, h)(n', h') = (n \, \theta_h(n'), hh'),$$

where $h \mapsto \theta_h \in \mathrm{Aut}\, N$ can be any group homomorphism. It's a good exercise to verify that $N \rtimes_\theta H$ is a group for any such θ, and to relate the internal and external semi-direct products. Note that $N \cong \{(n, e_H)\}, H \cong \{(e_N, h)\} \cong G/N$. Also, choosing the trivial homomorphism $\theta_h = id.$ recovers the (external) direct product. The semi-direct product is also called the 'inhomogeneous extension'.

For example, the dihedral group is a semi-direct product of $\mathbb{Z}_n$ with $\mathbb{Z}_2$. The group of isometries (distance-preserving maps) in 3-space is $\mathbb{R}^3 \rtimes (\{\pm I\} \times \mathrm{SO}_3)$, where $\mathbb{R}^3$ denotes the additive subgroup of translations, $-I$ denotes the reflection $x \mapsto -x$ through the origin, and SO_3 is the group of rotations. This continuous group is an example of a

Lie group (Section 1.4.2). Closely related is the *Poincaré group*, which is the semi-direct product of translations $\mathbb{R}^4$ with the Lorentz group $SO_{3,1}$.

Finally and most generally, if N is a normal subgroup of G then we say that G is an (internal) extension of N by the quotient group G/N. Equivalently, we say a group G is an (external) extension of N by H if each element g in G can be identified with a pair (n, h), for $n \in N$ and $h \in H$, and where the group operation is

$$(n, h)(n', h') = (\text{stuff}, hh'),$$

provided only that $(n, e_H)(n', e_H) = (nn', e_H)$.

That irritating *carry* in base 10 addition, which causes so many children so much grief, is the price we pay for building up our number system by repeatedly extending by the group $\mathbb{Z}_{10}$ (one for each digit) (see Question 1.1.8(c)).

A group G is *abelian* if $gh = hg$ for all $g, h \in G$. So $\mathbb{Z}_n$ is abelian, but the symmetric group S_n for $n > 2$ is not. A group is *cyclic* if it has only one generator. The only cyclic groups are the abelian groups $\mathbb{Z}_n$ and $\mathbb{Z}$. The *centre* $Z(G)$ of a group is defined to be all elements $g \in G$ commuting with all other $h \in G$; it is always a normal abelian subgroup.

Theorem 1.1.1 (Fundamental theorem of finitely generated abelian groups) *Let G be a finitely generated abelian group. Then*

$$G \cong \mathbb{Z}^r \times \mathbb{Z}_{m_1} \times \cdots \times \mathbb{Z}_{m_h}$$

where $\mathbb{Z}^r = \mathbb{Z} \times \cdots \times \mathbb{Z}$ (r times), and m_1 divides m_2 which divides . . . which divides m_h. The numbers r, m_i, h are unique. The group G is finite iff $r = 0$.

The proof isn't difficult – for example, see page 43 of [**504**]. Theorem 1.1.1 is closely related to other classical decompositions, such as that of the Jordan canonical form for matrices.

1.1.2 Finite simple groups

Theorem 1.1.1 gives among other things the classification of all finite abelian groups. In particular, the number of abelian groups G of order $\|G\| = n = \prod_p p^{a_p}$ is $\prod_p P(a_p)$, where $P(m)$ is the partition number of m (the number of ways of writing m as a sum $m = \sum_i m_i, m_1 \geq m_2 \geq \cdots \geq 0$).

What can we say about the classification of arbitrary finite groups? This is almost certainly hopeless. All groups of order p or p^2 (for p prime) are necessarily abelian. The smallest non-abelian group is the symmetric group S_3 (order 6); next are the dihedral group $\mathcal{D}_4$ and the quaternion group $\mathcal{Q}_4 = \{\pm 1, \pm i, \pm j, \pm k\}$ (both order 8). Table 1.1 summarises the situation up to order 50 – for orders up to 100, see [**418**]. This can't be pushed that much further, for example the groups of order 128 (there are 2328 of them) were classified only in 1990. One way to make progress is to restrict the class of groups considered.

Every group has two trivial normal subgroups: itself and $\{e\}$. If these are the only normal subgroups, the group is called *simple*. It is conventional to regard the trivial

Table 1.1. *The numbers of non-abelian groups of order* < 50

$\|G\|$	6	8	10	12	14	16	18	20	21	22	24	26	27
#	1	2	1	3	1	9	3	3	1	1	12	1	2

$\|G\|$	28	30	32	34	36	38	39	40	42	44	46	48
#	2	3	44	1	10	1	1	11	5	2	1	47

group $\{e\}$ as not simple (just as '1' is conventionally regarded as not prime). An alternate definition of a simple group G is that if $\varphi : G \to H$ is any homomorphism, then either φ is constant (i.e. $\varphi(G) = \{e\}$) or φ is one-to-one.

The importance of simple groups is provided by the *Jordan–Hölder Theorem*. By a 'composition series' for a group G, we mean a nested sequence

$$G = H_0 > H_1 > H_2 > \cdots > H_k > H_{k+1} = \{e\} \qquad (1.1.2)$$

of groups such that H_i is normal in H_{i-1} (though not necessarily normal in H_{i-2}), and the quotient H_{i-1}/H_i (called a 'composition factor') is simple. An easy induction shows that any finite group G has at least one composition series. If $H'_0 > \cdots > H'_{\ell+1} = \{e\}$ is a second composition series for G, then the Jordan–Hölder Theorem says that $k = \ell$ and, up to a reordering π, the simple groups H_{i-1}/H_i and $H'_{\pi j-1}/H'_{\pi j}$ are isomorphic.

The cyclic group $\mathbb{Z}_n$ is simple iff n is prime. Two composition series of $\mathbb{Z}_{12} = \langle 1 \rangle$ are

$$\mathbb{Z}_{12} > \langle 2 \rangle > \langle 4 \rangle > \langle 12 \rangle,$$
$$\mathbb{Z}_{12} > \langle 3 \rangle > \langle 6 \rangle > \langle 12 \rangle,$$

corresponding to composition factors $\mathbb{Z}_2, \mathbb{Z}_2, \mathbb{Z}_3$ and $\mathbb{Z}_3, \mathbb{Z}_2, \mathbb{Z}_2$. This is reminiscent of $2 \cdot 2 \cdot 3 = 3 \cdot 2 \cdot 2$ both being prime factorisations of 12. When all composition factors of a group are cyclic, the group is called *solvable*. The deep *Feit–Thompson Theorem* tells us that any group of odd order is solvable, as are all abelian groups and any group of order < 60 (Question 1.1.2). The name 'solvable' comes from Galois theory (Section 1.7.2).

Finite groups are a massive generalisation of the notion of number. The number n can be identified with the cyclic group $\mathbb{Z}_n$. The divisor of a number corresponds to a normal subgroup, so a prime number corresponds to a simple group. The Jordan–Hölder Theorem generalises the uniqueness of prime factorisations. Building up any number by multiplying primes becomes building up a group by (semi-)direct products and, more generally, by group extensions. Note however that $\mathbb{Z}_6 \times \mathbb{Z}_2$ and $\mathcal{S}_3 \times \mathbb{Z}_2$ – both different from $\mathbb{Z}_{12}$ – also have $\mathbb{Z}_2, \mathbb{Z}_2, \mathbb{Z}_3$ as composition factors. The lesson: unlike for numbers, 'multiplication' here does not give a unique answer. The semi-direct product $\mathbb{Z}_3 \rtimes \mathbb{Z}_2$ can equal either $\mathbb{Z}_6$ or $\mathcal{S}_3$, depending on how the product is taken.

The composition series (1.1.2) tells us that the finite group G is obtained inductively from the trivial group $\{e\}$ by extending $\{e\}$ by the simple group H_k/H_{k+1} to get H_k, then extending H_k by the simple group H_{k-1}/K_k to get H_{k-1}, etc. In other words, any finite group G can be obtained from the trivial group by extending inductively by simple groups; those simple groups are its 'prime factors' = composition factors.

Thus simple groups have an importance for group theory approximating what primes have for number theory. One of the greatest accomplishments of twentieth-century mathematics is the classification of the finite simple groups. Of course we would have preferred the complete finite group classification, but the simple groups are a decent compromise! This work, completed in the early 1980s (although gaps in the arguments are continually being discovered and filled [**22**]), runs to approximately 15 000 journal pages, spread over 500 individual papers, and is the work of a whole generation of group theorists (see [**256**], [**512**] for historical remarks and some ideas of the proof). A modern revision is currently underway to simplify the proof and find and fill all gaps, but the final proof is still expected to be around 4000 pages long. The resulting list, probably complete, is:

- the cyclic groups $\mathbb{Z}_p$ (p a prime);
- the alternating groups $\mathcal{A}_n$ for $n \geq 5$;
- 16 families of Lie type;
- 26 sporadic groups.

The alternating group $\mathcal{A}_n$ consists of the even permutations in the symmetric group $\mathcal{S}_n$, and so has order(=size) $\frac{1}{2} n!$. The groups of Lie type are essentially Lie groups (Section 1.4.2) defined over the finite fields $\mathbb{F}_q$, sometimes 'twisted'. See, for example, chapter I.4 of [**92**] for an elementary treatment. The simplest example is $\mathrm{PSL}_n(\mathbb{F}_q)$, which consists of the $n \times n$ matrices with entries in $\mathbb{F}_q$, with determinant 1, quotiented out by the centre of $\mathrm{SL}_n(\mathbb{F}_q)$ (namely the scalar matrices $\mathrm{diag}(a, a, \ldots, a)$ with $a^n = 1$) ($\mathrm{PSL}_2(\mathbb{Z}_2)$ and $\mathrm{PSL}_2(\mathbb{Z}_3)$ aren't simple so should be excluded). The 'P' here stands for 'projective' and refers to this quotient, while the 'S' stands for 'special' and means determinant 1.

The determinant $\det(\rho(g))$ of any representation ρ (Section 1.1.3) of a noncyclic simple group must be identically 1, and the centre of any noncyclic simple group must be trivial (why?). Hence in the list of simple groups of Lie type are found lots of P's and S's.

The smallest noncyclic simple group is $\mathcal{A}_5$, with order 60. It is isomorphic to $\mathrm{PSL}_2(\mathbb{Z}_5)$ and $\mathrm{PSL}_2(\mathbb{F}_4)$, and can also be interpreted as the group of all rotational symmetries of a regular icosahedron (reflections have determinant -1 and so cannot belong to any simple group $\ncong \mathbb{Z}_2$). The simplicity of $\mathcal{A}_5$ is ultimately responsible for the fact that the zeros of a general quintic polynomial cannot be solved by radicals (see Section 1.7.2).

The smallest sporadic group is the Mathieu group M_{11}, order 7920, discovered in 1861.[2] The largest is the Monster $\mathbb{M}$,[3] conjectured independently by Fischer and Griess

---

[2] ... although his arguments apparently weren't very convincing. In fact some people, including the Camille Jordan of Jordan–Hölder fame, argued in later papers that the largest of Mathieu's groups, M_{24}, couldn't exist. We now know it does, for example an elegant realisation is as the automorphisms of Steiner system S(5,8,24).

[3] Griess also came up with the symbol for the Monster; Conway came up with the name. It's a little unfortunate (but perhaps inevitable) that the Monster is not named after its codiscoverers, Berndt Fischer and Robert Griess; the name 'Friendly Giant' was proposed in [**263**] as a compromise, but 'Monster' stuck.

in 1973 and finally proved to exist by Griess [**263**] in 1980. Its order is given in (0.2.2). 20 of the 26 sporadic groups are involved in (i.e. are quotients of subgroups of) the Monster, and play some role in Moonshine, as we see throughout Section 7.3. We study the Monster in more detail in Section 7.1.1. Some relations among $\mathbb{M}$, the Leech lattice Λ and the largest Mathieu group M_{24} are given in chapters 10 and 29 of [**113**]. We collect together some of the data of the sporadics in Table 7.1.

This work reduces the construction and classification of all finite groups to understanding the possible extensions by simple groups. Unfortunately, group extensions turn out to be technically quite difficult and lead one into group cohomology.

There are many classifications in mathematics. Most of them look like phone books, and their value is purely pragmatic: for example, as a list of potential counterexamples, and as a way to prove some theorems by exhaustion. And of course obtaining them requires *at least* one paper, and with it some breathing space before those scoundrels on the grant evaluation boards. But when the classification has structure, it can resemble in ways a tourist guide, hinting at new sites to explore. The 18 infinite families in the finite simple group classification are well known and generic, much like the chain of MacDonald's restaurants, useful and interesting in their own ways. But the eye skims over them, and is drawn instead to the 26 sporadic groups and in particular to the largest: the Monster.

1.1.3 Representations

Groups typically arise as 'things that act'. This is their *raison d'être*. For instance, the symmetries of a square form the dihedral group $\mathcal{D}_4$ – that is, the elements of $\mathcal{D}_4$ act on the vertices by permuting them. When a group acts on a structure, you generally want it to preserve the essential features of the structure. In the case of our square, we want adjacent vertices to remain adjacent after being permuted.

So when a group G acts on a vector space V (over $\mathbb{C}$, say), we want it to act 'linearly'. The action $g.v$ of G on V gives V the structure of a G-*module*. In completely equivalent language, it is a *representation* ρ of G on $V \cong \mathbb{C}^n$, that is as a group homomorphism from G to the invertible matrices $\mathrm{GL}_n(\mathbb{C})$. So a representation ρ is a realisation of the group G by matrices, where multiplication in G corresponds to matrix multiplication:

$$\rho(gh) = \rho(g)\,\rho(h).$$

The identification of V with $\mathbb{C}^n$ is achieved by choosing a basis of V, so the module language is 'cleaner' in the sense that it is basis-independent, but this also tends to make it less conducive for practical calculations. The module action $g.v$ is now written $\rho(g)\mathbf{v}$, where $\mathbf{v}$ is the column vector consisting of the components of $v \in V$ with respect to the given basis. If $\rho(g)$ are $n \times n$ matrices, we say ρ is an n-dimensional representation.

For a practise example, consider the symmetric group

$$\mathcal{S}_3 = \{(1), (12), (23), (13), (123), (132)\}. \tag{1.1.3}$$

These cycles multiply as $(13)(123) = (12)$. One representation of $\mathcal{S}_3$ is one-dimensional, and sends all six elements of $\mathcal{S}_3$ to the 1×1 identity matrix:

$$\rho_1(\sigma) = (1), \qquad \forall \sigma \in \mathcal{S}_3.$$

Obviously (1.1.3) is satisfied, and so this defines a representation. But it's trivial, projecting away all structure in the group $\mathcal{S}_3$. Much more interesting is the defining representation ρ_3, which assigns to each $\sigma \in \mathcal{S}_3$ a 3×3 permutation matrix by using σ to permute the rows of the identity matrix I. For example

$$(12) \mapsto \begin{pmatrix} 0 & 1 & 0 \\ 1 & 0 & 0 \\ 0 & 0 & 1 \end{pmatrix}, \quad (13) \mapsto \begin{pmatrix} 0 & 0 & 1 \\ 0 & 1 & 0 \\ 1 & 0 & 0 \end{pmatrix}, \quad (123) \mapsto \begin{pmatrix} 0 & 0 & 1 \\ 1 & 0 & 0 \\ 0 & 1 & 0 \end{pmatrix}.$$

This representation is faithful, that is different permutations σ are assigned different matrices $\rho_3(\sigma)$. From this defining representation ρ_3, we get a second one-dimensional one – called the sign representation ρ_s – by taking determinants. For example, $(1) \mapsto (+1)$, $(12) \mapsto (-1)$, $(13) \mapsto (-1)$ and $(123) \mapsto (+1)$.

The most important representation associated with a group G is the *regular representation* given by the group algebra $\mathbb{C}G$. That is, consider the $\|G\|$-dimensional vector space (over $\mathbb{C}$, say) consisting of all formal linear combinations $\sum_{h \in G} \alpha_h h$, where $\alpha_h \in \mathbb{C}$. This has a natural structure of a G-module, given by $g.(\sum \alpha_h h) = \sum_h \alpha_h gh$.

When G is infinite, there will be convergence issues and hence analysis since infinite sums $\sum \alpha_h h$ are involved. The most interesting possibility is to interpret $h \mapsto \alpha_h$ as a $\mathbb{C}$-valued function $\alpha(h)$ on G. Suppose we have a G-invariant measure $\mathrm{d}\mu$ on this space of functions $\alpha : G \to \mathbb{C}$ – this means that the integral $\int_{gU} \alpha(h)\,\mathrm{d}\mu(h)$ will exist and equal $\int_U \alpha(gh)\,\mathrm{d}\mu(h)$ whenever the latter exists. For example, if G is discrete, define '$\int_G \alpha\,\mathrm{d}\mu$' to be $\sum_{h \in G} \alpha(h)$, while if G is the additive group $\mathbb{R}$, $\mathrm{d}\mu(x)$ is the Lebesgue measure (see Section 1.3.1). Looking at the g-coefficient of the product $(\sum_h \alpha_h h)(\sum_k \beta_k k)$, we get the formula $g \mapsto \sum \alpha_h \beta_{h^{-1}g}$, which we recognise as the convolution product (recall $(\alpha * \beta)(x) = \int \alpha(x) \beta(x - y)\,\mathrm{d}y$) in, for example, Fourier analysis. In this context, the regular representation of G becomes the Hilbert space $L^2(G)$ of square-integrable functions (i.e. $\int |\alpha|^2 \mathrm{d}\mu < \infty$); the convolution product defines an action of $L^2(G)$ on itself. Note however that the $L^2(\mathbb{R})$-module $L^2(\mathbb{R})$, for a typical example, doesn't restrict to an $\mathbb{R}$-module: the action of $\mathbb{R}$ on $\alpha \in L^2(\mathbb{R})$ by $(x.\alpha)(y) = \alpha(x + y)$ corresponds to the convolution product of α with the 'Dirac delta' distribution δ centred at x. We return to $L^2(G)$ in Section 1.5.5.

Two representations ρ, ρ' are called *equivalent* if they differ merely by a change of coordinate axes (basis) in the ambient space $\mathbb{C}^n$, that is if there exists a matrix U such that $\rho'(g) = U\rho(g)U^{-1}$ for all g. The *direct sum* $\rho' \oplus \rho''$ of representations is given by

$$(\rho' \oplus \rho'')(g) = \begin{pmatrix} \rho'(g) & 0 \\ 0 & \rho''(g) \end{pmatrix}. \tag{1.1.4a}$$

The tensor product $\rho' \otimes \rho''$ of representations is given by $(\rho' \otimes \rho'')(g) = \rho'(g) \otimes \rho''(g)$, where the Kronecker product $A \otimes B$ of matrices is defined by the following block form:

$$A \otimes B = \begin{pmatrix} a_{11}B & a_{12}B & \cdots \\ a_{21}B & a_{22}B & \cdots \\ \vdots & \vdots & \ddots \end{pmatrix}. \tag{1.1.4b}$$

The *contragredient* or *dual* ρ^* of a representation is given by the formula

$$\rho^*(g) = (\rho(g^{-1}))^t, \tag{1.1.4c}$$

so called because it's the natural representation on the space V^* dual to the space V on which ρ is defined. For any finite group representation defined over a subfield of $\mathbb{C}$, the dual ρ^* is equivalent to the complex conjugate representation $g \mapsto \overline{\rho(g)}$.

Returning to our $\mathcal{S}_3$ example, the given matrices for ρ_3 were obtained by having $\mathcal{S}_3$ act on coordinates with respect to the standard basis $\{(1, 0, 0), (0, 1, 0), (0, 0, 1)\}$. If instead we choose the basis $\{(1, 1, 1), (1, -1, 0), (0, 1, -1)\}$, these matrices become

$$(12) \mapsto \begin{pmatrix} 1 & 0 & 0 \\ 0 & -1 & 1 \\ 0 & 0 & 1 \end{pmatrix}, \quad (13) \mapsto \begin{pmatrix} 1 & 0 & 0 \\ 0 & 0 & -1 \\ 0 & -1 & 0 \end{pmatrix}, \quad (123) \mapsto \begin{pmatrix} 1 & 0 & 0 \\ 0 & 0 & -1 \\ 0 & 1 & -1 \end{pmatrix}.$$

It is manifest here that ρ_3 is a direct sum of ρ_1 (the upper-left 1×1 block) with a two-dimensional representation ρ_2 (the lower-right 2×2 block) given by

$$(12) \mapsto \begin{pmatrix} -1 & 1 \\ 0 & 1 \end{pmatrix}, \quad (13) \mapsto \begin{pmatrix} 0 & -1 \\ -1 & 0 \end{pmatrix}, \quad (123) \mapsto \begin{pmatrix} 0 & -1 \\ 1 & -1 \end{pmatrix}.$$

An *irreducible* or *simple* module is a module that contains no nontrivial submodule. 'Submodule' plays the role of divisor here, and 'irreducible' the role of prime number. A module is called *completely reducible* if it is the direct sum of finitely many irreducible modules. For example, the $\mathcal{S}_3$ representations ρ_1, ρ_s and ρ_2 are irreducible, while $\rho_3 \cong \rho_1 \oplus \rho_2$ is completely reducible.

A representation is called *unitary* if it is equivalent to one whose matrices $\rho(g)$ are all unitary (i.e. their inverses equal their complex conjugate transposes). A more basis-independent definition is that a G-module V is unitary if there exists a Hermitian form $\langle u, v \rangle \in \mathbb{C}$ on V such that

$$\langle g.u, g.v \rangle = \langle u, v \rangle.$$

By definition, a *Hermitian form* $\langle u, v \rangle : V \times V \to \mathbb{C}$ is linear in v and anti-linear in u, i.e.

$$\langle au + a'u', bv + b'v' \rangle = \overline{a}b\langle u, v \rangle + \overline{a}b'\langle u, v' \rangle + \overline{a'}b\langle u', v \rangle + \overline{a'}b'\langle u', v' \rangle,$$

for all $a, a', b, b' \in \mathbb{C}$, $u, u', v, v' \in V$, and finally $\langle u, u \rangle > 0$ for all nonzero $u \in V$. When V is finite-dimensional, a basis can always be found in which its Hermitian form looks like $\langle x, y \rangle = \sum_i \overline{x}_i y_i$. Most representations of interest in quantum physics are unitary. Unitary representations are much better behaved than non-unitary ones.

For instance, an easy argument shows that finite-dimensional unitary representation is completely reducible.

An *indecomposable* module is one that isn't the direct sum of smaller ones. An indecomposable module may be *reducible*: its matrices could be put into the form

$$\rho(g) = \begin{pmatrix} A(g) & B(g) \\ 0 & D(g) \end{pmatrix},$$

where for some g the submatrix $B(g)$ isn't the 0-matrix (otherwise we would recover (1.1.4a)). Then $A(g)$ is a subrepresentation, but $D(g)$ isn't. For finite groups, however, irreducible=indecomposable:

Theorem 1.1.2 (Burnside, 1904)   *Let G be finite and the field be $\mathbb{C}$. Any G-module is unitary and will be completely reducible if it is finite-dimensional. There are only finitely many irreducible G-modules; their number equals the number of conjugacy classes of G.*

The *conjugacy classes* are the sets $K_g = \{h^{-1}gh \mid h \in G\}$. This fundamental result fails for infinite groups. For example, take G to be the additive group $\mathbb{Z}$ of integers. Then there are uncountably many one-dimensional representations of G, and there are representations that are reducible but indecomposable (see Question 1.1.6(a)). Theorem 1.1.2 is proved using a projection defined by certain averaging over G, as well as:

Lemma 1.1.3 (Schur's Lemma)   *Let G be finite and ρ, ρ' be representations.*
(a) *ρ is irreducible iff the only matrices A commuting with all matrices $\rho(g)$, $g \in G$ – that is $A\rho(g) = \rho(g)A$ – are of the form $A = aI$ for $a \in \mathbb{C}$, where I is the identity matrix.*
(b) *Suppose both ρ and ρ' are irreducible. Then ρ and ρ' are isomorphic iff there is a nonzero matrix A such that $A\rho(g) = \rho'(g)A$ for all $g \in G$.*

Schur's Lemma is an elementary observation central to representation theory. It's proved by noting that the kernel (nullspace) and range (column space) of A are G-invariant.

The character[4] ch_ρ of a representation ρ is the map $G \to \mathbb{C}$ given by the trace:

$$\mathrm{ch}_\rho(g) = \mathrm{tr}\,(\rho(g)). \tag{1.1.5}$$

We see that equivalent representations have the same character, because of the fundamental identity $\mathrm{tr}(AB) = \mathrm{tr}(BA)$. Remarkably, for finite groups (and $\mathbb{C}$), the converse is also true: inequivalent representations have different character. That trace identity also tells us that the character is a 'class function', i.e. $\mathrm{ch}_\rho(hgh^{-1}) = \mathrm{tr}(\rho(h)\,\rho(g)\,\rho(h)^{-1}) = \mathrm{ch}_\rho(g)$ so ch_ρ is constant on each conjugacy class K_g. Group characters are enormously simpler than representations: for example, the smallest nontrivial representation of the Monster

---

[4] Surprisingly, characters were invented before group representations, by Frobenius in 1868. He defined characters indirectly, by writing the 'class sums' C_j in terms of the idempotents of the centre of the group algebra. It took him a year to realise they could be reinterpreted as the traces of matrices.

Table 1.2. *The character table of S_3*

ch\σ	(1)	(12)	(123)
ch_1	1	1	1
ch_s	1	−1	1
ch_2	2	0	−1

$\mathbb{M}$ consists of about 10^{54} matrices, each of size $196\,883 \times 196\,883$, while its character consists of 194 complex numbers. The reason is that the representation matrices have a lot of redundant, basis-dependent information, to which the character is happily oblivious.

The Thompson trick mentioned in Section 0.3 tells us: *A dimension can (and should) be twisted; that twist is called a character.* Indeed, $ch_\rho(e) = \dim(\rho)$, where the dimension of ρ is defined to be the dimension of the underlying vector space V, or the size n of the $n \times n$ matrices $\rho(g)$. When we see a positive integer, we should try to interpret it as a dimension of a vector space; if there is a symmetry present, then it probably acts on the space, in which case we should see what significance the other character values may have.

Algebra searches for structure. What can we say about the set of characters? First, note directly from (1.1.4) that we can add and multiply characters:

$$ch_{\rho \oplus \rho'}(g) = ch_\rho(g) + ch_{\rho'}(g), \tag{1.1.6a}$$

$$ch_{\rho \otimes \rho'}(g) = ch_\rho(g)\, ch_{\rho'}(g), \tag{1.1.6b}$$

$$ch_{\rho^*}(g) = \overline{ch_\rho(g)}. \tag{1.1.6c}$$

Therefore the complex span of the characters forms a (commutative associative) *algebra*. For G finite (and the field algebraically closed), each matrix $\rho(g)$ is separately diagonalisable, with eigenvalues that are roots of 1 (why?). This means that each character value $ch_\rho(g)$ is a sum of roots of 1.

By the *character table* of a group G we mean the array with rows indexed by the characters ch_ρ of irreducible representations, and the columns by conjugacy classes K_g, and with entries $ch_\rho(g)$. An example is given in Table 1.2. Different groups can have identical character tables: for instance, for any n, the dihedral group $\mathcal{D}_{4n}$ has the same character table as the quaternionic group $\mathcal{Q}_{4n}$ defined by the presentation

$$\mathcal{Q}_{4n} = \langle a, b \,|\, a^2 = b^{2n},\ abab = e \rangle. \tag{1.1.7}$$

In spite of this, the characters of a group G tell us much about G – for example, its order, all of its normal subgroups, whether or not it's simple, whether or not it's solvable ... In fact, the character table of a finite simple group determines the group uniquely [**100**] (its order alone usually distinguishes it from other simple groups). This suggests:

Problem *Suppose G and H have identical character tables (up to appropriate permutations of rows and columns). Must they have the same composition factors?*

After all, the answer is certainly yes for solvable G (why?).

It may seem that 'trace' is a fairly arbitrary operation to perform on the matrices $\rho(g)$ – certainly there are other invariants we can attach to a representation ρ so that equivalent representations are assigned equal numbers. For example, how about $g \mapsto \det\rho(g)$? This is too limited, because it is a group homomorphism (e.g. what happens when G is simple?). But more generally, choose an independent variable x_g for each element $g \in G$, and for any representation ρ of G define the group determinant of ρ

$$\Theta_\rho = \det\left(\sum_{g \in G} x_g \, \rho(g)\right).$$

This is a multivariable polynomial Θ_ρ, homogeneous of degree $n = \dim(\rho)$. The character $\mathrm{ch}_\rho(g)$ can be obtained from the group determinant Θ_ρ: it is the coefficient of the $x_g x_e^{n-1}$ term. In fact, the group G is uniquely determined by the group determinant of the regular representation $\mathbb{C}G$. See the review article [**315**].

One use of characters is to identify representations. For this purpose the orthogonality relations are crucial: given any characters ch, ch' of G, define the Hermitian form

$$\langle \mathrm{ch}, \mathrm{ch}' \rangle = \frac{1}{\|G\|} \sum_{g \in G} \mathrm{ch}(g)\overline{\mathrm{ch}'(g)}. \tag{1.1.8a}$$

Write ρ_i for the irreducible representations, and ch_i for the corresponding traces. Then

$$\langle \mathrm{ch}_i, \mathrm{ch}_j \rangle = \delta_{ij}, \tag{1.1.8b}$$

that is the irreducible characters ch_i are an orthonormal basis with respect to (1.1.8a). If the rows of a matrix are orthonormal, so are the columns. Hence (1.1.8b) implies

$$\sum_i \mathrm{ch}_i(g) \, \overline{\mathrm{ch}_i(h)} = \frac{\|G\|}{\|K_g\|} \delta_{K_g, K_h}. \tag{1.1.8c}$$

The decomposition of $\mathbb{C}G$ into irreducibles is now immediate:

$$\mathbb{C}G \cong \bigoplus_i (\dim \rho_i) \, \rho_i,$$

that is each irreducible representation appears with multiplicity given by its dimension. Taking the dimension of both sides, we obtain the useful identity

$$\|G\| = \sum_i (\dim \rho_i)^2.$$

The notion of vector space and representation can be defined over any field $\mathbb{K}$. One thing that makes representations over, for example, the finite field $\mathbb{K} = \mathbb{Z}_p$ much more difficult is that characters no longer distinguish inequivalent representations. For instance, take $G = \{e\}$ and consider the representations

$$\rho(1) = (1) \quad \text{and} \quad \rho'(1) = \begin{pmatrix} 1 & 0 & 0 \\ 0 & 1 & 0 \\ 0 & 0 & 1 \end{pmatrix}.$$

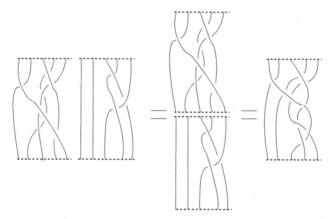

Fig. 1.1 Multiplication in the braid group $\mathcal{B}_5$.

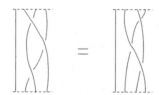

Fig. 1.2 The relation $\sigma_2\sigma_3\sigma_2 = \sigma_3\sigma_2\sigma_3$ in $\mathcal{B}_4$.

These are certainly different representations – their dimensions are different. But over the field $\mathbb{Z}_2$, their characters ch and ch′ are identical. Theorem 1.1.2 also breaks down here. Unless otherwise stated, in this book we restrict to characteristic 0 (but see *modular Moonshine* in Section 7.3.5).

1.1.4 Braided #1: the braid groups

Fundamental to us are the braid groups, especially $\mathcal{B}_3$. By an *n-braid* we mean n non-intersecting strands as in Figure 1.1. We are interested here in how the strands interweave, and not how they knot, and so we won't allow the strands to double-back on themselves. We regard two n-braids as equivalent if they can be deformed continuously into each other – we make this notion more precise in Section 1.2.3. The set of equivalence classes of n-braids forms a group, called the *braid group* $\mathcal{B}_n$, with multiplication given by vertical concatenation, as in Figure 1.1.

Artin (1925) gives a very useful presentation of $\mathcal{B}_n$:

$$\mathcal{B}_n = \langle \sigma_1, \ldots, \sigma_{n-1} \mid \sigma_i\sigma_j = \sigma_j\sigma_i, \; \sigma_i\sigma_{i+1}\sigma_i = \sigma_{i+1}\sigma_i\sigma_{i+1}, \; \text{whenever } |i - j| \geq 2 \rangle.$$
$$(1.1.9)$$

Here σ_i denotes the braid obtained from the identity braid by interchanging the ith and $(i + 1)$th strands, with the ith strand on top. See Figure 1.2 for an illustration.

Of course $\mathcal{B}_1$ is trivial and $\mathcal{B}_2 \cong \mathbb{Z}$, but the other $\mathcal{B}_n$ are quite interesting. Any non-trivial element in $\mathcal{B}_n$ has infinite order. Let $\sigma = \sigma_1 \sigma_2 \cdots \sigma_{n-1}$; then $\sigma \sigma_i = \sigma_{i+1} \sigma$, so the generators are all conjugate and the braid $Z = \sigma^n$ lies in the centre of $\mathcal{B}_n$. In fact, for $n \geq 2$ the centre $Z(\mathcal{B}_n) \cong \mathbb{Z}$, and is generated by that braid Z. We're most interested in $\mathcal{B}_3$: then $Z = (\sigma_1 \sigma_2)^3$ generates the centre, and we will see shortly that

$$\mathcal{B}_3/\langle Z^2 \rangle \cong SL_2(\mathbb{Z}), \tag{1.1.10a}$$

$$\mathcal{B}_3/\langle Z \rangle \cong PSL_2(\mathbb{Z}). \tag{1.1.10b}$$

There is a surjective homomorphism $\phi : \mathcal{B}_n \to \mathcal{S}_n$ taking a braid α to the permutation $\phi(\alpha) \in \mathcal{S}_n$, where the strand of α starting at position i on the top ends on the bottom at position $\phi(\alpha)(i)$. For example, $\phi(\sigma_i)$ is the transposition $(i, i+1)$. The kernel of ϕ is called the *pure braid group* $\mathcal{P}_n$. A presentation for $\mathcal{P}_n$ is given in lemma 1.8.2 of [**59**]. We find that $\mathcal{P}_2 = \langle \sigma_1^2 \rangle \cong \mathbb{Z}$ and

$$\mathcal{P}_3 = \langle \sigma_1^2, \sigma_2^2, Z \rangle \cong \mathcal{F}_2 \times \mathbb{Z}. \tag{1.1.10c}$$

Another obvious homomorphism is the degree map deg : $\mathcal{B}_n \to \mathbb{Z}$, defined by $\deg(\sigma_i^{\pm 1}) = \pm 1$. It is easy to show using (1.1.9) that 'deg' is well defined and is the number of signed crossings in the braid. Its kernel is the commutator subgroup $[\mathcal{B}_n \mathcal{B}_n]$ (see Question 1.1.7(a)).

The most important realisation of the braid group is as a fundamental group (see (1.2.6)). It is directly through this that most appearances of $\mathcal{B}_n$ in Moonshine-like phenomena arise (e.g. Jones' braid group representations from subfactors, or Kohno's from the monodromy of the Knizhnik–Zamolodchikov equation).

The relation of $\mathcal{B}_3$ to modularity in Moonshine, however, seems more directly to involve the faithful action of $\mathcal{B}_n$ on the free group $\mathcal{F}_n = \langle x_1, \ldots, x_n \rangle$ (see Question 6.3.5). This action allows us to regard $\mathcal{B}_n$ as a subgroup of Aut $\mathcal{F}_n$.

As is typical for infinite discrete groups, $\mathcal{B}_n$ has continua of representations. For instance, there is a different one-dimensional for every choice of nonzero complex number $w \neq 0$, namely $\alpha \mapsto w^{\deg \alpha}$. It seems reasonable to collect these together and regard them as different specialisations of a single one-dimensional $\mathbb{C}[w^{\pm 1}]$-representation, which we could call $w^{\deg}$, where $\mathbb{C}[w^{\pm 1}]$ is the (Laurent) polynomial algebra in w and w^{-1}.

The *Burau representation* (Burau, 1936) of $\mathcal{B}_n$ is an n-dimensional representation with entries in the Laurent polynomials $\mathbb{C}[w^{\pm 1}]$, and is generated by the matrices

$$\sigma_i \mapsto I_{i-1} \oplus \begin{pmatrix} 1-w & w \\ 1 & 0 \end{pmatrix} \oplus I_{n-i-1}, \tag{1.1.11a}$$

where I_k here denotes the $k \times k$ identity matrix. $\mathbb{C}[w^{\pm 1}]$ isn't a field, but checking determinants confirms that all matrices $\rho(\sigma_i)$ are invertible over it. The Burau representation is reducible – in particular the column vector $v = (1, 1, \ldots, 1)^t$ is an eigenvector with eigenvalue 1, for all the matrices in (1.1.11a), and hence $\mathcal{B}_n$ acts trivially on the subspace $\mathbb{C}v$. The remaining $(n-1)$-dimensional representation is the *reduced Burau representation.* For example, for $\mathcal{B}_3$ it is

$$\sigma_1 \mapsto \begin{pmatrix} -w & 1 \\ 0 & 1 \end{pmatrix}, \qquad \sigma_2 \mapsto \begin{pmatrix} 1 & 0 \\ w & -w \end{pmatrix}, \qquad (1.1.11b)$$

and so the centre-generator Z maps to the scalar matrix $w^3 I$. Note that the specialisation $w = -1$ has image $\mathrm{SL}_2(\mathbb{Z})$ – in fact it gives the isomorphism (1.1.10a) – while $w = 1$ has image $\mathcal{S}_3$ and is the representation ρ_2.

There are many natural ways to obtain the representation (1.1.11a). The simplest uses derivatives $\frac{\partial}{\partial x_i}$ acting in the obvious way on the group algebra $\mathbb{C}\mathcal{F}_n$. To any n-braid $\alpha \in \mathcal{B}_n$ define the $n \times n$ matrix whose (i, j)-entry is given by

$$w^{\deg} \frac{\partial}{\partial x_j}(\alpha.x_i),$$

where $\alpha.x_i$ denotes the action of $\mathcal{B}_n$ on $\mathcal{F}_n$ and where $w^{\deg}$ is the obvious representation of $\mathcal{F}_n$, extended linearly to $\mathbb{C}\mathcal{F}_n$. Then this recovers (1.1.11a).

All irreducible representations of $\mathcal{B}_3$ in dimension ≤ 5 are found in [**531**]. Most are non-unitary. For example, any two-dimensional irreducible representation is of the form

$$\sigma_1 \mapsto \begin{pmatrix} \lambda_1 & \lambda_2 \\ 0 & \lambda_2 \end{pmatrix}, \qquad \sigma_2 \mapsto \begin{pmatrix} \lambda_2 & 0 \\ -\lambda_1 & \lambda_1 \end{pmatrix},$$

for some nonzero complex numbers λ_1, λ_2 (compare (1.1.11b)). This representation will be unitary iff both $|\lambda_1| = |\lambda_2| = 1$ and $\lambda_1/\lambda_2 = e^{it}$ for $\pi/3 < t < 5\pi/3$. Not all representations of $\mathcal{B}_3$ are completely reducible, however (Question 1.1.9).

Question 1.1.1. Identify the group $\mathrm{PSL}_2(\mathbb{Z}_2)$ and confirm that it isn't simple.

Question 1.1.2. If G and H are any two groups with $\|G\| = \|H\| < 60$, explain why they will have the same composition factors.

Question 1.1.3. Verify that the dihedral group $\mathcal{D}_n$ (1.1.1) has order $2n$. Find its composition factors. Construct $\mathcal{D}_n$ as a semi-direct product of $\mathbb{Z}_2$ and $\mathbb{Z}_n$.

Question 1.1.4. (a) Using the methods and results given in Section 1.1.3, compute the character table of the symmetric group $\mathcal{S}_4$.
(b) Compute the tensor product coefficients of $\mathcal{S}_4$. That is, if $\rho_1, \rho_2, \ldots$ are the irreducible representations of $\mathcal{S}_4$, compute the multiplicities T_{ij}^k defined by

$$\rho_i \otimes \rho_j \cong \oplus_k T_{ij}^k \rho_k.$$

Question 1.1.5. Prove that $\mathrm{ch}(g^{-1}) = \overline{\mathrm{ch}(g)}$. Can you say anything about the relation of $\mathrm{ch}(g^\ell)$ and $\mathrm{ch}(g)$, for other integers ℓ?

Question 1.1.6. (a) Find a representation over the field $\mathbb{C}$ of the additive group $G = \mathbb{Z}$, which is indecomposable but not irreducible. Hence show that inequivalent (complex) finite-dimensional representations of $\mathbb{Z}$ can have identical characters.
(b) Let p be any prime dividing some $n \in \mathbb{N}$. Find a representation of the cyclic group $G = \mathbb{Z}_n$ over the field $\mathbb{K} = \mathbb{Z}_p$, which is indecomposable but not irreducible.

Question 1.1.7. (a) Let G be any group, and define the *commutator subgroup* $[G, G]$ to be the subgroup generated by the elements $ghg^{-1}h^{-1}$, for all $g, h \in G$. Prove that $[G, G]$ is a normal subgroup of G, and that $G/[G, G]$ is abelian. (In fact, $G/[G, G]$ is isomorphic to the group of all one-dimensional representations of G.)
(b) Show that the free groups $\mathcal{F}_n \cong \mathcal{F}_m$ iff $n = m$, by using Theorem 1.1.2.

Question 1.1.8. (a) Explicitly show how the semi-direct product $\mathbb{Z}_3 \rtimes_\theta \mathbb{Z}_2$ can equal $\mathbb{Z}_6$ or $\mathcal{S}_3$, depending on the choice of θ.
(b) Show that $\mathbb{Z}_2 \rtimes_\theta H \cong \mathbb{Z}_2 \times H$, for any group H and homomorphism θ.
(c) Hence $\mathbb{Z}_4$ can't be written as a semi-direct product of $\mathbb{Z}_2$ with $\mathbb{Z}_2$. Explicitly construct it as an external group extension of $\mathbb{Z}_2$ by $\mathbb{Z}_2$.

Question 1.1.9. Find a two-dimensional representation of the braid group $\mathcal{B}_3$ that is not completely reducible.

1.2 Elementary geometry

Geometry and algebra are opposites. We inherited from our mammalian ancestors our subconscious facility with geometry; to us geometry is intuitive and has implicit meaning, but because of this it's harder to generalise beyond straightforward extensions of our visual experience, and rigour tends to be more elusive than with algebra. The power and clarity of algebra comes from the conceptual simplifications that arise when content is stripped away. But this is equally responsible for algebra's blindness. Although recently physics has inspired some spectacular developments in algebra, traditionally geometry has been the most reliable star algebraists have been guided by. We touch on geometry throughout this book, though for us it adds more colour than essential substance.

1.2.1 Lattices

Many words in mathematics have multiple meanings. For example, there are vector *fields* and number *fields*, and *modular* forms and *modular* representations. 'Lattice' is another of these words: it can mean a 'partially ordered set', but to us a lattice is a discrete maximally periodic set – a toy model for everything that follows.

Consider the real vector space $\mathbb{R}^{m,n}$: its vectors look like $x = (x_+; x_-)$, where x_+ and x_- are m- and n-component vectors, respectively, and inner-products are given by $x \cdot y = x_+ \cdot y_+ - x_- \cdot y_-$. The inner-products $x_\pm \cdot y_\pm$ are given by the usual $\sum_i (x_\pm)_i (y_\pm)_i$. For example, the familiar Euclidean (positive-definite) space is $\mathbb{R}^n = \mathbb{R}^{n,0}$, while the Minkowski space-time of special relativity is $\mathbb{R}^{3,1}$.

Now choose any basis $\beta = \{b^{(1)}, \dots, b^{(m+n)}\}$ in $\mathbb{R}^{m,n}$. If we consider all possible linear combinations $\sum_i a_i b^{(i)}$ over the real numbers $\mathbb{R}$, then we recover $\mathbb{R}^{m,n}$; if instead we consider linear combinations over the integers only, we get a *lattice*.

Definition 1.2.1 *Let V be any n-dimensional inner-product space, and let $\{b^{(1)}, \dots, b^{(n)}\}$ be any basis. Then $L(\beta) := \mathbb{Z}b^{(1)} + \cdots + \mathbb{Z}b^{(n)}$ is called a* lattice.

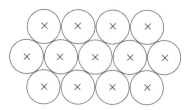

Fig. 1.3 Part of the A_2 disc packing.

A lattice is discrete and closed under sums and integer multiples. For example, $\mathbb{Z}^{m,n}$ is a lattice (take the standard basis in $\mathbb{R}^{m,n}$). A more interesting lattice is the hexagonal lattice (also called A_2), given by the basis $\beta = \{(\frac{\sqrt{2}}{2}, \frac{\sqrt{6}}{2}), (\sqrt{2}, 0)\}$ of $\mathbb{R}^2$ – try to plot several points. If you wanted to slide a bunch of identical coins on a table together as tightly as possible, their centres would form the hexagonal lattice (Figure 1.3). Another important lattice is $II_{1,1} \subset \mathbb{R}^{1,1}$, given by $\beta = \{(\frac{1}{\sqrt{2}}; \frac{1}{\sqrt{2}}), (\frac{1}{\sqrt{2}}; \frac{-1}{\sqrt{2}})\}$; equivalently, it can be thought of as the set of all pairs $(a, b) \in \mathbb{Z}^2$ with inner-product

$$(a, b) \cdot (c, d) = ad + bc. \tag{1.2.1}$$

Different bases may or may not result in a different lattice. For a trivial example, consider $\beta = \{1\}$ and $\beta' = \{-1\}$ in $\mathbb{R} = \mathbb{R}^{1,0}$: they both give the lattice $\mathbb{Z} = \mathbb{Z}^{1,0}$. Two lattices $L(\beta) \subset V$ and $L(\beta') \subset V'$ are called *equivalent* or *isomorphic* if there is an orthogonal transformation $T : V \to V'$ such that the lattices $T(L(\beta))$ and $L(\beta')$ are identical as sets, or equivalently if $b'_i = T \sum_j c_{ij} b_j$, for some integer matrix $C = (c_{ij}) \in GL_n(\mathbb{Z})$ with determinant ± 1.

This notion of lattice equivalence is important in that it emphasises the essential properties of a lattice and washes away the unpleasant basis-dependence of Definition 1.2.1. In particular, the ambient space V in which the lattice lives, and the basis β, are non-essential. The transformation T tells us we can change V, and C is a change-of-basis matrix for which both C and C^{-1} are defined over $\mathbb{Z}$.

For example, $\beta = \{(\frac{1}{\sqrt{2}}, \frac{1}{\sqrt{2}}), (\frac{1}{\sqrt{2}}, \frac{-1}{\sqrt{2}})\}$ in $\mathbb{R}^2$ yields a lattice equivalent to $\mathbb{Z}^2$. The basis $\beta' = \{(-1, 1, 0), (0, -1, 1)\}$ for the plane $a + b + c = 0$ in $\mathbb{R}^3$ yields the lattice $L(\beta') = \{(a, b, c) \in \mathbb{Z}^3 \mid a + b + c = 0\}$, equivalent to the hexagonal lattice A_2.

The *dimension* of the lattice is the dimension $\dim(V)$ of the ambient vector space. The lattice is called *positive-definite* if it lies in some $\mathbb{R}^m$ (i.e. $n = 0$), and *integral* if all inner-products $x \cdot y$ are integers, for $x, y \in L$. A lattice L is called *even* if it is integral and in addition all norm-squareds $x \cdot x$ are *even* integers. For example, $\mathbb{Z}^{m,n}$ is integral but not even, while A_2 and $II_{1,1}$ are even. The *dual* L^* of a lattice L consists of all vectors $x \in V$ such that $x \cdot L \subset \mathbb{Z}$. A natural basis for the dual $L(\beta)^*$ is the *dual basis* β^*, consisting of the vectors $c_j \in V$ obeying $b_i \cdot c_j = \delta_{ij}$ for all i, j. A lattice is integral iff $L \subseteq L^*$. A lattice is called *self-dual* if $L = L^*$. The lattices $\mathbb{Z}^{m,n}$ and $II_{1,1}$ are self-dual, but A_2 is not. We are most interested in even positive-definite lattices.

To any n-dimensional lattice $L(\beta)$, define an $n \times n$ matrix A (called a *Gram matrix*) by $A_{ij} = b_i \cdot b_j$. Two lattices with identical Gram matrices are necessarily equivalent, but the converse is not true. Note that the Gram matrix of $L(\beta^*)$ is the inverse of the Gram

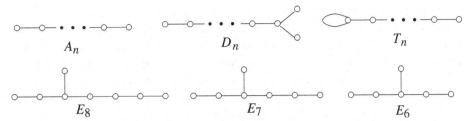

Fig. 1.4 The graphs with largest eigenvalue < 2.

matrix for $L(\beta)$. The *determinant* $|L|$ of a lattice is the determinant of the Gram matrix; geometrically, it is the volume-squared of the fundamental parallelepiped of L defined by the basis. This will always be positive if L is positive-definite. The determinant of a lattice is independent of the specific basis β chosen; equivalent lattices have equal determinant, though the converse isn't true. An integral lattice L is self-dual iff $|L| = \pm 1$. If $L' \subseteq L$ are of equal dimension, then the quotient L/L' is a finite abelian group of order

$$\|L/L'\| = \sqrt{|L'|/|L|} \in \mathbb{N}. \tag{1.2.2}$$

Given two lattices L, L', their (orthogonal) *direct sum* $L \oplus L'$ is defined to consist of all pairs (x, x'), for $x \in L$, $x' \in L'$, with inner-product defined by $(x, x') \cdot (y, y') = x \cdot y + x' \cdot y'$. The dimension of $L \oplus L'$ is the sum of the dimensions of L and L'. The direct sum $L \oplus L'$ will be integral (respectively self-dual) iff both L and L' are integral (respectively self-dual).

An important class of lattices are the so-called *root lattices* A_n, D_n, E_6, E_7, E_8 associated with simple Lie algebras (Section 1.5.2). They can be defined from the graph ('Coxeter–Dynkin diagram') in Figure 1.4 (but ignore the 'tadpole' T_n for now): label the nodes of such a graph from 1 to n, put $A_{ii} = 2$ and put $A_{ij} = -1$ if nodes i and j are connected by an edge. Then this matrix A (the *Cartan matrix* of Definition 1.4.5) is the Gram matrix of a positive-definite integral lattice. Realisations of some of these are given shortly; bases can be found in table VII of [**214**], or planches I–VII of [**84**]. Of these, E_8 is the most interesting as it is the even self-dual positive-definite lattice of smallest dimension.

The following theorem characterises norm-squared 1,2 vectors.

Theorem 1.2.2   *Let L be an n-dimensional positive-definite integral lattice.*
(a) *Then L is equivalent to the direct sum $\mathbb{Z}^m \oplus L'$, where L has precisely $2m$ unit vectors and L' has none.*
(b) *If L is spanned by its norm-squared 2 vectors, then L is a direct sum of root lattices.*

Theorem 1.2.2(b) gives the point-of-contact of Lie theory and lattices. The densest packing of circles in the plane (Figure 1.3) is A_2, in the sense that the centres of these circles are the points of A_2. The obvious pyramidal way to pack oranges is also the densest, and likewise gives the A_3 root lattice. The densest known sphere packings in dimensions 4, 5, 6, 7, 8 are the root lattices D_4, D_5, E_6, E_7, E_8, respectively.

The Leech lattice Λ is one of the most distinguished lattices, and like E_8 is directly related to Moonshine. It can be constructed using 'laminated lattices'

([**113**], chapter 6). Start with the zero-dimensional lattice $L_0 = \{0\}$, consisting of just one point. Use it to construct a one-dimensional lattice L_1, with minimal (nonzero) norm 2, built out of infinitely many copies of L_0 laid side by side. The result of course is simply the even integers $2\mathbb{Z}$. Now construct a two-dimensional lattice L_2, of minimal norm 2, built out of infinitely many copies of L_1 stacked next to each other. There are lots of ways to do this, but choose the densest lattice possible. The result is the hexagonal lattice A_2 rescaled by a factor of $\sqrt{2}$. Continue in this way: L_3, L_4, L_5, L_6, L_7 and L_8 are the root lattices A_3, D_4, D_5, E_6, E_7 and E_8, respectively, all rescaled by $\sqrt{2}$.

The 24th repetition of this construction yields uniquely the Leech lattice $\Lambda = L_{24}$. It is the unique 24-dimensional even self-dual lattice with no norm-squared 2-vectors, and provides among other things the densest known packing of 23-dimensional spheres S^{23} in $\mathbb{R}^{24}$. It is studied throughout [**113**]. After dimension 24, chaos reigns in lamination (23 different 25-dimensional lattices have an equal right to be called L_{25}, and over 75 000 are expected for L_{26}). So lamination provides us with a sort of no-input construction of the Leech lattice. Like the Mandelbrot set, the Leech lattice is a subtle structure with an elegant construction – a good example of the mathematical meaning of 'natural'.

Question 1.2.1 asks you to come up with a definition for the automorphism group of a lattice. An automorphism is a symmetry, mapping the lattice to itself, preserving all essential lattice properties. It is how group theory impinges on lattice theory.

Most (positive-definite) lattices have trivial automorphism groups, consisting only of the identity and the reflection $x \mapsto -x$ through the origin. But the more interesting lattices tend to have quite large groups. The reflection through the hyperplane orthogonal to a norm-squared 2-vector in an integral lattice defines an automorphism; together, these automorphisms form what Lie theory calls a Weyl group.

Typically the Weyl group has small index in the full automorphism group, though a famous counterexample is the Leech lattice (which, as we know, has trivial Weyl group). Its automorphism group is denoted Co_0 and has approximately 8×10^{18} elements. The automorphism $x \mapsto -x$ lies in its centre; if we quotient by this 2-element centre we get a sporadic simple group Co_1. Define Co_2 and Co_3 to be the subgroups of Co_0 consisting of all $g \in Co_0$ fixing some norm-squared 4-vector and some norm-squared 6-vector, respectively. These three groups Co_1, Co_2, Co_3 are all simple. In fact, a total of 12 sporadic finite simple groups appear as subquotients in Co_0, and can best be studied geometrically in this context. Gorenstein [**256**] wrote:

> … if Conway had studied the Leech lattice some 5 years earlier, he would have discovered a total of 7 new simple groups! Unfortunately he had to settle for 3. However, as consolation, his paper on .0[=Co_0] will stand as one of the most elegant achievements of mathematics.

1.2.2 Manifolds

On what structures do lattices act naturally? An obvious place is on their ambient space ($\mathbb{R}^n$, say). They act by addition. Quotient out by this action. Topologically, we have

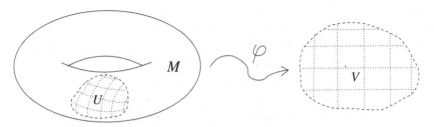

Fig. 1.5 A coordinate patch.

created a *manifold* (to be defined shortly); to each point on this manifold corresponds an orbit in $\mathbb{R}^n$ of our lattice action.

Consider first the simplest case. The number $n \in \mathbb{Z}$ acts on $\mathbb{R}$ by sending $x \in \mathbb{R}$ to $x + n$. The orbits are the equivalence classes of the reals mod 1. We can take as representatives of these equivalence classes, that is as points of $\mathbb{R}/\mathbb{Z}$, the half-open interval $[0, 1)$. This orbit space inherits a *topology* (i.e. a qualitative notion of points being close; for basic point-set topology see e.g. [**481**], [**104**]), from that of $\mathbb{R}$, and this is almost captured by the interval $[0, 1)$. The only problem is that the orbit of $0.999 \equiv -0.0001$ is pretty close to that of 0, even though they are at opposite ends of the interval. What we should do is identify the two ends, i.e. glue together 0 and 1. The result is a circle.

We say that $\mathbb{R}/\mathbb{Z}$ is topologically the circle S^1. The same argument applies to $\mathbb{R}^n/L$, and we get the n-torus $S^1 \times \cdots \times S^1$ (see Question 1.2.2).

The central structure in geometry is a manifold – geometries where calculus is possible. Locally, a manifold looks like a piece of $\mathbb{R}^n$ (or $\mathbb{C}^n$), but these pieces can be bent and stitched together to create more interesting shapes. For instance, the n-torus is an n-dimensional manifold. The definition of manifold, due to Poincaré at the turn of the century, is a mathematical gem; it explains how flat patches can be sewn together to form smooth and globally interesting shapes.

Definition 1.2.3 *A C^∞ manifold M is a topological space with a choice of open sets $U_\alpha \subset M$, $V_\alpha \subset \mathbb{R}^n$ and homeomorphisms $\varphi_\alpha : U_\alpha \to V_\alpha$, as in Figure 1.5, such that the U_α cover M (i.e. $M = \cup_\alpha U_\alpha$) and whenever $U_\alpha \cap U_\beta$, the map $\varphi_\alpha \circ \varphi_\beta^{-1}$ is a C^∞ map from some open subset (namely $\varphi_\beta(U_\alpha \cap U_\beta)$) of V_β to some open subset (namely $\varphi_\alpha(U_\alpha \cap U_\beta)$) of V_α.*

A homeomorphism means an invertible continuous map whose inverse is also continuous. By a C^∞ map f between open subsets of $\mathbb{R}^n$, we mean that

$$f(x) = (f_1(x_1, \ldots, x_n), \ldots, f_n(x_1, \ldots, x_n))$$

is continuous, and all partial derivatives $\frac{\partial^k}{\partial x_{i_1} \cdots \partial x_{i_k}} f_j$ exist and are also continuous.

This is the definition of a *real* manifold; a *complex* manifold is similar. An n-dimensional complex manifold is a $2n$-dimensional real one. A one-dimensional manifold is called a *curve*, and a two-dimensional one a *surface*. 'Smooth' is often used for C^∞.

Using φ_α, each 'patch' $U_\alpha \subset M$ inherits the structure of $V_\alpha \subset \mathbb{R}^n$. For instance, we can coordinatise V_α and do calculus on it, and hence we get coordinates for, and can do calculus on, U_α. The overlap condition for $\varphi_\alpha \circ \varphi_\beta^{-1}$ guarantees compatibility. For example, the familiar latitude/longitude coordinate system comes from covering the Earth with two coordinate patches V_i – one centred on the North pole and the other on the South, and both stretching to the Equator – with polar coordinates chosen on each V_i.

More (or less) structure can be placed on the manifold, by constraining the overlap functions $\varphi_\alpha \circ \varphi_\beta^{-1}$ more or less. For example, a 'topological manifold' drops the C^∞ constraint; the result is that we can no longer do calculus on the manifold, but we can still speak of continuous functions, etc. A *conformal manifold* requires that the overlap functions preserve angles in $\mathbb{R}^n$ – the angle between intersecting curves in $\mathbb{R}^n$ is defined to be the angle between the tangents to the curves at the point of intersection. Conformal manifolds inherit the notion of angle from $\mathbb{R}^n$. Stronger is the notion of *Riemannian manifold*, which also enables us to speak of length.

It is now easy to compare structures on different manifolds. For instance, given two manifolds M, M', a function $f : M \to M'$ is 'C^∞' if each composition $\varphi'_\beta \circ f \circ \varphi_\alpha^{-1}$ is a C^∞ map from some open subset of V_α to V'_β; M and M' are C^∞-*diffeomorphic* if there is an invertible C^∞-function $f : M \to M'$ whose inverse is defined and is also C^∞.

Note that our definition doesn't assume the manifold M is embedded in some ambient space $\mathbb{R}^m$. Although it is true (Whitney) that any n-dimensional real manifold M can be embedded in Euclidean space $\mathbb{R}^{2n}$, this embedding may not be natural. For example, we are told that we live in a 'curved' four-dimensional manifold called space-time, but its embedding in $\mathbb{R}^8$ presumably has no physical significance.

Much effort in differential geometry has been devoted to questions such as: Given some topological manifold M, how many inequivalent differential structures (compatible with the topological structure) can be placed on M? It turns out that for any topological manifold of dimension ≤ 3, this differential structure exists and is unique. Moreover, $\mathbb{R}^n$ has a unique differential structure as well in dimensions ≥ 5. Remarkably, in four (and only four) dimensions it has uncountably many different differential structures (see [**195**])! Could this have anything to do with the appearance of macroscopic space-time being $\mathbb{R}^4$? Half a century before that discovery, the physicist Dirac prophesied [**139**]:

> ... as time goes on it becomes increasingly evident that the rules which the mathematician finds interesting are the same as those which Nature has chosen ... only four-dimensional space is of importance in physics, while spaces with other dimensions are of about equal interest in mathematics. It may well be, however, that this discrepancy is due to the incompleteness of present-day knowledge, and that future developments will show four-dimensional space to be of far greater mathematical interest than all the others.

Given any open set U in a manifold M, write $C^\infty(U)$ for the space of C^∞-functions $f : U \to \mathbb{R}$. When $U \subseteq U_\alpha$, we can use local coordinates and write $f(x^1, \ldots, x^n)$ (local coordinates are often written with superscripts). A fundamental lesson of geometry (perhaps learned from physics) is that one studies the manifold M through the (local) smooth

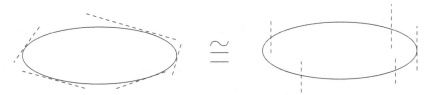

Fig. 1.6 The tangent bundle of S^1.

functions $f \in C^\infty(U)$ that live on it. This approach to geometry has been axiomatised into the notion of *sheaf* (see e.g. [537]), to which we return in Section 5.4.2.

For example, identifying S^1 with $\mathbb{R}/\mathbb{Z}$, the space $C^\infty(S^1)$ consists of the smooth period-1 functions $f : \mathbb{R} \to \mathbb{R}$, i.e. $f(\theta + 1) = f(\theta)$. Or we can identify S^1 with the locus $x^2 + y^2 = 1$, in which case $C^\infty(S^1)$ can be identified with the algebra $C^\infty(\mathbb{R}^2)$ of smooth functions in two variables, quotiented by the subalgebra (in fact ideal) consisting of all smooth functions $g(x, y)$ vanishing on all points satisfying $x^2 + y^2 = 1$; when $f(x, y), g(x, y)$ are polynomials, then they are identical functions in $C^\infty(S^1)$ iff their difference $f(x, y) - g(x, y)$ is a polynomial multiple of $x^2 + y^2 - 1$.

Fix a point $p \in M$ and an open set U containing p. In Section 1.4.2 we need the notion of tangent vectors to a manifold M. An intuitive approach starts from the set $S(U, p)$ of curves passing through p, i.e. $\sigma : (-\epsilon, \epsilon) \to U_\alpha$ is smooth and $\sigma(0) = p$. Call curves $\sigma_1, \sigma_2 \in S(U, p)$ equivalent if they touch each other at p, that is

$$\sigma_1 \approx_p \sigma_2 \quad \text{iff} \quad \frac{\mathrm{d}}{\mathrm{d}t} f(\sigma_1(t))|_{t=0} = \frac{\mathrm{d}}{\mathrm{d}t} f(\sigma_2(t))|_{t=0}, \qquad \forall f \in C^\infty(U, p). \quad (1.2.3a)$$

This defines an equivalence relation; the equivalence class $\langle \sigma \rangle_p$ consisting of all curves equivalent to σ is an infinitesimal curve at p. Equivalently, define a tangent vector to be a linear map $\xi : C^\infty(M) \to \mathbb{R}$ that satisfies the Leibniz rule

$$\xi(fg) = \xi(f) g(p) + f(p) \xi(g). \quad (1.2.3b)$$

In local coordinates $\xi = \sum_{i=1}^{n} \alpha_i \frac{\partial}{\partial x^i}|_{x=p}$, where the α_i are arbitrary real numbers. The bijection between these two definitions associates with any infinitesimal curve $v = \langle \sigma \rangle_p$ the tangent vector called the *directional derivative* $D_v : C^\infty(M) \to \mathbb{R}$, given by

$$D_v(f) = \frac{\mathrm{d}}{\mathrm{d}t} f(\sigma(t))|_{t=0}. \quad (1.2.3c)$$

The *tangent space* $T_p(M)$ *at* p is the set of all tangent vectors. Equation (1.2.3b) shows that $T_p(M)$ has a natural vector space structure; its dimension equals that of M. These tangent spaces can be glued together into a $2n$-dimensional manifold called the *tangent bundle* TM. Figure 1.6 shows why TS^1 is the cylinder $S^1 \times \mathbb{R}$. However, this is exceptional: although *locally* the tangent bundle TM of any manifold is trivial – that is, each TU_α is diffeomorphic to the direct product $U_\alpha \times \mathbb{R}^n$ – *globally* most tangent bundles TM are different from $M \times \mathbb{R}^n$.

A *vector field* is an assignment of a tangent vector to each point on the manifold, a smooth map $X : M \to TM$ such that $X(p) \in T_p(M)$. Equivalently, we can regard it

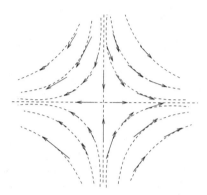

Fig. 1.7 The flow of a vector field.

as a *derivation* $X : C^\infty(M) \to C^\infty(M)$, i.e. a first-order differential operator acting on functions $f : M \to \mathbb{R}$ and obeying $X(fg) = X(f)\,g + f\,X(g)$. For example, the vector fields on the circle consist of the operators $g(\theta)\frac{\mathrm{d}}{\mathrm{d}\theta}$ for any smooth period-1 function $g(\theta)$.

Let Vect(M) denote the set of all vector fields on a manifold M. Of course this is an infinite-dimensional vector space, but we see in Section 1.4.1 that it has a much richer algebraic structure: it is a Lie algebra. Vect(S^1) is central to Moonshine, and in Section 3.1.2 we start exploring its properties.

A vector field X on M can be interpreted as being the instantaneous velocity of a fluid confined to M. We can 'integrate' this, by solving a first-order ordinary differential equation, thus covering M with a family of non-intersecting curves. Each curve describes the motion, or *flow*, of a small particle dropped into the fluid at the given point $p \in M$. The tangent vector to the curve at the given point p equals $X(p)$ – see Figure 1.7. Equivalently, corresponding to a vector field X is a continuous family $\varphi_t : M \to M$ of diffeomorphisms of M, one for each 'time' t, obeying $\varphi_t \circ \varphi_s = \varphi_{t+s}$, where $\varphi_t(p)$ is defined to be the position on M where the point p flows to after t seconds.

So it is natural to ask, what can we do with a diffeomorphism α of M? Clearly, α gives rise to an automorphism of the algebra $C^\infty(M)$, defined by $f \mapsto f^\alpha = f \circ \alpha$. Using this, we get an automorphism of Vect(M), $X \mapsto X^\alpha$, given by $X^\alpha(f) = (X(f^\alpha))^{\alpha^{-1}}$, or more explicitly, $X^\alpha(f)(p) = X(f \circ \alpha)(\alpha^{-1}(p))$. We return to this in Section 1.4.2.

One thing you can do with a *continuous* family of diffeomorphisms is construct a derivative for the algebras $C^\infty(M)$, Vect(M), etc. Defining a derivative of, say, a vector field X requires that we compare tangent vectors $X(p)$, $X(p')$ at neighbouring points on the manifold. This can't be done directly, since $X(p) \in T_p M$ and $X(p') \in T_{p'} M$ lie in different spaces. Given a vector field X, and corresponding flow φ_t, define the Lie derivative $\mathcal{L}_X(Y) \in$ Vect(M) of any vector field $Y \in$ Vect(M) by

$$\mathcal{L}_X(Y)(p) = \lim_{t \to 0} \frac{Y^{\varphi_t}(p) - Y(p)}{t} \in T_p M.$$

The Lie derivative $\mathcal{L}_X(f)$ of a function $f \in C^\infty(M)$ is defined similarly, and equals $X(f)$.

Dual to the tangent vectors are the *differential 1-forms*. Just as the tangent spaces $T_p(M)$ together form the $2n$-dimensional tangent bundle TM, so their duals $T_p^*(M)$ form the $2n$-dimensional *cotangent bundle* T^*M. At least for finite-dimensional manifolds, the vector spaces $T_p^*(M)$ and $T_p(M)$, as well as the manifolds T^*M and TM, are homeomorphic, but without additional structure on M this homeomorphism is not canonical (it is basis- or coordinate-dependent). If $x = (x^1, \ldots, x^n) \mapsto M$ is a coordinate chart for manifold M, then $\partial_i := \frac{\partial}{\partial x^i}|_p$ is a basis for the tangent space $T_p M$, and its dual basis is written $dx^i \in T_p^*(M)$: by definition they obey $dx^i(\partial_j) = \delta_{ij}$.

Changing local coordinates from x to $y = y(x)$, the chain rule tells us

$$\frac{\partial}{\partial y^i} = \sum_{j=1}^n \frac{\partial x^j}{\partial y^i} \frac{\partial}{\partial x^j}, \tag{1.2.4a}$$

and hence the 1-form basis changes by the inverse formula:

$$dy^i = \sum_{j=1}^n \frac{\partial y^i}{\partial x^j} dx^j. \tag{1.2.4b}$$

The main purpose of differential forms is integration (hence their notation). If we regard the integrand of a line-integral as a 1-form field (i.e. a choice of 1-form for each point $p \in M$), we make manifest the choice of measure. Rather than saying the ambiguous 'integrate the constant function "$f(p) = 1$" along the manifold S^1', we say the unambiguous 'integrate the 1-form "$\omega_p = d\theta$" along the manifold S^1'. Likewise, the integrands of double-, triple-, etc. integrals are 2-forms, 3-forms, etc., dual to tensor products of tangent spaces. We can evaluate these integrals by introducing coordinate patches and thus reducing them to usual $\mathbb{R}^n$ integrals over components of the differential form. The spirit of manifolds is to have a coordinate-free formalism; changing local coordinates (e.g. when moving from one coordinate patch to an overlapping one) changes those components as in (1.2.4b) in such a way that the value of the integral won't change.

A standard example of a 1-form field is the *gradient* df of a function $f \in C^\infty(M)$, defined at each point $p \in M$ by the rule: given any tangent vector $D_v \in T_p(M)$, define the number $(df)(D_v)$ to be the value of the directional derivative $D_v(f)(p)$ at p.

A familiar example of a *2-form* $g_p \in T_p^*(M) \otimes T_p^*(M)$ is the *metric tensor* on $T_p(M)$. Given two vectors $u, v \in T_p$, the number $g_p(u, v)$ is to be thought of as their inner-product. A *Riemannian manifold* is a manifold M together with a 2-form field g, which is symmetric and nondegenerate (usually positive-definite).[5] Given a local coordinate about $p \in M$, a basis for the tangent space $T_p M$ is $\frac{\partial}{\partial x^i}$ and we can describe the metric tensor g_p using an $n \times n$ matrix whose ij-entry is $g_{ij}(p) := g_p(\frac{\partial}{\partial x^i}, \frac{\partial}{\partial x^j})$, or in infinitesimal language as $ds^2 = \sum_{i,j=1}^n g_{ij} dx^i dx^j$, a form more familiar to most physicists.

---

[5] Whitney's aforementioned embedding of M into Euclidean space implies that any manifold can be given a Riemannian structure, since a submanifold of a Riemannian manifold naturally inherits the Riemannian structure. The *Beautiful Mind* of John Nash proved that any Riemannian structure on a given n-dimensional manifold M can likewise be inherited from its embedding into some sufficiently large-dimensional Euclidean space.

Much structure comes with this metric tensor field g. Most important, of course, we can define lengths of curves and the angles with which they intersect. In particular, the arc-length of the curve $\gamma : [0, 1] \to M$ is the integral

$$\int_0^1 \sqrt{g_{\gamma(t)}\left(\frac{d\gamma}{dt}, \frac{d\gamma}{dt}\right)}\, dt,$$

a quantity independent of the specific parametrisation $t \mapsto \gamma(t)$ chosen (verify this).

Also, we can use the metric to identify each T_p^* with T_p, just as the standard inner-product in $\mathbb{R}^n$ permits us to identify a column vector $u \in \mathbb{R}^n$ with its transpose $u^t \in \mathbb{R}^{n*}$. Moreover, given any curve $\sigma : [0, 1] \to M$ connecting $\sigma(0) = p$ to $\sigma(1) = q$, we can identify the tangent spaces $T_p(M)$ and $T_q(M)$ by *parallel-transport*. Using this, we can define a derivative (the so-called 'covariant derivative') that respects the metric, and a notion of geodesic (a curve that parallel-transports its own tangent vector, and which plays the role of 'straight line' here). In short, on a Riemannian manifold geometry in its fullest sense is possible. See, for example, [104] for more details.

Many manifolds locally look like a Cartesian product $A \times B$. A *fibre bundle* $p : E \to B$ locally (i.e. on small open sets U of E) looks like $F \times V$, where $F \cong p^{-1}(b)$ (for any $b \in B$) is called the *fibre*, and V is an open set in the *base* B. For example, the (open) cylinder and Möbius stip are both fibre bundles with base S^1 and fibre $(0, 1) \subset \mathbb{R}$. A *section* $s : B \to E$ obeys $p \circ s = id.$, that is for each small open set V of B it is a function $V \to F$. A *vector bundle* is a fibre bundle with fibre a vector space $F = V$, for example the tangent bundle TM is a vector bundle with base M and fibre $\cong T_p M$. We write $\Gamma(E)$ for the space of sections of a vector bundle E. A *line bundle* is a vector bundle with one-dimensional fibre (so the sections of a line bundle locally look like complex- or real-valued functions on the base). A *connection* on a vector bundle $E \to B$ is a way to differentiate sections (the *covariant derivative*). An example is a Riemannian structure on the tangent bundle $E = TB$. See, for example, [104] for details and examples.

Felix Klein's *Erlangen Programm* (so called because he announced it there) is a strategy relating groups and geometry. Geometry, it says, consists of a manifold (the space of points) and a group of automorphisms (transformations) of the manifold preserving the relevant geometric structures (e.g. length, angle, lines, etc.). Conversely, given a manifold and a group of automorphisms, we should determine the invariants relative to the group. Several different geometries are possible on the same manifold, distinguished by their preferred transformations.

For example, Euclidean geometry in its strongest sense (i.e. with lengths, angles, lines, etc.) has the group of symmetries generated by rotations, reflections and translations – that is any transformation of the form $x \mapsto xA + a$, where $x, a \in \mathbb{R}^n$ (regarded as row vectors, say) and A is an orthogonal $n \times n$ matrix. If our context is scale-independent (e.g. when studying congruent triangles), we can allow A to obey $AA^t = \lambda I$ for any $\lambda \in \mathbb{R}$.

More interesting is *projective geometry*. Here, angles and lengths are no longer invariants, but lines are. Projective geometry arose from the theory of perspective in art. The transformations of projective n-geometry come from projections $\mathbb{R}^{n+1} \to \mathbb{R}^n$.

More precisely, consider real projective n-space $\mathbb{P}^n(\mathbb{R})$. We coordinatise it using *homogeneous coordinates*: $\mathbb{P}^n(\mathbb{R}) = \mathbb{R}^{n+1\prime}/\sim$ consists of $(n+1)$-tuples of real numbers, where we identify points with their multiples. The origin $(0,0,0,\ldots,0)$ in $(n+1)$-space is excluded from projective space (hence the prime), as it belongs to all such lines. A projective 'point' consists of points on the same line through the origin; a projective 'line' consists of planes through the origin; etc. By convention, any equation in homogeneous coordinates is required to be homogeneous (so that a point satisfies an equation iff its whole line does). Complex projective space $\mathbb{P}^n(\mathbb{C})$ is defined similarly.

To see what projective geometry is like, consider first the projective line $\mathbb{P}^1(\mathbb{R})$. Take any point in $(x,y) \in \mathbb{P}^1(\mathbb{R})$. If $y \neq 0$ we may divide by it, and we get points of the form $(x', 1)$. These are in one-to-one correspondence with the points in the real line. If, on the other hand, $y = 0$, then we know $x \neq 0$ and so we should divide by x: what we get is the point $(1, 0)$, which we can think of as the infinite point $(\frac{1}{0}, 1)$. Thus the real projective line $\mathbb{P}^1(\mathbb{R})$ consists of the real line, together with a point 'at infinity'. Similarly, the complex projective line consists of the complex plane $\mathbb{C}$ together with a point at infinity; topologically, this is a sphere named after Riemann.

More generally, $\mathbb{P}^n(\mathbb{R})$ consists of the real space $\mathbb{R}^n$, together with a copy of $\mathbb{P}^{n-1}(\mathbb{R})$ as the hyperplane of infinite points. These points at infinity are where parallel lines meet. Intuitively, projective geometry allows us to put 'finite' and 'infinite' points on an equal footing; we can see explicitly how, for example, curves look at infinity.

For example, the 'parallel' lines $x = 0$ and $x = 1$ in $\mathbb{P}^2(\mathbb{R})$ correspond to the homogeneous equations $x = 0$ and $x = z$, and so to the points with homogeneous coordinates $(0, y, z)$ and (x, y, x). They intersect at the 'infinite' point $(0, y, 0) \sim (0, 1, 0)$. The parabola $y = x^2$ has only one infinite point (namely $(0,1,0)$), the hyperbola $xy = 1$ has two infinite points $((1,0,0)$ and $(0,1,0))$, while the circle $x^2 + y^2 = 1$ doesn't have any. Intuitively, the parabola is an ellipse tangent to the line (really, circle) at infinity, while the hyperbola is an ellipse intersecting it transversely.

Klein's group of transformations here is the projective linear group $\text{PGL}_{n+1}(\mathbb{R})$, that is all invertible $(n+1) \times (n+1)$ matrices A where we identify A with λA for any nonzero number λ. It acts on the homogeneous coordinates in the usual way: $x \mapsto xA$. This group mixes thoroughly the so-called infinite points with the finite ones, and emphasises that infinite points in projective geometry are completely on a par with finite ones. For example, the transformation $A = \begin{pmatrix} 0 & 0 & 1 \\ 0 & 1 & 0 \\ 1 & 0 & 0 \end{pmatrix}$ maps the parabola $y = x^2$ to the hyperbola $xy = 1$, indicating that these are projectively identical curves.

Projective geometry is central to modern geometry. The projective plane can be axiomatised, for example one axiom says that any two lines intersect in exactly one point. A remarkable property of projective geometry is that any theorem remains a theorem if the words 'line' and 'point' are interchanged.

In summary, there are many different geometries. Which geometry to use (e.g. Euclidean, projective, conformal) in a given context depends on the largest possible group of transformations that respect the basic quantities.

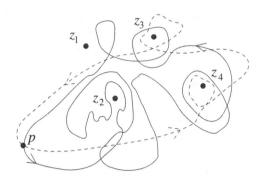

Fig. 1.8 Two homotopic loops in $\pi_1 \cong \mathcal{F}_4$.

1.2.3 Loops

The last subsection used curves to probe the infinitesimal neighbourhood of any point $p \in M$. We can also use curves to probe global features of manifolds.

Let M be any manifold, and put $I = [0, 1]$. A *loop* at $p \in M$ is any continuous curve $\sigma : I \to M$ with $\sigma(0) = \sigma(1) = p$. So σ starts and ends at the point p. Let $\Omega(M, p)$ be the set of all such loops. Loops $\sigma_0, \sigma_1 \in \Omega(M, p)$ are *homotopic* if σ_0 can be continuously deformed into σ_1, that is if there is a continuous map $F : I \times I \to M$ with $\sigma_i(\star) := F(\star, i) \in \Omega(M, p)$, for $i = 0, 1$. This defines an equivalence relation on $\Omega(M, p)$. For instance all loops in $M = \mathbb{R}^n$ are homotopic, while the homotopy equivalence classes for the circle $M = S^1$ are parametrised by their winding number $n \in \mathbb{Z}$, that is by the contour integral $\frac{1}{2\pi i} \int_{\sigma(I)} \frac{dz}{z}$.

Let $\pi_1(M, p)$ denote the set of all homotopy equivalence classes for $\Omega(M, p)$. It has a natural group structure: $\sigma \sigma'$ is the curve that first goes from p to p following σ, and then from p to p following σ'. More precisely,

$$(\sigma \sigma')(t) = \begin{cases} \sigma(2t) & \text{if } 0 \leq t \leq \frac{1}{2} \\ \sigma'(2t - 1) & \text{if } \frac{1}{2} \leq t \leq 1 \end{cases}. \tag{1.2.5}$$

For instance, the inverse σ^{-1} is given by the curve traversed in the opposite direction: $t \mapsto \sigma(1 - t)$. The identity is the constant curve $\sigma(t) = p$. With this operation $\pi_1(M, p)$ is called the *fundamental group* of M (the subscript '1' reminds us that a loop is a map from S^1; likewise π_k considers maps from the k-sphere S^k to M). As long as any two points in M can be connected with a path, then all $\pi_1(M, p)$ will be isomorphic and we can drop the dependence on 'p'. When $\pi_1(M) = \{e\}$, we say M is *simply connected*.

For example, $\pi_1(\mathbb{R}^n) \cong 1$ and $\pi_1(S^1) \cong \mathbb{Z}$. The complex plane $\mathbb{C}$ with n points removed has fundamental group $\pi_1(\mathbb{C}\backslash\{z_1, \ldots, z_n\}) \cong \mathcal{F}_n$, the free group – Figure 1.8 gives two paths homotopic to $x_4 x_3^{-1} \in F_4$. The torus $S^1 \times S^1$ has $\pi_1 \cong \mathbb{Z} \oplus \mathbb{Z}$.

The braid group (1.1.9), as with any group, also has a realisation as a fundamental group. Let $\mathfrak{C}_n$ be $\mathbb{C}^n$ with all diagonals removed:

$$\mathfrak{C}_n = \{(z_1, \ldots, z_n) \in \mathbb{C}^n \mid z_i \neq z_j \text{ whenever } i \neq j\}. \tag{1.2.6}$$

Then it is easy to see that the pure braid group $\mathcal{P}_n$ is isomorphic to $\pi_1(\mathfrak{C}_n)$ – indeed, given any braid $\alpha \in \mathcal{B}_n$, the value of the ith coordinate $\sigma(t)_i$ of the corresponding loop

Fig. 1.9 Some trivial knots.

Fig. 1.10 The trefoil.

Fig. 1.11 A wild knot.

$\sigma \in \pi_1(\mathfrak{C}_n)$ will be the position of the ith strand when we take a slice at t through our braid ($t = 0$ is the top of the braid, $t = 1$ the bottom). Now, the symmetric group $\mathcal{S}_n$ acts freely (i.e. without fixed points) on $\mathfrak{C}_n$ by permuting the coordinates: $\pi.z = (z_{\pi 1}, \ldots, z_{\pi n})$. The space $\mathfrak{C}_n/\mathcal{S}_n$ of orbits under this action has fundamental group $\pi_1(\mathfrak{C}_n/\mathcal{S}_n) \cong \mathcal{B}_n$.

Note that if $f : M' \to M$ is a homeomorphism, then it induces a group homomorphism $f_* : \pi_1(M') \to \pi_1(M)$. We return to this in Section 1.7.2.

By a *link* we mean a diffeomorphic image of $S_1 \cup \cdots \cup S^1$ into $\mathbb{R}^3$. A *knot* is a link with one strand – see Figures 1.9 and 1.10. Since S^1 comes with an orientation, so does each strand of a link. The reason for requiring the embedding $f : S^1 \cup \cdots \cup S^1 \to \mathbb{R}^3$ to be *differentiable* is that we want to avoid 'wild knots' (see Figure 1.11); almost every *homeo*morphic image of $S^1 \cup \cdots \cup S^1$ will be wild at almost every point.

Two links are equivalent, i.e. *ambient isotopic*, if continuously deforming one link yields the other. The word 'ambient' is used because the isotopy is applied to the ambient space $\mathbb{R}^3$. This is the intuitive notion of equivalent knots in a string, except that we glue the two ends of the string together (we can trivially untie any knotted open string by slipping the knot off an end). By a trivial knot or the *unknot* we mean any knot homotopic to (say) the unit circle in the xy-plane in $\mathbb{R}^3$.

We choose $\mathbb{R}^3$ for the ambient space because any link in $\mathbb{R}^n$, for $n \geq 4$, is trivial, and the Jordan Curve Theorem tells us that there are only two different 'knots' in $\mathbb{R}^2$

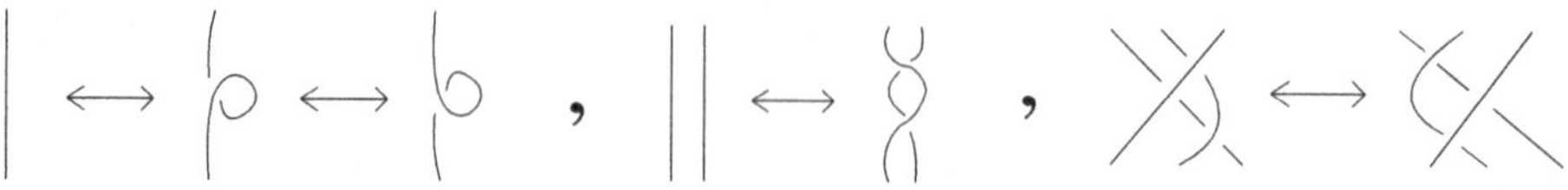

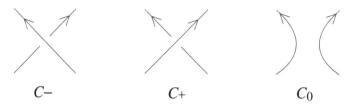

Fig. 1.12 The Reidemeister moves I, II, III, respectively.

$C-$ $C+$ C_0

Fig. 1.13 The possible (non)crossings.

(distinguished by their orientation). More generally, knotted k-spheres S^k in $\mathbb{R}^n$ are nontrivial only when $n = k + 2$ [**478**].

It isn't difficult to show [**478**] that two links are ambient isotopic iff their diagrams can be related by making a finite sequence of moves of the form given in Figure 1.12. The Reidemeister moves are useless at deciding directly whether two knots are equivalent, or even whether a given knot is trivial. Indeed, this seems difficult no matter which method is used, although a finite algorithm (by Häken and Hemion [**283**]) apparently exists. A very fruitful approach has been to assign to a link a quantity (called a *link invariant*), usually a polynomial, in such a way that ambient isotopic links get the same quantity. One of these is the *Jones polynomial* J_L, which can be defined recursively by a *skein relation*. Start with any (oriented) link diagram and choose any crossing; up to a rotation it will either look like the crossing C_+ or C_- in Figure 1.13. There are two things we can do to this crossing: we can pass the strings through each other (so the crossing of type $C_\pm$ becomes one of type $C_\mp$); or we can erase the crossing as in C_0. In this way we obtain three links: the original one (which we could call $L_\pm$ depending on the orientation of the chosen crossing) and the two modified ones ($L_\mp$ and L_0). The skein relation is

$$t^{-1} J_{L_+}(t) - t J_{L_-}(t) + \left(t^{-\frac{1}{2}} - t^{\frac{1}{2}}\right) J_{L_0}(t) = 0. \tag{1.2.7}$$

We also define the polynomial $J(t)$ of the unknot to be identically 1.

For a link with an odd number of components, $J_L(t) \in \mathbb{Z}[t^{\pm 1}]$ is a Laurent polynomial in t, while for an even number $J_L(t) \in \sqrt{t}\mathbb{Z}[t^{\pm 1}]$. For example, applying (1.2.7) twice, we get that the Jones polynomial of the trefoil in Figure 1.10 is $J(t) = -t^4 + t^3 + t$.

Are the trefoil and its mirror image ambient isotopic? The easiest argument uses the Jones polynomial: taking the mirror image corresponds to replacing t with t^{-1}, and we see that the Jones polynomial of the trefoil is not invariant under this transformation.[6]

---

[6] More generally, a knot with odd crossing number will be inequivalent with its mirror image (the crossing number is the minimum number of crossings needed in a diagram of the knot).

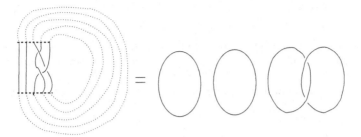

Fig. 1.14 The link associated with a braid.

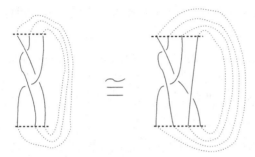

Fig. 1.15 A Markov move of type II.

The Reidemeister moves quickly prove $J_L(t)$ is a knot invariant, i.e. equivalent knots have the same polynomial, although inequivalent knots can also have the same one. But it was the first new knot polynomial in 56 years. It triggered discoveries of several other invariants while making unexpected connections elsewhere (Section 6.2.6), and secured for Jones a Fields medal. The problem then became that there were too many link invariants. We explain how we now organise them in Section 1.6.2.

Braids and links are directly related by theorems of Alexander (1923) and Markov (1935). Given any braid α we can define a link by connecting the ith spot on the bottom of the braid with the ith spot on the top, as in Figure 1.14. Alexander's theorem tells us that all links come from a braid in this way. Certainly though, different braids can correspond to the same link – for example, take any $\alpha, \beta \in \mathcal{B}_n$, then the links of α and $\beta\alpha\beta^{-1}$ are the same (slide the braid β^{-1} counterclockwise around the link until it is directly above, and hence cancels, β). This is called a Markov move of type I. A Markov move of type II changes the number of strands in a braid by ± 1, in a simple way – see Figure 1.15. Markov's theorem [59] says that two braids $\alpha \in \mathcal{B}_n$, $\beta \in \mathcal{B}_m$ correspond to equivalent links iff they are related by a finite sequence of Markov[7] moves. In Section 6.2.5, we explain how to use these two theorems to construct link invariants.

Question 1.2.1. Come up with a reasonable definition for the automorphism group of a lattice. Prove that the automorphism group of a positive-definite lattice is always finite.

---

[7] His father is the Markov of Markov chains.

Question 1.2.2. Let $x = (x_1, x_2)$ be any vector with nonzero coordinate x_2. Write $L(x)$ here for the lattice $\mathbb{Z}(1, 0) + \mathbb{Z}x$, and $T(x)$ for the torus $\mathbb{R}^2/L(x)$. Which x's give pointwise identical lattices (i.e. given x, find all y such that $L(x) = L(y)$)? Verify that all tori are diffeomorphic. Which tori $T(x)$ are obviously conformally equivalent?

Question 1.2.3. If we drop the requirement in Definition 1.2.1 that the $x^{(i)}$ be a basis, does anything really bad happen?

Question 1.2.4. Prove Theorem 1.2.2.

Question 1.2.5. Let L be an integral lattice. What is special about the reflection r_α through a vector $\alpha \in L$ with norm-squared $\alpha \cdot \alpha = 2$? (The formula for the reflection r_α is $r_\alpha(x) = x - \frac{2x \cdot \alpha}{\alpha \cdot \alpha} \alpha$.)

Question 1.2.6. Prove from (1.2.7) that the Jones polynomial for a link and its mirror image can be obtained from each other by the switch $t \leftrightarrow t^{-1}$. Prove that the Jones polynomial of a link is unchanged if the orientation of any component (i.e. the arrow on any strand) is reversed.

Question 1.2.7. Find the Jones polynomial of the disjoint union of n circles.

1.3 Elementary functional analysis

Moonshine concerns the occurrence of modular forms in algebra and physics, and care is taken to avoid analytic complications as much as possible. But spaces here are unavoidably infinite-dimensional, and through this arise subtle but significant points of contact with analysis. For example, the $q^{1/24}$ prefactor in the Dedekind eta (2.2.6b), and the central extension of loop algebras (3.2.2a), are analytic fingerprints. Lie group representations usually involve functional analysis (see e.g. Section 2.4.2 where we relate the Heisenberg group to theta functions). Much of functional analysis was developed to address mathematical concerns in quantum theory, and perhaps all of the rich subtleties of quantum field theory can be interpreted as functional analytic technicalities. For example, *anomalies* (which for instance permit derivations of the Atiyah–Singer Index Theorem from super Yang–Mills calculations) can be explained through a careful study of domains of operators [172]. Moreover, the natural culmination of the Jones knot polynomial is a deep relation between subfactors and conformal field theories (Section 6.2.6). The necessary background for all this is supplied in this section.

In any mature science such as mathematics, the division into branches is a convenient lie. In this spirit, analysis can be distinguished from, say, algebra by the central role played in the former by numerical inequalities. For instance, inequalities appear in the definition of derivatives and integrals as limits. Functional analysis begins with the reinterpretation of derivatives and integrals as linear operators on vector spaces. These spaces, which consist of appropriately restricted functions, are infinite-dimensional. The complexity and richness of the theory comes from this infinite-dimensionality.

Section 1.3.1 assumes familiarity with elementary point-set topology, as well as the definition of Lebesgue measure. All the necessary background is contained in standard textbooks such as [**481**].

1.3.1 Hilbert spaces

By a vector space $\mathcal{V}$, we mean something closed under *finite* linear combinations $\sum_{i=1}^{n} a_i v^{(i)}$. Here we are primarily interested in infinite-dimensional spaces over the complex numbers (i.e. the scalars a_i are taken from $\mathbb{C}$), and the vectors v are typically functions f. By a (complex) *pre-Hilbert space* we mean a vector space $\mathcal{V}$ with a Hermitian form $\langle f, g \rangle \in \mathbb{C}$ ('Hermitian form' is defined in Section 1.1.3). All complex n-dimensional pre-Hilbert spaces are isomorphic to $\mathbb{C}^n$ with Hermitian form

$$\langle u, v \rangle = \overline{u_1} v_1 + \cdots + \overline{u_n} v_n.$$

The analogue of $\mathbb{C}^n$ in countably many dimensions is $\ell^2(\infty)$, which consists of all sequences $u = (u_1, u_2, \ldots)$ with finite sum $\sum_{i=1}^{\infty} |u_i|^2 < \infty$. The reader can verify that it is closed under sums and thus forms a pre-Hilbert space. Another example consists of the C^{∞}-functions $f : \mathbb{R}^n \to \mathbb{C}$, say, with 'compact support' (that means that the set of all $x \in \mathbb{R}^n$ for which $f(x) \neq 0$ is bounded). The Hermitian form here is

$$\langle f, g \rangle = \int_{-\infty}^{\infty} \cdots \int_{-\infty}^{\infty} \overline{f(x)} \, g(x) \, \mathrm{d}^n x \; ; \tag{1.3.1}$$

this pre-Hilbert space is denoted $C_{cs}^{\infty}(\mathbb{R}^n)$. For instance, the function defined by

$$f(x) = \begin{cases} \exp[\frac{1}{x^2 - 1}] & \text{for } -1 < x < 1 \\ 0 & \text{otherwise} \end{cases}$$

lies in $C_{cs}^{\infty}(\mathbb{R})$. A larger space, arising for instance in quantum mechanics, is denoted $\mathcal{S}(\mathbb{R}^n)$ and consists of all functions $f \in C^{\infty}(\mathbb{R}^n)$ that, together with their derivatives, decrease to 0 faster than any power of $|x|^{-1}$, as $|x| \to \infty$. The space $\mathcal{S}$ is a pre-Hilbert space, again using (1.3.1). It contains functions such as $\mathrm{poly}(x_1, \ldots, x_n) \, e^{-x_1^2 - \cdots - x_n^2}$.

A pre-Hilbert space has a notion of distance, or *norm* $\|f\|$, given by $\|f\|^2 = \langle f, f \rangle$. Using this we can define limits, Cauchy sequences, etc. in the usual way [**481**]. We call a subset X of $\mathcal{V}$ *dense* in $\mathcal{V}$ if for any $f \in \mathcal{V}$ there is a sequence $f_n \in X$ that converges to f. For instance, the rationals $\mathbb{Q}$ are dense in the reals $\mathbb{R}$, but the integers aren't. Any convergent sequence is automatically Cauchy; a pre-Hilbert space $\mathcal{V}$ is called *complete* if conversely all Cauchy sequences in it converge.

Definition 1.3.1 *A Hilbert space $\mathcal{H}$ is a complete pre-Hilbert space.*

For example, each $\mathbb{C}^n$ is Hilbert, as is $\ell^2(\infty)$. Most pre-Hilbert spaces aren't Hilbert, for example neither $C_{cs}^{\infty}(\mathbb{R}^n)$ nor $\mathcal{S}(\mathbb{R}^n)$ are. However, given any pre-Hilbert space $\mathcal{V}$, there is a Hilbert space $\mathcal{H}$ that contains $\mathcal{V}$ as a dense subspace. This Hilbert space $\mathcal{H}$ is called the completion $\overline{\mathcal{V}}$ of $\mathcal{V}$, and is unique up to isomorphism. The construction of $\mathcal{H}$ from $\mathcal{V}$ is analogous to the construction of $\mathbb{R}$ from $\mathbb{Q}$, obtained by defining an equivalence relation on the Cauchy sequences.

The Hilbert space completion $\overline{\mathcal{C}_{cs}^{\infty}(\mathbb{R}^n)} = \overline{\mathcal{S}(\mathbb{R}^n)}$ is defined using the 'Lebesgue measure' μ, which is an extension of the usual notion of length to a much more general class of subsets $X \subset \mathbb{R}$ than the intervals, and the 'Lebesgue integral' $\int f(x)\,\mathrm{d}\mu(x)$, which is an extension of the usual Riemann integral to a much more general class of functions than the piecewise continuous ones. For example, what is the length of the set X consisting of all rational numbers between 0 and 1? This isn't defined, but its Lebesgue measure is easily seen to be 0. We won't define Lebesgue measures and integrals here, because we don't really need them; a standard account is [481]. The completion of $\mathcal{C}_{cs}^{\infty}(\mathbb{R}^n)$ is the Hilbert space $L^2(\mathbb{R}^n)$ consisting of all square-integrable functions $f : \mathbb{R}^n \to \mathbb{C} \cup \{\infty\}$. The Hermitian form is given by $\langle f, g \rangle = \int_{\mathbb{R}^n} \overline{f(x)}\,g(x)\,\mathrm{d}\mu(x)$. By f 'square-integrable' we mean that f is 'measurable' (e.g. any piecewise continuous function is measurable) and $\langle f, f \rangle < \infty$. We must identify two functions f, g if they agree *almost everywhere*, that is the set X of all $x \in \mathbb{R}^n$ at which $f(x) \neq g(x)$ has Lebesgue measure 0. This is because any two such functions have the property that $\langle f, h \rangle = \langle g, h \rangle$ for all h.

All Hilbert spaces we will consider, such as $L^2(\mathbb{R}^n)$, are *separable*. This means that there is a countable orthonormal set X of vectors $e_n \in \mathcal{H}$ (so $\langle e_n, e_m \rangle = \delta_{nm}$) such that the pre-Hilbert space span(X) consisting of all *finite* linear combinations $\sum a_m e_m$ is dense in $\mathcal{H}$. That is, given any $f \in \mathcal{H}$, $f = \lim_{n \to \infty} \sum_{i=1}^{n} \langle e_i, f \rangle\, e_i$ – we say that the *topological span* of X is $\mathcal{H}$. All infinite-dimensional separable Hilbert spaces are isomorphic to $\ell^2(\infty)$. The easy proof sends $f \in \mathcal{H}$ to the sequence $(\langle e_1, f \rangle, \langle e_2, f \rangle, \ldots) \in \ell^2(\infty)$.

We are interested in linear maps. The first surprise is that continuity is not automatic. In fact, let $T : \mathcal{V}_1 \to \mathcal{V}_2$ be a linear map between pre-Hilbert spaces. Then T is continuous at one point iff it's continuous at all points, iff it is *bounded* – that is, iff there exists a constant C such that $\|Tf\| \leq C\,\|f\|$, for all $f \in \mathcal{V}_1$. If $\mathcal{V}_1$ is finite-dimensional, then it is easy to show that any linear T is bounded. But in quantum mechanics, for example, most operators of interest are unbounded.

Another complication of infinite-dimensionality is that in practise we're often interested in linear operators whose domain is only a (dense) subspace of $\mathcal{H}$. For example, the domains of the operators $f(x) \mapsto xf(x)$ or $f(x) \mapsto \frac{\mathrm{d}}{\mathrm{d}x} f(x)$ (the 'position' and 'momentum' operators of quantum mechanics – see Section 4.2.1) are proper subspaces of $L^2(\mathbb{R})$. Those operators are well-defined though on $\mathcal{S}(\mathbb{R})$ (indeed, this is precisely why the space $\mathcal{S}$ is so natural for quantum mechanics). Once again bounded operators are simpler: if T is a bounded linear operator on some dense subspace $\mathcal{V}$ of a Hilbert space $\mathcal{H}$, then there is one and only one way to continuously extend the domain of T to all of $\mathcal{H}$.

The *dual* (or *adjoint*) $\mathcal{V}^*$ of a pre-Hilbert space $\mathcal{V}$ is defined as the space of all continuous linear maps (*functionals*) $\mathcal{V} \to \mathbb{C}$. In general, $\mathcal{V}$ can be regarded as a subspace of $\mathcal{V}^*$, with $f \in \mathcal{V}$ being identified with the functional $g \mapsto \langle \overline{f}, g \rangle$; when $\mathcal{V}$ is a Hilbert space $\mathcal{H}$, this identification defines an isomorphism $\mathcal{H}^* \cong \mathcal{H}$.

The functionals for $\mathcal{C}_{cs}^{\infty}$ are called *distributions*, while those for $\mathcal{S}$ are *tempered distributions*. For example, the Dirac delta '$\delta(x - a)$' is defined as the element of $\mathcal{S}(\mathbb{R})^*$ sending functions $\varphi \in \mathcal{S}(\mathbb{R})$ to the number $\varphi(a) \in \mathbb{C}$. (Tempered) distributions F can all be realised (non-uniquely) as follows: given $a \in \mathbb{N}$ and a continuous function $f(x)$ of polynomial growth, we get a functional $F \in \mathcal{S}(\mathbb{R})^*$ by

$$F(\varphi) = \int_{\mathbb{R}} f(x) \frac{\mathrm{d}^a \varphi}{\mathrm{d} x^a} \, \mathrm{d}x. \tag{1.3.2}$$

A similar realisation holds for the spaces $\mathcal{S}(\mathbb{R}^n)$ and $C_{cs}^{\infty}(\mathbb{R}^n)$. Of course distributions are not functions, and we cannot rewrite (1.3.2) as $\int g(x) \varphi(x) \, \mathrm{d}^n x$ for some function g. Note that the Dirac delta is not well-defined on the completion $L^2(\mathbb{R})$ of $\mathcal{S}$, since the elements $f \in L^2(\mathbb{R})$ are equivalence classes of functions and hence have ambiguous function values $f(a)$. This beautiful interpretation of distributions like δ as linear functionals is due to Sobolev and was developed by Schwartz, the 1950 Fields medalist. Another interpretation, using formal power series, is given in Section 5.1.2.

Distributions can be differentiated arbitrary numbers of times, and their partial derivatives commute (something not true of all differentiable functions). However, they usually cannot be multiplied together and thus form only a vector space, not an algebra. For more on distributions, see chapter 2 of [67] or chapter I of [244].

We're most interested in unitary and self-adjoint operators. First, let's define the adjoint. Let $T : \mathcal{V} \to \mathcal{H}$ be linear, where $\mathcal{V}$ is a subspace of $\mathcal{H}$. Let $\mathcal{U}$ be the set of all $g \in \mathcal{H}$ for which there is a unique vector $g^* \in \mathcal{H}$ such that for all $f \in \mathcal{V}$, $\langle g^*, f \rangle = \langle g, Tf \rangle$. Define the map (*adjoint*) $T^* : \mathcal{U} \to \mathcal{H}$ by $T^*(g) = g^*$. The adjoint T^* exists (i.e. its domain $\mathcal{U}$ is non-empty) iff $\mathcal{V}$ is dense in $\mathcal{H}$. In particular, T^{**} need not equal T. When $\mathcal{V}$ is dense in $\mathcal{H}$, $\mathcal{U}$ is a vector space and T^* is linear. When T is bounded, so is T^*, and its domain $\mathcal{U}$ is all of $\mathcal{H}$. Note that $\langle g, Tf \rangle = \langle T^*g, f \rangle$ for all $f \in \mathcal{V}$, $g \in \mathcal{U}$, but that relation doesn't uniquely specify T^*.

We call T *self-adjoint* if $T = T^*$ (so in particular this implies that their domains $\mathcal{V}, \mathcal{U}$ are equal). This implies $\langle Tf, g \rangle = \langle f, Tg \rangle$, but as before the converse can fail. If T is self-adjoint and unbounded, then its domain cannot be all of $\mathcal{H}$.

A linear map $T : \mathcal{H}_1 \to \mathcal{H}_2$ between Hilbert spaces $\mathcal{H}_1, \mathcal{H}_2$ is *unitary* if it is both onto and obeys $\langle Tf, Tg \rangle = \langle f, g \rangle$. Equivalently, $T^*T = TT^* = 1$. The surjectivity assumption is not redundant in infinite dimensions (Question 1.3.2). A unitary map is necessarily bounded. A famous example of a unitary operator is the Fourier transform $f \mapsto \hat{f}$, which, as usually defined, maps $\mathcal{S}(\mathbb{R}^n)$ onto itself; it extends to a unitary operator on $L^2(\mathbb{R}^n)$.

To define limits, etc., one needs only a topology. This need not come from a norm, and in general many different topologies can naturally be placed on a space. For an artificial example, consider the real line $\mathbb{R}$ endowed with the discrete topology (in which any subset of $\mathbb{R}$ is open): then any function $f : \mathbb{R} \to \mathbb{R}$ will be continuous, a sequence $x_n \in \mathbb{R}$ will converge iff there is some N such that $x_N = x_{N+1} = x_{N+2} = \cdots$, and $\mathbb{R}$ with this topology is again complete. In the topology coming from the Hermitian form (1.3.1), $\mathcal{S}(\mathbb{R}^n)$ is incomplete, however it is common to refine that topology somewhat. In this new topology, a sequence $f_m \in \mathcal{S}(\mathbb{R})$ converges to 0 iff for every $a, b \in \mathbb{N}$ we have

$$\lim_{m \to \infty} \sup_{x \in \mathbb{R}} |x|^b \left| \frac{\mathrm{d}^a f_m(x)}{\mathrm{d} x^a} \right| = 0.$$

This topology comes from interpreting $\mathcal{S}$ as the intersection of countably many Hilbert spaces; with it, $\mathcal{S}$ is complete. When we speak of $\mathcal{S}(\mathbb{R}^n)$ elsewhere in this book, we always take its topology to be this one (or its higher-dimensional analogue). Similar comments can be made for $C_{cs}^{\infty}(\mathbb{R}^n)$ – see chapter I of [**244**] for details. With these new topologies, both $\mathcal{S}(\mathbb{R}^n)$ and $C_{cs}^{\infty}(\mathbb{R}^n)$ are examples of *nuclear spaces*;[8] although they are not themselves Hilbert spaces (the completeness in Definition 1.3.1 must be in terms of the norm topology), they behave in a more finite-dimensional way, as is indicated by the Spectral Theorem given below. See, for example, [**244**] for more on nuclear spaces.

The Spectral Theorem tells us in which sense we can diagonalise self-adjoint and unitary operators. To state it precisely, we need a small generalisation of the construction of $\ell^2(\infty)$. Consider any measure space (e.g. $X = \mathbb{R}$ or S^1 with Lebesgue measure μ). Fix $n = 1, 2, \ldots, \infty$, and suppose that for each $x \in X$ there is associated a copy $\mathcal{H}_n$ of $\mathbb{C}^n$ or (if $n = \infty$) $\ell^2(\infty)$. We want to define the (orthogonal) *direct integral* over $x \in X$ of these $\mathcal{H}_n$'s. Consider all functions $h : X \to \mathcal{H}_n, x \mapsto h_x$ that aren't too wild and that obey the finiteness condition $\int_X \|h_x\|^2 \, d\mu < \infty$. As usual, we identify two such functions h, g if they agree everywhere except on a subset of X of μ-measure 0. Defining a Hermitian form by $\langle h, g \rangle = \int_X \langle h_x, g_x \rangle \, d\mu$, the set of all such (equivalence classes of) h constitutes a Hilbert space denoted $\int_X \mathcal{H}_n \, d\mu$ (completeness is proved as for $\ell^2(\infty)$). It is trivial to drop the requirement that the separable space $\mathcal{H}_n$ be fixed – see, for example, chapter 2 of [**67**] for details of the direct integral $\int_X \mathcal{H}(x) \, d\mu$.

In finite dimensions any self-adjoint operator is diagonalisable. This fails in infinite dimensions, for example both the 'momentum operator' $i\frac{\partial}{\partial x}$ and the 'position operator' $f(x) \mapsto x f(x)$ are self-adjoint on the dense subspace $\mathcal{S}(\mathbb{R})$ of $L^2(\mathbb{R})$, but neither have any eigenvectors anywhere in $L^2(\mathbb{R})$. So we need to generalise eigen-theory.

The statement of the Spectral Theorem simplifies when our operators act on $\mathcal{S}$. So let $T : \mathcal{S}(\mathbb{R}^n) \to \mathcal{S}(\mathbb{R}^n)$ be linear. Diagonalising T would mean finding a basis for $\mathcal{S}$ consisting of eigenvectors of T. We can't do that, but we get something almost as good. By a *generalised eigenvector* corresponding to the *generalised eigenvalue* $\lambda \in \mathbb{C}$, we mean a tempered distribution $F \in \mathcal{S}^*$ such that $F(T\varphi) = \lambda F(\varphi)$ for all $\varphi \in \mathcal{S}$. For each λ, let $E_\lambda \subset \mathcal{S}^*$ be the generalised eigenspace consisting of all such F. We say that the set of all generalised eigenvectors $\cup_\lambda E_\lambda$ is *complete* if they distinguish all vectors in $\mathcal{S}$, i.e. if, for any $\varphi, \varphi' \in \mathcal{S}$, we have $F(\varphi) = F(\varphi')$ for all generalised eigenvectors $F \in \cup_\lambda E_\lambda$ iff $\varphi = \varphi'$.

Theorem 1.3.2 (Spectral Theorem)

(a) *Let $U : \mathcal{S}(\mathbb{R}^n) \to \mathcal{S}(\mathbb{R}^n)$ be unitary. Then U extends uniquely to a unitary operator on all of $L^2(\mathbb{R}^n)$. All generalised eigenvalues λ lie on the unit circle $|\lambda| = 1$. We can express $L^2(\mathbb{R}^n)$ as a direct integral $\int_{|\lambda|=1} \mathcal{H}(\lambda) \, d\mu(\lambda)$ of Hilbert spaces $\mathcal{H}(\lambda) \subseteq E_\lambda$, so*

---

[8] Nuclear spaces were first formulated by Grothendieck, who began his mathematical life as a functional analyst before revolutionising algebraic geometry. The term 'nuclear' comes from 'noyau' (French for both 'nucleus' and 'kernel'), since the *Kernel Theorem* is a fundamental result holding for them. The 'L' in both ℓ^2 and L^2 is in honour of Lebesgue, and the symbol $\mathcal{S}$ honours Schwartz.

that U *sends the function* $h \in L^2(\mathbb{R}^n)$ *to the function* Uh *with* λ*-component* $(Uh)_\lambda = \lambda h_\lambda \in \mathcal{H}(\lambda)$. *Moreover, the generalised eigenvectors are complete.*

(b) *Suppose* $A : \mathcal{S}(\mathbb{R}^n) \to \mathcal{S}(\mathbb{R}^n)$ *is self-adjoint. Then all generalised eigenvalues* λ *lie on the real line* $\mathbb{R}$. *We can express* $L^2(\mathbb{R}^n)$ *as a direct integral* $\int_{-\infty}^{\infty} \mathcal{H}(\lambda) \, d\mu(\lambda)$ *of Hilbert spaces* $\mathcal{H}(\lambda) \subseteq E_\lambda$, *so that for each* $h \in \mathcal{S}(\mathbb{R}^n)$, Ah *has* λ*-component* $(Ah)_\lambda = \lambda h_\lambda$. *Moreover, the generalised eigenvectors are complete.*

For a simple example, consider the linear map $U : L^2(\mathbb{R}) \to L^2(\mathbb{R})$ acting by translation: $(Uf)(x) = f(x + 1)$. This is unitary, but it has no true eigenvectors in L^2. On the other hand, each point $\lambda = e^{iy}$ on the unit circle is a generalised eigenvalue, corresponding to generalised eigenvector F_λ given by $F_\lambda(\varphi) = \int_{-\infty}^{\infty} e^{-iyx} \varphi(x) \, dx$. The direct integral interpretation of L^2 corresponds to the association of any $f(x) \in L^2$ with its Fourier transform $f_\lambda = \widehat{f}(y) = \int_{-\infty}^{\infty} e^{iyx} f(x) \, dx$. The completeness of the generalised eigenvectors is implied by the Plancherel identity

$$\int |f(x)|^2 dx = \frac{1}{2\pi} \int |\widehat{f}(y)|^2 dy. \qquad (1.3.3)$$

The Spectral Theorem as formulated also holds for C_{cs}^{∞} in place of $\mathcal{S}$, and more generally for any *rigged* (or *equipped*) *Hilbert space* $\mathcal{V} \subset \mathcal{H} \subset \mathcal{V}^*$, where $\mathcal{H}$ is separable and $\mathcal{V}$ is nuclear (chapter I of [**244**]). They help provide a mathematically elegant formulation of quantum theories.

1.3.2 Factors

von Neumann algebras (see e.g. [**319**], [**177**]) can be thought of as symmetries of a (generally infinite) group. Their building blocks are called *factors*. Vaughn Jones initiated the combinatorial study of *subfactors* N of M (i.e. inclusions $N \subseteq M$ where M, N are factors), relating it to, for example, knots, and for this won a Fields medal in 1990. In Section 6.2.6 we describe Jones's work and the subsequent developments; this subsection provides the necessary background. Our emphasis is on accessibility.

Let $\mathcal{H}$ be a (separable complex) Hilbert space. By $\mathcal{L}(\mathcal{H})$ we mean the algebra of all bounded operators on $\mathcal{H}$ (we write '1' for the identity). For example, $\mathcal{L}(\mathbb{C}^n)$ is the space $M_n(\mathbb{C})$ of all $n \times n$ complex matrices. Let '$*$' be the adjoint (defined in the last subsection). Given a set S of bounded operators, denote by S' its *commutant*, that is the set of all bounded operators $x \in \mathcal{L}(\mathcal{H})$ that commute with all $y \in S$: $xy = yx$. We write $S'' := (S')'$ for the commutant of the commutant – clearly, $S \subseteq S''$.

Definition 1.3.3 *A von Neumann algebra* M *is a subalgebra of* $\mathcal{L}(\mathcal{H})$ *containing the identity* 1, *which obeys* $M = M^*$ *and* $M = M''$.

This is like defining a group by a representation. A von Neumann algebra can also be defined abstractly, which is equivalent except that (as we will see shortly) the natural notions of isomorphism are different in the concrete and abstract settings (just as the same group can have non-isomorphic representations).

Of course $\mathcal{L}(\mathcal{H})$ is a von Neumann algebra. Given any subset $S \subset \mathcal{L}(\mathcal{H})$ with $S^* = S$, the double-commutant S'' is a von Neumann algebra, namely the smallest one containing S. The space $L^\infty(\mathbb{R})$ of bounded functions $f : \mathbb{R} \to \mathbb{C}$ forms an abelian von Neumann algebra on the Hilbert space $\mathcal{H} = L^2(\mathbb{R})$ by pointwise multiplication. More generally (replacing $\mathbb{R}$ with any other measure space X and allowing multiple copies of the Hilbert space $L^2(X)$), all abelian von Neumann algebras are of that form.

The centre $Z(M) = M \cap M'$ of a von Neumann algebra M is an abelian one. Using the above characterisation $Z(M) = L^\infty(X)$, we can write M as a direct integral $\int_X M(\lambda) \, d\lambda$ of von Neumann algebras $M(\lambda)$ with trivial centre: $Z(M(\lambda)) = \mathbb{C}1$. The direct integral, discussed last subsection, is a continuous analogue of direct sum.

Definition 1.3.4   *A factor M is a von Neumann algebra with centre $Z(M) = \mathbb{C}I$.*

Thus the study of von Neumann algebras is reduced to that of factors – the simple building blocks of any von Neumann algebra. $\mathcal{L}(\mathcal{H})$ is a factor. In finite dimensions, any (concrete) factor is of the form $M_n(\mathbb{C}) \otimes \mathbb{C}I_m$ acting in the Hilbert space $\mathbb{C}^n \otimes \mathbb{C}^m$ ('I_m' is the $m \times m$ identity matrix). Whenever the factor is (abstract) isomorphic to some $\mathcal{L}(\mathcal{H})$, its concrete realisation will have a similar tensor product structure, which is the source of the name 'factor'. In quantum field theory, where von Neumann algebras arise as algebras of operators (Section 4.2.4), a factor means there is no observable that can be measured simultaneously (with infinite precision) with all others.

The richness of the theory is because there are other factors besides $\mathcal{L}(\mathcal{H})$. In particular, factors fall into different families:

Type I_n: the factors (abstract) isomorphic to $\mathcal{L}(\mathcal{H})$ ($n = \dim \mathcal{H}$).
Type II_1: infinite-dimensional but it has a *trace* (i.e. a linear functional tr $: M \to \mathbb{C}$ such that $\mathrm{tr}(xy) = \mathrm{tr}(yx)$).
Type II_∞: the factors isomorphic to $II_1 \otimes \mathcal{L}(\mathcal{H})$.
Type III: everything else.

Choosing the normalisation $\mathrm{tr}(1) = 1$, the type II_1 trace will be unique. This is a very coarse-grained breakdown, and in fact the complete classification of factors is not known. There are uncountably many inequivalent type II_1 factors. Type III is further subdivided into families III_λ for all $0 \le \lambda \le 1$. von Neumann regarded the type III factors as pathological, but this was unfair (see Section 6.2.6). Almost every factor is isomorphic to type III_1 (i.e. perturbing an infinite-dimensional factor typically gives you one of type III_1). *Hyperfinite* factors are limits in some sense of finite-dimensional factors. There is a unique (abstract) hyperfinite factor of type II_1, II_∞ and III_λ for $0 < \lambda \le 1$; we are interested in the hyperfinite II_1 and III_1 factors. Incidentally, the von Neumann algebras arising in quantum field theory are always of type III_1.

Discrete groups impinge on the theory through the *crossed-product construction* of factors. Start with any von Neumann algebra $M \subset \mathcal{L}(\mathcal{H})$, and let G be a discrete group acting on M (so $g.(xy) = (g.x)(g.y)$ and $g.x^* = (g.x)^*$). Let $\mathcal{H}_G = \mathcal{H} \otimes \ell^2(G)$ be the Hilbert space consisting of all column vectors $\zeta = (\zeta_g)_{g \in G}$ with entries $\zeta_g \in \mathcal{H}$ and obeying $\sum_{g \in G} \|\zeta_g\|^2 < \infty$. M acts on $\mathcal{H}_G$ by $\zeta \mapsto \pi(x)(\zeta)$, where π is defined by

$$(\pi(x)(\zeta))_g := (g^{-1}.x)\zeta_g. \tag{1.3.4a}$$

In (1.3.4a), $g^{-1}.x$ is the action of G on M, and $g^{-1}.x \in M \subset \mathcal{L}(\mathcal{H})$ acts on $\zeta_g \in \mathcal{H}$ by definition. Likewise, G acts on $\mathcal{H}_G$ by $\zeta \mapsto \lambda(h)(\zeta)$, where λ is defined by

$$(\lambda(h)(\zeta))_g := \zeta_{h^{-1}g}. \tag{1.3.4b}$$

We can regard π and λ as embedding M and G in $\mathcal{L}(\mathcal{H}_G)$. The crossed-product is simply the smallest von Neumann algebra containing both these images:

$$M \rtimes G := (\pi(M) \cup \lambda(G))''. \tag{1.3.4c}$$

More explicitly (using the obvious orthonormal basis), any bounded operator $\tilde{y} \in \mathcal{L}(\mathcal{H}_G)$ is a matrix $\tilde{y} = (\tilde{y}_{g,h})$ with entries $\tilde{y}_{g,h} \in \mathcal{L}(\mathcal{H})$ for $g, h \in G$, and where $(\tilde{y}\zeta)_g = \sum_{h \in G} \tilde{y}_{g,h} \zeta_h$ (defining the infinite sum on the right appropriately [**319**]). Then for all $x \in M$ and $g, h, k \in G$, we get the matrix entries

$$\pi(x)_{g,h} = \delta_{g,h} \, h^{-1}.x,$$
$$\lambda(k)_{g,h} = \delta_{g,kh} \, 1.$$

The crossed-product is now a space of functions $y : G \to M$:

$$M \rtimes G \cong \{ y : G \to M \mid \exists \tilde{y} \in \mathcal{L}(\mathcal{H}_G) \text{ such that } \tilde{y}_{g,h} = h^{-1}.(y_{gh^{-1}}) \, \forall g, h \in G \} \tag{1.3.4d}$$

(see lemma 1.3.1 of [**319**]). In this notation the algebra structure of $M \rtimes G$ is given by

$$(xy)(g) = \sum_{h \in G} h^{-1}.(x_{gh^{-1}} \, y_h), \tag{1.3.4e}$$

$$(y^*)_g = g^{-1}.(y_{g^{-1}})^*. \tag{1.3.4f}$$

Crossed-products allow for elegant constructions of factors. For example, the (von Neumann) group algebra $\mathbb{C} \rtimes G$ is type II_1, for any discrete group G acting trivially on $\mathbb{C}$ and with the property that all of its conjugacy classes (apart from $\{e\}$) are infinite (examples of such G are the free groups $\mathcal{F}_n$ or $\mathrm{PSL}_2(\mathbb{Z})$). Also, any type III_1 factor is of the form $M \rtimes \mathbb{R}$, where M is type II_∞ and the $\mathbb{R}$ action scales the trace.

A proper treatment of factors (which this subsection is not) would involve *projections* onto closed subspaces, that is elements $p \in M$ satisfying $p = p^* = p^2$. These span (in the appropriate sense) the full von Neumann algebra. In the case of $M = M_n(\mathbb{C})$, the projections are precisely the orthogonal projections onto subspaces of $\mathbb{C}^n$, and thus have a well-defined dimension (namely the dimension of that subspace, so some integer between 0 and n). Remarkably, the same applies to any projection in any factor. For type II_1 this 'dimension' $\dim(p)$ is the trace $\tau(p)$, which we can normalise so that $\tau(1) = 1$. Then we get that $\dim(p)$ continuously fills out the interval $[0, 1]$. For type II_∞, the dimensions fill out $[0, \infty]$. For type III, every nonzero projection is equivalent (in a certain sense) to the identity and so the (normalised) dimensions are either 0 or 1.

Finally, one can ask for the relation between the abstract and concrete definitions of M – in other words, given a factor M, what are the different representations (= modules) of M, that is realisations of M as bounded operators on a Hilbert space $\mathcal{H}$. For example, for M of type I_n, these are of the form $M \otimes \mathbb{C}^M = M \oplus \cdots \oplus M$ (m times) for m finite, as well as $M \otimes \ell^\infty$. We see the type I_n modules are in one-to-one correspondence with the 'multiplicity' $m \in \{0, 1, \ldots, \infty\}$, which we can denote $\dim_M(\mathcal{H})$ and think of as $\dim(\mathcal{H})/\dim(M)$, at least when M is finite-dimensional. There is a similar result for type II: for each choice $d \in [0, \infty]$ there is a unique module $\mathcal{H}_d$, and any module $\mathcal{H}$ is equivalent to a unique $\mathcal{H}_d$. Finally, any two nontrivial representations of a type III factor will be equivalent. For a general definition of $\dim_M(\mathcal{H})$ and a proof of this representation theory, see theorem 2.1.6 in [**319**].

For type II_1, this parameter $d =: \dim_M(\mathcal{H})$ is sometimes called by von Neumann's unenlightening name 'coupling constant'. Incidentally, $\mathcal{H}_1$ is constructed in Question 1.3.6.

Question 1.3.1. (a) Verify explicitly that the position $f(x) \mapsto xf(x)$ and momentum $i\frac{d}{dx}$ operators are neither bounded nor continuous, for the Hilbert space $L^2(\mathbb{R})$.
(b) Verify explicitly that the position operator of (a) is not defined everywhere.

Question 1.3.2. Consider the shift operator $S(x_1, x_2, \ldots) = (0, x_1, x_2, \ldots)$ in $\ell^2(\infty)$. Verify that $S^*S = 1$ but $SS^* \neq 1$.

Question 1.3.3. Apply the Spectral Theorem to the momentum operator $i\frac{d}{dx}$.

Question 1.3.4. Let $\mathcal{V} = \{f \in C^\infty(S^1) \mid f(0) = 0\}$.
(a) Verify that $\mathcal{V}$ is dense in $\mathcal{H} = L^2(S^1)$.
(b) Verify that $D = i\frac{d}{d\theta}$ obeys $\langle Df, g \rangle = \langle f, Dg \rangle$ for all $f, g \in \mathcal{V}$.
(c) Construct the adjoint D^* of $D : \mathcal{V} \to \mathcal{H}$. Is D self-adjoint?
(d) For each $\lambda \in \mathbb{C}$, define $\mathcal{V}_\lambda$ to be the extension of $\mathcal{V}$ consisting of all functions smooth on the interval $[0, 2\pi]$ and with $f(0) = \lambda f(2\pi)$. Extend D in the obvious way to $\mathcal{V}_\lambda$. For which λ is D now self-adjoint?

Question 1.3.5. Let the free group $\mathcal{F}_2$ act trivially on $\mathbb{C}$. Find a trace for $\mathbb{C} \rtimes \mathcal{F}_2$. What is the centre of $\mathbb{C} \rtimes \mathcal{F}_2$?

Question 1.3.6. Let M be type II_1. Prove M is a pre-Hilbert space by defining $\langle x, y \rangle$ appropriately (*Hint*: use the trace). Let $L^2(M)$ be its completion. Show that $L^2(M)$ is a module over M.

1.4 Lie groups and Lie algebras

Undergraduates are often disturbed (indeed, reluctant) to learn that the vector-product $u \times v$ really only works in three dimensions. Of course, there are several generalisations to other dimensions: for example an antisymmetric $(N - 1)$-ary product (a determinant) in N dimensions, or the wedge product of k-forms in $2k + 1$ dimensions. Arguably the most fruitful generalisation is that of a Lie algebra, defined below. They are the tangent spaces of those differential manifolds whose points can be 'multiplied' together.

As we know, much of algebra is developed by analogy with elementary properties of integers. For a finite-dimensional Lie algebra, a divisor is called an ideal; a prime is called simple; and multiplying corresponds to semi-direct sum (Lie algebras behave simpler than groups but not as simple as numbers). In particular, simple Lie algebras are important for similar reasons that simple groups are, and can also be classified (with *much* less effort). One non-obvious discovery is that they are rigid: the best way to capture the structure of a simple Lie algebra is through a graph. We push this thought further in Section 3.3. For an elementary introduction to Lie theory, [**92**] is highly recommended.

1.4.1 Definition and examples of Lie algebras

An *algebra* is a vector space with a way to multiply vectors that is compatible with the vector space structure (i.e. the vector-valued product is required to be *bilinear*: $(au + a'u') \times (bv + b'v') = ab\,u \times v + ab'\,u \times v' + a'b\,u' \times v + a'b'\,u' \times v')$. For example, the complex numbers $\mathbb{C}$ form a two-dimensional algebra over $\mathbb{R}$ (a basis is 1 and i $= \sqrt{-1}$; the *scalars* here are real numbers and the *vectors* are complex numbers). The quaternions are four-dimensional over $\mathbb{R}$ and the octonions are eight-dimensional over $\mathbb{R}$. Incidentally, these are the only finite-dimensional normed algebras over $\mathbb{R}$ that obey the cancellation law: $u \neq 0$ and $u \times v = 0$ implies $v = 0$ (does the vector-product of $\mathbb{R}^3$ fail the cancellation law?). This important little fact makes several unexpected appearances [**29**]. For instance, imagine a ball (i.e. S^2) covered in hair. No matter how you comb it, there will be a part in the hair, or at least a point where the hair leaves in all directions, or some such problem. More precisely, there is no continuous nowhere-zero vector field on S^2. On the other hand, it is trivial to comb the hair on the circle S^1 without singularity: just comb it clockwise, for example. More generally, the even spheres S^{2k} can never be combed. Now try something more difficult: place k wigs on S^k, and try to comb all k of them so that at each point on S^k the k hairs are linearly independent. This is equivalent to saying that the tangent bundle $T\,S^k$ equals $S^k \times \mathbb{R}^k$. The only k-spheres S^k that can be 'k-combed' in this way (i.e. for which there exist k linearly independent continuous vector fields) are for $k = 1, 3$ and 7. This is intimately connected with the existence of $\mathbb{C}$, the quaternions and octonions (namely, S^1, S^3 and S^7 are the length 1 complex numbers, quaternions and octonions, respectively) [**104**].

Definition 1.4.1 *A Lie algebra* $\mathfrak{g}$ *is an algebra with product (usually called a 'bracket' and written $[xy]$) that is both 'anti-commutative' and 'anti-associative':*

$$[xy] + [yx] = 0; \tag{1.4.1a}$$

$$[x[yz]] + [y[zx]] + [z[xy]] = 0. \tag{1.4.1b}$$

Like most other identities in mathematics, (1.4.1b) is named after Jacobi (although he died years before Lie theory was created). Usually we consider Lie algebras over $\mathbb{C}$, but sometimes over $\mathbb{R}$. Note that (1.4.1a) is equivalent to demanding $[xx] = 0$ (except for fields of characteristic 2).

A homomorphism $\varphi : \mathfrak{g}_1 \rightarrow \mathfrak{g}_2$ between Lie algebras must preserve the linear structure as well as the bracket – i.e. φ is linear and $\varphi[xy] = [\varphi(x)\varphi(y)]$ for all $x, y \in \mathfrak{g}_1$. If φ is in addition invertible, we call $\mathfrak{g}_1$, $\mathfrak{g}_2$ *isomorphic*.

One important consequence of bilinearity is that it is enough to know the values of all the brackets $[x^{(i)}x^{(j)}]$ for $i < j$, for any basis $\{x^{(1)}, x^{(2)}, \ldots\}$ of the vector space $\mathfrak{g}$. (The reader should convince himself of this before proceeding.)

A trivial example of a Lie algebra is a vector space $\mathfrak{g}$ with a bracket identically 0: $[xy] = 0$ for all $x, y \in \mathfrak{g}$. Any such Lie algebra is called *abelian*, because in any representation (i.e. realisation by matrices) its matrices will commute. Abelian Lie algebras of equal dimension are isomorphic.

In fact, the only one-dimensional Lie algebra (for any choice of field $\mathbb{F}$) is the abelian one $\mathfrak{g} = \mathbb{F}$. It is straightforward to find all two- and three-dimensional Lie algebras (over $\mathbb{C}$) up to isomorphism: there are precisely two and six of them, respectively (though one of the six depends on a complex parameter). Over $\mathbb{R}$, there are two and nine (with two of the latter depending on real parameters). This exercise cannot be continued much further – for example, not all seven-dimensional Lie algebras (over $\mathbb{C}$ say) are known. Nor is it obvious that this would be a valuable exercise. We should suspect that our definition of Lie algebra is probably too general for anything obeying it to be automatically interesting. Most commonly, a classification yields a stale and useless list – a phone book more than a tourist guide.

Two of the three-dimensional Lie algebras are important in what follows. One of them is well known to the reader: the vector-product in $\mathbb{C}^3$. Taking the standard basis $\{e_1, e_2, e_3\}$ of $\mathbb{C}^3$, the bracket can be defined by the relations

$$[e_1 e_2] = e_3, \qquad [e_1 e_3] = -e_2, \qquad [e_2 e_3] = e_1. \qquad (1.4.2a)$$

This algebra, denoted A_1 or $\mathfrak{sl}_2(\mathbb{C})$, deserves the name 'mother of all Lie algebras' (Section 1.4.3). Its more familiar realisation uses a basis $\{e, f, h\}$ with relations

$$[ef] = h, \qquad [he] = 2e, \qquad [hf] = -2f. \qquad (1.4.2b)$$

The reader can find the change-of-basis (valid over $\mathbb{C}$ but not $\mathbb{R}$) showing that equations (1.4.2) define isomorphic *complex* (though not *real*) Lie algebras.

Another important three-dimensional Lie algebra is the Heisenberg algebra[9] $\mathfrak{Heis}$, the algebra of the canonical commutation relations in quantum mechanics, defined by

$$[xp] = h, \qquad [xh] = [ph] = 0. \qquad (1.4.3)$$

The most basic source of Lie algebras are the $n \times n$ matrices with *commutator*:

$$[AB] = [A, B] := AB - BA \qquad (1.4.4)$$

(the reader can verify that the commutator always obeys (1.4.1)). Let $\mathfrak{gl}_n(\mathbb{R})$ (respectively $\mathfrak{gl}_n(\mathbb{C})$) denote the Lie algebra of all $n \times n$ matrices with coefficients in $\mathbb{R}$ (respectively

---

[9] There actually is a family of 'Heisenberg algebras', with (1.4.3) being the one of least dimension.

$\mathbb{C}$), with Lie bracket given by (1.4.4). More generally, if $\mathcal{A}$ is any associative algebra, then $\mathcal{A}$ becomes a Lie algebra by defining the bracket $[xy] = xy - yx$.

Another general construction of Lie algebras starts with any (not necessarily associative or commutative) algebra $\mathcal{A}$. By a *derivation* of $\mathcal{A}$, we mean any linear map $\delta : \mathcal{A} \to \mathcal{A}$ obeying the Leibniz rule $\delta(ab) = \delta(a) b + a \delta(b)$. We can compose derivations, but in general the result $\delta_1 \circ \delta_2$ won't be a derivation. However, an easy calculation verifies that the commutator $[\delta_1 \delta_2] = \delta_1 \circ \delta_2 - \delta_2 \circ \delta_1$ of derivations is also a derivation. Hence the vector space of derivations is naturally a Lie algebra. If $\mathcal{A}$ is finite-dimensional, so will be its Lie algebra of derivations.

In particular, vector fields $X \in \text{Vect}(M)$ are derivations. We can compose them $X \circ Y$, but this results in a second-order differential operator. Instead, the natural 'product' is their commutator $[X, Y] = X \circ Y - Y \circ X$, as it always results in a vector field. $\text{Vect}(M)$ with this bracket is an infinite-dimensional Lie algebra. For example, recall $\text{Vect}(S^1)$ from Section 1.2.2 and compare

$$\left(f(\theta) \frac{\mathrm{d}}{\mathrm{d}\theta} \right) \circ \left(g(\theta) \frac{\mathrm{d}}{\mathrm{d}\theta} \right) = f(\theta) g(\theta) \frac{\mathrm{d}^2}{\mathrm{d}\theta^2} + f(\theta) g'(\theta) \frac{\mathrm{d}}{\mathrm{d}\theta},$$
$$\left[f(\theta) \frac{\mathrm{d}}{\mathrm{d}\theta}, g(\theta) \frac{\mathrm{d}}{\mathrm{d}\theta} \right] = (f(\theta) g'(\theta) - f'(\theta) g(\theta)) \frac{\mathrm{d}}{\mathrm{d}\theta}.$$

Incidentally, another natural way to multiply vector fields X, Y of vector fields, the Lie derivative $\mathcal{L}_X(Y)$ defined in Section 1.2.2, equals the commutator $[X, Y]$ and so gives the same Lie algebra structure on $\text{Vect}(M)$.

1.4.2 Their motivation: Lie groups

From Definition 1.4.1 it is far from clear that Lie algebras, as a class, should be natural and worth studying. After all, there are infinitely many possible axiomatic systems: why should this one be anything special *a priori*? Perhaps the answer could have been anticipated by the following line of reasoning.

Axiom *Groups are important and interesting.*
Axiom *Manifolds are important and interesting.*

Definition 1.4.2 *A Lie group G is a manifold with a compatible group structure.*

This means that 'multiplication' $\mu : G \times G \to G$ (which sends the pair (a, b) to ab) and 'inverse' $\iota : G \to G$ (which sends a to a^{-1}) are both differentiable maps. The manifold structure (Definition 1.2.3) of G can be chosen as follows: fix any open set U_e about the identity $e \in G$; then the open set $U_g := gU_e$ will contain $g \in G$. The real line $\mathbb{R}$ is a Lie group under addition: obviously, μ and ι defined by $\mu(a, b) = a + b$ and $\iota(a) = -a$ are both differentiable. A circle is also a Lie group: parametrise the points with the angle θ defined mod 2π; the 'product' of the point at angle θ_1 with the point at angle θ_2 is the point at angle $\theta_1 + \theta_2$. Surprisingly, the only other k-sphere that is a Lie group is S^3

(the product can be defined using quaternions of unit length,[10] or by identifying S^3 with the matrix group $\mathrm{SU}_2(\mathbb{C})$). This is because it is always possible to 'n-comb the hair' on an n-dimensional Lie group (Section 1.4.1) – more precisely, the tangent bundle TG of any Lie group is trivial $G \times \mathbb{R}^n$, something easy to see using the charts U_g.

A *complex* Lie group G is a complex manifold with a compatible group structure. For example, the only one-dimensional compact real Lie group is S^1, whereas there are infinitely many compact one-dimensional complex Lie groups, namely the tori or 'elliptic curves' $\mathbb{C}/L$, for any two-dimensional lattice L in the plane $\mathbb{C}$. Thought of as real Lie groups (i.e. forgetting their complex structure), elliptic curves all are real-diffeomorphic to $S^1 \times S^1$; they differ in their complex-differential structure. We largely ignore the complex Lie groups; unless otherwise stated, by 'Lie group' we mean 'real Lie group'.[11]

Many but not all Lie groups can be expressed as matrix groups whose operation is matrix multiplication. The most important are GL_n (invertible $n \times n$ matrices) and SL_n (ones with determinant 1).

Incidentally, Hilbert's 5th problem[12] asked how important the differentiability hypothesis is here. It turns out it isn't (see [**569**] for a review): if a group G is a topological manifold, and μ and ι are merely continuous, then it is possible to endow G with a differentiable structure in one and only one way so that μ and ι are differentiable.

In any case, a consequence of the above axioms is surely:

Corollary   *Lie groups should be important and interesting.*

Indeed, Lie groups appear throughout mathematics and physics, as we will see again and again. For example, the Lie groups of relativistic physics (Section 4.1.2) come from the group $\mathrm{O}_{3,1}(\mathbb{R})$ consisting of all 4×4 matrices Λ obeying $\Lambda G \Lambda^t = G$, where $G = \mathrm{diag}(1, 1, 1, -1)$ is the Minkowski metric. Any such Λ must have determinant ± 1, and has $|\Lambda_{44}| \geq 1$; these 2×2 possibilities define the four connected components of $\mathrm{O}_{3,1}(\mathbb{R})$. The (restricted) *Lorentz group* $\mathrm{SO}_{3,1}^+(\mathbb{R})$ consists of the determinant 1 matrices Λ in $\mathrm{O}_{3,1}(\mathbb{R})$ with $\Lambda_{44} \geq 1$. It describes rotations in 3-space, as well as 'boosts' (changes of velocity). $\mathrm{SO}_{3,1}^+(\mathbb{R})$ has a double-cover (i.e. an extension by $\mathbb{Z}_2$) isomorphic to $\mathrm{SL}_2(\mathbb{C})$, which is more fundamental. Finally, the *Poincaré group* is the semi-direct product of $\mathrm{SO}_{3,1}^+(\mathbb{R})$ with $\mathbb{R}^4$, corresponding to adjoining to $\mathrm{SO}_{3,1}^+(\mathbb{R})$ the translations in space-time $\mathbb{R}^4$. The Lorentz group is six-dimensional, while the Poincaré group is 10-dimensional.

As said in Section 1.2.2, the tangent spaces of manifolds are vector spaces of dimension equal to that of the manifold. The space structure is easy to see for Lie groups: choose any infinitesimal curves $u = \langle g(t) \rangle_e, v = \langle h(t) \rangle_e \in T_eG$, so $g(0) = h(0) = e$, and let $a, b \in \mathbb{R}$. Then $au + bv$ corresponds to the curve $t \mapsto g(at)h(bt)$.

Not surprisingly, G acts on the tangent vectors: let $u \in T_hG$ correspond to curve $h(t)$, with $h(0) = h$, and define gu for any $g \in G$ to be the vector in $T_{gh}G$ corresponding

---

[10] Similarly, the 7-sphere inherits from the octonions a *non-associative* (hence nongroup) product, compatible with its manifold structure.

[11] Our vector spaces (e.g. Lie algebras) are usually complex; our manifolds (e.g. Lie groups) are usually real.

[12] In the International Congress of Mathematicians in 1899, David Hilbert announced several problems chosen to anticipate (and direct) major areas of study. His list was deeply influential.

to the curve $t \mapsto g(h(t))$. This means that conjugating gug^{-1} for any element $u \in T_eG$ gives another element of T_eG, that is T_eG carries a representation of the group G called the adjoint representation.

This is all fine. However, we have a rich structure on our manifold – namely the group structure – and it would be deathly disappointing if this adjoint representation were the high-point of the theory. Fortunately we can go *much* further. Consider any $u, v \in T_eG$, where $v = \langle g(t) \rangle_e$. Then $g(t) u\, g(t)^{-1}$ lies in the vector space T_eG for all t, and hence so will the derivative. It turns out that the quantity

$$[uv] := \frac{\mathrm{d}}{\mathrm{d}t}(g(t) u\, g(t)^{-1})|_{t=0} \qquad (1.4.5)$$

depends only on u and v (hence the notation). A little work shows that it is bilinear, anti-symmetric, and anti-associative. That is, T_eG is a Lie algebra!

In the last subsection we indicated that Vect(M) carries a Lie algebra structure, for any manifold M. It is tempting to ask: when M is a Lie group G, what is the relation between the infinite-dimensional Lie algebra Vect(G), and the finite-dimensional Lie algebra T_eG? Note that G acts on the space Vect(M) by 'left-translation', that is if X is a vector field, which we can think of as a derivation of the algebra $C^\infty(G)$ of real-valued functions on G, and $g \in G$, then $g.X$ is the vector field given by $(g.X)(f)(h) = X(f)(gh)$. Then the Lie algebra T_eG is isomorphic to the subalgebra of Vect(G) consisting of the 'left-invariant vector fields', that is those X obeying $g.X = X$. Given any manifold M, the Lie algebra Vect(M) corresponds to the infinite-dimensional Lie group $\mathrm{Diff}^+(M)$ of orientation-preserving diffeomorphisms of M; when M is itself a Lie group, the left-invariant vector fields correspond in $\mathrm{Diff}^+(M)$ to a copy of M given by left-multiplication.

Fact *The tangent space of a Lie group is a Lie algebra. Conversely, any (finite-dimensional real or complex) Lie algebra is the tangent space T_eG to some Lie group.*

For example, consider the Lie group $G = \mathrm{SL}_n(\mathbb{R})$. Let $A(t) = (A_{ij}(t))$ be any curve in G with $A(0) = I_n$. We see that only one term in the expansion of $\det A(t)$ can contribute to its derivative at $t = 0$, namely the diagonal term $A_{11}(t) \cdots A_{nn}(t)$, so differentiating $\det(A(t)) = 1$ at $t = 0$ tells us that $A'_{11}(0) + \cdots + A'_{nn}(0) = 0$. Thus the tangent space $T_{I_n}G$ consists of all trace-zero $n \times n$ matrices, since the algebra like the group must be $(n^2 - 1)$-dimensional. We write it $\mathfrak{sl}_n(\mathbb{R})$. Now choose any matrices $U, V \in \mathfrak{sl}_n(\mathbb{R})$, and let $A(t)$ be the curve in $\mathrm{SL}_n(\mathbb{R})$ corresponding to V. Differentiating $A(t) A(t)^{-1} = I_n$, we see that $(A^{-1})' = -A^{-1} A' A^{-1}$ and thus (1.4.5) becomes

$$[VU] = A'(0) U\, I_n^{-1} + I_n U \left(-I_n^{-1} A'(0) I_n^{-1}\right).$$

In other words, the bracket in $\mathfrak{sl}_n(\mathbb{R})$ – as with any other matrix algebra – is given by the commutator (1.4.4).

Given the above fact, a safe guess would be:

Conjecture *Lie algebras are important and interesting.*

From this line of reasoning, it should be expected that historically Lie groups arose first. Indeed that is the case: Sophus Lie introduced them in 1873 to try to develop a Galois theory for ordinary differential equations. Galois theory can be used for instance to show that not all fifth degree (or higher) polynomials can be explicitly 'solved' using radicals (Section 1.7.2). Lie wanted to study the explicit solvability (integrability) of differential equations, and this led him to develop what we now call Lie theory. The importance of Lie groups, however, has grown well beyond this initial motivation.

A Lie algebra, being a linearised Lie group, is much simpler and easier to handle. The algebra preserves the local properties of the group, though it loses global topological properties (like compactness). A Lie group has a single Lie algebra, but a Lie algebra corresponds to many different Lie groups. The Lie algebra corresponding to both $\mathbb{R}$ and S^1 is $\mathfrak{g} = \mathbb{R}$ with trivial bracket. The Lie algebra corresponding to both $S^3 = SU_2(\mathbb{C})$ and $SO_3(\mathbb{R})$ is the vector-product algebra (1.4.2a) (usually called $\mathfrak{so}_3(\mathbb{R})$).

We saw earlier that many (but not all) examples of Lie groups are matrix groups, that is subgroups of $GL_n(\mathbb{R})$ or $GL_n(\mathbb{C})$. The *Ado–Iwasawa Theorem* (see e.g. chapter VI of [**314**]) says that all finite-dimensional Lie algebras (over any field) are realisable as Lie subalgebras of $\mathfrak{gl}_n(\mathbb{R})$ or $\mathfrak{gl}_n(\mathbb{C})$. This is analogous to Cayley's Theorem, which says any finite group is a subgroup of some symmetric group $\mathcal{S}_n$. Now, choose any Lie algebra $\mathfrak{g} \subseteq \mathfrak{gl}_n(\mathbb{C})$. Let G be the topological closure of the subgroup of $GL_n(\mathbb{C})$ generated by all matrices e^A for $A \in \mathfrak{g}$, where e^A is defined by the Taylor expansion

$$e^A = \sum_{k=0}^{\infty} \frac{1}{k!} A^k.$$

Then the Lie group G has Lie algebra $\mathfrak{g}$. Remarkably, the group operation on G (at least close to the identity) can be deduced from the bracket: the first few terms of the *Baker–Campbell–Hausdorff formula* read

$$\exp(X) \exp(Y) = \exp\left(X + Y + \frac{1}{2}[XY] + \frac{1}{12}[[XY]X] + \frac{1}{12}[[XY]Y] + \cdots \right).$$
(1.4.6)

See, for example, [**475**] for the complete formula and some of its applications.

We saw earlier that the condition 'determinant $= 1$' for matrix groups translates to the Lie algebra condition 'trace $= 0$'. This also follows from the identity $\det(e^A) = e^{\operatorname{tr} A}$, which follows quickly from the Jordan canonical form of A.

Of course all undergraduates are familiar, at least implicitly, with exponentiating operators. Taylor's Theorem tells us that for any analytic function f and any real number a, the operator $e^{a\frac{d}{dx}}$ sends $f(x)$ to $f(x+a)$. Curiously, the operator $\log(\frac{d}{dx})$ also has a meaning, in the context of, for example, affine Kac–Moody algebras [**344**].

The definition of a Lie algebra makes sense over any field $\mathbb{K}$. However, the definition of Lie groups is much more restrictive, because they are analytic rather than merely linear and hence require fields like $\mathbb{C}$, $\mathbb{R}$ or the p-adic rationals $\widehat{\mathbb{Q}}_p$. A good question is: which Lie-like group structures do Lie algebras correspond to, for the other fields? A good answer is: algebraic groups, which are to algebraic geometry what Lie groups are to differential geometry. See, for example, part III of [**92**] for an introduction.

The main relationship between real Lie groups and algebras is summarised by:

Theorem 1.4.3 *To any finite-dimensional real Lie algebra $\mathfrak{g}$, there is a unique connected simply-connected Lie group $\widetilde{G}$, called the universal cover group. If G is any other connected Lie group with Lie algebra $\mathfrak{g}$, then there exists a discrete subgroup H of the centre of $\widetilde{G}$, such that $G \cong \widetilde{G}/H$ and $H \cong \pi_1(G)$, the fundamental group of G.*

The definitions of simply-connected and π_1 are given in Section 1.2.3. The universal cover $\widetilde{\mathbb{R}}$ of the Lie algebra $\mathbb{R}$ is the additive group $\mathbb{R}$; the circle $G = S^1$ has the same Lie algebra and can be written as $S^1 \cong \mathbb{R}/\mathbb{Z}$. The real Lie groups $\mathrm{SU}_2(\mathbb{C})$ and $\mathrm{SO}_3(\mathbb{R})$ both have Lie algebra $\mathfrak{so}_3(\mathbb{R})$; $\mathrm{SU}_2(\mathbb{C}) \cong S^3$ is the universal cover, and $\mathrm{SO}_3(\mathbb{R}) \cong \mathrm{SU}_2(\mathbb{C})/\{\pm I_2\}$ is the 3-sphere with antipodal points identified. $\pi_1(\mathrm{SL}_2(\mathbb{R})) \cong S^1$, and its universal cover (see Question 2.4.4) is an example of a Lie group that is not a matrix group.

So the classification of (connected) Lie groups reduces to the much simpler classification of Lie algebras, together with the classification of discrete groups in the centre of the corresponding $\widetilde{G}$. The condition that G be connected is clearly necessary, as the direct product of a Lie group with any discrete group leaves the Lie algebra unchanged.

Lie group structure theory is merely a major generalisation of linear algebra. The basic constructions familiar to undergraduates have important analogues valid in many Lie groups. For instance, in our youth we were taught to solve linear equations and invert matrices by reducing a matrix to row-echelon form using row operations. This says that any matrix $A \in \mathrm{GL}_n(\mathbb{C})$ can be factorised $A = BPN$, where N is upper-triangular with 1's on the diagonal, P is a permutation matrix and B is an upper-triangular matrix. This is essentially the Bruhat decomposition of the Lie group $\mathrm{GL}_n(\mathbb{C})$. More generally (where it applies to any 'reductive' Lie group G), P will be an element of the so-called Weyl group of G, and B will be in a 'Borel subgroup'. For another example, everyone knows that any nonzero real number x can be written uniquely as $x = (\pm 1) \cdot |x|$, and many of us remember that any invertible matrix $A \in \mathrm{GL}_n(\mathbb{R})$ can be uniquely written as a product $A = OP$, where O is orthogonal and P is positive-definite. More generally, this is called the Cartan decomposition for a real semi-simple Lie group. This encourages us to interpret a linear algebra theorem as a special case of a Lie group theorem . . . a squirrel.

1.4.3 Simple Lie algebras

The reader already weary of such algebraic tedium won't be surprised to read that the typical algebraic definitions can be imposed on Lie theory. The analogue of direct product of groups here is direct sum $\mathfrak{g}_1 \oplus \mathfrak{g}_2$, with bracket $[(x_1, x_2), (y_1, y_2)] = ([x_1 y_1]_1, [x_2 y_2]_2)$. Semi-direct sum is defined as usual. The analogue of normal subgroup here is called an *ideal*: a subspace $\mathfrak{h}$ of $\mathfrak{g}$ such that $[\mathfrak{g}\mathfrak{h}] := \mathrm{span}\{[xy] \mid x \in \mathfrak{g}, y \in \mathfrak{h}\}$ is contained in $\mathfrak{h}$. A Lie group N is a normal subgroup of Lie group G iff the Lie algebra of N is an ideal of that of G. Given an ideal $\mathfrak{h}$ of a Lie algebra $\mathfrak{g}$, the quotient space $\mathfrak{g}/\mathfrak{h}$ has a natural Lie algebra structure; if $\varphi : \mathfrak{g}_1 \to \mathfrak{g}_2$ is a Lie algebra homomorphism, then the kernel $\ker(\varphi)$ is an ideal of $\mathfrak{g}_1$ and the image $\varphi(\mathfrak{g}_1)$ is a subalgebra of $\mathfrak{g}_2$ isomorphic to $\mathfrak{g}_1/\ker(\varphi)$. The name 'ideal' comes from number theory (Section 1.7.1). The centre $Z(\mathfrak{g}) := \{x \in \mathfrak{g} \mid [x\mathfrak{g}] = 0\}$ of $\mathfrak{g}$ always forms an ideal, as does $[\mathfrak{g}\mathfrak{g}]$.

A *simple* Lie algebra is one with no proper ideals. It is standard though to exclude the one-dimensional Lie algebras, much like is often done with the cyclic groups $\mathbb{Z}_p$. A *semi-simple* Lie algebra is defined as any $\mathfrak{g}$ for which $[\mathfrak{g}\mathfrak{g}] = \mathfrak{g}$; it turns out that $\mathfrak{g}$ is semi-simple iff $\mathfrak{g}$ is the (Lie algebra) direct sum $\oplus_i \mathfrak{g}_i$ of simple Lie algebras $\mathfrak{g}_i$. A *reductive* Lie algebra $\mathfrak{g}$ is defined by the relation $[\mathfrak{g}\mathfrak{g}] \oplus Z(\mathfrak{g}) = \mathfrak{g}$; $\mathfrak{g}$ is reductive iff $\mathfrak{g}$ is the direct sum of a semi-simple Lie algebra with an abelian one. Of course simple Lie algebras are more important, but semi-simple and reductive ones often behave similarly.

The finite-dimensional simple Lie algebras constitute an important class of Lie algebras. Although it is doubtful the reader has leapt out of his chair with surprise at this pronouncement, it is good to see explicit indications of this importance.

Simple Lie algebras serve as building blocks for all other finite-dimensional Lie algebras, in the following sense (called *Levi decomposition* – see, for example, chapter III.9 of [**314**] for a proof): any finite-dimensional Lie algebra $\mathfrak{g}$ over $\mathbb{C}$ or $\mathbb{R}$ can be written *as a vector space* in the form $\mathfrak{g} = \mathfrak{r} \oplus \mathfrak{h}$, where $\mathfrak{h}$ is the largest semi-simple Lie subalgebra of $\mathfrak{g}$, and $\mathfrak{r}$ is called the *radical* of $\mathfrak{g}$ and is by definition the maximal 'solvable' ideal of $\mathfrak{g}$. This means $\mathfrak{g}$ is the semi-direct sum of $\mathfrak{r}$ with $\mathfrak{h} \cong \mathfrak{g}/\mathfrak{r}$. A *solvable* Lie algebra is the repeated semi-direct sum by one-dimensional Lie algebras; more concretely, it is isomorphic to a subalgebra of the upper-triangular matrices in some $\mathfrak{gl}_n$. Levi decomposition is the Lie theoretic analogue of the Jordan–Hölder Theorem of Section 1.1.2.

It is reassuring that we can also see the importance of simple Lie algebras geometrically: given any finite-dimensional real Lie group that is 'compact' as a manifold (i.e. bounded and contains all its limit points), its Lie algebra is reductive. Conversely, any reductive real Lie algebra is the Lie algebra of a compact Lie group.

In our struggle to understand a structure, it is healthy to find new ways to capture old information. Let us begin with a canonical way to associate linear endomorphisms (which the basis-hungry of us can regard as square matrices) to elements of the Lie algebra $\mathfrak{g}$. Define the 'adjoint operator' $\mathrm{ad}\, x : \mathfrak{g} \to \mathfrak{g}$ to be the linear map given by $(\mathrm{ad}\, x)(y) = [xy]$. In this language, anti-associativity of the bracket translates to the facts that: (i) for each $x \in \mathfrak{g}$, $\mathrm{ad}\, x$ is a derivation of $\mathfrak{g}$; and (ii) the assignment $x \mapsto \mathrm{ad}\, x$ defines a 'representation' of $\mathfrak{g}$, called the *adjoint representation* (more on this next section).

The point is that there are basis-independent ways to get numbers out of matrices. The *Killing form* $\kappa : \mathfrak{g} \times \mathfrak{g} \to \mathbb{C}$ of a (complex) Lie algebra $\mathfrak{g}$ is defined by

$$\kappa(x|y) := \mathrm{tr}(\mathrm{ad}\, x \circ \mathrm{ad}\, y), \qquad \forall x, y \in \mathfrak{g}. \tag{1.4.7a}$$

By 'trace' we mean to choose a basis, get matrices, and take the trace in the usual way; the answer is independent of the basis chosen. The Killing form is symmetric, respects the linear structure of $\mathfrak{g}$ (i.e. is bilinear) and respects the bracket in the sense that

$$\kappa([xy]|z) = \kappa(x|[yz]), \qquad \forall x, y, z \in \mathfrak{g}. \tag{1.4.7b}$$

This property of κ is called *invariance* (Question 1.4.6(b)).

Table 1.3. *Freudenthal's Magic Square: the Lie algebra* $\mathfrak{g}(\mathcal{A}_1, \mathcal{A}_2)$

$\mathcal{A}_i$	$\mathbb{R}$	$\mathbb{C}$	quat	oct
$\mathbb{R}$	$\mathfrak{so}_3(\mathbb{R})$	$\mathfrak{su}_3(\mathbb{R})$	$\mathfrak{sp}_3(\mathbb{R})$	F_4
$\mathbb{C}$	$\mathfrak{su}_3(\mathbb{R})$	$\mathfrak{su}_3(\mathbb{R}) \oplus \mathfrak{su}_3(\mathbb{R})$	$\mathfrak{su}_6(\mathbb{R})$	E_6
quat	$\mathfrak{sp}_3(\mathbb{R})$	$\mathfrak{su}_6(\mathbb{R})$	$\mathfrak{so}_{12}(\mathbb{R})$	E_7
oct	F_4	E_6	E_7	E_8

Let A, B be two $n \times n$ real matrices; then

$$\text{tr}(AB) = \sum_{i=1}^{n} A_{ii} B_{ii} + \sum_{1 \le i < j \le n} (A_{ij} B_{ji} + A_{ji} B_{ij}),$$

which can be interpreted as an indefinite inner-product on $\mathbb{R}^{n^2}$. Thus the Killing form $\kappa(x|y)$ should be thought of as an inner-product on the vector space $\mathfrak{g}$. It arose historically by expanding the characteristic polynomial $\det(\text{ad } x - \lambda I)$ (Question 1.4.6(c)).

An inner-product on a complex space V has only one invariant: the dimension of the subspace of null vectors. More precisely, define the radical of the Killing form to be

$$\mathfrak{s}(\kappa) := \{x \in \mathfrak{g} \mid \kappa(x|y) = 0 \; \forall y \in \mathfrak{g}\}.$$

By invariance of κ, $\mathfrak{s}$ is an ideal. It is always solvable.

Theorem 1.4.4 (Cartan's criterion) *Let $\mathfrak{g}$ be a (complex or real) finite-dimensional Lie algebra. Then $\mathfrak{g}$ is semi-simple iff κ is nondegenerate, i.e. $\mathfrak{s}(\kappa) = 0$.*

Moreover, $\mathfrak{g}$ is solvable iff $[\mathfrak{g}\mathfrak{g}] \subseteq \mathfrak{s}(\kappa)$. The nondegeneracy of the Killing form plays a crucial role in the theory of semi-simple $\mathfrak{g}$. For instance, it is an easy orthogonality argument that a semi-simple Lie algebra is the direct sum of its simple ideals.

The classification of simple finite-dimensional Lie algebras over $\mathbb{C}$ was accomplished at the turn of the century by Killing and Cartan. There are four infinite families A_r ($r \ge 1$), B_r ($r \ge 3$), C_r ($r \ge 2$) and D_r ($r \ge 4$), and five exceptionals E_6, E_7, E_8, F_4 and G_2. A_r can be thought of as $\mathfrak{sl}_{r+1}(\mathbb{C})$, the $(r+1) \times (r+1)$ matrices with trace 0. The orthogonal algebras B_r and D_r can be identified with $\mathfrak{so}_{2r+1}(\mathbb{C})$ and $\mathfrak{so}_{2r}(\mathbb{C})$, respectively, where $\mathfrak{so}_n(\mathbb{C})$ is all $n \times n$ anti-symmetric matrices $A^t = -A$. The symplectic algebra C_r is $\mathfrak{sp}_{2r}(\mathbb{C})$, i.e. all $2r \times 2r$ matrices A obeying $A\Omega = -\Omega A^t$, where $\Omega = \begin{pmatrix} 0 & I_r \\ -I_r & 0 \end{pmatrix}$ and I_r is the identity. In all these cases the bracket is the commutator (1.4.4). The exceptional algebras can be constructed using, for example, the octonions. For instance, G_2 is the algebra of derivations of octonions. In fact, given any pair $\mathcal{A}_1$, $\mathcal{A}_2$ of normed division rings (so $\mathcal{A}_i$ are $\mathbb{R}$, $\mathbb{C}$, the quaternions or the octonions), there is a general construction of a simple Lie algebra $\mathfrak{g}(\mathcal{A}_1, \mathcal{A}_2)$ (over $\mathbb{R}$) – see, for example, section 4 of [**29**]. The results are summarised in *Freudenthal's Magic Square* (Table 1.3). The interesting thing here is the uniform construction of four of the five exceptional Lie algebras. In Sections 1.5.2 and 1.6.2 we give further reasons for thinking of the exceptional Lie algebras as fitting into a sequence – a nice paradigm whenever multiple exceptional structures are present.

To verify that (1.4.2b) truly is $\mathfrak{sl}_2(\mathbb{C})$, put

$$e = \begin{pmatrix} 0 & 1 \\ 0 & 0 \end{pmatrix}, \qquad f = \begin{pmatrix} 0 & 0 \\ 1 & 0 \end{pmatrix}, \qquad h = \begin{pmatrix} 1 & 0 \\ 0 & -1 \end{pmatrix}. \qquad (1.4.8)$$

The names A, B, C, D have no significance: since the four series start at $r = 1, 2, 3, 4$, they were called A, B, C, D, respectively. Unfortunately, misfortune struck: at random $B_2 \cong C_2$ was called orthogonal, although the affine Coxeter–Dynkin diagrams (Figure 3.2) reveal that it is actually symplectic and only accidentally looks orthogonal. In hindsight the names of the B- and C-series really should have been switched.

For reasons we explain in Section 1.5.2, all semi-simple finite-dimensional Lie algebras over $\mathbb{C}$ have a presentation of the following form.

Definition 1.4.5 (a) *A Cartan$_{ss}$ matrix A is an $n \times n$ matrix with integer entries a_{ij}, such that:*

c1. *each diagonal entry $a_{ii} = 2$;*
c2. *each off-diagonal entry a_{ij}, $i \ne j$, is a nonpositive integer;*
c3. *the zeros in A are symmetric about the main diagonal (i.e. $a_{ij} = 0$ iff $a_{ji} = 0$); and*
c4. *there exists a positive diagonal matrix D such that the product AD is positive-definite (i.e. $(AD)^t = AD$ and $x^t ADx > 0$ for any real column vector $x \ne 0$).*

(b) *Given any Cartan$_{ss}$ matrix A, define a Lie algebra $\mathfrak{g}(A)$ by the following presentation. It has $3n$ generators e_i, f_i, h_i, for $i = 1, \ldots, n$, and obeys the relations*

R1. $[e_i f_j] = \delta_{ij} h_i$, $[h_i e_j] = a_{ij} e_j$, $[h_i f_j] = -a_{ij} f_j$, *and* $[h_i h_j] = 0$, *for all i, j; and*
R2. $(\operatorname{ad} e_i)^{1 - a_{ij}} e_j = (\operatorname{ad} f_i)^{1 - a_{ij}} f_j = 0$ *whenever $i \ne j$.*

'ss' stands for 'semi-simple'; it is standard to call these matrices A 'Cartan matrices', but this can lead to terminology complications when in Section 3.3.2 we doubly generalise Definition 1.4.5(a). As always, $\operatorname{ad} e : \mathfrak{g} \to \mathfrak{g}$ is defined by $(\operatorname{ad} e)f = [ef]$, so if $a_{ij} = 0$ then $[e_i e_j] = 0$, while if $a_{ij} = -1$ then $[e_i [e_i e_j]] = 0$. It is a theorem of Serre (1966) that $\mathfrak{g}(A)$ is finite-dimensional semi-simple, and any complex finite-dimensional semi-simple Lie algebra $\mathfrak{g}$ equals $\mathfrak{g}(A)$ for some Cartan$_{ss}$ matrix A.

The terms 'generators' and 'basis' are sometimes confused. Both build up the whole algebra; the difference lies in which operations you are permitted to use. For a basis, you are only allowed to use linear combinations (i.e. addition of vectors and multiplication by numbers), while for generators you are also permitted multiplication of vectors (the bracket here). 'Dimension' refers to basis, while 'rank' usually refers to generators. For instance, the (commutative associative) algebra of polynomials in one variable x is infinite-dimensional, but the single polynomial x is enough to generate it (so its rank is 1). Although $\mathfrak{g}(A)$ has $3r$ generators, its dimension will usually be far greater.

The entries of Cartan$_{ss}$ matrices are mostly zeros, so it is more transparent to realise them with a graph, called the Coxeter–Dynkin diagram.[13] The diagram corresponding

---

[13] The more common name 'Dynkin diagram' is historically inaccurate. Coxeter was the first to introduce these graphs, originally in the context of reflection groups, but in 1934 he applied them also to Lie algebras. Dynkin's involvement with them occurred over a decade later.

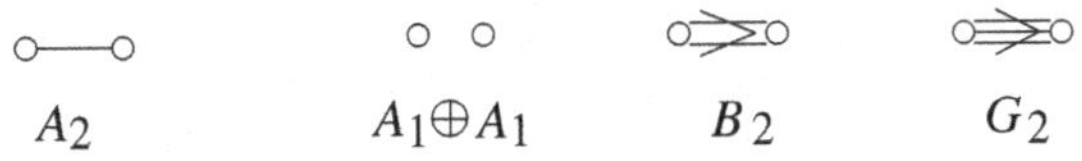

A_2 $A_1 \oplus A_1$ B_2 G_2

Fig. 1.16 The rank 2 Coxeter–Dynkin diagrams.

to matrix A has r nodes; the ith and jth nodes are connected with $a_{ij}a_{ji}$ edges, and if $a_{ij} \ne a_{ji}$, we put an arrow over those edges pointing to i if $a_{ij} < a_{ji}$.

For example, the 2×2 Cartan$_{ss}$ matrices are

$$\begin{pmatrix} 2 & -1 \\ -1 & 2 \end{pmatrix}, \quad \begin{pmatrix} 2 & 0 \\ 0 & 2 \end{pmatrix}, \quad \begin{pmatrix} 2 & -2 \\ -1 & 2 \end{pmatrix}, \quad \begin{pmatrix} 2 & -1 \\ -3 & 2 \end{pmatrix}.$$

The third and fourth matrices can be replaced by their transposes, which correspond to isomorphic algebras. Their Coxeter–Dynkin diagrams are given in Figure 1.16.

To get a better feeling for relations R1, R2, consider a fixed i. The generators $e = e_i, f = f_i, h = h_i$ obey (1.4.2b). In other words, every node in the Coxeter–Dynkin diagram corresponds to a copy of the A_1 Lie algebra. The lines connecting these nodes tell how these r copies of A_1 intertwine. For instance, the first Cartan matrix given above corresponds to the Lie algebra $A_2 = \mathfrak{sl}_3(\mathbb{C})$. The two A_1 subalgebras that generate it (one for each node) can be chosen to be the trace-zero matrices of the form

$$\begin{pmatrix} \star & \star & 0 \\ \star & \star & 0 \\ 0 & 0 & 0 \end{pmatrix}, \quad \begin{pmatrix} 0 & 0 & 0 \\ 0 & \star & \star \\ 0 & \star & \star \end{pmatrix}.$$

The Lie algebra corresponding to a disjoint union $\cup_i \mathcal{D}_i$ of diagrams is the direct sum $\oplus_i \mathfrak{g}_i(\mathcal{D}_i)$ of algebras. Thus we may require the matrix A to be *indecomposable*, or equivalently that the Coxeter–Dynkin diagram be connected, in which case the Lie algebra $\mathfrak{g}(\mathcal{D})$ will be simple. Of the four in Figure 1.16, only the second is decomposable.

Theorem 1.4.6 (a) *The complete list of indecomposable Cartan$_{ss}$ matrices, or equivalently the connected Coxeter–Dynkin diagrams, is given in Figure 1.17. The series A_r, B_r, C_r, D_r are defined for $r \ge 1, r \ge 3, r \ge 2, r \ge 4$, respectively.*
(b) *The complete list of finite-dimensional simple Lie algebras over $\mathbb{C}$ are $\mathfrak{g}(\mathcal{D})$ for each of the Coxeter–Dynkin diagrams in Figure 1.17.*

This classification changes if the field – the choice of scalars – is changed. As always, $\mathbb{C}$ is better behaved than $\mathbb{R}$ because every polynomial can be factorised completely over $\mathbb{C}$ (we say $\mathbb{C}$ is *algebraically closed*). This implies every matrix has an eigenvector over $\mathbb{C}$, something not true over $\mathbb{R}$. Over $\mathbb{C}$, each simple algebra has its own symbol $X_r \in \{A_r, \ldots, G_2\}$; over $\mathbb{R}$, each symbol corresponds to a number of inequivalent algebras. See section VI.10 of [**348**] or chapter 8 of [**214**] for details. For example, 'A_1' corresponds to three different real simple Lie algebras, namely the matrix algebras $\mathfrak{sl}_2(\mathbb{R})$, $\mathfrak{sl}_2(\mathbb{C})$ (interpreted as a *real* vector space) and $\mathfrak{su}_2(\mathbb{C}) \cong \mathfrak{so}_3(\mathbb{R})$. The simple Lie algebra classification is known in any characteristic $p > 7$ (see e.g. [**559**]). Smaller primes usually behave poorly, and the classification for characteristic 2 is probably hopeless.

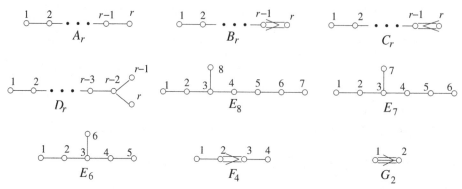

Fig. 1.17 The Coxeter–Dynkin diagrams of the simple Lie algebras.

Simple Lie algebras need not be finite-dimensional. An example is the *Witt algebra* 𝔚itt, defined (over $\mathbb{C}$) by the basis[14] ℓ_n, $n \in \mathbb{Z}$, and relations

$$[\ell_m \ell_n] = (m - n)\ell_{m+n}. \tag{1.4.9}$$

Using the realisation $\ell_n = -ie^{-in\theta}\frac{d}{d\theta}$, 𝔚itt is seen to be the polynomial subalgebra of the complexification $\mathbb{C} \otimes \text{Vect}(S^1)$ (i.e. the scalar field of $\text{Vect}(S^1)$ is changed from $\mathbb{R}$ to $\mathbb{C}$). Incidentally, infinite-dimensional Lie algebras need not have a Lie group: for example, the real algebra $\text{Vect}(S^1)$ has the Lie group $\text{Diff}(S^1)$ of diffeomorphisms $S^1 \to S^1$, but its complexification $\mathbb{C} \otimes \text{Vect}(S^1)$ has no Lie group (Section 3.1.2). The Witt algebra is fundamental to Moonshine. We study it in Section 3.1.2.

Question 1.4.1. Let G be a finite group, and $\mathbb{C}G$ be its group algebra (i.e. all formal linear combinations $\sum_g a_g g$ over $\mathbb{C}$). Verify that $\mathbb{C}G$ becomes a Lie algebra when given the bracket $[g, h] = gh - hg$ (extend linearly to all of $\mathbb{C}G$). Identify this Lie algebra.

Question 1.4.2. Let $\mathbb{K}$ be any field. Find all two-dimensional Lie algebras over $\mathbb{K}$, up to (Lie algebra) isomorphism.

Question 1.4.3. Prove the Witt algebra (1.4.9) is simple.

Question 1.4.4. Prove the Lie algebraic analogue of the statement that any homomorphism $f : G \to H$ between simple groups is either constant or a group isomorphism.

Question 1.4.5. The nonzero quaternions $a1 + bi + cj + dk$, for $a, b, c, d \in \mathbb{R}$, form a Lie group by multiplication (recall that $i^2 = j^2 = k^2 = -1$, $ij = -ji = k$, $jk = -kj = i$ and $ki = -ik = j$). Find the Lie algebra.

Question 1.4.6. (a) Verify that $\text{ad}\,[xy] = \text{ad}\,x \circ \text{ad}\,y - \text{ad}\,y \circ \text{ad}\,x$, for any elements x, y in a Lie algebra 𝔤.
(b) Verify that the Killing form is invariant (i.e. obeys (1.4.7b)) for any Lie algebra.

---

[14] In order to avoid convergence complications, only finite linear combinations of basis vectors are typically permitted in algebra. Infinite linear combinations would require taking some completion.

(c) Let $\mathfrak{g}$ be n-dimensional and semi-simple. Choose any $x \in \mathfrak{g}$. Verify that the coefficient of λ^{n-2} in the characteristic polynomial $\det(\operatorname{ad} x - \lambda I)$ is proportional to $\kappa(x|x)$.

Question 1.4.7. Consider the complex Lie algebra $\mathfrak{g}(A)$, for $A = \begin{pmatrix} 2 & -1 \\ -1 & 2 \end{pmatrix}$, defined in Definition 1.4.5(b).

(a) Prove that a basis for $\mathfrak{g}$ is $\{e_i, f_i, h_i, [e_1 e_2], [f_1 f_2]\}$ and thus that $\mathfrak{g}$ is eight-dimensional. Prove from first principles that $\mathfrak{g}$ is simple.

(b) Verify that the following generates a Lie algebra isomorphism of $\mathfrak{g}$ with $\mathfrak{sl}_3(\mathbb{C})$:

$$e_1 \mapsto \begin{pmatrix} 0 & 1 & 0 \\ 0 & 0 & 0 \\ 0 & 0 & 0 \end{pmatrix}, \quad f_1 \mapsto \begin{pmatrix} 0 & 0 & 0 \\ 1 & 0 & 0 \\ 0 & 0 & 0 \end{pmatrix}, \quad h_1 \mapsto \begin{pmatrix} 1 & 0 & 0 \\ 0 & -1 & 0 \\ 0 & 0 & 0 \end{pmatrix},$$

$$e_2 \mapsto \begin{pmatrix} 0 & 0 & 0 \\ 0 & 0 & 1 \\ 0 & 0 & 0 \end{pmatrix}, \quad f_2 \mapsto \begin{pmatrix} 0 & 0 & 0 \\ 0 & 0 & 0 \\ 0 & 1 & 0 \end{pmatrix}, \quad h_2 \mapsto \begin{pmatrix} 0 & 0 & 0 \\ 0 & 1 & 0 \\ 0 & 0 & -1 \end{pmatrix}.$$

Question 1.4.8. Show that property c3 can be safely dropped. That is, given a $\mathbb{Z}$-matrix A obeying c1, c2 and c4, show that there is a Cartan matrix A' such that the Lie algebra $\mathfrak{g}(A)$ (defined as in Definition 1.4.5(b)) is isomorphic to $\mathfrak{g}(A')$.

Question 1.4.9. Are $\operatorname{Vect}(\mathbb{R})$ and $\operatorname{Vect}(S^1)$ isomorphic as Lie algebras?

1.5 Representations of simple Lie algebras

The representation theory of the simple Lie algebras can be regarded as an enormous generalisation of trigonometry. For instance, the facts that $\frac{\sin(nx)}{\sin(x)}$ can be written as a polynomial in $\cos(x)$ for any $n \in \mathbb{Z}$, and that

$$\frac{\sin(mx)\sin(nx)}{\sin(x)} = \sin((m+n)x) + \sin((m+n-2)x) + \cdots + \sin((m-n)x)$$

for any $m, n \in \mathbb{N}$ are both easy special cases of the theory. Representation theory is vital to the classification and structure of simple Lie algebras, and leads to the beautiful geometry and combinatorics of root systems. The relevance of Lie algebras to Moonshine and conformal field theory – which is considerable – is through their representations. The book [219] is a standard treatment of Lie representation theory; it is presented with more of a conformal field theoretic flavour in [214].

1.5.1 Definitions and examples

Although we have learned over the past couple of centuries that commutativity can be dropped without losing depth and usefulness, most interesting algebraic structures obey some form of associativity. In fact, true associativity (as opposed to, for example, anti-associativity) really simplifies the arithmetic. Given the happy accident that the commutator $[x, y] := xy - yx$ in any associative algebra obeys anti-associativity, it is tempting to seek ways in which associative algebras $\mathfrak{A}$ can 'model' or *represent* a given

Lie algebra. That is, we would like a map $\rho : \mathfrak{g} \to \mathfrak{A}$ that preserves the linear structure (i.e. ρ is linear) and sends the bracket $[xy]$ in $\mathfrak{g}$ to the commutator $[\rho(x), \rho(y)]$ in $\mathfrak{A}$.

In practise groups often appear as symmetries, and algebras as their infinitesimal generators. These symmetries often act linearly. In other words, the preferred associative algebras are usually matrix algebras, and so we are interested in Lie algebra homomorphisms $\rho : \mathfrak{g} \to \mathfrak{gl}_n$. The *dimension* of this representation is the number n.

Completely equivalent to a representation is the notion of '$\mathfrak{g}$-module M', as is the case for finite groups (Section 1.1.3). A $\mathfrak{g}$-*module* is a vector space M on which $\mathfrak{g}$ acts (on the left) by product $x.v$, for $x \in \mathfrak{g}$, $v \in M$. This product must be bilinear, and must obey $[xy].v = x.(y.v) - y.(x.v)$. We use 'module' and 'representation' interchangeably.

Lie algebra modules behave much like finite group modules. Let $\rho_i : \mathfrak{g} \to \mathfrak{gl}(V_i)$ be two representations of $\mathfrak{g}$. We define their *direct sum* $\rho_1 \oplus \rho_2 : \mathfrak{g} \to \mathfrak{gl}(V_1 \oplus V_2)$ as usual by

$$(\rho_1 \oplus \rho_2)(x)(v_1, v_2) = (\rho_1(x)(v_1), \rho_2(x)(v_2)), \qquad \forall x \in \mathfrak{g}, \ v_i \in V_i. \qquad (1.5.1a)$$

Lie algebras are special in that (like groups) we can *multiply* their representations: define the *tensor product representation* $\rho_1 \otimes \rho_2 : \mathfrak{g} \to \mathfrak{gl}(V_1 \otimes V_2)$ through

$$(\rho_1 \otimes \rho_2)(x)(v_1 \otimes v_2) = (\rho_1(x)v_1) \otimes v_2 + v_1 \otimes (\rho_2(x)v_2), \qquad \forall x \in \mathfrak{g}_1, \ v_i \in V_i. \tag{1.5.1b}$$

Recall that the vector space $V_1 \otimes V_2$ is defined to be the span of all $v_1 \otimes v_2$, so the value $(\rho_1 \otimes \rho_2)(x)(v)$ on generic vectors $v \in V_1 \otimes V_2$ requires (1.5.1b) to be extended linearly. It is easy to verify that (1.5.1b) defines a representation of $\mathfrak{g}$; the obvious but incorrect attempt $(\rho_1(x)v_1) \otimes (\rho_2(x)v_2)$ would lose linear dependence on x. As usual, the dimension of $\rho_1 \oplus \rho_2$ is $\dim(\rho_1) + \dim(\rho_2)$, while $\dim(\rho_1 \otimes \rho_2)$ is $\dim(\rho_1) \dim(\rho_2)$.

A rich representation theory requires in addition a notion of *dual* or *contragredient*. Recall that the dual space V^* is the space of all linear functionals $v^* : V \to \mathbb{C}$. Given a $\mathfrak{g}$-module V, the natural module structure on V^* is the contragredient, defined by

$$(x.v^*)(u) = -v^*(x.u), \qquad \forall x \in \mathfrak{g}, \ v^* \in V^*, \ u \in V. \qquad (1.5.1c)$$

This defines $\rho^*(x)v^* \in V^*$ by its value at each $u \in V$. In terms of matrices, (1.5.1c) amounts to choosing $\rho^*(x)$ to be $-\rho(x)^t$, the negative of the transpose of $\rho(x)$. The negative sign is needed for the Lie brackets to be preserved.

The definition of unitary representation ρ for finite groups says each $\rho(g)$ should be a unitary matrix. Since the exponential of a Lie algebra representation should be a Lie group representation, we would like to say that a unitary representation ρ of a Lie algebra should obey $\rho(x)^\dagger = -\rho(x)$ for any $x \in \mathfrak{g}$, where '$\dagger$' is the adjoint (complex conjugate-transpose), that is to say all matrices $\rho(x)$ should be anti-self-adjoint. This works for real Lie algebras, but not for complex ones: if $\rho(x)$ is anti-self-adjoint, then $\rho(\mathrm{i}x) = \mathrm{i}\rho(x)$ will be self-adjoint!

The correct notion of unitary representation $\rho : \mathfrak{g} \to \mathfrak{gl}(V)$ for complex Lie algebras is that there is an anti-linear map $\omega : \mathfrak{g} \to \mathfrak{g}$ obeying $\omega[xy] = -[\omega x, \omega y]$, such that $\rho(x)^\dagger = \rho(\omega x)$. 'Anti-linear' means $\omega(ax + y) = \overline{a}\omega(x) + \omega(y)$. Equivalently, ρ is unitary if the complex vector space V has a Hermitian form $\langle u, v \rangle \in \mathbb{C}$ on it, such that

$$\langle u, \rho(x)v \rangle = \langle \rho(\omega x)u, v \rangle. \tag{1.5.2}$$

For the case of real Lie algebras, $\omega x = -x$ works. For the complex semi-simple Lie algebra $\mathfrak{g}(A)$ of Definition 1.4.5, the most common choice is $\omega e_i = f_i, \omega f_i = e_i, \omega h_i = h_i$ (this is the negative of the so-called *Chevalley involution*).

A *submodule* of a $\mathfrak{g}$-module V is a subspace $U \subseteq V$ obeying $\mathfrak{g}.U \subseteq U$. The obvious submodules are $\{0\}$ and V; an *irreducible module* is one whose only submodules are those trivial ones. Schur's Lemma (Lemma 1.1.3) holds verbatim, provided G is replaced with a finite-dimensional Lie algebra $\mathfrak{g}$, and ρ, ρ' are also finite-dimensional.

Finding all possible modules, even for the simple Lie algebras, is probably hopeless. For example, all simple Lie algebras have uncountably many irreducible ones. However, it is possible to find all of their *finite-dimensional* modules.

Theorem 1.5.1 *Let $\mathfrak{g}$ be a complex finite-dimensional semi-simple Lie algebra of rank r. Then any finite-dimensional $\mathfrak{g}$-module is completely reducible into a direct sum of irreducible modules. Moreover, there is a unique unitary irreducible module $L(\lambda)$ for each r-tuple $\lambda = (\lambda_1, \ldots, \lambda_r)$ of nonnegative integers, and all irreducible ones are of that form.*

Let $P_+ = P_+(\mathfrak{g})$ denote the set of all r-tuples λ of nonnegative integers; $\lambda \in P_+$ are called *dominant integral weights*. The module $L(\lambda)$ is called the irreducible module with *highest weight* λ. We explain how to prove Theorem 1.5.1 and construct $L(\lambda)$ in Section 1.5.3, but to get an idea of what $L(\lambda)$ looks like, consider A_1 from (1.4.2b). For any $\lambda \in \mathbb{C}$, define $x_0 \neq 0$ to formally obey $h.x_0 = \lambda x_0$ and $e.x_0 = 0$. Define inductively $x_{i+1} := f.x_i$ for $i = 0, 1, \ldots$ The span of all x_i, call it $M(\lambda)$, is an infinite-dimensional A_1-module: the calculations $h.x_{i+1} = h.(f.x_i) = ([hf] + fh).x_i = (-2f + fh).x_i$ and $e.x_{i+1} = e.(f.x_i) = ([ef] + fe).x_i = (h + fe).x_i$ show inductively that $h.x_m = (\lambda - 2m)x_m$ and $e.x_m = (\lambda - m + 1)m, x_{m-1}$. The linear independence of the x_i follow from these. $M(\lambda)$ is called a *Verma module* with highest weight λ, and x_0 its highest-weight vector.

Is $M(\lambda)$ unitary? Here, ω interchanges e and f, and fixes h. The calculation

$$\langle x_i, x_i \rangle = \langle f.x_{i-1}, x_i \rangle = \langle x_{i-1}, e.x_i \rangle = (\lambda - i + 1)\langle x_{i-1}, x_{i-1} \rangle \tag{1.5.3}$$

tells us that the norm-squares $\langle x_i, x_i \rangle$ and $\langle x_{i-1}, x_{i-1} \rangle$ can't both be positive, if i is sufficiently large. Thus no Verma module $M(\lambda)$ is unitary.

Now specialise to $\lambda = n \in \mathbb{N} := \{0, 1, 2, \ldots\}$. Since $e.x_{n+1} = 0$ and $h.x_{n+1} = (-n - 2)x_{n+1}$, $M(n)$ contains a *submodule* with highest-weight vector x_{n+1}, isomorphic to $M(-n - 2)$. x_{n+1} is called a *singular* or *null vector*, because by (1.5.3) it has norm-squared $\langle x_{n+1}, x_{n+1} \rangle = 0$. In other words, we could set $x_{n+1} := 0$ and still have an A_1-module – a *finite-dimensional* module $L(n) := M(n)/M(-n - 2)$ with basis $\{x_0, x_1, \ldots, x_n\}$ and dimension $n + 1$. This basis is orthogonal and $L(n)$ is unitary.

For example, the basis $\{x_0, x_1\}$ of $L(1)$ recovers the representation $\mathfrak{sl}_2(\mathbb{C})$ of (1.4.8). The *adjoint representation* of Section 1.5.2 is $L(2)$.

The situation for the other simple Lie algebras X_r is similar (Section 1.5.3). On the other hand, non-semi-simple Lie algebras have a much more complicated representation theory. They have finite-dimensional modules that aren't completely reducible. For example, given any finite-dimensional representation $\rho : \mathfrak{g} \to \mathfrak{gl}(V)$ of any *solvable* Lie algebra $\mathfrak{g}$, a basis can be found for V such that every matrix $\rho(x)$ will be upper-triangular (i.e. the entries $\rho(x)_{ij}$ will equal 0 when $i > j$) – see Lie's Theorem in section 4.1 of [**300**]. This implies that any finite-dimensional irreducible module of a solvable $\mathfrak{g}$ is one-dimensional, and thus a finite-dimensional representation ρ will be completely reducible iff all matrices $\rho(x)$ are simultaneously diagonalisable. See Question 1.5.2.

1.5.2 The structure of simple Lie algebras

Representation theory is important in the structure theory of the Lie algebra itself, and as such is central to the classification of simple Lie algebras. In particular, any Lie algebra $\mathfrak{g}$ is itself a $\mathfrak{g}$-module with action $x.y := (\mathrm{ad}\,x)(y) = [xy]$ – the so-called adjoint representation. In this subsection we use this representation to associate a Cartan matrix to each semi-simple $\mathfrak{g}$.

Consider for concreteness the $\mathfrak{g} = \mathfrak{sl}_n(\mathbb{C})$, the Lie algebra of all trace-0 $n \times n$ matrices, for $n \geq 2$. Let $\mathfrak{h}$ be the set of all diagonal trace-0 matrices. Then the matrices in $\mathfrak{h}$ commute with themselves, so $\mathfrak{h}$ is an abelian Lie subalgebra of $\mathfrak{g}$. Restricting the adjoint representation of $\mathfrak{g}$, we can regard $\mathfrak{g}$ as an $(n^2 - 1)$-dimensional $\mathfrak{h}$-module. Unlike most $\mathfrak{h}$-modules, this one is completely reducible.

In particular, let $E_{(ab)}$ be the $n \times n$ matrix with entries $(E_{(ab)})_{ij} = \delta_{ai}\delta_{bj}$, that is with 0's everywhere except for a '1' in the ab entry. Since $E_{(ab)}E_{(cd)} = \delta_{bc}E_{(ad)}$, we get

$$[E_{(ab)}, E_{(cd)}] = \delta_{bc}E_{(ad)} - \delta_{ad}E_{(cb)}. \tag{1.5.4a}$$

Now, a basis for $\mathfrak{h}$ is $A_a = E_{(a,a)} - E_{(a+1,a+1)}$ for $a = 1, \ldots, n - 1$. Thus

$$[A_a, E_{(cd)}] = (\delta_{ad} + \delta_{a+1,c} - \delta_{ac} - \delta_{a+1,d})E_{(cd)} \tag{1.5.4b}$$

and the basis $\{E_{(cd)}\}_{1 \leq c \neq d \leq n} \cup \{A_a\}_{1 \leq a < n}$ of $\mathfrak{g}$ simultaneously diagonalises all endomorphisms $\mathrm{ad}\,A_a$. In other words, this representation $\mathrm{ad}\,\mathfrak{h}$ decomposes into a direct sum of one-dimensional $\mathfrak{h}$-modules. Define functionals $\alpha_{(cd)} \in \mathfrak{h}^*$ by

$$\alpha_{(cd)}(A_a) = \delta_{ad} + \delta_{a+1,c} - \delta_{ac} - \delta_{a+1,d}.$$

Then we can write

$$\mathfrak{g} = \oplus_{1 \leq c \neq d \leq n}\mathbb{C}E_{(cd)} \oplus \mathrm{span}\{A_a\}_{1 \leq a < n} = \oplus_{\alpha \in \Phi}\mathfrak{g}_\alpha \oplus \mathfrak{h}, \tag{1.5.4c}$$

where $\Phi = \{\alpha_{(cd)}\}_{1 \leq c \neq d \leq n}$ and $\mathfrak{g}_{\alpha_{(cd)}} = \mathbb{C}E_{(cd)}$. The functional $\alpha = \alpha_{(cd)}$ is called a *root* because $\alpha(A)$ is the eigenvalue of the operator $\mathrm{ad}\,A$ on the eigenspace $\mathbb{C}E_{(cd)}$ and thus is a *root* of the characteristic polynomial of $\mathrm{ad}\,A$. We avoid calling 0 (the functional for $\mathfrak{h}$) a root because it behaves differently, for example $\mathfrak{g}_0 = \mathfrak{h}$ has dimension $n - 1$ but all other $\mathfrak{g}_\alpha$ have dimension 1. In Section 3.3.1 we identify 0 though as a precursor to the so-called imaginary roots of Kac–Moody algebras.

From the identity

$$(\operatorname{ad} A)[xy] = [(\operatorname{ad} A)x, y] + [x, (\operatorname{ad} A)y]$$

(which holds in any Lie algebra), or more concretely from (1.5.4a), we see that the decomposition (1.5.4c) defines a grading $[\mathfrak{g}_\alpha, \mathfrak{g}_\beta] \subseteq \mathfrak{g}_{\alpha+\beta}$, for any roots $\alpha, \beta \in \Phi$, where we put $\mathfrak{g}_{\alpha+\beta} = \{0\}$ if $\alpha + \beta \notin \Phi$. In fact, a little more care verifies that equality always holds:

$$[\mathfrak{g}_\alpha, \mathfrak{g}_\beta] = \mathfrak{g}_{\alpha+\beta}, \qquad \forall \alpha, \beta \in \Phi. \tag{1.5.4d}$$

In Question 1.5.3 you compute the Killing form (1.4.7a). We find that $\kappa(E_{(ab)}|E_{(cd)}) = 0$ unless $(d, c) = (a, b)$, and that κ is positive-definite when restricted to the real $(n-1)$-dimensional space $\mathfrak{h}_\mathbb{R}$ spanned over $\mathbb{R}$ by $A_1, \ldots, A_{n-1}$.

The roots $\alpha_1 = \alpha_{(1,2)}, \ldots, \alpha_{n-1} = \alpha_{(n-1,n)}$ form a basis Δ for the dual space $\mathfrak{h}^*$, and are called *simple roots*. Explicitly, the root $\alpha_{(cd)} \in \Phi$ is

$$\alpha_{(cd)} = \begin{cases} \alpha_c + \alpha_{c+1} + \cdots + \alpha_{d-1} & \text{if } c < d \\ -\alpha_d - \alpha_{d+1} - \cdots - \alpha_{c-1} & \text{if } c > d \end{cases}.$$

Note that for each root $\alpha = \alpha_{(ab)}$, the elements $e_\alpha := E_{(ab)}$, $f_\alpha := E_{(ba)}$, $h_\alpha := A_a - A_b$ span a copy of $\mathfrak{sl}_2$. In particular, the $\mathfrak{sl}_2$'s coming from the simple roots α_i generate all of $\mathfrak{sl}_n(\mathbb{C})$, thanks to the grading (1.5.4d). For each $\alpha_i, \alpha_j \in \Delta$, let

$$a_{ij} = \alpha_i(h_{\alpha_j}) = \begin{cases} 2 & \text{if } i = j \\ -1 & \text{if } |i - j| = 1 \\ 0 & \text{otherwise} \end{cases}.$$

This defines a Cartan matrix A. To verify that $\mathfrak{g}(A)$ is $\mathfrak{sl}_n(\mathbb{C})$, do calculations such as

$$[e_{\alpha_i}[e_{\alpha_i} e_{\alpha_{i\pm1}}]] \in \mathfrak{g}_{2\alpha_i + \alpha_{i\pm1}} = \{0\}.$$

This analysis continues to hold for any semi-simple $\mathfrak{g}$. The space $\mathfrak{h}$ of diagonal matrices becomes any subalgebra of $\mathfrak{g}$, all of whose elements x have diagonalisable operator $\operatorname{ad} x$. Any maximal such Lie subalgebra is called a *Cartan subalgebra*. Since almost every polynomial has distinct roots, almost every matrix is diagonalisable; for semi-simple $\mathfrak{g}$, almost every $\operatorname{ad} x$ is diagonalisable. A Cartan subalgebra is necessarily abelian.

Given a Cartan subalgebra $\mathfrak{h}$, we get a *root-space decomposition*

$$\mathfrak{g} = \oplus_{\alpha \in \Phi} \mathfrak{g}_\alpha \oplus \mathfrak{h} \tag{1.5.5a}$$

as in (1.5.4c), by simultaneously diagonalising all $\operatorname{ad} \mathfrak{h}$. The $\alpha \in \Phi \subset \mathfrak{h}^*$ are called roots as before; the *root spaces* $\mathfrak{g}_\alpha$ are defined to be the simultaneous eigenspaces

$$\mathfrak{g}_\alpha := \{x \in \mathfrak{g} \mid [hx] = \alpha(h)x\}. \tag{1.5.5b}$$

The $\mathfrak{g}_\alpha$ are always one-dimensional and define a grading as in (1.5.4d). The Killing form κ is a nondegenerate inner-product, with $\kappa(\mathfrak{g}_\alpha|\mathfrak{g}_\beta) = 0$ unless $\beta = -\alpha$. The finite set Φ of roots is called the *root system*; the full algebra $\mathfrak{g}$ can be reconstructed directly from Φ.

Each $\mathfrak{g}$ has uncountably many possible Cartan subalgebras. They are related by auto-morphisms of $\mathfrak{g}$ – in fact 'inner automorphisms' $\exp(\mathrm{ad}\,x)$ (Section 1.5.4) – so they yield equivalent root systems Φ. Let $N(\mathfrak{h})$ denote the set of all inner automorphisms that map the space $\mathfrak{h}$ onto itself, and let $C(\mathfrak{h}) = \exp(\mathrm{ad}\,\mathfrak{h})$ denote the set of all inner automor-phisms that fix $\mathfrak{h}$ pointwise. Then $C(\mathfrak{h})$ is a normal subgroup of $N(\mathfrak{h})$, and the quotient $N(\mathfrak{h})/C(\mathfrak{h})$ of these continuous groups is a finite group called the *Weyl group* W. It is a symmetry of the data of $\mathfrak{g}$, as we will see.

The Killing form identifies $\mathfrak{h}$ and its dual (this is the raising/lowering of indices familiar to any physicist, or transpose familiar to everyone else). We thus get an inner-product on the dual space $\mathfrak{h}^*$, positive-definite on the real span of the roots. For increased readability, we write $(\beta|\beta')$ in place of $\kappa(\beta|\beta')$, for $\beta, \beta' \in \mathfrak{h}^*$. The Weyl group W acts on $\mathfrak{h}^*$; in particular it is generated by the reflections

$$r_\alpha(\beta) = \beta - 2\frac{(\beta|\alpha)}{(\alpha|\alpha)}\alpha \tag{1.5.5c}$$

through each root $\alpha \in \Phi$ (recall Question 1.2.5). The Weyl group W permutes the roots and preserves the Killing form. Each reflection r_α fixes the hyperplane orthogonal to α. Removing those hyperplanes decomposes $\mathfrak{h}^*$ into connected components, one for every element of W. Choose one at random and call it the *positive chamber C*.

The $\mathbb{Z}$-span of the roots $\alpha \in \Phi$ is called the root lattice of $\mathfrak{g}$; it is positive-definite, the orthogonal direct sum of copies of $\mathbb{Z}$ and the lattices A_n, D_n, E_6, E_7, E_8 of Section 1.2.1, all appropriately scaled. The Weyl group is a group of automorphisms of the root lattice, normal and of small index in the full automorphism group.

Let $\alpha_1, \ldots, \alpha_r$ be the roots orthogonal to the walls of the positive chamber C, with the sign of each α_i chosen so that $(\alpha_i|C)$ is positive. Then those α_i form a basis Δ for $\mathfrak{h}^*$, called a *base*; the α_i are called *simple roots*. Moreover, given any root $\alpha \in \Phi$, either α or $-\alpha$ lies in $\mathbb{N}\alpha_1 + \cdots + \mathbb{N}\alpha_r$ – we say α is *positive* or *negative*, respectively. The root-space decomposition (1.5.5a) can be written in the form

$$\mathfrak{g} = \eta_+ \oplus \mathfrak{h} \oplus \eta_-, \tag{1.5.5d}$$

called a *triangular decomposition*, where $\eta_\pm$ is the sum of the positive (negative) root spaces. The grading implies $[\mathfrak{h}\mathfrak{h}] = 0$, $[\eta_\pm\eta_\pm] \subseteq \eta_\pm$, $[\mathfrak{h}\eta_\pm] \subseteq \eta_\pm$. Any Lie algebra with a triangular decomposition has Verma modules, as we will see [**432**].

Once we have a base Δ, we get a Cartan matrix A (and hence a Coxeter–Dynkin diagram) through the formula

$$a_{ij} = 2\frac{(\alpha_i|\alpha_j)}{(\alpha_j|\alpha_j)}.$$

For each simple root $\alpha_i \in \Delta$, we get elements $e_i \in \mathfrak{g}_{\alpha_i}$, $f_i \in \mathfrak{g}_{-\alpha_i}$, $h_i \in \mathfrak{h}$ that span a copy of $\mathfrak{sl}_2(\mathbb{C})$, and together these $3r$ elements generate all of $\mathfrak{g}$. In fact, these are the elements referred to in Definition 1.4.5(b), and $\mathfrak{g}$ is isomorphic to that Lie algebra $\mathfrak{g}(A)$. The cardinality r of any base is called the *rank* of $\mathfrak{g}$. Incidentally, an arrow between vertices i, j in a diagram always points towards the simple root of smaller norm.

Thus we get a Coxeter–Dynkin diagram from $\mathfrak{g}$ by making two arbitrary choices: a Car-tan subalgebra $\mathfrak{h}$ and a positive chamber C. Different choices are related by symmetries

Table 1.4. *The simple roots and fundamental weights for the classical algebras*

Algebra	Simple root α_i	Fundamental weight ω_i
A_r	$e_i - e_{i+1}$, $1 \le i \le r$	$\sum_{j=1}^{i} e_j - \frac{i}{r+1} \sum_{j=1}^{r+1} e_j$
B_r	$e_i - e_{i+1}$, $1 \le i < r$	$e_1 + \cdots + e_i$, $1 \le i < r$
	$2e_r$	$\frac{1}{2}(e_1 + \cdots + e_r)$
C_r	$\sqrt{2}(e_i - e_{i+1})$, $1 \le i < r$	$\frac{1}{\sqrt{2}}(e_1 + \cdots + e_i)$, $1 \le i \le r$
	$\sqrt{2}e_r$	
D_r	$e_i - e_{i+1}$, $1 \le i < r$	$e_1 + \cdots + e_i$, $1 \le i < r - 1$
	$e_{r-1} + e_r$	$\frac{1}{2}(e_1 + e_2 + \cdots + e_{r-2} + e_{r-1} - e_r)$, $i = r - 1$
		$\frac{1}{2}(e_1 + e_2 + \cdots + e_r)$, $i = r$

(inner automorphisms) of $\mathfrak{g}$, and the resulting diagram is uniquely determined. This is a powerful paradigm: to understand and classify a rigid structure, find and study a combinatorial characterisation. Later we apply this strategy to conformal field theories.

These choices though should disturb the mathematician in us. Perhaps the presence of the Weyl group in the following is a hint that we are doing Lie theory badly. Just as the vector space 'symmetry' GL_n is the artificial consequence of choosing a basis, so is the Weyl group the bad karma caused by selecting one positive chamber over all others. Probably an approach based on Vogel's universal Lie algebra (Section 1.6.2) will ultimately be preferable.

In any case, we are most interested in the Killing form and Weyl group restricted to $\mathfrak{h}^*$. Given simple roots α_i, define *fundamental weights* $\omega_i \in \mathfrak{h}^*$ to be the dual basis $(\omega_i | \alpha_j) = \delta_{ij}$. They lie on the edges of the chamber C. Their $\mathbb{Z}$-span is the lattice dual to the root lattice, called the *weight lattice*. Denote by P_+ the intersection of the weight lattice with C, so $\lambda \in P_+$ if and only if $\lambda = \sum_{i=1}^{r} \lambda_i \omega_i$ where each *Dynkin label* λ_i lies in $\mathbb{N}$. These $\lambda \in \mathbb{N}$, called *dominant integral weights*, are the r-tuples of Theorem 1.5.1.

Table 1.4 gives the α_i and ω_i for the classical algebras, using an orthonormal basis of $\mathbb{R}^r$ ($\mathbb{R}^{r+1}$ for A_r). Nodes are labelled as in Figure 1.17 – this is the labelling used in, for example, [328] but not by all other authors. The table makes manifest the Killing form on $\mathfrak{h}^*$, and is useful in the study of affine Kac–Moody algebras (Section 3.2). More data for the simple Lie algebras, including the exceptional ones (avoided here for reasons of brevity), can be found in section 6.7 of [328], chapter 7 of [214], and especially pages 265–90 of [84].

The Weyl group of $\mathfrak{g} = \mathfrak{sl}_n(\mathbb{C})$ is the symmetric group $\mathcal{S}_n$ and acts on $\mathfrak{h}^*$ by permuting the subscripts: $\sigma \sum_i h_i \omega_i = \sum_i h_i \omega_{\sigma i}$. Figure 1.18 gives the root systems of the semi-simple Lie algebras of rank 2. A choice of simple roots is indicated by the numerals '1' and '2'. In Figure 1.19 a portion of the weight lattices of $\mathfrak{g} = \mathfrak{sl}_2(\mathbb{C})$ and $\mathfrak{g} = \mathfrak{sl}_3(\mathbb{C})$ are displayed, along with simple roots and fundamental weights, and the Weyl reflections $r_i = r_{\alpha_i}$ through the simple roots. Note the $\mathcal{S}_2 \cong \{\pm 1\}$ symmetry of the A_1 weight lattice, and the $\mathcal{S}_3$ symmetry of the A_2 weight lattice.

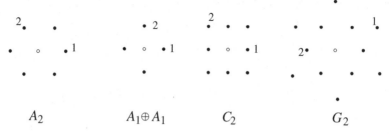

$$A_2 \qquad A_1 \oplus A_1 \qquad C_2 \qquad G_2$$

Fig. 1.18 The root systems of the rank 2 algebras.

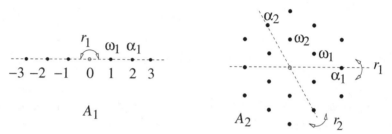

$$A_1 \qquad\qquad A_2$$

Fig. 1.19 Some of the weights of A_1 and A_2.

								$A_1, 2$
							$u_1, 1$	$A_2, 3$
						$0, 1$	$A_1, 3$	$G_2, 7$
					$0, 1$	$2u_1, 2$	$3A_1, 4$	$D_4, 8$
				$0, 2$	$A_1, 5$	$A_2, 8$	$C_3, 14$	$F_4, 26$
			$0, 1$	$2u_1, 3$	$A_2, 2$	$A_2, 9$	$A_5, 15$	$E_6, 27$
		$u_1, 2$	$A_1, 4$	$3A_1, 8$	$C_3, 14$	$A_5, 20$	$D_6, 32$	$E_7, 56$
	$A_1, 3$	$A_2, 8$	$G_2, 14$	$D_4, 28$	$F_4, 52$	$E_6, 78$	$E_7, 133$	$E_8, 248$

Fig. 1.20 Cvitanović's Magic Triangle.

The first hint that the exceptional Lie algebras are not especially exceptional (i.e. that they fall into a common series) is Freudenthal's Magic Square (Table 1.3). A second is *Cvitanović's Magic Triangle* [**126**], [**129**] (Figure 1.20). The clearest example of a family of Lie algebras is A_n, where in fact the representation rings of smaller A_n embed in those of the larger (the characters are the Schur polynomials in infinitely many variables, appropriately restricted). For example, the formulae $L(\omega_1) \otimes L(\omega_k) = L(\omega_1 + \omega_k) \oplus L(\omega_{k+1})$ and $\dim L(\omega_k) = \binom{n+1}{k}$ hold for all k and A_n, although, for example, $L(\omega_2) = L(0)$ and $L(\omega_3) = 0$ for A_1. Something similar (though more complicated) happens for the 'exceptional series', i.e. the Lie algebras in the bottom row of the Magic Triangle. For instance, the decomposition of various powers $\mathfrak{g}^{\otimes k}$ of the adjoint modules into irreducibles take the same form (e.g. $\mathfrak{g} \otimes \mathfrak{g} = L(0) \oplus Y_2 \oplus Y_2^* \oplus \mathfrak{g} \oplus X_2$, where for, for example, $\mathfrak{g} = G_2, F_4, E_8$, respectively we have $Y_2 = L(2\omega_1), L(2\omega_1), L(2\omega_7), Y_2^* = L(2\omega_2), L(2\omega_4), L(\omega_1), \mathfrak{g} = L(\omega_1), L(\omega_1), L(\omega_7)$ and $X_2 = L(3\omega_2), L(\omega_2), L(\omega_6))$, and the dimension of the adjoint representation is given by the uniform equation $\dim \mathfrak{g} = 2(5h^\vee - 6)(h^\vee + 1)/(h^\vee + 6)$, where $h^\vee$ is the dual Coxeter number of Section 3.2.3. For more examples, see [**126**], [**129**] and references therein.

Note that the exceptional series is nested:

$$A_1 \subset A_2 \subset G_2 \subset D_4 \subset F_4 \subset E_6 \subset E_7 \subset E_8.$$

Taking any pair $\mathfrak{h} \subset \mathfrak{g}$, the corresponding entry in the Magic Triangle is the centraliser $\mathfrak{c}$ of $\mathfrak{h}$ in $\mathfrak{g}$, and the number there is the dimension of an irreducible module of $\mathfrak{c}$, unique up to outer automorphism, defined by the decomposition of $\mathfrak{g}$ as a $\mathfrak{c} \oplus \mathfrak{h}$-module. For simplicity Figure 1.20 is watered-down by using Lie algebras in place of Lie groups (e.g. the 0's along the top diagonal are really finite groups) – see [**129**] for details. This exceptional series is explained by Vogel's universal Lie algebra (Section 1.6.2).

1.5.3 Weyl characters

Let $\mathfrak{g}$ be any complex finite-dimensional semi-simple Lie algebra. The analysis of the last subsection on the adjoint representation can be generalised to the other finite-dimensional $\mathfrak{g}$-modules. Recall the notation introduced last subsection. Let Φ^+ be the positive roots. For each $\alpha \in \Phi^+$, choose $e_\alpha \in \mathfrak{g}_\alpha$, $f_\alpha \in \mathfrak{g}_{-\alpha}$ and $h_\alpha \in \mathfrak{h}$ as before, and write e_i, f_i, h_i for these corresponding to the simple root $\alpha_i \in \Delta$. Let ω_i be the fundamental weights, as before.

For all representations $\rho : \mathfrak{g} \to \mathfrak{gl}(V)$ of interest to us, in particular all of the finite-dimensional ones, the matrices $\rho(h)$ for $h \in \mathfrak{h}$ will be simultaneously diagonalisable. The analogue of (1.5.4c) is the *weight-space decomposition*

$$V = \oplus_{\beta \in \Omega(\rho)} V_\beta, \tag{1.5.6a}$$

where these functionals $\beta \in \Omega(\rho) \subset \mathfrak{h}^*$ are called the *weights* of ρ. For example, the non-zero weights of the adjoint representation ad $\mathfrak{g}$ are the roots. For any finite-dimensional ρ, the β all lie in the weight lattice $\mathbb{Z}\omega_1 + \cdots + \mathbb{Z}\omega_r$. These *weight spaces*

$$V_\beta := \{v \in V \mid h.v = \beta(h)v \,\forall h \in \mathfrak{h}\} \tag{1.5.6b}$$

will no longer be one-dimensional in general – the dimension $\dim V_\beta$ is called the multiplicity of β in ρ. The grading (1.5.4d) now becomes

$$f_\alpha V_\beta \subseteq V_{\beta+\alpha}, \qquad e_\alpha V_\beta \subseteq V_{\beta-\alpha}. \tag{1.5.6c}$$

The weight-space decomposition, or equivalently the weights $\beta \in \Omega(\rho)$ and their multiplicities, uniquely determines any finite-dimensional module (up to equivalence). The Weyl group W acts on weights via (1.5.5c), and preserves multiplicities:

$$\dim V_\beta = \dim V_{w\beta}, \qquad \forall w \in W, \ \beta \in \Omega(\rho). \tag{1.5.6d}$$

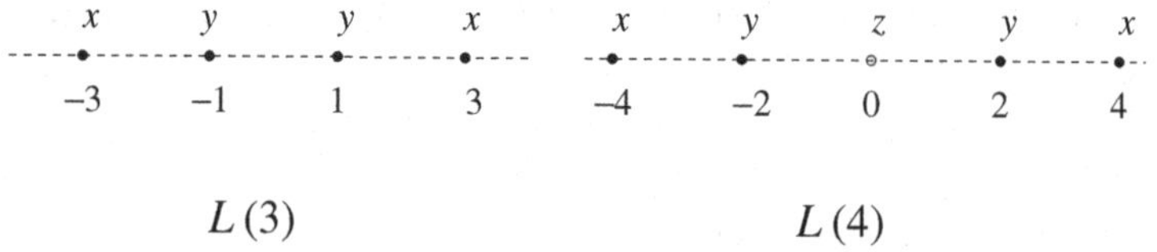

$L(3)$ $L(4)$

Fig. 1.21 The weights of representations of A_1.

In Section 3.2.3 we learn that this innocent symmetry (1.5.6d) is a key to the appearance of modularity in affine Kac–Moody algebras.

For an example, recall the Verma module $M(\lambda)$ for $\mathfrak{sl}_2(\mathbb{C})$ constructed in Section 1.5.1. Strictly speaking we should write $\lambda\omega_1$ for the highest weight λ. This representation has weights $(\lambda - 2j)\omega_1$ for $j = 0, 1, 2, \ldots$, all with multiplicity 1. Moreover, the unitary module $L(n) = L(n\omega_1)$ has weights $(n - 2j)\omega_1$ for $j = 0, 1, \ldots, n$, again all with multiplicity 1. The weight-spaces $L(n)_m$ are $\mathbb{C}x_{(n-m)/2}$. The Weyl group $W \cong \mathbb{Z}_2$ acts here by sending $i\omega_1$ to $-i\omega_1$. See Figure 1.21 for the weights of A_1-representations $L(3\omega_1)$ and $L(4\omega_1)$. We label weights in the same Weyl orbit with the same letter.

Given any functional $\lambda = \sum_{i=1}^{r} \lambda_i \omega_i \in \mathfrak{h}^*$, a *highest-weight module M with highest weight λ* is a $\mathfrak{g}$-module generated by a nonzero vector $v \in M$ obeying

$$e_\alpha.v = 0, \quad \forall \alpha \in \Phi^+, \tag{1.5.7a}$$

$$h_i.v = \lambda_i v, \quad 1 \le i \le r. \tag{1.5.7b}$$

Of course by linearity (1.5.7b) implies that $h_\alpha.v = (\lambda|\alpha)v$ for all positive roots α (not just the simple ones), and more generally $h.v = \lambda(h)v \; \forall h \in \mathfrak{h}$. The module M is generated by v in the sense that M is the span of all vectors of the form

$$x_{(1)} \cdots x_{(m)}.v := x_{(1)}.(\cdots(x_{(m)}.v)\cdots),$$

as the vectors $x_{(j)}$ range over all of $\mathfrak{g}$. This v is called the *highest-weight vector*. The $\mathfrak{g}$-modules of greatest interest to us are the highest-weight ones. The name comes from the fact that for all $\mu \in \Omega(\lambda)$ except $\mu = \lambda$, $\lambda - \mu$ lies in the positive chamber C.

By the *Verma module $M(\lambda)$* we mean the largest or universal or free $\mathfrak{g}$-module with highest weight λ. Any other $\mathfrak{g}$-module with highest weight λ can be constructed from this. To make this more precise, we first define the analogue here of the group algebra $\mathbb{C}G$.

As we know, a basis for $\mathfrak{g}$ is e_α, f_α for all positive roots $\alpha \in \Phi^+$, together with the elements h_i. The *universal enveloping algebra $U(\mathfrak{g})$* is the largest associative algebra generated by those $\|\Phi\| + \|\Delta\|$ symbols e_α, f_α, h_i, which obey all identities of the form $xy - yx = [xy]$ for all $x, y \in \mathfrak{g}$. More precisely, $U(\mathfrak{g})$ is the quotient of the free associative algebra on those $\|\Phi\| + \|\Delta\|$ symbols, with the ideal generated by all elements $xy - yx - [xy]$. The starting point for the theory of $U(\mathfrak{g})$ is:

Theorem 1.5.2 (Poincaré – Birkhoff–Witt)   *A basis for $U(\mathfrak{g})$ is the set of monomials*

$$\left(\prod_\alpha f_\alpha^{m_\alpha}\right)\left(\prod_\alpha e_\alpha^{n_\alpha}\right)\left(\prod_{i=1}^{r} h_i^{p_i}\right),$$

for all choices of integers $m_\alpha \ge 0$, $n_\alpha \ge 0$, $p_i \ge 0$.

The basis element corresponding to $m_\alpha = n_\alpha = p_i = 0$ is denoted 1. The associative algebra $U(\mathfrak{g})$ is not commutative, so to define the products $\prod_\alpha$ we must make some arbitrary ordering of the positive roots Φ^+ – it doesn't matter how we do this. The Poincaré–Birkhoff–Witt Theorem holds for any Lie algebra (not necessarily semi-simple). See the proof and discussion in chapter III of [**348**]. That those monomials span $U(\mathfrak{g})$ is clear; more difficult is to show that they are linearly independent.

In Section 6.2.3 we use $U(\mathfrak{g})$ to construct *quantum groups*. Here what is significant is that its representation theory is identical to that of $\mathfrak{g}$. This isn't deep: the matrices $\rho(x)$, for $x \in \mathfrak{g}$, generate an associative (matrix) algebra. Thus we have replaced the task of finding modules of the *non-associative* algebra $\mathfrak{g}$ with the simpler but equivalent task of finding modules of the *associative* (though infinite-dimensional) algebra $U(\mathfrak{g})$. The relation between $\mathfrak{g}$ and $U(\mathfrak{g})$ is quite analogous to that between G and $\mathbb{C}G$, except that $\mathbb{C}G$ is somewhat simpler due to G already having an associative product.

Let $J(\lambda)$ be the left-ideal of $U(\mathfrak{g})$ generated by all e_α and all $h_i - \lambda_i 1$. This means

$$J(\lambda) = \left\{ \sum_\alpha x_\alpha e_\alpha + \sum_i y_i (h_i - \lambda_i 1) \,|\, x_\alpha, y_i \in U(\mathfrak{g}) \right\}.$$

The Verma module $M(\lambda)$ can now be defined to be the quotient of $U(\mathfrak{g})$ by $J(\lambda)$. It is a (left) $U(\mathfrak{g})$-module, and hence a $\mathfrak{g}$-module. By the Poincaré–Birkhoff–Witt Theorem, the infinite set of elements of the form

$$v_{\{m\}} := \left(\prod_\alpha f_\alpha^{m_\alpha} \right) v, \tag{1.5.8a}$$

for all integers $m_\alpha \geq 0$, forms a basis for $M(\lambda)$. The action of $e_\alpha, f_\alpha, h_i \in \mathfrak{g}$ on these vectors $v_{\{m\}}$ is obtained using the commutation relations of $\mathfrak{g}$ together with (1.5.7). In particular, we find that $v_{\{m\}}$ is an eigenvector for all operators h_i, and corresponds to weight $\lambda - \sum_\alpha m_\alpha \alpha$. Thus the weight-space decomposition of the Verma module $M(\lambda)$ is

$$M(\lambda) = \bigoplus_{\alpha' \in \mathbb{N}\alpha_1 + \cdots + \mathbb{N}\alpha_r} M(\lambda)_{\lambda - \alpha'}, \tag{1.5.8b}$$

where $M(\lambda)_{\lambda - \alpha'}$ has basis consisting of all $v_{\{m\}}$ with $\alpha' = \sum_{\alpha \in \Delta^+} m_\alpha \alpha$.

The Verma module $M(\lambda)$ is indecomposable but may or may not be reducible (see Question 1.5.5). The general way to handle modules that aren't completely reducible is to use quotients, exactly as we did with the composition series for finite groups. In particular, $M(\lambda)$ always has a unique maximal submodule $K(\lambda) \neq M(\lambda)$, and for it the quotient $L(\lambda) := M(\lambda)/K(\lambda)$ is irreducible. More generally, every $U(\mathfrak{g})$-module with highest weight λ can be obtained by quotienting $M(\lambda)$ by some submodule; the quotient $L(\lambda)$ can thus be regarded as the smallest $U(\mathfrak{g})$-module with highest weight λ, and is the module in which we are primarily interested. In particular, the finite-dimensional irreducible modules named in Theorem 1.5.1 are precisely these quotients $L(\lambda)$.

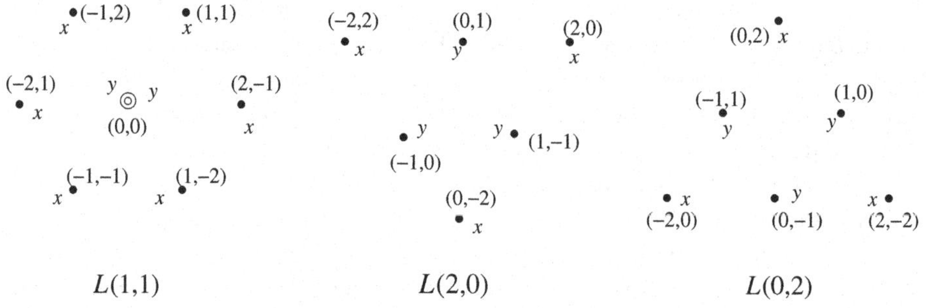

$L(1,1)$ $L(2,0)$ $L(0,2)$

Fig. 1.22 The weights of modules of A_2.

This maximal submodule $K(\lambda)$ is the space of all null-vectors. For dominant integral weights $\lambda \in P_+$, it is the span of all vectors of the form

$$\left(\prod_\alpha f_\alpha^{c_\alpha}\right) (f_i)^{\lambda_i+1} v,$$

for any choice of integers $c_\alpha \in \mathbb{N}$, and any i.

Figure 1.22 gives the weights for A_2-modules $L(\omega_1 + \omega_2)$, $L(2\omega_1)$ and $L(2\omega_2)$. We denote a weight $\beta = \sum_i \beta_i \omega_i$ by its *Dynkin labels* $\beta_i \in \mathbb{Z}$. All multiplicities in Figures 1.21 and 1.22 are 1 except for $L(\omega_1 + \omega_2)_{(0,0)}$, which has multiplicity 2. Incidentally, $L(\omega_1 + \omega_2)$ is the adjoint representation, while $L(2\omega_1)$ and $L(2\omega_2)$ are contragredient.

As usual, it is hard to compare modules directly: ρ and ρ' could be equivalent (i.e. differ merely by a change-of-basis) but look very different. Or given some module, we may wish to decompose it into the direct sum of irreducible modules $L(\lambda^{(i)})$. For finite groups, we use characters to clarify their representation theory, projecting away the extraneous basis-dependent details; Weyl showed that something similar works here.

The *character* of a $\mathfrak{g}$-module V, with weight-space decomposition (1.5.6a), is

$$\mathrm{ch}_V(z) := \sum_{\beta \in \Omega(V)} \dim V_\beta \, e^{\beta(z)}, \tag{1.5.9a}$$

for any $z \in \mathfrak{h}$. If we coordinatise $\mathfrak{h}$ and $\mathfrak{h}^*$ by $z = \sum_i z_i h_i$ and $\beta = \sum_i \beta_i \omega_i$, we can use

$$\beta(z) = \sum_{i=1}^r \beta_i z_i \frac{2}{(\alpha_i|\alpha_i)}. \tag{1.5.9b}$$

For example, for the A_1-module $L(n\omega_1)$ we find

$$\mathrm{ch}_{L(n\omega_1)}(zh) = \sum_{i=0}^n e^{(n-2i)z} = \frac{e^{(n+1)z} - e^{-(n+1)z}}{e^z - e^{-z}}, \tag{1.5.10a}$$

where we obtained the formula on the right by summing the geometric series. Note that its numerator and denominator are alternating sums over the Weyl group $\mathcal{S}_2$ of A_1. By comparison, the character for the Verma module $M(\lambda\omega_1)$ is

$$\mathrm{ch}_{M(\lambda\omega_1)}(zh) = \sum_{i=0}^\infty e^{(\lambda-2i)z} = \frac{e^{\lambda z}}{1 - e^{-2z}}. \tag{1.5.10b}$$

More generally, the Verma module $M(\lambda)$ for $\mathfrak{g}$ has character

$$\text{ch}_{M(\lambda)}(z) = \frac{e^{\lambda(z)}}{\prod_{\alpha \in \Delta^+}(1 - e^{-\alpha(z)})}. \tag{1.5.10c}$$

The *Weyl character formula* expresses the character of any finite-dimensional irreducible module $L(\lambda)$ for any semi-simple $\mathfrak{g}$ as a fraction: the numerator is an alternating sum over the Weyl group W, and the denominator is a product over positive roots $\alpha \in \Delta^+$. More precisely,

$$\text{ch}_\lambda(z) := \text{ch}_{L(\lambda)}(z) = e^{-\rho \cdot z} \frac{\sum_{w \in W} \det(w) e^{w(\lambda + \rho) \cdot z}}{\prod_{\alpha \in \Delta^+}(1 - e^{-\alpha(z)})}, \tag{1.5.11}$$

where $\rho = \sum_{i=1}^r \omega_i$ here is the *Weyl vector*. For a proof see, for example, chapter 14 of [214]. This formula and its generalisations have profound consequences (see Section 3.4.2).

Finite groups have only finitely many irreducible modules, while Lie algebras have infinitely many. Otherwise their theory is quite analogous, and in particular Lie algebra characters work as effectively as finite group characters.

Theorem 1.5.3 *Let $\mathfrak{g}$ be a finite-dimensional semi-simple Lie algebra, and M, N two finite-dimensional modules. Then $\text{ch}_M(z) = \text{ch}_N(z)$ for all $z \in \mathfrak{h}$ iff M and N are equivalent as $\mathfrak{g}$-modules. Moreover, $\text{ch}_{M \oplus N}(z) = \text{ch}_M(z) + \text{ch}_N(z)$, $\text{ch}_{M \otimes N}(z) = \text{ch}_M(z) \text{ch}_N(z)$ and $\text{ch}_{M^*}(z) = \text{ch}_M(\bar{z})$.*

As before, the characters are also enormously simpler than the modules themselves: for example, the smallest nontrivial representation of $\mathfrak{g} = E_8$ is a map from $\mathbb{C}^{248}$ to the space of 248×248 matrices, while its character is a function $\mathbb{C}^8 \to \mathbb{C}$. But why is Weyl's definition (1.5.9a) natural? How did he come up with it?

He used the relation with groups. Consider for concreteness $\mathfrak{g} = A_r$. Given any representation ρ, the map $e^x \mapsto e^{\rho(x)}$ is a representation of the Lie group $G = \text{SL}_{r+1}(\mathbb{C})$ corresponding to $\mathfrak{g}$ (the exponential e^A of a matrix is defined by the usual power series, and always converges). The trace of the matrix $e^{\rho(x)}$ is the *group* character value at $e^x \in G$, so we define it to be the *algebra* character value at $x \in \mathfrak{g}$. Again, it suffices to restrict to representatives of each conjugacy class of G, because the character is a class function. Now, almost every matrix is diagonalisable (since almost any $n \times n$ matrix has n distinct eigenvalues), and so we shouldn't lose much by restricting $x \in \mathfrak{g}$ to *diagonalisable* matrices. Hence we may take our conjugacy class representatives to be *diagonal* matrices $x \in \mathfrak{g}$, i.e. to $x \in \mathfrak{h}$. So the Lie algebra character can be chosen to be a function of $z \in \mathfrak{h}$. Finally, the trace of the matrix $e^{\rho(x)}$ is the sum (with multiplicities) of its eigenvalues, which gives us (1.5.9a). This is the intuition behind Weyl's definition (1.5.9a) of character.

However, different diagonal matrices can be conjugate. For instance in A_1,

$$\begin{pmatrix} 0 & -1 \\ 1 & 0 \end{pmatrix} \begin{pmatrix} a & 0 \\ 0 & b \end{pmatrix} \begin{pmatrix} 0 & -1 \\ 1 & 0 \end{pmatrix}^{-1} = \begin{pmatrix} b & 0 \\ 0 & a \end{pmatrix},$$

so e^{zh} and e^{-zh} lie in the same $G = \mathrm{SL}_2(\mathbb{C})$ conjugacy class and $\mathrm{ch}_M(z) = \mathrm{ch}_M(-z)$. This $z \mapsto -z$ symmetry is the Weyl group action on the Cartan subalgebra $\mathfrak{h} = \mathbb{C}h$. Each character $\mathrm{ch}_{L(\lambda)}$ of any semi-simple $\mathfrak{g}$ is similarly invariant under the Weyl group of $\mathfrak{g}$, thanks to (1.5.6d).

At first glance, it may seem that the Weyl character formula (1.5.9a) is not very practical, at least for large rank. For instance, the numerator of (1.5.9a) for E_8 would involve an alternating sum over the Weyl group, which has about 700 million elements! On the other hand, one alternating sum is very easy to compute: a determinant is an alternating sum over the symmetric group (the Weyl group of the A-series). Since all Weyl groups have symmetric subgroups of relatively small index, the numerators and denominators of (1.5.9a) actually can be computed quite effectively.

It is common practise in physics to use dimensions to specify irreducible modules. For example, the defining representation of $\mathfrak{sl}_3(\mathbb{C})$ is denoted $\mathbf{3}$, and its contragredient by $\overline{\mathbf{3}}$. This is a terrible habit, as many unrelated modules can have identical dimension. For instance, $\mathfrak{sl}_5$ has six different irreducible modules with dimension 175: namely $L(\lambda)$ with $\lambda = (1, 2, 0, 0), (1, 1, 0, 1), (0, 3, 0, 0)$ and their contragredients $(0,0,2,1), (1,0,1,1),$ $(0,0,3,0)$. The practise should rather be to use highest weights, not dimensions, when labelling finite-dimensional modules.

1.5.4 Twisted #1: automorphisms and characters

A fundamental theme of this book is twisting by automorphisms. As we see later, it is central to conformal field theory and string theory, as well as vertex operator algebras, and is implicit in the definition of the McKay–Thompson series T_g. Its role in finite-dimensional Lie theory is more elementary, but can be regarded as a toy model for several of this book's most important subsections.

Let $\mathfrak{g}$ be any Lie algebra over $\mathbb{C}$. An automorphism γ of $\mathfrak{g}$ is an endomorphism (i.e. an invertible linear map $\gamma : \mathfrak{g} \to \mathfrak{g}$) that obeys

$$\gamma[x, y] = [\gamma x, \gamma y], \qquad \forall x, y \in \mathfrak{g}.$$

Write $\mathrm{Aut}(\mathfrak{g})$ for the group of automorphisms of $\mathfrak{g}$. When $\gamma \in \mathrm{Aut}(\mathfrak{g})$ has order $n < \infty$, it is diagonalisable on the space $\mathfrak{g}$ (why?). Hence we can write $\mathfrak{g}$ as a direct sum

$$\mathfrak{g} = \oplus_{k=0}^{n-1} \mathfrak{g}_k \tag{1.5.12a}$$

of eigenspaces of γ, where

$$\gamma x = \xi_n^k x, \qquad \forall x \in \mathfrak{g}_k \tag{1.5.12b}$$

(as always, ξ_n denotes the root of unity $\exp[2\pi \mathrm{i}/n]$). Because γ is an automorphism, (1.5.12a) defines a $\mathbb{Z}_n$-grading on $\mathfrak{g}$, in the sense that

$$[\mathfrak{g}_k, \mathfrak{g}_\ell] \subseteq \mathfrak{g}_{k+\ell}. \tag{1.5.12c}$$

The γ-invariant space $\mathfrak{g}_0$ is a subalgebra of $\mathfrak{g}$, and the other subspaces $\mathfrak{g}_k$ are $\mathfrak{g}_0$-modules.

For example, let $\mathfrak{g} = \mathfrak{sl}_3(\mathbb{C})$ and choose the usual basis e_1, e_2, $e_{12} := [e_1 e_2]$, f_1, f_2, $f_{12} := [f_1 f_2]$, h_1, h_2. There is an order-2 automorphism γ of $\mathfrak{sl}_3$, corresponding to the left–right symmetry of the A_2 Coxeter–Dynkin diagram. It exchanges e_1 and e_2; therefore

$$\gamma e_{12} = [\gamma e_1, \gamma e_2] = [e_2 e_1] = -e_{12}.$$

Continuing in this way, we find that γ exchanges f_1 and f_2, as well as h_1 and h_2, and sends f_{12} to $-f_{12}$. Thus

$$\mathfrak{g}_0 = \mathrm{span}\{e_1 + e_2, f_1 + f_2, h_1 + h_2\},$$
$$\mathfrak{g}_1 = \mathrm{span}\{e_1 - e_2, e_{12}, f_1 - f_2, f_{12}, h_1 - h_2\}.$$

The reader can verify that the Lie subalgebra $\mathfrak{g}_0$ is isomorphic to $\mathfrak{sl}_2(\mathbb{C})$, while $\mathfrak{g}_1$ is the irreducible five-dimensional A_1-module.

Every Lie algebra has nontrivial automorphisms. For instance, let $x \in \mathfrak{g}$ be such that the operator $\mathrm{ad}\, x$ on $\mathfrak{g}$ is nilpotent, that is there is some integer k such that

$$(\mathrm{ad}\, x)^k y := [x[x \cdots [xy] \cdots]] = 0, \qquad \forall y \in \mathfrak{g}.$$

For instance, any $x = e_i$ or $x = f_j$ works when $\mathfrak{g}$ is semi-simple. Then $\exp(\mathrm{ad}\, x)$ (defined by the usual power series expansion) is a well-defined invertible operator on $\mathfrak{g}$ and is in fact an automorphism. These automorphisms $\exp(\mathrm{ad}\, x)$ together generate a normal subgroup of $\mathrm{Aut}(\mathfrak{g})$ called the *inner automorphisms* of $\mathfrak{g}$. The quotient of $\mathrm{Aut}(\mathfrak{g})$ by the inner automorphisms defines a group called the *outer automorphisms*.

For example, for $\mathfrak{g} = \mathfrak{sl}_n(\mathbb{C})$ the inner automorphisms form a group isomorphic to $\mathrm{PGL}_n(\mathbb{C}) \cong \mathrm{GL}_n(\mathbb{C})/\{\mathbb{C}^\times I_n\}$, and the group of outer automorphisms is $\mathbb{Z}_2$ for $n > 2$ (and $\{1\}$ for $n = 2$). The outer automorphism takes a matrix $x \in \mathfrak{sl}_n(\mathbb{C})$ to $-x^t$.

As an aside, the group of inner automorphisms of a simple Lie algebra over $\mathbb{C}$ is always a simple group (though infinite). It could be hoped that the same would be true if instead we consider Lie algebras over a finite field $\mathbb{F}_q$. Indeed this is the case (except for five small counterexamples, involving the fields $\mathbb{F}_2$ and $\mathbb{F}_3$). This gives rise to nine of the infinite families of finite simple groups of Lie type (Section 1.1.2); the seven remaining ones are various twists of these groups.

Given any two Cartan subalgebras $\mathfrak{h}_1$, $\mathfrak{h}_2$ of a simple algebra $\mathfrak{g}$, an inner automorphism can be found mapping $\mathfrak{h}_1$ to $\mathfrak{h}_2$ (we say the inner automorphisms *act transitively* on the set of Cartan subalgebras). Moreover, for any choice of Cartan subalgebra $\mathfrak{h}$, if we take the subgroup of inner automorphisms mapping $\mathfrak{h}$ to itself, and quotient it by the subgroup of inner automorphisms fixing $\mathfrak{h}$ pointwise, then we get the Weyl group of $\mathfrak{g}$. This means that (modulo an inner automorphism) an automorphism of $\mathfrak{g}$ permutes the simple roots; conversely, this permutation uniquely determines it. In other words, the outer automorphisms for semi-simple $\mathfrak{g}$ are in a natural one-to-one correspondence with the symmetries of the Coxeter–Dynkin diagram. These are the most important choices of automorphisms, for our purposes, as the fixed-point subalgebras $\mathfrak{g}_0$ are maximally large.

In particular, the fixed-point subalgebra $\mathfrak{g}_0$ for $\mathfrak{g} = \mathfrak{sl}_{2n}$, when γ is taken to be the outer automorphism permuting e_i and e_{2n-i} (the order-2 diagram symmetry), is isomorphic to $\mathfrak{sp}_{2n} \cong C_n$. Likewise, taking $\mathfrak{g}$ to be (respectively) A_{2n+1}, D_n, D_4 and E_6 and taking γ to be the diagram symmetry of order 2, 2, 3 and 2, yields a fixed-point subalgebra $\mathfrak{g}_0$ isomorphic to $\mathfrak{so}_{2n+1} = B_n$, $\mathfrak{so}_{2n-1} \cong B_{n-1}$, G_2 and F_4, respectively.

The automorphism group $\mathrm{Aut}(\mathfrak{g})$ permutes the $\mathfrak{g}$-modules, through the formula

$$\rho^\gamma(x) = \rho(\gamma x).$$

Sometimes ρ^γ and ρ are isomorphic as $\mathfrak{g}$-modules. In this case there is a matrix $A \in \mathrm{GL}(V)$, where V is the underlying space of ρ and ρ^γ, such that

$$\rho(\gamma x) = A^{-1}\rho(x)A, \qquad \forall x \in \mathfrak{g}.$$

Let us assume for convenience that ρ is irreducible. Then by Schur's Lemma this matrix A will be well defined up to a scalar multiple.

In fact, for $\mathfrak{g}$ semi-simple and γ corresponding to a diagram symmetry, and ρ the module $L(\lambda)$, ρ^γ will be the module $L(\gamma\lambda)$, where γ acts on weights by permuting Dynkin labels. Thus $\rho^\gamma \cong \rho$ iff $\gamma\lambda = \lambda$. In this case there is a canonical choice of matrix A, sending weight-space $L(\lambda)_\beta$ to weight-space $L(\lambda)_{\gamma\beta}$, given by

$$e_{m_1} \cdots e_{m_k}.v_\lambda \mapsto e_{\gamma m_1} \cdots e_{\gamma m_k}.v_\lambda.$$

Recall Thompson's trick: twisting the graded dimension (0.3.2) to get the McKay–Thompson series T_g of (0.3.3). Here, this becomes the γ-*twisted* or *twining character*

$$\mathrm{ch}_\lambda^\gamma(h) := \mathrm{tr}_V\, A \exp[\rho(h)] = \sum_{\beta = \gamma\beta} \mathrm{tr}(A_\beta) \exp[\beta(h)], \qquad (1.5.13)$$

where we can restrict the sum to all weights $\beta \in \Omega(L(\lambda))$ that are fixed by γ, and where A_β is the restriction of A to the weight-space $L(\lambda)_\beta$. The term 'twining', introduced in [213], is short for 'intertwining'. In terms of the basis (1.5.8a) for the weight-spaces $L(\lambda)_\beta$, A_β is a permutation matrix when β is fixed by γ (only these β survive in (1.5.13)).

For example, consider first $\mathfrak{g} = D_4$ and γ the diagram automorphism interchanging the third and fourth nodes. The dominant weight $\lambda = (1, 0, 0, 0)$ is invariant under γ. The D_4-representation $L(\lambda)$ is eight-dimensional, with all weight-spaces $L(\lambda)_\beta$ having dimension 1. It is thus easy to compute the twisted character $\mathrm{ch}_\lambda^\gamma$: it has a term with coefficient 1 for each γ-invariant weight $\beta = (\pm 1, 0, 0, 0)$, $(0, \pm 1, 0, 0)$. For a more complicated example, consider D_4 again, but with the order-3 automorphism ('triality') and the invariant dominant weight $\lambda = (0, 1, 0, 0)$: this D_4-representation is 28-dimensional but only its weights $\beta = (0, \pm 1, 0, 0)$, $\pm(1, -1, 1, 1)$, $\pm(1, -2, 1, 1)$, $(0, 0, 0, 0)$ are triality-invariant. Of those, the weight-spaces are all one-dimensional except for $L(0, 1, 0, 0)_{(0,0,0,0)}$, which is four-dimensional. A basis for that weight-space consists of

$$f_3 f_2 f_4 f_1 f_2.v, \quad f_4 f_2 f_3 f_1 f_2.v, \quad f_4 f_2 f_3 f_4 f_2.v, \quad f_2 f_3 f_4 f_1 f_2.v.$$

The map $A_{(0,0,0,0)}$ cyclically permutes the first three basis vectors, but fixes the fourth. Thus the twisted character has seven terms, each with coefficient 1. For similar calculations with small-rank algebras, the concrete bases given in [383] are useful.

If we restrict to h in the Cartan subalgebra of the fixed-point subalgebra $\mathfrak{g}_0$, the result will lie in the character ring of $\mathfrak{g}_0$. Thus the twisted character $\mathrm{ch}_\lambda^\gamma$ is a *virtual character* for the fixed-point subalgebra $\mathfrak{g}_0$, that is, it is a linear combination over $\mathbb{Z}$ of true characters. However, $\mathrm{ch}_\lambda^\gamma$ itself need not be a true character of $\mathfrak{g}_0$.

For example, recall the example $\mathfrak{g} = D_4$, weight $\lambda = (1, 0, 0, 0)$, and γ interchanging nodes 3 and 4. Then the fixed-point subalgebra $\mathfrak{g}_0$ is B_3 and the twisted character $\mathrm{ch}_\lambda^\gamma$ is the virtual B_3 character $\mathrm{ch}_{(1,0,0)}^B - \mathrm{ch}_{(0,0,0)}^B$ (Question 1.5.8(a)). On the other hand, the other D_4 example has fixed-point subalgebra G_2, and the twisted character equals the true character $L(0, 1)$.

Surprisingly, $\mathrm{ch}_\lambda^\gamma$ is always a *true* character for the algebra $\mathfrak{g}_0^{\mathrm{op}}$ obtained by reversing the arrows in the diagram of $\mathfrak{g}_0$. $\mathfrak{g}^{\mathrm{op}}$ is called the *orbit Lie algebra* in [213].

For example, when $\mathfrak{g} = D_4$ and $\lambda = (1, 0, 0, 0)$, we find (Question 1.5.8(b)) that the twisted character $\mathrm{ch}_{(1,0,0,0)}^\gamma$ equals the character $\mathrm{ch}_{(1,0,0)}^C$ of the orbit Lie algebra $\mathfrak{g}_0^{\mathrm{op}}$.

More generally, we find:

Theorem 1.5.4 [213] *Let $\mathfrak{g}$ be semi-simple and finite-dimensional, and let γ be the automorphism of $\mathfrak{g}$ corresponding to a Coxeter–Dynkin diagram symmetry. Let $\lambda \in P_+(\mathfrak{g})$ be any dominant integral weight fixed by γ. Then the twisted character $\mathrm{ch}_\lambda^\gamma$ defined in (1.5.13), restricted to the Cartan subalgebra of the fixed-point subalgebra $\mathfrak{g}_0$, is a virtual character of $\mathfrak{g}_0$ and a true character $\chi_{\overline{\lambda}}$ of the orbit Lie algebra $\mathfrak{g}_0^{\mathrm{op}}$, for some $\overline{\lambda} \in P_+(\mathfrak{g}_0^{\mathrm{op}})$.*

A weight $\lambda \in P_+(A_{2n})$ fixed by the order-two diagram symmetry looks like $\lambda = (\lambda_1, \ldots, \lambda_n, \lambda_n, \ldots, \lambda_1)$; likewise, $\lambda \in P_+(A_{2n-1})$ fixed by the order-two diagram symmetry looks like $\lambda = (\lambda_1, \ldots, \lambda_{n-1}, \lambda_n, \lambda_{n-1}, \ldots, \lambda_1)$; while a weight $\lambda \in P_+(D_{n+1})$ fixed by the $n - 1 \leftrightarrow n$ diagram symmetry looks like $\lambda = (\lambda_1, \ldots, \lambda_n, \lambda_n)$. The orbit Lie algebra $\mathfrak{g}_0^{\mathrm{op}}$ here is C_n for A_{2n} or D_{n+1}, and B_n for A_{2n-1}. In all three cases, $\overline{\lambda}$ has Dynkin labels $(\lambda_1, \ldots, \lambda_n)$.

The proof of Theorem 1.5.4 follows that of the Weyl character formula. Although Theorem 1.5.4 is not itself important for us, the obvious generalisation holds for affine algebras (Theorem 3.4.1), and provides a striking special case of the important orbifold construction in string theory and vertex operator algebras.

In hindsight it is easy to see that $\mathfrak{g}_0^{\mathrm{op}}$ is the more natural algebra: for modules, $\mathfrak{h}^*$ is more relevant than $\mathfrak{h}$ since that is where the weights live. Consider, for example, D_4 again, with diagram symmetry $3 \leftrightarrow 4$. Then a γ-invariant weight looks like $\beta_1 \omega_1 + \beta_2 \omega_2 + \beta_3(\omega_3 + \omega_4)$. Using Table 1.4, we see that these vectors $\{\omega_1, \omega_2, \omega_3 + \omega_4\}$ have the same inner-products with each other that the fundamental weights of C_3 have (up to a global factor of 2, which is merely conventional).

Incidentally, some version of these remarks holds for finite groups. Let γ be an automorphism of a finite group G; then γ permutes the irreducible representations of G, $\rho \mapsto \rho \circ \gamma$, as before. Choose any irreducible representation $\rho \cong \rho \circ \gamma$ and let A be the isomorphism. The γ-twisted character of ρ is the trace $\mathrm{ch}_\rho^\gamma(g) := \mathrm{tr}\, A\, \rho(g)$. It won't be a class function of G – for example, for the inner automorphism $g \mapsto h^{-1}gh$, $\mathrm{ch}_\rho^h(g) = \mathrm{ch}_\rho(hg)$. But this calculation shows that it suffices to consider outer automorphisms. In particular, diagram automorphisms of finite reductive groups should be interesting in this context.

1.5.5 Representations of Lie groups

We are more interested in (complex) Lie algebras, but (real) Lie groups do occasionally arise. Once again, it is their representation theory that is of greatest interest to us.

Let G be a real finite-dimensional Lie group, and let $\mathcal{H}$ be a complex Hilbert space. Let $\mathcal{B}(\mathcal{H})$ be the group of bounded linear operators with bounded inverse – boundedness is equivalent to continuity (Section 1.3.1). A *representation* or *module* of G on $\mathcal{H}$ is a homomorphism $\pi : G \to \mathcal{B}(\mathcal{H})$ such that the map $G \to \mathcal{H}$, defined by $g \mapsto \pi(g)v$, is continuous for every $v \in \mathcal{H}$. We call two modules π, π' *equivalent* if there is a bounded operator $A : \mathcal{H} \to \mathcal{H}'$, with bounded inverse, such that $A^{-1}\pi'(g)A = \pi(g)$ for all $g \in G$. The module π is *unitary* if each operator $\pi(g)$ is unitary, that is surjective and

$$\langle \pi(g)v, \pi(g)v' \rangle = \langle v, v' \rangle, \qquad \forall v, v' \in \mathcal{H}.$$

The module π is *irreducible* if there is no closed nontrivial subspace V, such that $\pi(g)V \subseteq V$ for all $g \in G$. Most important are the irreducible unitary modules, and these together form a topological space called the *unitary dual* $\widehat{G}$ of G.

For example, all one-dimensional modules of the additive group $G = \mathbb{R}$ are of the form $x \mapsto e^{i\alpha x}$ for any $\alpha \in \mathbb{C}$; it will be unitary iff $\alpha \in \mathbb{R}$. The map $x \mapsto \begin{pmatrix} 1 & x \\ 0 & 1 \end{pmatrix}$ is a representation of $\mathbb{R}$ that is not irreducible (consider $V = \mathbb{C} \times \{0\} \subset \mathbb{C}^2 = \mathcal{H}$). The one-dimensional modules of the group $G = S^1$ are $e^{i\theta} \mapsto e^{in\theta}$ for $n \in \mathbb{Z}$, and all are unitary. The unitary duals of $\mathbb{R}$ and S^1 are $\mathbb{R}$ and $\mathbb{Z}$, respectively.

Continuity is an important requirement. For instance, let $\{b_\beta\}_{\beta \in B}$ be a basis for $\mathbb{R}$ treated as a vector space over $\mathbb{Q}$ (so B is uncountable). Then for any choice of complex numbers α_β, the assignment $\sum_\beta r_\beta b_\beta \mapsto \prod_\beta e^{i r_\alpha \alpha_\beta}$ defines a (rather chaotic) group (for $r_\beta \in \mathbb{Q}$) homomorphism $\mathbb{R} \to \mathbb{C}^\times$. Continuity of π is needed in order to obtain from π a module of the Lie algebra $\mathfrak{g}$ of G.

Call a vector $v \in \mathcal{H}$ *smooth* if $g \mapsto \pi(g)v$ is a smooth function from G to $\mathcal{H}$. The space $\mathcal{H}_\infty$ of smooth vectors forms a dense G-invariant subspace of $\mathcal{H}$; if $\mathcal{H}$ is finite-dimensional, $\mathcal{H}_\infty$ equals $\mathcal{H}$. Recall that the Lie algebra $\mathfrak{g}$ is the tangent space T_eG, and the exponential map exp sends $\mathfrak{g}$ to G. For any $v \in \mathcal{H}_\infty$ and $x \in \mathfrak{g}$, define

$$\delta\pi(x)v = \frac{\mathrm{d}}{\mathrm{d}t}\left(\pi(e^{tx})v\right)_{t=0}. \tag{1.5.14}$$

This defines a $\mathfrak{g}$-module on $\mathcal{H}_\infty$ called the *derived module*. Of course, a (complex) module of the real Lie algebra $\mathfrak{g}$ lifts to a complex module of its complexification $\mathfrak{g}_\mathbb{C} := \mathbb{C} \otimes_\mathbb{R} \mathfrak{g}$.

The theory simplifies enormously if G is compact (for simplicity we also assume connectivity). Then G is a subgroup of the unitary group $U_n(\mathbb{C})$. Moreover:

Theorem 1.5.5 (Peter–Weyl) *Let G be a connected compact finite-dimensional Lie group. Any module π of G is equivalent to a unitary one, is completely reducible, and*

$\delta\pi$ *is a module of the reductive Lie algebra* $\mathfrak{g}_{\mathbb{C}} = \mathbb{C} \otimes \mathfrak{g}$. *Any irreducible G-module is finite-dimensional, and the derived module for* $\mathfrak{g}_{\mathbb{C}}$ *is also irreducible as a Lie algebra module.*

The unitary dual $\widehat{G}$ is thus a countable discrete space. The key to proving Theorem 1.5.5 is that it is possible to average (integrate) over the group. This G-invariant *Haar measure* plays the role here of the ubiquitous $\sum_{g \in G}$ in the finite group theory. For example, a G-invariant Hermitian form on $\mathcal{H}$ is obtained by averaging any given Hermitian form over its translates – compactness of G is needed to show that integral converges. See, for example, chapter II.9 of [92] for an elementary proof of Theorem 1.5.5.

If G is simply-connected as well as compact and connected, then any irreducible module of $\mathfrak{g}_{\mathbb{C}}$ lifts to one of G. Otherwise, $G = \widetilde{G}/Z$, where $\widetilde{G}$ is the universal cover and Z is some discrete subgroup of $\widetilde{G}$ (Theorem 1.4.3), and a $\mathfrak{g}_{\mathbb{C}}$-module will lift to one on G iff, once it is lifted to $\widetilde{G}$, it is trivial on Z. If it isn't trivial on Z, it would be a *projective representation* for G (Section 3.1.1).

An elementary example of this is provided by the modules of $\mathbb{R} \cong \widetilde{S^1}$ and $S^1 \cong \mathbb{R}/\mathbb{Z}$, given earlier. More interesting is to compare the universal cover $SU_2(\mathbb{C})$ of the group $SO_3(\mathbb{R}) \cong SU_2(\mathbb{C})/\left\langle \begin{pmatrix} -1 & 0 \\ 0 & -1 \end{pmatrix} \right\rangle$. Their complexified Lie algebra $\mathfrak{g}_{\mathbb{C}}$ is $\mathfrak{sl}_2(\mathbb{C})$, whose irreducible modules correspond to highest weights $\lambda \in P_+ = \{0, \omega_1, 2\omega_1, \ldots\}$. Each of these exponentiates to an irreducible module of $SU_2(\mathbb{C})$. In particular, the $SU_2(\mathbb{C})$-module corresponding to highest weight $\lambda = n\omega_1$ can be realised as the space of homogeneous polynomials $p(z_1, z_2)$ of degree n, with $SU_2(\mathbb{C})$ action given by

$$\begin{pmatrix} a & b \\ c & d \end{pmatrix} \cdot p(z_1, z_2) = p(az_1 + cz_2, bz_1 + dz_2). \tag{1.5.15}$$

This will be a module of $SO_3(\mathbb{R})$ iff $\begin{pmatrix} -1 & 0 \\ 0 & -1 \end{pmatrix}$ acts trivially, i.e. iff $p(z_1, z_2) = p(-z_1, -z_2)$ for all p, i.e. iff n is even. Physicists call $n/2$ the 'spin', and the modules with n odd are called 'spinors'. See, for example, chapter 20 of [214] for more on this. More generally, the dominant weight $\lambda = \sum_{i=1}^{n-1} \lambda_i \omega_i \in P_+$ gives a module of $PSL_n(\mathbb{C}) \cong SL_n(\mathbb{C})/\mathbb{Z}_n$ iff n divides $\sum i\lambda_i$.

Let G be any compact simply-connected connected Lie group, and $\mathfrak{g}$ its (real) Lie algebra. The simply-connected connected complex Lie group associated with the complex Lie algebra $\mathfrak{g}_{\mathbb{C}}$ is called the *complexification* $G_{\mathbb{C}}$ of G. For example, the complexification of $SU_n(\mathbb{C})$ is $SL_n(\mathbb{C})$. *Weyl's unitary trick* says that the irreducible modules of G, $\mathfrak{g}$, $\mathfrak{g}_{\mathbb{C}}$ and $G_{\mathbb{C}}$ are all in natural bijection, using the derived module, complexification of the algebra module, 'exponentiation' of an algebra module to a simply-connected Lie group and restriction. Depending on the context, it is sometimes more convenient to look at the modules of G, $\mathfrak{g}_{\mathbb{C}}$ or $G_{\mathbb{C}}$.

All of the irreducible modules of a compact connected Lie group G are constructed explicitly by the *Borel–Weil Theorem*. It suffices of course to consider simply-connected G. Take $G = SU_n(\mathbb{C})$ for concreteness. Let B be the upper-triangular matrices in $G_{\mathbb{C}} = SL_n(\mathbb{C})$. It is called the *Borel subgroup* and is a maximal solvable subgroup in $G_{\mathbb{C}}$. Given a dominant integral weight $\lambda = \sum \lambda_i \omega_i$, put $t = \lambda_1 + 2\lambda_2 + \cdots + (n-1)\lambda_{n-1}$ and

$$\mu = \left(\sum_{i=1}^{n-1} \lambda_i - \frac{1}{n} t, \sum_{i=2}^{n-1} \lambda_i - \frac{1}{n} t, \ldots, \lambda_{n-1} - \frac{1}{n} t, -\frac{1}{n} t \right) \in \mathbb{R}^n.$$

Let $\Gamma(\lambda)$ be the space of holomorphic functions $f(g)$ on $G_{\mathbb{C}}$ (regarded as a complex manifold) such that

$$f(gb) = b_1^{\mu_1} \cdots b_n^{\mu_n} f(g), \qquad \forall g \in G_{\mathbb{C}}, \ b = \begin{pmatrix} b_1 & * & * \\ 0 & \ddots & * \\ 0 & 0 & b_n \end{pmatrix}. \tag{1.5.16}$$

Then this is a $G_{\mathbb{C}}$-module (namely, one induced from a one-dimensional B-module), and it is easy to identify its weights since the maximal torus T (the exponentiation of the Cartan subalgebra $\mathfrak{h}$ of $\mathfrak{g}_{\mathbb{C}}$, i.e. the diagonal determinant-1 matrices) is contained in B: we find that $\Gamma(\lambda)$ is the contragredient of the highest-weight representation $V(\lambda)$. From this picture, the Weyl character formula arises through fixed-point formulae for the $G_{\mathbb{C}}$-action on $G_{\mathbb{C}}/B$ [83].

The geometry of this construction is quite pretty (see e.g. section 23.3 of [219] or [83]). Geometrically, the space $G_{\mathbb{C}}/B \cong G/T$ is a flag variety whose points are the various choices $0 \subset V_1 \subset \cdots \subset V_{n-1} \subset \mathbb{C}^n$ of subspaces, where $\dim V_i = i$. Then $\Gamma(\lambda)$ is the space of holomorphic sections of a line bundle $G_{\mathbb{C}} \times_B \mathbb{C}$ on $G_{\mathbb{C}}/B$ naturally associated with λ. Similar comments apply to any other G. Something similar happens for the Virasoro algebra, where the flag manifold is replaced by the moduli space of curves (Section 3.1.2).

As discussed in Section 1.1.3, the natural analogue of the group algebra for a Lie group G is the space $L^2(G)$ of functions $f : G \to \mathbb{C}$, with convolution product. The main importance of these spaces of functions is that they are natural G-modules, using right translation: $(h.f)(g) := f(gh)$. For example, consider $G = S^1$, so $f \in L^2(S^1)$ can be regarded as a function $f(x)$ with period 2π. We find that $L^2(S^1)$ decomposes into the infinite direct sum

$$L^2(S^1) = \oplus_{n \in \mathbb{Z}} V(n)$$

of irreducible one-dimensional modules $V(n)$. More precisely, $L^2(S^1)$ will be a completion of this algebraic direct sum. This means that any 'vector' $f \in L^2(S^1)$ can be written as $\sum_{n \in \mathbb{Z}} f_n$ where each summand $f_n \in V(n)$. Now, $V(n)$ consists of those functions f_n on which $e^{iy} \in S^1$ acts as $(e^{iy}.f_n)(x) := f_n(x+y) = e^{iyn} f_n(x)$ – in other words $f_n(x) = c_n e^{inx}$ for some complex number c_n. Using the orthogonality of the e^{inx}, we can explicitly construct the projection operator $L^2(S^1) \to V(n)$, and we find

$$c_n = \frac{1}{2\pi} \int_0^{2\pi} f(x) e^{-inx} \, \mathrm{d}x,$$

which we recognise as the Fourier transform $\widehat{f}(n)$ of f.

More generally, for arbitrary compact G, the Peter–Weyl Theorem tells us that the matrix entries $\pi(g)_{ij}$ of the irreducible representations of G are dense in the space of functions on G. More precisely, the *Fourier transform* associates with a function $f \in L^2(G)$, a matrix-valued function $\widehat{f}(\pi)$ on the unitary dual $\widehat{G}$, defined by

$$\widehat{f}(\pi) = \int_G f(g)\,\pi(g)\,\mathrm{d}g,$$

where as usual we're using the Haar measure on G, normalised so that the volume of G is 1. Then for any $f \in L^2(G)$,

$$f(g) = \sum_{\pi \in \widehat{G}} \dim \pi \, \mathrm{tr}\left(\widehat{f}(\pi)\,\pi(g)^\dagger\right).$$

As is familiar from the abelian case, the convolution product is sent to the ordinary (matrix) product: $\widehat{f_1 * f_2}(\pi) = \widehat{f_1}(\pi)\,\widehat{f_2}(\pi)$. We also get a unitary isomorphism between $L^2(G)$ and what we can call $L^2(\widehat{G})$ (the space of these matrix-valued $\widehat{f}$), called the Plancherel formula:

$$\int_G |f(g)|^2\,\mathrm{d}g = \sum_{\pi \in \widehat{G}} (\dim \pi)\,\mathrm{tr}\left(\widehat{f}(\pi)\widehat{f}(\pi)^\dagger\right).$$

The representation theory of noncompact Lie groups is completely different. This can already be seen for the additive group $G = \mathbb{R}$, which has a continuum of irreducible unitary modules (namely $e^{i\alpha x}$ for all $\alpha \in \mathbb{R}$). The unitary dual $\widehat{G}$ can involve both continuous and discrete parts, and can have a wild topology. Once again, a unitary module is completely reducible into irreducible unitary ones, but for a general noncompact G a direct integral (Section 1.3.1), rather than a direct sum, will be needed, and for wild groups the uniqueness of this decomposition will be lost.

Any connected Lie group is (up to central extensions) the semi-direct product of a solvable Lie group with a semi-simple Lie group – this is the Levi decomposition (see e.g. appendix B in [**348**]). The representation theory of solvable groups is quite well understood, using the *orbit method*. It relates the unitary dual to certain orbits of G on the dual $\mathfrak{g}^*$ of the Lie algebra $\mathfrak{g}$ of G (see [**346**] for an excellent introduction, although section 2 of [**563**] may be more accessible to physicists). Physically, this is just geometric quantisation: G is a symmetry of a physical system; the classical phase space is a symplectic manifold on which G acts (these are essentially the coadjoint orbits); quantum mechanically we would like this to correspond to a Hilbert space carrying a unitary representation of G. Geometric quantisation tries to do for quantum theories what the symplectic geometry of Hamiltonian mechanics does for classical ones: provide an elegant and natural mathematical formulation.

The effect of the semi-direct product on the unitary dual is also under control. However, the representation theory of the (noncompact real) semi-simple groups is poorly understood. See [**349**] for a modern review.

For example, the Heisenberg group H consisting of all matrices

$$\begin{pmatrix} 1 & a & b \\ 0 & 1 & c \\ 0 & 0 & 1 \end{pmatrix}, \qquad \forall a, b, c \in \mathbb{R},$$

is simply-connected and solvable. Its irreducible unitary modules are given in Theorem 2.4.2 below, and we can naturally identify its unitary dual with the xy-plane in $\mathbb{R}^3$ together with the z-axis. On the other hand, $SL_2(\mathbb{R})$ is a semi-simple noncompact group, topologically equivalent to the interior of a solid torus; its unitary irreducible modules are described in Section 2.4.1, and its unitary dual consists of three one-dimensional families (the principal, spherical principal, and complementary series) and a countable family (the discrete series).

Question 1.5.1. Interpret the trigonometric identities given at the beginning of this section, in terms of the character theory of A_1.

Question 1.5.2. Classify all two-dimensional representations of the abelian Lie algebra $\mathfrak{g} = \mathbb{C}^2$. Which of these are completely reducible?

Question 1.5.3. Let $\mathfrak{g} = \mathfrak{sl}_n(\mathbb{C})$. From first principles, compute the Killing form $\kappa(A_a|E_{cd}), \kappa(A_a|A_b), \kappa(E_{ab}|E_{cd})$.

Question 1.5.4. In effect, Question 1.4.7 defines a representation $\mathfrak{g}$ of $\mathfrak{sl}_3(\mathbb{C})$.
(a) Find the weight-space decomposition of this representation of $\mathfrak{sl}_3(\mathbb{C})$, as well as the corresponding character.
(b) Find the root-space decomposition of $\mathfrak{sl}_3(\mathbb{C})$, i.e. the weight-space decomposition of the adjoint representation of $\mathfrak{sl}_3(\mathbb{C})$. Also compute the character.

Question 1.5.5. Recall the Verma modules $M(\lambda)$ for A_1 constructed in Section 1.5.1.
(a) Prove that each $M(\lambda)$ is indecomposable (i.e. cannot be written as the direct sum of two submodules).
(b) When $\lambda \notin \mathbb{N}$, prove that $M(\lambda)$ is irreducible. Thus $L(\lambda) = M(\lambda)$ for these λ.
(c) When $\lambda = n \in \mathbb{N}$, find all submodules. Verify that the maximal one has highest weight vector x_{n+1}.

Question 1.5.6. Let $\mathfrak{g} = \mathfrak{sl}_2(\mathbb{C})$.
(a) Set $C := ef + fe + \frac{1}{2}h^2 \in U(\mathfrak{g})$. Show that C is in the centre of $U(\mathfrak{g})$. (C is called the *quadratic Casimir* of $\mathfrak{g}$; there is an analogue for any semi-simple $\mathfrak{g}$.)
(b) Given any irreducible module π of $\mathfrak{g}$, prove that $Z := 2\pi(f)\pi(e) + \pi(h) + \frac{1}{2}\pi(h)^2$ is a scalar multiple of the identity.

Question 1.5.7. Let $G = SU(2)$. Then $\mathfrak{g} = \mathfrak{sl}_2(\mathbb{C})$ (which is the complexification of the Lie algebra of G) acts naturally on the space $\mathbb{C}^\infty(G)$ of all smooth complex-valued functions on G. In particular, $\mathfrak{g}$ can be identified as the space of all left-invariant first-order differential operators. Prove that $U(\mathfrak{g})$ can be identified with the space of all left-invariant finite-order differential operators on $C^\infty(G)$.

Question 1.5.8. (a) Verify the claim in Section 1.5.4 that $\mathfrak{g} = D_4$ with $L(1, 0, 0, 0)$ has twisted character restricting to B_3-character $\mathrm{ch}_{(1,0,0)} - \mathrm{ch}_{(0,0,0)}$ and C_3-character $\mathrm{ch}_{(1,0,0)}$.
(b) Repeat this calculation for $\mathfrak{g} = A_4$ and $\lambda = (1, 0, 0, 1)$.

Question 1.5.9. (a) In Section 1.5.5 we gave a module of $SU_2(\mathbb{C})$ using degree n polynomials. Find the derived module for the Lie algebra $\mathfrak{sl}_2(\mathbb{C})$, find its weight-spaces, and prove the equivalence with $L(n\omega)$.
(b) Work out the Borel–Weil representation $\Gamma(\lambda)$ for $SU_2(\mathbb{C})$, for any $\lambda = n\omega_1, n \in \mathbb{N}$.

1.6 Category theory

The only difficulty in understanding categories is in realising that they have no real content. They're just a language, highly abstract like the more familiar set theory, but one that can be both natural and suggestive. It tries to deflect some of our instinctive infatuation with objects (nouns), to the mathematically more fruitful one with structure-preserving maps between objects (verbs).

Category theory is intended as a universal language of mathematics, so all concepts should be translated into it. Much as beavers, who as a species hate the sound of running water, plaster a creek with mud and sticks until alas that cursed tinkle stops, so do category theorists devise elaborate and obscure definitions in an attempt to capture a concept that to most of us seemed perfectly clear before they got to it. But at least sometimes this works admirably – for instance no one can be immune to the charm of treating knot invariants with braided monoidal categories.

1.6.1 General philosophy

A *category* **C** consists of two kinds of things. One are the *objects*, and the other are the *arrows* (or *morphisms*). An arrow, written $f : A \to B$, has an initial and a final object (A and B, respectively). We let Hom(A, B) denote all arrows $A \to B$ in the category. Arrows f, g can be composed to yield a new arrow $f \circ g$, if the final object of g equals the initial object of f. Maps between categories are called *functors* if they take each object (respectively, arrow) of one to the objects (respectively, arrows) of the other, and preserve composition. A gentle introduction to the mathematics of categories is [**370**]; the standard reference is [**397**].

The standard category is called **Set**, where the 'objects' are sets, and the arrows from A to B are functions $f : A \to B$. Many algebraic categories are of that form, with objects being sets with certain structure, and the arrows being structure-preserving maps. A typical example is **Vect**, where the objects are vector spaces over some fixed field and the arrows are linear maps. A rather trivial example of a functor $\mathcal{F} :$**Vect** $\to$ **Set** sends a vector space to its underlying set – $\mathcal{F}$ simply 'forgets' the vector space structure on V and ignores the fact that the arrows f in **Vect** are linear.

Geometric categories often employ the idea of cobordism. For instance, fix a manifold M; let the objects be points $p \in M$, and the arrows $p \to q$ be homotopy equivalence classes of paths σ in M from p to q. Composition of arrows is given by (1.2.5). This category is called the *fundamental groupoid* of M – note that Hom(p, p) $= \pi_1(M, p)$. A higher-dimensional example is called **Riem**: its objects are disjoint unions of (parametrised) circles S^1, and the arrows are (conformal equivalence classes of) *cobordisms*, that is (Riemann) surfaces whose boundaries are those circles. Composition of arrows in **Riem** amounts to sewing the surfaces along the appropriate boundary circles. A final example of a geometric category is **Braid**: its objects are any finite number (possibly 0) of 'hooks', Hom(m, n) is empty unless $m = n$, in which case the arrows are the n-braids $\beta \in \mathcal{B}_n$. Such categories, where arrows consist of equivalence classes, are called *quotient categories* [**397**].

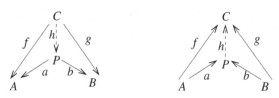

Fig. 1.23 The definition of product and sum.

For a baby example of the translation of the familiar into category theory, consider the usual definition of a one-to-one function: $f(x) = f(y)$ only when $x = y$. Category theory replaces this with the right cancellation law: call an arrow $f : A \to B$ 'one-to-one' if for any object C and any arrows $g, h \in \mathrm{Hom}(C, A)$, $f \circ g = f \circ h$ implies $g = h$. The reader can easily verify that in **Set** this agrees with the usual definition. What does this redefinition gain us? It certainly doesn't seem any simpler. But it does change the focus from the *argument* of f to the *global* functional behaviour of f, and a change of perspective can never be bad. It allows us to transport the idea of one-to-one-ness to arbitrary categories. For instance, in the category **Riem**, all arrows are 'one-to-one'.

Or consider the notion of *product*. In category theory, we say that the triple (P, a, b) is a product of objects A, B if $a : P \to A$ and $b : P \to B$ are arrows, and if for any $f : C \to A, g : C \to B$, there is a unique arrow $h : C \to P$ such that $f = a \circ h$ and $g = b \circ h$. See the left diagram in Figure 1.23. This notion unifies several constructions (each of which is the 'product' in an appropriately chosen category): Cartesian product of sets; intersection of sets; multiplication of numbers; the logical operator 'and'; direct product; infimum in a partially ordered set; etc. *Sum* can be defined similarly, by reversing the orientation of all the arrows in the diagram for product (see the right diagram in Figure 1.23). This unifies the constructions of disjoint union, 'or', addition, tensor product, direct sum, supremum, etc. Of course the specific construction of sum and product depends sensitively on the category. For example, in the category **Ab-Group**, where objects are abelian groups and arrows are homomorphisms, the sum of the cyclic groups $\mathbb{Z}_2$ and $\mathbb{Z}_3$ is their direct product $\mathbb{Z}_2 \times \mathbb{Z}_3 \cong \mathbb{Z}_6$, while in the category **Group**, where objects are groups and arrows homomorphisms, the direct sum of $\mathbb{Z}_2$ and $\mathbb{Z}_3$ is $\mathrm{PSL}_2(\mathbb{Z})$! See Question 1.6.3.

This generality of course comes with a price: it can wash away all of the endearing special features of a favourite theory or structure. There certainly are contexts where, for example, all human beings should be considered equal, but there are other contexts where the given human is none other than your mother and must be treated as such.

1.6.2 Braided monoidal categories

This book tries to identify the natural context for Moonshine. Categories more than sets provide the most appealing language for this context. The starting point for this formulation is braided monoidal categories. Standard references include chapter 1 of [**534**], chapter 1 of [**32**] and chapter XIII of [**338**].

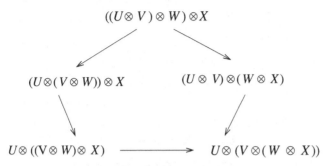

Fig. 1.24 The associativity pentagon.

Let us try to translate the vector space tensor product into category theoretic language. The result, called a *monoidal* or *tensor category*, was obtained by MacLane (1963).

Let $U_i, V_i, i = 1, 2, 3$, be vector spaces, and choose any linear maps $f_j : U_j \to U_{j+1}$, $g_j : V_j \to V_{j+1}$, $j = 1, 2$. Then the composition of the tensor product maps $f_j \otimes g_j : U_j \otimes V_j \to U_{j+1} \otimes V_{j+1}$ is given by $(f_2 \otimes g_2) \circ (f_1 \otimes g_1) = (f_2 \circ f_1) \otimes (g_2 \circ g_1)$. This is exactly the same as saying that '$\otimes$' is a functor between the categories **Vect** $\times$ **Vect** and **Vect**, where the Cartesian product of categories has the obvious meaning.

The tensor product should be associative up to isomorphism: for any objects U, V, W, there should be an isomorphism $a_{UVW} : (U \otimes V) \otimes W \to U \otimes (V \otimes W)$ (called the *associativity constraint*). It should obey a consistency condition coming from the iso-morphism $((U \otimes V) \otimes W) \otimes X \cong U \otimes (V \otimes (W \otimes X))$; that is, there are two ways of computing that isomorphism in terms of associativity, and the resulting isomorphisms should agree:

$$(id_V \otimes a_{VWX}) \circ a_{U,V \otimes W,X} \circ (a_{UVW} \otimes id_X) = a_{U,V,W \otimes X} \circ a_{U \otimes V,W,X}. \qquad (1.6.1)$$

This is called the *pentagon axiom*, thanks to its depiction in Figure 1.24.

Moreover, tensoring any object V with the one-dimensional vector space (call it '1') must give back V, so there are isomorphisms $l_V : 1 \otimes V \to V, r_V : V \otimes 1 \to V$. These are required to be consistent with the associativity constraint, by requiring the *triangle axiom*

$$r_V \otimes id_W = (id_V \otimes l_W) \circ a_{V1W}. \qquad (1.6.2)$$

A monoidal category [**397**] is any category **C** possessing such a functor $\otimes$, with unit 1 and invertible arrows l_V, r_V, a_{UVW} satisfying (1.6.1) and (1.6.2). Of course **Vect** with tensor products is monoidal, as is **Set** with disjoint union. **Braid** is monoidal; the tensor product of an n-braid with an m-braid is the $(n + m)$-braid obtained by placing the two braids side-by-side. There are numerous other examples. The word 'monoidal' comes from 'monoid', meaning a group-like structure without inverses.

MacLane proved two things. The first is *coherence*, which says that (1.6.1) and (1.6.2) are sufficient. Remarkably, any other consistency condition we may care to write down will be redundant. To give a random example, the identity involving a's, l's and r's

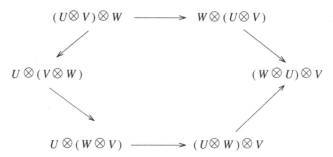

Fig. 1.25 The hexagon equation.

saying that the isomorphisms coming from $U \otimes ((V \otimes W) \otimes (1 \otimes (X \otimes Y))) \cong (U \otimes (V \otimes W)) \otimes (X \otimes Y)$ must agree can be derived from the pentagon and the triangle.

Secondly, MacLane proved that any monoidal category $\mathbf{C}$ is (monoidally) equivalent to a monoidal category $\mathbf{C}^{strict}$ where the associativity constraints are identity maps. Such a monoidal category is called *strict*; in it we can drop all associativity constraints as trivial, and with them all braces '(' and ')' in our tensor products.

Now that we've handled associativity of the tensor product, let's turn next to commutativity. We can't expect anything like MacLane's strictness to apply here – although the vector spaces $U \otimes V$ and $V \otimes U$ are naturally isomorphic, they are not equal. We proceed though in the same way.

For any objects U, V, we have an invertible arrow (called a *commutativity constraint*) $c_{UV} : U \otimes V \rightarrow V \otimes U$. Some natural relations are

$$c_{U'V'} \circ (f \otimes g) = (g \otimes f) \circ c_{UV}, \qquad (1.6.3a)$$

$$c_{VU} \circ c_{UV} = id_{U \otimes V}, \qquad (1.6.3b)$$

$$c_{U, V \otimes W} = c_{UV} \circ c_{UW}. \qquad (1.6.3c)$$

The isomorphism $(U \otimes V) \otimes W \cong (W \otimes U) \otimes V$, or more explicitly the equation

$$(c_{UW} \otimes id_V) \circ (id_U \otimes c_{VW}) = c_{U \otimes V, W}, \qquad (1.6.3d)$$

is called the *hexagon axiom* (see Figure 1.25).

Any monoidal category with commutativity constraints c_{UV} obeying (1.6.3) is called a *symmetric monoidal category* (MacLane, 1965). **Vect** is an example. Another is the categories Rep $\mathfrak{g}$ or Rep G of finite-dimensional $\mathfrak{g}$- or G-modules, for a Lie algebra $\mathfrak{g}$ (or Lie group G), with tensor product. In fact, *Tannaka–Krein duality* states that a monoidal category with both product and sum, that looks like Rep G (e.g. it has a unit object 1, a contragredient, and all objects decompose intoa sum of simple ones), *is* Rep G for a unique such group G. See, for example, section 9.4 of [**398**] for details and a generalisation.

In 1985, Joyal and Street [**321**] suggested to drop the symmetry condition (1.6.3b). The resulting categories they call *braided monoidal*, for reasons that will be clear shortly. They also pointed out that there is a very convenient graphical calculus in such categories,

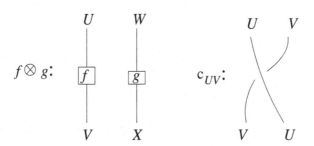

Fig. 1.26 The graphical calculus.

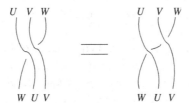

Fig. 1.27 The hexagon axiom revisited.

Fig. 1.28 The commutativity constraint c_{43} in **Braid**.

which elegantly keeps track of all relations. Namely, write arrows vertically and tensor products horizontally. Composition is given by vertical concatenation. The left-most diagram in Figure 1.26 represents the arrow $f \otimes g$ where $f \in \mathrm{Hom}(U, V)$ and $g \in \mathrm{Hom}(W, X)$, while the commutativity constraint c_{UV} is depicted as in the right-most. The associativity constraint a_{ABC} is ignored as we identify it with the identity. So we label strands with objects, which can change labels only at a box ('coupon'). The hexagon axiom takes the form of Figure 1.27, which we recognise as two equivalent braids. One immediate consequence is that the category **Braid** described last subsection is braided monoidal, provided we define c_{mn} as in Figure 1.28.

In terms of the graphical calculus, MacLane's symmetry condition (1.6.3b) would permit us to slip one strand through another, reducing the content of a braid (i.e. some combination of commutativity constraints) to that of its underlying permutation.

Joyal–Street also proved coherence for braided monoidal categories, that is equations (1.6.2a), (1.6.2c) and (1.6.3) are sufficient to establish the well-definedness of other

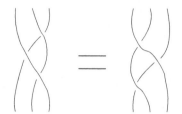

Fig. 1.29 The Yang–Baxter equation.

isomorphisms involving associativity and commutativity. For a famous example, $U \otimes V \otimes W \cong W \otimes V \otimes U$ yields the *Yang–Baxter equation*

$$(c_{VW} \otimes id_U) \circ (id_V \otimes c_{UW}) \circ (c_{UV} \otimes id_W) = (id_W \otimes c_{UV}) \circ (c_{UW} \otimes id_V) \circ (id_U \otimes c_{VW}),$$
$$(1.6.4)$$

which corresponds graphically to the braid equivalence of Figure 1.29 (compare Figure 1.2). We return to the Yang–Baxter equation in Section 6.2.3.

It's not a coincidence that Figure 1.29 is a braid equivalence – it *must* be, since **Braid** is a braided monoidal category. Conversely, any braid equivalence yields an equation holding in any braided monoidal category. **Braid** is the least-common divisor of all braided monoidal categories, the one with commutativity constraints and nothing else, obeying the minimum possible relations – it is *universal* or *free*. More precisely:

Theorem 1.6.1 [321] *Let **C** be any (strict) braided monoidal category, and A any object in it. Then there exists a unique braided monoidal functor $F :$ **Braid** $\to$ **C** with $F(1) = A$ and $F(c_{1,1}) = c_{A,A}$.*

A 'braided monoidal' functor is one preserving the braided monoidal structure in the obvious way. The object '1' of **Braid** denotes one hook, which generates via tensoring all other objects in **Braid**. This important theorem relates topology and algebra.

The simplest example (in fact too simple) of such universality is the freeness of $\mathbb{Z}$: given any group G with one generator g, there is a unique group homomorphism $\varphi : \mathbb{Z} \to G$ sending $1 \in \mathbb{Z}$ to $g \in G$. Any such G defines an *invariant* for $\mathbb{Z}$: the integer n is assigned the invariant $\varphi(n)$. We call it an invariant, because equal integers must get assigned the same G-value, even if they look different (e.g. 3 and $2 - 1 + 2$ superficially look different, but will be assigned the same G-value $\varphi(3) = \varphi(2 - 1 + 2)$). For example, the invariant φ for $G = \mathbb{Z}_2 = \{[0], [1]\}$ assigns $[0]$ to any even $n \in \mathbb{Z}$ and $[1]$ to any odd n. Because φ is structure-preserving, computing this invariant is relatively easy. Of course integer invariants are not terribly exciting, because it is so easy to determine if two integer expressions (involving arbitrary sums and subtractions) are equal.

Likewise, the universality of **Braid** means that, given any braided monoidal category **C** and any braid $\beta \in \mathcal{B}_n$, we get a braid-invariant $F(\beta) \in \text{Hom}_{\mathbf{C}}(A^{\otimes n}, A^{\otimes n})$. Here the object $A^{\otimes n}$ of **C** means $A \otimes \cdots \otimes A$ (n times). It is not so difficult to determine directly whether two braids are the same (ambient isotopic) – for example, by 'combing the braid' (see e.g. pages 24–5 of [59]) – and thus these braid invariants are also not intrinsically valuable. But they are a stepping stone to something that is.

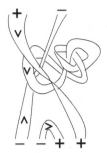

Fig. 1.30 A typical ribbon in Hom $((+, -), (-, -, +, +))$.

Fig. 1.31 Evaluation, coevaluation and twist.

Theorem 1.6.1 implies that, in any braided monoidal category **C**, the braid group $\mathcal{B}_n$ acts on both $\mathrm{Hom}_{\mathbf{C}}(U^{\otimes n}, V)$ and $\mathrm{Hom}_{\mathbf{C}}(U, V^{\otimes n})$ and the pure braid group $\mathcal{P}_n$ acts on both $\mathrm{Hom}_{\mathbf{C}}(U_1 \otimes \cdots \otimes U_n, V)$ and $\mathrm{Hom}_{\mathbf{C}}(U, V_1 \otimes \cdots \otimes V_n)$ (why?). Thus the groups governing braided monoidal categories are the braid groups $\mathcal{B}_n$ and $\mathcal{P}_n$, while those of symmetric monoidal categories are the symmetric groups $\mathcal{S}_n$ (hence their names).

If we continue with our project of categorising tensor product, we will be rewarded. We can introduce the notion of duals A^* of objects (in the sense of the dual vector space), duals of arrows f^* (the analogue of transpose of matrices), the evaluation map $A^* \otimes A \to 1$ (the evaluation $f(a)$ of a functional $f \in A^*$ on a vector $a \in A$), coevaluation $1 \to A^* \otimes A$ (let b_i be a basis of vector space A and $b_i^* \in A^*$ the dual basis, then the element $\sum_i b_i^* \otimes b_i \in A^* \otimes A$ is independent of the choice of basis). These obey the obvious relations (see, for example, chapter 1 of [**534**]) and the result is called a *ribbon category*; in place of the formal definition it suffices to give the universal ribbon category.

The objects of **Ribbon** are ordered n-tuples $A = (a_1, \ldots, a_n)$ of signs, $a_i = \pm$, for $n \geq 0$ ($n = 0$ is the empty object $\emptyset$). Hom(A, B) consists of isotopy classes of knotted linked twisted oriented strips, called ribbons. A strip can start at position i on the top (or position j on the bottom) only if $a_i = +1$ (or $b_j = -1$, respectively); similarly, it can end at i or j only if $a_i = -1$ or $b_j = +1$ – see Figure 1.30. Braiding is as before. The dual of $(a_1, \ldots, a_n)$ is $(-a_n, \ldots, -a_1)$, and the dual of a ribbon is given by rotation through $180°$. The evaluation and coevaluation are given in Figure 1.31.

We use ribbons (strips) rather than links (strands) because the $360°$ turn depicted on the right of Figure 1.31 cannot be straightened without introducing a twist in the strip. Up to isotopy, a ribbon can be thought of as braided knotted strands (the spine of each strip) together with an integer assigned to each strand (saying how much that strip is twisted).

As on the left of Figure 1.26, it is very useful to colour ribbons. Let S be any set; by **Ribbon**$_S$ we mean the category with objects $((A_1, s_1), \ldots, (A_k, s_k))$ for $A_i \in S, s_i \in \{\pm\}$. The arrows are as before except now they are coloured with $A \in S$; if the ribbon has endpoints, they must be of the form (A, s) and (A, s') where the signs s, s' are as before.

Two isotopic ribbons define an identity holding in any ribbon category:

Theorem 1.6.2 [473] *Let* **C** *be a (strict) ribbon category and let S be the set of its objects. Then there exists a unique ribbon functor $F :$* **Ribbon**$_S \to$ **C** *such that $F(A, +) = A$ and $F(A, -) = A^*$.*

By the usual arguments, any ribbon category **C** gives us (isotopy) invariants of braided knotted ribbons. The most interesting (because it is the simplest) special case concerns any ribbon $\mathcal{R} \in \text{Hom}_{\textbf{Ribbon}}(\emptyset, \emptyset)$ without ends: the invariant $F(\mathcal{R})$ will lie in $\text{Hom}_\mathbb{C}(F(\emptyset), F(\emptyset))$. This gives an invariant for any link, by drawing its ribbon with zero twist for each strip. Of course some ribbon categories give a complete link invariant (why?).

Unlike for braids, we have no effective way to determine if linked ribbons or links are ambient isotopic (but see [**283**]), so these invariants are topologically interesting. For example, they permit an easy proof that the trefoil and its mirror image are not ambient isotopic, something that took a clever argument from Dehn to do originally. The functoriality property of F makes them relatively easy to compute.

It is far from obvious that there are any nontrivial calculationally practical examples of ribbon categories, independent of **Ribbon**. Fortunately though there are: although **Ribbon** is geometric, there are several ribbon categories coming from algebra (namely representation theory). In fact, there are now so many that the main value of Theorem 1.6.2 is organisational, conceptually gathering together a plethora of link invariants that have been accumulating since the 1980s, starting with the Jones polynomial.

This treatment can and should be pushed much further, starting with the direct sum $U \oplus V$ of objects. See [**534**], [**398**], [**353**] for more details and developments. The refinement called *modular category* is the one of greatest relevance to the mathematics and physics related to Moonshine. We return to categories in Section 4.4.1.

Vogel [**547**] defined a monoidal category $\mathcal{D}'$, which looks like the category of modules of a Lie algebra. He calls it the *Universal Lie algebra*, since given any simple Lie (super)algebra $\mathfrak{g}$, there is a unique functor from $\mathcal{D}'$ to the category of $\mathfrak{g}$-modules satisfying certain natural properties. Roughly, Vogel assigns each such Lie (super)algebra a different point on the projective plane, from which much of its data can easily be computed. For example, the A-series corresponds to the projective coordinates $[n, 2, -2]$, while the exceptional series (the bottom row of Figure 1.20) falls on the line $[-2, a + 4, 2a + 4]$. The 'universal decompositions' and dimension formulae described in Section 1.5.2 arise because they hold for $\mathcal{D}'$.

Question 1.6.1. A variety is a solution set to a system of polynomial equations over some ring R. Interpret this as a functor from a category of rings to a category of sets.

Question 1.6.2. (a) Find what (if anything) product and sum (in the sense of Figure 1.23) are in the category **Set**.
(b) Same question for the category **Riem**.

Question 1.6.3. (a) Show that in the category **Ab-Group** (where objects are abelian groups and arrows are group homomorphisms), sum and product are identical.
(b) Show that in the category **Group**, product is direct product, but sum is not.

Question 1.6.4. Let L be any lattice (Section 1.2.1). Define a category whose objects are elements of L, with $\mathrm{Hom}(v, v) = \mathbb{C}$ and $\mathrm{Hom}(v, w) = \{0\}$ whenever $v \neq w$. Composition of arrows is multiplication. Complete the construction of a ribbon category for this category, where the braiding $c_{v,w}$ is $e^{iv \cdot w}$.

1.7 Elementary algebraic number theory

The coefficients of the McKay–Thompson series T_g are always integers, as are the fusion multiplicities $\mathcal{N}_{ab}^c$ in RCFT. But non-integers often lurk in the shadows, secretly watching their more arrogant brethren the integers strut. One of the consequences of their presence can be the existence of certain Galois symmetries. The Galois theory of cyclotomic fields plays a background role in Moonshine, much as it does for finite groups and modular forms. We sketch the basics in this section.

Galois automorphisms are a generalisation of complex conjugation. If in your problem complex conjugation seems interesting, then there is a good chance other Galois automorphisms will play a role.

1.7.1 Algebraic numbers

Euler and Lagrange were the first to show that 'weird' (complex) numbers could tell us about the integers, but it took Gauss (*c.* 1831) to do this with care and subtlety. For an example of this idea, suppose we are interested in the equation $n = a^2 + b^2$. Consider for concreteness $5 = 2^2 + 1^2$. We can write this as $5 = (2 + i)(2 - i)$, so we are led to consider complex numbers of the form $a + bi$, for $a, b \in \mathbb{Z}$. These are now called 'Gaussian integers'.

Fact *Let $p \in \mathbb{Z}$ be any prime number. Then p factorises (i.e. is composite) over the Gaussian integers iff $p = 2$ or $p \equiv 1 \pmod{4}$.*

Now suppose $p \not\equiv 3 \pmod 4$ is prime, and factorise it $p = (a + bi)(c + di)$. Then $p^2 = (a^2 + b^2)(c^2 + d^2)$, so $a^2 + b^2 = c^2 + d^2 = p$. Conversely, suppose $p = a^2 + b^2$, then $p = (a + bi)(a - bi)$. Thus:

Consequence[15] *Let $p \in \mathbb{Z}$ be any prime number. Then $p = a^2 + b^2$ for $a, b \in \mathbb{Z}$ iff $p = 2$ or $p \equiv 1 \pmod 4$.*

---

[15] This result was first stated by Fermat in one of his infamous margin notes (another is discussed shortly), and was finally proved a century later by Euler. For a one-line proof see Question 1.7.1.

Now we can answer the question: when can n be written as a sum of two squares $n = a^2 + b^2$? Write out the prime decomposition $n = \prod p^{a_p}$. Then $n = a^2 + b^2$ has a solution iff a_p is *even* for every $p \equiv 3 \pmod 4$. For instance, $60 = 2^2 \cdot 3^1 \cdot 5^1$ cannot be written as the sum of two squares, but $90 = 2^1 \cdot 3^2 \cdot 5^1 = \{(1+i)3(1+2i)\}\{(1-i)3(1-2i)\}$ can (e.g. $90 = (-3)^2 + 9^2$). In fact we can find and count all solutions.

More generally, let $\mathbb{K}$ be any subfield of $\mathbb{C}$ (usually we take $\mathbb{K} = \mathbb{Q}$) and $\alpha_1, \ldots, \alpha_n$ be any complex numbers. We discussed 'field' in Section 1.1.1. By $\mathbb{L} = \mathbb{K}(\alpha_1, \ldots, \alpha_n)$ we mean the smallest field containing $\mathbb{K}$ and all α_i. In other words, $\mathbb{L}$ consists of all rational functions poly/poly of the α_i, with coefficients in $\mathbb{K}$. Then $\mathbb{L}$ can be thought of as a vector space over $\mathbb{K}$; write $[\mathbb{L} : \mathbb{K}] \leq \infty$ for the dimension of that vector space. We say $\mathbb{L}$ is an *extension* of the *base-field* $\mathbb{K}$ of *degree* $[\mathbb{L} : \mathbb{K}]$. The most interesting case is when the degree $[\mathbb{L} : \mathbb{K}]$ is *finite*. In this case we can find a single number $\alpha \in \mathbb{L}$ such that $\mathbb{L} = \mathbb{K}[\alpha]$, where as always we write $R[x]$ for all polynomials in x with coefficients in R. Then α will be a zero of a monic polynomial $p(x) \in \mathbb{K}[x]$ and of degree $[\mathbb{L} : \mathbb{K}]$, called the *minimal polynomial* of α. Such α are called *algebraic*, and such extensions $\mathbb{K}[\alpha]$ are called *finite*. The finite extensions most relevant for this book are discussed in Section 1.7.3.

Numbers of course arise throughout science in their role as coordinates; less appreciated is that observing the specific kinds of numbers that arise can provide profound structural information. *This is very much how algebraic number theory impinges on the areas considered in this book.* For an elementary example, recall that Euclid's books are filled with geometric constructions, particularly those involving straight-edge (i.e. drawing the line passing through two points) and compass (i.e. drawing the circle with given centre and radius). The reader can discover for herself how to trisect line segments and double the area of a square, using only straight-edge and compass. But some problems weren't solved back then: for example, how to trisect an angle or double the volume of a cube. To solve these, consider coordinates. Suppose we start with N points (x_i, y_i). We can construct the line joining any two of those points, and the circle centred at some (x_i, y_i) with some radius $|(x_j, y_j) - (x_k, y_k)|$; we can construct new points only as intersections of these lines and circles. Now, if we let $\mathbb{K}$ denote the field generated from $\mathbb{Q}$ by all $2N$ coordinates x_i, y_i, then the equations of our lines and circles will have coefficients belonging to $\mathbb{K}$. The coordinates of the intersection of any two such lines will lie in $\mathbb{K}$, while that of the intersection of a line with a circle, or of two circles, will lie in an extension $\mathbb{L}_1$ of $\mathbb{K}$ of degree $[\mathbb{L}_1 : \mathbb{K}] = 2$. Continuing in this way, we see that any construction, no matter how involved, can only construct points whose coordinates lie in some extension $\mathbb{L}$ of $\mathbb{K}$ of degree a power of 2. Now, given an angle θ, defined by points $(0,0)$, $(1,0)$ and $(\cos(\theta), \sin(\theta))$, trisecting θ means constructing the point $(\cos(\theta/3), \sin(\theta/3))$. But $\alpha = \cos(\theta/3)$ obeys $\cos(\theta) = 4\alpha^3 - 3\alpha$, i.e. $\cos(\theta/3)$ lies (generically) in a degree-3 extension of $\mathbb{K} = \mathbb{Q}[\cos(\theta), \sin(\theta)]$. Thus we cannot trisect that angle, using only a compass and straight-edge, for most θ (e.g. $\theta = 60°$).

The degree $[\mathbb{L} : \mathbb{K}]$ is a (rather crude) invariant of the field extension $\mathbb{L} \supset \mathbb{K}$. We have just seen the power of this simple invariant; in the next subsection we refine it considerably, giving it a group structure.

Consider 'Fermat's Last Theorem', which asserts that there are no positive integer solutions to the equation $x^n + y^n = z^n$, for $n > 2$. It is tempting, as Fermat himself probably did, to factorise this into

$$\prod_{j=0}^{n-1} \left(x + \xi_{2n}^{2j+1} y\right) = z^n,$$

where $\xi_m = \exp[2\pi i/m]$, and to try to show from this that each $a + \xi_{2n}^{2j+1} b$ has an 'integral' nth root, if $x = a$, $y = b$, $z = c$ is an integral solution. We return to Fermat's Last Theorem in Section 2.2.1.

These examples should give the reader some appreciation for the value of using non-integers to study integers, and also provide some impetus for extending the tools and notions of high school number theory (primes, divisibility, etc.) to complex numbers. The result is *algebraic number theory*. A classic introduction is [282]; the book [515] is filled with concrete examples.

Euler worked with numbers of the form $\ell + m\sqrt{n}$, for $\ell, m, n \in \mathbb{Z}$, and regarded them as generalised integers, carrying over (without proof) their divisibility laws, etc. from the usual integers. However, it was soon learned that care must be taken. For a simple example, the factorisation $2 = (n - \sqrt{n^2 - 2})(n + \sqrt{n^2 - 2})$ holds for all $n \in \mathbb{Z}$, so what should the 'unique prime factorisation' of 2 be?

The basic theory was developed in the nineteenth century, by Kummer, Dedekind, Frobenius and others. Take the base field $\mathbb{K}$ to be $\mathbb{Q}$ for convenience, and fix a finite extension $\mathbb{L} = \mathbb{Q}[\alpha]$. Any $z \in \mathbb{L}$ is algebraic, i.e. satisfies $a_m z^m + a_{m-1} z^{m-1} + \cdots + a_0 = 0$ for some $a_i \in \mathbb{Z}$ (not all zero). The $\mathbb{L}$-*integers* are those numbers $z \in \mathbb{L}$ that satisfy $z^m + a_{m-1} z^{m-1} + \cdots + a_0 = 0$ for some $a_i \in \mathbb{Z}$ (i.e. $a_m = 1$). The sum and products of $\mathbb{L}$-integers are $\mathbb{L}$-integers, and so we call the set $R_\mathbb{L}$ of all these $\mathbb{L}$-integers the *ring of integers*. For example, when $\mathbb{L} = \mathbb{Q}$, $\mathbb{Q}[i]$, $\mathbb{Q}[\sqrt{2}]$ and $\mathbb{Q}[\sqrt{5}]$, respectively, the ring of integers are $\mathbb{Z}$, $\mathbb{Z} + i\mathbb{Z}$, $\mathbb{Z} + \sqrt{2}\mathbb{Z}$, and

$$\{(m + n\sqrt{5})/2 \,|\, m, n \in \mathbb{Z}, \ m - n \in 2\mathbb{Z}\},$$

respectively. All elements of $\mathbb{L}$ are quotients of $\mathbb{L}$-integers, just as all $r \in \mathbb{Q}$ equal a/b for $a, b \in \mathbb{Z}$.

What should prime mean here? The obvious guess would be any number $\gamma \in R_\mathbb{L}$ whose only divisors β are trivial, i.e. the only $\beta \in R_\mathbb{L}$ with $\gamma/\beta \in R_\mathbb{L}$ are units or γ times units. Units are the analogue here of ± 1: an $\mathbb{L}$-integer u is a unit iff u^{-1} is also an $\mathbb{L}$-integer. The only problem with this definition of prime is that unique factorisation is usually lost. For example, in $\mathbb{L} = \mathbb{Q}[\sqrt{-26}]$, the $\mathbb{L}$-integers are $\mathbb{Z} + \sqrt{-26}\mathbb{Z}$; we have the equation

$$3^3 = 27 = (1 + \sqrt{-26})(1 - \sqrt{-26})$$

and yet, as the reader can easily verify, both 3 and $1 \pm \sqrt{-26}$ are primes by our definition. Incidentally, most finite extensions $\mathbb{L}$ have infinitely many $\mathbb{L}$-units (e.g. $(1 + \sqrt{2})^n$ is a unit of $\mathbb{Q}[\sqrt{2}]$ for any $n \in \mathbb{Z}$).

The correct definition of prime (Dedekind, 1871) is a gem. Replace the single $\mathbb{L}$-integer $\gamma \in R_\mathbb{L}$ with the set of all multiples $R_\mathbb{L}\gamma =: (\gamma)$ of that number. This washes away the irritating ambiguity due to units. Any subset $I \subseteq R_\mathbb{L}$ closed under $R_\mathbb{L}$-linear combinations (i.e. for which $\sum a_i z_i \in I$ for all $z_i \in I$ and $a_i \in R_\mathbb{L}$) is called an *ideal* of $R_\mathbb{L}$. For example, (γ) is always an ideal, though for typical rings $R_\mathbb{L}$, most ideals won't have a single generator. Consider any ideals I, J of $R_\mathbb{L}$. By the product of ideals we mean

$$IJ = \left\{ \sum a_i b_i \mid a_i \in I, \ b_i \in J \right\}.$$

A prime ideal is defined to be any nonzero ideal $P \neq R_\mathbb{L}$ such that $IJ = P$ for ideals I, J only if $I = R_\mathbb{L}$ or $J = R_\mathbb{L}$. In $R_\mathbb{L}$, any prime ideal P is maximal (and conversely): the only ideals I satisfying $P \subset I \subset R_\mathbb{L}$ are $I = P, R_\mathbb{L}$. Although unique factorisation usually won't hold for $\mathbb{L}$-integers, it always holds for ideals: any nonzero ideal I of the ring $R_\mathbb{L}$ of integers can be written uniquely as a product of prime ideals.

For example, the prime ideals of $\mathbb{Z}$ are (p) for p prime, and this reduces to the usual unique factorisation of integers. The unique factorisation of the ideal (27) in the field $\mathbb{Q}[\sqrt{-26}]$ is $(27) = P_+^3 P_-^3$, where $P_\pm := (3, 1 \pm \sqrt{-26}) = (3) \cap (1 \pm \sqrt{-26})$. Thus neither $(3) = P_+ P_-$ nor $(1 \pm \sqrt{-26}) = P_\pm^3$ are prime.

We are thus led to picture $\mathbb{L}$-integers as ideals of the ring $R_\mathbb{L}$. In fact the name 'ideal', now standard in algebra, was chosen because it corresponds to an *ideal* – as opposed to *true* – number.

This reinterpretation of integers as ideals has a striking geometric parallel. We are taught to study a geometric space X through the functions $f \in \mathbb{C}[X]$ that live on it. In this language, what should play the role of a point $x \in X$? Given any point $a \in X$, we can evaluate these functions $f(x)$ at $x = a$. Algebraically, this corresponds to a homomorphism $\mathbb{C}[X] \to \mathbb{C}$. Those homomorphisms, via their kernels, are essentially in one-to-one correspondence with ideals of the ring $\mathbb{C}[X]$, and thus we should identify points $x \in X$ with certain ideals in $\mathbb{C}[X]$. Looking at concrete examples such as $X = \mathbb{C}^n$, we find that ideals correspond more generally to submanifolds (subvarieties) in X, and that maximal ideals correspond to points. This unexpected and deep connection between number theory and geometry is a great illustration of the effectiveness of abstract algebra.

1.7.2 Galois

Evariste Galois was a brilliantly original French mathematician. Born shortly before Napoleon's ill-fated invasion of Russia, he died shortly before the ill-fated 1832 uprising in Paris. His last words: 'Don't cry, I need all my courage to die at 20'.

Galois grew up in a time and place confused and excited by revolution. He was known to say 'if only I were sure that a body would be enough to incite the people to revolt, I would offer mine'. On 2 May 1832, after frustration over failure in love and failure to convince the Paris mathematical establishment of the depth of his ideas, he made his decision. A duel was arranged with a friend, but only his friend's gun would be loaded. Galois died the day after that bullet perforated his intestine. At his funeral it was discovered that a famous general had also just died, and the revolutionaries decided to use the general's death rather than Galois' as a pretext for an armed uprising. A few days later the streets of Paris were blocked by barricades, but not because of Galois' sacrifice: his death had been pointless [**529**].[16]

Galois theory in its most general form is the study of relations between objects defined implicitly by some conditions.[17] For example, the objects could be the solutions to a given differential equation. Or the objects could be the different points $\pi^{-1}(p) \subset Y$ sitting above a given point $p \in X$ in a cover $\pi : Y \to X$. In the most familiar incarnation of Galois theory, the objects are the zeros of certain polynomials.

Look at complex conjugation: $\overline{wz} = \overline{w}\,\overline{z}$ and $\overline{w+z} = \overline{w} + \overline{z}$. Also, $\overline{x} = x$ for any $x \in \mathbb{R}$. So we can say that $z \mapsto \overline{z}$ is a structure-preserving map $\mathbb{C} \to \mathbb{C}$ (called an *automorphism* of $\mathbb{C}$) fixing the reals. We say that complex conjugation belongs to the Galois group $\mathrm{Gal}(\mathbb{C}/\mathbb{R})$ of $\mathbb{C}$ over $\mathbb{R}$; apart from complex conjugation, it contains only the identity automorphism.

A way of thinking about the automorphism $\overline{z}$ is that it says that, as far as the real numbers are concerned, i and $-$i are identical twins. Algebra alone can't tell that i is in the upper half-plane, or that going from 1 to i is going counterclockwise about 0, while 1 to $-$i is clockwise.

Let $\mathbb{L}$ be any field containing $\mathbb{Q}$. The *Galois group* $\mathrm{Gal}(\mathbb{L}/\mathbb{Q})$ is the set of all automorphisms=symmetries of $\mathbb{L}$ that fix all rationals.

For example, $\mathbb{L} = \mathbb{Q}[\sqrt{5}]$ is the field of all numbers of the form $a + b\sqrt{5}$, where $a, b \in \mathbb{Q}$. Let's try to find its Galois group. Let $\sigma \in \mathrm{Gal}(\mathbb{F}/\mathbb{Q})$. Then $\sigma(a + b\sqrt{5}) = \sigma(a) + \sigma(b)\sigma(\sqrt{5}) = a + b\sigma(\sqrt{5})$, so once we know what σ does to $\sqrt{5}$, we know everything about σ. But $5 = \sigma(5) = \sigma(\sqrt{5}^2) = (\sigma(\sqrt{5}))^2$, so $\sigma(\sqrt{5}) = \pm\sqrt{5}$ and again there are precisely two possible Galois automorphisms here (one is the identity). As far as the arithmetic of $\mathbb{Q}$ is concerned, $\pm\sqrt{5}$ are interchangeable.

Consider more generally any extension $\mathbb{L}$ of the base field $\mathbb{K}$ of degree $n = [\mathbb{L} : \mathbb{K}] < \infty$. As mentioned in the last subsection, these are always of the form $\mathbb{L} = \mathbb{K}[\alpha]$, where α is the root of a monic polynomial $p(x)$ of degree n with coefficients in $\mathbb{K}$. This means any $z \in \mathbb{L}$ is expressible as a polynomial in α with coefficients in $\mathbb{K}$, of degree $< n$. Hence, any automorphism $\sigma \in \mathrm{Gal}(\mathbb{L}/\mathbb{K})$ is uniquely specified by the value $\sigma(\alpha) \in \mathbb{L}$. Since $\sigma(p(x)) = p(\sigma x)$, σ must send α to one of the n roots of $p(x)$. Thus $\|\mathrm{Gal}(\mathbb{L}/\mathbb{K})\| \leq [\mathbb{L} : \mathbb{K}]$. Extensions $\mathbb{L}$ for which $\mathrm{Gal}(\mathbb{L}/\mathbb{K})$ is maximally large (i.e. of order n) are the most interesting and are called *Galois*: they are the extensions for which all roots of $p(x)$ are in $\mathbb{L}$.

---

[16] Apparently this treatment of Galois' life has been disputed. But surely the main purposes of history are for supplying a context and motivation, for its sheer entertainment value, and for drawing Lofty Morals. And at least when they are successful, it is probably wisest if neither motivation nor entertainment nor Morality be investigated too closely...

[17] This is the dynamic point of view, but the reader should be warned that there is an alternate interpretation. Abstracting out the more structural side of Galois theory, many authors regard Galois theory as ultimately a contravariant functorial correspondence associating to some objects $A, B, \ldots$ (e.g. groups) other objects $K, L, \ldots$ (e.g. fields invariant under the group action) in such a way that $A \subset B$ corresponds to $K \supset L$.

Let $\mathbb{L} \supset \mathbb{K}$ be a finite Galois extension, and write $G := \mathrm{Gal}(\mathbb{L}/\mathbb{K})$. The classical Galois Theorem sets up a natural bijection between fields $\mathbb{J}$, $\mathbb{L} \supset \mathbb{J} \supset \mathbb{K}$, and subgroups H of G. In particular, to the field $\mathbb{J}$ associate the subgroup $H = \mathrm{Gal}(\mathbb{L}/\mathbb{J})$, and to the subgroup H associate the space (in fact field) $\mathbb{J} = \mathbb{L}^H$ of all elements $z \in \mathbb{L}$ fixed by all $\sigma \in H$. Then $[\mathbb{J} : \mathbb{K}] = \|G/H\|$, and the extension $\mathbb{J} \supset \mathbb{K}$ is Galois iff H is normal in G, in which case $\mathrm{Gal}(\mathbb{J}/\mathbb{K}) \cong G/H$.

We saw earlier the power of the numerical invariant $[\mathbb{L} : \mathbb{K}]$. We should think of $\mathrm{Gal}(\mathbb{L}/\mathbb{K})$ as a group-valued refinement of degree. For an application, suppose for contradiction that we have a general formula for the zeros of any polynomial $a_n x^n + a_{n-1} x^{n-1} + \cdots + a_0$ of degree n. For $n = 2$ we have the quadratic formula (which involves square-roots), and we've all seen the formula for $n = 3$ (which involves square-roots and cube-roots). Does there exist a formula for any n, involving taking arbitrary nested roots of rational expressions in the coefficients a_i? Let $\mathbb{K} = \mathbb{Q}[a_0, \ldots, a_n, \xi_1, \xi_2, \ldots]$ – we include in $\mathbb{K}$ all roots of unity so that all extensions below will be Galois. Then the first kth root we come to in our formula will move us into a Galois extension $\mathbb{K}_1$ of $\mathbb{K}$, with Galois group $\mathrm{Gal}(\mathbb{K}_1/\mathbb{K}) \cong \mathbb{Z}_k$. If the hypothetical formula involves a second radical, requiring us to take say an ℓth root of a rational expression in $\mathbb{K}_1$, then this takes us into a Galois extension $\mathbb{K}_2$ of $\mathbb{K}_1$, with Galois group $\mathrm{Gal}(\mathbb{K}_2/\mathbb{K}_1) \cong \mathbb{Z}_\ell$ – that is, $\mathrm{Gal}(\mathbb{K}_2/\mathbb{K})$ is an extension of the cyclic group $\mathbb{Z}_\ell$ by $\mathbb{Z}_k$. Continuing in this way until all roots in our hypothetical formula are exhausted, we would find that the zeros of the general degree-n polynomial would lie in a Galois extension $\mathbb{L}$ of $\mathbb{K}$ whose Galois group is obtained by repeatedly extending by cyclic groups. Such a group is called *solvable* (Section 1.1.3) for this reason. It is easy to see that $\mathrm{Gal}(\mathbb{L}/\mathbb{K})$ here is in fact the symmetric group $\mathcal{S}_n$, and that $\mathcal{S}_n$ is solvable iff $n \leq 4$ (recall that $\mathcal{A}_5$ is simple!). Thus a general formula for the roots of a general polynomial of degree n, involving nested radicals, can exist only for $n \leq 4$.

Every area of mathematics has a Galois-type theory. In geometry, for instance, covers $f : M \to N$ of a fixed manifold N are in one-to-one correspondence with subgroups $H \cong \pi_1(M)$ of the fundamental group $G := \pi_1(N)$; $\gamma \in \pi_1(N)$ belongs to H iff γ lifts to a closed loop in M. When the subgroup H is normal, G/H is naturally isomorphic to the group of all homeomorphisms $\alpha : M \to M$ satisfying $f \circ \alpha = f$ (these α are called *covering transformations*). See the beautiful book [**363**]. The question 'What is the Galois theory for von Neumann algebras?' led Jones to subfactor theory $M \supset N$ – for instance, his index $[M : N] \in \mathbb{R} \cup \{\infty\}$ plays the role of the degree $[\mathbb{L} : \mathbb{K}] \in \mathbb{Z} \cup \{\infty\}$. Just as the degree $[\mathbb{L} : \mathbb{K}]$ can be refined into the Galois group $\mathrm{Gal}(\mathbb{L}/\mathbb{K})$, the Jones index can be refined into a topological field theory (see Section 6.2.6).

Galois theory is reminiscent, at least qualitatively, of Gödel's Incompleteness Theorem. In mathematics we generally start with a model (e.g. Euclidean geometry or the natural numbers) that we try to capture implicitly by an axiomatic system. Gödel's Theorem tells us that there are infinitely many different models compatible with the given axiomatic system, regardless of how many axioms we include. Each of these is obtained by realising in incompatible ways the undefined terms of the axiomatic system.

Of course it is the *model* and not the *axiomatic system* in which most mathematics occurs. For example, we don't criticise Wiles' work on Fermat's Last Theorem on the grounds that his proof assumes $\mathbb{N}$ is embedded in $\mathbb{C}$, even though this transcendental interpretation of $\mathbb{N}$ surely is not a consequence of Peano's axioms (the axiomatic system describing the natural numbers). Likewise, [**459**] gives a simple statement about $\mathbb{N}$; it is easy to prove using standard arguments involving $\mathbb{R}$, but neither it nor its negation can be proved using only Peano's axioms.

1.7.3 Cyclotomic fields

We are primarily interested in a simple class of numbers: those in the *cyclotomic extensions* of $\mathbb{Q}$. These are the fields $\mathbb{Q}[\xi_n]$, consisting of all polynomials $a_m \xi_n^m + a_{m-1} \xi_n^{m-1} + \cdots + a_0$ in the root of unity $\xi_n := \exp[2\pi i/n]$, for all $a_i \in \mathbb{Q}$. For instance, $\cos(\pi r)$, $\sin(\pi r)$ and $\sqrt{r}$ are cyclotomic numbers for any $r \in \mathbb{Q}$. In particular,

$$\cos\left(2\pi \frac{m}{n}\right) = \frac{\xi_n^m + \xi_n^{-m}}{2}, \tag{1.7.1a}$$

$$\sin\left(2\pi \frac{m}{n}\right) = \frac{\xi_n^m - \xi_n^{-m}}{2i}, \tag{1.7.1b}$$

$$\sqrt{p} = c_p \sum_{n=0}^{p-1} \xi_p^{n^2}, \tag{1.7.1c}$$

for any nonzero $m, n \in \mathbb{Z}$, and any odd prime p, where $c_p = 1$ or $-i$ for $p \equiv \pm 1$ (mod 4), respectively ((1.7.1c) is called a *Gauss sum*). Only countably many complex numbers are cyclotomic, i.e. lie in $\cup_{n=1}^{\infty} \mathbb{Q}[\xi_n]$, so almost every complex number is not cyclotomic.

Cyclotomic numbers are the numbers in the character tables of finite groups, the values of Lie group characters at elements of finite order, the values of quantum-dimensions in RCFT, and the matrix entries in the $SL_2(\mathbb{Z})$-representation coming from rational VOAs. The theory is deeply entwined with that of modular forms and functions, as we see in Section 2.3.3. The key property of cyclotomic numbers, which accounts for their ubiquity, has to do with their Galois groups.

As usual, an automorphism $\sigma \in \mathrm{Gal}(\mathbb{Q}[\xi_n]/\mathbb{Q})$ is uniquely determined by what it does to the generator ξ_n. Since $\xi_n^n = 1$, we see that σ must send ξ_n to another nth root of 1, ξ_n^ℓ say; in fact $\sigma(\xi_n)$ must be another 'primitive' nth root of 1, that is ℓ must be coprime to n. So $\mathrm{Gal}(\mathbb{Q}[\xi_n]/\mathbb{Q})$ is isomorphic to the multiplicative group $\mathbb{Z}_n^\times$ of numbers between 1 and n coprime to n. To see what σ does to some $z \in \mathbb{Q}[\xi_n]$, we find the $\ell \in \mathbb{Z}_n^\times$ corresponding to σ and write z as a $\mathbb{Q}$-polynomial $p(\xi_n)$: then $\sigma z = p(\xi_n^\ell)$. For example,

$$\sigma\left(\cos(2\pi a/n)\right) = \sigma\left(\frac{\xi_n^a + \xi_n^{-a}}{2}\right) = \frac{\xi_n^{a\ell} + \xi_n^{-a\ell}}{2} = \cos(2\pi a\ell/n).$$

The defining property of cyclotomic numbers is a central result of classical number theory:

Theorem 1.7.1 (Kronecker–Weber) *Let $\mathbb{L}$ be a finite Galois extension of $\mathbb{Q}$ with abelian Galois group $\mathrm{Gal}(\mathbb{L}/\mathbb{Q})$. Then $\mathbb{L}$ is contained in some cyclotomic extension $\mathbb{Q}[\xi_n]$.*

The proof is quite complicated. Conversely, any cyclotomic extension $\mathbb{Q}[\xi_n]$ of $\mathbb{Q}$ is finite Galois and has abelian Galois group. In fact, the degree of $\mathbb{Q}[\xi_n]$ is given by Euler's ϕ-function:

$$[\mathbb{Q}[\xi_n] : \mathbb{Q}] = \phi(n) := n \prod_{p|n} \frac{p-1}{p}.$$

The minimal polynomial of ξ_n is called the nth *cyclotomic polynomial*; a manifestly integral construction for it is given in [64]. Its zeros are ξ_n^i for each i coprime to n.

The ring of *cyclotomic integers* $R_{\mathbb{Q}[\xi_n]}$ is simply $\mathbb{Z}[\xi_n]$. For all $n \neq 1, 2, 4$, $\mathbb{Q}[\xi_n]$ has infinitely many units: for example, $(\xi_n^i - 1)/(\xi_n - 1)$ is a unit of infinite order, for any $1 < i < n - 1$ coprime to n. Unique factorisation at the level of numbers (as opposed to ideals, which always holds) fails in all but 30 cyclotomic fields ($\mathbb{Q}[\xi_{23}]$ is the first field for which it fails).

Kronecker's *Jungentraum* ('dream of youth') [546] proposes that just as all abelian extensions of $\mathbb{Q}$ are obtained by adjoining to $\mathbb{Q}$ the values of a transcendental function (namely $\exp[2\pi i z]$) at certain algebraic numbers (namely $z \in \mathbb{Q}$), something similar should happen for abelian extensions of other finite extensions $\mathbb{K}$. This is still far from understood in general, but we know that any abelian extension of $\mathbb{K} = \mathbb{Q}[\sqrt{-d}]$ is contained in an extension of $\mathbb{K}$ by a root of unity, square-roots of integers, and the j-function (0.1.8) evaluated at $(a + \sqrt{b})/2$ for some $a, b \in \mathbb{Z}$.

Question 1.7.1. [572] (a) Show that a prime $p \equiv 3 \pmod 4$ cannot be written in the form $a^2 + b^2$ for integers a, b.
(b) Let $p \equiv 1 \pmod 4$ be prime. Define

$$S_p = \{(x, y, z) \in \mathbb{Z}^3 \mid x > 0, y > 0, z > 0, \ x^2 + 4yz = p\}.$$

Verify that for any $(x, y, z) \in S_p$, both $x \neq y - z$ and $x \neq 2y$. Define a map L on S_p by

$$L(x, y, z) = \begin{cases} (x + 2z, z, y - x - z) & \text{if } x < y - z \\ (2y - x, y, x - y + z) & \text{if } y - z < x < 2p \\ (x - 2y, x - y + z, y) & \text{if } x > 2y \end{cases}$$

Verify that L is an involution (i.e. $L(L(x, y, z)) = (x, y, z)$), and that L has exactly one fixed point. Show that this implies that the cardinality $\|S_p\|$ must be odd, and thus that the involution $(x, y, z) \mapsto (x, z, y)$ must also have a fixed point. Conclude that any prime $p \equiv 1 \pmod 4$ has a solution $p = a^2 + b^2$.

Question 1.7.2. Suppose we are given two points P, Q in the plane, distance 1 apart. Determine whether it is possible, using only a straight-edge and compass, to construct a point R collinear with P and Q such that the distance between P and R is $2^{-1/3}$. What if the distance between P and R is instead required to be $2^{-1/4}$?

Question 1.7.3. Let $\mathbb{K} = \mathbb{Q}[2^{1/3}]$. Show that $\text{Gal}(\mathbb{K}/\mathbb{Q})$ is trivial.

Question 1.7.4. Let $\mathbb{L} = \mathbb{Q}[\sqrt{2}, \sqrt{3}]$.
(a) Find an α such that $\mathbb{L} = \mathbb{Q}[\alpha]$.
(b) Find $\text{Gal}(\mathbb{L}/\mathbb{Q})$. Is $\mathbb{L}$ Galois?
(c) For each subgroup H of $\text{Gal}(\mathbb{Q}[\sqrt{2}, \sqrt{3}]/\mathbb{Q})$, find the corresponding extension $\mathbb{J}$.

Question 1.7.5. (a) Show that the values $\text{ch}(g)$ of characters are always cyclotomic integers. After reading this section, can you add anything to your answer to Question 1.1.5?
(b) Let G be any finite group. Prove: G is simple iff for all irreducible characters ch of G, $\text{ch}(a) = \text{ch}(e)$ only when $a = e$.

Question 1.7.6. Find all rational numbers r such that $\cos(2\pi r) \in \mathbb{Q}$.

2

Modular stuff

This chapter introduces modular functions and forms, a subject central to the remainder of the book. Some earlier parts of this chapter are beautifully covered in [**414**].

Section 2.1 supplies the underlying geometry, but can be skimmed on a first reading. In spite of this background material, the theory of modular forms and functions discussed in Sections 2.2 and 2.3 will probably appear as somewhat arbitrary to the uninitiated reader. Section 2.4.1 addresses some of this apparent artificiality, by developing the broader context of automorphic forms.

As explained in the introductory chapter, Moonshine involves unexpected occurrences of modularity. The modularity of Moonshine functions follows from Zhu's Theorem (Theorem 5.3.8). However, the complexity of the underlying mathematics begs the question: Can modularity be established in a more elementary way? The simplest example of Moonshine involves theta functions. Hence we explore the limits and potentials of four classical strategies for proving the modularity of theta functions: Poisson summation, Dirichlet series, the heat kernel and representations of Heisenberg groups (Sections 2.2.3, 2.3.1, 2.3.4 and 2.4.2, respectively).

Moonshine has really only been worked out in genus 1,[1] but conformal field theory tells us that there is an analogue for every genus (Section 6.3.1). It will be much more complicated, but it will be more rewarding because the number theoretic side is much less developed. In other words, we will find traces of, for example, the Monster in automorphic forms for the higher mapping class groups $\Gamma_{g,n}$ and $\mathrm{Sp}_{2n}(\mathbb{Z})$. We include Sections 2.1.4 and 2.3.5 in anticipation of this most natural and significant future development.

2.1 The underlying geometry

2.1.1 The hyperbolic plane

The birth of hyperbolic geometry is one of the most remarkable and instructive in the history of mathematics. Euclid's Fifth Postulate[2] was noticeably more complicated than the other axioms, looking more like a theorem than a self-evident proposal. Indeed, its converse was a theorem proved by Euclid. For example, compare it with Euclid's First

---

[1] There are two possible meanings of 'genus' in a phrase like 'higher genus Moonshine'. Ordinary Monstrous Moonshine is genus 0 in the sense that the j-function is a Hauptmodul, i.e. a function on a sphere. It is genus 1 in the sense that the argument τ of j parametrises different tori. In this paragraph we are anticipating Moonshine's extension to higher genus in this second sense.

[2] Also called the Parallel Postulate, it is equivalent to the simpler statement: *Given any line* L *and a point* p *not on* L, *there is a unique line parallel to* L *that passes through* p.

Fig. 2.1 Several parallel lines in the hyperbolic plane $\mathbb{H}$.

Postulate: *There is a unique line passing through any two points*, or Euclid's Fourth Postulate: *All right angles are equal.* For centuries, starting with Archimedes, mathematicians (both professional and amateur) tried to prove it from the other axioms. Finally in 1868 Beltrami established its independence by finding models for the hyperbolic plane, proving the conjecture of Gauss, Bolyai and Lobachevski as to the existence (i.e. internal consistency) of this non-Euclidean geometry. (More precisely, Beltrami's models reduced the question of the consistency of hyperbolic geometry to the consistency of Euclidean geometry.) Far from being an artificial construct, we've now learned that hyperbolic geometry is far more important than Euclidean geometry, at least in two and three dimensions.

In place of the Euclidean plane $\mathbb{R}^2$, consider the upper half-plane

$$\mathbb{H} := \{(x, y) \in \mathbb{R}^2 \mid y > 0\} = \{\tau \in \mathbb{C} \mid \mathrm{Im}\,\tau > 0\}. \tag{2.1.1}$$

The angles between intersecting curves in $\mathbb{H}$ are measured as in $\mathbb{R}^2$ (namely, take the angle between the two Euclidean lines tangent to the curves at the point of intersection). However, the hyperbolic lines consist of all half-lines perpendicular to the x-axis, together with all semi-circles with centre on the x-axis (see Figure 2.1). All axioms of Euclidean geometry hold here (e.g. between any two distinct points there passes a unique line), except for the Parallel Postulate: there are always infinitely many hyperbolic lines parallel to a given hyperbolic line L and passing through a given point $p \notin L$.

It is possible to prove from the other axioms that the remaining possibility (namely that there are *no* lines parallel to line L through point p) cannot occur. Nevertheless, there is a second kind of non-Euclidean geometry, called *spherical* geometry. In place of $\mathbb{R}^2$ we have the sphere S^2, and lines now are great circles. If we identify antipodal points $\pm p \in S^2$, then we get a geometry satisfying most of Euclid's axioms. The exceptions are that we can't speak unambiguously of the portion of a line between two points, and the Parallel Postulate (there are no parallel lines). Spherical geometry is older than Euclid – we needed it, for example, in our study of the night sky.

In Euclidean $\mathbb{R}^2$ the metric (infinitesimal length-squared) is given by $\mathrm{d}s^2 = \mathrm{d}x^2 + \mathrm{d}y^2$, and so the arc-length of a curve $\gamma : [0, 1] \to \mathbb{R}^2$ is

$$\mathrm{length}(\gamma) := \int_0^1 \sqrt{\gamma_1'(t)^2 + \gamma_2'(t)^2}\,\mathrm{d}t.$$

On $\mathbb{H}$ the arc-length of a curve $\gamma : [0, 1] \to \mathbb{H}$ becomes

$$\mathrm{length}_{\mathbb{H}} := \int_0^1 \frac{\sqrt{\gamma_1'(t)^2 + \gamma_2'(t)^2}}{\gamma_2(t)}\,\mathrm{d}t = \int_0^1 \frac{|\gamma'(t)|}{\mathrm{Im}\,\gamma(t)}\,\mathrm{d}t. \tag{2.1.2}$$

Define the hyperbolic distance $\text{dist}_{\mathbb{H}}(p, q)$ between two points $p, q \in \mathbb{H}$ to be the infimum $\inf_\gamma \text{length}_{\mathbb{H}}(\gamma)$ of the arc-lengths of all paths γ between $p = \gamma(0)$ and $q = \gamma(1)$. Just as the shortest path (geodesic) between two points in Euclidean geometry is the line segment between them, so in hyperbolic geometry it is the hyperbolic line segment.

The 'boundary' for $\mathbb{R}^2$ can be thought of as the circular horizon of 'points at infinity', parametrised by angle, and every line touches this circle at two points. Likewise, the boundary of $\mathbb{H}$ can be thought of as the circle $\mathbb{R} \cup \{\infty\}$, and again every line touches this circle at two points. This circle will appear as the infinitely distant horizon to beings living in $\mathbb{H}$. The point '∞' here is often written $i\infty$ to emphasise its relation to the vertical lines. The difference is that in $\mathbb{R}^2$, all parallel lines share the same two points at infinity; in $\mathbb{H}$, parallel lines share at most one point at infinity.

The most compelling model of the hyperbolic plane is perhaps the Poincaré disc

$$\mathbb{D} := \{z \in \mathbb{C} \mid |z| < 1\}.$$

Here, angles are again as in $\mathbb{R}^2$, but lines consist of diameters of the boundary circle $|z| = 1$, together with the intersection of $\mathbb{D}$ with circles hitting the boundary $|z| = 1$ at right angles. The metric is $|dz|^2/(1 - |z|^2)^2$, and the 'points at infinity' form the boundary circle $|z| = 1$. The equivalence with $\mathbb{H}$ is given by the isometry $\tau \mapsto \frac{\tau - i}{\tau + i}$ taking $\mathbb{H}$ onto $\mathbb{D}$.

It may seem strange that both models $\mathbb{H}$ and $\mathbb{D}$ of hyperbolic geometry have a distorted notion of length and line. Is there any way to realise hyperbolic geometry, using a surface embedded in $\mathbb{R}^3$ inheriting the usual metric and angle of $\mathbb{R}^3$? Hilbert proved the answer is No: *There is no complete surface in $\mathbb{R}^3$ with constant negative curvature* (see e.g. page 51 of [**527**]). Nash's Theorem (footnote 5 in chapter 1) implies though that there will be an embedding of the hyperbolic plane in some $\mathbb{R}^n$ ($n = 5$ works). 'Complete' means that any Cauchy sequence converges, so there aren't any points missing. To find the curvature of a surface at a point, first find the smallest and largest circles hugging the surface the closest at that point; the curvature is the inverse product $r^{-1}R^{-1}$ of their radii. For example, a sphere of radius r has constant curvature r^{-2}. A surface with 0 curvature is (locally) flat in one direction – for example, a cylinder or torus has constant curvature 0. The small and large circles for a surface Σ with negative curvature have centres on opposite sides of the tangent plane $T_p\Sigma$, like a saddle curving up from front to back, but curving down from side to side. The hyperbolic plane has constant negative curvature (Theorem 2.1.4(b)).

What is the significance of the word 'hyperbolic' here? It was chosen by Klein, partly because sinh and cosh appear in many formulae, but also because of another model of $\mathbb{H}$. Consider the hyperboloid $x_1^2 + x_2^2 - x_3^2 = -1$, embedded in Minkowski space $\mathbb{R}^{2,1}$ (so it is a Minkowski sphere of radius i). It consists of two sheets; let's focus on the upper one (where $x_3 \geq 1$). As a surface in $\mathbb{R}^{2,1}$, it inherits its notions of angle and metric $ds^2 = dx_1^2 + dx_2^2 - dx_3^2$ – in particular this induced geometry is equivalent to the hyperbolic plane. The lines here consist of the intersection of planes through the origin with the upper sheet (when those intersections are non-empty). Stereographic projection from the point $(0, 0, -1)$ conformally maps the upper sheet onto the Poincaré disc $\mathbb{D} \times \{0\}$.

Just as the area of a region $R \subset \mathbb{R}^2$ is given by the double integral $\int_R \mathrm{d}x\, \mathrm{d}y$, so is the hyperbolic area of region $R \subset \mathbb{H}$ given by

$$\mathrm{area}_{\mathbb{H}}(R) := \int_R \frac{\mathrm{d}x\, \mathrm{d}y}{y^2}. \tag{2.1.3a}$$

This just says that the hyperbolic area of the infinitesimal rectangle $[x, x + \mathrm{d}x] \times [y, y + \mathrm{d}y]$ is the product $\frac{\mathrm{d}x}{y} \times \frac{\mathrm{d}y}{y}$ of hyperbolic length with hyperbolic height. This area formula fails for macroscopic rectangles, if for no other reason than that there are *no* macroscopic rectangles! In fact, one of the most remarkable formulae of geometry must be the expression, originally due to Lambert (1766),[3] for the area of a triangle T in terms of its interior angles $\alpha_1, \alpha_2, \alpha_3$:

$$\mathrm{area}_{\mathbb{H}}(T) = \pi - \alpha_1 - \alpha_2 - \alpha_3. \tag{2.1.3b}$$

More generally, the area of an n-sided hyperbolic polygon is $(n - 2)\pi - \sum_i \alpha_i$. From this we obtain the non-existence of rectangles. These formulae apply even in the limiting case where some vertices lie on the boundary $\mathbb{R} \cup \{i\infty\}$. In particular, the area of any hyperbolic triangle is bounded above (even though $\mathbb{H}$ itself has infinite area)!

Klein proposed to study geometry using the group of symmetries of whichever geometric quantities are important to the context (Section 1.2.2). The group $\mathrm{Isom}(\mathbb{R}^2)$ of isometries (i.e. distance-preserving maps) of $\mathbb{R}^2$ consists of all translations $x \mapsto x + a$, all orthogonal maps (rotations and reflections) $x \mapsto xA$ where $AA^t = I$, and all combinations $xA + b$ thereof. Likewise, the group $\mathrm{Isom}(\mathbb{H})$ of hyperbolic isometries consists of all *Möbius*, or *fractional linear*, transformations

$$z \mapsto \frac{az + b}{cz + d}, \qquad \forall a, b, c, d \in \mathbb{R} \text{ with } ad - bc = 1, \tag{2.1.4a}$$

together with the reflection $z \mapsto -\bar{z}$, and all combinations thereof. As in the Euclidean case, $\mathrm{Isom}(\mathbb{H})$ is a three-dimensional real Lie group, with two connected components; the component $\mathrm{Isom}^+(\mathbb{H})$ containing the identity consists of (2.1.4a), and is isomorphic to

$$\mathrm{PSL}_2(\mathbb{R}) := \mathrm{SL}_2(\mathbb{R}) / \left\{ \pm \begin{pmatrix} 1 & 0 \\ 0 & 1 \end{pmatrix} \right\}. \tag{2.1.4b}$$

As in the Euclidean case, isometries preserve the absolute value $|\theta|$ of angles; maps $\alpha \in \mathrm{Isom}^+(\mathbb{H})$ preserve the angles themselves and so are conformal. Isometries preserve area and send hyperbolic lines to hyperbolic lines. $\mathrm{PSL}_2(\mathbb{R})$ preserves everything of geometric significance and is thus the group of symmetries of the hyperbolic plane.

Likewise, the group $\mathrm{Isom}^+(S^2)$ of symmetries of spherical geometry is $\mathrm{PSL}_2(\mathbb{C})$, acting on the Riemann sphere $\mathbb{P}^1(\mathbb{C})$ by Möbius transformations. The symmetries $\mathrm{PSL}_2(\mathbb{R})$ of $\mathbb{H}$ are precisely those transformations in $\mathrm{PSL}_2(\mathbb{C})$ that send $\mathbb{H}$ to itself. The only reason this action by Möbius transformations of the 2×2 matrices on $\mathbb{P}^1(\mathbb{C})$ or $\mathbb{H} \cup \{i\infty\}$ might not look strange to us, is because familiarity breeds numbness. Much more natural is

---

[3] This is the same Lambert who proved the irrationality of π and e.

the action of $n \times n$ matrices on $\mathbb{C}^n$, and this induces their action on $\mathbb{C}^{n-1}$ (together with a codimension-2 set of 'points at infinity') by interpreting $\mathbb{C}^n$ as homogeneous coordinates for $\mathbb{C}^{n-1}$ (Section 1.2.2). Specialising to $n = 2$ gives us the action (2.1.4a). In Section 2.4.1 we interpret (2.1.4a) using the multiplication of matrices in $\mathrm{SL}_2(\mathbb{R})$.

A model for n-dimensional hyperbolic geometry is the upper half-space $\mathbb{H}^n := \{(x_i) \in \mathbb{R}^n \mid x_n > 0\}$, which is conformally equivalent to the interior of the unit n-ball, or to the upper (i.e. $x_{n+1} > 0$) sheet of the hyperboloid $x_1^2 + \cdots + x_n^2 - x_{n+1}^2 = -1$. Euclidean angle is used, but the metric is $ds^2 = (dx_1^2 + \cdots + dx_n^2)/x_n^2$. Hyperbolic lines consist of half-lines and semi-circles perpendicular to the boundary hyperplane $x_0 = 0$; hyperbolic planes in $\mathbb{H}^n$ consist of half-planes and half-spheres perpendicular to the boundary hyperplane $x_0 = 0$. The hyperboloid model makes it clear that the isometries $\mathrm{Isom}(\mathbb{H}^n)$ of hyperbolic n-space is isomorphic to the group of those matrices $A \in \mathrm{O}_{n,1}(\mathbb{R})$ with $A_{n+1,n+1} \geq 1$. The group $\mathrm{Isom}^+(\mathbb{H}^n)$ of *conformal* isometries is the Lorentz group $\mathrm{SO}_{n,1}(\mathbb{R})^+$, obeying in addition the condition $\det(A) = 1$. Of course the Lorentz group $\mathrm{SO}_{3,1}(\mathbb{R})^+$ is more famous in its incarnation as the symmetry of special relativity (Section 4.1.2). By identifying the boundary plane of $\mathbb{H}^3$ with $\mathbb{C}$, the group $\mathrm{Isom}^+(\mathbb{H}^3) \cong \mathrm{SO}_{3,1}(\mathbb{R})^+$ can be naturally identified with the Möbius transformations $\mathrm{PSL}_2(\mathbb{C})$.

Recall Hilbert's theorem from a few paragraphs ago. Although no surface embedded in $\mathbb{R}^3$ can provide a model of the full hyperbolic plane, they can provide a model of a piece of that plane (i.e. be 'incomplete'). This is accomplished by any surface of constant negative curvature. For example, consider the 'tractrix' – the path traced by a stone, initially placed at $(0,1)$, pulled ('tractored') by a string of length 1 as we walk along the x-axis. Take the tractrix in the xy-plane and rotate it about the x-axis; the result is called the 'pseudo-sphere', and is a surface of constant negative curvature in $\mathbb{R}^3$. More generally, by a *hyperbolic surface* we mean a surface that is also a metric space (i.e. it has a notion of distance between points, and of arc-length), which is locally isometric to $\mathbb{H}$ (i.e. the open sets V_α in Definition 1.2.3 are taken to be in $\mathbb{H} \subset \mathbb{R}^2$, and the transition functions $\varphi_{\alpha\beta}$ are in $\mathrm{Isom}(\mathbb{H})$). The pseudo-sphere is an example of a hyperbolic surface different from the hyperbolic plane; crocheting constructs several other examples [284]. Similarly, we can define hyperbolic manifolds of arbitrary dimension. We conclude this subsection with the classification of all hyperbolic surfaces. But first we need the notion of a Fuchsian group.

As was discussed in Section 1.2.2, tori $S^1 \times S^1$ arise from the quotient $\mathbb{R}^2/L$ of the plane by a two-dimensional lattice. This construction is equivalent to the familiar depiction of a torus as a parallelogram with opposite sides identified. We discuss the Riemann surfaces in more detail next subsection, but a genus-g surface can be depicted by identifying appropriate sides in a $4g$-gon (see Figure 2.2 for the situation with a genus 2 surface). This arises from making $2g$ circular cuts into the surface and flattening it out. But can we also interpret that $4g$-gon as corresponding to some quotient of $\mathbb{R}^2$, generalising the $\mathbb{R}^2/L$ construction of a torus? The answer is no – the group $\mathrm{Isom}(\mathbb{R}^2)$ doesn't have a rich enough supply of discrete subgroups. We *can* interpret the $4g$-gon as a quotient, but of the *hyperbolic* plane and not the Euclidean one.

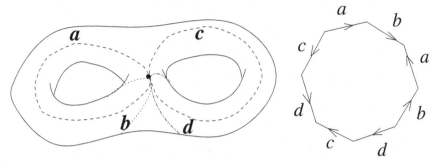

Fig. 2.2 A genus 2 surface and its octagon.

Definition 2.1.1 *A Fuchsian group is a discrete subgroup* Γ *of* $SL_2(\mathbb{R})$*, i.e. one with* $\inf\{(a-1)^2 + b^2 + c^2 + (d-1)^2\} > 0$*, where the infimum is over all* $\begin{pmatrix} a & b \\ c & d \end{pmatrix} \neq I$ *in* Γ.

We identify a subgroup Γ of $SL_2(\mathbb{R})$ with its canonical projection $\overline{\Gamma}$ into $PSL_2(\mathbb{R})$, since these give rise to identical surfaces. Examples of Fuchsian subgroups are

$$G_N = \left\{ \begin{pmatrix} \cos(\pi k/N) & \sin(\pi k/N) \\ -\sin(\pi k/N) & \cos(\pi k/N) \end{pmatrix} \mid 0 \leq k < N \right\}, \qquad \forall N = 1, 2, \ldots,$$

$$G_{\mathbb{Z}} = \left\{ \begin{pmatrix} 1 & k \\ 0 & 1 \end{pmatrix} \mid k \in \mathbb{Z} \right\},$$

and the modular group $SL_2(\mathbb{Z})$. The latter is certainly the most interesting of these.

Let Γ be a Fuchsian group. Most points $z \in \mathbb{H}$ (i.e. all but at most countably many) are fixed only by the identity in Γ (why?). Let $z_0 \in \mathbb{H}$ be any of those generic points. Define the set

$$D_{\Gamma}(z_0) := \{w \in \mathbb{H} \mid \text{dist}_{\mathbb{H}}(z_0, w) < \text{dist}_{\mathbb{H}}(\gamma.z_0, w) \text{ for all } \gamma \in \Gamma, \ \gamma \neq \pm I\}.$$

So $D_{\Gamma}(z_0)$ is the intersection of a number of hyperbolic half-planes. This set $D = D_{\Gamma}(z_0)$ is called a *fundamental domain* of Γ, as it satisfies the following properties: (i) it is open; (ii) each orbit $\Gamma.z$ intersects D in at most one point, and every orbit intersects the closure of D in at least one point; (iii) the boundary ∂D of D in $\mathbb{H}$ consists of at most countably many hyperbolic line segments. (In fact, as long as Γ is finitely generated, D can be chosen with boundary consisting of only finitely many segments.)

For example, a fundamental domain for G_N consists of the points lying between any pair of hyperbolic lines intersecting at i with angle $2\pi/N$. A fundamental domain for $G_{\mathbb{Z}}$ is $\{z \in \mathbb{H} \mid -\frac{1}{2} < \text{Re } z < \frac{1}{2}\}$. Choosing $z_0 = 2i$, we get the fundamental domain D for $SL_2(\mathbb{Z})$ depicted in Figure 2.3: the vertical sides are $\text{Re } z = \pm\frac{1}{2}$, and the circle is $|z| = 1$.

Applying Γ to a fundamental domain D will tile the hyperbolic plane – see Escher's *Circle Limit I,II, . . .* for examples. Since $\Gamma \subset \text{Isom}^+(\mathbb{H})$, each of these tiles is an identical copy (a congruent translate) of D. All this holds as well in hyperbolic n-space – for example, an analogue of $SL_2(\mathbb{Z})$ for $\mathbb{H}^3$ is $SL_2(\mathbb{Z} + i\mathbb{Z})$.

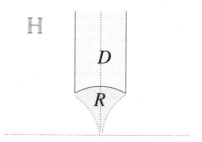

Fig. 2.3 Two fundamental domains for $SL_2(\mathbb{Z})$.

Just as we constructed the torus by identifying opposite sides of the parallelogram, so we can obtain a surface by identifying the appropriate sides of the fundamental domain of a Fuchsian group Γ. This surface will be a realisation of the orbit space $\Gamma\backslash\mathbb{H}$ (we write Γ on the left because it acts on the left). Provided no $\gamma \in \Gamma$ has fixed points in $\mathbb{H}$ (except for the trivial maps $\gamma = \pm I$), the orbit space $\Gamma\backslash\mathbb{H}$ will inherit the hyperbolic geometry of $\mathbb{H}$ and be a hyperbolic surface.

Theorem 2.1.2   *Any complete hyperbolic surface Σ is isometric to a surface of the form $\overline{\Gamma}\backslash\mathbb{H}$ where $\overline{\Gamma}$ is a torsion-free Fuchsian subgroup of $PSL_2(\mathbb{R})$. Two such subgroups $\overline{\Gamma_1}, \overline{\Gamma_2}$ define isometric surfaces $\overline{\Gamma_1}\backslash\mathbb{H}$ and $\overline{\Gamma_2}\backslash\mathbb{H}$ iff $\alpha\overline{\Gamma_1}\alpha^{-1} = \overline{\Gamma_2}$ for some $\alpha \in PSL_2(\mathbb{R})$.*

'Torsion-free' means that all nontrivial elements of $\overline{\Gamma}$ have infinite order – see Question 2.1.2(b). Almost all surfaces with a conformal or metric or complex structure are $\Gamma\backslash\mathbb{H}$ for some Fuchsian subgroup Γ. An unexpected revelation of Thurston's Programme is that something similar happens in three dimensions – see the review [**497**]. Any surface of genus $g \geq 2$ supports uncountably many different hyperbolic structures. By contrast, the Mostow Rigidity Theorem (1973) tells us that a connected compact oriented manifold of dimension $n \geq 3$ supports only one.

2.1.2 Riemann surfaces

Manifolds M, N are homeomorphic if there is a continuous map $M \to N$ with continuous inverse. Compact connected orientable surfaces are characterised, up to homeomorphism, by the *genus* $g \in \mathbb{N}$. A sphere is genus 0, a torus genus 1, and the double-torus of Figure 2.2 is genus 2. The surface of a wine glass or fork is topologically a sphere, while coffee mugs and keys are (usually) tori. A ladder with n rungs has genus $n - 1$. The surface of a pair of pants is genus 2, while that of a sweater is genus 3.

A torus can be realised in many different ways. One is the Cartesian product $S^1 \times S^1$ of circles (lay one circle horizontally, then from each point on it place a vertical circular rib perpendicular to it, filling out the torus's surface). A complex curve of the form $y^2 = ax^3 + bx^2 + cx + d$ is a torus (at least if the points at infinity are included), as is the quotient $\mathbb{C}/L$ of the complex plane with a two-dimensional lattice L (Section 1.2.1).

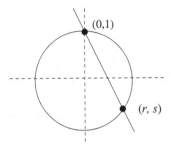

Fig. 2.4 Diophantus' argument.

If we drop the requirement that our surface be *compact*, then up to homeomorphism it is uniquely specified by two numbers: the genus g as above, and the number of punctures (or boundary components) n. For instance, a sphere with one puncture is homeomorphic to an open disc or equivalently the plane $\mathbb{C}$. We see this when we pop a balloon: the sphere becomes a rather jagged-edged disc. A sphere with two punctures is a cylinder or annulus.

The *non-orientable* surfaces have a very similar classification. For example, if we could create a $\mathbb{P}^2(\mathbb{R})$-shaped balloon, then popping it would create a jagged-edged Möbius band. We always require orientability in this book.

The surfaces we encounter have more structure than mere topology. If the surface Σ is in fact *smooth* (Section 1.2.2), then we are interested in their classification up to *diffeomorphism*. In this case though nothing changes, the surface is again parametrised by the genus and number of punctures: any surface Σ has a unique differential structure compatible with its topology. In order to obtain a finer distinction between the surfaces, we need to further enrich their structure. The easiest way to do this is by introducing a metric onto the tangent spaces, or give the surface a complex or conformal structure. More on the resulting *Riemann surfaces* shortly. Nevertheless, the genus remains the single most important invariant distinguishing Riemann surfaces. There are many qualitative differences captured by genus – we will give three of them.

Diophantus [45] was a mathematical giant who lived in Alexandria in the second or third century A.D. He seems to have been the first Greek to regard fractions as legitimate numbers, and he was the first to use negative numbers (though only in intermediate arithmetical calculations, so probably didn't believe their ontological reality), and the first to invent an abstract symbolism for algebra. The following (expressed in modern language) is how Diophantus found all Pythagorean triples, that is the integer solutions to $a^2 + b^2 = c^2$.

First, it's enough to look for all rational solutions to the circle $x^2 + y^2 = 1$. Then the integers a, b, c can be recovered by clearing denominators. Consider a line through the point $(0, 1)$ that intersects the circle at another rational point (r, s) (see Figure 2.4). Clearly this line must have rational (or infinite) slope $\frac{s-1}{r}$. Conversely, consider any line through $(0,1)$ with rational slope u: its equation will be $y = ux + 1$. Where does it intersect the circle? We get $1 = x^2 + (ux + 1)^2 = (u^2 + 1)x^2 + 2ux + 1$, i.e. $x\left((u^2 + 1)x + 2u\right) = 0$.

So apart from our original point $(0, 1)$, it will also intersect the circle at

$$(x, y) = \left(\frac{-2u}{u^2 + 1}, \frac{1 - u^2}{u^2 + 1} \right).$$

As long as u is rational, so will be this point. Thus Diophantus found a parametrisation of all rational points on the circle, and hence all Pythagorean triples.

His method is far more general than this, as he knew. In fact, consider any nondegenerate conic. To find all rational points on it, we first find one rational point, and then consider all lines with rational slope through that point. This will exhaust all rational points on the curve. Thus if a conic has one rational point (it might have none), then it will have infinitely many, and all can be found explicitly.

Why won't this trick work for other equations of this sort? For example, Fermat's Last Theorem challenges us to find a nontrivial rational solution to $x^n + y^n = 1$, for $n > 2$. If we draw a line through the obvious solution $(x, y) = (0, 1)$, we simply get a mess. What's so special, geometrically, about conics?

The modern way (due to Bezout in the eighteenth century) to think about this is to regard the given equation, say $x^2 + y^2 = 1$, as an equation relating two complex numbers $(x, y) \in \mathbb{C}^2$. The result will be a complex curve, that is a real surface. To which complex curve does $x^2 + y^2 = 1$ correspond? The *real* curve (a circle) is parametrised by $x = \cos \theta$ and $y = \sin \theta$, and a moment's deliberation will convince oneself that permitting θ to take complex values will exhaust all points on the *complex* curve. So write $x = \frac{1}{2}(w + w^{-1})$ and $y = \frac{i}{2}(w - w^{-1})$ for any $w \in \mathbb{C}$ except $w = 0$; this identifies the complex curve $x^2 + y^2 = 1$ with the complex plane punctured at 0, that is a cylinder. The unit circle in $\mathbb{R}^2$ is merely the slice of this cylinder in $\mathbb{C}^2$ by the plane passing through the two real axes of $\mathbb{C}^2$. A different slice will produce, for instance, an hyperbola.

More generally, any polynomial in x, y defines a noncompact surface in $\mathbb{C}^2$. For example, a nondegenerate cubic $y^2 = x^3 + ax^2 + bx + c$ is a once-punctured torus – explicitly, the quotient $\mathbb{C}'/(\mathbb{Z} + \tau\mathbb{Z})$, where $\mathbb{C}'$ means deleting from $\mathbb{C}$ the lattice points $\mathbb{Z} + \tau\mathbb{Z}$, is equivalent in every sense one could want (e.g. conformally) to the cubic

$$y^2 = 4x^3 - 60G_4(\tau)x - 140G_6(\tau),$$

where the Eisenstein series $G_k(\tau)$ is defined in (0.1.5). Similarly, the complex curve $x^3 + y^3 = 1$ corresponds to the torus $\mathbb{C}/(\mathbb{Z} + \tau\mathbb{Z})$ with three points removed.

In any case, we can now answer our question: What is so special geometrically about the conics, that Diophantus' method works for them? The answer: They are (punctured) spheres, that is have genus 0.

It will always seem that some points 'at infinity' are missing from these complex curves. Kepler back in 1604 knew that adding such points simplifies the geometry. We do this by *projectifying* the given equation (Section 1.2.2). For example, $x^2 + y^2 = 1$ corresponds to the homogeneous equation $x^2 + y^2 = z^2$, where we identify (x, y, z) and $(\lambda x, \lambda y, \lambda z)$ for $\lambda \neq 0$. The two 'infinite' points, that is the points with $z = 0$, are then $(1, \pm 1, 0)$. Similarly, the three missing points on the Fermat curve $x^3 + y^3 = 1$ have homogeneous coordinates $(x, y, z) = (1, -\xi, 0)$ for any third root of unity ξ. We see that in homogeneous coordinates the 'infinite points' don't look so bad.

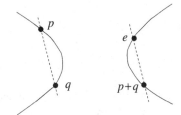

Fig. 2.5 Addition of points on a hyperbola.

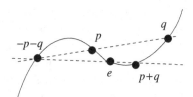

Fig. 2.6 Addition of points on a cubic.

Another special property of conics (avoiding the infinite points) is that they are additive groups. Fix any point e on the conic C (it will be the identity); given any two finite points p, q on the conic, the sum $p + q \in C$ is defined to be the intersection with C of the line through e parallel to the line through p and q (Figure 2.5). Associativity follows from Pascal's Theorem concerning hexagons inscribed in conics. For example, choosing the identity $e = (1, 0)$ and the parametrisation $(x(t), y(t)) = (\cos(t), \sin(t))$ of the circle $x^2 + y^2 = 1$, this addition of points corresponds to addition of angle t. The same conclusion holds for the hyperbola $x^2 - y^2 = 1$, with $e = (1, 0)$ and parametrisation $t \mapsto (\cosh(t), \sinh(t))$ of the $x > 0$ branch. See Question 2.1.3.

Better known is the addition of points on a nondegenerate (projective) cubic C. Fix any $e \in C$ (again it will play the role of identity), and choose any points $p, q \in C$. Let $r \in C$ be the intersection with C of the line through p, q; the sum $p + q$ is defined to be $-r$, that is the intersection with C of the line through r and e (see Figure 2.6). This also is commutative and associative, provided we include the points at infinity. Addition continues to work when the cubic is complexified, and that's how to make sense of it: the resulting surface is a torus, equivalent to one of the form $\mathbb{C}/(\mathbb{Z} + \tau\mathbb{Z})$ for some $\tau \in \mathbb{C}$, and this addition on the cubic lifts to ordinary addition on $\mathbb{C}$. Incidentally, the addition of points is only one of a number of senses in which conics are toy models for the much richer theory of elliptic curves (i.e. cubics with a marked point e) [372].

The simplest quantitative distinction between surfaces of different homeomorphism type (g, n) is the fundamental group π_1, defined in Section 1.2.3. For example, $\pi_1(S^2) = 1$ since S^2 is simply connected, and π_1 of a torus is $\mathbb{Z} \oplus \mathbb{Z}$. Let Σ_g be a compact genus $g > 0$ surface. Then $\pi_1(\Sigma_g)$ has presentation

$$\pi_1(\Sigma_g) \cong \left\langle \alpha_1, \ldots, \alpha_g, \beta_1, \ldots, \beta_g \mid \alpha_1\beta_1\alpha_1^{-1}\beta_1^{-1} \cdots \alpha_g\beta_g\alpha_g^{-1}\beta_g^{-1} = 1 \right\rangle. \quad (2.1.5a)$$

The generators α_i, β_j are chosen as in Figure 2.2 ($\alpha_1 = a, \beta_1 = b$, etc.). The easiest way to read off the genus from (2.1.5a) is to compute the abelianisation $\pi_1/[\pi_1, \pi_1]$ (which equals incidentally the first homology group $H_1(\Sigma_g, \mathbb{Z})$); as is clear from (2.1.5a), it is the abelian group $\mathbb{Z}^{2g}$ generated by α_i, β_j. On the other hand, the fundamental group of a genus-g surface $\Sigma_{g,n}$ with $n > 0$ punctures is free (see e.g. page 64 of [103]):

$$\pi_1\left(\Sigma_{g,n}\right) \cong \mathcal{F}_{2g+n-1}. \tag{2.1.5b}$$

The preceding discussion indicates the significance of genus. Now let's impose more structure. A *Riemann surface* is a connected orientable surface with a *conformal structure*, together with a choice of orientation. Equivalently, a Riemann surface can be defined as a complex analytic curve: any polynomial equation in x, $y \in \mathbb{C}$ inherits the conformal and differential structure of $\mathbb{C}$. This is because locally the conformal maps in $\mathbb{R}^2$ are precisely the locally holomorphic maps in $\mathbb{C}$ with nonvanishing derivative (theorem 14.2 of [481]). A third possible definition is that Riemann surfaces consist of those connected 2-manifolds with a complete metric with constant curvature. As mentioned above, its homeomorphism class is given by its genus g and number of punctures n, and the surface is compact iff $n = 0$. We are primarily interested in compact Riemann surfaces.

Any topological surface can be made into a Riemann surface, usually in a continuum of inequivalent ways (Section 2.1.4). We identify two Riemann surfaces if they are conformally equivalent, or holomorphically equivalent, or isometric. In Section 2.1.4 we discuss the classification of Riemann surfaces up to conformal equivalence.

The basic example of a Riemann surface is the complex plane $\mathbb{C}$. Also important is the complex projective line $\mathbb{P}^1(\mathbb{C}) = \mathbb{C} \cup \{\infty\}$; stereographic projection verifies that it is topologically a sphere, called the *Riemann sphere*. Now, a *meromorphic function* $f : D \to \mathbb{C}$ by definition is holomorphic everywhere except for isolated poles; if f has poles at z_i, then defining $f(z_i) = \infty$ gives a *conformal* map $f : D \to \mathbb{P}^1(\mathbb{C})$ between Riemann surfaces (perhaps it is this picture, in which z_i is sent to the 'north pole' ∞, which is the origin of the term 'pole'). Likewise, we can extend the *domain* of a function f on $\mathbb{C}$ to $\mathbb{P}^1(\mathbb{C})$, provided it is meromorphic at ∞. For example, if p is a polynomial of degree n, then p has a pole of degree n at ∞, and we obtain a holomorphic map $p : \mathbb{P}^1(\mathbb{C}) \to \mathbb{P}^1(\mathbb{C})$. By comparison, the functions e^z and $\cos(z)$ have essential singularities at ∞ and so cannot be extended to $\mathbb{P}^1(\mathbb{C})$.

Historically, Riemann surfaces were introduced by Riemann to supply the maximal domain (via analytic continuation) of a holomorphic function. The problem is that many of the most natural complex functions are multivalued, for example $f(z) = \sqrt{z}$ or $g(z) = \log z$ or other inverses of nice functions. As we move counterclockwise along the unit circle $|z| = 1$, starting at $z = 1$, the value $f(z) = \sqrt{z}$ changes continuously from $f(1) = 1$ to $f(1) = -1$, and the value of $g(z) = \log z$ changes continuously from $g(1) = 0$ to $g(1) = 2\pi \mathrm{i}$. To Riemann, we should regard $f(z)$ as a holomorphic function on a double cover $D = D^b \cup D^t$ of the complex plane, and $g(z)$ is holomorphic on a helix. As we move along the circle, the argument z of $f(z)$ moves from the bottom sheet $D^b \cong \mathbb{C}$ to the top sheet $D^t \cong \mathbb{C}$, and if we continue a second time around the circle, we return from the sheet D^t to D^b. To identify D homeomorphically, cut both D^b and D^t from 0 to ∞, and glue the $\theta = 0^+$ slit of D^b to the $\theta = 0^-$ slit of D^t and vice versa. The result is homeomorphic to a sphere with one puncture, corresponding to the point at infinity.

Note that $f : D \to \mathbb{C}$ is well-defined and holomorphic; it is an example of what we will shortly call a cover of $\mathbb{C}$, ramified at $z = 0$.

The remainder of this subsection describes an important realisation (called *uniformisation*) of any Riemann surface. The idea is simple. There are two different connected real curves, up to homeomorphism, and they are the line $\mathbb{R}$ and the circle S^1. The circle can be realised as $S^1 \cong \mathbb{R}/\mathbb{Z}$. We call $\mathbb{R}$ the 'universal cover' $\widetilde{S^1}$ of S^1, because it is simply-connected; $\mathbb{Z}$ here is the fundamental group $\pi_1(S^1)$. See also Theorem 1.4.3.

The same works with surfaces. For example, the sphere with two punctures (a cylinder) and a torus both have universal cover homeomorphic to $\mathbb{C}$; the cylinder itself is homeomorphic to $S^1 \times \mathbb{R}$ and the torus to $\mathbb{C}/(\mathbb{Z} + i\mathbb{Z})$, where $\mathbb{Z}$ and $\mathbb{Z} + i\mathbb{Z}$ are isomorphic to their fundamental groups. Let's make these ideas more precise, and incorporate as well the conformal structure.

Definition 2.1.3 *Let Σ^*, Σ be two Riemann surfaces. We say that Σ^* covers Σ by f if $f : \Sigma^* \to \Sigma$ is a holomorphic map from Σ^* onto Σ. If in addition f is locally conformal, we call f a* conformal *or* unramified cover. *If $f : \Sigma^* \to \Sigma$ is a conformal cover, and Σ^* is simply-connected, then we call Σ^* a* universal cover *of Σ.*

Let $U_\alpha \subset \Sigma$, $\varphi_\alpha : U_\alpha \to V_\alpha \subset \mathbb{C}$ be a family of coordinate charts for Σ (Definition 1.2.3); by *local coordinates* we mean the complex numbers $z \in V_\alpha$. In local coordinate z about point $p^* \in \Sigma^*$, a cover f sends a neighbourhood of p^* to one of $f(p^*) \in \Sigma$ with local coordinates $a + cz^n +$ higher terms, for some constants a and $c \neq 0$. To be conformal, this order n must always be 1 (otherwise we say f is ramified at p^*).

If $f : \Sigma^* \to \Sigma$ is a conformal cover, then the fundamental group $\pi_1(\Sigma^*)$ is naturally isomorphic to a subgroup of $\pi_1(\Sigma)$ (Section 1.7.2). In this way, the covers Σ^* of Σ (up to homeomorphism) are in one-to-one correspondence with conjugacy classes of subgroups of $\pi_1(\Sigma)$. A universal cover $\widetilde{\Sigma}$ is the 'largest' and most important cover, and is unique up to conformal equivalence. It can be identified as the space of all homotopy-equivalence classes of paths on Σ with fixed initial point $p \in \Sigma$. For example, visualise a 'point' $\tilde{p}$ on $\widetilde{S^1}$ as a curve starting at $1 \in S^1$ and ending at $e^{i\theta}$ ($0 \leq \theta < 2\pi$), and wrapping around the circle (i.e. crossing $1 \in S^1$) n times; the identification of $\widetilde{S^1}$ with $\mathbb{R}$ comes from identifying this path with the number $\theta + 2\pi n \in \mathbb{R}$.

We are now ready to state the basic result of this subsection.

Theorem 2.1.4 (Uniformisation Theorem)

(a) *Up to conformal equivalence, the only simply-connected Riemann surfaces (i.e. the only candidates for a universal cover) are the sphere $S^2 = \mathbb{P}^1(\mathbb{C}) = \mathbb{C} \cup \{\infty\}$, the plane $\mathbb{C}$ and the upper half-plane $\mathbb{H}$.*

(b) *Let Σ be any Riemann surface, and let $\widetilde{\Sigma}$ be its universal cover. Then Σ is conformally equivalent to $\widetilde{\Sigma}/\Gamma$, where $\Gamma \cong \pi_1(\Sigma)$ is a subgroup of the automorphisms of $\widetilde{\Sigma}$ that act on $\widetilde{\Sigma}$ without fixed points. A metric can be chosen for Σ with constant curvature $+1, 0, -1$, respectively, if $\widetilde{\Sigma} = S^2, \mathbb{C}, \mathbb{H}$, respectively. Two surfaces $\widetilde{\Sigma}/\Gamma, \widetilde{\Sigma'}/\Gamma'$ are conformally equivalent iff the universal covers $\widetilde{\Sigma}$ and $\widetilde{\Sigma'}$ are the same, and Γ and Γ' are conjugate subgroups in $\mathrm{Aut}(\widetilde{\Sigma})$.*

Table 2.1. *The universal covers of the genus g surfaces with n punctures*

$g\backslash n$	0	1	2	≥ 3
0	S^2	$\mathbb{C}, \mathbb{H}$	$\mathbb{C}, \mathbb{H}$	$\mathbb{H}$
1	$\mathbb{C}$	$\mathbb{H}$	$\mathbb{H}$	$\mathbb{H}$
≥ 2	$\mathbb{H}$	$\mathbb{H}$	$\mathbb{H}$	$\mathbb{H}$

Of course $\mathbb{H}$ and $\mathbb{C}$ are homeomorphic, but they aren't conformally equivalent (replacing $\mathbb{H}$ with the disc $\mathbb{D}$, this follows from *Liouville's Theorem*: a bounded holomorphic function on $\mathbb{C}$ must be constant). Part (a) is due to Klein, Poincaré and Koebe. These three possibilities for $\widetilde{\Sigma}$ correspond respectively to the three geometries: spherical, Euclidean and hyperbolic. The group of automorphisms of $\widetilde{\Sigma}$ is just Isom$^+$. The condition that Γ acts without fixed points (apart from the identity in Γ) is significant – fixed points change the geometry. A famous example of an orbit space with fixed points is $SL_2(\mathbb{Z})\backslash\mathbb{H}$, which has conical singularities at i and $e^{\pi i/3}$.

Table 2.1 gives the universal cover of any Riemann surface, as a function of the genus and number of punctures. We see there that almost every surface is hyperbolic: *the generic geometry in two dimensions is hyperbolic*.

The Uniformisation Theorem easily proves *Picard's Theorem* ('the range $f(\mathbb{C})$ of any holomorphic nonconstant function $f : \mathbb{C} \to \mathbb{C}$ can omit at most one point from $\mathbb{C}$'). The proof, which the reader can fill in, uses Liouville's Theorem together with the fact that the universal cover of the twice-punctured plane is $\mathbb{D}$.

2.1.3 Functions and differential forms

The last subsection gives several equivalent notions of a Riemann surface. Here we see that any compact Riemann surface is the locus of a homogeneous polynomial $f(a, b, c) = 0$ in the complex projective plane $\mathbb{P}^2(\mathbb{C})$.

We study a manifold through the functions living on it. Two manifolds differing merely by a single point can have a completely different family of functions. For instance, we all know many examples of holomorphic functions on $\mathbb{C}$. But the only functions holomorphic on $\mathbb{C}$ and also holomorphic at ∞ are the constants. More generally, any noncompact Riemann surface Σ has several functions $f : \Sigma \to \mathbb{C}$ holomorphic everywhere, while if Σ is compact, the only holomorphic functions $f : \Sigma \to \mathbb{C}$ are the constants. We are more interested in compact Σ.

Given any Riemann surface Σ, let $\mathcal{K}(\Sigma)$ denote all the meromorphic functions $f : \Sigma \to \mathbb{C}$ – equivalently, all holomorphic functions $f : \Sigma \to \mathbb{P}^1(\mathbb{C})$ (by convention we discard the constant function $f \equiv \infty$). Let $U_\alpha \subset \Sigma$, $\varphi_\alpha : U_\alpha \to V_\alpha \subset \mathbb{C}$ be a family of coordinate charts for Σ. Then $f \in \mathcal{K}(\Sigma)$ iff each $f \circ \varphi_\alpha^{-1}$ is a meromorphic function of the local coordinate $z \in V_\alpha$.

For example, $\mathcal{K}(\mathbb{P}^1(\mathbb{C}))$ consists of all rational functions $f(z) = \frac{\text{poly}(z)}{\text{poly}(z)}$, while $\mathcal{K}(\mathbb{C})$ is much larger. This space $\mathbb{K}(\Sigma)$ is in fact always a field; its algebraic structure determines the surface Σ (up to conformal equivalence) and naturally mirrors all aspects of Σ. A compact Riemann surface Σ has genus 0 iff $\mathbb{K}(\Sigma) \cong \mathbb{C}(z)$, the field of rational functions in some variable z. For positive genus, two generators are needed.

Theorem 2.1.5 *Let Σ be a compact Riemann surface of genus $g > 0$. Choose any nonconstant function $f \in \mathcal{K}(\Sigma)$. Then there exists another nonconstant function $g \in \mathcal{K}(\Sigma)$, such that $\mathcal{K}(\Sigma) = \mathbb{C}(f)[g]$, i.e. for some $n \in \mathbb{N}$, any $h \in \mathcal{K}(\Sigma)$ can be written in the form $h = \sum_{i=0}^{n-1} a_i(f) g^i$, where $a_i(z)$ are rational. Moreover, there is an irreducible polynomial $P(z, w)$ such that $P(f, g) = 0$, and such that $\mathcal{K}(\Sigma)$ is isomorphic as a field to the quotient $\mathbb{C}(z, w)/(P(z, w))$ of the algebra of rational functions in z, w by the ideal generated by polynomial P. Moreover, writing P as a homogeneous polynomial in three variables, Σ is conformally equivalent to the complex curve $P = 0$ in the complex projective plane $\mathbb{P}^2(\mathbb{C})$.*

For a proof and more material on Riemann surfaces, see [**180**]. It is nontrivial that we can embed any Riemann surface into the complex projective plane. In fact, most complex n-tori $\mathbb{C}^n/L$ (where $L \subset \mathbb{C}^n$ is a $2n$-dimensional lattice), for $n > 1$, *cannot* be embedded in *any* projective space (Section 6.3.2). The plane curve $P = 0$ will typically have 'singularities', that is points where all three partial derivatives vanish, where the curve self-intersects transversely. These singularities can be 'blown up', that is the two intersecting 'complex strands' (i.e. open discs in $\mathbb{C}$) can be separated, but this requires the complex curve to be embedded in $\mathbb{P}^3$, not $\mathbb{P}^2$.

Every geometric feature (except the choice of orientation) of the surface Σ has an algebraic analogue in $\mathcal{K}(\Sigma)$, and hence the geometry of Σ can be studied via algebra. For example, a $\mathbb{C}$-algebra homomorphism $F : \mathcal{K}(\Sigma') \to \mathcal{K}(\Sigma)$ lifts to a holomorphic map $\widetilde{F} : \Sigma \to \Sigma'$. This general observation is the starting point of both algebraic geometry and noncommutative geometry. For example, the space of smooth complex-valued functions on a manifold M will be an infinite-dimensional commutative algebra, since the target $\mathbb{C}$ is a commutative algebra. Connes suggests that we study a noncommutative algebra as if it too is the algebra of functions on some manifold. The hope is that this should be directly relevant to quantum theories, since we access space-time only indirectly, via the functions ('quantum fields') living on it. We seem to get into problems in quantum field theory when we take too literally the (naive and improbable) intuition that space-time is anything like a manifold. In any case calculus in noncommutative geometry formally resembles quantum mechanics (e.g. the role of coordinates is played by self-adjoint operators – observables – and infinitesimal distance ds by the fermion propagator).

For a concrete example of Theorem 2.1.5, consider the torus $T_\tau = \mathbb{C}/(\mathbb{Z} + \tau\mathbb{Z})$. A meromorphic function $f : T_\tau \to \mathbb{C}$ lifts to a meromorphic function (which we also call f) on $\mathbb{C}$, with periods 1 and τ. That is, $f \in \mathcal{K}(T_\tau)$ iff $f : \mathbb{C} \to \mathbb{C}$ is meromorphic and $f(z + m + n\tau) = f(z)$ $\forall z \in \mathbb{C}$, $\forall m, n \in \mathbb{Z}$. Any such doubly-periodic meromorphic function is called an *elliptic function*, for fairly obscure reasons.[4] We know

---

[4] One of the more carefree creative outlets for mathematicians is through their happy role as nomenclators. Elliptic functions first arose historically as the functional inverse of a certain class of integrals called 'elliptic integrals'. This class got its name since it included the integral computing arc-lengths of ellipses. Likewise, the name 'elliptic curve' for a genus-1 complex curve arose since the functions living on it are those elliptic functions. There is however no direct relation between ellipses and elliptic curves.

any nonconstant $f \in \mathcal{K}(T_\tau)$ must have at least one pole in the 'fundamental parallelogram' P_τ with corners at $0, \tau, 1, 1 + \tau$. Moreover, the contour integral $\int_C f$ about the parallelogram $C = \partial P_\tau$ vanishes by periodicity, so the sum of residues of f inside P_τ must vanish. Hence any nonconstant elliptic function must have at least two poles in P_τ.

We can construct an elliptic function by averaging $f(z) = \sum_{m,n} g(z + m + n\tau)$ for any function g over each orbit $z + \mathbb{Z} + \tau\mathbb{Z}$. As the simplest possibility for a nonconstant elliptic function would have a single pole of order 2 at the lattice points, it is tempting to take $g(z) = z^{-2}$. Unfortunately, for large m, n, $(z + m + n\tau)^{-2}$ is close to $(m + n\tau)^{-2}$, and so its sum over all m, n won't converge. Thus we are led to consider its 'regularisation'

$$\wp(z) := z^{-2} + \sum_{m,n=-\infty}^{\infty}{}' \{(z + m + n\tau)^{-2} - (m + n\tau)^{-2}\} \qquad (2.1.6a)$$

function, called the *Weierstrass function* (although Eisenstein knew of it years earlier), where $\sum'$ here means to avoid $m = n = 0$. Its derivative

$$\wp'(z) = -2 \sum_{m,n=-\infty}^{\infty} (z + m + n\tau)^{-3} \qquad (2.1.6b)$$

is also elliptic. Being meromorphic functions on a compact Riemann surface, $\wp$ and $\wp'$ must be polynomially related: we find

$$\wp'(z)^2 = 4(\wp(z) - e_1)(\wp(z) - e_2)(\wp(z) - e_3), \qquad (2.1.6c)$$

where $e_1 = \wp(1/2)$, $e_2 = \wp(\tau/2)$ and $e_3 = \wp((1 + \tau)/2)$. This is shown by verifying that $(\wp - e_1)(\wp - e_2)(\wp - e_3)/\wp'$ has no poles and hence must be constant. Together, $\wp$ and $\wp'$ generate $\mathcal{K}(T_\tau)$: we can write any elliptic function $f \in \mathcal{K}(T_\tau)$ as $R_1(\wp) + \wp' R_2(\wp)$, where $R_1(\wp(z))$ is the even part $(f(z) + f(-z))/2$ of f and $\wp'(z) R_2(\wp(z))$ the odd part. T_τ is conformally equivalent to the projective curve with 'finite' points $(\wp(z), \wp'(z), 1) \in \mathbb{P}^2(\mathbb{C})$, together with the 'infinite' point $(0, 1, 0)$ corresponding to the pole of $\wp$ and $\wp'$ at $z = 0$.

One way to embed Riemann surfaces into projective space uses *theta functions*:

$$\theta_{r,s}(\tau, z) := \sum_{m \in \mathbb{Z}} \exp[\pi \mathrm{i}\tau (m + r)^2 + 2\pi \mathrm{i} (m + r)(z + s)], \qquad (2.1.7a)$$

for any $r, s \in \mathbb{Q}$. These functions and their generalisations are central to Moonshine, but for now note that they converge for all $(\tau, z) \in \mathbb{H} \times \mathbb{C}$ to a function holomorphic in both τ and z. These $\theta_{r,s}$ are nearly doubly-periodic in z: if $r, s \in \frac{1}{N}\mathbb{Z}$ then

$$\theta_{r,s}(\tau, z + Nm + \tau Nn) = \exp[-\pi \mathrm{i}N^2 n^2 \tau - 2\pi \mathrm{i}Nnz] \theta_{r,s}(\tau, z), \qquad (2.1.7b)$$

for all $m, n \in \mathbb{Z}$. Apart from a constant root of unity, $\theta_{r,s}$ depends only on the values of r and s mod 1. Enumerate the N^2 pairs $(r_i, s_i) \in \frac{1}{N}\mathbb{Z}_N \times \frac{1}{\mathbb{Z}_N}\mathbb{Z}_N$. Then for any N and any

$\tau \in \mathbb{H}$, the map from T_τ to $\mathbb{P}^{N^2-1}(\mathbb{C})$ defined in homogeneous coordinates by

$$z \mapsto (\theta_{r_1,s_1}(\tau, Nz), \theta_{r_2,s_2}(\tau, Nz), \ldots) \in \mathbb{P}^{N^2-1}(\mathbb{C})$$

is well defined (to see that this N^2-tuple can never be the 0-vector, find explicitly the zeros of $\theta_{r,s}$). This map is one-to-one, that is it embeds the torus T_τ as a complex submanifold of $\mathbb{P}^{N^2-1}(\mathbb{C})$. We can specify this submanifold more explicitly (in the simplest case, namely $N = 2$) by the homogeneous polynomials

$$\theta_{0,0}(\tau)^2 z_1^2 = \theta_{0,1/2}(\tau)^2 z_2^2 + \theta_{1/2,0}(\tau)^2 z_3^2, \qquad \theta_{0,0}(\tau)^2 z_4^2 = \theta_{1/2,0}(\tau)^2 z_2^2 - \theta_{0,1/2}(\tau)^2 z_3^2,$$

where $(z_1, z_2, z_3, z_4) \in \mathbb{P}^3(\mathbb{C})$ are homogeneous coordinates and $\theta_{r,s}(\tau) = \theta_{r,s}(\tau, 0)$. The fact that the image of T_τ satisfies those equations follows from the Riemann theta identities. Moreover, any elliptic function $f : T_\tau \to \mathbb{C}$ can be written in the form

$$f(z) = c \prod_{1 \le i \le \ell} \frac{\theta_{0,0}(\tau, z - a_i)}{\theta_{0,0}(\tau, z - b_i)},$$

for arbitrary complex numbers a_i, b_i, c subject to the relation $\sum_i a_i = \sum_i b_i$. The Weierstrass $\wp$-function can be written

$$\wp(z) = -\frac{d^2}{dz^2}\theta_{1/2,1/2}(\tau, z) - \frac{\pi^2}{3}.$$

For any $k \in \mathbb{Z}$, a holomorphic (respectively meromorphic) k-form ω (Section 1.2.2) on a complex curve Σ looks like $f\, dz^k$ in local coordinates, where f is holomorphic (respectively meromorphic). If we change local coordinates $z_1 \mapsto \varphi_2(\varphi_1^{-1}(z_1))$, then (1.2.4b) becomes

$$f_\beta(z_\beta) = \frac{d^k z_\alpha}{dz_\beta^k} f_\alpha(z_\alpha). \tag{2.1.8}$$

For example, dz is a meromorphic (but not holomorphic) 1-differential on $\mathbb{P}^1(\mathbb{C})$ (it has a pole of order 2 at ∞). Let $\mathcal{H}^k(\Sigma)$ be the vector space of holomorphic k-forms, and $\mathcal{M}^k(\Sigma)$ be the space of meromorphic ones. Given any $\omega, \omega' \in \mathcal{M}^k(\Sigma)$, ω' not identically 0, the ratio ω/ω' lies in the function field $\mathcal{K}(\Sigma)$. Of course, as vector spaces $\mathcal{M}^0(\Sigma) = \mathcal{K}(M)$. For any surface Σ and integer k, $\mathcal{M}^k(\Sigma)$ is infinite-dimensional, but for any compact surface Σ and any integer k, the Riemann–Roch theorem implies that $\mathcal{H}^k(\Sigma)$ is always finite-dimensional and may be 0.

2.1.4 Moduli

In physics, the *phase space* lets us consider all possible states of a physical system; the actual time-evolution of a given instance of that system will be a curve in phase space. Likewise, we often want to consider simultaneously families of manifolds, rather than fix a single manifold. For example, last subsection we treated all tori T_τ simultaneously. The role of phase space is played by a *moduli space*, the space of orbits of a group of diffeomorphisms of a geometric structure placed on a manifold. A path on the moduli space connecting orbits $[p]$ and $[q]$ is a continuous deformation from the geometric structure on p to that on q.

The notion of moduli space for surfaces is due to Riemann, who also computed its dimension. The idea is to consider the space $\mathfrak{M}(\Sigma_0)$ of all conformal equivalence classes of Riemann surfaces homeomorphic to a given surface Σ_0. As Σ_0 is completely characterised by the genus g and number n of punctures, we also denote this by $\mathfrak{M}_{g,n}$. With a few exceptions mentioned shortly, $\mathfrak{M}_{g,n}$ has complex dimension $3g - 3 + n$. However, these moduli spaces usually aren't manifolds (they have conical singularities). It was for this reason that Teichmüller introduced a cover, now called the *Teichmüller space* $\mathfrak{T}_{g,n}$. The moduli space is recovered by the quotient $\mathfrak{M}_{g,n} = \mathfrak{T}_{g,n}/\Gamma_{g,n}$, where $\Gamma_{g,n}$ is a discrete group called the *mapping class group* (see Definition 2.1.6). Teichmüller space is much better behaved than the moduli space – it is a complex manifold (except for certain small g, n), and as a real manifold is diffeomorphic to $\mathbb{R}^{6g-6+2n}$.

As we shall see, there's a small number of pairs (g, n) that don't behave completely generically for one reason or another: namely, $(0, 0)$, $(0, 1)$, $(0, 2)$, $(0, 3)$, $(0, 4)$, $(1, 0)$, $(1, 1)$ and $(2,0)$. We mention some of their individual peculiarities below.

In order to anticipate the definitions, consider a torus T (so $g = 1, n = 0$). For concreteness (this doesn't lose any generality), restrict to tori coming from a parallelogram in the complex plane $\mathbb{C}$, with one pair of opposite sides labelled '1', and the other pair labelled '2'; the torus is recovered by first identifying the opposite sides labelled '1', and then identifying the opposite sides labelled '2' (changing this order changes the shape – though not the conformal class – of the torus). By translating, rotating and rescaling this parallelogram, we can put the vertices at $0, 1, \tau$ and $\tau + 1$, for some $\tau \in \mathbb{H}$, where the horizontal sides are labelled '1', which continuously deforms the torus without changing its conformal equivalence class. This is the best we can do, if we restrict to continuous deformations. The resulting parameter space, namely the upper half-plane $\mathbb{H}$, is the Teichmüller space $\mathfrak{T}_{1,0}$ for the torus. The torus corresponding to $\tau \in \mathbb{H}$ is $T_\tau = \mathbb{C}/(\mathbb{Z} + \mathbb{Z}\tau)$.

However, different points τ in $\mathbb{H}$ can correspond to conformally equivalent tori. For example, we can cut the torus open along the seam '2', twist the open arm m complete turns, and then sew it back up. This amounts to replacing parameter τ with $\tau + m$. As long as m is an integer, this is a conformal diffeomorphism of the torus (if m isn't an integer, this map isn't even continuous). Thus the points $\tau + \mathbb{Z}$ all correspond to the same conformal structure. Similarly, cutting open seam '1' and giving the upper cap n complete twists before resewing corresponds to replacing the parallelogram $0, 1, \tau$ and $\tau + 1$ with the parallelogram $0, 1 + n\tau, \tau$ and $(n + 1)\tau + 1$ – after putting it into canonical form, this replaces τ with $\tau/(n\tau + 1)$. Both these twists are called *Dehn twists*. We can also switch the roles of sides '1' and '2', which replaces τ with $-1/\tau$ (why?). More generally, the tori corresponding to parameters τ and $\frac{a\tau+b}{c\tau+d}$ are conformally equivalent, for any $\begin{pmatrix} a & b \\ c & d \end{pmatrix} \in \mathrm{SL}_2(\mathbb{Z})$. This accounts for all redundancies in the parametrisation by $\mathbb{H}$ of the conformal equivalence classes of tori. The orbit space $\mathrm{SL}_2(\mathbb{Z})\backslash\mathbb{H}$ is the 'moduli space' $\mathfrak{M}_{1,0}$ for the torus. Note that $\mathfrak{M}_{1,0}$ has conical singularities at the orbits

$[\tau] = [\mathrm{i}]$ and $[e^{2\pi\mathrm{i}/3}]$, corresponding to those tori with additional automorphisms. This happens in higher genus too. Indeed, any finite group G is the automorphism group of some surface of sufficiently high genus. For example, there will be a compact Riemann surface with automorphism group exactly the Monster $\mathbb{M}$, though it will have genus at least 9.6×10^{51}.

Definition 2.1.6 *Let Σ_0 be a fixed Riemann surface. Consider all pairs (Σ, f), where f is an orientation-preserving homeomorphic map of Σ_0 onto Σ. Write $(\Sigma, f) \sim (\Sigma', f')$ if there exists a conformal homeomorphism $h : \Sigma \to \Sigma'$ such that the homeomorphism $f'^{-1} \circ h \circ f : \Sigma_0 \to \Sigma_0$ is homotopic to the identity. The set of these equivalence classes is the* Teichmüller space $\mathfrak{T}(\Sigma_0)$. *The* mapping class group $\Gamma(\Sigma_0)$ *is the quotient* $\mathrm{Homeo}_+(\Sigma_0)/\mathrm{Homeo}_0(\Sigma_0)$ *of the group of orientation-preserving self-homeomorphisms f of Σ_0, by the (normal) subgroup consisting of those homotopic to the identity.*

For example, $\Gamma_{1,0} = \mathrm{SL}_2(\mathbb{Z})$ and $\mathfrak{T}_{1,0} = \mathbb{H}$; as we explain in Section 2.2.4, the moduli space $\mathfrak{M}_{1,0}$ is a punctured sphere. Because $\mathbb{C}/(\mathbb{Z} + \tau\mathbb{Z})$ can also be interpreted as a torus with a special point, namely the additive identity 0, we also have $\mathfrak{T}_{1,1} = \mathbb{H}$ and $\Gamma_{1,1} = \mathrm{SL}_2(\mathbb{Z})$. For a different reason, we also have $\mathfrak{T}_{0,4} = \mathbb{H}$ and $\Gamma_{0,4} = \mathrm{SL}_2(\mathbb{Z})$.

The basic idea, illustrated above, is that the Teichmüller space $\mathfrak{T}_{g,n}$ accounts for 'continuous' conformal equivalences, while the mapping class group $\Gamma_{g,n}$ contains the left-over 'discontinuous' ones. To help make this important but abstract definition more accessible, consider the following artificial example. Let $X = \mathbb{R}^2$, and suppose the additive group $G = \mathbb{Z} \times \mathbb{R}$ acts on X by addition. Then G is a disconnected Lie group with connected components $G_n := \{n\} \times \mathbb{R}$ for each $n \in \mathbb{Z}$; the component G_0 is the one containing the identity $(0, 0)$. The group $\pi_0 = G/G_0 \cong \mathbb{Z}$ interchanges the components in the obvious way. We can mod out first by the continuous part G_0 of G (which should be relatively easy), then by the discontinuous π_0: the orbit space X/G is then $(X/G_0)/\pi_0 = \mathbb{R}/\mathbb{Z} = S^1$. Of course, here X plays the role of the infinite-dimensional space of all conformal structures, G plays the role of all conformal homeomorphisms, and X/G is the moduli space. The identity component G_0 corresponds to the homeomorphisms homotopic to the identity, π_0 is the mapping class group and X/G_0 is the Teichmüller space.

The mapping class groups are central to our story, so we'll try to make them more accessible. More details and proofs are provided in [56], [270], [60] and chapter 4 of [59]. A simple presentation of the mapping class group $\Gamma_{g,n}$ for $n = 0, 1$ – the cases of greatest interest to us – is given in [550].

$\Gamma_{g,n}$ acts like a braid group. For example, any $f \in \mathrm{Homeo}_+(\Sigma)$ permutes the n punctures, so the same is true of $\gamma \in \Gamma_{g,n}$; the 'pure' mapping class group $\mathrm{P}\Gamma_{g,n}$ consists of those $\gamma \in \Gamma_{g,n}$ that fix each puncture. Then $\mathrm{P}\Gamma_{g,n}$ is normal in $\Gamma_{g,n}$ and has quotient $\Gamma_{g,n}/\mathrm{P}\Gamma_{g,n} = \mathcal{S}_n$.

A braid group $\mathcal{B}_n(\Sigma)$ can be associated with any surface Σ in the obvious way [59]. For genus $g \geq 2$ and any $n \geq 0$, the group $\Gamma_{g,n}$ is an extension of $\mathcal{B}_n(\Sigma_g)$, by the group $\Gamma_{g,0}$. For genus $g = 1$ and $n \geq 2$, $\Gamma_{1,n}$ is an extension of the quotient $\mathcal{B}_n(\Sigma_1)/Z(\mathcal{B}_n(\Sigma_1))$ by $\mathrm{PSL}_2(\mathbb{Z})$, where the centre $Z(\mathcal{B}_n(\Sigma_1)) \cong \mathbb{Z}^2$. For genus $g = 0$ and $n \geq 3$, the group

$\Gamma_{0,n}$ is isomorphic to the quotient $\mathcal{B}_n(S^2)/Z(\mathcal{B}_n(S^2))$, where $Z(\mathcal{B}_n(S^2)) \cong \mathbb{Z}_2$. For any n, $\Gamma_{0,n}$ is a homomorphic image (i.e. a quotient) of the braid group $\mathcal{B}_n$.

Let Σ be a compact Riemann surface. To any simple closed loop γ in Σ, we can define the *Dehn twist* about γ, by cutting out from Σ a neighbourhood of the loop homeomorphic to a cylinder, giving one end of this cylinder an integral twist, and gluing it back. The Dehn twists about the $2g$ elementary loops a_i, b_j defined in Section 2.1.2 generate the mapping class group of Σ.

Teichmüller space need not be connected. In particular, there are three different kinds of twice-punctured spheres: one is flat and has conformal structure given by the cylinder $\mathbb{C}/\mathbb{Z}$; one is the punctured disc $0 < |z| < 1$ and corresponds to the half-cylinder $\mathbb{H}/\langle z \mapsto z + 1\rangle$; and finally, we have the family of annuli $A_r := \{r < |z| < 1\}$, which are all of the form $\mathbb{H}/\langle z \mapsto \lambda z\rangle$ for $\lambda > 1$. Thus $\mathfrak{T}_{0,2}$ and $\mathfrak{M}_{0,2}$ consist of two isolated points and an open line segment $(0, 1)$ say. $\Gamma_{0,2} \cong \mathbb{Z}_2$ consists of the identity, and the inversion through 0 that exchanges the two boundary circles. Similarly, both $\mathfrak{T}_{0,1}$ and $\mathfrak{M}_{0,1}$ consist of two isolated points.

The mapping class group usually (but not always) acts *faithfully* on Teichmüller space (a faithful action means that the only group element that acts trivially is the identity element). $\Gamma_{1,0} = \Gamma_{1,1} = \Gamma_{0,4}$ are exceptions: $-I \in \mathrm{SL}_2(\mathbb{Z})$ acts trivially on $\mathbb{H}$. Also, consider the thrice-punctured sphere $\mathbb{P}^1(\mathbb{C})/\{z_1, z_2, z_3\}$. As is well known, $\mathrm{Aut}(S^2) \cong \mathrm{PSL}_2(\mathbb{C})$ can conformally move any three points to any other three points, so we can send $z_1, z_2, z_3 \in \mathbb{P}^1(\mathbb{C})$ respectively to 0, 1, ∞. Thus $\mathfrak{T}_{0,3}$ consists of a single point. However, we could have moved, for example, z_2, z_1, z_3 instead to 0, 1, ∞, respectively. A total of six different choices could have been made, corresponding to the mapping class group $\Gamma_{0,3} = \mathcal{S}_3$, which acts trivially on Teichmüller space.

$\mathfrak{M}_{g,n}$ is simultaneously the moduli space of: (i) conformal equivalence classes of real surfaces; (ii) complete Riemannian metrics of constant negative curvature on real surfaces; and (iii) complex-analytic structures on complex curves. This is an accident of small dimensions, for example the Mostow Rigidity Theorem says that in three dimensions the moduli space of (ii) consists of a single point.

A different approach to moduli spaces ties in with Sections 2.3.5 and 6.3.2. First, by the *Siegel upper half-space* $\mathbb{H}_g$ we mean the space of all symmetric $g \times g$ complex matrices Ω whose imaginary part $\mathrm{Im}(\Omega)$ is positive-definite – that is, $v^t \, \mathrm{Im}(\Omega) \, v > 0$ for any nonzero column vector $v \in \mathbb{R}^g$. $\mathbb{H}_g$ is a higher-genus generalisation of $\mathbb{H}$. The role of the group $\mathrm{SL}_2(\mathbb{Z})$ here is played by the symplectic group $\mathrm{Sp}_{2g}(\mathbb{Z})$, that is the group of all determinant 1 $2g \times 2g$ matrices M satisfying $M^t \begin{pmatrix} 0 & I \\ -I & 0 \end{pmatrix} M = \begin{pmatrix} 0 & I \\ -I & 0 \end{pmatrix}$, where $I = I_g$ and 0 are, respectively, the $g \times g$ identity and $g \times g$ zero matrices. The familiar action $\begin{pmatrix} a & b \\ c & d \end{pmatrix} . \tau = \frac{a\tau + b}{c\tau + d}$ is replaced by the action

$$\begin{pmatrix} A & B \\ C & D \end{pmatrix} . \Omega = (A\Omega + B)(C\Omega + D)^{-1}, \qquad \forall \begin{pmatrix} A & B \\ C & D \end{pmatrix} \in \mathrm{Sp}_{2g}(\mathbb{R}), \ \forall \Omega \in \mathbb{H}_g.$$

$$(2.1.9a)$$

The generalisation of the Jacobi theta function (2.1.7a) is *Siegel's theta function*

$$\theta(\Omega, z) = \sum_{n \in \mathbb{Z}^g} \exp(\pi \mathrm{i}\, n^t \Omega n + 2\pi \mathrm{i}\, n \cdot z), \qquad (2.1.9\mathrm{b})$$

which converges for all $\Omega \in \mathbb{H}_g$ and $z \in \mathbb{C}^g$.

Where does $\mathbb{H}_g$ come from? Associate with a compact genus-g surface Σ_g its Jacobian variety, as follows. The space $\mathcal{H}^1(\Sigma_g)$ of holomorphic 1-forms is g-dimensional, so let $\{\omega_1, \ldots, \omega_g\}$ be a basis. Fix any base-point $p \in \Sigma_g$; then we get a map from $\Sigma_g \times \cdots \times \Sigma_g$ to $\mathbb{C}^g$ by integrating:

$$(q_1, \ldots, q_g) \mapsto \sum_{i=1}^{g} \left(\int_{C_i} \omega_1, \int_{C_i} \omega_2, \ldots, \int_{C_i} \omega_g \right),$$

where C_i is any path on Σ_g from p to q_i. Of course the result depends on which paths C_i are chosen, and so isn't well defined as a function of q_i's alone. However, consider the set L of all possible values $(\int_C \omega_1, \ldots, \int_C \omega_g) \in \mathbb{C}^g$, where C runs over all possible closed loops in Σ_g passing through P. Then our ill-defined map $\Sigma_g \times \cdots \times \Sigma_g \to \mathbb{C}^g$ will become well-defined (i.e. independent of the choice of path C_i) if we replace the target $\mathbb{C}^g$ with $\mathbb{C}^g/L$. It isn't hard to show that L is a $2g$-dimensional lattice (in fact a basis is given by the values on the $2g$ loops we call α_i, β_j in (2.1.5a)), and so $\mathbb{C}^g/L$ is a $2g$-dimensional torus, called the *Jacobian variety* $\mathrm{Jac}(\Sigma_g)$. This map $\Sigma_g \times \cdots \times \Sigma_g \to \mathbb{C}^g/L$ is holomorphic and surjective ('Jacobi Inversion'). Restricting it to the diagonal embedding $q \mapsto (q, \ldots, q) \in \Sigma_g \times \cdots \times \Sigma_g$, we get a one-to-one conformal embedding $q \mapsto F(C, \ldots, C)$ of Σ_g into $\mathrm{Jac}(\Sigma_g)$. When $g = 1$, Σ_1 and $\mathrm{Jac}(\Sigma_1)$ are identical; when $g > 1$ the embedding is into a proper submanifold of the Jacobian (check dimensions).

Now, we can select our basis ω_i of 1-forms so that the integral $\int_{\alpha_i} \omega_j$ equals δ_{ij}. This choice means that our lattice L contains $\mathbb{Z}^g$. The remaining basis vectors of L are $(\int_{\beta_i} \omega_1, \ldots, \int_{\beta_i} \omega_g) \in \mathbb{C}^g$, and it can be shown (the 'Riemann bilinear relations') that these basis vectors will be column vectors of a symmetric $g \times g$ matrix Ω whose imaginary part is positive-definite – that is, the *period matrix* Ω lies in $\mathbb{H}_g$. So the lattice L becomes $\mathbb{Z}^g + \Omega\mathbb{Z}^g$ and the Jacobian becomes $T_\Omega := \mathbb{C}^g/(\mathbb{Z}^g + \Omega\mathbb{Z}^g)$, where we regard vectors in $\mathbb{Z}^g$ and $\mathbb{C}^g$ as column vectors. Different choices of bases correspond to the $\mathrm{Sp}_{2g}(\mathbb{Z})$-orbit of Ω.

So every surface Σ_g corresponds to an $\mathrm{Sp}_{2g}(\mathbb{Z})$-orbit in $\mathbb{H}_g$. The *Schottky Problem* asks which points in $\mathbb{H}_g$ arise as period matrices. Call this subset $\mathfrak{C}_g$. Our moduli space $\mathfrak{M}_{g,0}$ can be identified with $\mathfrak{C}_g/\mathrm{Sp}_{2g}(\mathbb{Z})$ and $\mathrm{Sp}_{2g}(\mathbb{Z})$ is a homomorphic image (or quotient) of $\Gamma_{g,0}$. Since the symplectic group $\mathrm{Sp}_{2g}(\mathbb{Z})$ is much more accessible than the mapping class group $\Gamma_{g,0}$, the main difficulty is to find a nice characterisation of $\mathfrak{C}_g$ and the kernel of $\Gamma_{g,0} \to \mathrm{Sp}_{2g}(\mathbb{Z})$. For a formal solution to the Schottky problem, see e.g. [**12**].

The moduli space $\mathfrak{M}_{g,n}$ is rarely compact. A very natural way to compactify $\mathfrak{M}_{g,n}$, due to Deligne and Mumford, is fundamental to conformal field theory. Consider first the complex curve $w^2 = (z - 2)(z + 1 - \alpha)(z - 1 - \alpha)$, where α is a parameter. Provided $\alpha \neq 0, \pm 1$, this is a genus-1 nonsingular curve, conformally equivalent to the torus

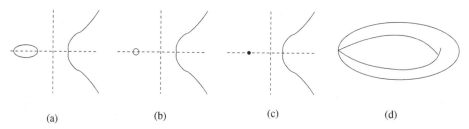

(a)   (b)   (c)   (d)

Fig. 2.7 The surface $w^2 = (z - 2)(z + 1 - \alpha)(z + 1 + \alpha)$.

$\mathbb{C}/(\mathbb{Z} + \tau\mathbb{Z})$ where

$$j(\tau) = \frac{(\alpha^2 + 3)^2 - (2\alpha^2 - 2)^3}{(2 - 1 + \alpha)^2(2 + 1 - \alpha)^2(1 - \alpha - 1 + \alpha)^2}.$$

We know that $\mathfrak{M}_{1,0}$ is real-diffeomorphic to a sphere with one point removed. As we vary α, we move through $\mathfrak{M}_{1,0}$, and as $\alpha \to 0$ we approach the missing point. What happens to the curve in that limit? In Figure 2.7(a)–(c) we intersect our curve, for $\alpha = 1/2, 1/4, 0$, respectively, with the plane $\mathbb{R}^2 \subset \mathbb{C}^2$. Figure 2.7(d) gives a picture of the complex curve at $\alpha = 0$: it is a pinched torus. We call the nonsmooth point $(z, w) = (-1, 0)$ a *node*. This is the surface to which the boundary point of $\mathfrak{M}_{1,0}$ corresponds. Including it, compactifies $\mathfrak{M}_{1,0}$ to $\overline{\mathfrak{M}_{1,0}} \cong S^2$.

More generally, we add to each moduli space $\mathfrak{M}_{g,n}$ the surfaces Σ with nodes. These are connected compact spaces where the neighbourhood of any point either looks like $\mathbb{C}$ (i.e. Σ is smooth there) or like $zw = 0$ at $(0, 0)$ (these are the nodes). We say Σ has type (g, n) if unpinching each node results in a genus-g surface with n punctures – for example, Figure 2.7(d) has type $(1,0)$. We require these surfaces to have the following property: when you delete all nodes and the surface falls into connected pieces, none of those pieces is a sphere with one or two punctures (the only exception is that we also allow a torus with one node). These surfaces are called *stable*, because they have a finite automorphism group (this terminology is explained by visualising a marble versus a dice on a tabletop). As we know, the larger the automorphism group, the worse the singularity is in moduli space.

The moduli space $\mathfrak{M}_{g,n}$ is compactified if we include the conformal equivalence classes of stable type (g, n) surfaces with nodes. The resulting space $\overline{\mathfrak{M}_{g,n}}$ is called the *moduli space of stable surfaces*. A nice review is given in [447]. For example, the moduli space $\mathfrak{M}_{0,4}$ is also a sphere with one missing point. That missing point corresponds to pinching a sphere with four punctures into two spheres, each with two punctures.

Moduli spaces of curves seem first to have been introduced into string theory and conformal field theory by Polyakov in 1981, and have played an important role there ever since. We are actually more interested in an enhanced moduli space, obtained by decorating Riemann surfaces with additional structure. Many more or less equivalent alternatives have appeared in the literature. In particular, let Σ be a compact genus-g surface, possibly with nodes, with n marked points $p_i \in \Sigma$ (none of which are at a node). About each point p_i is chosen a local coordinate z_i, vanishing at p_i, identifying a neighbourhood

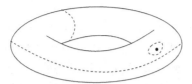

Fig. 2.8 The Dehn twists on the torus with one marked point.

of p_i with a neighbourhood of 0 in $\mathbb{C}$ (see section 2.1 of [**530**] for details). We call this data $(\Sigma, \{p_i\}, \{z_i\})$ an *enhanced surface* of type (g, n). It is essentially equivalent to removing a disc from Σ about p_i and choosing a parametrisation about the boundary circle. We call two enhanced surfaces $(\Sigma, \{p_i\}, \{z_i\})$ and $(\Sigma', \{p_i'\}, \{z_i'\})$ equivalent if there is a conformal equivalence $h : \Sigma \to \Sigma'$ such that $h(p_i) = p_i'$ and $z_i'(hx) = z_i(x)$ locally about p_i. The resulting moduli space $\widehat{\mathfrak{M}}_{g,n}$ will be infinite-dimensional, but the mapping class group $\widehat{\Gamma}_{g,n}$ will be an extension of the usual $\Gamma_{g,n}$ by $\mathbb{Z}^n$.

These groups $\widehat{\Gamma}_{g,n}$ are of great interest to us – for example, a rational conformal field theory gives a projective finite-dimensional representation of each of them. This yields the braid group representations in quantum groups or Jones subfactor theory, as well as the modularity of Moonshine. They are discussed, with many examples, in section 5.1 of [**32**] (where they are denoted $\Gamma_{g,n}$, and what we call $\Gamma_{g,n}$ is denoted there Γ_g^n). For example, $\widehat{\Gamma}_{1,1}$ is the braid group $\mathcal{B}_3$, a central extension of $\mathrm{SL}_2(\mathbb{Z})$ by $\mathbb{Z}$. It is generated by the Dehn twists depicted in Figure 2.8. We return to this in Sections 4.3.3, 5.3.4 and 7.2.4.

The main reason we prefer extended surfaces to ordinary Riemann surfaces is that there are canonical ways to sew them together. This sewing operation is fundamental in conformal field theory, because it permits us to decompose a higher-genus surface into discs and 'pairs-of-pants' (Section 4.4.1).

Question 2.1.1. How would a hyperbolic mathematician model the Euclidean plane?

Question 2.1.2. (a) Let $\gamma = \begin{pmatrix} a & b \\ c & d \end{pmatrix} \in \mathrm{SL}_2(\mathbb{R})$, $\gamma \neq \pm I$. We can regard γ as a map from the extended upper-half plane $\mathbb{H} \cup \mathbb{R} \cup \{\infty\}$ to itself. Show that:

(i) $|a + d| = 2$ iff γ has exactly one fixed point on the boundary $\mathbb{R} \cup \{\infty\}$, iff γ can be conjugated in $\mathrm{SL}_2(\mathbb{R})$ to the translation $z \mapsto z + t$;
(ii) $|a + d| > 2$ iff γ has exactly two distinct fixed points on the boundary $\mathbb{R} \cup \{\infty\}$, iff γ can be conjugated in $\mathrm{SL}_2(\mathbb{R})$ to the dilation $z \mapsto \lambda z$;
(iii) $|a + d| < 2$ iff γ has exactly one fixed point in $\mathbb{H}$, iff γ can be conjugated in $\mathrm{SL}_2(\mathbb{R})$ to the rotation $z \mapsto \frac{\cos(\theta)z + \sin(\theta)}{-\sin(\theta)z + \cos(\theta)}$ about i with fixed point i.

(b) Suppose Γ is a Fuchsian group. Prove that $\gamma \in \Gamma$ has a fixed point in $\mathbb{H}$ iff γ has finite order.

Question 2.1.3. Explain how the addition of points on a conic is a degenerate case of the addition of points on a cubic.

Question 2.1.4. Find all rational solutions (r, s) to $r^2 - 2rs + r + 2s - s^2 = 0$. Verify that, for the choice of identity $e = (0, 0)$ and addition defined as in Figure 2.6, the rational points form a subgroup. As an abstract group, what is this subgroup isomorphic to?

Question 2.1.5. Using the conformal map $z \mapsto (x, y) = (\wp(z), \wp'(z))$ between $\mathbb{C}/(\mathbb{Z} + \tau\mathbb{Z})$ and the cubic $y^2 = 4(x - e_1)(x - e_2)(x - e_3)$, verify that the addition of points on the cubic corresponds to the addition $z_1 + z_2 \pmod{\mathbb{Z} + \tau\mathbb{Z}}$ in $\mathbb{C}$.

Question 2.1.6. Identify $\mathcal{M}_{0,4}$ with a space of $\mathcal{S}_3$-orbits in $\mathbb{C} \setminus \{0, 1\}$.

Question 2.1.7. Let G be a finite group. Define

$$K(g, h) = \frac{1}{\|G\|} \sum_\rho \dim(\rho) \operatorname{ch}_\rho(gh^{-1}),$$

for $g, h \in G$, where the sum is over all irreducible representations ρ of G.
(a) Verify that $K(g, h) = \delta_{g,h}$.
(b) For any $\gamma \in \mathbb{N}$, take $f : G^{2\gamma} \to G$ by

$$f(g_1, h_1, \ldots, g_\gamma, h_\gamma) = g_1 h_1 g_1^{-1} h_1^{-1} g_2 h_2 g_2^{-1} h_2^{-1} \cdots g_\gamma h_\gamma g_\gamma^{-1} h_\gamma^{-1}.$$

Define $I = \sum_{(g_i, h_i) \in G^{2\gamma}} K(f(g_i, h_i), e)$. By evaluating I in two ways, obtain the formula

$$\|\operatorname{Hom}(\pi_1(\Sigma_\gamma), G)\| = \|G\|^{2\gamma - 1} \sum_\rho \dim(\rho)^{2 - 2\gamma},$$

where Σ_γ is a compact genus-γ surface.

2.2 Modular forms and functions

Number theory, at its most elemental level, is concerned with finding integer solutions to various (systems of) equations. It is truly remarkable how this seemingly pedestrian pursuit has resulted in the creation of the richest and deepest mathematics. Indeed, it is tempting to suspect that beneath any spot on the mathematical turf, no matter how remote or seemingly barren, is a gemstone merely requiring hard work and discerning fingertips to unearth.

2.2.1 Definition and motivation

As we saw in several different contexts in Section 2.1, the group $\mathrm{SL}_2(\mathbb{R})$ of 2×2 matrices with real entries and determinant 1 acts on the upper half-plane $\mathbb{H} = \{\tau \in \mathbb{C} \mid \operatorname{Im}(\tau) > 0\}$ by Möbius transformations (2.1.4a). For example, the matrices $s := \begin{pmatrix} 0 & -1 \\ 1 & 0 \end{pmatrix}$ and $t := \begin{pmatrix} 1 & 1 \\ 0 & 1 \end{pmatrix}$ correspond to the functions $\tau \mapsto -1/\tau$ and $\tau \mapsto \tau + 1$, respectively.

Consider $\Gamma = \mathrm{SL}_2(\mathbb{Z})$, the subgroup of $\mathrm{SL}_2(\mathbb{R})$ consisting of the matrices with integer entries. It is generated by s and t:

$$\mathrm{SL}_2(\mathbb{Z}) = \left\langle \begin{pmatrix} 0 & -1 \\ 1 & 0 \end{pmatrix}, \begin{pmatrix} 1 & 1 \\ 0 & 1 \end{pmatrix} \right\rangle = \langle s, t \mid s^4 = e, (st)^3 = s^2 \rangle. \tag{2.2.1a}$$

Because $-I \in SL_2(\mathbb{Z})$ yields the trivial map in $\mathbb{H}$, we are also interested in the group

$$PSL_2(\mathbb{Z}) = SL_2(\mathbb{Z})/\{\pm I\} = \langle s, t \mid s^2 = (st)^3 = e \rangle =: \mathbb{Z}_2 * \mathbb{Z}_3, \qquad (2.2.1b)$$

the free product of $\mathbb{Z}_2$ with $\mathbb{Z}_3$. Groups like Γ act on the extended upper half-plane $\overline{\mathbb{H}} := \mathbb{H} \cup \{i\infty\} \cup \mathbb{Q}$ in the obvious way (e.g. s interchanges 0 and $i\infty$). The extra points $\{i\infty\} \cup \mathbb{Q}$ are called *cusps* because of the hyperbolic triangle R in Figure 2.3, which points at one of them. Cusps correspond to tori with a single node (Figure 2.7(d)), and compactify the moduli space $\mathfrak{M}_{1,0}$.

Recall Definition 0.1: a modular function for Γ is a meromorphic function $f : \overline{\mathbb{H}} \to \mathbb{C}$, symmetric with respect to Γ. A related definition is:

Definition 2.2.1 *A modular form f for $\Gamma = SL_2(\mathbb{Z})$ of weight $k \in \mathbb{Q}$ and multiplier $\mu : \Gamma \to \mathbb{C}$, $|\mu| = 1$ is a holomorphic function $f : \mathbb{H} \to \mathbb{C}$, which is also holomorphic at the cusps $\mathbb{Q} \cup \{i\infty\}$ and obeys the transformation law*

$$f\left(\frac{a\tau + b}{c\tau + d}\right) = \mu\begin{pmatrix} a & b \\ c & d \end{pmatrix} (c\tau + d)^k f(\tau), \qquad \forall \begin{pmatrix} a & b \\ c & d \end{pmatrix} \in \Gamma. \qquad (2.2.2)$$

For fractional k we choose the branch of the kth power to be the principal one (so $x^k > 0$ when $x > 0$). For number-theoretic purposes, we require the values of μ to be roots of unity. Writing $\mu(t) = e^{2\pi i h}$, we can expand f in powers of q: $f(\tau) = q^h \sum_{n=-\infty}^{\infty} a_n q^n$. By 'meromorphic at $i\infty$' we mean that all but finitely many negative n have $a_n = 0$, so f has a pole of finite order at $q = 0$; by 'holomorphic at $i\infty$' we mean $h \geq 0$ and $a_n = 0$ for all negative n. Meromorphicity or holomorphicity at the other cusps is implied by that at $i\infty$, because of (2.2.2) and the fact that all cusps lie in the same $SL_2(\mathbb{Z})$-orbit.

For the significance, which is considerable, of the condition that f be meromorphic at the cusps, see Question 2.2.1. The moduli spaces $\overline{\mathfrak{M}_{1,0}}, \overline{\mathfrak{M}_{1,1}}$ and $\overline{\mathfrak{M}_{0,4}}$ all are $SL_2(\mathbb{Z})\backslash\overline{\mathbb{H}}$. The cusps $\mathbb{Q} \cup \{i\infty\}$ of $\mathbb{H}$ correspond to pinched tori or spheres (Section 2.1.4). Meromorphicity at the cusps says f respects this surface degeneration in the appropriate way.

If the weight k is an integer, the multiplier μ will necessarily be a one-dimensional representation of Γ; when k is rational, μ will be a projective representation. We define projective representations, and explain what to do with them, in Section 3.1.1. An intriguing implication for fractional k is described in Section 2.4.3.

The function f is called a modular *form* because $f(\tau)\, d^{-k/2}\tau$ is a holomorphic $(-k/2)$-form on the space $SL_2(\mathbb{Z})\backslash\mathbb{H}$; by contrast, a modular *function* f is a meromorphic function on the space $SL_2(\mathbb{Z})\backslash\mathbb{H}$.

The easiest examples of modular forms of weight $k \geq 4$ (k even) are the Eisenstein series G_k defined in equation (0.1.5). It is conventional to normalise them as follows:

$$E_k(\tau) := \frac{1}{2\zeta(k)} G_k(\tau) = 1 - \frac{2k}{B_k} \sum_{n=1}^{\infty} \sigma_{k-1}(n)\, q^n \in \mathbb{Z}[q], \qquad (2.2.3a)$$

where B_k are the Bernoulli numbers, defined by the generating function $\frac{x}{e^x-1} = \sum_{k=0}^{\infty} B_k \frac{x^k}{k!}$, and where $\sigma_{k-1}(n)$ and the Riemann zeta function $\zeta(s)$ are defined by

$$\sigma_m(n) = \sum_{d|n} d^m, \tag{2.2.3b}$$

$$\zeta(s) = \sum_{n=1}^{\infty} n^{-s} = \prod_{p \text{ prime}} (1 - p^{-s})^{-1} \tag{2.2.3c}$$

(see Section 2.3.1). The E_k and G_k have multiplier $\mu \equiv 1$.

Indeed, the Eisenstein series generate all modular forms for $SL_2(\mathbb{Z})$ with trivial multiplier μ. More specifically, the span of all such modular forms (over all k) is a ring graded by k (i.e. the product of modular forms of weight k and k' is one of weight $k + k'$). This ring is generated (over $\mathbb{C}$) by the Eisenstein series $E_4(\tau)$ and $E_6(\tau)$ – that is, any level k modular form f can be written as a polynomial (homogeneous in the obvious sense) in E_4 and E_6. Moreover, E_4 and E_6 are algebraically independent, so that polynomial is unique. Using this we can readily compute the dimension of (and find a basis for) the space of weight k modular forms. For instance, a basis for the weight 24 modular forms is $\{E_6^4, E_6^2 E_4^3, E_4^6\}$.

The definition of modular forms seems fairly arbitrary. For example, one may ask how significant the upper half-plane $\mathbb{H}$ is, or where the factor $(c\tau + d)^k$ in (2.2.2) comes from. We confront this in Section 2.4.1. But for now note that Definition 2.2.1 (like Definition 0.1 before it) also makes perfect sense if $SL_2(\mathbb{Z})$ is replaced by any Fuchsian group Γ that sends the cusps $\mathbb{Q} \cup \{i\infty\}$ to themselves. The only (minor) complication is that the cusps may not lie in the same orbit. See, for example, [**352**] for the proper definition. We are interested in Γ *commensurable* with $SL_2(\mathbb{Z})$, that is, $\Gamma \cap SL_2(\mathbb{Z})$ has finite index in both Γ and $SL_2(\mathbb{Z})$. Typical choices for Γ are the *congruence subgroups*

$$\Gamma(N) := \left\{ \begin{pmatrix} a & b \\ c & d \end{pmatrix} \in SL_2(\mathbb{Z}) \,\middle|\, \begin{pmatrix} a & b \\ c & d \end{pmatrix} \equiv \pm \begin{pmatrix} 1 & 0 \\ 0 & 1 \end{pmatrix} \pmod{N} \right\}, \tag{2.2.4a}$$

$$\Gamma_0(N) := \left\{ \begin{pmatrix} a & b \\ c & d \end{pmatrix} \in SL_2(\mathbb{Z}) \,\middle|\, c \equiv 0 \pmod{N} \right\}, \tag{2.2.4b}$$

for any $N \in \mathbb{N}$. Incidentally, for $N > 1$, $\Gamma(N)/\{\pm 1\}$ is always free (i.e. isomorphic to some $\mathcal{F}_m$), while $\Gamma_0(N)$ may or may not be free.

It is not at all obvious that modular forms and functions should be interesting, but in fact they are unavoidable in modern number theory. For example, consider the question of writing numbers as sums of squares. We can write $5 = 1^2 + (-2)^2 = (-1)^2 + 1^2 + 0^2 + 1^2 + (-1)^2$, to give a couple of trivial examples. Let $N_n(k)$ be the number of ways we can write the integer n as a sum of k squares, counting order and signs. For example, $N_5(1) = 0$ (since 5 is not a perfect square), $N_5(2) = 8$ (since $5 = (\pm 1)^2 + (\pm 2)^2 = (\pm 2)^2 + (\pm 1)^2$), $N_5(3) = 24$, etc. Their generating functions are:[5]

$$\sum_{n=0}^{\infty} N_n(k) x^n = \theta(x)^k,$$

---

[5] A fundamental principle in mathematics is: whenever you have a subscript with an infinite range, make a power series (called a *generating function*) out of it.

where

$$\theta(x) = 1 + 2x + 2x^4 + \cdots = \sum_{n \in \mathbb{Z}} x^{n^2}$$

is called a *theta function*. It turns out that θ transforms nicely with respect to $SL_2(\mathbb{Z})$, once we make the change-of-variables $x = \exp[\pi i \tau]$ (what we usually call $\sqrt{q}$). Write $\theta_3(\tau)$ for $\theta(x)$. Then θ_3 is clearly invariant under the action of $\begin{pmatrix} 1 & 2 \\ 0 & 1 \end{pmatrix}$, and a little work (done next subsection) shows that $\begin{pmatrix} 0 & -1 \\ 1 & 0 \end{pmatrix}$ takes $\theta_3(\tau)$ to $\sqrt{\frac{\tau}{i}}\, \theta_3(\tau)$. Together those two modular transformations generate the group

$$\Gamma_\theta := \left\langle \begin{pmatrix} 0 & -1 \\ 1 & 0 \end{pmatrix}, \begin{pmatrix} 1 & 2 \\ 0 & 1 \end{pmatrix} \right\rangle = \left\{ \begin{pmatrix} a & b \\ c & d \end{pmatrix} \in SL_2(\mathbb{Z}) \,|\, ac \equiv bd \equiv 0 \;(\mathrm{mod}\; 2) \right\}.$$

$$(2.2.5)$$

θ_3 is a modular form of weight $\frac{1}{2}$ and nontrivial multiplier for Γ_θ.

Jacobi introduced that important change-of-variables $x = \exp[\pi i \tau]$ two centuries ago, in his analysis of elliptic integrals. His theory is poorly remembered today, which is very disheartening considering how much of modern mathematics is touched by it. Have a look at the book [**94**], written over a century ago; the style of mathematics in our time is rather different from that in Jacobi's, and we've lost a little in innocence what we've gained in power. See also the beautiful book [**414**]. Let's briefly sketch Jacobi's theory.

Just as we could develop a theory of 'circular functions' (i.e. sine, etc.) starting from the integral $s(a) = \int_0^a \frac{dx}{\sqrt{1-x^2}}$, so we can develop a theory of 'elliptic functions' starting from the elliptic integral $F(k, a) = \int_0^a \frac{dx}{\sqrt{(1-x^2)(1-k^2 x^2)}}$. Inverting $s(a)$ gives a function both more useful and with nicer properties than $s(a)$: we call it $\sin(u)$. Similarly, for any k the elliptic function $\mathrm{sn}(k, u)$ is defined by $u = F(k, \mathrm{sn}(k, u))$. Just as we can define a numerical constant π by $\sin(\frac{1}{2}\pi) = 1$ (i.e. $\frac{1}{2}\pi = \int_0^1 \frac{dx}{\sqrt{1-x^2}}$), we get a function $K(k) = \int_0^1 \frac{dx}{\sqrt{(1-x^2)(1-k^2 x^2)}}$. Just as $\sin(u)$ has period $4(\frac{1}{2}\pi)$, so sn has u-period $4K(k)$. sn also turns out to have u-period $4i K(k')$ where $k' = \sqrt{1 - k^2}$ – today we take this as the starting point and define an elliptic function to be doubly periodic or, what is the same thing, to be a function on a torus (Section 2.1.3).

Theta functions aren't elliptic functions, but they are closely related, as we see in Section 2.1.3. In Jacobi's language, we have

$$\theta_3 \left(\frac{i K(k')}{K(k)} \right) = \sqrt{\frac{2K(k)}{\pi}}.$$

The 'modular transformation' $\tau \mapsto \frac{-1}{\tau}$ interchanges the 'modulus' k with the 'complementary modulus' k', and is completely natural in Jacobi's theory. The important formula $\theta_3(\frac{-1}{\tau}) = \sqrt{\frac{\tau}{i}}\, \theta_3(\tau)$ is trivial here. Closely related to this is Poincaré's remarkable path to modular functions (Section 3.2.4).

Surprisingly, many seemingly innocent questions can be dragged (usually with effort) into the richly developed realm of elliptic curves and modular forms, where they are often solved. For instance, we all know the ancient Greeks were interested in Pythagorean triples: integer solutions a, b, c to $a^2 + b^2 = c^2$, or equivalently right-angle triangles with rational side-lengths (Section 2.1.1).

There are two ways of pushing this. One is to ask which $n \in \mathbb{Z}$ can arise as areas of these rational right-angle triangles. It turns out $n = 5$ is the smallest one: $a = \frac{3}{2}, b = \frac{20}{3}, c = \frac{41}{6}$ works ($5 = \frac{1}{2}(\frac{3}{2})(\frac{20}{3})$ and $(\frac{3}{2})^2 + (\frac{20}{3})^2 = (\frac{41}{6})^2$). This is a hard problem – just try to show $n = 1$ cannot work. The number $n = 157$ works, though the simplest triangle has a and b as quotients of integers of size around 10^{25}, and c as the quotient of integers around 10^{47}. Although this problem was studied by the ancient Greeks and also by the Arabs in the tenth century, it was finally cracked in the 1980s by first translating it into the question of whether the elliptic curve $y^2 = x^3 - n^2 x$ has infinitely many rational points.

The other extension of Pythagorean triples is more famous: find all integer solutions to $a^n + b^n = c^n$. 350 years ago Fermat wrote in the margin of a book he was reading (the book was describing at that point Diophantus'classification of Pythagorean triples) that he had found a 'truly marvelous' proof that for $n > 2$ there are no nontrivial solutions, but that the margin was too narrow to contain it. This result came to be known as 'Fermat's Last Theorem'[6] and despite considerable effort no one has succeeded in rediscovering his proof. Most people believe that Fermat soon realised his 'proof' wasn't valid, otherwise he would have alluded to it in later letters. In any case, a very long and complicated proof was finally achieved in the 1990s: the 'Taniyama–Shimura conjecture' says that a certain function associated with any elliptic curve over $\mathbb{Q}$ will be modular; if $a^n + b^n = c^n$ for some $n > 2$ and nonzero integers a, b, c, then the elliptic curve $y^2 = x^3 + (a^n - b^n)x^2 - a^n b^n$ will violate that conjecture; finally, Wiles proved the Taniyama–Shimura conjecture.

A certain interpretation of modular functions also indicates their usefulness, and explains the adjective 'modular'. The moduli space of tori is $SL_2(\mathbb{Z})\backslash\mathbb{H}$ (Section 2.1.4). So if we have a complex-valued function F on the set of all tori, which associates the same value to conformally equivalent tori (an example is the genus-1 partition function (4.3.8b) in conformal field theories), then F is a function $F : \mathbb{H} \to \mathbb{C}$, symmetric with respect to $SL_2(\mathbb{Z})$.

Likewise, suppose we are interested in meromorphic functions $f : \Sigma \to \mathbb{C}$ living on some surface Σ. We know from the last section that almost every surface Σ is a quotient $\Sigma = \Gamma\backslash\mathbb{H}$, for some Fuchsian group Γ. Then f can be lifted to a meromorphic function on $\mathbb{H}$ with symmetry Γ.

What is the meaning of the Fourier expansion? Think of the parameter q as the local coordinate about the cusp i∞. The Fourier expansion is simply the local expansion of f about that cusp. There is a similar expansion about any other cusp $x \in \mathbb{Q}$. In the case of $SL_2(\mathbb{Z})$, all cusps are equivalent, but for smaller groups the cusps typically fall into

---

[6] It was called his 'Last Theorem' because it was the last of his 48 margin notes to be proved by other mathematicians – a different margin note is discussed in Section 1.7. The story of Fermat's Last Theorem is a fascinating one, but alas this footnote is too small to do it credit. See for instance the excellent book [508].

several distinct orbits, and the corresponding expansions carry independent information. These coefficients are often quite interesting (e.g. they may give the numbers of solutions to various equations, or the dimensions of certain subspaces). The modular form f is a holomorphic interpolation between this local information.

2.2.2 Theta and eta

Two modular forms that appear throughout the following are the Jacobi theta function θ_3 and the Dedekind eta function η:

$$\theta_3(\tau) := 1 + 2 \sum_{m=1}^{\infty} q^{n^2/2} = \prod_{n=1}^{\infty} \left(1 + q^{(2n-1)/2}\right)^2 \prod_{n=1}^{\infty} (1 - q^n), \qquad (2.2.6a)$$

$$\eta(\tau) := q^{1/24} \prod_{n=1}^{\infty} (1 - q^n) = q^{1/24} \sum_{m=-\infty}^{\infty} (-1)^m q^{(3m^2+m)/2}. \qquad (2.2.6b)$$

The equality in (2.2.6a) comes from the denominator identity (3.4.5b) for $A_1^{(1)}$, while that in (2.2.6b) comes from Euler's pentagonal identity; in both cases the first expressions are more important. We saw θ_3 last subsection. Unlike the Eisenstein series, its modularity is not obvious. It can be established though in a number of ways, the most familiar perhaps being *Poisson summation*. This says that for any rapidly decreasing smooth function $g : \mathbb{R} \to \mathbb{C}$ (g is in the Schwartz space $\mathcal{S}(\mathbb{R})$ of Section 1.3.1),

$$\sum_{n \in \mathbb{Z}} g(n) = \sum_{m \in \mathbb{Z}} \widehat{g}(m), \qquad (2.2.7a)$$

where $\widehat{g}$ is the Fourier transform of g:

$$\widehat{g}(y) = \int_{-\infty}^{\infty} e^{-2\pi i x y} g(x) \, dx. \qquad (2.2.7b)$$

Choose $g(x) = e^{-\pi t x^2}$ with $t \in \mathbb{R}$, so $\tau = it \in \mathbb{H}$; then $\widehat{g}(y) = \sqrt{1/t} \, e^{-\pi y^2/t}$ and we obtain (by analytic continuation to all $\tau \in \mathbb{H}$) the transformation formula for θ_3 under $\tau \mapsto -1/\tau$:

$$\theta_3\left(\frac{-1}{\tau}\right) = \sqrt{\frac{\tau}{i}} \, \theta_3(\tau). \qquad (2.2.7c)$$

θ_3 is a modular form for Γ_θ (2.2.5) of weight $1/2$ and nontrivial multiplier. Both Poisson summation and its application to (2.2.7c) are due to Gauss. In Question 2.2.4 you are asked to prove Poisson summation, and next subsection we try to understand what it is saying. In Sections 2.3.1, 2.3.4 and 2.4.2 we give alternate proofs of (2.2.7c).

The modularity of η can be summarised by

$$\eta(\tau + 1) = \xi_{24} \, \eta(\tau), \qquad (2.2.8a)$$

$$\eta\left(\frac{-1}{\tau}\right) = \sqrt{\frac{\tau}{i}} \, \eta(\tau), \qquad (2.2.8b)$$

where $\xi_{24} = \exp[2\pi i/24]$.

More generally, we get the complicated transformation law

$$\eta\left(\frac{a\tau + b}{c\tau + d}\right) = \mu(a, b, c, d)\sqrt{c\tau + d}\,\eta(\tau), \qquad \forall \begin{pmatrix} a & b \\ c & d \end{pmatrix} \in SL_2(\mathbb{Z}), \qquad (2.2.8c)$$

where, for $c > 0$, $\mu(a, b, c, d) = \exp(\pi i\,(\frac{a+d}{12c} - \frac{1}{2} - s(d, c)))$ for the *Dedekind sum*

$$s(d, c) = \sum_{i=1}^{c-1} \frac{i}{c}\left(\frac{di}{c} - \left\lfloor\frac{di}{c}\right\rfloor - \frac{1}{2}\right). \qquad (2.2.8d)$$

For $c = 0$, the transformation follows immediately from (2.2.8a), while for $c < 0$ an analogue to (2.2.8c) holds. The denominator of the rational number $s(d, c)$ will always divide $6c$; μ will always be a 24th root of 1. Although Dedekind sums have many special properties [**468**], we find in Section 2.4.3 a much cleaner way to write (2.2.8c). In any case, η is a modular form for $SL_2(\mathbb{Z})$ of weight $\frac{1}{2}$ and nontrivial multiplier.

Once again, (2.2.8a) is immediate from the definition (2.2.6b) and isn't deep. There are several arguments in the literature that establish (2.2.8b), including Poisson summation applied to the series in (2.2.6b). Here is another, which is instructive for other reasons. In the following paragraph, let's not be distracted by mere analytic concerns, like convergence or interchanging integrals and infinite sums.

Fix $\tau = it$, $t > 0$. The expression

$$-\frac{1}{4}\int (\theta_3(ist) - 1)(\theta_3(is/t) - 1)\,ds \qquad (2.2.9a)$$

is manifestly invariant under the transformation $t \mapsto 1/t$. Applying the transformation (2.2.7c) to $\theta_3(is/t)$ and expanding out both θ_3's, we get

$$-\frac{1}{4}\int \left(\sum_{\ell=1}^{\infty} 2e^{-\pi st\ell^2}\right)\left(\sqrt{\frac{t}{s}}\left(1 + 2\sum_{n=1}^{\infty} e^{-\pi tn^2/s}\right) - 1\right)ds \qquad (2.2.9b)$$

$$= -\sum_{\ell=1}^{\infty}\sum_{n=1}^{\infty}\int \sqrt{\frac{t}{s}}e^{-\pi st\ell^2 - \pi tn^2/s}\,ds + \frac{1}{2}\sum_{\ell=1}^{\infty}\int e^{\pi st\ell^2}\,ds - \frac{1}{2}\sum_{\ell=1}^{\infty}\int \sqrt{\frac{t}{s}}e^{-\pi st\ell^2}\,ds.$$

Now, replace the indefinite integral here with $\int_0^\infty$. The third term in the right-side of (2.2.9b) is independent of t (to see this, change variables: $y = ts$) and so is a constant. The second term can be evaluated explicitly:

$$\frac{1}{2}\sum_{\ell=1}^{\infty}\int_0^\infty e^{-\pi st\ell^2}\,ds = \frac{1}{2}\sum_{\ell=1}^{\infty}\frac{1}{\pi t\ell^2} = \frac{1}{2\pi t}\frac{\pi^2}{6} = \frac{\pi}{12t}. \qquad (2.2.9c)$$

To simplify the first term of (2.2.9b), replace s with x^2 and apply the identity

$$e^{-2\sqrt{ab}} = 2\sqrt{\frac{a}{\pi}}\int_0^\infty e^{-ax^2 - bx^{-2}}\,dx$$

(this is identity 3.325 of [**258**]) with $a = \pi t\ell^2$, $n = \pi tn^2$. The first term becomes

$$-\sum_{\ell=1}^{\infty}\sum_{n=1}^{\infty}\frac{1}{\ell}e^{-2\pi t\ell n} = -\sum_{n=1}^{\infty}\log(1 - e^{-2\pi tn}).$$

Putting these together, we get

$$-\frac{1}{4}\int_0^\infty (\theta_3(\mathrm{i}st) - 1)(\theta_3(\mathrm{i}s/t) - 1)\,\mathrm{d}s = \log\eta(\mathrm{i}t) + \frac{\pi t}{12} + \frac{\pi}{12t} + C$$

for some constant C.

Two unfortunate remarks should probably be made regarding this calculation. First, it would imply (2.2.8b) holds without the prefactor $\sqrt{\tau/\mathrm{i}}$. Second, the constant C diverges, as does the integral in (2.2.9a). Calculations like this mellow somewhat one's disdain for analysis. The way to proceed is to 'regularise' (2.2.9a) by subtracting from the integrand near $s = 0$ the term s^{-1} responsible for the divergence. This results in the identity

$$\log\eta(\mathrm{i}t) = -\frac{1}{4}\int_1^\infty (\theta_3(\mathrm{i}st) - 1)(\theta_3(\mathrm{i}s/t) - 1)\,\mathrm{d}s - \frac{1}{4}\int_0^1 (\theta_3(\mathrm{i}st) - 1)(\theta_3(\mathrm{i}s/t) - 1)$$

$$-s^{-1}\,\mathrm{d}s - \frac{\pi t}{12} - \frac{\pi}{12t} - \frac{1}{4}\log t. \tag{2.2.9d}$$

In Question 2.2.5 the reader is asked to fill in the details, proving (2.2.9d) and thus (2.2.8b). We see from this argument that the mysterious power $1/24$ in (2.2.6b), required for the modularity of η, in fact equals $\zeta(2)/(2\pi)^2$.

At least in spirit, this calculation is reminiscent of the regularisation of Feynman integrals in quantum field theory (Section 4.2.3). For example, the Dedekind eta arises in the calculation of the one-loop partition function of a boson compactified on a circle (see e.g. section 8 of [**246**]). The normalisation factor there involves the product of the nonzero eigenvalues of the Laplacian $\frac{\partial^2}{\partial x^2} + \frac{\partial^2}{\partial y^2}$ on the torus $\mathbb{C}/(\mathbb{Z} + \tau\mathbb{Z})$: namely the modulus-squared $|D|^2$ of

$$D(\tau) = \prod_{(m,n)\neq(0,0)} \frac{\pi}{\tau_2}(n - \tau m), \tag{2.2.10a}$$

where $\tau_2 = \mathrm{Im}(\tau) > 0$. This expression diverges enthusiastically, but it is to be interpreted using the substitutions (*zeta-function regularisation*)

$$\prod_{n=1}^\infty a = a^{\zeta(0)} = a^{-\frac{1}{2}}, \quad \prod_{n=1}^\infty n^\alpha = e^{-\alpha\zeta'(0)} = (2\pi)^{\alpha/2}, \quad \prod_{n=1}^\infty a^n = a^{\zeta(-1)} = a^{-\frac{1}{12}},$$

$$\tag{2.2.10b}$$

where ζ here is the Riemann zeta function (2.2.3c). It is found that

$$D(\tau) = 2\tau_2\,\eta(\tau)^2. \tag{2.2.10c}$$

In this 'derivation' of η, the exponent $1/24$ in (2.2.6b) equals $-\zeta(-1)/2$. Since the values $\zeta(-1)$ and $\zeta(2)$ are related by the functional equation (2.3.2), they are indeed equivalent. Also, note that (2.2.10a) obeys $D(\tau + 1) = D(\tau)$ and $D(-1/\tau) = D(\tau)/\tau$, while (2.2.10c) obeys $D(\tau + 1) = e^{\pi\mathrm{i}/6}D(\tau)$ and $D(-1/\tau) = -\mathrm{i}D(\tau)/\overline{\tau}$. Thus the identifications (2.2.10b) don't preserve modular behaviour. It is somewhat reminiscent of the $-s^{-1}$ regularisation in (2.2.9), which breaks the $t \leftrightarrow 1/t$ symmetry.

Prefactors q^m as in (2.2.6b) are very common, as we shall see later with the characters of Kac–Moody algebras or vertex algebras. In Monstrous Moonshine, this is the q^{-1} with which the j-function begins. These factors are a little mysterious – for example, why start the grading in (0.3.1) at -1 rather than 0 – and there are several explanations (Sections 3.1.2, 3.2.3 and 5.3.4). The point of our little digression into string theory is to introduce its term *conformal anomaly* for this factor q^m. In physics, an anomaly is a symmetry of a classical system that is broken in its quantisation. Here, the $\tau \mapsto \tau + 1$ symmetry (an aspect of conformal invariance) of $D(\tau)$ is broken by regularisation, an anomaly.

We see in (2.2.3a) that the coefficients of the q-expansion of Eisenstein series are interesting. In fact, we are usually more interested in the coefficients of a modular form than in the function itself. A classic example of this is the theta series of a lattice. Let $L \subset \mathbb{R}^n$ be any n-dimensional positive-definite lattice (Section 1.2.1), and choose any vector $t \in \mathbb{R}^n$. Define

$$\Theta_{t+L}(\tau) := \sum_{x \in t+L} q^{x \cdot x/2}. \tag{2.2.11a}$$

In words, the coefficient of q^r is the number of vectors in $t + L$ with length $\sqrt{2r}$. For example, $\Theta_{\mathbb{Z}} = \theta_3$. Let L be rational (i.e. for all $u, v \in L$ we have $u \cdot v \in \mathbb{Q}$) and t have finite order m in L (i.e. $mt \in L$). Then Poisson summation again yields

$$\Theta_{t+L}\left(\frac{-1}{\tau}\right) = \frac{(\tau/\mathrm{i})^{n/2}}{\sqrt{|L|}} \sum_{k=0}^{m-1} \xi_m^k \, \Theta_{ks+L_0}(\tau), \tag{2.2.11b}$$

where (as always) $\xi_m := \exp[2\pi \mathrm{i}/m]$, $s \in L^*$ satisfies $s \cdot t \equiv \frac{1}{m} \pmod 1$ (why must such a vector s always exist?) and where $L_0 = \{u \in L^* \mid u \cdot t \in \mathbb{Z}\}$. In particular,

$$\Theta_L\left(\frac{-1}{\tau}\right) = \frac{(\tau/\mathrm{i})^{n/2}}{\sqrt{|L|}} \Theta_{L^*}(\tau). \tag{2.2.11c}$$

Definition 2.2.2 *Let $\mathcal{I}$ be a finite set, and suppose for each $i \in \mathcal{I}$ we have a function $f_i(\tau)$ meromorphic in $\mathbb{H}$ and with q-expansion $f_i(\tau) = \sum_{r \in \mathbb{Q}} a_{r,i} q^r$, such that for each N only finitely many $r < N$ have nonzero coefficients $a_{r,i}$. We call the set $\{f_i(\tau)\}_{i \in \mathcal{I}}$ a vector-valued modular function for $\mathrm{SL}_2(\mathbb{Z})$ with multiplier $\rho : \mathrm{SL}_2(\mathbb{Z}) \to \mathrm{GL}_{\mathcal{I}}(\mathbb{C})$ if, for each $A \in \mathrm{SL}_2(\mathbb{Z})$ and $i \in \mathcal{I}$, we have*

$$f_i\left(\frac{a\tau + b}{c\tau + d}\right) = \sum_{j \in \mathcal{I}} \rho(A)_{ij} \, f_j(\tau).$$

The strange condition on the $a_{r,i}$ simply says that each f_i is meromorphic at $\tau = \mathrm{i}\infty$. Vector-valued modular forms are studied in, for example, [350]. By the usual argument, ρ will be a $\|\mathcal{I}\|$-dimensional representation of $\mathrm{SL}_2(\mathbb{Z})$. We are interested in the case when the matrices $\rho(A)$ are unitary. In this case, at least when the functions $f_i(\tau)$ are linearly independent, a vector-valued modular function for $\mathrm{SL}_2(\mathbb{Z})$ defines a flat, holomorphic, Hermitian vector bundle over $\mathfrak{M}_{1,0}$: namely, the diagonal quotient ($\mathbb{H} \times \mathrm{span}\{f_i(\tau)\})/\mathrm{PSL}_2(\mathbb{Z})$. The fibre above any point in $\mathfrak{M}_{1,0}$ will be $\|\mathcal{I}\|$-dimensional, except possibly for the singular points [i] and $[e^{\pi \mathrm{i}/3}]$. The f_i are holomorphic sections of this bundle.

A classical property of theta functions, apparently due in this generality to Hecke in 1940, anticipates beautifully what we see later in this book in more and more generality.

Theorem 2.2.3 *Let $L \subset \mathbb{R}^n$ be any n-dimensional positive-definite lattice.*
(a) *Suppose for all $v \in L$ that $v \cdot v \in \mathbb{Q}$. Let $t \in \mathbb{R}^n$ be any vector with finite order in L: i.e. $mt \in L$ for some nonzero $m \in \mathbb{Z}$. Then the theta series $\Theta_{L+t}(\tau)$, divided by $\eta(\tau)^n$, is a modular function for some $\Gamma(N)$.*
(b) *Suppose further that L is an even lattice (i.e. all $v \cdot v$ lie in $2\mathbb{Z}$), and let L^* be its dual. Write $t_i + L$, $i = 1, \ldots, M$, for the finitely many cosets in L^*/L. Define a column vector $\vec{\chi}_L(\tau)$ with ith component $\Theta_{t_i+L}(\tau)/\eta(\tau)^n$. Then $\vec{\chi}_L$ forms a vector-valued modular function for $SL_2(\mathbb{Z})$.*

For the proof of part (a), see theorem 20 of [**456**]. Part (b) follows quickly from (2.2.11b) and (2.2.8). This theorem can be interpreted as being a special case of Theorem 3.2.3 below, when $\mathfrak{g}$ is the affinisation of the reductive (abelian) Lie algebra $\mathbb{C}^n$. Note, however, that the functions in (2.2.11a) are linearly dependent, and so the matrices $\rho(A)$ are not uniquely defined by (b). The easiest way to get linear independence is by adding variables (Section 2.3.2).

The Leech lattice Λ (Section 1.2.1) is to lattices much as the Moonshine module $V^\natural$ is to VOAs (see Section 7.2.1 below). It has no length-squared 2-vectors, and has precisely 196 560 length-squared 4-vectors – a number remarkably close to the monstrous 196 883. Indeed its theta function $\Theta_\Lambda(\tau)$, when divided by $\eta(\tau)^{24}$, equals $J(\tau) + 24$. Is this another example of Moonshine, on par with McKay's equation (0.2.1a)?

Indeed it is. However, for the Leech lattice Λ, we can quickly identify $\Theta_\Lambda(\tau)$ in terms of $J(\tau)$ (see Question 2.2.7). Although the $196\,560 \approx 196\,884$ coincidence is thus easy to explain, it nevertheless turns out to be an instructive example of Moonshine.

2.2.3 Poisson summation

Theta series (2.2.11a) are sums, over periodic sets, of the exponential of a quadratic polynomial. According to the argument given last subsection, two ingredients go into their modularity: together with Poisson summation (2.2.7a), we also needed the fact that the Fourier transform of the Gaussian e^{-tx^2} is essentially itself. Poisson summation requires the infinite periodic sum. There are many other simple functions f that are likewise nearly invariant under Fourier transform: for example, the Fourier transform over $\mathbb{R}^2$ of $f(x, y) = e^{ix^3/y}y^{-2/3}\text{sign}(y)$ is $if(x, -y/27)$. For several other examples, see [**176**]. To see how to use this to get 'cubic' analogues of theta functions (which will transform nicely with respect to $SL_3(\mathbb{Z})$), as well as possible applications to physics, see the intriguing review [**462**] and references therein.

What is the other ingredient, Poisson summation, really saying? Meaning arises from a natural embedding of the particular into a more general context, so let's try to generalise Poisson summation.

First, let G be a group – we require it to be a topological group (separable and locally compact). As defined in Section 1.5.5, its unitary dual $\widehat{G}$ consists of all unitary irreducible representations. For example, the unitary duals of $\mathbb{R}$ and $\mathbb{Z}$ can be identified with $\mathbb{R}$ and S^1, respectively, while the unitary dual of compact groups (like finite G or $G = S^1$) consists of a discrete set of points. When the group is abelian, the representations $\pi \in \widehat{G}$ are all one-dimensional; the dual $\widehat{G}$ itself forms an abelian group, and *Pointrjagin duality* says that the double-dual $\widehat{\widehat{G}}$ is isomorphic to G. For example, the representations in $\widehat{\mathbb{R}}$ look like $\psi_\lambda(x) = e^{2\pi i \lambda x}$ for each $\lambda \in \mathbb{R}$, so $\widehat{\mathbb{R}} \cong \mathbb{R}$. When G is non-abelian, Pointrjagin duality becomes the more abstract Tannaka–Krein duality of Section 1.6.2.

Let us begin with abelian groups. Let Γ be a (discrete) subgroup of an abelian group G, such that the quotient $\Gamma \backslash G$ is compact. The theta series modularity arguments last subsection correspond to the choices $G = \mathbb{R}$ and $\Gamma = \mathbb{Z}$ and, more generally, $G = \mathbb{R}^n$ and $\Gamma = L$; of course the circle $\mathbb{Z} \backslash \mathbb{R}$ and the n-torus $L \backslash \mathbb{R}^n$ are compact.

The Fourier transform $f \mapsto \widehat{f}$ for the group G – explicitly, $\widehat{f}(\psi) = \int_G f(x) \overline{\psi(x)} \, dx$ – is a unitary map taking Schwartz functions on G to Schwartz functions on the dual $\widehat{G}$. Incidentally, the integrals here and below are with respect to the invariant Haar measure (Section 1.5.4). Then the classical Poisson summation (2.2.7a) becomes

$$\int_\Gamma f(\gamma) \, d\gamma = \int_{\Gamma^\perp} \widehat{f}(\psi) \, d\psi, \tag{2.2.12}$$

where $\Gamma^\perp$ consists of all $\psi \in \widehat{G}$ such that $\psi(\gamma) = 1$ for all $\gamma \in \Gamma$. The integrals here reduce to sums, thanks to discreteness. It is through $\Gamma^\perp$ that the dual lattice L^* enters into (2.2.11c). Since $\mathbb{Z}^\perp = \mathbb{Z}$, we find that (2.2.12) is indeed a generalisation of (2.2.7a).

(2.2.12) is too easy a generalisation to help us much. The meaning of Poisson summation, and of (2.2.12), becomes a little clearer when we generalise to non-abelian groups. Let Γ now be an arbitrary discrete closed subgroup of a separable locally compact group G. G and Γ may or may not be abelian. For simplicity we assume that the coset space $\Gamma \backslash G$ is compact. Then $\Gamma \backslash G$ has a finite invariant measure, and the space $L^2(\Gamma \backslash G)$ of square-integrable functions forms a Hilbert space (Section 1.3.1). The regular representation R of G on $L^2(\Gamma \backslash G)$ is defined by $(R(x)f)(y) = f(yx)$, as usual, and is unitary. This representation decomposes as a direct sum of irreducible unitary representations:

$$L^2(\Gamma \backslash G) = \oplus_{\pi \in \widehat{G}} m_\pi \pi,$$

where the numbers $m_\pi \geq 0$ are the (finite) multiplicities.

Even though R is infinite-dimensional, we can define a character for it as follows. For any sufficiently nice function ϕ on G (e.g. ϕ smooth and of compact support), define the operator $R(\phi) = \int_G \phi(y) R(y) \, dy$ on $L^2(\Gamma \backslash G)$ by

$$(R(\phi)f)(x) = \int_G \phi(y) \, f(xy) \, dy.$$

This assignment $\phi \mapsto R(\phi)$ forms a representation of the algebra of smooth functions with compact support, with multiplication given by convolution $\phi * \phi'$. The trace of an operator is defined to be the sum of its eigenvalues. It can be shown that the trace tr $R(\phi)$

exists, and in fact equals

$$\sum_{\pi \in \widehat{G}} m_\pi \operatorname{tr} \pi(\phi) = \sum_{\gamma \in T} \operatorname{vol}(\Gamma_\gamma \backslash G_\gamma) \int_{G_\gamma \backslash G} \phi(x^{-1} \gamma x) \, dx, \qquad (2.2.13)$$

where T is a set of conjugacy class representatives in Γ, and Γ_γ and G_γ are the stabilisers of γ in Γ and G, respectively (e.g. $\Gamma_\gamma = \{g \in \Gamma \mid g\gamma g^{-1} = \gamma\}$). The left side of (2.2.13) is obviously spectral, that is involves eigenvalues. The right side is geometric; the integral over $G_\gamma \backslash G$ is called an 'orbital integral'. Equation (2.2.13) has an immediate generalisation: replace the regular representation R of G on $L^2(\Gamma \backslash G)$ with any representation of G induced from a finite-dimensional unitary representation ρ of Γ. The trivial representation of Γ yields the regular representation R. [20] gives the straightforward proof of (2.2.13) as well as other generalisations.

In the abelian case (e.g. $G = \mathbb{R}^n$, $\Gamma = L$), all $m_\pi = 0$ or 1 and $\Gamma^\perp$ consists of all $\pi \in \widehat{G}$ with $m_\pi = 1$, and (2.2.13) reduces to (2.2.12). In effect we have reinterpreted the Fourier transform $\widehat{f}(\psi)$ by fixing $\psi \in \widehat{G}$ and varying the function f, as a sort of character value for the (possibly infinite-dimensional) irreducible representation ψ. Another special case of (2.2.13) is to take the group G to be finite, in which case it reduces to Frobenius reciprocity. Interesting finite group applications are described in chapters 22–25 of [522].

Equation (2.2.13) is called the *Selberg trace formula*; there is a more complicated version (due in fuller generality to Arthur) when $\Gamma \backslash G$ is noncompact (in which case there are continuous parts to the spectrum). Selberg (a 1950 Fields medalist) was most interested in the case where $G = \mathrm{SL}_2(\mathbb{R})$ and, for example, $\Gamma = \mathrm{SL}_2(\mathbb{Z})$, which has noncompact quotient. For this G he found explicit expressions for the orbital integrals, and the resulting trace formula has powerful consequences.

The Selberg trace formula (2.2.13) can be thought of as an expression for the character of the regular representation of G on $L^2(\Gamma \backslash G)$. This expression is geometric in the sense that for typical groups, the quantities on the right-side typically have geometric interpretations (e.g. for $G = \mathrm{SL}_2(\mathbb{R})$, and Γ a Fuchsian group acting without fixed points, the orbital integrals can be expressed using lengths of closed geodesics on the compact Riemann surface $\Gamma \backslash \mathbb{H}$). Of course these orbital integrals, and hence much of the potential geometry, are trivial in the abelian group case used last section.

Although Poisson summation, and its generalisations like the Selberg trace formula, play a central role in the theory of automorphic forms and Langlands programme, they have only played sporadic roles so far in Moonshine and conformal field theory. For example, [130] applies the Selberg trace formula to string theory, to find the trace of the heat kernel. Orbital integrals also play a fundamental role in the approach [346] to understand group representations via coadjoint orbits; I. Frenkel extended this method to express the characters of affine Kac–Moody algebras as orbital integrals [198], and in this way obtained new proofs of the Macdonald identities. It seems unlikely though that Poisson's and Selberg's formulae can provide a unified explanation of all modularity proofs in Moonshine. A rigorous proof in mathematics may be too slick, much as a painting can be too photographic. It seems to this author that, although Poisson summation permits a quick proof of theta function modularity, it doesn't tell us *why* it's true. A conceptual proof should open the door to natural generalisations of the given theorem, by underscoring the confluence of properties needed for that theorem to hold.

2.2.4 Hauptmoduls

Let's identify the orbit space $\mathrm{SL}_2(\mathbb{Z})\backslash\overline{\mathbb{H}}$, by studying the fundamental domain D of Figure 2.3. Apart from the boundary of D, every $\mathrm{SL}_2(\mathbb{Z})$-orbit will intersect D in one and only one point. But what should we do about the boundary? Well, the edge $\mathrm{Re}(\tau) = -\frac{1}{2}$ gets mapped by the translation $T : \tau \mapsto \tau + 1$ to the edge $\mathrm{Re}(\tau) = \frac{1}{2}$, so we should identify these, i.e. glue them together. The result is a cylinder running off to infinity, with a strange lip at the bottom. The inversion $S : \tau \mapsto -1/\tau$ tells us how we should close that lip: identify $ie^{i\theta}$ and $ie^{-i\theta}$. This seals the bottom of the cylinder, so we get an infinitely tall cup with a strangely puckered base. In fact the top of this cup is also capped off, by the cusp $i\infty$. So what we have (topologically speaking) is a *sphere*. It inherits the smoothness of $\mathbb{H}$ except for conical singularities at the fixed points i and $e^{\pi i/3}$. The cusps are responsible for compactness. This interpretation of $\mathrm{SL}_2(\mathbb{Z})\backslash\overline{\mathbb{H}}$ means that a modular function for $\mathrm{SL}_2(\mathbb{Z})$ can be reinterpreted as a meromorphic complex-valued function on this sphere. There is a canonical sphere in complex analysis, namely the Riemann sphere $\mathbb{P}^1(\mathbb{C}) = \mathbb{C} \cup \{\infty\}$. The meromorphic functions on the Riemann sphere must be *rational*, that is of the form $f(w) = \frac{\text{some polynomial } P(w)}{\text{some polynomial } Q(w)}$, where w is the complex parameter on the Riemann sphere. So a modular function $f(\tau)$ for $\mathrm{SL}_2(\mathbb{Z})$ is simply some rational function P/Q evaluated at the change-of-local-parameters, or at the uniformising function $w = c(\tau)$ that maps us from our sphere $\Gamma\backslash\overline{\mathbb{H}}$ to the Riemann sphere. There are many different choices for this function $c(\tau)$, but the standard one is the j-function:[7]

$$j(\tau) := \frac{\left(1 + 240\sum_{n=1}^{\infty}\sigma_3(n)\,q^n\right)^3}{q\prod_{n=1}^{\infty}(1-q^n)^{24}} = \frac{\Theta_{E_8}(\tau)^3}{\eta(\tau)^{24}} = q^{-1} + 744 + 196\,884\,q + \cdots$$

(2.2.14)

(see also (0.1.8)), where σ_3 is in (2.2.3b), Θ_{E_8} is the theta series of the E_8 root lattice (2.2.11a) and η is the Dedekind eta (2.2.6b). Thus, any modular function for $\mathrm{SL}_2(\mathbb{Z})$ can be written as a rational function $f(\tau) = P(j(\tau))/Q(j(\tau))$ in the j-function. Conversely, any such function is modular.

This is analogous to (and much stronger than) saying that any function $g(x)$ periodic under $x \mapsto x + 1$ is really a function on the unit circle $S^1 \subset \mathbb{C}$ evaluated at the uniformising function $x \mapsto e^{2\pi ix}$, and hence has a Fourier expansion $\sum_n g_n \exp[2\pi inx]$.

We can generalise the argument that led to j. Recall (2.2.4).

Definition 2.2.4 *Call a discrete subgroup Γ of $\mathrm{SL}_2(\mathbb{R})$ a congruence subgroup if it contains some $\Gamma(N)$. Call it of moonshine-type if it contains some $\Gamma_0(N)$, and obeys*

$$\begin{pmatrix} 1 & t \\ 0 & 1 \end{pmatrix} \in \Gamma \;\Rightarrow\; t \in \mathbb{Z}.$$

(2.2.15)

---

[7] Historically, j was the standard choice, but in Monstrous Moonshine the preferred choice would be the function $J = j - 744$ with zero constant term.

The congruence subgroups are relatively rare among finite index subgroups of $SL_2(\mathbb{Z})$, but their theory is much better developed. Let f be a modular function for a congruence subgroup Γ. Then we can expand f as a Laurent series in $q^{1/N}$. We analyse this as before: look at the orbit space $\Sigma = \Gamma \backslash \overline{\mathbb{H}}$; because Γ is not too big, Σ will be a Riemann surface; because Γ is not too small, Σ will be compact.

We call Γ 'genus g' if its surface Σ has genus g. If Γ is a subgroup of $\Gamma(1) = SL_2(\mathbb{Z})$, and without loss of generality we have $-I \in \Gamma$, then the genus is given by

$$g = 1 + \frac{n}{12} - \frac{n_2}{4} - \frac{n_3}{3} - \frac{n_\infty}{2}, \qquad (2.2.16)$$

where n is the index $\|\Gamma(1)/\Gamma\|$ of Γ in $\Gamma(1)$, and where n_k ($k = 2, 3, \infty$) is the number of Γ-orbits of order-$2k$ fixed points. For the easy proof from the Hurwitz formula, see proposition 1.40 of [505]. Note that n_∞ is the number of punctures of $\Gamma \backslash \mathbb{H}$. For example, for $\Gamma = SL_2(\mathbb{Z})$ we have $n = 1 = n_2 = n_3 = n_\infty$ and we recover our result that the genus is 0. The values n, n_2, n_3, n_∞ for all $\Gamma(N)$ and $\Gamma_0(N)$ are given in Section 1.6 of [505].

For example, $\Gamma = \Gamma_0(2)$ and $\Gamma = \Gamma_0(25)$ are both genus 0 (with 2, respectively 6, punctures), while $\Gamma_0(50)$ is genus 2 with 12 punctures and $\Gamma_0(24)$ is genus 1 with 7 punctures. Once again, we are interested here in the genus-0 case. As before, this means that there is a uniformising function J_Γ that is a modular function for Γ, and all other modular functions for Γ can be written as a rational function in it. Because of (2.2.15), we can choose J_Γ to look like

$$J_\Gamma(\tau) = q^{-1} + a_1(\Gamma)q + a_2(\Gamma)q^2 + \cdots$$

So J_Γ, the *Hauptmodul* for Γ, plays exactly the same role for Γ that $J := j - 744$ plays for $SL_2(\mathbb{Z})$. For example, $\Gamma_0(2)$, $\Gamma_0(13)$ and $\Gamma_0(25)$ are all genus 0, with Hauptmoduls

$$J_2(\tau) = q^{-1} + 276\,q - 2048\,q^2 + 11202\,q^3 - 49152\,q^4 + 184024\,q^5 + \cdots,$$
$$\qquad (2.2.17a)$$
$$J_{13}(\tau) = q^{-1} - q + 2\,q^2 + q^3 + 2\,q^4 - 2\,q^5 - 2\,q^7 - 2\,q^8 + q^9 + \cdots, \quad (2.2.17b)$$
$$J_{25}(\tau) = q^{-1} - q + q^4 + q^6 - q^{11} - q^{14} + q^{21} + q^{24} - q^{26} + \cdots \qquad (2.2.17c)$$

The smaller (sparser) the modular group, the smaller the coefficients of the Hauptmodul. In this sense, the j-function is optimally bad among the Hauptmoduls: for example, for it $a_{23} \approx 10^{25}$.

In Theorem 2.1.5 we see what happens in genus > 0: two generators, not one, are needed, although they will be polynomially related.

As is mentioned in Chapter 0, Monstrous Moonshine is interested directly in genus-0 groups. We construct certain functions associated with the Monster, and it turns out unexpectedly that these functions are actually Hauptmoduls.

An obvious question is, how many genus-0 groups (equivalently, how many Hauptmoduls) are there? It turns out that $\Gamma_0(p)$ is genus 0, for a prime p, iff $p - 1$ divides 24. Thompson [526] proved that for any g, there are only finitely many genus-g groups of moonshine type. Cummins [121] has shown that there are in fact exactly 6486 genus-0 groups of moonshine type. 616 of these have Hauptmoduls with integer coefficients $a_i(\Gamma)$, and all of the remainder have q-coefficients in some cyclotomic field.

Question 2.2.1. How important are the conditions at the cusps for the definition of modular functions or forms? For example, describe all functions f holomorphic on $\mathbb{C}$, symmetric with respect to $SL_2(\mathbb{Z})$ (i.e. $f(\gamma.\tau) = f(\tau)$ for all $\gamma \in SL_2(\mathbb{Z})$), but which need not be holomorphic or even meromorphic at the cusps (i.e. f may have an essential singularity there).

Question 2.2.2. Show that if f is a modular form of weight k, and 3 doesn't divide k, then $f(e^{2\pi i/3}) = 0$.

Question 2.2.3. Suppose f is a modular form, not identically 0, for some Γ, with multiplier μ and *integral* weight k. Prove that μ must be a one-dimensional representation of Γ. Where does the proof go wrong if k is fractional?

Question 2.2.4. Prove Poisson summation (2.2.7a). (*Hint:* $x \mapsto \widetilde{f}(x) = \sum_{n\in\mathbb{Z}} f(n+x)$ is periodic, so can be Fourier expanded. Compute $\widetilde{f}(0)$ in two different ways.)

Question 2.2.5. By modifying slightly the argument beginning with (2.2.9a), prove (2.2.9d) and thus (2.2.8b).

Question 2.2.6. Let L be any self-dual positive-definite lattice. Then $\Theta_L(\tau)$ is a polynomial in $\theta_3(\tau)$ and $\Theta_{E_8}(\tau)$ (you can assume this, which is proved for instance in [503]). Using this fact, show that the theta function for any self-dual positive-definite lattice of dimension < 24 is uniquely determined by the numbers N_1, N_2 of norm-squared 1- and 2-vectors.

Question 2.2.7. Let L be a positive-definite 24-dimensional even self-dual lattice. Prove that $\Theta_L(\tau)/\eta(\tau)^{24} = J(\tau) + c_L$ for some constant c_L. Find that constant.

Question 2.2.8. Find the genus of $\Gamma(2)$, using (2.2.16).

2.3 Further developments

2.3.1 Dirichlet series

One of the most remarkable formulae in science is surely

$$1 + 2 + 3 + 4 + \cdots = -\frac{1}{12}. \tag{2.3.1}$$

Of course the right side is the value at $s = -1$ of the Riemann zeta function (2.2.3c). The expressions in (2.2.3c) converge absolutely when $\mathrm{Re}(s) > 1$, where ζ is then holomorphic, and ζ has a unique holomorphic extension to all of $\mathbb{C}$, except for a simple pole at $s = 1$ (the harmonic series). Equation (2.3.1) is used in quantum field theory in the context of zeta function regularisation (2.2.10); it is related to the $q^{1/24}$ in the Dedekind eta function (2.2.6b) and the normalisation $C/12$ in Lie brackets (3.1.5a) of the Virasoro algebra.

The equality of the infinite sum and product in (2.2.3c) is merely an analytic reformulation of unique factorisation in $\mathbb{Z}$, but it shows crucially the relation between $\zeta(s)$ and the primes. For a trivial example, taking logs of (2.2.3c) quickly gives the divergence of $\sum 1/p$.

As important as analytic continuation and the product expansion are, more important for us is the *functional equation*

$$\Lambda(1-s) = \Lambda(s), \tag{2.3.2}$$

where $\Lambda(s) := \pi^{-s/2}\Gamma(s/2)\,\zeta(s)$, using the Gamma function

$$\Gamma(s) := (2\pi)^s \int_0^\infty e^{-2\pi y} y^{s-1}\,dy.$$

Indeed, Hecke discovered that (2.3.2) is equivalent to modularity (2.2.7c).

Theorem 2.3.1 (Hecke, 1936) *Let $f(\tau) = \sum_{n=0}^\infty a_n e^{2\pi i n\tau/d}$ and $\phi(s) = \sum_{n=1}^\infty a_n n^{-s}$, where $|a_n| < Cn^c$ for some constants d, C, c. Define $\Phi(s) = (2\pi/d)^{-s}\Gamma(s)\phi(s)$. Then the following two statements are equivalent:*
(i) $f(\frac{-1}{\tau}) = \left(\frac{\tau}{i}\right)^k f(\tau)$;
(ii) $\Phi(k-s) = \Phi(s)$, and $\Phi(s) + \frac{a_0}{s} + \frac{a_0}{k-s}$ is holomorphic and bounded in each vertical strip in $\mathbb{H}$.

Proof: The key idea of the proof is that $\Phi(s)$ and $f(\tau)$ are related by the Mellin transform:

$$\Phi(s) = \int_0^\infty x^{s-1}\,(f(ix) - a_0)\,dx, \tag{2.3.3a}$$

$$f(ix) - a_0 = \frac{1}{2\pi i}\int_{\mathrm{Re}(s)=a} x^{-s}\,\Phi(s)\,ds, \tag{2.3.3b}$$

for any constant $a > 0$ sufficiently large.

To prove (ii) from (i), write $\int_0^\infty = \int_0^1 + \int_1^\infty$ in (2.3.3a), so we get the sum $\Phi(s) = \Phi_0 + \Phi_\infty$. Note that $\Phi_\infty(s)$ is clearly holomorphic everywhere, and $\phi_0(s)$ is holomorphic when $\mathrm{Re}(s)$ is sufficiently large. Then, using (i), for those s

$$\Phi_0(s) = \int_0^1 x^{s-1}\,(f(ix) - a_0)\,dx = \int_1^\infty x^{-s-1}x^k f(ix)\,dx - \frac{a_0}{s}$$
$$= \Phi_\infty(k-s) - \frac{a_0}{s} - \frac{a_0}{k-s}.$$

Therefore $\Phi_0(s)$ extends holomorphically everywhere, except for simple poles at $s = 0$ and $s = k$, and $\Phi_0(s) = \Phi_\infty(k-s) - a_0 s^{-1} - a_0(k-s)^{-1}$ holds $\forall\, s \neq 0, k$. Thus

$$\Phi(k-s) = \Phi_0(k-s) + \Phi_\infty(k-s)$$
$$= \left(\Phi_\infty(s) - \frac{a_0}{k-s} - \frac{a_0}{s}\right) + \left(\Phi_0(s) + \frac{a_0}{s} + \frac{a_0}{k-s}\right) = \Phi(s).$$

To prove (i) from (ii), shift the vertical contour $\text{Re}(s) = a > 0$ in (2.3.3b) to the left, to $\text{Re}(s) = b < 0$, and pick up residues $-a_0$ at $s = 0$ and $x^{-k}a_0$ at $s = k$:

$$f(\mathrm{i}x) - a_0 x^{-k} = \frac{1}{2\pi\mathrm{i}} \int_{\text{Re}(s)=b} x^{-s}\Phi(s)\,\mathrm{d}s = \frac{1}{2\pi\mathrm{i}} \int_{\text{Re}(s)=k-b} x^{-(k-s)}\Phi(s)\,\mathrm{d}s$$
$$= x^{-k}(f(\mathrm{i}/x) - a_0).$$

Therefore $f(\mathrm{i}/x) = x^k f(\mathrm{i}x)$, and (i) follows by analytic continuation.   ∎

When f is a modular form, we call ϕ the *Dirichlet series* or *L-function* corresponding to f (the term L-function is usually reserved for those ϕ which also have product expansions as in (2.2.3c)). The modular form corresponding to the Riemann zeta function $\zeta(s)$ is $f(\tau) = \frac{1}{2}\theta_3(\tau)$. Theorem 2.3.1 applies with $k = \frac{1}{2}, d = 2$ and $\Lambda(2s) = \Phi(s)$, and relates (2.3.2) directly to (2.2.7c). Another famous example, due to Ramanujan, is $f = \eta^{24}$. Its Φ is holomorphic everywhere and its ϕ has a product form $\prod_p (1 - \tau(p)p^{-s} + p^{11-2s})^{-1}$, where τ here is the so-called *Ramanujan tau-function* (see e.g. (3.4.6)).

Mysteriously, we can associate Dirichlet series to many of the basic objects of arithmetic – modular forms, number fields, algebraic varieties, etc. – in such a way that basic operations performed on, and relations between, the Dirichlet series correspond to natural operations on, and relations between, the arithmetic objects. In its most general form, this is Langlands functoriality. For a famous special case, given an elliptic curve E defined over $\mathbb{Q}$, its L-function keeps track of the number of points on E as we vary its field of definition from $\mathbb{Q}$ to the finite fields. The Taniyama–Shimura Conjecture states that E is modular, i.e. that this L-function is the Dirichlet series of a modular form of weight 2. As we know, Wiles *et al.* proved Taniyama–Shimura and hence Fermat's Last Theorem.

See [**456**] for a clear treatment of the material of this subsection. We have been hurried since there is at this point no evidence for its direct relevance to Moonshine. There are many generalisations of Theorem 2.3.1. Let us mention one. Generators for the groups $\text{SL}_2(\mathbb{Z})$ and Γ_θ are given in (2.2.1a) and (2.2.5), so Theorem 2.3.1 gives a Dirichlet series characterisation for f to be a modular form for those groups. When Γ is smaller (say $\Gamma = \Gamma(N)$), to which Dirichlet series conditions does the modularity of f translate? The list of generators is far more complicated. An answer is provided by Weil's Converse Theorem (Section 2.3.3).

2.3.2 Jacobi forms

The general quadratic polynomial in one variable x looks like $ax^2 + bx + c$, so we might try to generalise $\theta_3(\tau)$ by replacing $n^2\tau$ with $an^2\tau + bnz + cu$. Consider then the function

$$\theta_3(\tau, z, u) = \sum_{n\in\mathbb{Z}} e^{\pi\mathrm{i}\tau n^2 + 2\pi\mathrm{i}zn + 2\pi\mathrm{i}u}, \tag{2.3.4}$$

where $\tau, z, u \in \mathbb{C}$. We've seen these kinds of functions before in (2.1.7a). The $2\pi\mathrm{i}$'s in front of z and u are conventional. As before, convergence requires $\tau \in \mathbb{H}$. Obviously, the u-dependence is rather trivial and is retained only for book-keeping.

Fix $\tau \in \mathbb{H}$ and $u \in \mathbb{C}$, and consider this as a function of $z \in \mathbb{C}$. It has period 1 and quasi-period τ:

$$\theta_3(\tau, z + m\tau + \ell, u + mz + m^2\tau/2) = \theta_3(\tau, z, u), \qquad \forall m, \ell \in \mathbb{Z}, \qquad (2.3.5a)$$

and thus is a function living (projectively) on the torus $\mathbb{C}/(\mathbb{Z} + \tau\mathbb{Z})$.

Next, fix $z, u \in \mathbb{C}$ and consider θ_3 as a function of $\tau \in \mathbb{H}$. Completing the square $\tau n^2 + 2nz = \tau (n + \frac{z}{\tau})^2 - \frac{z^2}{\tau}$ and restricting τ, z to the imaginary axis, Poisson summation (2.2.7a) and analytic continuation gives us

$$\theta_3(\tau, z, u) = \sqrt{\frac{i}{\tau}} \theta_3 \left(\frac{-1}{\tau}, \frac{z}{\tau}, u - \frac{z^2}{2\tau} \right), \qquad (2.3.5b)$$

valid for all $\tau \in \mathbb{H}$ and $z, u \in \mathbb{C}$.

Definition 2.3.2 [170] *By a Jacobi form for $SL_2(\mathbb{Z})$ of weight k and index m we mean a holomorphic function $f : \mathbb{H} \times \mathbb{C} \to \mathbb{C}$ satisfying*

$$f \left(\frac{a\tau + b}{c\tau + d}, \frac{z}{c\tau + d} \right) = (c\tau + d)^k \exp \left[2\pi i \frac{mcz}{c\tau + d} \right] f(\tau, z), \qquad (2.3.6a)$$

$$f(\tau, z + \ell\tau + n) = \exp[-2\pi i m (\ell^2\tau + 2\ell z)] f(\tau, z), \qquad (2.3.6b)$$

for all $\begin{pmatrix} a & b \\ c & d \end{pmatrix} \in SL_2(\mathbb{Z})$ *and* $\ell, n \in \mathbb{Z}$. *Moreover, f must have a Fourier expansion of the form*

$$f(\tau, z) = \sum_{n \in \mathbb{N}} \sum_{r \in \mathbb{Z}, r^2 \le 4mn} c_{n,r} e^{2\pi i (n\tau + rz)}. \qquad (2.3.6c)$$

Similarly, we call $\theta_3(\tau, z, 0)$ a Jacobi form of weight $\frac{1}{2}$ and index 0 for Γ_θ. The Weierstrass $\wp$-function $\wp(\tau, z)$ in (2.1.6a) is a Jacobi form for $SL_2(\mathbb{Z})$ of level 2 and index 1 (Question 2.3.1). A Jacobi form is a natural blend of the notions of modular form and elliptic function: the parameter $\tau \in \mathbb{H}$ tells us where on the moduli space of tori we are, and the parameter z lives on that torus. Given such classical examples, it is hard to understand why their theory was developed only in the 1980s. The introduction of the index m in Definition 2.3.2 may be somewhat unexpected, but is explained in Section 2.4.1.

We can generalise the example (2.3.4) to lattices (and in fact to translates of lattices). Let L be an n-dimensional lattice in $\mathbb{R}^n$. Define

$$\Theta_L(\tau, z, u) = \sum_{v \in L} \exp[\pi i \tau \, v \cdot v + 2\pi i z \cdot v + 2\pi i u], \qquad (2.3.7)$$

where $z \in \mathbb{C}^n$, $u \in \mathbb{C}$ and $\tau \in \mathbb{H}$. The z-periods of Θ_L fill out the dual lattice L^*, and the z-quasi-periods fill out τL^*. Provided L is a rational lattice, we get the obvious analogue of (2.2.11c), again from Poisson summation. To make Θ_L into a Jacobi form for some $\Gamma(N)$ at weight n and index 0, it suffices to embed $z \in \mathbb{C}$ into $\mathbb{C}^n$ along any nonzero dual weight vector $u^* \in L^*$: i.e. $\Theta_L(\tau, zu^*, 0)$ will be a Jacobi form.

As any string theorist knows, there are several different lattices L, L' that have the same theta function: $\Theta_L(\tau) = \Theta_{L'}(\tau)$. Perhaps the most famous example of this is the pair of even self-dual lattices of dimension 16 (namely, D_{16}^+ and $E_8 \oplus E_8$ [113]). Actually there are lattice examples in every dimension ≥ 3 [108]. However, their Jacobi forms are unique in the strongest form possible (see Question 2.3.2).

Writing theta functions as Jacobi forms is crucial to their interpretation as heat kernels, or using Heisenberg groups, as we see in Sections 2.3.4 and 2.4.2. In Theorem 3.2.3 we find that the characters of affine Kac–Moody algebras are Jacobi forms of weight and index 0. Indeed, they are rational functions of lattice Jacobi forms (2.3.7).

An obvious question to ask is, to any modular form $f(\tau)$, is there a Jacobi form $f(\tau, z)$ for the same group and at the same weight such that $f(\tau, 0) = f(\tau, z)$? And if so, is this Jacobi form unique? It turns out that every weight-k modular form f, at least for $SL_2(\mathbb{Z})$, *can* be lifted to a Jacobi form for the same weight and group, at index $m = 1$. This Jacobi form is far from unique, even at $m = 1$. In fact, the redundancy has the same dimension as the space of weight-$k + 2$ cusp forms for $SL_2(\mathbb{Z})$. This fact is a consequence of theorem 3.5 in [170].

2.3.3 Twisted #2: shifts and twists

Recall the classical Jacobi theta functions $\theta_1 = \theta_{\frac{1}{2}, \frac{1}{2}}, \theta_2 = \theta_{\frac{1}{2}, 0}, \theta_3 = \theta_{0,0}, \theta_4 = \theta_{0, \frac{1}{2}}$, using the notation of (2.1.7a). These obey simple modular transformation rules, most concisely stated in vector notation as

$$\begin{pmatrix} \theta_1 \\ \theta_2 \\ \theta_3 \\ \theta_4 \end{pmatrix}(\tau + 1, z) = \begin{pmatrix} e^{\pi i/4} & 0 & 0 & 0 \\ 0 & e^{\pi i/4} & 0 & 0 \\ 0 & 0 & 0 & 1 \\ 0 & 0 & 1 & 0 \end{pmatrix} \begin{pmatrix} \theta_1 \\ \theta_2 \\ \theta_3 \\ \theta_4 \end{pmatrix}(\tau, z), \qquad (2.3.8a)$$

$$\begin{pmatrix} \theta_1 \\ \theta_2 \\ \theta_3 \\ \theta_4 \end{pmatrix}\left(\frac{-1}{\tau}, \frac{z}{\tau}\right) = e^{\pi i z^2/\tau} \sqrt{\frac{\tau}{i}} \begin{pmatrix} 1 & 0 & 0 & 0 \\ 0 & 0 & 0 & 1 \\ 0 & 0 & 1 & 0 \\ 0 & 1 & 0 & 0 \end{pmatrix} \begin{pmatrix} \theta_1 \\ \theta_2 \\ \theta_3 \\ \theta_4 \end{pmatrix}(\tau, z). \qquad (2.3.8b)$$

That is, these θ_i define a vector-valued Jacobi form for $SL_2(\mathbb{Z})$ (Definition 2.2.2). The q-expansions of θ_1 and θ_4 have negative coefficients; we can make 'positive' combinations of these theta functions that have almost as nice transformations under $SL_2(\mathbb{Z})$:

$$\theta_{[0]}(\tau, z) = \frac{\theta_3(\tau, z) + \theta_4(\tau, z)}{2} = 1 + q^2(r^2 + r^{-2}) + q^8(r^4 + r^{-4}) + \cdots,$$

$$(2.3.9a)$$

$$\theta_{[1]}(\tau, z) = \frac{\theta_1(\tau, z) + \theta_2(\tau, z)}{2} = q^{1/8} r^{1/2}(1 + qr^{-2} + q^3 r^2 + q^6 r^{-4} + \cdots),$$

$$(2.3.9b)$$

$$\theta_{[2]}(\tau, z) = \frac{\theta_3(\tau, z) - \theta_4(\tau, z)}{2} = q^{1/2}((r + r^{-1}) + q^4(r^3 + r^{-3}) + \cdots),$$

$$(2.3.9c)$$

$$\theta_{[3]}(\tau, z) = \frac{\theta_2(\tau, z) - \theta_1(\tau, z)}{2} = q^{1/8} r^{1/2}(r^{-1} + qr + q^3 r^{-3} + q^6 r^3 + \cdots),$$

$$(2.3.9d)$$

where $r = e^{2\pi i z}$. Note that $\theta_{[i]}$ has the geometric interpretation as the theta series (2.2.11a) of the translate $2\mathbb{Z} + \frac{i}{2}$.

We regard $\theta_1, \theta_2, \theta_4$ as $\mathbb{Z}_2$-twists and -shifts of θ_3. More generally, the parameter $r \in \frac{1}{N}\mathbb{Z}$ in $\theta_{r,s}$ corresponds to a $\mathbb{Z}_N$-shift, and $s \in \frac{1}{N}\mathbb{Z}$ to a $\mathbb{Z}_N$-twist. A far-reaching generalisation of this simple construction is studied in Section 5.3.6; the analogue there of the positive combinations (2.3.9) is the characters for a vertex operator algebra. In Monstrous Moonshine the twists of $J(\tau)$ are the McKay–Thompson series $J_g(\tau)$, and its more general shifts and twists are the Norton series of Maxi-Moonshine (Section 7.3.2). Physically, this corresponds to the orbifold construction (Section 4.3.4). There, the positive linear combinations have the direct interpretation as graded dimensions of sectors of the conformal field theory.

As always, the clearest example is provided by lattices (Section 1.2.1). Let L be an integral positive-definite lattice and let r, s be two vectors in $\mathbb{Q} \otimes L$. As in (2.1.7a), write

$$\Theta_{L;r,s}(\tau, z) = \sum_{x \in L} e^{\pi i \tau (x+r) \cdot (x+r)} e^{\pi i (z+s) \cdot (2x+r)}, \tag{2.3.10a}$$

where as before $z \in \mathbb{C} \otimes L$. Then $\Theta_{L;r,s}$ will be a Jacobi form for some subgroup of $SL_2(\mathbb{Z})$, as is (2.3.7). In fact, if L is even and self-dual, we can be much more explicit. For any $r, s \in \mathbb{Q} \otimes L$, and any $\begin{pmatrix} a & b \\ c & d \end{pmatrix} \in SL_2(\mathbb{Z})$, we have

$$\Theta_{L;r,s} \left(\frac{a\tau + b}{c\tau + d}, \frac{z}{c\tau + d} \right) = (c\tau + d)^{n/2} \exp \left[\pi i \frac{cz}{c\tau + d} \right] \Theta_{L;ar+cs,br+ds}(\tau, z), \tag{2.3.10b}$$

where n is the dimension of L.

As usual, certain positive combinations of these $\Theta_{L;r,s}$ have a direct (geometric) interpretation. Again let L be self-dual, and suppose the vector $s \in \mathbb{Q} \otimes L$ has order m in L (so $ms \in L$). Then there will be a vector $s' \in L$ such that $s \cdot s' \equiv \frac{1}{m} \pmod{1}$. For any integer k, and vector $r \in \mathbb{Q} \otimes L$, we get this generalisation of (2.3.9):

$$\frac{1}{m} \sum_{j=0}^{m-1} \exp \left[2\pi i j \left(s \cdot r - \frac{k}{m} \right) \right] \Theta_{L;r,js} = \Theta_{L_0+r+ks'}, \tag{2.3.10c}$$

the theta series of a translate of the lattice $L_0 = \{v \in L \mid v \cdot s \in \mathbb{Z}\}$.

In the orbifold construction of vertex operator algebras and *chiral* conformal field theory, the role of vectors r, s is played by automorphisms g, h in some group G, and the role of the sublattice L_0 in (2.3.10c) is played by the vertex operator subalgebra $\mathcal{V}^G$ fixed by G. However, as we see in Section 4.3, *full* conformal field theory or string theory involves the interplay of two vertex operator algebras; the orbifold construction there involves in addition a reconstruction of a new full conformal field theory from $\mathcal{V}^G$. We address this further in Sections 4.3.4 and 5.3.6.

This reconstruction is again beautifully illustrated by lattices. Let L be any rational lattice and $T = \{t_i\}$ be a finite set of vectors in $\mathbb{Q} \otimes L$. Then by $L\{T\}$ we mean the set

$$L\{T\} = \left\{ x + \sum_i \ell_i t_i \mid \ell_i \in \mathbb{Z}, \ x \in L, \ \left(x + \sum_i \ell_i t_i \right) \cdot t_j \in \mathbb{Z} \ \forall j \right\}.$$

Then $L\{T\}$ is a lattice *rationally equivalent* to L (i.e. there is an orthogonal transformation $T : \mathbb{Q} \otimes L\{T\} \to \mathbb{Q} \otimes L$). Conversely, if L_1 and L_2 are rationally equivalent integral lattices, then there is a finite set $T = \{t_1, \ldots, t_m\} \subset \mathbb{Q} \otimes L_1$ such that $L_1\{T\}$ is isomorphic to L_2 [238]. Clearly the theta series of $L\{T\}$ is the average of $\Theta_{L;r,s}$ for a finite number of r, s in the $\mathbb{Z}$-span of T. The important special case is when L is self-dual; then $L\{T\}$ will also be self-dual provided all $t_i \cdot t_j \in \mathbb{Z}$. In this case,

$$L\{T\} = \bigcup_{\ell_i \in \mathbb{Z}} \left(L_0 + \sum_i \ell_i t_i \right),$$

where $L_0 = \{x \in L \mid x \cdot t_i \in \mathbb{Z}\}$. Call two self-dual lattices L_1, L_2 *neighbours* if there is some vector t with integer length-square $t \cdot t$ such that $2t \in L_1$, and L_2 and $L_1\{t\}$ are isomorphic. Then any two self-dual lattices, with equal dimensions $n_+ + n_-$ and signature $n_+ - n_-$, will be neighbours of neighbours of $\cdots$ of neighbours of each other [238].

Another way to collect some of these results is through *Dirichlet characters*, which are important in the classical theory of modular forms. A Dirichlet character is a function $\chi : \mathbb{Z} \to \mathbb{C}$, with some period N, such that $\chi(a) \neq 0$ iff a is coprime to N, and for all $a, b \in \mathbb{Z}$ $\chi(ab) = \chi(a)\chi(b)$. Dirichlet introduced these χ in his proof that there are infinitely many primes in any arithmetic series $a, a + b, a + 2b, \ldots$, provided only that a and b are coprime (clearly a necessary condition). He proved this by twisting the Riemann zeta function (2.2.3c) by χ:

$$L(\chi, s) = \sum_{i=1}^{\infty} \chi(n) n^{-s} = \prod_p (1 - \chi(p)p^{-s})^{-1}. \tag{2.3.11}$$

Given the lesson of Section 2.3.1, it should also be interesting to Dirichlet-twist modular forms.

Modular forms and functions for the principal congruence subgroup $\Gamma(N)$ can be defined as in Definitions 2.2.1 and 0.1, except now there are several orbits of cusps, and we have invariance under only $\begin{pmatrix} 1 & N \\ 0 & 1 \end{pmatrix}$, so the q-expansion takes the form

$$f(\tau) = \sum_{n \in \mathbb{Z}} a_n e^{2\pi i n\tau/N} = \sum_{n \in \mathbb{Z}} a_n q^{n/N}. \tag{2.3.12}$$

Given any Dirichlet character χ, we can twist this function f and obtain

$$f_\chi(\tau) = \sum_{n \in \mathbb{Z}} \chi(n) a_n q^{n/N}. \tag{2.3.13}$$

Then if f is a modular form for $\Gamma(N)$, f_χ will be a modular form of the same weight for some $\Gamma(M)$. It isn't very deep that modularity should be preserved – see Question 2.3.4 for one such argument. Theorem 14 in [456] provides a generalisation. The Dirichlet twist takes on a clear algebraic significance in the context of automorphic representations (Section 2.4.1).

A deeper use of Dirichlet twists is *Weil's Converse Theorem* (see e.g. theorem 17 of [456] or page 64 of [90]), which characterises modular forms for $\Gamma(N)$ by generalising

Theorem 2.3.1, using infinitely many Dirichlet twists. It is a 'converse' in that it gener-alises the converse of (i) $\Rightarrow$ (ii). Applications of this are given in sections 1.9 and 1.10 of [89].

A more surprising example of twisting is by Galois automorphisms. Let F_N be the space (in fact field) of all modular functions for $\Gamma(N)$, with q-expansion as in (2.3.12), where each coefficient a_i lies in the cyclotomic field $\mathbb{Q}[\xi_N]$ (recall Section 1.7.3). This field F_N is explicitly constructed in section 6.2 of [505]. Clearly, $j(\tau)$ lies in each F_N. It can be shown that F_N is a Galois extension over $\mathbb{Q}(j)$, with Galois group

$$\operatorname{Gal}(F_N/\mathbb{Q}(j(\tau))) \cong \operatorname{GL}_2(\mathbb{Z}_N)/\{\pm 1\} \qquad (2.3.14a)$$

(see Section 1.7.2 for definitions). For any matrix $\overline{A} \in \operatorname{GL}_2(\mathbb{Z}_N)$, we can find an integer $\ell \in \mathbb{Z}_N^\times$ (namely $\ell = \det(A)$) and a matrix $B = \begin{pmatrix} a & b \\ c & d \end{pmatrix} \in \operatorname{SL}_2(\mathbb{Z})$ such that $\overline{A} = B \begin{pmatrix} 1 & 0 \\ 0 & \ell \end{pmatrix} \pmod{N}$. Then the action of $\overline{A} \in \operatorname{GL}_2(\mathbb{Z}_N)$ on a modular function $f(\tau)$ is given by

$$\overline{A}.f(\tau) = (\sigma_\ell f)\left(\frac{a\tau + c}{b\tau + d}\right), \qquad (2.3.14b)$$

$$\sigma_\ell \sum_{n\in\mathbb{Z}} a_n q^{n/N} = \sum_{n\in\mathbb{Z}} \sigma_\ell(a_n) q^{n/N}, \qquad (2.3.14c)$$

where $\sigma_\ell \in \operatorname{Gal}(\mathbb{Q}[\xi_N]/\mathbb{Q})$ sends ξ_N to ξ_N^ℓ. This Galois action plays a technical but important role in both Moonshine (e.g. Question 7.3.3) and rational conformal field theory (e.g. Section 6.1); see Section 6.3.3 for some speculation.

2.3.4 The remarkable heat kernel

Various topological proofs of modularity, inspired by conformal field theory, have arisen in recent years. For instance [24], [203] and section 6 of [502] all provide proofs for $\eta(\tau)$. These suggest the thought that, more generally, modularity – hence Moonshine – may be a topological effect (Section 7.2.4). The oldest and perhaps most fundamental observation along these lines is the relation between theta function modularity and the heat kernel.

Fourier determined that the rate of flow of heat energy in a material is proportional to the gradient of the temperature, and thus wrote down the *diffusion* or *heat equation*, which in one dimension looks like

$$\frac{\partial}{\partial t} u(t, x) = \frac{1}{4\pi} \frac{\partial^2}{\partial x^2} u(t, x), \qquad \forall x \in \mathbb{R}, \forall t > 0 \qquad (2.3.15a)$$

(the harmless normalisation $1/4\pi$ is introduced for later convenience). Suppose that the initial distribution of heat in the infinite rod is $f(x) = \lim_{t\to 0} u(t, x)$. Then Fourier analysis tells us how to find a solution $u(t, x)$ for all times t. Letting

$$\widehat{u}(t, \alpha) = \frac{1}{2\pi} \int_{-\infty}^{\infty} u(t, y) e^{-i\alpha y} \, dy, \qquad \widehat{f}(\alpha) = \frac{1}{2\pi} \int_{-\infty}^{\infty} f(y) e^{-i\alpha y} \, dy,$$

the equation to be solved has been transformed to $\partial \widehat{u}/\partial t = -\alpha^2 \widehat{u}/4\pi$, with initial condition $\widehat{f}$, which has the solution $\widehat{u}(t, \alpha) = \widehat{f}(\alpha) e^{-\alpha^2 t/4\pi}$. We can now find u by using the inverse transform:

$$u(t, x) = \frac{1}{2\pi} \int_{-\infty}^{\infty} e^{i\alpha x - \alpha^2 t/4\pi} \int_{-\infty}^{\infty} f(y) e^{-i\alpha y} \, dy \, d\alpha.$$

But

$$\frac{1}{2\pi} \int_{-\infty}^{\infty} e^{i\alpha z - \alpha^2 t/4\pi} \, d\alpha = t^{-1/2} e^{-\pi z^2/t} =: K(t, z).$$

Thus $u(t, x)$ is given by the convolution

$$u(t, x) = \int_{-\infty}^{\infty} K(t, x - y) f(y) \, dy. \tag{2.3.15b}$$

We see that $K(t, x)$ is itself a solution to the heat equation, with initial condition $f(x) = \delta(x)$, the Dirac delta. Physically, K corresponds to an infinitely hot spot placed at position $x = 0$ at time $t = 0$, on an otherwise uniform, infinitely long rod. This fundamental solution $K(t, x)$ is called the *heat kernel* or *propagator* for $\mathbb{R}$.

What has this to do with the theta function? Consider the specialisation $\theta_3(it, x)$, where $t, x \in \mathbb{R}$, $t > 0$. Note that

$$\frac{\partial}{\partial t} \theta_3(it, x) = \frac{1}{4\pi} \frac{\partial^2}{\partial x^2} \theta(it, x),$$

so θ_3 is a solution to the heat equation. Also, in the $t \to 0$ limit, $\theta_3(0, x)$ becomes the distribution $\sum_{n=-\infty}^{\infty} \delta(x - n)$ (this is proved by evaluating $\lim_{t\to 0} \int_0^1 \theta_3(it, x) f(x) \, dx$, but is merely the statement that $\sum_n e^{2\pi imx} = \sum_m \delta(x - m)$). Thus θ_3 plays the same role on the circle $\mathbb{R}/\mathbb{Z}$ that $K(t, x)$ played on the line $\mathbb{R}$: θ_3 is the heat kernel for the circle. But we can obtain this kernel in another way, by averaging the heat kernel $K(t, x)$ for $\mathbb{R}$:

$$\sum_{n=-\infty}^{\infty} t^{-1/2} e^{-\pi (x-n)^2/t} = t^{-1/2} e^{-\pi x^2/t} \theta_3 \left(\frac{i}{t}, \frac{x}{it} \right).$$

Equating this to $\theta_3(it, x)$ recovers (2.3.15b).

As with Poisson summation, the notion of heat kernel can be generalised considerably. For example, let M be a compact n-dimensional Riemannian manifold and let Δ be the Laplacian. In local coordinates,

$$\Delta(x) = -\sum_{i,j=1}^{n} g^{ij}(x) \frac{\partial^2}{\partial x^i \partial x^j},$$

where $g^{ij}(x)$ is the metric. The heat equation on M is

$$\frac{\partial}{\partial t} u(t, x) = -\Delta u(t, x), \quad x \in M, \ t > 0,$$

with initial condition $f(x) = \lim_{t\to 0} u(t, x)$. This can be solved formally by the expression $u(t, x) = e^{-t\Delta} f(x)$. In fact $e^{-t\Delta}$ makes sense as an operator on $L^2(M)$, for any $t \in \mathbb{C}$

with $\text{Re}(t) > 0$. By the heat kernel $K(t, x, y)$ for M we mean as before the solution to the heat equation with initial condition $\delta(x, y)$, or equivalently $K(t, x, y)$ generates the solution $(e^{-t\Delta} f)(x) = \int_M K(t, x, y) f(y) \, dy$ to the heat equation with arbitrary initial condition f. The heat kernel always exists and is unique, and is analytic for $t > 0$. In fact the heat kernel can be expressed as

$$K(t, x, y) = \sum_n e^{-\lambda_n t} \phi_n(x) \overline{\phi_n(y)},$$

where $\lambda_n \geq 0$ are the (discrete) eigenvalues of the Laplacian Δ with (orthonormal) eigenfunctions $\phi_n \in C^\infty(M) \subset L^2(M)$. Incidentally, K is the kernel of the operator $e^{-t\Delta}$ in the sense of the Schwartz kernel theorem. For t small,

$$K(t, x, y) = (4\pi t)^{-n/2} e^{-d(x,y)^2/4t} \sum_{i=0}^\infty t^i f_i(x, y)$$

where $d(x, y)$ is the distance between $x, y \in M$, and f_i are certain functions. In the language of quantum field theory, the heat kernel $K(t, x, y)$ equals $\langle x | e^{-t\Delta} | y \rangle$. The heat kernel stores geometric information on M, and interpolates between the identity operator of $L^2(M)$ at $t = 0$ and the projection onto the kernel of Δ as $t \to \infty$.

For example, for $M = \mathbb{R}^n$ the heat kernel is $K(t, x, y) = (4\pi t)^{-n/2} \exp[-|x - y|^2/4t]$, so for any n-dimensional lattice $L \subset \mathbb{R}^n$ the heat kernel of the n-torus $\mathbb{R}^n/L$ is

$$(4\pi t)^{-n/2} \sum_{v \in L} \exp[-|x - y - v|^2/4t].$$

But it also equals (normalising the arguments appropriately) $\frac{1}{\sqrt{|L|}} \Theta_{L^*}$, and so we recover the modularity of (2.3.7).

The natural generalisation of the $M = \mathbb{R}^n$ calculation is performed by [**231**]. In particular, let G be a connected, noncompact reductive Lie group, let K be a maximal compact subgroup, and let Γ be a discrete subgroup of G such that the quotient $\Gamma \backslash G$ is compact. Then two expressions for the heat kernel, and its trace, on the space $\Gamma \backslash G / K$ are obtained. In the special case of $G = \mathbb{R}^n$ and Γ being a lattice, the trace formula reduces to the usual formula expressing $\Theta_L(-1/\tau)$. The naturality of this construction $\Gamma \backslash G / K$ will be clear after reading Section 2.4.1. Moreover, [**181**] proves the Macdonald identities using the heat equation on compact Lie groups.

Further generalisations are possible (see e.g. [**52**]). For example, degree 1 and 0 terms can be added to the Laplacian Δ, and we can consider more generally differential operators on sections of line bundles over M, rather than on M. Heat kernel techniques can be used to prove various formulations of the Atiyah–Singer Index Theorem, and equivariant analogues of the theory yield the Atiyah–Bott fixed-point theorem. The strategy typically followed by these applications is to consider the integral $I(t) = \int_M K(t, f(y), x) \, dy$ for some map $f : M \to N$, where K is the heat kernel on N. The $t \to 0$ limit collapses the integral to an integral or sum over $f^{-1}(x)$. But a global expression for $I(t)$ can often be found, for example using representation theory or geometry; taking its $t \to 0$ limit yields an identity between the local integral $\int_{f^{-1}(x)}$ and some global data of M and N. See, for example, [389]. Question 2.1.7 is essentially an example of this strategy – what we call $K(g, h)$ there is the heat kernel at $t = 0$ of the finite group G.

Some of the many applications and occurrences of the heat kernel are collected in [320]. But can the heat kernel be directly relevant to Moonshine? This seems very possible. After all, the Atiyah–Bott fixed-point theorem yields an elegant proof of the Weyl character formula for compact Lie groups. In the conformal field theories associated with Lie groups (namely, the Wess–Zumino–Witten models), the heat kernel is used to explicitly construct the flat Knizhnik–Zamolodchikov connection on spaces of chiral blocks [288] (more on this starting in Section 3.2.4). This is significant because, according to conformal field theory, it is the monodromy of the Knizhnik–Zamolodchikov equation that is responsible (in genus 1) for the modularity of the affine algebra characters.

To this author's knowledge, heat kernel methods have never been used directly in the context of Monstrous Moonshine, but surely they can be used to prove at minimum the modularity of the McKay–Thompson series, and to help us understand a little better the geometry of Monstrous Moonshine. It seems possible that equivariant heat kernel methods could provide a geometric umbrella under which herd the more interesting examples of Moonshine.

2.3.5 Siegel forms

Vaughn Jones considered how one von Neumann algebra can be embedded in another (e.g. itself), and the result – subfactor theory – is profoundly interesting. This success suggests the following analogue of Galois theory:

The Jones Programme *Study the ways in which one infinite beast can be embedded in another.*

Let's probe this thought with the simplest infinite beast this author can think of: lattices (Section 1.2.1). Let $L \subset \mathbb{R}^n$, $L' \subset \mathbb{R}^{n'}$ be lattices of dimension n and n', respectively. Fix bases $\{x^{(1)}, \ldots, x^{(n)}\}$, $\{y^{(1)}, \ldots, y^{(n')}\}$ and construct the $n \times n$ matrix M, whose columns are the $x^{(i)}$. An embedding of L' into L is a linear map $\varphi : L' \to L$ that preserves all inner-products. It is determined by the values $\varphi(y^{(j)}) = \sum_i \varphi_{ji} x^{(i)}$. The coefficients φ_{ji} all lie in $\mathbb{Z}$ and form an $n' \times n$ matrix (φ). Now, φ preserves all inner-products, iff $\varphi(y^{(i)}) \cdot \varphi(y^{(j)}) = y^{(i)} \cdot y^{(j)} \ \forall i, j$, iff

$$(\varphi) M^t M (\varphi)^t = M^t M. \tag{2.3.16}$$

Let $N(L', L)$ be the number of these embeddings, i.e. the number of $n' \times n$ $\mathbb{Z}$-matrices (φ) satisfying (2.3.16). This number will be 0 unless $n' \leq n$.

For example, $N(\mathbb{Z}, L)$ equals the number of unit vectors in L. Thus, if L is integral, the generating function $\sum_{k=0}^{\infty} N(\sqrt{k}\mathbb{Z}, L) x^k$ is the theta function $\Theta_L(\tau)$, for $x = e^{\pi i \tau}$. We might hope that the numbers $N(L', L)$ are coefficients of some other modular-like function.

Construct a multi-variable generating function as follows. Fix an n-dimensional integral lattice L. Let x_{ij}, $1 \le i, j \le n$, be variables. Consider

$$\mathrm{Th}_L(x_{ij}) := \sum_{n'=0}^{n} \sum_{[L']} \sum_{\mathbb{Z}\{\beta_1,\dots,\beta_n\}=L'} \frac{N(L',L)}{\mathrm{Aut}(L')} \prod_{1 \le i,j \le n} x_{ij}^{\beta_i \cdot \beta_j}. \tag{2.3.17a}$$

The sum over $[L']$ is of all isomorphism classes of n'-dimensional even lattices. For each of these classes, fix a representative $L' \subset \mathbb{R}^n$. The $\{\beta_i\}$ run over all possible ordered n-tuples of lattice vectors that span L'. There is an equivalent but cleaner way to write (2.3.17a). Let $\mathcal{A}_n$ be the set of all $n \times n$ positive semidefinite matrices A with integer entries and even integers down the diagonal. These are precisely the matrices $A_{ij} = \beta_i \cdot \beta_j$. Then

$$\mathrm{Th}_L(x_{ij}) = \sum_{A' \in \mathcal{A}_n} N(L',L) \prod_{1 \le i,j \le n} x_{ij}^{A'_{ij}}, \tag{2.3.17b}$$

where L' is any lattice realising the matrix A' of inner-products.

In any case, this generating function Th_L, after making the change-of-variables $x_{ij} = e^{\pi i T_{ij}}$, is a *Siegel modular form*! We return to it shortly.

Let's try to find a version of modular forms where $\mathbb{H}$ is replaced by a higher-dimensional space. Start with $\Theta_L(\tau, z)$ in equation (2.3.7), but reinterpret this as a function of the complex matrix $T := \tau A$, with entries $A_{ij} = b^{(i)} \cdot b^{(j)}$ for a basis $b^{(i)}$ of the lattice L. We thus get

$$\Theta(T, z) := \sum_{n \in \mathbb{Z}^n} \exp[\pi i n \cdot T n + 2 \pi i n \cdot z]. \tag{2.3.18}$$

How far can we extend the domain T? We may as well restrict to symmetric matrices T. For which symmetric matrices T does (2.3.18) converge to a holomorphic function? We know from (2.3.7) that it does whenever $T = xA + iA$ for any positive-definite matrix A and real number x, but there is no need to restrict to such T. Indeed, it is straightforward to obtain that (2.3.18) converges to a holomorphic function for any $z \in \mathbb{C}^n$ and any T in the Siegel upper half-space $\mathbb{H}_n$ defined in Section 2.1.4.

Of course, (2.3.18) is quasi-periodic in the z variable:

$$\Theta(T, z + m) = \Theta(T, z), \qquad \forall m \in \mathbb{Z}^n \tag{2.3.19a}$$

$$\Theta(T, z + Tm) = \exp[-\pi i m \cdot T m - 2 \pi i m \cdot z] \, \Theta(T, z), \qquad \forall m \in \mathbb{Z}^n. \tag{2.3.19b}$$

The Siegel theta function $\Theta(T, z)$ is an easy generalisation of the Jacobi theta function (2.3.7). What makes it so remarkable is its symmetries as a function of T:

$$\Theta((AT + B)(CT + D)^{-1}, (CT + D)^{t-1}z)$$
$$= \xi_\gamma \det(CT + D)^{\frac{1}{2}} \exp[\pi i z \cdot (CT + D)^{-1}z] \, \Theta(T, z) \tag{2.3.20}$$

for all $\gamma = \begin{pmatrix} A & B \\ C & D \end{pmatrix} \in \mathrm{Sp}_{2n}(\mathbb{Z})$ for which all diagonal entries of $A^t C$ and $B^t D$ are even. Call this subgroup Γ_θ^n, in analogy with (2.2.5). The numbers $\xi_\gamma \in \mathbb{C}$ are certain eighth roots of unity.

We defined $\text{Sp}_{2n}(\mathbb{Z})$ in Section 2.1.4. The modularity of $\Theta(T, z)$ is proved much the way that modularity of θ_3 was proved. The analogue of (2.2.1a) is

$$\text{Sp}_{2n}(\mathbb{Z}) = \left\langle \begin{pmatrix} I & A \\ 0 & I \end{pmatrix}, \begin{pmatrix} B & 0 \\ 0 & B^{t-1} \end{pmatrix}, \begin{pmatrix} 0 & -I \\ I & 0 \end{pmatrix} \mid \forall A \in M_{n \times n}(\mathbb{Z}), \right.$$

$$\left. A = A^t, \forall B \in \text{GL}_n(\mathbb{Z}) \right\rangle. \tag{2.3.21}$$

If we insist the matrices A in (2.3.21) have even diagonals, then we generate Γ_θ^n. Verifying invariance of $\Theta(T, z)$ under $\begin{pmatrix} I & A \\ 0 & I \end{pmatrix}$ and $\begin{pmatrix} B & 0 \\ 0 & B^{t-1} \end{pmatrix}$ is routine; use Poisson summation for $\begin{pmatrix} 0 & -I \\ I & 0 \end{pmatrix}$. The argument is given in detail in chapter 2.5 of [**439**].

Section 2.1.4 relates $\mathbb{H}_n$ to Riemann surfaces of genus n. As we recall, the possible period matrices Ω of a given surface form an $\text{Sp}_{2n}(\mathbb{Z})$-orbit in $\mathbb{H}_n$. The Jacobian of the surface is $\mathbb{C}^n/(\mathbb{Z}^n + \Omega\mathbb{Z}^n)$. Quasi-periodicity (2.3.19) embeds these Jacobians into projective space. Most points in $\mathbb{H}_n$ (at least for $n > 2$) aren't period matrices of surfaces, and as we recall the moduli space $\mathfrak{M}_{n,0}$ can be identified with $\mathfrak{C}_n/\text{Sp}_{2n}(\mathbb{Z})$ for some subset $\mathfrak{C}_n$ in $\mathbb{H}_n$.

We should thus regard $\Theta(T, z)$, $\text{Sp}_{2n}(\mathbb{Z})$ and $\mathbb{H}_n$ as the genus n versions of θ_3, $\text{SL}_2(\mathbb{Z}) \cong \text{Sp}_2(\mathbb{Z})$ and $\mathbb{H}$, where τ becomes an $n \times n$ matrix. The hyperbolic geometry of $\mathbb{H}$ becomes symplectic geometry on $\mathbb{H}_n$ (see e.g. section 4 of [**395**]). As mentioned in footnote 1 of this chapter, the future will find Moonshine expanding into higher genus. The calculations will be far more complicated, and this is presumably the reason for the delay. One of the only explicit works in this direction is [**533**], which looks at the lattice $\leftrightarrow$ theta function example of Figure 0.1 (or equivalently the bosonic string compactified on a torus) at genus 2. As expected, Siegel modular forms play a dominant role. See also [**9**] for some calculations with multi-loop heterotic strings, which heavily involve Siegel theta functions.

Definition 2.3.3 *Let* $\Gamma \subset \text{Sp}_{2n}(\mathbb{Z})$ $(n > 1)$ *have finite index. Then a* Siegel modular form *of weight k and level Γ is a holomorphic function f on $\mathbb{H}_n$ such that*

$$f((AT + B)(CT + D)^{-1}) = \det(CT + D)^k f(T), \qquad \forall \begin{pmatrix} A & B \\ C & D \end{pmatrix} \in \Gamma.$$

A growth condition at the cusps (requiring holomorphicity) is automatically satisfied when $n > 1$. Another simplification of higher genus is that any subgroup $\Gamma \subset \text{Sp}_{2n}(\mathbb{Z})$ of finite index includes some congruence group $\Gamma^n(N) := \{A \in \text{Sp}_{2n}(\mathbb{Z}) \mid A \equiv I \pmod{N}\}$ with finite index.

For example, $\Theta(T, z)^2$ is a modular form of weight 1 and level $\Gamma^n(4)$. Eisenstein series for $\text{Sp}_{2n}(\mathbb{Z})$ can be defined in the obvious way, as a sum of $\det(CT + D)^{-2k}$ over appropriately defined pairs $\{C, D\}$ of matrices (see e.g. section 14 of [**395**] for details). A final example plays the same role for $\Theta(T, z)$ that $\Theta_L(\tau)$ played for $\theta_3(\tau)$: let L be any m-dimensional rational lattice and let A be its Gram matrix, then

$$\Theta_L(T, Z) := \sum_N \exp[\pi i \, \text{tr}(N^t T N A) + 2\pi i \text{tr}(N^t Z)],$$

where $T \in \mathbb{H}_n$, Z is an $n \times m$ complex matrix, and the sum is over all $n \times m$ $\mathbb{Z}$-matrices. This is a specialisation of Θ for $Sp_{2nm}(\mathbb{Z})$, and is a Siegel modular form of weight $m/2$ for some $\Gamma^n(M)$ (see e.g. chapter 2.6 of [**439**]). We met Θ_L in (2.3.17).

Finally, let us describe the analogue of Fourier expansion here. For convenience take Γ to be $Sp_{2n}(\mathbb{Z})$. Then a modular form f for Γ obeys the periodicity $f(T+B) = f(T)$ for all $n \times n$ $\mathbb{Z}$-matrices B. Together with holomorphicity, this means f has an expansion

$$f(T) = \sum_{M \geq 0} a(M) \exp[2\pi \mathrm{i}\, \mathrm{tr}(TM)], \qquad (2.3.22)$$

where the sum is over all positive-semidefinite symmetric $n \times n$ matrices M with entries $M_{ii} \in \mathbb{Z}$ and $M_{ij} \in \frac{1}{2}\mathbb{Z}$. These numbers $a(M)$ play the role of Fourier coefficients here. For example, (2.3.17b) gives the Fourier expansion of $\Theta_L(T)$.

Question 2.3.1. Prove that the Weierstrass $\wp$ function (2.1.6a) is a Jacobi form for $SL_2(\mathbb{Z})$ with weight $k = 2$ and index $m = 1$.

Question 2.3.2. Let L, L' be two n-dimensional rational lattices in $\mathbb{R}^n$, and let $u, u' \in \mathbb{R}^n$ be vectors of finite order for L and L', respectively.
(a) *Prove:* If $\Theta_{L+u}(\tau, z) = \Theta_{L'+u'}(\tau, z)$ for all $\tau \in \mathbb{H}$, $z \in \mathbb{C}^n$, then $L + u = L' + u'$ as sets.
(b) Prove that L and L' are isomorphic (Section 1.2.1) iff there exists an orthogonal map $T \in O_n(\mathbb{R})$ such that $\Theta_L(\tau, z) = \Theta_{L'}(\tau, Tz)$ for all $\tau \in \mathbb{H}$, $z \in \mathbb{C}^n$.

Question 2.3.3. Let L be any integral lattice of dimension n. For each $m = 0, 1, 2, \ldots$, let $L_{(m)}$ denote all the vectors $u \in L$ with norm-squared $u \cdot u = m$. Each automorphism ω of L permutes the vectors in $L_{(m)}$, so for each m we get a $\|L_{(m)}\|$-dimensional representation $\alpha_{(m)}$ of $\mathrm{Aut}(L)$ by permutation matrices. Thus, for each $\omega \in \mathrm{Aut}(L)$, we can *twist* Θ_L as follows: define

$$\Theta_L^{(\omega)}(\tau) := \sum_{m=0}^{\infty} \chi_{(m)}(\omega) \exp[\pi \mathrm{i}\tau m],$$

where $\chi_{(m)}$ is the character of the representation $\alpha_{(m)}$. For example, $\Theta_L^{(id)} = \Theta_L$ and $\Theta_L^{(-id)}(\tau) = 1$. Prove that, for each $\omega \in \mathrm{Aut}(L)$, $\Theta_L^{(\omega)}$ will be a modular form for some $\Gamma(N)$ and some weight $0 \leq k \leq n/2$, and that $k = n/2$ iff $\omega = id$.

Question 2.3.4. Let f be a modular form of weight k, for some $\Gamma(N)$.
(a) Prove that, for each choice of $r \in \mathbb{Q}$, the function $g(\tau) := f(\tau + r)$ is a modular form of level k, for some $\Gamma(M)$ (M depending on r).
(b) For any field $\mathbb{F}$, prove that $SL_2(\mathbb{F})$ is generated by the matrices $\begin{pmatrix} 1 & r \\ 0 & 1 \end{pmatrix}$, for $r \in \mathbb{F}$, together with $\begin{pmatrix} 0 & -1 \\ 1 & 0 \end{pmatrix}$. From this, prove that if f is a modular form for $\Gamma(N)$ of weight k, then for any $\begin{pmatrix} a & b \\ c & d \end{pmatrix} \in SL_2(\mathbb{Q})$, the function $h(\tau) := f(\frac{a\tau+b}{c\tau+d})$ will be a modular form of weight k for some $\Gamma(M)$ (M depending on a, b, c, d).

2.4 Representations and modular forms

According to I. M. Gel'fand, mathematics of any kind is representation theory.[8]
This section applies this beautiful strategy to modular forms.

There are at least formal similarities between quantum theory and modular forms. Wigner taught that a particle should be identified with a unitary representation of $SL_2(\mathbb{C})$ or $SL_2(\mathbb{R})$, in $(3+1)$- or $(2+1)$-dimensional space-time, respectively. In this section we associate modular forms to unitary representations of $SL_2(\mathbb{R})$, and the picture generalises naturally to, for example, $SL_2(\mathbb{C})$. Could there be some cross-fertilisation between the methods and ideas of quantum field theory and modular forms?

In the 1962 International Congress of Mathematicians, I. M. Gel'fand remarked somewhat cryptically that there is an intriguing analogy between the scattering matrix of quantum mechanics and zeta functions. Ten years later the idea was exploited and clarified by Faddeev and Pavlov, who applied the Lax–Phillips scattering theory to the theory of automorphic forms. For example, poles of the scattering matrix (which in quantum field theory would correspond to particles) correspond to zeros of the Riemann zeta function. Their work is generalised in [**371**], where we find for instance a new proof of the Selberg Trace Formula for SL_2. These applications are significant, and hopefully a small hint of things to come. See also [**562**].

2.4.1 Automorphic forms

Definitions 0.1 and 2.2.1 of modular functions and forms for $SL_2(\mathbb{Z})$ should seem very arbitrary. In mathematics we attack arbitrariness through generalisation. A good generalisation helps us to see the meaning of each feature, and puts the whole theory into a broader perspective. Of course we can generalise these definitions by replacing $SL_2(\mathbb{Z})$ with other Fuchsian groups $\Gamma < SL_2(\mathbb{R})$, but this is too obvious to be helpful.

Much more valuable is to understand the relation between $\mathbb{H}$ and $G = SL_2(\mathbb{R})$. In particular, an easy calculation shows that our action of G on $\mathbb{H}$ is *transitive*. That is, any point in $\mathbb{H}$ can get mapped to any other point in $\mathbb{H}$ by a matrix in G. In particular, $\gamma_{x+\mathrm{i}y} = \begin{pmatrix} \sqrt{y} & x/\sqrt{y} \\ 0 & 1/\sqrt{y} \end{pmatrix} \in G$ sends i to $x + \mathrm{i}y$. We call $\mathbb{H}$ a *homogeneous space* for G. Moreover, the subgroup of G fixing i $\in \mathbb{H}$, say, is $K = SO_2(\mathbb{R})$. Thus

$$\mathbb{H} \cong SL_2(\mathbb{R})/SO_2(\mathbb{R}) = G/K. \qquad (2.4.1a)$$

More precisely, we have the Iwasawa decomposition

$$\begin{pmatrix} a & b \\ c & d \end{pmatrix} = y^{-1/2} \begin{pmatrix} y & x \\ 0 & 1 \end{pmatrix} \begin{pmatrix} \cos\theta & \sin\theta \\ -\sin\theta & \cos\theta \end{pmatrix}, \qquad (2.4.1b)$$

$$x + \mathrm{i}y = \frac{a\mathrm{i} + b}{c\mathrm{i} + d}, \quad e^{\mathrm{i}\theta} = \frac{d - \mathrm{i}c}{|d - \mathrm{i}c|}. \qquad (2.4.1c)$$

In fact $SO_2(\mathbb{R})$ is the unique (up to conjugation) maximal compact subgroup of G.

---

[8] See the quotation on page 840 of *Proc. ICM* (American Mathematical Society, Providence 1987), edited by A. M. Gleason.

In mathematics we try to find hidden structure, and that is the spirit in which (2.4.1a) should be read. The key here was the transitive action: an expression like (2.4.1a) arises whenever one has a homogeneous space. Note that the action $\gamma.\tau$ of G on $\mathbb{H}$ now reduces to matrix multiplication: $\gamma\gamma_\tau K$.

Do modular forms respect (2.4.1a)? Can we lift modular forms $f : \mathbb{H} \to \mathbb{C}$ into functions $\phi_f : G \to \mathbb{C}$? Yes, and in fact we gain something in the process. Use (2.4.1b):

$$\phi_f \begin{pmatrix} a & b \\ c & d \end{pmatrix} = f\left(\frac{ai+b}{ci+d}\right)(ci+d)^{-k} = f(x+iy)\,y^{k/2}\,e^{i\theta k}, \qquad (2.4.2a)$$

where k is the weight of f. Then for any $A \in \mathrm{SL}_2(\mathbb{Z})$ and $\alpha \in \mathbb{R}$, we get

$$\phi_f\left(A\begin{pmatrix} a & b \\ c & d \end{pmatrix}\begin{pmatrix} \cos\alpha & \sin\alpha \\ -\sin\alpha & \cos\alpha \end{pmatrix}\right) = \phi_f\begin{pmatrix} a & b \\ c & d \end{pmatrix}e^{-ik\alpha}. \qquad (2.4.2b)$$

The point of multiplication by $(ci+d)^{-k}$ is now clear: it makes ϕ_f left-invariant with respect to $\mathrm{SL}_2(\mathbb{Z}) = \Gamma$. Thus we've sacrificed K-invariance and Γ-covariance, for K-covariance and Γ-invariance. This is significant, because compact Lie groups like K are much easier to handle than infinite discrete groups like $\mathrm{SL}_2(\mathbb{Z})$.

In particular, we find that the right multiplication in (2.4.2b) defines a one-dimensional representation of K on $\mathbb{C}\phi_f$. We know that the finite-dimensional irreducible K-representations are parametrised by a nonnegative integer, and all are one-dimensional. Thus we get an algebraic interpretation for the parameter k in Definition 2.2.1: it is the highest weight of a representation of the maximal compact subgroup $\mathrm{SO}_2(\mathbb{R})$ of $\mathrm{SL}_2(\mathbb{R})$.

We also get a representation of $\mathrm{SL}_2(\mathbb{R})$ on the left side, given by $\phi_f \mapsto \phi_f \circ \gamma^{-1}$. The vector space here is the infinite-dimensional function space given by the $\mathbb{C}$-span of the $\mathrm{SL}_2(\mathbb{R})$-orbit of ϕ_f. The result is an irreducible representation of $\mathrm{SL}_2(\mathbb{R})$, which is constant on $\Gamma = \mathrm{SL}_2(\mathbb{Z})$. This representation is unitary – in fact it is a subrepresentation of the regular representation of G on the Hilbert space $L^2(\Gamma\backslash G)$.

As an aside, note that everything generalises very naturally to Siegel modular forms. There, G is $\mathrm{Sp}_{2n}(\mathbb{R})$, Γ is $\mathrm{Sp}_{2n}(\mathbb{Z})$ or a similar discrete group like Γ_θ^n, and $K = \mathrm{SO}_{2n}(\mathbb{R}) \cap \mathrm{Sp}_{2n}(\mathbb{R}) \cong U_n(\mathbb{C})$. Once again, $\mathbb{H}_n \cong G/K$. For Jacobi forms, G is a semi-direct product of $\mathrm{SL}_2(\mathbb{R})$ with the Heisenberg group (it is constructed next subsection), and K is $\mathrm{SO}_2(\mathbb{R}) \times S^1$: once again $G/K \cong \mathbb{H} \times \mathbb{C}$, as it should. The weight k and index m in Definition 2.3.2 parametrise the irreducible one-dimensional representations of $\mathrm{SO}_2(\mathbb{R})$ and S^1, that is to say K. Thus the index of a Jacobi form has a natural algebraic interpretation, as it should.

So the generalisation of modular forms and functions is starting to be clearer. We are looking for functions on the space $\Gamma\backslash G$, for discrete subgroups Γ of real Lie groups G, and we should study them via the representation of G they generate. The relation between modular forms and representation theory was accomplished in the 1950s by Gel'fand and Fomin. Let's make it more precise.

The unitary irreducible representations of $G = \mathrm{SL}_2(\mathbb{R})$ were classified by Bargmann [44]. His motivation was physics (the Lorentz group). Of course there is the one-dimensional identity representation. The remaining irreducible unitary representations are all infinite-dimensional, and fall into three series: the principal series $\mathcal{P}_s^\pm$ for $s \in \mathbb{R}$,

the complementary series $\mathcal{C}_s$ for $0 < s < 1$, and the discrete series $\mathcal{D}_n^{\pm}$ for $n = 2, 3, \ldots$
In addition, G has many irreducible non-unitary representations. See, for example, chapter 1.3 of [**243**] for explicit realisations of all the unitary representations. For example, the discrete series $\mathcal{D}_n^+$ consists of holomorphic functions f on $\mathbb{H}$, with Peterssen Hermitian form $\langle f, g \rangle = \int_{\mathbb{H}} f(\tau)\overline{g(\tau)} y^{n-1} \mathrm{d}x \, \mathrm{d}y$, and action $f \mapsto (-c\tau + a)^{-n} f \left(\frac{d\tau - b}{-c\tau + a} \right)$.
Obviously our G-representation associated with Φ_f is isomorphic to $\mathcal{D}_k^+$. What f's come from the other G-representations?

Associated with the principal series are functions such as this analogue of the Eisenstein series, called a *Maass form*:

$$E(\tau, s) = \sum_{m,n \in \mathbb{Z}} ' \frac{y^s}{|m\tau + n|^{2s}}, \ s \in \mathbb{C}.$$

This may look less strange when one considers the formula $\mathrm{Im}(\gamma.\tau) = y/|c\tau + d|^2$.
For fixed $\tau \in \mathbb{H}$, the Maass form is absolutely convergent for $\mathrm{Re}(s) > 1$ and has a meromorphic extension to all $s \in \mathbb{C}$. For fixed $s \in \mathbb{C}$, it is invariant under $\mathrm{SL}_2(\mathbb{Z})$. It is not a holomorphic function of τ, and so cannot be a modular form in the usual sense, but holomorphicity in Definitions 0.1 and 2.2.1 is a feature we must be prepared to lose, since most real Lie groups G aren't complex manifolds. In fact we lost the holomorphicity of f when we wrote (2.4.2a). What takes its place?

What is holomorphicity, other than the solution to differential equations (the Cauchy–Riemann equations, or the Laplacian $\frac{\partial^2}{\partial x^2} + \frac{\partial^2}{\partial y^2}$ on $\mathbb{R}^2$)? The Maass forms aren't holomorphic, but they are eigenfunctions of the Laplacian on $\mathbb{H}$, namely $-y^2 \left(\frac{\partial^2}{\partial x^2} + \frac{\partial^2}{\partial y^2} \right)$. By the Laplacian on $\mathbb{H}$ we mean a second-order differential operator that is invariant under all isometries $\mathrm{SL}_2(\mathbb{R})$.

We are thus led to the role of differential operators. These can be understood as follows. Whenever we have a Lie group representation, we also get an associated action of the Lie algebra (the derived module of Section 1.5.5). The Lie algebra will typically act as first-order differential operators; on $L^2(G)$ it acts by Lie derivatives. More precisely, to $X \in \mathfrak{sl}_2(\mathbb{R})$ we get the action $f(g) \mapsto \frac{\mathrm{d}}{\mathrm{d}t} f(ge^{tX})|_{t=0}$. For example, $\begin{pmatrix} 0 & 1 \\ -1 & 0 \end{pmatrix} \in \mathfrak{sl}_2(\mathbb{R})$
corresponds to $\frac{\partial}{\partial \theta}$, using the parametrisation of (2.4.1a). An action of $\mathfrak{sl}_2(\mathbb{R})$ implies an action of the universal enveloping algebra $U(\mathfrak{sl}_2(\mathbb{R}))$, in our case simply by composing differential operators to get ones of higher order. As always, the centre $Z(U(\mathfrak{sl}_2(\mathbb{R})))$ naturally plays a fundamental role. Here, it is generated by the second-order operator

$$y^2 \left(\frac{\partial^2}{\partial x^2} + \frac{\partial^2}{\partial y^2} \right) - y \frac{\partial^2}{\partial x \, \partial \theta}.$$

This is how the Laplacian arises, algebraically. By definition, it commutes with all operators, so studying its eigenspaces helps decompose $L^2(\Gamma \backslash G)$ – we used a similar idea in decomposing Lie algebra modules into weight-spaces. Understanding that decomposition is essentially equivalent to understanding the space of modular forms for Γ, and can be called the harmonic analysis of automorphic forms.

We have only scratched the surface, but this discussion and the following definition should give the reader a glimpse of the resulting theory.

Definition 2.4.1   *Let Γ be a discrete subgroup of a real semi-simple Lie group G, and let K be a maximal compact subgroup of G. Let χ be a one-dimensional representation of K. We call a smooth function $f : G \to \mathbb{C}$ an* automorphic form *for Γ if:*

(i) *$f(\gamma g k) = \chi(k) f(g)$ for all $\gamma \in \Gamma, g \in G, k \in K$;*

(ii) *f is an eigenfunction of every operator in $Z(U(\mathfrak{g}))$;*

(iii) *f obeys a certain growth condition.*

The term 'automorphic form' (going back to Klein in 1890) is much older than this definition. Here, $\mathfrak{g}$ is the Lie algebra of G and $Z(U(\mathfrak{g}))$ is the centre of its universal enveloping algebra, which will be isomorphic to a polynomial algebra in r variables, where r is the rank of $\mathfrak{g}$. As mentioned above, the differential equations in (ii) take the place of holomorphicity. The growth condition is too technical to give here, but for $SL_2(\mathbb{Z})$ it reduces to holomorphicity at the cusps. For more on the relation between automorphic forms and representations, see, for example, [**89**].

All modern material on automorphic functions uses the language of *adèles* and *idèles*,[9] which unify and simplify the theory (at the expense of making it more abstract). However, since they have no role in the remaining material of this book, we only sketch their motivation, and remain true here to the spirit of this not-completely-self-contained subsection.

Projective or *inverse limits* are the way algebra 'integrates' an infinite tower of structures into a single structure. A classic – and relevant – example is divisibility by powers of primes. We say that a given integer n is divisible by p^a if the canonical projection $\mathbb{Z} \to \mathbb{Z}_{p^a}$ ('reduce mod p^a') sends n to 0. Now, the rings $\mathbb{Z}_{p^a}$ and $\mathbb{Z}_{p^b}$ are related by a homomorphism $\mathbb{Z}_{p^a} \to \mathbb{Z}_{p^b}$, provided $a \geq b$. So we get a tower

$$\cdots \to \mathbb{Z}/p^3\mathbb{Z} \to \mathbb{Z}/p^2\mathbb{Z} \to \mathbb{Z}/p\mathbb{Z} \to 0.$$

The corresponding integrated structure is the *projective limit* $\lim_{\leftarrow}\mathbb{Z}_{p^a} =: \widehat{\mathbb{Z}}_p$, the p-adic integers, which can be realised as formal power series $\sum_{a=0}^{\infty} a_n p^n$, $a_i \in \mathbb{Z}/p\mathbb{Z}$. Doing arithmetic on them amounts to treating all $\mathbb{Z}/p^a\mathbb{Z}$ simultaneously – in this sense it is the integration of all $\mathbb{Z}_{p^a}$. For example,

$$\sqrt{2} = 3 + 1 \cdot 7 + 2 \cdot 7^2 + 6 \cdot 7^3 + \cdots$$

in $\widehat{\mathbb{Z}}_7$. The p-adic rationals $\widehat{\mathbb{Q}}_p$ are the field of fractions of $\widehat{\mathbb{Z}}_p$, or equivalently the formal Laurent series $\sum_{i=-N}^{\infty} a_i p^i$, $p_i \in \mathbb{Z}/p\mathbb{Z}$. They are to the ordinary rationals much as $\mathbb{R} =: \widehat{\mathbb{Q}}_\infty$ is: a completion, on which calculus can be defined. For a readable introduction to the p-adics, see [**257**]. Projective limits play a huge role in Section 6.3.3.

The more intuitive notion of limit, namely the *injective* or *direct limit*, arises when all arrows are reversed (i.e. when we have a sequence of embeddings rather than projections),

---

[9] Idèles were introduced by Chevalley in 1935 to remove some of the analysis being used with L-functions, etc. The word comes from 'ideal'. Adèles were introduced in 1945 as an additive version of idèles.

and is the algebraic analogue of taking derivatives. The prototypical example is the space of smooth functions $F_M(U)$ on an open patch of a manifold M: the direct limit $\lim_{\to} F_M(U)$, as $U \to \{p\}$, is isomorphic to the space of germs at p.

The modern theory of automorphic forms collects together the $\widehat{\mathbb{Q}}_p$ into the additive group of adèles $\mathbb{A}$ and multiplicative group of idèles $\mathbb{A}^\times$. The adèles are defined to be the group of all sequences $(x_\infty, x_2, x_3, x_5, \ldots, x_p, \ldots)$, where $x_\infty \in \mathbb{R}$, $x_p \in \widehat{\mathbb{Q}}_p$, and for all but finitely many p, $x_p \in \widehat{\mathbb{Z}}_p$. The idèles are defined similarly, and we obtain

$$\mathbb{A}^\times \cong \mathbb{Q}^\times \times \mathbb{R}_>^\times \times \prod \widehat{\mathbb{Z}}_p^\times,$$

where $\sum_{i=0}^\infty a_i p^i \in \mathbb{Z}_p^\times$ if $a_0 \neq 0$. The rationals $\mathbb{Q}$ embed in each $\widehat{\mathbb{Q}}_p$, and so embed diagonally in $\mathbb{A}$ ($r \mapsto \widehat{\mathbb{Z}}_p$ for any prime p not dividing the denominator of r). There are many generalisations of $\mathbb{A}$ and $\mathbb{A}^\times$, for example we can replace $\mathbb{Q}$ by other number fields. But what good are they? What have they to do with modular forms?

There are many situations where the level of a modular form is variable. For example, any $A \in \mathrm{SL}_2(\mathbb{Q})$ takes a modular form for $\Gamma(N)$ to one for some other $\Gamma(N')$ (see Question 2.3.4). We have natural maps from the surface $\Gamma(n)\backslash\mathbb{H}$ to any $\Gamma(d)\backslash\mathbb{H}$, when d divides n. Collecting together this tower of surfaces $\Gamma(n)\backslash\mathbb{H}$ into a single structure amounts to taking the limit space $\widehat{\mathbb{H}} := \lim_{\leftarrow} \Gamma(n)\backslash\mathbb{H}$. Functions on $\widehat{\mathbb{H}}$ include ratios f/g of modular forms of the same weight but different levels. Much as

$$\lim_{\leftarrow} \mathbb{R}/n\mathbb{Z} \cong \mathbb{A}/\mathbb{Q}$$

as topological groups, we get

$$\widehat{\mathbb{H}} \cong \mathrm{SL}_2(\mathbb{Q})\backslash\mathrm{SL}_2(\mathbb{A})/K_\infty, \tag{2.4.3}$$

where K_∞ consists of all sequences of matrices $(A, I_2, I_2, \ldots)$ where $A \in \mathrm{SO}_2(\mathbb{R}) \subset \mathrm{SL}_2(\mathbb{R})$ and the I_2's are the identity matrices in each $\mathrm{SL}_2(\widehat{\mathbb{Q}}_p)$. In Section 4.3.3 we discover $\widehat{\mathbb{H}}$ naturally in nonperturbative string theory.

Similarly, a Dirichlet character (see Section 2.3.3) can be thought of as a continuous one-dimensional representation on $\mathbb{Q}^\times\backslash\mathbb{A}^\times$, and the Galois group of a finite abelian extension of $\mathbb{Q}$ can be thought of as a subgroup of $\mathbb{Q}^\times\backslash\mathbb{A}^\times$.

The Langlands conjectures suggest that the n-dimensional representations of the absolute Galois group $\mathrm{Gal}(\overline{\mathbb{K}}/\mathbb{K})$ of a field $\mathbb{K}$ (such as $\mathbb{Q}$) correspond to 'automorphic representations' of $\mathrm{GL}_n(\mathbb{A})$, where $\mathbb{A}$ here is the group of adèles of $\mathbb{K}$. This correspondence can be seen through the corresponding L-functions. For GL_1 and $\mathbb{K} = \mathbb{Q}$, this correspondence involves the Kronecker–Weber Theorem and Dirichlet characters. For GL_2 this relates two-dimensional representations of Galois groups to modular forms. A recent accessible introduction to the Langlands Programme is [90]. Although there are hints of some sort of relation between the Langlands conjectures and Moonshine in its more general sense, these are still too speculative to go into here. However, Section 6.3.3 may whet one's appetite.

2.4.2 Theta functions as matrix entries

The relationship between representation theory and modular forms discussed last section is quite democratic in the sense that it exists at the level of the vector space of modular forms. Democracy is all well and good, but we are not equally interested in all modular forms – some have names!

The Jacobi theta function $\theta_3(\tau, z)$ is the unique quasi-periodic entire function, in the sense that any entire function $f : \mathbb{C} \to \mathbb{C}$ obeying $f(z + 1) = f(z)$ and $f(z + \tau) = a \, e^{-2\pi i z}$ for some constants $\tau \in \mathbb{H}$ and $a \in \mathbb{C}$ is a constant multiple of the function

$$f(z) = 1 + \sum_{n=-\infty}^{\infty} e^{\pi i (n^2 - n)\tau} a^{-n} e^{2\pi i n z}.$$

For an elementary analytic proof see section 1.1 of [**439**]. From this uniqueness, all properties of θ_3 can be quickly derived. In this section we sketch a striking algebraic version of this argument.

Starting in the 1960s, theta functions were interpreted as matrix entries in a representation of the Heisenberg group. The motivation was pure Moonshine:

> A force d'habitude, le fait que les séries thêta définissent des fonctions modulaires a presque cessé de nous étonner. Mais l'apparition du groupe symplectique comme un *deus ex machina* dans les célèbres travaux de Siegel sur les formes quadratiques n'a rien perdu encore de son caractère mystérieux. Le but de ce mémoire, et de ceux qui lui feront suite, n'est pas, bien entendu, d'élucider définitivement la question, mais de jeter un peu de lumière sur certains aspects de cette théorie qui étaient restés dans l'ombre jusqu'à présent. [**555a**][10]

The resulting explanation of the transformation $\theta_3(-1/\tau) = \sqrt{\frac{\tau}{i}}\,\theta_3(\tau)$ can be extended to many other functions arising in Moonshine. First let us sketch the basic idea, before giving details and generalisations.

The starting point is the thought of realising special functions as matrix entries of Lie group representations. An elementary example of this involves the representation of $S^1 = U_1(\mathbb{R})$ as rotations in $\mathbb{R}^2$:

$$\theta \mapsto \begin{pmatrix} \cos \theta & \sin \theta \\ -\sin \theta & \cos \theta \end{pmatrix}. \tag{2.4.4}$$

The basic properties of $\sin(\theta)$ and $\cos(\theta)$ (e.g. angle-sum formulae, or even-oddness) can quickly be derived from this. We want to do something similar with θ_3.

Begin by recalling the full variable dependence of $\theta_3(\tau, z, u)$, given in (2.3.4). For fixed u we get a Jacobi form, and for fixed τ and u we get an elliptic function for the

---

[10] 'By force of habit, the fact that theta series define modular forms has nearly ceased to amaze us. But the appearance of the symplectic group as a *deus ex machina* in the famous work of Siegel on quadratic forms has still lost none of its mysterious character. The goal of this paper, and of those which follow it, is not of course to clarify definitively the question, but rather to shed a little light on certain aspects of this theory which have remained in the dark up to now.'

torus $\mathbb{C}/(\mathbb{Z} + \mathbb{Z}\tau)$. This leads us to consider two translation operators on the space of (say) entire functions $f : \mathbb{C} \to \mathbb{C}$, as follows. Fix $\tau \in \mathbb{H}$ and define

$$(S_b f)(z) = f(z + b), \tag{2.4.5a}$$

$$(T_a f)(z) = \exp[\pi i a^2 \tau + 2\pi i a z] \, f(z + a\tau), \tag{2.4.5b}$$

for any $a, b \in \mathbb{R}$. In this way, for each fixed $\tau \in \mathbb{H}$, $\mathbb{R}^2$ acts on the space of entire functions – the role of τ being primarily to parametrise different isomorphisms between the additive groups $\mathbb{R}^2$ and $\mathbb{C}$. However, an easy calculation shows that T_a and S_b don't commute, rather $S_b \circ T_a = \exp[2\pi i a b] \, T_a \circ S_b$. So the group $\langle T_a, S_b \rangle$ generated by all T_a's and S_b's is the semi-direct product of S^1 with $\mathbb{R}^2$, consisting of all pairs $[\lambda, x]$ for $\lambda \in \mathbb{C}$, $|\lambda| = 1$ and $x = (x_1, x_2) \in \mathbb{R}^2$, and operation

$$[\lambda, x] \cdot [\mu, y] = [\lambda \mu \exp[2\pi i x_2 y_1], x + y].$$

This group is called the *Heisenberg group* H. Then (2.4.5) says that θ_3 is a vector in a space carrying a representation of H. Now, it turns out that all irreducible representations $(\pi, \mathcal{H})$ of H are essentially isomorphic. A more natural and useful way to see θ_3 in any such representation $(\pi, \mathcal{H})$ is by defining a vector $f_\tau \in \mathcal{H}$ and distribution $\mu_\mathbb{Z}$ such that the Hermitian product

$$\langle \pi_{[1,x]} f_\tau, \mu_\mathbb{Z} \rangle = c \, e^{\pi i x_1 (\tau x_1 + x_2)} \, \theta_3 \, (\tau, x_1 \tau + x_2) \tag{2.4.6}$$

for some nonzero constant c. The exponential factor on the right side of (2.4.6) simplifies the quasi-periodicity of the right side.

We will see that $\mathrm{SL}_2(\mathbb{R})$ acts as automorphisms on the Heisenberg group H. Hence for any $\gamma \in \mathrm{SL}_2(\mathbb{R})$, we get a new representation π_γ of H by $[\lambda, x] \mapsto \pi_{\gamma.[\lambda,x]}$. This representation must be isomorphic to π, so there is a (unitary) operator R_γ on $\mathcal{H}$ such that $\pi_{\gamma.[\lambda,x]} = R_\gamma \circ \pi_{[\lambda,x]} \circ R_\gamma^{-1}$. The assignment $\gamma \mapsto R_\gamma$ defines a projective representation of $\mathrm{SL}_2(\mathbb{R})$ on $\mathcal{H}$. Modularity of θ_3 now follows from the calculation

$$\langle \pi_{[1,x]} f_\tau, \mu_\mathbb{Z} \rangle = \langle R_\gamma \pi_{[1,x]} f_\tau, R_\gamma \mu_\mathbb{Z} \rangle = \langle \pi_{\gamma.[1,x]} R_\gamma f_\tau, R_\gamma \mu_\mathbb{Z} \rangle, \tag{2.4.7}$$

together with the computation of $R_\gamma f_\tau$ and $R_\gamma \mu_\mathbb{Z}$ for the $\gamma \in \Gamma_\theta < \mathrm{SL}_2(\mathbb{Z})$. Let us now fill in the details.

For reasons that will be clear shortly, it is preferable to work instead of $[\lambda, x]$ with the realisation of the group H given by all pairs (λ, x) with operation

$$(\lambda, x) \cdot (\mu, y) = (\lambda \mu \exp [\pi i (x_1 y_2 - x_2 y_1)], x + y).$$

The isomorphism between these realisations of H is given by the correspondence

$$(\lambda, x) \longleftrightarrow [\lambda^{-1} \exp[\pi i x_1 x_2], x].$$

This group H is a three-dimensional real Lie group corresponding to the Heisenberg Lie algebra $\mathfrak{Heis}$ defined in (1.4.3). It is a quotient by $\mathbb{Z} \cong \left\langle \begin{pmatrix} 1 & 0 & 1 \\ 0 & 1 & 0 \\ 0 & 0 & 1 \end{pmatrix} \right\rangle$ of the group $\widetilde{H}$

of upper-triangular matrices

$$
\begin{pmatrix} 1 & a & c \\ 0 & 1 & b \\ 0 & 0 & 1 \end{pmatrix} \in \mathrm{SL}_3(\mathbb{R}).
$$

$\widetilde{H}$ is the (unique) simply-connected Lie group with Lie algebra $\mathfrak{Heis}$; it isn't important that we're focusing on H rather than its universal cover $\widetilde{H}$. The group H and its $(2n+1)$-dimensional versions (the obvious extension of $\mathbb{R}^{2n}$ by S^1) were studied originally in the context of quantum mechanics, hence their name.

The representation theory of these groups was established around 1930. Let π be a unitary irreducible representation of H, in a Hilbert space $\mathcal{H}$. Recall from Section 1.5.5 that this means π is a homomorphism from H into the group of unitary operators of $\mathcal{H}$; moreover, for each $f \in \mathcal{H}$, the map from H to $\mathcal{H}$ given by $(\lambda, x) \mapsto \pi_{(\lambda, x)} f$ is continuous. First note that by Schur's Lemma (the analogue here of Lemma 1.1.3), the central element $(\lambda, 0) \in H$ will act in $\mathcal{H}$ by a scalar multiple λ^n for some $n \in \mathbb{Z}$.

Theorem 2.4.2 (Stone–von Neumann) *Let π be a unitary irreducible representation of H, obeying $\pi_{(\lambda, 0)}(f) = \lambda^n f$.*
(i) *If $n \ne 0$, then π is infinite-dimensional and any other unitary irreducible representation π' of H obeying $\pi'_{(\lambda, 0)}(f) = \lambda^n f$ will be unitarily equivalent to π.*
(ii) *If $n = 0$, then π is one-dimensional and unitarily equivalent to $(\lambda, x) \mapsto e^{i a \cdot x} \in \mathbb{C}$ for some vector $a \in \mathbb{R}^2$.*

We're interested in the case $n = 1$; see, for example, theorem 1.2 in [**440**] for a proof of this special case. There are many different realisations for this unique irreducible representation. The simplest (sometimes called the Schrödinger representation) uses the Hilbert space $\mathcal{H} = L^2(\mathbb{R})$. The action of $(\lambda, x) \in H$ on $f \in L^2(\mathbb{R})$ is given by the unitary operator $U_{(\lambda, x)}$ defined by

$$
(U_{(\lambda, x)} f)(y) = \lambda \exp\left[\pi i \left(2 y x_2 + x_1 x_2\right)\right] f(y + x_1).
$$

This is (essentially) the exponential of the defining representation (4.2.5) of $\mathfrak{Heis}$. Incidentally, the action of S_b, T_a in (2.4.5) on entire functions extends to an $n = -1$ representation of H; this representation is anti-linearly equivalent to the Schrödinger representation.

We want to recover the theta function naturally from the $n = 1$ representation. As always, 'natural' means free of arbitrary choices, such as a specific realisation of the $n = 1$ representation, or a specific basis of the underlying Hilbert space. Begin with any realisation $(\pi, \mathcal{H})$ of the $n = 1$ representation of H.

As we see in Section 1.5.5, a unitary representation U of a Lie group G on a space $\mathcal{H}$ induces a representation δU (the derived module) of the corresponding Lie algebra $\mathfrak{g}$ on a dense subspace $\mathcal{H}_\infty$ of $\mathcal{H}$ by anti-Hermitian operators. For example, the representation (2.4.4) of $U_1(\mathbb{R})$ acts on the Hilbert space $\mathcal{H} = L^2(S^1) \oplus L^2(S^1)$ of all pairs $\left(\begin{smallmatrix} f(\theta) \\ g(\theta) \end{smallmatrix}\right)$. To see how the corresponding Lie algebra $u_1(\mathbb{R}) = \mathbb{R}$ acts, decompose (2.4.4) into irreducibles (i.e. diagonalise):

$$\begin{pmatrix} \cos\theta & \sin\theta \\ -\sin\theta & \cos\theta \end{pmatrix} = \begin{pmatrix} 1 & i \\ i & 1 \end{pmatrix}^{-1} \begin{pmatrix} e^{i\theta} & 0 \\ 0 & e^{-i\theta} \end{pmatrix} \begin{pmatrix} 1 & i \\ i & 1 \end{pmatrix}.$$

Thus the Lie algebra $\mathfrak{u}_1(\mathbb{R})$ acts as

$$x \mapsto \begin{pmatrix} 1 & i \\ i & 1 \end{pmatrix}^{-1} \begin{pmatrix} ix\frac{d}{d\theta} & 0 \\ 0 & -xi\frac{d}{d\theta} \end{pmatrix} \begin{pmatrix} 1 & i \\ i & 1 \end{pmatrix} = \begin{pmatrix} 0 & -x\frac{d}{d\theta} \\ x\frac{d}{d\theta} & 0 \end{pmatrix}.$$

The domain of these operators isn't the whole of the Hilbert space $\mathcal{H}$, but it does contain the dense subspace consisting of the infinitely differentiable functions.

Similarly, our representation π of H on $\mathcal{H}$ induces an anti-Hermitian representation $\delta\pi$ of $\mathfrak{Heis}$ on a dense subspace $\mathcal{H}_\infty$ of $\mathcal{H}$. If we write $e^{x_1 A} = (1, (x_1, 0))$, $e^{x_2 B} = (1, (0, x_2))$ and $e^{tC} = (e^{2\pi it}, 0)$, then using the Baker–Campbell–Hausdorff formula (1.4.6), these generators obey $[A, B] = C, [A, C] = [B, C] = 0$. As an example, in the Schrödinger $n = 1$ representation on space $\mathcal{H} = L^2(\mathbb{R})$, these become the 'momentum operator' $\delta U_A f = \frac{df}{dx}$, the 'position operator' $(\delta U_B f)(x) = 2\pi ix f(x)$ and the central term $\delta U_C f = 2\pi if$. In this example, the dense subspace $\mathcal{H}_\infty$ is the Schwartz space $S(\mathbb{R})$ (Section 1.3.1) consisting of infinitely differentiable, rapidly decreasing functions.

We are now ready to define the two vectors $f_\tau, e_{\mathbb{Z}}$ in (2.4.6). Consider the subspace W_τ consisting of all $f \in \mathcal{H}$ for which $(\delta\pi_A - \tau\delta\pi_B)f$ is defined, and equals 0. This can be thought of as a holomorphicity condition $\frac{\partial}{\partial\bar{z}}f = 0$ (recall τ corresponds to $\sqrt{-1}$). We know that W_τ will be one-dimensional for our choice of π, since it manifestly is for the Schrödinger representation U: there, $W_\tau = \mathbb{C}e^{\pi i\tau y^2}$. Choose any nonzero $f_\tau \in W_\tau$.

The map $\sigma(n) := ((-1)^{n_1 n_2}, n)$ defines a homomorphism $\mathbb{Z}^2 \to H$, and obeys $(\rho \circ \sigma)(n) = n$ for the obvious projection $\rho : H \to \mathbb{R}^2$ – we say ρ 'splits over $\mathbb{Z}^2$'. Define V to be the common 1-eigenspace of all $U_{\sigma(n)}$. More precisely, let V consist of all (tempered) distributions $\mu \in \mathcal{H}_0^*$ with the property that, for all $n \in \mathbb{Z}^2$ and all $f \in \mathcal{H}_\infty$, $\langle \pi_{\sigma(n)} f, \mu \rangle = \langle f, \mu \rangle$. For example, in the Schrödinger representation, we must have $e^{2\pi in_2 y}\mu(y + n_1) = \mu(y)$ for all $n \in \mathbb{Z}^2$. Note that $\mu(y) = \Sigma_{n\in\mathbb{Z}}\delta(y + n)$ satisfies that, and using test functions $f(y) = e^{2\pi imy}$ it quickly follows that this μ is unique up to scalar multiplication. Therefore, for our representation π, V will also be one-dimensional. Choose any nonzero $\mu_{\mathbb{Z}} \in V$. It encodes quasi-periodicity.

Thus we obtain, in the Schrödinger representation,

$$\langle U_{(1,x)} f_\tau, \mu_{\mathbb{Z}} \rangle = \left\langle e^{\pi i(2yx_2 + x_1 x_2)} e^{\pi i\tau(y + x_1)^2}, \sum_n \delta(y + n) \right\rangle,$$

which simplifies to the right side of (2.4.6) with $c = 1$. Therefore, by uniqueness of π and basis independence of the Hermitian product $\langle\ ,\ \rangle$, we get that (2.4.6) holds regardless of the realisation $(\pi, \mathcal{H})$ and vectors $f_\tau, \mu_{\mathbb{Z}}$ we choose.

The reader can verify that quasi-periodicity is automatic (Question 2.4.3). The modularity is of course more difficult (and more interesting). To do this, we need to describe the action of $SL_2(\mathbb{R})$ on the space $\mathcal{H}$ (which we can take to be $L^2(\mathbb{R})$).

Any $\gamma \in \mathrm{SL}_2(\mathbb{R})$ defines an automorphism of H by $(\lambda, x) \mapsto (\lambda, \gamma.x)$, by

$$\gamma.x = \begin{pmatrix} 0 & -1 \\ 1 & 0 \end{pmatrix}^{-1} \begin{pmatrix} a & b \\ c & d \end{pmatrix} \begin{pmatrix} 0 & -1 \\ 1 & 0 \end{pmatrix} \begin{pmatrix} x_1 \\ x_2 \end{pmatrix} = \begin{pmatrix} dx_1 - cx_2 \\ -bx_1 + ax_2 \end{pmatrix}. \tag{2.4.8}$$

The precise form of this action is chosen so that (2.4.10a) below will involve the usual Möbius action of $\mathrm{SL}_2(\mathbb{R})$ on $\mathbb{H}$. We can twist by γ and thus get a new representation $(\pi', \mathcal{H})$ of H, defined by $\pi'_{(\lambda,x)} f = \pi_{(\lambda,\gamma.x)} f$. Obviously π' is also irreducible and has central parameter $n = 1$, so by the Stone–von Neumann Theorem must be unitarily equivalent to π. That is, there exists a unitary operator R_γ on the Hilbert space $\mathcal{H}$ that intertwines π and π': $R_\gamma \pi = \pi' R_\gamma$. The assignment $\gamma \mapsto R_\gamma$ is only defined up to a constant, and so we get a projective representation of $\mathrm{SL}_2(\mathbb{R})$ on $\mathcal{H}$. As we learn in Section 3.1.1, projective representations become true representations when we centrally extend. In particular, we get a true representation when we replace $\mathrm{SL}_2(\mathbb{R})$ with a double-cover called the *metaplectic group* $\mathrm{Mp}_2(\mathbb{R})$.

The metaplectic group is the unique connected double-cover of $\mathrm{SL}_2(\mathbb{R})$. It can be thought of as a way of keeping track of which branch of the square-root we're on in equations like (2.3.5b), and this provides its easiest realisation. Define $\mathrm{Mp}_2(\mathbb{R})$ to be the set of all pairs (γ, s), where $\gamma \in \mathrm{SL}_2(\mathbb{R})$ and $s = s(\tau)$ is a choice of holomorphic square-root of $c\tau + d$. Since there are two choices for s (differing by a sign), this is indeed a double-cover. The group operation is

$$(\gamma, s(\tau))(\gamma', s'(\tau)) = (\gamma\gamma', s(\gamma'.\tau) s'(\tau)), \tag{2.4.9}$$

as can be seen by calculating from (2.3.6) with $k = 1/2$.

Returning to the γ-twist π' of the representation π of H, it is possible to choose unitary operators $R_{(\gamma,s)}$, for each $(\gamma, s) \in \mathrm{Mp}_2(\mathbb{R})$, such that $R_{(\gamma,s)} \pi = \pi' R_{(\gamma,s)}$ and $(\gamma, s) \mapsto R_{(\gamma,s)}$ defines a representation of the metaplectic group $\mathrm{Mp}_2(\mathbb{R})$.

Recalling the definition of f_τ and $\mu_\mathbb{Z}$ as eigenvectors, it isn't difficult to see that

$$R_{(\gamma,s)} f_\tau = s(\tau)^{-1} f_{\gamma.\tau}, \qquad \forall (\gamma, s) \in \mathrm{Mp}_2(\mathbb{R}), \tag{2.4.10a}$$

$$R_{(\gamma,s)} \mu_\mathbb{Z} = \mu_{(\gamma,s)} e_\mathbb{Z}, \qquad \forall (\gamma, s) \in \widetilde{\Gamma}_\theta = \{(\gamma, s) \in \mathrm{Mp}_2(\mathbb{R}) \mid \gamma \in \Gamma_\theta\}, \tag{2.4.10b}$$

where $\gamma.\tau$ is the usual action (2.1.4a) and where $\mu : \widetilde{\Gamma}_\theta \to \mathbb{C}^*$ is some one-dimensional representation (with values in eighth roots of unity). See chapter 8 of [**440**] for the detailed calculation. We now immediately obtain from (2.4.6) and (2.4.7) that

$$c\, e^{\pi i x_1 (\tau x_1 + x_2)} \theta_3 (\tau, x_1 \tau + x_2) = \langle \pi_{(1,\gamma.x)} R_\gamma f_\tau, R_\gamma \mu_\mathbb{Z} \rangle = s(\tau)^{-1} \langle \pi_{(1,\gamma.x)} f_{\gamma.\tau}, \mu_{(\gamma,s)} e_\mathbb{Z} \rangle$$

$$= c\, s(\tau)^{-1} \mu_{(\gamma,s)} \exp \left[\pi i (dx_1 - cx_2) \left(\frac{a\tau + b}{c\tau + d} (dx_1 - cx_2) + (-bx_1 + ax_2) \right) \right]$$

$$\times \theta_3 \left(\frac{a\tau + b}{c\tau + d}, (dx_1 - cx_2) \frac{a\tau + b}{c\tau + d} + (-bx_1 + ax_2) \right), \tag{2.4.11}$$

for all $\gamma = \begin{pmatrix} a & b \\ c & d \end{pmatrix} \in \Gamma_\theta$, which simplifies down to the desired modularity (2.3.5b).

Last subsection we learned that $\mathrm{SL}_2(\mathbb{R})$ acts transitively on $\mathbb{H}$. Using this and (2.4.10a), we can refine (2.4.6) and write θ_3 as a matrix entry of a unitary representation of the obvious semi-direct product of $\mathrm{Mp}_2(\mathbb{R})$ with H. We obtain

$$c\, e^{\pi i x_1 (\tau x_1 + x_2)} \theta_3 (\tau, x_1 \tau + x_2) = \sqrt{c\mathrm{i} + d} \, \langle \pi_{(1,x)} R_{(\gamma,s)} f_\tau, \mu_{\mathbb{Z}} \rangle, \qquad (2.4.12)$$

where $\tau = \frac{bd+ac+\mathrm{i}}{c^2+d^2}$, for $\gamma = \begin{pmatrix} a & b \\ c & d \end{pmatrix} \in \mathrm{SL}_2(\mathbb{R})$.

This argument is far longer and more technically difficult than the other proofs of theta function modularity given in this chapter, and it is easy to get lost in the details. But it is a remarkable argument, and much more conceptual than, for example, Poisson summation. The modular group $\mathrm{SL}_2(\mathbb{Z})$ (or rather its subgroup Γ_θ) arises here as a group of automorphisms of H transforming in a controlled way the vectors f_τ and $\mu_{\mathbb{Z}}$. The intrinsically algebraic nature of the argument means it generalises easily, and with little extra effort we could have given the proof for Siegel theta functions. (Nonholomorphic) Eisenstein series can also be constructed and studied in a similar way (by first lifting to $\mathrm{SL}_2(\mathbb{R})$). But as with the previous modularity proofs, new ideas would be needed to generalise it beyond these classical functions into a general device providing uniform proofs of modularity for Moonshine functions. In the next subsection though we explain why it might after all have something to do with Moonshine.

2.4.3 Braided #2: from the trefoil to Dedekind

The decomposition (2.4.1b) says that $\mathrm{SL}_2(\mathbb{R})$ is topologically homeomorphic to $\mathbb{R}^2 \times S^1$, i.e. the interior of a solid torus (or if one prefers, the complement of S^1 in $\mathbb{R}^3$). In remarkable work in the context of computing $k_2(\mathbb{Z})$ (see Section 2.5.1), Quillen showed that the space $\mathrm{SL}_2(\mathbb{Z})\backslash\mathrm{SL}_2(\mathbb{R})$ is naturally diffeomorphic to the complement of the trefoil knot in the sphere S^3 (see pages 84–5 of [419] for the elementary argument). Namely, the Eisenstein series $a = G_4$, $b = G_6$ in (0.1.5) identify the space $\mathrm{GL}_2(\mathbb{Z})\backslash\mathrm{GL}_2(\mathbb{R})$ of two-dimensional lattices with the complement of the complex curve $20a^3 - 49b^2 = 0$ (which corresponds to degenerate lattices); the intersection of $20a^3 - 49b^2 = 0$ with the sphere $|a|^2 + |b|^2 = 1$ in $\mathbb{C}^2$ (to get instead $\mathrm{SL}_2(\mathbb{Z})\backslash\mathrm{SL}_2(\mathbb{R})$) is then identified with the trefoil (the (2,1)-torus knot, drawn in Figure 1.10). Now, in Section 2.4.1 we lift modular forms for $\mathrm{SL}_2(\mathbb{Z})$ to the space $L^2(\mathrm{SL}_2(\mathbb{Z})\backslash\mathrm{SL}_2(\mathbb{R}))$: thus, for example, the j-function is a complex-valued function on the complement of the trefoil. More generally, as we will see later, the characters of an affine algebra, or vertex operator algebra, or rational conformal field theory, are vector-valued functions on the complement of the trefoil. The cusps of $\mathbb{H}$ can be interpreted as rational points on the trefoil. Can modular forms and functions somehow see this topological trefoil? The answer is yes!

First, the fundamental group of the complement of the trefoil is easy to compute using the Wirtinger presentation (Section 6.2.5), and is naturally isomorphic to the braid group $\mathcal{B}_3$. This suggests the following picture. Write G for $\mathrm{SL}_2(\mathbb{R})$, $\widetilde{G}$ for its universal cover and Γ for $\mathrm{SL}_2(\mathbb{Z})$. Then

$$\widetilde{G} \xrightarrow{\pi} G \xrightarrow{q} \Gamma\backslash G. \qquad (2.4.13)$$

Of course π is surjective and has kernel $\pi_1(G) \cong \mathbb{Z}$. $\widetilde{G}$ is also the universal cover of the trefoil-complement $\Gamma \backslash G$, and the kernel of this surjective map $q \circ \pi$ is the central extension $\pi_1(\Gamma \backslash G) \cong \mathcal{B}_3$ of the modular group $\mathrm{SL}_2(\mathbb{Z})$. The map $\mathcal{B}_3 \to \mathrm{SL}_2(\mathbb{Z})$ is simply the reduced Burau representation (1.1.11b) specialised to $w = -1$ (recall (1.1.10a)).

So what does this mean for modular forms? Recall from Section 2.2.1 that modular forms for $\mathrm{SL}_2(\mathbb{Z})$ have multiplier μ that carries a *projective* representation of $\mathrm{SL}_2(\mathbb{Z})$ – it will be a true representation only when the weight k is an integer. As we emphasise in Section 3.1.1, *projective* representations become *true* representations when one centrally extends. Especially when the weight is fractional, the role of $\mathrm{SL}_2(\mathbb{R})$ really should be played by the more fundamental Lie group $\widetilde{\mathrm{SL}_2(\mathbb{R})}$, and likewise the modular group $\mathrm{SL}_2(\mathbb{Z})$ should be replaced by its central extension $\mathcal{B}_3$.

For a good example, recall the Dedekind eta function $\eta(\tau)$ of (2.2.6b). As we see in (2.2.8), it is a modular form for $\mathrm{SL}_2(\mathbb{Z})$ of weight $\frac{1}{2}$, whose multiplier μ is quite complicated as a function on $\mathrm{SL}_2(\mathbb{Z})$. But $\mathcal{B}_3$ is the more fundamental transformation group underlying $\eta(\tau)$. Indeed, in terms of $\mathcal{B}_3$, the multiplier is trivial to describe:

$$\mu(\beta) = \exp\left[\frac{2\pi i}{24} \deg \beta\right], \tag{2.4.14}$$

where the degree of a braid is the length of its word in σ_1, σ_2 (Section 1.1.4). More generally, the multiplier for any modular form for $\mathrm{SL}_2(\mathbb{Z})$ will be similar, with '24' replaced by some other rational. Surely this algebraic interpretation of Dedekind sums in terms of $\mathcal{B}_3$ is related to the topological interpretation of Dedekind sums reviewed and explored in [24]; see also [23], [43].

Of course the multiplier of η is almost as trivial if we write $\begin{pmatrix} a & b \\ c & d \end{pmatrix} \in \mathrm{SL}_2(\mathbb{Z})$ as a monomial in the generators S, T, but finding that monomial isn't easy. On the other hand, finding 'deg β' by looking at the braid β is easy: just count the crossings in β, with signs. The multiplier, as a function of β, is far simpler than as a function of a, b, c, d. Our topological considerations have been rewarded!

Likewise, the multiplier in the vector-valued Jacobi form (2.3.8) (again of weight $\frac{1}{2}$) defines a four-dimensional projectivere presentation of $\mathrm{SL}_2(\mathbb{Z})$, given by the tensor product of the one-dimensional representation $\exp[2\pi i \deg \beta / 8]$ of $\mathcal{B}_3$, with a true four-dimensional representation of $\mathrm{SL}_2(\mathbb{Z})$.

Of course the metaplectic group was introduced last subsection for essentially the same reason $(\mathrm{Mp}_2(\mathbb{R})$ is also a quotient of $\widetilde{\mathrm{SL}_2(\mathbb{R})})$. Indeed, since most modular forms arising in the literature have weight in $\frac{1}{2}\mathbb{Z}$, the metaplectic group is a large enough central extension, and $\widetilde{\mathrm{SL}_2(\mathbb{R})}$ may seem like overkill. But modular forms with fractional weight exist in abundance for arbitrarily large denominator (see e.g. [303] for examples). The important 'one-point functions on a torus' (Section 4.3.2) in conformal field theory (CFT), to which family the Moonshine functions naturally belong, can form vector-valued modular forms of arbitrary rational weight. We will see in Section 7.2.4 how nicely the CFT machinery accommodates this universal $\mathcal{B}_3$ action, and also how other considerations in (Monstrous) Moonshine are trying to focus our attention on the relation of $\mathcal{B}_3$ to modular functions.

The braid group $\mathcal{B}_3$ is at least as relevant for the nonholomorphic automorphic forms of $SL_2(\mathbb{Z})$, alluded to in Section 2.4.1. For a simple example, [**379**] studies the Maass cusp forms $u(\tau)$ (with weight 0), identifying them with 'period functions' $\psi(z)$; the exact symmetry $u(-1/\tau) = u(\tau)$ becomes $\psi(1/z) = z^{2s}\psi(z)$, where s is the 'spectral parameter' of u. This transformation of the ψ's, with the factor z^{2s}, is what one would expect from the braid group (compare (7.2.4)).

We should regard $\mathcal{B}_3$ as the universal symmetry of (not necessarily holomorphic) modular forms for $SL_2(\mathbb{Z})$. If instead we have modular forms for some subgroup Γ of $SL_2(\mathbb{Z})$, then the role of $\mathcal{B}_3$ is replaced by its subgroup that projects (via the reduced Burau representation (1.1.11b) specialised to $w = -1$) to Γ. For instance, the principal congruence subgroup $\Gamma(2)$ corresponds to the pure braid group $\mathcal{P}_3$. It would be interesting to find the topological interpretation of $\Gamma_0(p)+$ in (7.1.5) and the other modular groups appearing in Monstrous Moonshine.

The lesson of Section 2.4.1 is that, whenever we have some sort of modularity for, for example, $SL_2(\mathbb{Z})$, we should lift the domain to that of the relevant Lie group (e.g. $SL_2(\mathbb{R})$). This should be especially valuable for providing perspective and clarity when we are investigating a new modular-like phenomenon. To give one example among many, [**519**] introduces nonholomorphic deformations of familiar modular forms relevant to strings on a pp-wave background (a 1-parameter deformation of flat space-time). Of more direct relevance to us is the question: *Is it natural to regard the modular functions (characters) of RCFT, VOAs and Moonshine as functions on $SL_2(\mathbb{R})$?*

The lesson of this subsection is that an $SL_2(\mathbb{Z})$-action may become simpler when lifted to its central extension $\mathcal{B}_3$. The braid group provides a clean universal formulation especially appropriate when metaplectic groups or other central extensions of $SL_2(\mathbb{Z})$ arise. Mathematics thrives on having alternate interpretations for the same phenomenon: here we replace the matrix group $SL_2(\mathbb{Z})$ (or its subgroups) with the topologically defined $\mathcal{B}_3$ (or its subgroups). Some things will be easier in one formalism, and presumably other things in the other (e.g. the multipliers μ are much easier for $\mathcal{B}_3$). It is tempting to apply this to the so-called S-duality of superstrings (Section 3.2.5). *Are there other ways modular forms for $SL_2(\mathbb{Z})$ see the trefoil?*

The modularity argument of Section 2.4.2 has never been applied to Monstrous Moonshine, to this author's knowledge. But one hint that it might be the shadow of such a device is that the braid group lurks here. In particular, there is an action of $\mathcal{B}_3$ on $G \times G$, for any group G (Question 2.4.4); the action (2.4.8) of $SL_2(\mathbb{Z})$ on H is really this action of $\mathcal{B}_3$ on $\mathbb{R}^2$ – it factors through to $SL_2(\mathbb{Z})$ because $\mathbb{R}^2$ is abelian. In Section 7.3.3 we use this same action, this time applied to $\mathbb{M} \times \mathbb{M}$, to identify the group-theoretic property of the Monster $\mathbb{M}$ that could be responsible for the genus-0 properties of the McKay–Thompson series T_g.

Another hint, perhaps more substantial, of its relevance to Moonshine-like phenomena is the repeated appearance of Maslov indices in the study of gluing anomalies in three-dimensional topological field theory (see chapter IV of [**534**]). This suggested to Turaev an intimate relation of topological field theory with the Segal–Shale–Weil representations of the metaplectic groups. These representations also appear in the context of braids and subfactors [**252**] – metaplectic representations arise naturally there when constructing knot invariants from braids. Much of the mathematical background is developed in [**387**], where we also learn that the universal cover $\widetilde{SL}_2(\mathbb{R})$ can easily be expressed using Maslov indices.

Question 2.4.1. Use the decomposition (2.4.1b) to find a (noncanonical) group structure on $\mathbb{H}$, inherited from that of $SL_2(\mathbb{R})$.

Question 2.4.2. Show that uniqueness of the representation in Theorem 2.4.2 fails if H is replaced with infinitely many coupled Heisenberg groups. (This is a major complication for quantum field theory, as we see in Section 4.2.2 in the context of Haag's Theorem.)

Question 2.4.3. Verify that any function of the form $F(x) = \langle \pi_{(1,x)} f, \mu_{\mathbb{Z}} \rangle$, for any f for which F is defined, necessarily obeys $F(x + n) = (-1)^{n_1 n_2} e^{\pi i (n_1 x_2 - n_2 x_1)} F(x)$. Hence $\mu_{\mathbb{Z}}$ is responsible for the quasi-periodicity (2.3.5a) of θ_3.

Question 2.4.4. (a) Let G be a finite group. Verify that we obtain a right braid group $\mathcal{B}_3$ action on the Cartesian product $G \times G \times G$, by defining

$$(g, h, k).\sigma_1 = (ghg^{-1}, g, k), \ (g, h, k).\sigma_2 = (g, hkh^{-1}, h), \qquad (2.4.15a)$$

where σ_i are the usual generators of $\mathcal{B}_3$ (recall (1.1.9)). Also, verify that there is a right $\mathcal{B}_3$-action on $G \times G$, generated by

$$(g, h).\sigma_1 = (g, gh), \ (g, h).\sigma_2 = (gh^{-1}, h). \qquad (2.4.15b)$$

(b) Let $C \subseteq G \times G$ consist of all pairs (g, h) where $gh = hg$. Show that this $\mathcal{B}_3$ action takes C to itself, and that its restriction to C actually defines an action of $SL_2(\mathbb{Z})$ on C.
(c) Extend the $\mathcal{B}_3$ actions of (a) to $\mathcal{B}_n$ actions on G^n and G^{n-1}.

Question 2.4.5. (a) Show that $SL_2(\mathbb{R})$ is isomorphic to the group

$$SU_{1,1}(\mathbb{C}) := \left\{ \begin{pmatrix} \alpha & \bar{\beta} \\ \beta & \bar{\alpha} \end{pmatrix} \in SL_2(\mathbb{C}) \right\},$$

by showing they are conjugate in $SL_2(\mathbb{C})$.
(b) Verify that $SU_{1,1}(\mathbb{C})$ is isomorphic to the set of all pairs (γ, θ), where $|\gamma| < 1$ and $-1 < \theta \leq 1$, with group operation $(\gamma, \theta)(\gamma', \theta') = (\theta'', \theta'')$ where

$$\gamma'' = \frac{\gamma + \gamma' e^{-2\pi i \theta}}{1 + \bar{\gamma} \gamma' e^{-2\pi i \theta}}, \qquad \theta'' = \theta + \theta' + \frac{1}{2\pi i} \log \frac{\gamma + \gamma' e^{-2\pi i \theta}}{1 + \bar{\gamma} \gamma' e^{-2\pi i \theta}} \ (\mathrm{mod} \ 2).$$

(c) Using (b), realise the universal cover $\widetilde{SL}_2(\mathbb{R})$ of $SL_2(\mathbb{R})$.
(d) Realise $\mathcal{B}_3$ as a subgroup of $\widetilde{SL}_2(\mathbb{R})$.

2.5 Meta-patterns in mathematics

2.5.1 Twenty-four

There are lots of 'meta-patterns' in mathematics, i.e. collections of seemingly different problems that have similar answers, or structures that appear more often than we would have expected. Once one of these meta-patterns is identified it is always helpful to understand what is responsible for it, to see what simple structure or basic lemma underlies it. Why are groups so important in mathematics and science? Because they are the devices through which we 'act' on sets, spaces, etc. Mathematics is not above metaphysics; like any area it grows by asking questions, and changing one's perspective – even to a metaphysical one – should suggest new questions.

To give a trivial example, years ago while the author was writing up his PhD thesis he noticed in several places the numbers 1, 2, 3, 4 and 6. For instance $\cos(2\pi r) \in \mathbb{Q}$ for $r \in \mathbb{Q}$ iff the denominator of r is 1, 2, 3, 4 or 6. Likewise, the theta function $\Theta_{\mathbb{Z}+r}(\tau)$ for $r \in \mathbb{Q}$ can be written as $\sum a_i \theta_3(b_i \tau)$ for some $a_i, b_i \in \mathbb{R}$ iff the denominator of r is 1, 2, 3, 4 or 6. This pattern is easy to explain: they are precisely those positive integers n with Euler totient $\phi(n) \leq 2$, that is there are at most two positive numbers less than n coprime to n. The various incidences of these numbers can usually be reduced to this $\phi(n) \leq 2$ property. For example, the number field $\mathbb{Q}[\cos(2\pi \frac{a}{b})]$ (see Section 1.7.1), considered as a vector space over $\mathbb{Q}$, has dimension $\phi(b)/2$.

A more interesting meta-pattern involves the number 24 and its divisors (especially 8). One sees 24 wherever modular forms naturally appear. For instance, we see it in the critical dimensions in string theory: the bosonic string lives in a background space-time of dimension $24 + 2$, while the fermionic string lives in $8 + 2$ dimensions. Another example: the dimensions of even self-dual positive-definite lattices must be a multiple of 8 (e.g. the E_8 root lattice has dimension 8, while the Leech lattice has dimension 24). The meta-pattern 24 is also easy to understand: the fundamental problem for which it is the answer is the following one. Fix n, and consider the congruence $x^2 \equiv 1 \pmod{n}$. Certainly in order to have a chance of satisfying this, x and n must be coprime. The extreme situation[11] is when *every* number x coprime to n satisfies this congruence: that is,

$$\gcd(x, n) = 1 \quad \Longleftrightarrow \quad x^2 \equiv 1 \pmod{n}. \tag{2.5.1}$$

The reader can try to verify the following simple fact: n obeys this extreme situation (2.5.1) iff n divides 24. What does this congruence property have to do with these other occurrences of 24? The elementary argument for even self-dual positive-definite lattices involves the construction $L\{T\}$ of Section 2.3.3 and is sketched in Question 2.5.1.

The '24' appearing in the $q^{1/24}$ of η is the same as the 24 in $c/24$ appearing in, for example, (3.1.10); in both cases they come from $\zeta(-1) = -1/12$ or equivalently

---

[11] This is a standard trick in mathematics: when some sort of bound is established, look at the extremal cases that realise that bound. If your bound is a good one, it should be possible to say something about those extremal cases, and having something to say is always of paramount importance. This strategy is used, for instance, in the definition of normal subgroup in Section 1.1.1 and of simple-currents in Section 6.1.1.

$\zeta(2) = \pi^2/6$. Are these the same as the 24 in (2.5.1)? Note from the right side of (2.2.6b) that

$$\eta(\tau) = \Theta_{\mathbb{Z}+\frac{1}{12}}(12\tau) - \Theta_{\mathbb{Z}+\frac{5}{12}}(12\tau).$$

Using this identity, the fact that $\eta(\tau + 1)$ is a constant multiple of $\eta(\tau)$ is indeed related to (2.5.1). Moreover, this '1/24' is directly related to the abelianisation

$$\mathrm{SL}_2(\mathbb{Z})/[\mathrm{SL}_2(\mathbb{Z}), \mathrm{SL}_2(\mathbb{Z})] \cong \mathbb{Z}_{12} \qquad (2.5.2)$$

of $\mathrm{SL}_2(\mathbb{Z})$: writing $\eta(-1/\tau)^2 = a\tau\eta(\tau)^2$ and $\eta(\tau+1)^2 = b\eta(\tau)^2$, the multiplier $s \mapsto a, t \mapsto b$ must define a one-dimensional representation of $\mathrm{SL}_2(\mathbb{Z})$, since η^2 has weight 1 (recall Question 2.2.3); for any group G, and in particular $\mathrm{SL}_2(\mathbb{Z})$, the abelianisation $G/[GG]$ is isomorphic to the group of all one-dimensional representations of G. This argument forces b to be some 12th root of unity, and a to be b^3.

Perhaps the most intriguing '24' occurs as a K-theoretic invariant of the integers. *K-theory* is a generalised (co)homology theory, and as such associates a sequence of abelian groups $K_i(X)$ to the object X, which can capture some subtle aspects of X. When X is a ring, the definition of these invariants $K_i(X)$ is quite involved, and their calculation is very difficult (see e.g. [**419**] – for example, for $X = \mathbb{Z}$ the groups are known only for $0 \le i \le 5$, where they equal $\mathbb{Z}, \mathbb{Z}_2, \mathbb{Z}_2, \mathbb{Z}_{48}, 0, \mathbb{Z}$, respectively. $K_0(\mathbb{Z}) \cong \mathbb{Z}$ says that the projective $\mathbb{Z}$-modules are the free $\mathbb{Z}$-modules $\mathbb{Z}^n$, while $K_1(\mathbb{Z}) \cong \mathbb{Z}_2$ tells us that the Euclidean domain $\mathbb{Z}$ has only two units (namely, ± 1). The first interesting group in this list is $\mathbb{Z}_{48}$, which arises naturally here as an extension of $\mathbb{Z}_{24}$. Thus 24 (or 48) is a number intimately associated with $\mathbb{Z}$. This author knows no direct connection with our definition (2.5.1) of 24, but there is a conjectural relation of $\|K_{4n-2}(\mathbb{Z})_{torsion}\|/\|K_{4n-1}(\mathbb{Z})_{torsion}\|$ with values $\zeta(1-2n)$ of the Riemann zeta function (see e.g. [**230a**]). In particular, $K_3(\mathbb{Z}) \cong \mathbb{Z}_{48}$ is related to $\zeta(-1) = -\frac{1}{12}$, which in turn is related to our 24.

2.5.2 A–D–E

A much deeper and still not-completely-understood meta-pattern is called *A–D–E* (see [**16**] for a discussion and examples). The name comes from the *simply-laced Lie algebras*, i.e. the simple finite-dimensional Lie algebras whose Coxeter–Dynkin diagrams – see Figure 1.17 – contain only single edges (i.e. no arrows). These are the $A_\star$- and $D_\star$-series, along with the E_6, E_7 and E_8 exceptionals. The observation is that many other problems, which don't have anything directly in common with simple Lie algebras, have a solution that falls into this *A–D–E* pattern. Of course, for an object to be meaningfully labelled X_ℓ at least some of the data associated with the algebra X_ℓ should reappear in some form in that object. Let's look at some examples.

Consider any even positive-definite integral lattice L (Section 1.2.1). The smallest possible nonzero length-squareds in L will be 2, and the vectors of length-squared 2 are special and are called *roots* (Question 1.2.5). It is important in lattice theory to know the lattices that are spanned by their roots; it turns out these are precisely the orthogonal direct sums of lattices called A_n, D_n and E_6, E_7 and E_8 (Theorem 1.2.2). They carry those names for a number of reasons. For example, the lattice called X_n has a basis $\{\alpha_1, \ldots, \alpha_n\}$ with the property that the Gram matrix $A_{ij} := \alpha_i \cdot \alpha_j$ is the Cartan matrix (see Section 1.4.5) for the Lie algebra X_n. Also, the group generated by reflections in the roots of the lattice X_n is naturally isomorphic to the Weyl group of the Lie algebra X_n. Moreover, to any simple Lie algebra there is canonically associated a lattice called the root lattice; for the simply-laced algebras, these are isomorphic to the lattice of the same name. Incidentally, the root lattices for the non-simply-laced simple Lie algebras are (up to rescalings) orthogonal direct sums of the simply-laced root lattices.

A famous A–D–E example is due to McKay.[12] Consider any finite subgroup G of the Lie group $SU_2(\mathbb{C})$ (i.e. the 2×2 unitary matrices with determinant 1). For example, there is the cyclic group $\mathbb{Z}_n$ of n elements generated by the matrix

$$M_n = \begin{pmatrix} \exp[2\pi i/n] & 0 \\ 0 & \exp[-2\pi i/n] \end{pmatrix}.$$

There are also the (doubles of) dihedral groups $\mathcal{D}_n$, and the binary tetrahedral, binary octahedral and binary icosahedral groups of orders 24, 48 and 120, respectively. Let R_i be the irreducible representations of G. For instance, for $\mathbb{Z}_n$, there are precisely n of these, all one-dimensional, given by sending the generator M_n to $\exp[2\pi ik/n]$ for each $k = 1, 2, \ldots, n$. Now consider the tensor product $G \otimes R_i$, where we interpret $G \subset SU_2(\mathbb{C})$ here as a two-dimensional representation. By Theorem 1.1.2 we can decompose that product into a direct sum $\oplus_j m_{ij} R_j$ of irreducibles (the m_{ij} here are multiplicities). Now create a graph with one node for each R_i, and with the ith and jth nodes ($i \neq j$) connected with precisely m_{ij} directed edges $i \to j$. If $m_{ij} = m_{ji}$, we agree to erase the double arrows from the m_{ij} edges. Then McKay [**411**] observed that this graph, for any of these finite $G < SU_2(\mathbb{C})$, is a distinct extended Coxeter–Dynkin diagram of A–D–E type (these are all listed in Figure 3.2). For instance, the cyclic group with n elements yields the extended graph of A_{n-1}.

How was McKay led to his remarkable correspondence? He knew that the sum of the labels $a_i = 1, 2, 3, 4, 5, 6, 4, 2, 3$ associated with each node of the extended E_8 diagram (Figure 3.2) equals 30, the Coxeter number of E_8. So what do their *squares* add to? 120, which he recognised as the cardinality of one of the exceptional finite subgroups of $SU_2(\mathbb{C})$, and that got him thinking . . .

A deep example of A–D–E, due to Arnol'd, are the *simple singularities*. A *singularity* or *critical point* of a smooth function $f : \mathbb{C}^n \to \mathbb{C}$ is a point $z \in \mathbb{C}^n$ where all first partial derivatives $\partial_i f$ vanish. For example, $f(z) = z^{k+1}$ has a singularity at $z = 0$ for any integer $k \geq 1$. We identify singularities if locally they merely differ by a change-of-coordinates – see, for example, [**19**] for details. For example, any singularity of $f : \mathbb{C} \to \mathbb{C}$ is equivalent to one of the form $f(z) = z^{k+1}$. A simple singularity is an isolated singularity and behaves like the poles $f(z) = z^{-n}$ of usual complex analysis – again see [**19**] for the precise definition. For example, $z_1^2 + z_2^{k+1}$ is simple but $z_1^4 + 3z_1^2 z_2^2 + z_2^4$ is not (the coefficient '3' can be deformed, yielding a continuum of inequivalent singularities).

---

[12] He is the same John McKay we celebrated in Chapter 0.

Table 2.2. *The simple singularities in* $\mathbb{C}^2$

Name	A_k	D_k	E_6	E_7	E_8
Representative	$x^2 + y^{k+1}$	$x^2 y + y^{k-1}$	$x^3 + y^4$	$x^3 + xy^3$	$x^3 + y^5$

Table 2.2 lists the simple singularities in $\mathbb{C}^2$ up to equivalence. In higher dimensions we get the same list, with the extra variables coming in as $z_3^2 + \cdots + z_n^2$. These singularities can be related to McKay's A–D–E as follows. The group $SU_2(\mathbb{C})$ acts on $\mathbb{C}^2$ in the obvious way (matrix multiplication). If G is a discrete subgroup of $SU_2(\mathbb{C})$, then consider the ring of polynomials in two variables w_1, w_2 invariant under G. It turns out it will have three generators $x(w_1, w_2)$, $y(w_1, w_2)$, $z(w_1, w_2)$, which are connected by one polynomial relation (syzygy). For instance, take G to be the cyclic group $\mathbb{Z}_n$, then we're interested in polynomials $p(w_1, w_2)$ invariant under $w_1 \mapsto \exp[2\pi i/n]w_1, w_2 \mapsto \exp[-2\pi i/n]w_2$. Any such invariant $p(w_1, w_2)$ is clearly generated by (i.e. can be written as a polynomial in) $w_1 w_2$, w_1^n and w_2^n. Choosing instead the generators $x = \frac{w_1^n - w_2^n}{2}$, $y = w_1 w_2$, $z = i\frac{w_1^n + w_2^n}{2}$, we get the syzygy $y^n = -(x^2 + z^2)$. For any G, generators x, y, z can always be found so that the syzygy will be one of the polynomials in Table 2.2 (with '$+z^2$' appended), and this will give the equation of the algebraic surface $\mathbb{C}^2/G$ as a two-dimensional complex surface in $\mathbb{C}^3$. For example, the complex surfaces $\mathbb{C}^2/\mathbb{Z}_n$ and $\{(x, y, z) \in \mathbb{C}^3 \mid x^2 + y^2 + z^n = 0\}$ are equivalent.

There are other ways these singularities can be associated with A–D–E. Given a surface $\Sigma \subset \mathbb{C}^3$ with a single singularity, a *resolution* $\tilde{\Sigma}$ is a smooth surface without singularities that agrees with Σ away from the singularity (again see [19] for details). A *minimal resolution* is one through which any other resolution must factor. The minimal resolution exists and is unique. For example, the A_1 singularity $x^2 + y^2 + z^2 = 0$ has the resolution

$$\tilde{\Sigma} = \{(x, y, z, (a, b)) \in \mathbb{C}^3 \times \mathbb{P}^1(\mathbb{C}) \mid x^2 + y^2 + z^2 = 0, \ xb = ya\}.$$

For $(x, y) \neq (0, 0)$, $xb = ya$ uniquely determines the homogeneous coordinates (a, b), but the singularity $(x, y) = (0, 0)$ is blown up into the sphere $\mathbb{P}^1(\mathbb{C})$; the points on the sphere parametrise the different (complex) directions in which the singularity can be approached.

More generally, given a minimal resolution $\pi : \tilde{\Sigma} \to \Sigma$ of a simple singularity, $\pi^{-1}(0)$ will be a union of r spheres $\cup C_i$. duVal [165] noticed that these classes $[C_i]$ form a basis of the homology group $H_2(\tilde{\Sigma}, \mathbb{Z})$, on which there is defined a $\mathbb{Z}$-valued intersection form; this form makes $H_2(\tilde{\Sigma}, \mathbb{Z})$ into a negative-definite lattice isomorphic (up to a factor of $\sqrt{-1}$) to the root lattice of X_r, where $[C_i]$ map to a basis of simple roots. The Weyl group of X_r is isomorphic to the so-called monodromy group of the singularity (see [19] for details).

Incidentally, the *McKay correspondence* refers to the strategy of describing the geometry of the resolution of the orbifold singularities $\mathbb{C}^n/G$ for finite subgroups G of $SL_n(\mathbb{C})$,

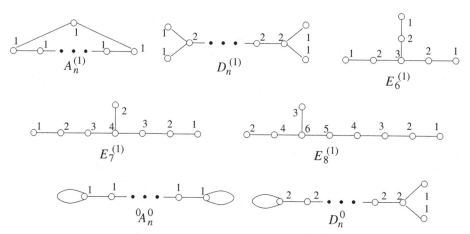

Fig. 2.9 The connected multigraphs with largest eigenvalue 2.

through the representation theory of G. See [**254**] for the $n = 2$ story (i.e. for the simple singularities) and [**471**] for fascinating speculations on what happens in dimension $n > 2$.

Arguably the first A–D–E classification goes back to Theaetetus, who classified the regular solids in 400 B.C. For instance, the tetrahedron can be associated with E_6 while the cube is matched with E_7. This A–D–E is only partial, as there are no regular solids assigned to the A-series, and to get the D-series one must look at 'degenerate regular solids', that is the regular polygons.

The closest we have to an explanation of the A–D–E meta-pattern would seem to be graphs of small eigenvalues. Consider any multigraph $\mathcal{G}$ – that is, we allow multiple edges (there can be more than one edge connecting two vertices) and loops (an edge running from a node to itself), but all edges are undirected. We can also assume without loss of generality that $\mathcal{G}$ is connected. Assign a positive number a_i to each node. If this assignment has the property that for each i, $2a_i = \sum a_j$ where the sum is over all nodes j adjacent to i (counting multiplicities of edges), then we call it 'PF2'. The column vector $(a_1, \ldots, a_n)^t$ will be a strictly positive eigenvector (called the Perron–Frobenius eigenvector) of the adjacency matrix of $\mathcal{G}$, with eigenvalue 2. A multigraph has a PF2 assignment iff the eigenvalue λ of its adjacency matrix with largest absolute value $|\lambda|$ is $\lambda = 2$ (see Theorem 2.5.1 below). For instance, for the multigraph $\circ\!\!=\!\!\circ$, corresponding to adjacency matrix $\begin{pmatrix} 0 & 2 \\ 2 & 0 \end{pmatrix}$, the assignment $a_1 = 1 = a_2$ is PF2 but the assignment $a_1 = 1, a_2 = 2$ is not. The question is, which multigraphs have a PF2 assignment? The answer is given in Figure 2.9. The names $A_n^{(1)}$ to $E_6^{(1)}$ there come from Figure 3.2; the names ${}^0A_n^0$ and D_n^0 are invented. We see that the PF2 multigraphs without loops are precisely the extended Coxeter–Dynkin diagrams of A–D–E type, and their PF2 assignments are unique (up to constant proportionality) and are given by the *labels* a_i of the corresponding affine algebra (i.e. the numbers attached to the graphs in Figures 2.9 and 3.2).

The unextended diagrams have a similar depiction. For them, we assign positive numbers a_i to each node so that $2a_i \geq \sum_j a_j$, where as before we sum over all adjacent j. We also require that for at least one vertex i, we don't get an equality. Call this a PF2$^-$ assignment. A multigraph $\mathcal{G}$ has a PF2$^-$ assignment iff the absolute value $|\lambda|$ of each eigenvalue λ of its adjacency matrix is < 2. In Figure 1.4 we list all multigraphs for which there is a PF2$^-$ assignment.

Perron–Frobenius theory studies the eigenvectors/eigenvalues of nonnegative matrices. We revisit this theory elsewhere in the book. The basic result is:

Theorem 2.5.1 (Perron–Frobenius) *Let A be an $n \times n$ matrix with real nonnegative entries $A_{ij} \geq 0$ ($1 \leq i, j \leq n$).*

(a) *Let $\rho(A) := \max_\lambda |\lambda|$ be the maximum of the absolute values of the eigenvalues of A. Then $\rho(A)$ is itself an eigenvalue of A, called the 'Perron–Frobenius eigenvalue', and it has an eigenvector $(a_1, \ldots, a_n)^t \geq 0$ (i.e. each $a_i \geq 0$), called a 'Perron–Frobenius eigenvector'.*

(b) *If it is not possible to simultaneously permute the rows and columns of A so that A takes the form*

$$A = \begin{pmatrix} B & C \\ 0 & D \end{pmatrix}$$

for submatrices B, C, D (such a matrix A is called 'irreducible'), then the Perron–Frobenius eigenvector is strictly positive and is unique up to scalar multiples.

(c) *Suppose A is irreducible in the sense of (b), and B is an $n \times n$ matrix obeying $0 \leq B_{ij} \leq A_{ij}$ $\forall i, j$. Then $\rho(B) \leq \rho(A)$, with equality iff $B = A$.*

See, for example, [420] for a proof and further results of this kind. In our case A is the adjacency matrix of a connected multigraph and so, being symmetric, is irreducible in the sense of (b). The classification of all PF2 and PF2$^-$ multigraphs follows by repeatedly applying Theorem 2.5.1(c) (see Question 2.5.2).

What do eigenvalues have to do with the other *A–D–E* classifications? Consider a finite subgroup G of SU$_2(\mathbb{C})$. Take the dimension of the equation $G \otimes R_i = \oplus_j m_{ij} R_j$: we get $2d_i = \sum_j m_{ij} d_j$, where $d_j = \dim(R_j)$. Hence the dimensions of the irreducible representations define a PF2 assignment for each of McKay's graphs, and hence those graphs must be of *A–D–E* type (provided we know $m_{ij} = m_{ji}$ and $m_{ii} = 0$).

Or consider lattices: let α_i be a basis of a positive-definite lattice, with all norm-squareds $\alpha_i \cdot \alpha_i = 2$. Then by the Cauchy–Schwarz inequality, $\alpha_i \cdot \alpha_j \in \{0, \pm 1\}$ for $i \neq j$. For $i < j$, if $\alpha_i \cdot \alpha_j = +1$ then replace α_j with $\alpha_j - \alpha_i$. What this means is that we can assume that each $\alpha_i \cdot \alpha_j \in \{0, -1\}$ for $i \neq j$. Put $A_{ij} = \alpha_i \cdot \alpha_j$ and $B = 2I - A$. Then B is a symmetric $\mathbb{N}$-matrix with zeros down the diagonal, and is easily seen to have Perron–Frobenius eigenvalue < 2. Thus B falls into the *A–D–E* pattern.

Suggestion *There are two different, though related, fundamental A–D–E patterns: namely, the PF2 and PF2$^-$ multigraph classifications. Any other instance of an A–D–E pattern reduces to one or the other of these.*

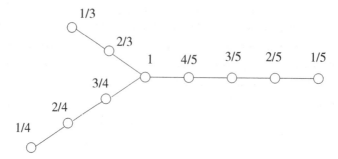

Fig. 2.10 The tree corresponding to $p = 3, q = 4, r = 5$.

This suggestion should be treated with some caution – as simple singularities illustrate, the same area may realise both types of A–D–E patterns, depending on the specific questions asked. In particular, duVal corresponds to Figure 1.4 and McKay to Figure 2.9. What relates these is that one of the nodes in the McKay graph (namely, that corresponding to the identity) is distinguished, and when it is deleted duVal's graph is recovered. We return to singularities in Section 3.2.5.

We encounter other A–D–E's later in this book. One of these (Theorem 6.2.2) is the only instance of A–D–E known to this author that hasn't yet been related to PF2 or PF2$^-$.

Incidentally, it is commonly suggested that a possible explanation for A–D–E may be the set of all triples $p, q, r \in \mathbb{N}$ for which

$$\frac{1}{p} + \frac{1}{q} + \frac{1}{r} > 1. \tag{2.5.3}$$

Then $(1, q, r)$, $(2, 2, r)$ and $(2, 3, 3), (2, 3, 4), (2, 3, 5)$ (corresponding to A_{q+r-1}, D_{r+2}, $E_{6,7,8}$, respectively) exhausts all solutions except for $p = 1, q \neq r$. However, this is not as fundamental as the graph explanation suggested above. In particular, given any triple obeying (2.5.3), construct the tree consisting of three strings leaving a common central vertex, of lengths $p - 1, q - 1, r - 1$, respectively (see Figure 2.10). Give this graph the assignment indicated in the figure – that is, label the ith vertex from the end of the first (respectively second, third) string $\frac{i}{p}$ (respectively $\frac{i}{q}, \frac{i}{r}$). Then inequality (2.5.3) is precisely the statement that this assignment is PF2$^-$, and thus that the graph will be of (unextended) A–D–E type. The reverse implication, showing that any PF2$^-$ graph $\mathcal{G}$ will necessarily correspond to a triple obeying (2.5.3), is much less elementary.

What comes after A–D–E? Natural candidates should be the graphs with largest eigenvalue $\rho = 3$, say. For the same reason that those with $\rho = 2$ arise in so many contexts, those with $\rho = 3$ surely will too. The difference is that the number and variety of graphs grows dramatically with the largest eigenvalue ρ. The list of graphs with $\rho = 2$ has such a simple and tight structure that different situations will automatically share a family resemblance, provided only that they depend critically on graphs with $\rho = 2$. For instance, the eigenvalues of any graph with $\rho = n$ must be character values of an n-dimensional representation of SU$_n$, if the graph is to have a chance at being the

McKay graph of a finite subgroup of SU_n; although this is automatic for $n = 2$, it is a severe constraint for $n \geq 3$. A different $\rho = 3$ situation can carry with it its own severe constraints, which would thus overwhelm the presence of the $\rho = 3$ graphs. We could say that $\rho = 2$ is a dominant gene, while $\rho = 3$ is recessive; this is why A–D–E is so ubiquitous, and why there seems to be no effective successor meta-pattern to A–D–E. (But see Section 6.3.2.)

For a final meta-pattern, consider 'modular functions'. After all, they appear in many places and disguises. Maybe we shouldn't regard their ubiquity as fortuitous. Instead, perhaps there's a deeper common 'situation' that is the source for that ubiquity. Two-dimensional lattices, perhaps? Riemann surfaces? The braid group $\mathcal{B}_3$?

Question 2.5.1. Let $L \subset \mathbb{R}^n$ be an even self-dual n-dimensional lattice. Assume there exists an orthonormal basis e_i of $\mathbb{R}^n$ and a number k such that the orthogonal lattice $2^k(\mathbb{Z}e_1 \oplus \cdots \oplus \mathbb{Z}e_n)$ is a sublattice of L (this is true for any self-dual L – see theorem 3.15 of [**238**]).

(a) Let L' be the orthonormal lattice $\mathbb{Z}e_1 \oplus \cdots \oplus \mathbb{Z}e_n$. Then the abelian group $L/(L \cap L')$ must be isomorphic to $\mathbb{Z}_{2^{k_1}} \times \cdots \times \mathbb{Z}_{2^{k_m}}$ for $0 < k_m \leq \cdots \leq k_1 \leq k$. Generators $\omega_1, \ldots, \omega_m \in L$ for it can be chosen so that $\omega_i = \frac{1}{2^{k_i}} \sum_j \omega_{ij} e_j$, where $\omega_{ij} \in \mathbb{Z}$, such that $\sum_i c_i \omega_i \in L'$ for $c_i \in \mathbb{Z}$ iff 2^{k_i} divides c_i for each i. Prove that there exist vectors $r_1, \ldots, r_j \in L'$ such that $r_i \cdot \omega_j \equiv \frac{1}{2^{k_i}} \delta_{ij}$ (mod 1).

(b) Let $x = \sum_i 2^{k_i - k_m} r_i^2 \omega_i = \frac{1}{2^{k_m}} \sum_j x_j e_j$, so $x \in L$ and $x_j \in \mathbb{Z}$. Prove that each x_j is odd (*Hint*: consider $\sum_i \omega_{ij} r_j - e_i$).

(c) Conclude from (2.5.1) that 8 must divide the dimension n.

Question 2.5.2. Using Theorem 2.5.1(c), prove that the multigraphs in Figures 1.4 and 2.9 exhaust all connected multigraphs whose eigenvalues λ all obey $|\lambda| \leq 2$.

Question 2.5.3. Why are there no loops in the McKay graph corresponding to any finite subgroup of $SL_2(\mathbb{C})$? Why don't these McKay graphs have directed edges?

Question 2.5.4. The classifications in Figures 1.4 and 2.9 depend on the requirement that the matrices be symmetric, i.e. that the multigraphs have no arrows. Find all 2×2 nonnegative integer matrices whose eigenvalues λ all obey $|\lambda| \leq 2$.

3

Gold and brass: affine algebras and generalisations

This chapter introduces the nontwisted affine algebras – infinite-dimensional Lie algebras of considerable mathematical and physical interest – and searches for generalisations that preserve and enhance those special features. The affine algebras supply classic examples of Moonshine, in that the characters of their integrable modules are vector-valued Jacobi functions for $SL_2(\mathbb{Z})$. They thread through the remainder of the book, guiding all subsequent mathematical developments. Their Lie groups are discussed in Section 3.2.6.

Algebraically, the affine algebras naturally generalise to the Kac–Moody algebras (Section 3.3.1), although that generalisation seems to lose some of their magic. In turn, the Kac–Moody algebras generalise naturally to the Borcherds–Kac–Moody algebras (Section 3.3.2), which play a significant role in Borcherds' proof of Monstrous Moonshine through their denominator identities (Section 3.4.2). Two other natural generalisations of affine algebras are described elsewhere in Section 3.3. In Section 3.4.1 we study an important special case of what we later call the orbifold construction, and in the final subsection we touch on a more recent and tangential development.

The Virasoro algebra (Section 3.1.2) plays a prominent structural role in conformal field theory (Chapter 4) and vertex operator algebras (Chapter 5); its relation to moduli spaces is a fundamental source of Moonshine itself.

3.1 Modularity from the circle

3.1.1 Central extensions

Let V be any (complex) vector space, and let $GL(V)$ denote the group of all invertible linear maps $V \to V$. A *projective representation* of a group G is a map $P : G \to GL(V)$ such that $P(e) = I$ (the identity), and given any elements $g, h \in G$, there is a nonzero complex number $\alpha(g, h)$ such that

$$P(g) P(h) = \alpha(g, h) P(gh). \qquad (3.1.1a)$$

We call P an *α-representation*. So just as a (true) representation is a group homomorphism $R : G \to GL(V)$, a projective representation defines a group homomorphism P from G into the projective group $PGL(V) := GL(V)/\{\mathbb{C}^\times I\}$ (hence the name); conversely, given a homomorphism $\pi : G \to PGL(V)$, arbitrarily choosing a 'section', that is a representative $P(g) \in GL(V)$ in each equivalence class $\pi(g) \in PGL(V)$, defines a projective representation of G. A projective representation P is a true representation iff $\alpha(g, h) = 1$ for all $g, h \in G$.

Projective representations are plentiful. For example, the multiplier μ in Definition 2.2.1 is a projective representation of $\mathrm{SL}_2(\mathbb{Z})$ whenever the weight k is rational. Quasi-periodicity (2.3.5a) is a projective representation of the abelian group $\mathbb{C}^2$ on the space of functions $f : \mathbb{C} \to \mathbb{C}$. In quantum physics (Section 4.2) the state of a system is completely described by a nonzero vector v in a Hilbert space. However, any nonzero multiple λv describes a physically identical state. Thus projective representations arise naturally also in quantum physics, where they are called 'ray representations'.

Note that associativity

$$\alpha(h, k)\,\alpha(g, hk)\,P(ghk) = P(g)\,(P(h)\,P(k)) = (P(g)\,P(h))\,P(k)$$
$$= \alpha(g, h)\,\alpha(gh, k)\,P(ghk)$$

tells us that

$$\alpha(h, k)\,\alpha(g, hk) = \alpha(gh, k)\,\alpha(g, h), \qquad \forall g, h, k \in G. \tag{3.1.1b}$$

This equation may remind the reader of a two-cocycle condition, hinting of the relevance of cohomology. Indeed, this function $\alpha : G \times G \to \mathbb{C}^\times$ is called a *2-cocycle* and group cohomology organises the projective representations.

Two projective representations $P_i : G \to \mathrm{GL}(V_i)$ are *(linearly) equivalent* if there is a vector space isomorphism $\varphi : V_1 \to V_2$ such that $\varphi^{-1} \circ P_1 \circ \varphi = P_2$. Equivalent projective representations must have the same 2-cocycle α. For a given α, the number of inequivalent irreducible α-representations of G equals the number of conjugacy classes of α-regular elements $g \in G$ (g is called α-*regular* if $\alpha(g, h) = \alpha(h, g)$ for all $h \in C_G(g)$). Hence this number is at most the number of inequivalent irreducible true G-representations.

We call projective representations $P_i : G \to \mathrm{GL}(V_i)$ *projectively equivalent* when there is a function $\beta : G \to \mathbb{C}^\times$ and a vector space isomorphism $\varphi : V_1 \to V_2$ such that

$$\varphi^{-1}(P_1(\varphi(g))) = \beta(g)\,P_2(g), \qquad \forall g \in G.$$

The 2-cocycles of projectively equivalent projective representations are related by

$$\alpha_2(g, h) = \alpha_1(g, h)\,\beta(gh)\,\beta^{-1}(g)\,\beta^{-1}(h).$$

β plays the role of a coboundary, so the 2-cocycles α_i of projectively equivalent projective representations lie in the same cohomology class $[\alpha] \in H^2(G, \mathbb{C}^\times)$, and $H^2(G, \mathbb{C}^\times)$ classifies the projectively inequivalent projective representations. $H^2(G, \mathbb{C}^\times)$ is an abelian group, called the *Schur multiplier*, and is finite when G is finite. The point of converting a problem into algebraic topology is that machinery (and experts!) are available to help compute these groups. For example, $H^2(\mathbb{Z}_n, \mathbb{C}^\times) = H^2(\mathrm{SL}_2(\mathbb{Z}), \mathbb{C}^\times) = H^2(\mathbb{M}, \mathbb{C}^\times) = \{0\}$ while $H^2(Co_1, \mathbb{C}^\times) \cong \mathbb{Z}_2$. This implies, for instance, that any projective representation of the Monster $\mathbb{M}$ is projectively equivalent to a true representation of $\mathbb{M}$.

Projective representations of Lie algebras are defined similarly: $P : \mathfrak{g} \to \mathrm{End}(V)$ is linear, and equations (3.1.1) become

$$[P(x), P(y)] = P([x, y]) + c(x, y) I, \tag{3.1.2a}$$

$$c(x, y) = -c(y, x), \tag{3.1.2b}$$

$$c([xy], z) = c([yz], x) = c([zx], y) = 0, \tag{3.1.2c}$$

where the 2-cocycle c is complex-valued and I is the identity endomorphism.

Geometrically, projective representations often arise from the following fundamental construction. Let $\mathcal{L} \to M$ be any line bundle with connection ∇ over some manifold M (Section 1.2.2). Let $\varphi : \mathfrak{g} \to \mathrm{Vect}(M)$ be a homomorphism from some Lie algebra $\mathfrak{g}$ to the Lie algebra of vector fields on M. The map $x \mapsto \nabla_{\varphi(x)}$, sending $x \in \mathfrak{g}$ to the covariant derivative in the direction $\varphi(x)$, associates with each $x \in \mathfrak{g}$ a differential operator on the space of sections of $\mathcal{L}$. Since

$$[\nabla_X, \nabla_Y] = \nabla_{[X,Y]} + R(X, Y) I$$

for each vector field X, Y, where R is the curvature of the connection, this map defines a projective representation of $\mathfrak{g}$ on the space $\Gamma(\mathcal{L})$ of sections of $\mathcal{L}$, with cocycle $c = R$. As we will see later this chapter, the central extensions of both the Witt and the loop algebras can be interpreted in this way [13]. This construction is well known in physics, where it falls under the slogan 'curvature is a local anomaly' (by contrast, global anomalies are monodromy effects like modularity).

A standard trick (central extensions) converts *projective* representations into *true* representations. Let G be any group, and let A be any abelian group. By a *central extension* $\widehat{G}$ of G by A, we mean that A can be identified with a subgroup of the centre of $\widehat{G}$, and the quotient $\widehat{G}/A$ is isomorphic to G. For example, the dihedral group $\mathcal{D}_4$ is a central extension (by $\mathbb{Z}_2$) of a central extension (by $\mathbb{Z}_2$) of a central extension (by $\mathbb{Z}_2$) of $\{e\}$.

Let P be a projective representation of a group G, and assume for simplicity that no operator $P(g)$ is a scalar multiple $a I$ of the identity. Let $\widehat{G}$ be the group consisting of all operators $a P(g)$, for $a \in \mathbb{C}^\times$ and $g \in G$. Then $\widehat{G}$ is a central extension of G by $\mathbb{C}^\times$, and $\widehat{G}$ is defined by a faithful representation in V. The projective representation of G has been transformed into a true representation of the larger group $\widehat{G}$. The specific situation for finite groups and the most common finite-dimensional Lie groups is simpler:

Theorem 3.1.1 (a) *Let G be a finite group. Then there is a central extension $\widetilde{G}$ of G by its Schur multiplier $H^2(G, \mathbb{C}^\times)$, with the following property: any projective representation $P : G \to \mathrm{GL}(V)$ of G lifts to a true representation $\widetilde{P} : \widetilde{G} \to \mathrm{GL}(V)$ of $\widetilde{G}$.*
(b) *Let G be a connected, finite-dimensional semi-simple Lie group over $\mathbb{R}$ or $\mathbb{C}$, and let $\widetilde{G}$ be its universal cover group (which is a central extension of G by the fundamental group $\pi_1(G)$). Then any continuous finite-dimensional projective representation $P : G \to \mathrm{GL}(V)$ of G lifts to a true representation $\widetilde{P} : \widetilde{G} \to \mathrm{GL}(V)$ of $\widetilde{G}$.*

Conversely, a true representation of $\widetilde{G}$ restricts to a projective representation of G. The central extension $\widetilde{G}$ in Theorem 3.1.1(a) is a finite group (e.g. for $\widetilde{Co}_1$ take the Conway group Co_0), and in (b) is a Lie group of the same dimension as G (see Theorem 1.4.3). For Lie groups there is a topological (π_1) as well as cohomological (H^2) obstacle to the trivialisation of projective representations. The assumption in (b) that G be semi-simple was made only to guarantee that the Schur multiplier of G would be trivial. The conclusion to Theorem 3.1.1(b) also holds for certain non-semi-simple Lie groups, such as the Poincaré group important to relativistic physics. On the other hand, the Galilei group, which plays the same role in pre-relativistic physics, has nontrivial Schur multiplier. In this case, the relevant cover will be a Lie group of higher dimension. The simplest example of this phenomenon is the additive group $\mathbb{C}^2$, and its central extension the three-dimensional Heisenberg group (see Question 3.1.3). It is through projective representations of $\mathbb{C}^2$ that the Heisenberg group and algebra arise in both theta functions (Section 2.4.2) and quantum physics (Section 4.2). Similarly, the Galilei group must act on nonrelativistic wave-functions (i.e. solutions to the Schrödinger equation (4.2.1)) *projectively* – this is a consequence of the nontriviality of the Schur multiplier of the Galilei group (Question 4.2.1).

Incidentally, the Schur multiplier $H^2(G, \mathbb{C}^\times)$ of a finite group G appears in another context. Consider any presentation of G, with say m generators and n relations. The finiteness of G requires that $m \leq n$. The Schur multiplier of G is a finite abelian group, so let h be its number of generators as in Theorem 1.1.1. Then $n - m \geq h$.

We are primarily interested in *one-dimensional central extensions* $\widehat{\mathfrak{g}}$ of Lie algebras $\mathfrak{g}$, that is a vector space $\widehat{\mathfrak{g}} = \mathfrak{g} \oplus \mathbb{C}C$ together with the brackets

$$[ab]_{\text{new}} = [ab]_{\text{old}} + c(a, b)\,C, \tag{3.1.3a}$$

$$[aC] = 0. \tag{3.1.3b}$$

The element C is called the *central term*. Equivalently, we have

$$0 \to \mathbb{C} \to \widehat{\mathfrak{g}} \to \mathfrak{g} \to 0, \tag{3.1.3c}$$

together with the requirement that the image $\mathbb{C}C$ of $\mathbb{C}$ in $\widehat{\mathfrak{g}}$ is in the centre of $\widehat{\mathfrak{g}}$. The short exact sequence (3.1.3c) says that there is an ideal in $\widehat{\mathfrak{g}}$ (namely the image of the second arrow) isomorphic as a Lie algebra to $\mathbb{C}$, and that when this ideal is projected out (by the third arrow) we recover $\mathfrak{g}$.

The exact sequence (3.1.3c) has the charm of not requiring an explicit *splitting* of $\widehat{\mathfrak{g}}$ into a $\mathfrak{g}$-part $\overline{\mathfrak{g}}$ (namely, a *lift* of the Lie algebra $\mathfrak{g}$ onto a subspace $\overline{\mathfrak{g}}$) and a $\mathbb{C}$-part $\mathbb{C}C$. The point is that there are many possible splittings: for example, given any such splitting $\widehat{\mathfrak{g}} = \overline{\mathfrak{g}} \oplus \mathbb{C}C$, choose a linear map $f : \overline{\mathfrak{g}} \to \mathbb{C}$; then a new splitting is obtained by replacing the subspace $\overline{\mathfrak{g}}$ with the span of the $a + f(a)C$, as a runs through $\overline{\mathfrak{g}}$. Modern mathematics abhors arbitrary choices, and so would encourage us to delay the choice of such a splitting as long as Good Fortune permits. Of course this is merely the current century-long fad, and there are advantages and disadvantages to it, and indeed physics prefers the opposite choice.

$\widehat{\mathfrak{g}}$ will be a Lie algebra iff the function $c : \mathfrak{g} \times \mathfrak{g} \to \mathbb{C}$ obeys (3.1.2b), (3.1.2c); as before, c is called the *2-cocycle* associated with the extension (3.1.3). The trivial 2-cocycle $c \equiv 0$ always works, in which case $\widehat{\mathfrak{g}}$ is merely the Lie algebra direct sum $\mathfrak{g} \oplus \mathbb{C}$.

We regard two extensions $\widehat{\mathfrak{g}}_1, \widehat{\mathfrak{g}}_2$ as equivalent if there is a Lie algebra isomorphism $\varphi : \widehat{\mathfrak{g}}_1 \to \widehat{\mathfrak{g}}_2$ that sends the ideal $\mathbb{C}C_1$ of $\widehat{\mathfrak{g}}_1$ onto $\mathbb{C}C_2 \subset \widehat{\mathfrak{g}}_2$. One way (but not the only way) to get equivalent extensions is to change the splitting $\widehat{\mathfrak{g}} = \overline{\mathfrak{g}} \oplus \mathbb{C}C$, as mentioned before. In the language of Lie algebra cohomology (see e.g. [183] for a mathematical treatment, or [27] for a physically motivated one), $f : \overline{\mathfrak{g}} \to \mathbb{C}$ is a 2-coboundary, and the resulting 2-cocycles c_1, c_2 define the same class in the cohomology space $H^2(\mathfrak{g})$. There are other ways though to obtain equivalent extensions – for example, the central term can be rescaled – so $H^2(\mathfrak{g})$ is in general too fine to serve as a 'moduli space' of one-dimensional central extensions of $\mathfrak{g}$, but it gives a very useful partial answer. For example, $H^2(\mathfrak{g})$ is trivial for any finite-dimensional semi-simple Lie algebra $\mathfrak{g}$, which means any such $\mathfrak{g}$ has only trivial central extensions (see Question 3.1.4).

For a concrete example, consider the n-dimensional abelian Lie algebra $\mathfrak{h} = \mathbb{C}^n$, with basis $\{e_1, \ldots, e_n\}$. A one-dimensional central extension $\widehat{\mathfrak{h}}$ of $\mathfrak{h}$ is uniquely determined by n^2 numbers $\alpha_{ij} \in \mathbb{C}$ defined by $[e_i e_j] = \alpha_{ij} C$, where $C \in \widehat{\mathfrak{h}}$ is central (all other brackets of $\widehat{h}$ are determined by bilinearity and $[e_i C] = 0$). Anti-commutativity requires $\alpha_{ij} = -\alpha_{ji}$, and anti-associativity is automatically satisfied. Thus each choice of an anti-symmetric $n \times n$ matrix $A = (\alpha_{ij})$ defines a one-dimensional central extension $\widehat{\mathfrak{h}}_A$ of $\mathfrak{h} = \mathbb{C}^n$, and conversely. The dependence of this argument on an arbitrary choice of basis e_i means there is redundancy here: in particular, two such central extensions $\widehat{\mathfrak{h}}_A$ and $\widehat{\mathfrak{h}}_{B'}$ define isomorphic Lie algebras iff there is an invertible matrix B such that $A' = B A B^t$. The reader can verify that any anti-symmetric matrix A is equivalent in this sense to the direct sum of k copies of $\begin{pmatrix} 0 & 1 \\ -1 & 0 \end{pmatrix}$, and $\ell = n - 2k$ copies of (0), where $2k$ is the rank of A. Thus we get a different one-dimensional central extension of $\mathbb{C}^n$, for each $k = 0, 1, \ldots, \lfloor n/2 \rfloor$. When A is invertible (i.e. $k = n/2$), we call $\widehat{\mathfrak{h}}$ a Heisenberg algebra; as simple a (non-simple!) Lie algebra as it is, it's one of the most important.

3.1.2 The Virasoro algebra

Recall the Witt algebra $\mathfrak{Witt}$ in (1.4.9). For each choice of $\alpha, \beta \in \mathbb{C}$, we get a module $V_{\alpha, \beta}$, with basis v_k, $k \in \mathbb{Z}$, given by

$$\ell_n . v_k = -(k + \alpha + \beta + \beta n) v_{k+n}. \qquad (3.1.4a)$$

This can be obtained from the derived module (Section 1.5.5) coming from the natural action of a subgroup of the diffeomorphism group $\mathrm{Diff}(S^1)$ on the space of differential 'forms' $p(z) z^\alpha (\mathrm{d}z)^\beta$, where $p(z) \in \mathbb{C}[z^{\pm 1}]$ are Laurent polynomials. Clearly, $V_{\alpha+m, \beta} \cong V_{\alpha, \beta}$ for any $m \in \mathbb{Z}$.

As usual, we are interested in unitary modules (Section 1.5.1), and for this we need an anti-homomorphism ω of $\mathfrak{Witt}$. Up to an automorphism of $\mathfrak{Witt}$, the unique choice is $\omega \ell_n = \ell_{-n}$. Then for this choice, $V_{\alpha,\beta}$ is unitary iff both $\text{Re}(\beta) = 1/2$ and $\alpha + \beta \in \mathbb{R}$ [**334**]. These modules are also irreducible.

The element $\ell_0 \in \mathfrak{Witt}$ is obviously special and plays the role of energy operator (Hamiltonian) in the application to physics. The most interesting $\mathfrak{Witt}$-modules are unitary ones with diagonalisable ℓ_0. In this case the eigenvalues of ℓ_0 will necessarily be real, and should have the physical interpretation of energy. Unfortunately, the only nontrivial unitary irreducible $\mathfrak{Witt}$-modules with ℓ_0 diagonalisable are those $V_{\alpha\beta}$. This is unfortunate because the eigenvalues of ℓ_0 in any $V_{\alpha\beta}$ have no upper or lower bound. For reasons of stability, physics wants energy to be bounded below. The space $V_{\alpha\beta}$ is infinite-dimensional, but ℓ_0 defines on it a natural grading into finite-dimensional subspaces, and so we are led to formally define its *graded-dimension* to be

$$\text{tr}_{V_{\alpha\beta}} q^{\ell_0} = \sum_{k \in \mathbb{Z}} q^{k+\alpha+\beta}. \tag{3.1.4b}$$

Unfortunately this never converges.

Central extensions are a common theme in infinite-dimensional Lie theory.[1] Their *raison d'être* is always the same: a richer supply of representations. The *Virasoro algebra* $\mathfrak{Vir}$ is the one-dimensional central extension $\mathfrak{Vir} = \mathfrak{Witt} \oplus \mathbb{C} C$ with brackets

$$[L_m L_n] = (m - n) L_{m+n} + \delta_{n,-m} \frac{m (m^2 - 1)}{12} C, \tag{3.1.5a}$$

$$[L_m C] = 0. \tag{3.1.5b}$$

As always, we avoid convergence issues by defining $\mathfrak{Vir}$ to consist of only finite linear combinations of these basis vectors. Incidentally, a common mistake in the physics literature is to regard C as a number: it is in fact a vector, though in most modules of interest to, for example, mathematical physics it is mapped to a scalar multiple cI of the identity.

The reason for the strange-looking (3.1.5a) is that we have little choice: $\mathfrak{Vir}$ is the unique nontrivial one-dimensional central extension of $\mathfrak{Witt}$ (Question 3.1.5). The factor $\frac{1}{12}$ there is conventional, but arises naturally in the realisations of $\mathfrak{Vir}$ by normal-ordered operators in Fock spaces (see (3.2.13), (3.2.14) for such a calculation). In fact, the normal-ordering prescription is somewhat arbitrary and actually we are much more interested in a slightly different basis of $\mathfrak{Vir}$, with L_0 replaced by $L_0 - C/24$. This is the combination appearing in almost every expression for characters from this point on. Where does this $-C/24$ come from? With this modified L_0, the brackets (3.1.5a) simplify (Question 3.1.8). According to conformal field theory or vertex operator algebras, this new basis corresponds to a change in topology (see Section 5.3.4), which can be calculated using the Atiyah–Singer Index Theorem [**8**], so physically the 'conformal anomaly' term $-c/24$ is a Casimir effect. But the best algebraic explanation for this $-c/24$ is given Section 3.2.3.

As before, $L_0 \in \mathfrak{Vir}$ is the energy operator, and so we want irreducible $\mathfrak{Vir}$-modules where L_0 is diagonalisable and its eigenvalues are bounded below. Let v be any eigenvector of L_0 in such a module, say $L_0 v = E v$, and suppose $L_n v \neq 0$ for some $n > 0$. Then

---

[1] On the other hand, the *finite-dimensional* simple Lie algebras do not have nontrivial central extensions.

$L_0(L_n v) = -nL_n v + L_n L_0 v = (E - n)L_n v$ and thus $(L_n)^\ell v$ will be an eigenvector of L_0 whose eigenvalue $E - n\ell$ has real part going to $-\infty$ as $\ell \to \infty$. Thus any $\mathfrak{Vir}$-module whose L_0-eigenvalues have real part bounded below must be a *highest-weight module*.

More precisely, because $\mathfrak{Vir}$ has a triangular decomposition (recall (1.5.5d))

$$\mathfrak{Vir}_- \oplus \mathfrak{Vir}_0 \oplus \mathfrak{Vir}_+ = \mathrm{span}\{L_n\}_{n<0} \oplus \mathrm{span}\{L_0, C\} \oplus \mathrm{span}\{L_n\}_{n>0},$$

we can mimic the construction of highest-weight modules in Section 1.5.3. In particular, for any $h, c \in \mathbb{C}$, the *Verma module* $M(c, h)$ is the universal $\mathfrak{Vir}$-module generated by a vector $v \neq 0$ obeying

$$L_0 v = h v, \quad C v = c v, \quad L_n v = 0, \qquad \forall n > 0.$$

The pair (c, h) is the highest weight; c is the *central charge* and h the *conformal weight*. As before, it can be more explicitly defined using the universal enveloping algebra, or equivalently by inducing the module from $\mathfrak{Vir}_0 \oplus \mathfrak{Vir}_+$ to all of $\mathfrak{Vir}$. By the Poincaré–Birkhoff–Witt Theorem 1.5.2, $M(c, h)$ has a basis given by all vectors

$$L_{-i_1} L_{-i_2} \cdots L_{-i_n} v,$$

for all integers $i_1 \geq i_2 \geq \cdots \geq i_n \geq 1$. Any other $\mathfrak{Vir}$-module with highest weight (c, h) is a homomorphic image of $M(c, h)$, or equivalently the quotient of $M(c, h)$ by some ideal.

Each Verma module $M(c, h)$ is indecomposable, but may not be irreducible. However, they all have a unique nontrivial irreducible quotient $V(c, h)$, which is then the unique irreducible $\mathfrak{Vir}$-module with highest weight (c, h).

The anti-linear anti-homomorphism ('adjoint') of $\mathfrak{Vir}$ sends L_n to L_{-n}, and fixes C. The only unitary irreducible $\mathfrak{Vir}$-modules where L_0 is diagonalisable and all its eigenspaces are finite-dimensional are certain $V_{\alpha,\beta}$ in (3.1.4a) (these are $\mathfrak{Vir}$-modules with C acting trivially), as well as certain highest-weight modules $V(c, h)$ and their duals, the lowest-weight modules $V(c, h)^*$. In fact, $V(c, h)$ (and $V(c, h)^*$) are unitary iff either: (i) both $c \geq 1$ and $h \geq 0$; or (ii) c and h fall into the *discrete series*, i.e. for $m, r, s \in \mathbb{N}$ with $1 \leq s \leq r \leq m + 1$,

$$c = c_m := 1 - \frac{6}{(m + 2)(m + 3)}, \quad h = h_{m;rs} := \frac{((m + 3)r - (m + 2)s)^2 - 1}{4(m + 2)(m + 3)}. \quad (3.1.6)$$

These $V(c, h)$ are called *positive-energy representations* since the spectrum of L_0 is positive. Thus the only unitary irreducible $\mathfrak{Vir}$-modules with L_0 diagonalisable, with finite-dimensional L_0-eigenspaces, and with the L_0-spectrum bounded below, are the $V(c, h)$ in (i) and (ii). They are the building blocks of the most interesting affine algebra representations, vertex operator algebra modules and conformal field theories.

For unitary $V(c, h)$, we have $V(c, h) = M(c, h)$ when both $c > 1$ and $h > 0$, or when $c = 1$ and $2\sqrt{h} \notin \mathbb{Z}$. In these cases, by analogy with (3.1.4b), $V(c, h)$ has graded-dimension

$$\dim_{V(c,h)}(q) := \mathrm{tr}_{V(c,h)} q^{L_0} = q^h \prod_{n=1}^{\infty}(1-q^n)^{-1}, \qquad (3.1.7\mathrm{a})$$

as the infinite product gives the generating function for the partition numbers:

$$\prod_{n=1}^{\infty}(1-q^n)^{-1} = \sum_{m=1}^{\infty} p(m) q^m \qquad (3.1.7\mathrm{b})$$

where $p(m)$ is the number of ways to write m as a sum $m = a_1 + a_2 + \cdots + a_k$ for positive integers $1 \le a_1 \le a_2 \le \cdots \le a_k$. Unlike (3.1.4b), this converges whenever $|q| < 1$. In fact, we recognise (3.1.7b) as (up to a factor of $q^{1/24}$) the reciprocal of the Dedekind eta $\eta(\tau)$ (2.2.6b), once we change variables by $q = e^{2\pi i \tau}$ – we saw last chapter that $\eta(\tau)$ is a modular form for $SL_2(\mathbb{Z})$. In fact we obtain

$$\dim_{V(c,h)}\left(e^{2\pi i(\tau+1)}\right) = e^{2\pi i(h-\frac{1}{24})}\dim_{V(c,h)}(e^{2\pi i\tau}), \qquad (3.1.7\mathrm{c})$$

$$\dim_{V(c,h)} e^{-2\pi i/\tau} = \sqrt{\frac{i}{\tau}} \int_{-\infty}^{\infty} \exp[2\pi i h h']\, \dim_{V(c,h)} e^{2\pi i\tau}\, \mathrm{d}h'. \qquad (3.1.7\mathrm{d})$$

This is our first glimpse of modularity from a graded dimension, though it certainly won't be our last. But $\eta(\tau)$ arises here through elementary combinatorics, so it is tempting to dismiss this modularity as accidental. This however would be an error.

What should be the characters of these $\mathfrak{Vir}$-modules? For simple Lie algebras, we define the character as a trace over formal exponentials of elements of the Cartan subalgebra. The analogue of the Cartan subalgebra here is $\mathfrak{Vir}_0 = \mathbb{C}L_0 \oplus \mathbb{C}C$, so the character of $V(c, h)$ should be

$$\mathrm{ch}_{c,h}(z_L, z_C) := \mathrm{tr}_{V(c,h)} e^{2\pi i z_L L_0 + 2\pi i z_C C}, \qquad (3.1.8)$$

which equals $e^{2\pi i c z_C}$ times the graded-dimension of $V(c, h)$ (with $q = e^{2\pi i z_L}$).

The characters of the discrete series (3.1.6) are calculated in [**477**], and again converge for $|e^{2\pi i z_L}| < 1$. Moreover, they obey a much more interesting modularity than do the graded-dimensions in (3.1.7): let $\begin{pmatrix} a & b \\ f & d \end{pmatrix} \in SL_2(\mathbb{Z})$ act on $\mathfrak{Vir}_0$ by

$$(z_L, z_C) \mapsto \left(\frac{a z_L + b}{f z_L + d}, z_C + \frac{f z_L^2 + (d-a) z_L - b}{24(f z_L + d)} \right); \qquad (3.1.9)$$

then $\mathrm{ch}_{c_m, h_{m;rs}}(z_L, z_C)$ is fixed by some $\Gamma(N)$ (recall (2.2.4a)), and for each fixed m (i.e. fixed central charge c), the span over all $1 \le s \le r \le m+1$ of the characters $\mathrm{ch}_{c_m, h_{m,rs}}$ is invariant under $SL_2(\mathbb{Z})$. They furnish a good example of *modular data* (Definition 6.1.6). This $SL_2(\mathbb{Z})$ action (3.1.9) is a little complicated; if instead we specialise to the variables $z_L = \tau$ and $z_C = -\tau/24$, then each

$$\mathrm{ch}_{c_m; h_{m;rs}}(\tau) := \mathrm{ch}_{c_m; h_{m;rs}}(\tau, -\tau/24) = e^{-2\pi i c/24}\, \mathrm{tr}_{V(c,h)} e^{2\pi i z_L L_0} \qquad (3.1.10)$$

is a modular function for some $\Gamma(N)$ for $\tau \in \mathbb{H}$, and for fixed m the characters $\mathrm{ch}_{c_m; h_{m;rs}}(\tau)$ form a vector-valued modular function for $SL_2(\mathbb{Z})$ (Definition 2.2.2).

The best explanation for the mysterious-looking discrete series (3.1.6) will probably come from the orbit method [563], but the analysis is still incomplete. At least part of the discrete series of the Virasoro algebra has been related to (co)homology theory of the universal cover of $SL_2(\mathbb{R})$, given a discrete topology [164]. This should be explored further.

The characters of the non-unitary $V(c, h)$, for $c, h \in \mathbb{R}$, have most of the properties of those of the unitary ones, and it is unfair to completely ignore them. For example, for $c, h \in \mathbb{R}$ the modules $V(c, h)$ have a contravariant nondegenerate Hermitian form $\langle \star, \star \rangle$, apart from the positive-definiteness condition. Lie algebras typically have too many representations and some criterion is needed that isolates the interesting ones, but unitarity is too restrictive here.

As we know, the Lie algebra $\mathrm{Vect}(S^1)$ of vector fields on the circle contains the real Witt algebra $\mathfrak{Witt}_{\mathbb{R}}$ (i.e. the span over $\mathbb{R}$ of the generators ℓ_n in (1.4.9)) as a dense 'Laurent polynomial' subalgebra. The connected real Lie group naturally associated with $\mathrm{Vect}(S^1)$ is the group $\mathrm{Diff}^+(S^1)$ of orientation-preserving diffeomorphisms $S^1 \to S^1$ of the circle. As a group, $\mathrm{Diff}^+(S^1)$ is simple [286] but as a manifold it is not simply connected: its universal cover $\widetilde{\mathrm{Diff}}(S^1)$ is the group of all diffeomorphisms $\phi : \mathbb{R} \to \mathbb{R}$ of the real line satisfying the periodicity condition $\phi(x + 2\pi) = \phi(x) + 2\pi$. The centre of the universal cover is $\mathbb{Z}$ (namely $\phi_n(x) = x + 2\pi n$) and $\widetilde{\mathrm{Diff}}(S^1)/\mathbb{Z} \cong \mathrm{Diff}^+(S^1)$.

Nontrivial central extensions of $\mathrm{Diff}^+(S^1)$ by a circle are explicitly constructed in, for example, section 6.8 of [465] and appendix D.5 of [295]; these all have a Lie algebra isomorphic to the real Virasoro algebra $\mathfrak{Vir}_{\mathbb{R}}$ (i.e. the $\mathbb{R}$-span of the generators L_m, C of (3.1.5)).

Lie theory for the Virasoro and Witt algebras (and more generally the Lie algebra $\mathrm{Vect}(M)$ of vector fields on any manifold M) is much more complicated than the finite-dimensional semi-simple theory described in Chapter 1. For example, although the 'exponential' map $\exp: \mathrm{Vect}(S^1) \to \mathrm{Diff}^+(S^1)$ is defined here (by first integrating the vector field to its flow), it is neither locally one-to-one nor locally onto (proposition 3.3.1 of [465]). By comparison, the exponential map of compact Lie groups is locally one-to-one and globally onto. Moreover, the complex Lie algebra $\mathbb{C} \otimes \mathrm{Vect}(S^1)$ does not have a corresponding Lie group. After all, although a vector field on S^1 corresponds to a path in the space of maps (in fact diffeomorphisms) $S^1 \to S^1$, and these form a group by composition, a *complex* vector field on S^1 corresponds to a path in the space of maps $S^1 \to \mathbb{C}$ and these won't form a group. Segal [502] suggests that the complex Lie semigroup $\mathbf{C}_{0,2}$ defined in Section 4.4.1 is the closest we can come to the complexification of $\mathrm{Diff}^+(S^1)$.

We have two fairly general frameworks in which to understand Lie group representations: Borel–Weil and the orbit method (*a.k.a.* geometric quantisation). There is, as we recall from Section 1.5.5, a general philosophy that says the representations of a group G (here $\mathrm{Diff}(S^1)$) are in one-to-one correspondence with certain orbits of the coadjoint action of G on the Lie algebra $\mathfrak{g}$ of G (here $\mathfrak{Witt}$). As mentioned earlier, Witten [563] explored this possible relation for the Virasoro algebra. For example, the homogeneous space $\mathrm{Diff}(S^1)/S^1$ appears as an orbit, and can be associated with ghosts in string theory.

The main motivation would be to find a new interpretation for the discrete series (3.1.6), which is a little mysterious from the algebraic point of view. Witten identified the orbits to which these should correspond, but couldn't quantise those orbits (this is a common curse of the orbit method).

The space $\mathrm{Diff}(S^1)/\mathrm{PSL}_2(\mathbb{R})$ is also a coadjoint orbit. Something special happens here when we replace $\mathrm{Diff}(S^1)$ with the larger group $\mathrm{QS}(S^1)$ of *quasi-symmetric* homeomorphisms of S^1: then $\mathrm{QS}(S^1)/\mathrm{PSL}_2(\mathbb{R})$ is called the *universal Teichmüller space* $\mathfrak{T}$. Every Teichmüller space $\mathfrak{T}_{g,n}$ (recall Section 2.1.4) is naturally contained in $\mathfrak{T}$. Likewise, $\mathrm{Diff}(S^1)/\mathrm{PSL}_2(\mathbb{R})$ naturally embeds in $\mathfrak{T}$ (every diffeomorphism of S^1 is quasi-symmetric), and intersects each $\mathfrak{T}_{g,n}$ transversely. See the reviews [460], [168] for definitions and references. Given this, an intriguing answer to the challenge suggested by Manin in Section 5.4.1 is to consider the reparametrisations of strings using quasi-symmetric homeomorphisms rather than diffeomorphisms; see [460] for some physical speculations.

Pursuing an analogue of Borel–Weil is at least as interesting. Recall that for G compact, we get an action of G on line bundles on the flag manifold $G_{\mathbb{C}}/B$, and this accounts for the special (i.e. finite-dimensional) representations of G. Manin [402] suggested that something similar happens to $\mathfrak{Vir}$, with now the moduli spaces of curves playing the role of the flag manifold. This thought was made much more precise in [357], [49], [13]. Consider the enhanced moduli space $\widehat{\mathfrak{M}}_{g,n}$ of Section 2.1.4, where each of the n marked points on the genus-g surface is given a local coordinate z_i. A copy of $\mathfrak{Witt}$ for each marked point acts naturally on $\widehat{\mathfrak{M}}_{g,n}$: the vector field $z_i^{\ell}\partial/\partial z_i$, for $\ell \geq 1$, changes the coordinate z_i; $\partial/\partial z_i$ moves the ith point; and finally $z_i^{\ell}\partial/\partial z_i$ for $\ell \leq -1$ can change the conformal structure of the surface. This action fills out the tangent space to any point on $\widehat{\mathfrak{M}}_{g,n}$, i.e. we get a surjective Lie algebra homomorphism from $\mathfrak{Witt}$ to the tangent space at any point on $\widehat{\mathfrak{M}}_{g,n}$, and from this we can derive the central extension geometrically by considering determinant line bundles (a nice introduction to this important object is [192]) over $\widehat{\mathfrak{M}}_{g,n}$.

Pushing this much further would force us into the complexities (and riches) of algebraic geometry and $\mathcal{D}$-modules (see [116] for a gentle introduction to the simplest $\mathcal{D}$-modules). A far-reaching generalisation of the Borel–Weil Theorem is the equivalence of categories established by Beilinson–Bernstein and Brylinski–Kashiwara: given a Lie group G with semi-simple Lie algebra $\mathfrak{g}$, their 'localisation functor' relates an algebraic category, whose objects include the Verma modules of $\mathfrak{g}$, with a topological category of $\mathcal{D}$-modules (i.e. sheaves of modules over a ring of differential operators over the flag manifold $G_{\mathbb{C}}/B$). Describing this would take us far afield (see [80], [417] for reviews and references). In conformal field theory, the Virasoro algebra, moduli spaces $\mathfrak{M}_{g,n}$, and mapping class groups $\Gamma_{g,n}$ take the place of $\mathfrak{g}$, $G_{\mathbb{C}}/B$ and the Weyl group [402], [530]. [49] relates Virasoro modules to $\mathcal{D}$-modules on the enhanced moduli space $\widehat{\mathfrak{M}}_{g,n}$.

In any case, this deep relation between moduli spaces of curves and $\mathfrak{Vir}$ is significant to Moonshine, because of its relation to the analogues of the *Knizhnik–Zamolodchikov (KZ) equations* in any conformal field theory at any genus. We elaborate on this elsewhere (starting in Section 3.2.4), but for now let us say that 'chiral blocks' are sections over the moduli spaces $\widehat{\mathfrak{M}}_{g,n}$, and satisfy a system of partial differential equations saying roughly that they respect this $\mathfrak{Vir}$ action. The monodromy of those equations gives rise to projective actions of the mapping class groups on the spaces of chiral blocks. Now, the chiral blocks of the space $\mathfrak{M}_{1,1}$ (or rather $\widehat{\mathfrak{M}}_{1,1}$) are vertex operator algebra characters (including for instance (3.1.10)), and $\Gamma_{1,1} \cong \mathrm{SL}_2(\mathbb{Z})$ (or rather its central extension $\widehat{\Gamma}_{1,1} \cong \mathcal{B}_3$) acts on them. This is conformal field theory's explanation for the modularity of these characters. Thus the Virasoro algebra, through its action on the $\widehat{\mathfrak{M}}_{g,n}$, lies at the heart of Moonshine.

Question 3.1.1. (a) Let $G = \mathbb{Z}_2 \times \mathbb{Z}_2$. Define a map $P : G \to \mathrm{GL}_2(\mathbb{C})$ by

$$P(0,0) = \begin{pmatrix} 1 & 0 \\ 0 & 1 \end{pmatrix}, \qquad P(1,0) = \begin{pmatrix} 0 & 1 \\ -1 & 0 \end{pmatrix},$$

$$P(0,1) = \begin{pmatrix} i & 0 \\ 0 & -i \end{pmatrix}, \qquad P(1,1) = \begin{pmatrix} 0 & -i \\ -i & 0 \end{pmatrix}$$

Verify that P is a projective representation of G.
(b) Let Q be the order 8 'quaternion group', given by the following relations:

$$Q = \{\pm 1, \pm i, \pm j, \pm k \mid -1 = (\pm i)^2 = (\pm j)^2 = (\pm k)^2,\ ij = k = -ji,\ -1 \text{ is in centre}\}.$$

Show that there is a homomorphism $\varphi : Q \to G$ with kernel $\{\pm 1\}$.
(c) Show that there is a true representation R of Q such that

$$P(x) = \delta(x)\, R(r(x)), \qquad \forall x \in G,$$

where $r(x) \in \varphi^{-1}(x)$, and where $\delta : G \to \mathbb{C}^\times$.

Question 3.1.2. Identify $G = S^1$ with $\mathbb{R}/\mathbb{Z}$, and for any class $[x] \in \mathbb{R}/\mathbb{Z}$, choose the unique representative $0 \le x < 1$. Verify that for any complex number α, the map $[x] \mapsto \alpha^x$ defines a one-dimensional projective representation of S^1. Find the corresponding true representation on the universal cover $\widetilde{G}$ of S^1.

Question 3.1.3. For this question, let G be the additive group $\mathbb{C}^2$. Define the function $\alpha : G \times G \to \mathbb{C}^\times$ by $\alpha(z, w) = \exp[z_2 w_1 - z_1 w_2]$. Verify that α obeys the 2-cocycle condition (3.1.1b), and construct the corresponding central extension.

Question 3.1.4. Find all one-dimensional central extensions of the Lie algebra A_1.

Question 3.1.5. Show that there are only two one-dimensional central extensions of the Witt algebra, up to isomorphism. (*Hint:* first show, changing basis if necessary, that $[L_0, L_n] = -nL_n$. Then consider anti-associativity of $[L_0[L_m L_n]]$.)

Question 3.1.6. (a) The group $\mathrm{PSL}_2(\mathbb{R})$ acts naturally on the unit disc $|z| < 1$ by Möbius transformations. Use this to embed $\mathrm{PSL}_2(\mathbb{R})$ naturally in $\mathrm{Diff}^+(S^1)$, and find the corresponding Lie subalgebra of $\mathrm{Vect}(S^1)$.
(b) The group $\mathrm{SL}_2(\mathbb{R})$ naturally acts on the space of semi-infinite rays $\mathbb{R}_\ge(x, y)$ in $\mathbb{R}^2$ with endpoint at the origin $(0, 0)$. Find this action, and use it to embed $\mathrm{SL}_2(\mathbb{R})$ in $\mathrm{Diff}^+(S^1)$. Find the corresponding Lie subalgebra of $\mathrm{Vect}(S^1)$.

Question 3.1.7. Prove that the Lie algebra of derivations of the algebra $\mathbb{C}[x^{\pm 1}]$ of Laurent polynomials is $\mathrm{Vect}(S^1)$.

Question 3.1.8. Find the constant $\alpha \in \mathbb{C}$ for which the new basis $L'_n = L_n + \alpha\, \delta_{n,0} C$ of $\mathfrak{Vir}$ has especially simple brackets $[L'_m, L'_n]$.

3.2 Affine algebras and their representations

The theory of nontwisted affine Kac–Moody algebras (usually called *affine algebras*) is very analogous to that of the finite-dimensional simple Lie algebras. Nothing infinite-dimensional tries harder to be finite-dimensional than affine algebras. Their construction is so trivial that it seems surprising anything interesting and new can happen here. But a certain 'miracle' happens. . .

Standard references for the theory of affine algebras are [328], [337], [214], [551]. We will ignore here an interesting part of the story: the KP hierarchy [423].

3.2.1 Motivation

Generalisations are too easy; they should be justified before they are endured. Here we describe the original justifications for the study of Kac–Moody algebras.

Each simple finite-dimensional Lie algebra has, as we know, a Weyl group, which is a symmetry of most of the data of the algebra (e.g. the weight multiplicities of finite-dimensional modules) and which encodes much (but not all) of the structure of the algebra. These Weyl groups are a very special sort of group: they are generated by reflections (namely those through the simple roots).

Associated with any vector $\alpha \in \mathbb{R}^n$, the reflection r_α through α, sending α to $-\alpha$ and fixing the hyperplane perpendicular to α is given by (1.5.5c). More abstractly, a reflection r is simply an *involution* (i.e. order 2: $r^2 = e$). A *finite reflection group* is a finite group generated by reflections. Coxeter studied these as symmetries of a regular solids.

For example, the dihedral group $\mathcal{D}_n$ (the group of symmetries of a regular n-gon) is a finite reflection group, consisting of n reflections and n rotations, and is generated by any two neighbouring reflections. The symmetric group $\mathcal{S}_n$ is a finite reflection group: it acts on an orthonormal basis e_i of $\mathbb{R}^n$ by permuting the subscripts, and is generated by the transpositions $(i, i + 1)$, which are reflections r_{α_i} through the vector $e_i - e_{i+1}$.

Finite reflection groups have remarkably simple presentations.

Definition 3.2.1 *A* Coxeter group *G is a group with a set R of generators, whose complete list of relations is*

$$(rr')^{m(r,r')} = e, \quad \forall r, r' \in R,$$

where $m(r, r) = 1$ and the other $m(r, r')$ all lie in $\{2, 3, \ldots, \infty\}$. (The value $m(r, r') = \infty$ means that rr' has infinite order.)

The geometry of Coxeter groups is quite pretty – see, for example, [301], [84]. In Section 7.1.1 we describe a generalisation due to Conway, and its relation to the Monster $\mathbb{M}$.

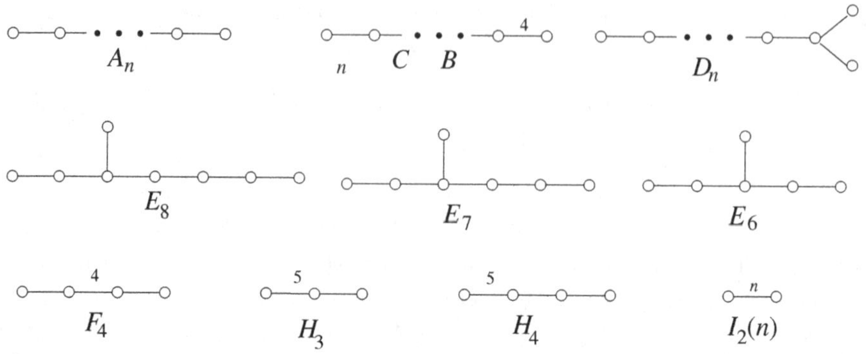

Fig. 3.1 The indecomposable finite Coxeter groups.

The list of finite Coxeter groups and finite reflection groups coincide. They are most easily described by the associated *Coxeter graph*: put a node for each generator $r \in R$, and connect two nodes with an edge labelled $m(r, r')$. To increase readability, erase the edge and label if $m(r, r') = 2$, and erase the label (but keep the edge) if $m(r, r') = 3$. The complete list of finite Coxeter groups (Coxeter, 1935) is given by arbitrary disjoint unions of the graphs of Figure 3.1. The group given by A_n is the symmetric group $\mathcal{S}_{n+1}$, and $I_2(n)$ is the dihedral group $\mathcal{D}_n$. The group H_3 is the symmetry group of the icosahedron, and is isomorphic to $\mathbb{Z}_2 \times \mathcal{A}_5$.

Figure 3.1 should remind us of Figure 1.17. Indeed, Figure 3.1 includes the Weyl groups of all simple finite-dimensional Lie algebras. More precisely, the Weyl groups consist of all finite Coxeter groups that obey the *crystallographic* condition: for all distinct $r, r' \in R, m(r, r') \in \{1, 2, 3, 4, 6\}$. Geometrically, the crystallographic condition says that the Coxeter group stabilises a lattice in $\mathbb{R}^n$ (see also Question 1.7.6). As we recall, the Weyl groups stabilise the corresponding root lattice.

Most Coxeter groups are infinite. As a graduate student, Robert Moody asked that, since the finite-dimensional semi-simple Lie algebras correspond to *finite crystallographic* Coxeter groups, what is the class of Lie algebras that correspond more generally to *any* Coxeter group? Presumably they should have a theory very similar to that of the semi-simple ones. The *partial* answer to Moody's beautiful question is that the Lie algebras corresponding to the (possibly infinite) *crystallographic* Coxeter groups are the Kac–Moody algebras! In fact, much of the interest in the affine algebras is due ultimately to their Weyl groups. We still don't know the Lie algebras corresponding to the noncrystallographic groups.

Victor Kac's road to these algebras was quite different. Let $\mathfrak{g}$ be a complex Lie algebra. By a $\mathbb{Z}$-*grading* we mean that we can write the vector space $\mathfrak{g}$ as $\mathfrak{g} = \bigoplus_{n=-\infty}^{\infty} \mathfrak{g}_n$, such that $[\mathfrak{g}_m, \mathfrak{g}_n] \subseteq \mathfrak{g}_{m+n}$ for all $m, n \in \mathbb{Z}$. We call $\mathfrak{g}$ a *simple $\mathbb{Z}$-graded* Lie algebra if, in addition, $\mathfrak{g}$ does not contain any nontrivial $\mathbb{Z}$-graded ideal.

It is probably hopeless to classify all simple $\mathbb{Z}$-graded Lie algebras – there are too many of them. However, decades earlier, Cartan had studied vector fields (i.e. derivations) on polynomial algebras, and found four infinite families that were simple $\mathbb{Z}$-graded, with the dimension $\dim(\mathfrak{g}_n)$ bounded above by some polynomial in n. We say that these $\mathbb{Z}$-graded algebras have *polynomial growth*. Kac conjectured, and Olivier Mathieu proved, the complete list of such algebras.

Theorem 3.2.2 [409] *The simple $\mathbb{Z}$-graded Lie algebras of polynomial growth are:*

(a) *the finite-dimensional simple Lie algebras;*

(b) *the loop algebras (possibly twisted);*

(c) *Cartan's four families; and*

(d) *the Witt algebra $\mathfrak{Witt}$.*

The proof is long and complicated. We've already met the finite-dimensional $\mathfrak{g}$ and the Witt algebra. Cartan's algebras are defined explicitly in, for example, [409]. The 'loop algebras' are constructed next subsection (there are six infinite families and seven exceptionals).

What we call the affine algebras – our main interest this chapter – are the central extensions of these loop algebras. Of course, such algebras cannot be simple because of their centres, and for this reason aren't in Mathieu's list. In any case, the affine algebras (together with the Virasoro algebra) answer a technical but natural algebraic question.

A couple of years after their mathematical introduction [325, 430], the nontwisted affine algebras were discovered independently in string theory [42], under the name *current algebras*.

The Lie algebras (a)–(d) in Mathieu's list are truly extraordinary, especially regarding their representation theory. The simplest of Cartan's families are the Weyl algebras, which are the differential operators on the algebra $\mathbb{C}[x_1, \ldots, x_n]$ of polynomials, generated by multiplication operators $x_1, \ldots, x_n$ and partial derivatives $\partial/\partial x_1, \ldots, \partial/\partial x_n$. Their modules are the simplest $\mathcal{D}$-modules and have deep connections throughout mathematics and physics (see [116], [80] for an introduction).

3.2.2 Construction and structure

Let $\overline{\mathfrak{g}}$ be any simple finite-dimensional Lie algebra. The affine algebra $\mathfrak{g} = \overline{\mathfrak{g}}^{(1)}$ is essentially the *(polynomial) loop algebra* $\mathcal{L}_{poly}\overline{\mathfrak{g}} = \mathbb{C}[t^{\pm 1}] \otimes \overline{\mathfrak{g}}$, defined to be all possible 'Laurent polynomials' $\sum_{n \in \mathbb{Z}} a_n t^n$ where each $a_n \in \overline{\mathfrak{g}}$ and all but finitely many $a_n = 0$. Treat t here as a formal variable. The bracket in $\mathcal{L}_{poly}\overline{\mathfrak{g}}$ is the obvious one: e.g. $[at^n, bt^m] = [ab]t^{n+m}$. Geometrically, $\mathcal{L}_{poly}\overline{\mathfrak{g}}$ is the Lie algebra of polynomial maps $S^1 \to \overline{\mathfrak{g}}$ (to see this realisation, think of $t = e^{2\pi i\theta}$). This explains the name, and also suggests several generalisations (e.g. take any manifold in place of S^1). But the loop algebra is simplest and best understood of these geometric Lie algebras, and the only one we consider in any depth (but see Section 3.3). Note that $\mathcal{L}_{poly}\overline{\mathfrak{g}}$ is infinite-dimensional. Its Lie groups are the *loop groups*, consisting of all maps of S^1 to a Lie group for $\overline{\mathfrak{g}}$ (Section 3.2.6).

We saw S^1 before, in the discussion of the Witt algebra, so we may expect the Virasoro and affine algebras to be related. In fact, the Witt algebra acts on the affine algebras as *derivations*. By definition, a derivation D is a linear map that obeys the product rule for derivatives: $D([xy]) = [(Dx)y] + [x(Dy)]$. The easiest examples are the 'inner derivations': $D = \text{ad}(x)$. All derivations of $\bar{\mathfrak{g}}$ are inner, but the loop algebra $\mathcal{L}_{poly}\bar{\mathfrak{g}}$ has several non-inner ones. In particular, because $\mathcal{L}_{poly}\bar{\mathfrak{g}}$ consists of all (polynomial) maps $S^1 \to \bar{\mathfrak{g}}$, the vector fields $\text{Vect}_{poly}(S^1)$, and hence the Witt algebra $\mathfrak{Witt}$, act on it. More precisely, using the realisation $\ell_j = -t^{j+1}d/dt$ of the basis vectors of (1.4.9), we get the action

$$\ell_j.(at^n) = -t^{j+1}\frac{d}{dt}(at^n) = -nat^{j+n}. \tag{3.2.1}$$

This relation between $\mathfrak{Witt}$ and $\mathcal{L}_{poly}\bar{\mathfrak{g}}$ plays an important role in the whole theory.

The loop algebra has a unique nontrivial one-dimensional central extension $\widehat{\mathcal{L}_{poly}\bar{\mathfrak{g}}} = \mathcal{L}_{poly}\bar{\mathfrak{g}} \oplus \mathbb{C}C$, defined by

$$[t^m x, t^n y] = t^{m+n}[x, y] + m\delta_{m,-n}\kappa(x|y)C \tag{3.2.2a}$$

for all $x, y \in \bar{\mathfrak{g}}$ and $m, n \in \mathbb{Z}$, where $\kappa(x|y)$ is the invariant bilinear form (Killing form) of $\bar{\mathfrak{g}}$. Thus $\widehat{\mathcal{L}_{poly}\bar{\mathfrak{g}}}$ has the same relation to $\mathcal{L}_{poly}\bar{\mathfrak{g}}$ that $\mathfrak{Vir}$ has to $\mathfrak{Witt}$. Incidentally, [**344**] relates the central extensions (3.2.2a) and (3.1.5) to logarithms of differential operators.

In addition, for a technical reason (namely, to make the simple roots linearly independent, so weight spaces can be finite-dimensional), a further noncentral one-dimensional extension is usually made. The result: by the affine algebra $\mathfrak{g} = \bar{\mathfrak{g}}^{(1)}$ we mean the extension of $\widehat{\mathcal{L}_{poly}\bar{\mathfrak{g}}}$ by the derivation $\ell_0 := t\frac{d}{dt}$. The Witt algebra also acts naturally on $\mathfrak{g}$ (Question 3.2.3). The superscript '(1)' denotes the fact that the loop algebra was twisted by an order-1 automorphism, in other words that it is nontwisted. It is called 'affine' because of its Weyl group, as we shall see.

For example, elements in $A_1^{(1)}$ are triples $(a(t), w, x)$ where $w, x \in \mathbb{C}$ and $a(t) = \sum_{n \in \mathbb{Z}} a_n t^n$, for all $a_n \in \mathfrak{sl}_2(\mathbb{C})$ and only finitely many $a_n \neq \begin{pmatrix} 0 & 0 \\ 0 & 0 \end{pmatrix}$. The Lie bracket is

$$[(a(t), w, x), (a'(t), w', x')] =$$
$$\left(\sum_{m,n} t^{m+n}[a_m, a'_n] + x\sum_n na'_n t^n - x'\sum_m ma_m t^m, \sum_m m\,\text{tr}(a_m a'_{-m}), 0 \right). \tag{3.2.2b}$$

Each object associated with $\bar{\mathfrak{g}}$ has an analogue here: Coxeter–Dynkin diagram, Weyl group, weights,... For instance, the affine Coxeter–Dynkin diagram (Figure 3.2) is obtained from that of $\bar{\mathfrak{g}}$ (Figure 1.17) by adding one node, labelled with an 'x'. We have included the labels a_i and (where different from a_i) colabels $a_i^\vee$, whose significance is given next subsection.

The Cartan subalgebra $\mathfrak{h}$ plays the same role here that it does in Chapter 1: decomposing modules into weight spaces. It can be chosen to be $\bar{\mathfrak{h}} \oplus \mathbb{C}C \oplus \mathbb{C}\ell_0$, where $\bar{\mathfrak{h}}$ is a Cartan subalgebra of the semi-simple algebra $\bar{\mathfrak{g}}$. In fact, $\mathfrak{g}$ has a triangular decomposition $\mathfrak{g} =$

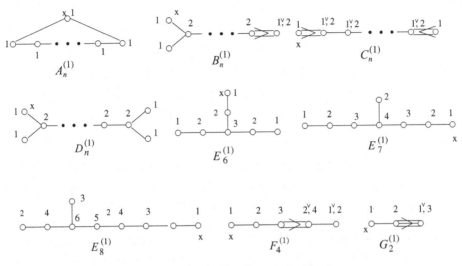

Fig. 3.2 The nontwisted affine Coxeter–Dynkin diagrams.

$\mathfrak{g}_+ \oplus \mathfrak{h} \oplus \mathfrak{g}_-$ (recall (1.5.5d)) where

$$\mathfrak{g}_\pm = \left(t^{\pm 1}\mathbb{C}[t^{\pm 1}] \otimes (\overline{\mathfrak{g}}_\mp \oplus \overline{\mathfrak{h}})\right) \oplus \mathbb{C}[t^{\pm 1}] \otimes \overline{\mathfrak{g}}_\pm \qquad (3.2.3a)$$

and $\overline{\mathfrak{g}} = \overline{\mathfrak{g}}_+ \oplus \overline{\mathfrak{h}} \oplus \overline{\mathfrak{g}}_-$ is a triangular decomposition of $\overline{\mathfrak{g}}$. Given $\mathfrak{h}$, we obtain the root-space decomposition of $\mathfrak{g}$, as in (1.5.5a):

$$\mathfrak{g} = \mathfrak{h} \oplus \bigoplus_{n\in\mathbb{Z}}\bigoplus_{\overline{\alpha}\in\overline{\Phi}} t^n\overline{\mathfrak{g}}_{\overline{\alpha}} \oplus \bigoplus_{n\in\mathbb{Z}\backslash 0} t^n\overline{\mathfrak{h}} \qquad (3.2.3b)$$

where $\overline{\mathfrak{g}} = \overline{\mathfrak{h}} \oplus \oplus_{\overline{\alpha}}\overline{\mathfrak{g}}_{\overline{\alpha}}$. We return to (3.2.3) when we study $\mathfrak{g}$-modules next subsection, but for now note that if $\overline{\mathfrak{g}}$ has rank r, then the root spaces $t^n\overline{\mathfrak{h}}$ of $\mathfrak{g}$ have dimension r while all $t^n\overline{\mathfrak{g}}_{\overline{\alpha}}$ have dimension 1. The latter, which act like root spaces in $\overline{\mathfrak{g}}$, are called *real*, while the former are called *imaginary*.

This loop algebra construction can be twisted. Let $\overline{\mathfrak{g}}$ again be any simple and finite-dimensional Lie algebra and let $\mathfrak{g}$ be the corresponding affine algebra. Choose any symmetry α of the Coxeter–Dynkin diagram of $\overline{\mathfrak{g}}$, of order N say, and extend this into an automorphism of $\overline{\mathfrak{g}}$ as in Section 1.5.4. We can further extend α to an automorphism of $\mathfrak{g}$, by requiring α to fix C and ℓ_0, and send at^n to $\alpha(a)\xi_N^{-n}t^n$. Then the fixed-point subalgebra $\mathfrak{g}_0$ of $\mathfrak{g}$ is

$$\mathfrak{g}_0 = \left\{\sum_n a_n t^n + wC + x\ell_0 \,\Big|\, a_n \in \overline{\mathfrak{g}}_{n \bmod N}\right\}, \qquad (3.2.4)$$

where $\overline{\mathfrak{g}}_i$ are the eigenspaces of α in $\overline{\mathfrak{g}}$ (recall (1.5.12)). This Lie algebra $\mathfrak{g}_0$ is called a *twisted affine algebra* and is denoted $\overline{\mathfrak{g}}^{(N)}$. All twisted affine algebras are listed in Figure 3.3, with their colabels. Twisted affine algebras behave very analogously to the nontwisted ones, and also have a significant role in the theory (Section 3.4.1).

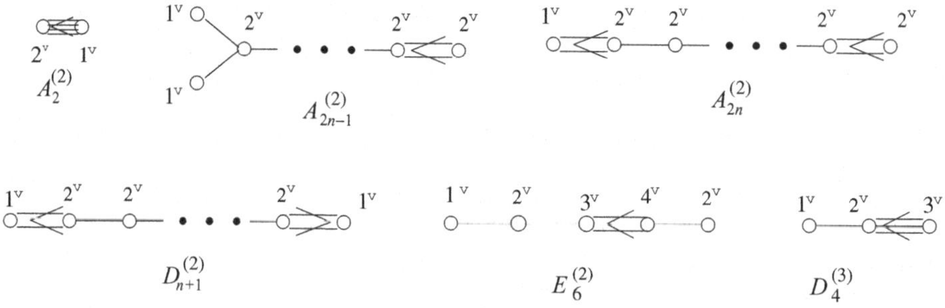

Fig. 3.3 The twisted affine Coxeter–Dynkin diagrams.

3.2.3 Representations

The loop algebra $\mathcal{L}_{poly}\bar{\mathfrak{g}}$ has no interesting modules, which is why we centrally extend it and introduce the affine algebras $\mathfrak{g} = \bar{\mathfrak{g}}^{(1)}$. No interesting $\mathfrak{g}$-module is finite-dimensional. However, $\mathfrak{g}$ has a triangular decomposition (3.2.3a), so highest-weight modules exist. Weights $\lambda \in \mathfrak{h}^*$ here are triples $(\bar{\lambda}, k, u) \in \bar{\mathfrak{h}}^* \times \mathbb{C}^2$; a weight vector v obeys

$$\bar{h}.v = \bar{\lambda}(h)\,v, \quad C.v = kv, \quad \ell_0.v = uv.$$

Define the Verma module $M(\bar{\lambda}, k, u)$ and the irreducible highest-weight module $L(\bar{\lambda}, k, u)$ – our greatest interest – as in Section 1.5.3. Given any highest-weight module M, the central term C acts as a multiple $k\,I$ of the identity; this constant k is a fundamental invariant of the representation called the *level* of M. On the other hand, the value of u is irrelevant (at least when the level is not 0) – see Question 3.2.4.

A highest-weight module M is infinite-dimensional but comes with a grading $M = \oplus_{n=0}^{\infty} M_{u+n}$ into eigenspaces of ℓ_0. Because ℓ_0 commutes with $\bar{\mathfrak{g}}$, these spaces M_{u+n} are all $\bar{\mathfrak{g}}$-modules, and the lowest, namely M_u, has highest weight $\bar{\lambda}$. Using this we can define the graded-dimension as in (3.1.4b). However, the ℓ_0-spaces of Verma modules will be infinite-dimensional, as will those of $L(\bar{\lambda}, k, u)$ unless $\bar{\lambda} \in P_+(\bar{\mathfrak{g}})$. There are two ways to proceed: either find a more suitable grading, or (more important) consider instead the *character*.

Defining these characters requires decomposing our modules into weight-spaces, and for this we should fix a basis for $\mathfrak{h}^*$. A basis for $\mathfrak{h}$ is $h_1, \ldots, h_r$ (the usual basis for $\bar{\mathfrak{h}}$) together with $h_0 := C - \sum_{i=1}^{r} a_i^\vee h_i$ and $-\ell_0$ ($a_i^\vee$ are the colabels of Figure 3.2). The reason for introducing h_0 will be clearer in Section 3.3.1. The dual basis for $\mathfrak{h}^*$, corresponding to $h_0, \ldots, h_r, -\ell_0$, is written $\omega_0, \ldots, \omega_r, \delta$. Recall from Sections 1.4.3 and 1.5.2 the Killing form $\kappa(\bar{h}|\bar{h}')$ and $(\bar{\lambda}|\bar{\mu})$ for $\bar{\mathfrak{g}}$; its analogue for affine algebras (Question 3.2.5) obeys

$$\kappa(\bar{z} + a\ell_0 + uC|\bar{z}' + a'\ell_0 + u'C) = \kappa(\bar{z}|\bar{z}') - au' - ua', \tag{3.2.5a}$$

$$\left(\sum_{i=0}^{r} \lambda_i \omega_i + b\delta \,\Big|\, \sum_{j=0}^{r} \mu_j \omega_j + d\delta\right) = \left(\sum_{i=1}^{r} \lambda_i \omega_i \,\Big|\, \sum_{j=1}^{r} \mu_j \omega_j\right) + \sum_{i=0}^{r} (d\lambda_i + b\mu_i). \tag{3.2.5b}$$

The level k is recovered from the weight λ by the formula

$$k = (\delta|\lambda) = \sum_{i=0}^{r} a_i^{\vee} \lambda_i. \tag{3.2.5c}$$

A useful formula gives the evaluation $\lambda(h)$:

$$\left(\sum_{i=0}^{r} \lambda_i \omega_i + b\delta\right)(\overline{z} + \tau\ell_0 + uC) = \left(\sum_{i=1}^{r} \lambda_i \omega_i\right)(\overline{z}) + ku - \tau b. \tag{3.2.5d}$$

In this notation, the roots of $\mathfrak{g}$ are $\overline{\alpha} - (\theta|\overline{\alpha})\omega_0 + n\delta$ for any root $\overline{\alpha}$ of $\overline{\mathfrak{g}}$ (these are the *real roots*, and have multiplicity 1), as well as $n\delta$ (the *imaginary roots*, with multiplicity equal to the rank r of $\overline{\mathfrak{g}}$). The root $\theta = \sum_{i=1}^{r} a_i \overline{\alpha}_i$ is called the *highest root* of $\overline{\mathfrak{g}}$, where a_i are the labels of $\mathfrak{g}$ (Figure 3.2). The positive roots are any of these with $n > 0$, together with $\overline{\alpha} - (\theta|\overline{\alpha})\omega_0$ for positive roots $\overline{\alpha}$. The simple roots are $\alpha_i := \overline{\alpha}_i - (\theta|\overline{\alpha}_i)\omega_0$ for $1 \leq i \leq r$, together with $\alpha_0 := \delta - \sum_{i=1}^{r} a_i \alpha_i$. Note that the adjoint representation of an affine algebra is not a highest-weight representation (why?). Many of these comments will make more sense when we associate a Coxeter–Dynkin diagram to $\mathfrak{g}$ in Section 3.3.1.

The weight-spaces for the Verma modules, and hence any highest-weight module M, are always finite-dimensional and so we can define their character ch_M as in (1.5.9a). For an easy example, the Verma module $M(\overline{\lambda}, k, 0) = M(\lambda)$ has character

$$\mathrm{ch}_{M(\lambda)}(h) = e^{\lambda(h)} \prod_{\alpha > 0} \left(1 - e^{-\alpha(h)}\right)^{-\mathrm{mult}\,\alpha}, \tag{3.2.6}$$

where 'mult α' denotes the dimension of the root-space $\mathfrak{g}_\alpha$ (which now may be > 1). We can obtain convergent graded dimensions by specialising this in any number of ways; the most obvious (called the *principal gradation*) chooses $h \in \mathfrak{h}$ so that $e^{\alpha_i(h)} = x$ for all simple roots α_i ($0 \leq i \leq r$), and $e^{\omega_0(h)} = 1$ (x is a formal variable). In other words, the principal grading of a vector with weight $\lambda - \sum_{i=0}^{r} n_i \alpha_i$ is $\sum_i n_i$ less than the grading of λ – this gradation keeps track of how many 'creation operators' f_i (using notation introduced in Section 3.3.1) are applied to the 'vacuum' v in order to create the given state.

For example, the affine algebra $A_1^{(1)}$ has positive roots $2\omega_1 - 2\omega_0 + n\delta$ (for $n \geq 0$) as well as $m\delta$ and $-2\omega_1 + 2\omega_0 + m\delta$ (for $m > 0$). All root multiplicities are 1. The simple roots are $\alpha_1 = 2\omega_1 - 2\omega_0$ and $\alpha_0 = \delta - \alpha_1$. A highest-weight λ looks like $\lambda_0\omega_0 + \lambda_1\omega_1$, with level $\lambda_0 + \lambda_1$. Applying the principal gradation to the $A_1^{(1)}$-Verma module, its character (3.2.6) specialises to the principally-graded dimension

$$\dim_{M(\lambda)}^{pg}(x) = x^{\lambda_1/2} \prod_{n=0}^{\infty} \left(1 - x^{-(1+2n)}\right)^{-1} \prod_{m=1}^{\infty} (1 - x^{-2m})^{-1} \prod_{m=1}^{\infty} \left(1 - x^{-(-1+2m)}\right)^{-1}$$

$$= e^{-\pi i \lambda_1 \tau} \eta(2\tau)/\eta(\tau)^{-2}, \tag{3.2.7}$$

where we write $x = e^{-2\pi i \tau}$ and recall the Dedekind eta function from (2.2.6b). Thus once again we find the remarkable fact that graded dimensions of Verma modules have something to do with the modular group $SL_2(\mathbb{Z})$ (compare (3.1.7)). Something similar happens for the highest-weight representations of any affine algebra!

Nothing particularly deep is happening here. The modularity of $\dim^{pg}$ arises here for free, simply from the combinatorics. Indeed, for any affine algebra, the specialised product of (3.2.6) is the generating function for some partition-like function as in (3.1.7b), and these have nice modular behaviour (by arguments like those used in Section 2.2.2).

More precisely, in the Verma module we get a free action of the creation operators of a Heisenberg subalgebra, coming from the central extension of the loop algebra of the Cartan subalgebra $\overline{\mathfrak{h}}$. Thus the modular group arises in affine algebra characters because of a Heisenberg algebra action. However, much as the discrete series (3.1.6) of $\mathfrak{Vir}$-modules behaves simpler than the other unitary $\mathfrak{Vir}$-modules, discretising the integral in (3.1.7d), certain families of $\mathfrak{g}$-modules have especially nice modular properties. What makes this work is the Weyl group. *It is this conjunction of the Heisenberg subalgebra with the affine Weyl group that makes affine algebras so special.*

The analogue for $\mathfrak{g}$ of the finite-dimensional modules of $\overline{\mathfrak{g}}$ are called the *integrable highest-weight modules*. Technically speaking, an integrable representation π is one where all $x \in \mathfrak{g}_{\pm}$ are locally nilpotent, that is, for each $v \in V$ there is a number $n_x(v)$ such that $\pi(x)^{n_x(v)} v = 0$. In particular, this means $e^{\pi(x)}$ is well-defined as an operator on the module by its Taylor series – in infinite dimensions most operators can't be exponentiated. These modules are called *integrable* because they are precisely those highest-weight modules that can be 'integrated' to a projective module of the corresponding loop group (Section 3.2.6). The integrable modules are precisely the unitary ones.

The *highest weight* $\lambda = \sum_{i=0}^{r} \lambda_i \omega_i$ is integrable iff each $\lambda_i \in \mathbb{N}$. Hence the set of all integrable level k highest weights is

$$P_+^k(\mathfrak{g}) := \left\{ \sum_{i=0}^{r} \lambda_i \omega_i \mid \lambda_i \in \mathbb{N}, \ k = \sum_{i=0}^{r} a_i^{\vee} \lambda_i \right\}. \tag{3.2.8}$$

Simple formulae for the cardinality $\| P_+^k(\mathfrak{g}) \|$ exist for all algebras (Question 3.2.6) – for example, for $A_r^{(1)}$ it is $\| P_+^k \| = \binom{k+r}{r}$. The most important weight in $P_+^k(\mathfrak{g})$ is $k\omega_0$, often denoted '0' in the literature. The module $L(k\omega_0)$ has a vertex operator algebra structure (Section 5.2.2) and corresponds to the vacuum sector in conformal field theory (Section 6.1.1).

The ℓ_0-eigenspaces of an integrable representation $L(\lambda)$ are all finite-dimensional representations of $\overline{\mathfrak{g}}$, and thus we can define its character $\mathrm{ch}_{L(\lambda)}$ as in (1.5.9a), although just as for the Virasoro algebra in (3.1.10) it proves to be more convenient to 'normalise' it:

$$\chi_\lambda(h) := e^{-(h_\lambda - c_\lambda/24)\,\delta(h)} \sum_{\beta \in \Omega(L(\lambda))} \dim L(\lambda)_\beta \, e^{\beta(h)}, \tag{3.2.9a}$$

where $L(\lambda) = \oplus L(\lambda)_\beta$ is the weight-space decomposition of $L(\lambda)$, $h \in \mathfrak{h}$, and

$$h_\lambda := \frac{(\lambda | \lambda + 2\rho)}{2(k + h^{\vee})}, \tag{3.2.9b}$$

$$c_\lambda := \frac{k}{k + h^{\vee}} \dim \overline{\mathfrak{g}} \tag{3.2.9c}$$

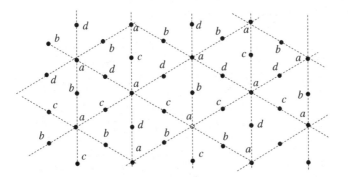

Fig. 3.4 The Weyl group of $A_2^{(1)}$ acting on level-2 weights.

are called the *conformal weight* and *central charge*, respectively, of $L(\lambda)$. The quantity $h^\vee = \sum_{i=0}^r a_i^\vee$ is called the *dual Coxeter number* and $\rho = \sum_{i=0}^r \omega_i$ the *Weyl vector*. The algebraic meaning of h_λ and c_λ involves the Virasoro algebra and is given shortly; $\delta(h)$ plucks out the coefficient $2\pi i \tau$ of ℓ_0 (recall (3.2.5d)). We are assuming in (3.2.9b) that the highest-weight component u has been set to 0 (Question 3.2.4). We discuss the normalisation (the exponential involving $h_\lambda - c_\lambda/24$) later in this subsection. As in (1.5.11), the character χ_λ can be written as an alternating sum over the Weyl group W, over a 'nice' denominator (namely the product in (3.2.6)). The difference is that W is now infinite.

See Figure 3.4 for the Weyl group of $A_2^{(1)}$ (projected to $\ddot{\mathfrak{h}}^*$), and Question 3.2.7 for some simple calculations. Much of the interest in affine algebras can be traced to the 'miracle' that their Weyl groups are a semi-direct product $Q^\vee \rtimes \ddot{W}$ of translations in a lattice $Q^\vee$ (the r-dimensional 'co-root lattice' of $\ddot{\mathfrak{g}}$) with the (finite) Weyl group $\ddot{W}$ of $\ddot{\mathfrak{g}}$. More precisely, for any root $\ddot{\alpha}$ of $\ddot{\mathfrak{g}}$ define the *co-root* $\ddot{\alpha}^\vee$ by $\ddot{\alpha}^\vee = 2\ddot{\alpha}/(\ddot{\alpha}|\ddot{\alpha})$; by the co-root lattice $Q^\vee \subset \ddot{\mathfrak{h}}^* \subset \mathfrak{h}^*$ of $\ddot{\mathfrak{g}}$ we mean the $\mathbb{Z}$-span of these co-roots. For any vector $\beta \in Q^\vee$, define the map

$$t_\beta(\mu) = \mu + (\mu|\delta)\,\beta - ((\mu|\beta) + (\beta|\beta)\,(\mu|\delta)/2)\,\delta, \qquad (3.2.10a)$$

$\forall \mu \in \mathfrak{h}^*$. It is straightforward to verify $t_\beta t_\gamma = t_{\beta+\gamma}$, and thus these deserve the name 'translations'. Any element of the Weyl group W of $\mathfrak{g}$ can be written uniquely as a pair (t_β, w) for some $\beta \in Q^\vee$ and some $w \in \ddot{W}$, and

$$(t_\beta, w) \circ (t_{\beta'}, w') = \left(t_\beta\, t_{w(\beta')},\, ww'\right). \qquad (3.2.10b)$$

As in (1.5.6d), weights $\mu \in \Omega(L(\lambda))$ in the same Weyl orbit of an integrable module have the same multiplicities. One thing this implies is that χ_λ will be of the form 'theta series'/denominator. In particular, the lattice is $Q^\vee$, and the '$(\beta|\beta)\delta$' term in (3.2.10a) provides the quadratic form in the lattice theta series. As we know from (2.3.10), theta series are modular forms, and this is the second complementary reason the modular group $SL_2(\mathbb{Z})$ makes an appearance (the first was the combinatorics of the free action of the Heisenberg subalgebra of creation operators). To make this more precise, consider the highest weight $\lambda = \lambda_0 \omega_0 + \lambda_1 \omega_1 \in P_+^k(A_1^{(1)})$. Then

$$\chi_\lambda(2\pi i (z + \tau \ell_0 + uC)) = \frac{\Theta_{\lambda_1+1}^{(k+2)}(\tau, z, u) - \Theta_{-\lambda_1-1}^{(k+2)}(\tau, z, u)}{\Theta_1^{(2)}(\tau, z, u) - \Theta_{-1}^{(2)}(\tau, z, u)}, \tag{3.2.11a}$$

$$\Theta_m^{(n)}(\tau, z, u) := e^{-2\pi i n u} \sum_{\ell \in \mathbb{Z}+\frac{m}{2n}} \exp[2\pi i n \tau \ell^2 - 2\sqrt{2} \pi i n \ell z]. \tag{3.2.11b}$$

Any $\mathfrak{g}$ has an analogue of (3.2.11), the *Weyl–Kac character formula*

$$\chi_\lambda(2\pi i (z + \tau \ell_0 + uC)) = \frac{\sum_{w \in \overline{W}} \epsilon(w) \Theta_{w(\lambda+\rho)}^{(k+h^\vee)}(\tau, z, u)}{\sum_{w \in \overline{W}} \epsilon(w) \Theta_{w(\rho)}^{(h^\vee)}(\tau, z, u)}, \tag{3.2.11c}$$

where both the numerator and denominator involve an alternating sum over the finite Weyl group $\overline{W}$ of $\overline{\mathfrak{g}}$, and where the theta series in (3.2.11c) involves a sum over the lattice $Q^\vee$ shifted by some weight and appropriately rescaled. For example, the Weyl group of A_1 is $\mathcal{S}_2$ and its co-root lattice $Q^\vee$ is $\sqrt{2}\mathbb{Z}$. The key variable in (3.2.11) is the modular one τ – the main role of the other variables is to ensure linear independence. The character χ_λ converges for any choice of $\tau \in \mathbb{H}$, $z \in \mathbb{C}^r$ and $u \in \mathbb{C}$.

Thus the denominator of the character of an irreducible integrable $\mathfrak{g}$-module $L(\lambda)$ is a modular form, by virtue of the combinatorics of Verma modules. The numerator is a modular form, by virtue of the structure and action of the affine Weyl group. Together they give a modular function.

Theorem 3.2.3 [333] *Let $\overline{\mathfrak{g}}$ be finite-dimensional and simple, and let $\mathfrak{g} = \overline{\mathfrak{g}}^{(1)}$ be the corresponding affine algebra. Define $\chi_\lambda(\tau, z, u) = \chi_\lambda(2\pi i (z + \tau \ell_0 + uC))$. Fix any level $k \in \mathbb{N}$. Then for any integrable weight $\lambda \in P_+^k(\mathfrak{g})$, $\chi_\lambda(\tau, 0, 0)$ is a modular function for some congruence subgroup $\Gamma(N)$. Moreover, define a column vector $\vec{\chi}(\tau, z, u)$ with entries $\chi_\lambda(\tau, z, u)$ for each $\lambda \in P_+^k(\mathfrak{g})$. Then there is a unitary representation ρ of $SL_2(\mathbb{Z})$ such that*

$$\vec{\chi}\left(\frac{a\tau+b}{f\tau+d}, \frac{z}{f\tau+d}, u - \frac{f(z|z)}{2(f\tau+d)}\right) = \rho\begin{pmatrix} a & b \\ f & d \end{pmatrix} \vec{\chi}(\tau, z, u),$$

for any $\begin{pmatrix} a & b \\ f & d \end{pmatrix} \in SL_2(\mathbb{Z})$.

We say that the characters χ_λ define a *vector-valued Jacobi form* for $SL_2(\mathbb{Z})$, with multiplier ρ (recall Definition 2.2.2). This modularity of affine characters is fundamental to this book, and a prototypical example of much of what follows. The complex matrices $\rho(A)$ here are examples of modular data (Sections 6.1.2 and 6.2.1). A $\Gamma(N)$ that uniformly works in Theorem 3.2.3 is to let N be the least common multiple of all denominators of $h_\lambda - c/24$ (these will always be rational), as λ runs through the finite set $P_+^k(\mathfrak{g})$.

We can now explain McKay's observation (0.5.1) that the coefficients of $j(\tau)^{\frac{1}{3}}$ are related to the E_8 Lie group. $j^{\frac{1}{3}}(\tau)$ equals the character $\chi_{\omega_0}(\tau, 0, 0)$ of the integrable $E_8^{(1)}$-module. The q-coefficients ($q = e^{2\pi i \tau}$) of $j^{\frac{1}{3}}(\tau)$ are thus dimensions of the

ℓ_0-eigenspaces of $L(\omega_0)$, which are automatically E_8-modules. Because $P_+^1(E_8^{(1)}) = \{\omega_0\}$, all modularity properties of the character $j^{\frac{1}{3}}$ are a direct consequence of Theorem 3.2.3.

All of this assumes the underlying finite-dimensional Lie algebra $\bar{\mathfrak{g}}$ is semi-simple. When it is merely reductive (i.e. the direct sum of copies of the one-dimensional abelian algebra $\mathfrak{u}_1$, with a number of simple Lie algebras), something different happens. For example, consider the affinisation of $\mathfrak{u}_1$ (the *oscillator algebra*). It has basis C, a_n ($n \in \mathbb{Z}$) and obeys relations

$$[C, a_n] = 0, \qquad [a_m, a_n] = m\delta_{m,-n}C. \tag{3.2.12a}$$

Its irreducible unitary modules are parametrised by a highest weight $\lambda \in \mathbb{R}$, and are Verma modules $M(\lambda)$. In particular, any $\lambda \in \mathbb{R}$ defines a different irreducible unitary module. They can be realised in the space of polynomials $\mathbb{C}[x_1, x_2, \ldots]$ by the operators $C.p(x) = p(x), a_0.p(x) = \lambda\, p(x)$, and for all $n \geq 1$

$$a_n p(x) = \frac{\partial}{\partial x_n} p(x), \qquad a_{-n} p(x) = n x_n p(x). \tag{3.2.12b}$$

Note that the level k here is 1 (why can we demand $k = 1$?). The reader can verify that this representation has (normalised) character

$$\chi_\lambda(\tau) = q^{\lambda^2/2}/\eta(\tau). \tag{3.2.12c}$$

These characters aren't linearly independent (since $\chi_{-\lambda} = \chi_\lambda$), but the reader can work out the usual remedy. Their modularity is discussed in Section 6.2.2. In the language of conformal field theory, the unitary modules of the oscillator algebra $\mathfrak{u}_1^{(1)}$ are *quasirational* while the integrable modules of affine algebras are *rational*. Nevertheless, the oscillator algebra (studied in detail in [**334**]) is a convenient toy model for the affine algebras.

Last subsection we saw that $\mathfrak{Witt}$ acts naturally on loop algebras by derivations. Does $\mathfrak{Witt}$ act on affine modules? Consider the oscillator algebra for simplicity. We will have a universal $\mathfrak{Witt}$ action on $\mathfrak{u}_1^{(1)}$-modules M if we can construct the basis ℓ_n of (1.4.9) out of the operators a_m of (3.2.12a), that is realise the ℓ_n in the universal enveloping algebra $U(\mathfrak{u}_1^{(1)})$ (or some completion thereof). We are led to consider quadratic combinations in the a_m, since that is the simplest after linear ones (which won't work), and also since ℓ_0 has the interpretation of a Hamiltonian, which always contains a quadratic part. Define

$$t_m = \sum_{i \in \mathbb{Z}} a_{-i} a_{m+i}. \tag{3.2.13a}$$

Being an infinite sum, convergence won't be automatic, but let's ignore that for now. Then

$$[t_m, a_n] = \sum_{i \in \mathbb{Z}} a_{-i}[a_{m+i}, a_n] + [a_{-i}, a_n]a_{m+i} = -n a_{m+n}C - nC a_{m+n} = -2n a_{m+n}C. \tag{3.2.13b}$$

In, for example, a highest-weight module, C acts as a scalar k, and so (at least for $k \neq 0$) $\ell_m := \frac{1}{2k} t_m$ mimics the action of the standard 𝔚itt action $\ell_m = -t^{m+1} d/dt$ on the loop algebra $\mathcal{L}_{poly} \mathfrak{u}_1$. This looks promising. We compute from (3.2.13b)

$$[\ell_m, \ell_n] = (2k)^{-2} \sum_{i \in \mathbb{Z}} [t_m, a_{-i}] a_{n+i} + (2k)^{-2} \sum_{j \in \mathbb{Z}} a_{-j} [t_m, a_{n-j}]$$

$$= (2k)^{-1} \sum_{i \in \mathbb{Z}} i a_{m-i} a_{n+i} + (2k)^{-1} \sum_{j \in \mathbb{Z}} (-n - j) a_{-j} a_{m+n+j} = (m - n) \ell_{m+n},$$

establishing that indeed the ℓ_m form a realisation of 𝔚itt in $U(\mathfrak{u}_1^{(1)})$.

Unfortunately, the sum in (3.2.13a) doesn't converge. Take M to have highest-weight vector v with highest weight (λ, k). Then

$$t_0.v = \sum_{i \leq -1} (a_i a_{-i} - iC).v + a_0 a_0.v + \sum_{j \geq 1} a_{-j} a_j.v = k^2 v + k \left(\sum_{j \geq 1} j \right) v, \quad (3.2.13c)$$

which diverges. This means (3.2.13a) must be modified. The simplest correction can be written $T_m := \sum_{i \in \mathbb{Z}} : a_{-i} a_{m+i} :$, where the *normal-ordering* $: a_m a_n :$ is defined to equal either $a_m a_n$ or $a_m a_n$, depending on whether or not $m \leq n$. For $m \neq 0$, $T_m = t_m$, but $T_0.v = k^2 v$. Indeed, each operator T_m will be defined on any Fock space. We find that

$$L_m := (2k)^{-1} \sum_{i=-\infty}^{\infty} : a_{-i} a_{m+i} : \qquad (3.2.14a)$$

satisfies both

$$[L_m, a_n] = -n a_{m+n}, \qquad (3.2.14b)$$

$$[L_m, L_n] = (m - n) L_{m+n} + \frac{m^3 - m}{12} \delta_{m,-n}. \qquad (3.2.14c)$$

Thus any highest-weight $\mathfrak{u}_1^{(1)}$-module is simultaneously a 𝔙ir-module with central charge $c = 1$. Thus this nonzero central charge arises as an analytic effect.

Using (3.2.12a), this normal-ordering (3.2.14) doesn't change $L_n = \ell_n$, for $n \neq 0$, but shifts the divergent ℓ_0 by the infinite multiple $\left(\sum_{i=1}^{\infty} i \right)$ of C. There is nothing particularly special about this normal-ordering; for example, for any fixed ℓ we could have replaced the condition '$m \leq n$' with '$m \leq n + \ell$', and nothing would have changed except L_0 would have been shifted by some other multiple of C. This is a clue to understanding what is so special about the $-c/24$ shifts in, for example, (3.1.10) or (3.2.9a). The arbitrariness of the normal-ordering can be removed by reinterpreting ('regularising') the divergent term in (3.2.13c) as $k\zeta(-1)$ (recall (2.3.1)). Equivalently, this amounts to replacing the normal-ordered L_0 with $L_0 - C/24$. This is the algebraic 'explanation' for the naturality of the shift, and hence the pervasive appearance of $-c/24$: simply put, algebra prefers $L_0 - C/24$ over all other combinations $L_0 + \alpha C$ (recall Question 3.1.8). It should thus not come as a complete surprise that so too does $\text{SL}_2(\mathbb{Z})$. Incidentally, this '24', $\zeta(-1)$, the special dimensions $8 + 2$ and $24 + 2$ in string theory and the 24 of Section 2.5.1 are all directly related.

More generally, Bloch [63] considered other algebras of differential operators on S^1. In particular, in place of $\ell_n = -t^{n+1} \mathrm{d}/\mathrm{d}t$ he considers

$$\ell_n^{(r)} = (-1)^{r+1}(t\mathrm{d}/\mathrm{d}t)^r t^n (t\mathrm{d}/\mathrm{d}t)^{r+1}.$$

He obtains a (projective) realisation of these $\ell_n^{(r)}$ by normal-ordering operators in a Fock (or highest-weight) module, exactly as we do here. The analogue of $(m^3 - m)/12$ in the bracket $[L_m^{(r)}, L_n^{(s)}]$ is a polynomial of degree $2r + 2s + 3$ in m. As before, we want to remove this arbitrary choice of normal-ordering. Naively dropping it introduces the divergence $1^{2r+1} + 2^{2r+1} + \cdots$, so as before replace it with the Riemann zeta value $\zeta(-1 - 2r)$, i.e. replace $L_0^{(r)}$ with $L_0^{(r)} + (-1)^r \zeta(-1 - 2r)C/2$. Then the polynomial in m becomes the monomial $(r + s + 1)!(r + s + 1)!m^{2r+2s+3}/(2(2r + 2s + 3)!)$. This appearance of 'zeta function regularisation' in algebra has been interpreted and generalised in the vertex operator algebra framework (see [375] for a review).

Identical comments hold for affine algebras. Choose a basis x_a of $\overline{\mathfrak{g}}$, orthonormal with respect to the Killing form: $\kappa(x_a|x_b) = \delta_{ab}$. Then for $\lambda \in P_+^k(\mathfrak{g})$, (3.2.14) become

$$L_m := \frac{1}{2(k + h^\vee)} \sum_{j \in \mathbb{Z}} \sum_a : (t^{-j}x_a)(t^{m+j}x_a) :, \tag{3.2.15a}$$

$$[L_m, xt^n] = -nxt^{m+n}, \qquad \forall x \in \overline{\mathfrak{g}}, \tag{3.2.15b}$$

$$[L_m, L_n] = (m - n)L_{m+n} + c_\lambda \frac{m^3 - m}{12}\delta_{m,-n}. \tag{3.2.15c}$$

Thus the $\mathfrak{g}$-module $L(\lambda)$ is also automatically a completely reducible $\mathfrak{Vir}$-module. Each irreducible $\mathfrak{Vir}$-submodule has central charge c_λ and conformal weight $h \in h_\lambda + \mathbb{N}$ (see (3.2.9)). In $L(\lambda)$, the Virasoro generator L_0 and the derivation ℓ_0 of $\mathfrak{g}$ are related by $L_0 = h_\lambda Id + \ell_0$. Equation (3.2.15a), known as the *Sugawara construction*, should remind us of the *quadratic Casimir* $\Omega := \frac{1}{2}\sum_a x_a x_a$ of $\overline{\mathfrak{g}}$, that is, the simplest nontrivial element in the centre of $U(\overline{\mathfrak{g}})$; it acts on the irreducible $\overline{\mathfrak{g}}$-module $L(\overline{\lambda})$ as multiplication by the scalar $(\lambda|\lambda + 2\rho)$ (recall (3.2.9b)). The shift by the dual Coxeter number $h^\vee$ in (3.2.15a) arises algebraically as the eigenvalue of Ω in the adjoint representation of $\overline{\mathfrak{g}}$; its physical significance is discussed in Section 6.2.1.

The integrable modules of twisted affine algebras $X_r^{(N)}$ (recall Figure 3.3) behave similarly. As we know from (3.2.4), $X_r^{(N)}$ is obtained from the nontwisted affine algebra $\mathfrak{g} = X_r^{(1)}$ and an order-N symmetry α of the Coxeter–Dynkin diagram of X_r. The integrable highest-weight $X_r^{(N)}$-modules $L(\lambda)$ are parametrised by $(r + 1)$-tuples $\lambda \in P_+^k$ as in (3.2.8), where the co-labels $a_i^\vee$ are now given in Figure 3.3. These modules also have weight-space decompositions as in (1.5.6a) and characters χ_λ as in (3.2.9a). Their characters are also modular (see theorem 13.9 of [328] for details).

Theorem 3.2.4 [333] *The characters χ_λ, $\lambda \in P_+^k(A_{2r}^{(2)})$ form a vector-valued Jacobi function for $SL_2(\mathbb{Z})$, as in Theorem 3.2.3. For $\mathfrak{g} = A_{2r-1}^{(2)}$, $D_{r+1}^{(2)}$, $E_6^{(2)}$ and $D_4^{(3)}$, respectively, define $\mathfrak{g}' = D_{r+1}^{(2)}$, $A_{2r-1}^{(2)}$, $E_6^{(2)}$, $D_4^{(3)}$ and $N = 2, 2, 2, 3$; then the characters χ_λ, $\lambda \in P_+^k(\mathfrak{g})$, form a vector-valued Jacobi function for $\Gamma_0(N)$ (recall (2.2.4b)),* *and for each* $\lambda \in P_+^k(\mathfrak{g})$,

$$\chi_\lambda\left(\frac{-1}{\tau}, \frac{z}{\tau}, u - \frac{(z|z)}{2\tau}\right) \in \text{span}_{\mu' \in P_+^k(\mathfrak{g}')} \chi_{\mu'}\left(\frac{\tau}{N}, \frac{z}{N}, u\right).$$

3.2.4 *Braided #3: braids and affine algebras*

According to conformal field theory, the modularity of, for example, affine algebra characters arises through the monodromy of a system of partial differential equations (the Knizhnik–Zamolodchikov equations for a torus with one puncture). In this subsection we anticipate this important idea by considering the simpler and better-known situation of a sphere. See also [355], [174]; the basic idea of differential equation monodromy is nicely described in [363].

Theorem 3.2.5 *Consider a simply-connected open region D in $\mathbb{C}$. Consider the differential equation*

$$\frac{d^2 w}{dz^2} + P(z)\frac{dw}{dz} + Q(z)\,w = 0, \tag{3.2.16a}$$

where $P(z)$ and $Q(z)$ are holomorphic in D. For any point $z_0 \in D$, and any $\alpha, \beta \in \mathbb{C}$, there is a unique function $w(z)$, holomorphic in D, satisfying the initial conditions

$$w(z_0) = \alpha, \tag{3.2.16b}$$

$$\frac{dw}{dz}(z_0) = \beta. \tag{3.2.16c}$$

Hence the solutions w to (3.2.16a) form a two-dimensional space, parametrised by $\alpha, \beta \in \mathbb{C}$. For a proof of this theorem, see, for example, chapter XII of [307].

What if D is not simply-connected? One way to proceed would be to make D simply-connected by cutting it. For example, if D is $\mathbb{C}$ with n points $z_1, \ldots, z_n$ removed, then we can cut D along a non-self-intersecting polygonal path connecting $z_1, \ldots, z_n$ and ∞, avoiding the point z_0. Call D' the resulting simply-connected subregion of D. Then a holomorphic function on D restricts to a holomorphic function on D'; however, most holomorphic functions on D' won't extend continuously to D.

The other way to proceed is to consider the (simply-connected) universal cover $\pi : \widetilde{D} \to D$ (recall Section 2.1.2). We can then identify D with $\widetilde{D}/G$ for some group G isomorphic to the fundamental group $\pi_1(D)$; each $\gamma \in G$ is an automorphism of $\widetilde{D}$ shuffling the points $\widetilde{z} \in \pi^{-1}(z)$ above each $z \in D$. Functions h holomorphic on D lift to functions $h \circ \pi$ holomorphic on $\widetilde{D}$, although a typical function $\widetilde{h}$ on $\widetilde{D}$ won't correspond to a well-defined function on D. However, $\pi^{-1}(D') \subset \widetilde{D}$ consists of several connected open components, one for each $\gamma \in \pi_1(D)$, and through this there is a many-to-one correspondence between the holomorphic functions on D' and those on $\widetilde{D}$.

Let's return to the situation of Theorem 3.2.5, except with D now being non-simply-connected (although still connected). Then there is a unique solution w to (3.2.16) in D'. Writing $\widetilde{P} = P \circ \pi$ and $\widetilde{Q} = Q \circ \pi$, and choosing any $\widetilde{z}_0 \in \pi^{-1}(z_0)$, we can lift the equations (3.2.16) to $\widetilde{D}$ and again we obtain a unique solution $\widetilde{w}$, this time holomorphic in $\widetilde{D}$. The space of solutions w on D', $\widetilde{w}$ on $\widetilde{D}$, are both two-dimensional. But we get more: both spaces carry naturally an action of the fundamental group $\pi_1(D)$, called the *monodromy representation*. More precisely, each automorphism $\gamma^* \in G \cong \pi_1(D)$ carries a solution $\widetilde{w}$ of (3.2.16a) to another solution $\widetilde{w} \circ \gamma^*$ – it preserves α, β but changes the choice $\widetilde{z}_0 \in \pi^{-1}(z_0)$. It corresponds to an analytic continuation of w across the polygonal path cut out from D, along closed paths γ corresponding to γ^*.

A simple example should make this clear. Consider

$$\frac{\mathrm{d}^2 w}{\mathrm{d}z^2} + z^{-1}\frac{\mathrm{d}w}{\mathrm{d}z} = 0. \tag{3.2.17a}$$

Here, D is the punctured plane $\mathbb{C} \setminus \{0\}$ so we can take D' to be $\mathbb{C}$ with the negative real axis removed. The fundamental group $\pi_1(D)$ is $\mathbb{Z}$, and the universal cover $\widetilde{D}$ is the infinite spiral staircase. Two solutions to (3.2.17a) in D' are $w = \log z$ and $w = 1$. Analytically extend $w(z) = \log z$ along the unit circle starting at $z_0 = 1$ and running counterclockwise: as we cross the negative real axis continuity requires the value of w to be shifted by $2\pi\mathrm{i}$ from its previous 'principal' value. More generally, the path $\gamma^* = n$, winding n times around the origin, would pick up a monodromy of $2\pi\mathrm{i}n$. On the other hand, the constant solution $w(z) = 1$ is of course unchanged under analytic continuation. In terms of our basis $\{\log z, 1\}$, we thus obtain the monodromy representation

$$n \mapsto \begin{pmatrix} 1 & 2\pi\mathrm{i}\,n \\ 0 & 1 \end{pmatrix}. \tag{3.2.17b}$$

We are interested here in a slightly more complicated situation than that of Theorem 3.2.5. Let $\overline{\mathfrak{g}}$ be any finite-dimensional semi-simple Lie algebra and choose n distinct points $z_1, \ldots, z_n$ in $\mathbb{C}$. Recall the space $\mathfrak{C}_n$ defined in (1.2.6). Choose a basis x_a of $\overline{\mathfrak{g}}$, orthonormal with respect to the Killing form κ. For each i, choose a finite-dimensional $\overline{\mathfrak{g}}$-representation R_i, acting on a space V_i. Fix some complex number $\gamma \neq 0$. By the *Knizhnik–Zamolodchikov* (or *KZ*) equations we mean

$$\frac{\partial w}{\partial z_i} = \gamma \sum_{j \neq i} \sum_a \frac{R_i(x_a) \otimes R_j(x_a)}{z_i - z_j}\, w, \qquad 1 \leq i \leq n, \tag{3.2.18a}$$

where $w : \mathfrak{C}_n \to V_1 \otimes \cdots \otimes V_n$, and where $R_i(x_a), R_j(x_a)$ act on the ith, jth components of the multilinear form w.

We recognise in (3.2.18a) the quadratic Casimir $\Omega = \sum_a x_a x_a$ discussed after (3.2.15). Physically (i.e. in the context of conformal field theory), w is a *chiral block* on the sphere $\mathbb{P}^1(\mathbb{C})$ with $n + 1$ distinct marked points (namely $z_1, \ldots, z_n$ and $z_{n+1} = \infty$) for a Wess–Zumino–Witten model (Section 4.3.2). Geometrically (see e.g. [**338**]),

$$\frac{1}{2}\mathrm{d} - \gamma \sum_{\substack{j,i \\ i \neq j}} \sum_a R_i(x_a) \otimes R_j(x_a)\, \frac{\mathrm{d}z_i - \mathrm{d}z_j}{z_i - z_j} \tag{3.2.18b}$$

defines a *connection* (Section 1.2.2) on the trivial vector bundle $\mathfrak{C}_n \times W$, for $W = V_1 \otimes \cdots \otimes V_n$. An easy calculation verifies this connection is flat (i.e. has 0 curvature). The partial differential equations (3.2.18a) say that w is a *horizontal* or *parallel section*.

In other words, restricting to a simply-connected subregion $\mathfrak{C}'_n$ of $\mathfrak{C}_n$, the unique solution $w(z_1, \ldots, z_n)$ to (3.2.18a) satisfying some initial condition $w(z^{(0)}) = w^{(0)}$ is obtained geometrically by parallel-transporting the vector $w^{(0)}$ along any path γ in $\mathfrak{C}'_n$ connecting $z^{(0)}$ to the desired point $(z_1, \ldots, z_n)$.

Our context here is thus analogous to that of Theorem 3.2.5: parallel transport plays the role of analytic continuation, and the flatness of $\mathfrak{C}_n$ corresponds to the Monodromy Theorem of complex analysis (e.g. theorem 16.15 of [481]). The result is that the space of solutions to (3.2.18a) carries a representation of the fundamental group $\pi_1(\mathfrak{C}_n)$, i.e. of the pure braid group $\mathcal{P}_n$. We get an action of the full braid group through 'half-monodromies': a braid $\beta \in \mathcal{B}_n$ will take a solution w of (3.2.18a) to a solution of (3.2.18a) with values in $V_{\beta 1} \otimes \cdots \otimes V_{\beta n}$, where β acts on the indices $\{1, \ldots, n\}$ through the natural homomorphism $\phi : \mathcal{B}_n \to \mathcal{S}_n$ described in Section 1.1.4. In particular, if all V_i are isomorphic, the space of solutions of (3.2.18a) will carry a representation of the full group $\mathcal{B}_n$.

The infinitely many irreducible finite-dimensional modules of a simple Lie algebra naturally span a symmetric monoidal category (recall Section 1.6.2 for definitions); its character ring is isomorphic to a polynomial ring in r variables, where r is the rank of the algebra. On the other hand, the finitely-many level-k irreducible integrable modules of a nontwisted affine algebra span a braided monoidal category (in fact ribbon and modular categories); the corresponding character ring is called a *fusion ring* and is described in Section 6.2.1. The key ingredient in this category – the braiding – comes from this braid group monodromy. In Section 6.2.2 we see that this braid group monodromy, and associated braided monoidal category, generalise to the modules of sufficiently nice vertex operator algebras, and this (or if you prefer, conformal field theories) serves as the natural context for the modularity in Moonshine.

There are many other occurrences of the braid group in the mathematics and physics neighbouring Moonshine, and most of these are directly related to this KZ monodromy on a sphere. For example, the knot invariants arising from subfactors and quantum groups come from braid group representations, and Drinfel'd and Kohno have proved that these representations are the same ones coming from KZ monodromy.

On the other hand, the relation of the braid group $\mathcal{B}_3$ to $\mathrm{SL}_2(\mathbb{Z})$ and its modular functions, which we have seen already in Section 2.4.3 and which we argue later plays a fundamental role in Monstrous Moonshine, does not have a *direct* relation to this braid group monodromy. But we will see later that modularity too is due to monodromy of a system of partial differential equations – the analogue of these KZ equations for a once-punctured torus – defining a flat connection on the extended moduli space $\widehat{\mathfrak{M}}_{1,1}$. The solutions of these equations are spanned by the affine algebra characters (or more generally the vertex operator algebra one-point functions). The associated monodromy group is the mapping class group of $\widehat{\mathfrak{M}}_{1,1}$, which is readily seen to be $\mathcal{B}_3$.

Intriguingly, this means that we've come full circle. Poincaré's 125-year-old path to modular functions (see [259] for a review) was differential equations of the form (3.2.16a). Let $f(z)$, $g(z)$ be a basis for the space of solutions, and write $\xi(z) = f(z)/g(z)$. Note that the monodromy group acts on ξ by Möbius transformations: $\xi \mapsto \frac{a\xi+b}{c\xi+d}$.

Poincaré found that, at least in some cases, when we invert $\xi(z)$ and write z as a function of ξ, then z will be a modular function for some discrete subgroup of $\mathrm{SL}_2(\mathbb{R})$, acting on $\xi \in \mathbb{H}$.

A simple example is Legendre's equation

$$\frac{\mathrm{d}^2 y}{\mathrm{d}k^2} + \frac{1 - 3k^2}{k(1 - k^2)}\frac{\mathrm{d}y}{\mathrm{d}k} - \frac{y}{1 - k^2} = 0.$$

This has the elliptic periods $K(k)$ and $K'(k) = K(k')$ as solutions (recall Section 2.2.1). It is more convenient to change variables to $z = k^2$, when this equation becomes

$$\frac{\mathrm{d}^2 w}{\mathrm{d}z^2} + \frac{1 - 2z}{z(1 - z)}\frac{\mathrm{d}w}{\mathrm{d}z} - \frac{w}{4z(1 - z)} = 0. \tag{3.2.19a}$$

Then $K'(z) = K(1 - z)$, since $k^2 + k'^2 = 1$. The domain D is the plane with $z = 0$ and $z = 1$ removed; its fundamental group π_1 is the free group $\mathcal{F}_2 = \langle \sigma_0, \sigma_1 \rangle$ generated by counter-clockwise loops σ_k about $z = k$. It turns out that $K(z)$ is holomorphic at $z = 0$, but $K'(z)$ has a logarithmic singularity there: $K'(z) + \frac{1}{\pi}K(z)\log z$ is holomorphic at $z = 0$. Thus as we go counter-clockwise in a small circle about $z = 0$, $K(z)$ is unchanged but $K'(z)$ becomes $K'(z) - 2\mathrm{i}K(z)$. Hence, as we go counter-clockwise in a small circle about $z = 1$, $K'(z)$ is unchanged but $K(z)$ becomes $K(z) + 2\mathrm{i}K'(z)$. Thus in terms of the basis $\{K(z), \mathrm{i}K'(z)\}$ of solutions to (3.2.19a), the monodromy representation becomes

$$\sigma_0 \mapsto \begin{pmatrix} 1 & 2 \\ 0 & 1 \end{pmatrix}, \qquad \sigma_1 \mapsto \begin{pmatrix} 1 & 0 \\ 2 & 1 \end{pmatrix}. \tag{3.2.19b}$$

For the details of this calculation, see chapter 14.5 of [**486**]. The image of (3.2.19b) is precisely the congruence subgroup $\Gamma(2)$, which indeed is isomorphic to $\mathcal{F}_2$. Now, Poincaré would have us invert the function $\mathrm{i}K'(z)/K(z)$. That ratio turns out to always be in $\mathbb{H}$, and so denote it $\tau(z)$. Expressing $z = k^2$ as a function of τ, we obtain

$$z(\tau) = \frac{\theta_2(\tau)^4}{\theta_3(\tau)^4}. \tag{3.2.19c}$$

Indeed, we know from (2.3.8) that (3.2.19c) is invariant under $\Gamma(2)$.

It is remarkable to recover in this way the group $\Gamma(2)$, its action on $\mathbb{H}$ and a modular function for $\Gamma(2)$ (in fact, $\Gamma(2)$ is genus-0 and θ_2^4/θ_3^4 generates all of its modular functions). There are many other examples of this kind, for example

$$w'' + z^{-1}w' + \left(\frac{31}{144}z - \frac{1}{36}\right)z^{-2}(z - 1)^{-2}w = 0$$

yields in this way the j-function. See [**516**] for more on the deep relation between modular forms and hypergeometric functions. The relations between affine algebras, the KZ equation and hypergeometric functions is explored in [**541**]. The Riemann–Hilbert problem asks that all linear representations of mapping class groups arise as monodromies; see the appendix of [**259**] for a history of this problem and chapter VIII of [**80**] for the modern treatment and generalisation using $\mathcal{D}$-modules.

Thus Poincaré, like conformal field theory over a century after him, finds it natural to interpret modularity using differential equation monodromy!

3.2.5 *Singularities and Lie algebras*

In this subsection we quickly review the geometry underlying the associations of singularities to simple Lie algebras (duVal) and affine Lie algebras (McKay), which are described in Section 2.5.2. This is related to mirror symmetry and provides a new explanation for the modularity of affine algebra characters.

Let Γ be a finite subgroup of $SU_2(\mathbb{C})$. Then the orbifold $\mathbb{C}^2/\Gamma$ has a critical point at the fixed point $(0, 0)$; the minimal resolution $X_\Gamma = \widetilde{\mathbb{C}^2/\Gamma}$ is a smooth noncompact real 4-manifold with an ALE ('asymptotically locally Euclidean') hyper-Kähler structure. An ALE manifold is Riemannian, with a metric tending quickly to the Euclidean one as $r \to \infty$. Physically, they correspond to positive-definite self-dual solutions to Einstein's gravitation equations in a vacuum ('gravitational instantons'). Conversely, any ALE hyper-Kähler manifold is diffeomorphic to some X_Γ for a unique Γ. The details are reviewed in [362].

Kronheimer–Nakajima [362] use the Atiyah–Singer Index Theorem to directly relate the duVal and McKay data associated with a simple singularity. Let X be an ALE hyper-Kähler manifold and $\Gamma < SU_2(\mathbb{C})$ the corresponding finite group. Then asymptotically at infinity, X is flat and in fact looks like $\mathbb{R}^4/\Gamma$. Given any vector bundle E over X, the fibre over ∞ defines a Γ-module R via monodromy. Kronheimer–Nakajima take E to be $\mathcal{R} \otimes \mathcal{R}^*$, where $\mathcal{R}$ is the tautological vector bundle, because its index vanishes. Then the monodromy representation R decomposes as $\sum_i \rho_i \otimes \overline{\rho_i}$, where ρ_i are the irreducible representations of Γ. The Index Theorem provides an expression for the numbers

$$\frac{1}{\|\Gamma\|} \sum_{\gamma \in \Gamma, \gamma \neq e} \frac{\mathrm{ch}_{\rho_i}(\gamma) \, \mathrm{ch}_{\rho_j^*}(\gamma)}{2 - \mathrm{ch}_\rho(\gamma)},$$

for $i, j = 0, 1, \ldots, n$, as an integral over X involving the intersection matrix, where ρ is the defining two-dimensional representation of $\Gamma < SL_2(\mathbb{C})$. From this they quickly establish the equivalence of duVal's observation that the intersection matrix is the negative of the $n \times n$ Cartan matrix, with McKay's interpretation of the $(n + 1) \times (n + 1)$ Cartan matrix as coefficients of the product $\rho \otimes \rho_i$.

The first direct relation between simple singularities and the Lie algebras A_r, D_r, E_r was established by Brieskorn [86]. Let $\overline{\mathfrak{g}}_\Gamma$ be the finite-dimensional simple Lie algebra associated with Γ, and G_Γ the corresponding Lie group. Let W be its (finite) Weyl group, and choose any Cartan subalgebra $\overline{\mathfrak{h}}$. Then Brieskorn obtained the singularity $\mathbb{C}^2/\Gamma$ and its resolution by studying the map $\overline{\mathfrak{g}}_\Gamma \to \overline{\mathfrak{h}}/W$, sending $x = x_s + x_n \in \overline{\mathfrak{g}}_\Gamma$ (this decomposition of x is just the Jordan canonical form [300]) to the orbit of the semi-simple part x_s under the adjoint action of G_Γ – these orbits are parametrised by $\overline{\mathfrak{h}}/W$ (Section 1.5.2).

More relevant for us is Nakajima's geometric realisation of affine algebras and their integrable representations (see e.g. his review [**445**]). Let E be an anti-self-dual Yang–Mills instanton over X with gauge group $U_k(\mathbb{C})$. These bundles E are associated with three discrete invariants: the monodromy representation R as above; the first Chern class $c_1(E)$; and the instanton number $ch_2(E) \in \mathbb{N}$.

The monodromy R is a k-dimensional representation of Γ. Decompose R into irreducibles: $R = \sum_i \lambda_i \rho_i$, where the multiplicities $\lambda_i \in \mathbb{N}$. Then taking dimensions we obtain $k = \sum_{i=0}^n a_i \lambda_i$, where $a_i = \dim \rho_i$. According to the McKay correspondence, a_i are the labels of the corresponding nontwisted affine algebra $\mathfrak{g}_\Gamma$, and so $\lambda = \sum_i \lambda_i \omega_i$ is a level-k integrable highest weight of $\mathfrak{g}_\Gamma$.

Nakajima proceeds to construct not only $\mathfrak{g}_\Gamma$ from the geometric data, but also the $\mathfrak{g}_\Gamma$-module $L(\lambda)$. The singularity at $(0,0)$ of $\mathbb{C}^2/\Gamma$ resolves locally into n copies of the sphere $\mathbb{P}^1(\mathbb{C})$. These give a basis of $H_2(X, \mathbb{Z})$; Nakajima identifies them with the usual basis h_i of a Cartan subalgebra of the finite-dimensional algebra $\overline{\mathfrak{g}_\Gamma}$ and their intersection form with the Killing form. Thus the dual vectors $c_1(E)$ are weights. The number $ch_2(E)$ is identified with an eigenvalue of the derivation $\delta = L_0$. The other generators e_i, f_i of $\mathfrak{g}_\Gamma$ can be interpreted likewise. The moduli space $\mathcal{M}(k)$ of $U_k(\mathbb{C})$-instantons on X has a finite-dimensional connected component $\mathcal{M}(k)_{\lambda,\mu,n}$ for every choice of monodromy λ, $c_1 = \mu$ and $ch_2 = n$. The infinite-dimensional cohomology space $H^*(\mathcal{M}(k))$ carries a natural though reducible module of the affine algebra $\mathfrak{g}_\Gamma$. However, the middle-dimensional cohomology

$$\oplus_{\mu,n} H^d(\mathcal{M}(k)_{\lambda,\mu,n}), \qquad d = \frac{1}{2}\dim(\mathcal{M}(k)_{\lambda,\mu,n}) \qquad (3.2.20)$$

is isomorphic to $L(\lambda)$, with each summand being a weight-space (the middle-dimensional cohomology spaces are generally the most interesting – for example, the pairing defines a bilinear form, here the Killing form, on them).

This construction generalises considerably [**445**]. It also has a natural interpretation in string theory. The *Bogomol'nyi–Prasad–Sommerfeld ('BPS') states* generally form an algebra closely related to Borcherds–Kac–Moody algebras (Section 3.3.2) [**276**]. Inside this BPS algebra for the heterotic string on the torus T^4 is the associated affine algebra. This string theory is dual to that of a type IIA string on a K3 surface (X_Γ is essentially a noncompact K3), where Nakajima's construction is very natural. So string theory interprets Nakajima's cohomological construction of affine algebras as a manifestation of mirror symmetry [**276**]. In this context, Vafa–Witten suggested that the modularity of affine algebra characters may have to do with S-duality [**540**], an $SL_2(\mathbb{Z})$-symmetry of the heterotic string. It seems unlikely though that this can account for the modularity in arbitrary RCFT. We revisit mirror symmetry [**291**] in Section 7.3.8.

Physically, instantons are configurations for which the classical action (4.1.3) has a local minima. This means that in the corresponding quantum theory, we should perturb about them just as we do about the vacua. See the review [**159**] on instantons in supersymmetric theories. It turns out that (not necessarily holomorphic) modular forms appear naturally in this context, with the modular group arising again through S-duality.

Recalling Section 2.4.3, we can ask: *Can S-duality sometimes be extended naturally into a $\mathcal{B}_3$ symmetry?* This may provide a universal simplification, for example, for fractional instantons.

3.2.6 Loop groups

This brief subsection introduces the Lie groups of the affine algebras, by translating the previous subsections into this geometric language. See the book [**465**] for more details. Loop groups appear directly in Wess–Zumino–Witten string theory, and in the study of certain differential equations (solitons), but otherwise the affine algebra is mathematically prior. From our (limited) perspective, the geometric insight gained isn't obviously worth the analytic subtleties.

Choose any compact Lie group G, and let $\bar{\mathfrak{g}}$ be its Lie algebra. By the *loop group* $\mathcal{L}G$ we mean all smooth maps $S^1 \to G$, and by the *loop algebra* $\mathcal{L}\bar{\mathfrak{g}}$ we mean all smooth maps $S^1 \to \bar{\mathfrak{g}}$. The loop group $\mathcal{L}G$ has a group structure given by pointwise product, and in fact it forms an infinite-dimensional Lie group with Lie algebra $\mathcal{L}\bar{\mathfrak{g}}$.

Think of G as a subgroup of $U_n(\mathbb{C})$, as we can. The *polynomial loop group* $\mathcal{L}_{poly}G$ is the set of all loops $\gamma \in \mathcal{L}G$ that can be written in the form

$$\gamma(z) = \sum_{m=-\infty}^{\infty} a_m z^m,$$

i.e. as a matrix-valued function, where $z \in S^1$ and each a_m is an $n \times n$ complex matrix, with all but finitely many $a_m = 0$. Note that $\mathcal{L}_{poly}G$ is indeed a group – for example, inverse is given by $\gamma(z)^{-1} = \sum_m a_m^\dagger z^{-m} \in \mathcal{L}_{poly}G$. However, note that $\mathcal{L}_{poly}S^1$ consists of the monomials az^m for some constants $m \in \mathbb{Z}$ and $a \in S^1 \subset \mathbb{C}$ (to see this, multiply $\gamma(z)$ by $\gamma(z)^\dagger$; the result is a Laurent polynomial in z with coefficients in $\mathbb{C}$, which identically equals 1 for uncountably many $z \in \mathbb{C}$). Thus $\mathcal{L}_{poly}S^1$ has Lie algebra $i\mathbb{R} \neq \mathcal{L}_{poly}S^1$. For semi-simple G, however, $\mathcal{L}_{poly}G$ has Lie algebra $\mathcal{L}_{poly}\bar{\mathfrak{g}}$, as we'd like.

The loop group $\mathcal{L}G$ is generally better behaved than $\mathcal{L}_{poly}G$. For example, we know the exponential map exp: $\bar{\mathfrak{g}} \to G$ is onto and locally one-to-one. The exponential map $\mathcal{L}\bar{\mathfrak{g}} \to \mathcal{L}G$ is defined in the obvious way (as the exponential of a matrix-valued function), and it is locally (but not globally) both one-to-one and onto. On the other hand, the exponential of a Laurent polynomial will usually not be a Laurent polynomial, and so the exponential map doesn't exist for polynomial loops. By way of comparison, as we mentioned in Section 3.1.2, exp: $\mathrm{Vect}(S^1) \to \mathrm{Diff}(S^1)$ is neither locally one-to-one nor locally onto (in fact its image is nowhere dense).

$\mathrm{Diff}(S^1)$ acts naturally on $\mathcal{L}G$, by changing the parametrisation of the loop (for simple G, the only other automorphisms of $\mathcal{L}G$ come from the loop group of $\mathrm{Aut}(G)$).

To enrich the representation theory of $\mathcal{L}G$, we centrally extend $\mathcal{L}G$ by S^1. For simple G, $\mathcal{L}G$ has an inequivalent central extension $\widetilde{\mathcal{L}G}_n$ for each $n = 0, 1, 2, \ldots$, and these exhaust all of them. $\widetilde{\mathcal{L}G}_0 \cong S^1 \times \mathcal{L}G$ is the trivial extension; $\widetilde{\mathcal{L}G}_1$ is the unique simply-connected such extension. $\widetilde{\mathcal{L}G}_n$ is obtained from $\widetilde{\mathcal{L}G}_1$ by quotienting by the order-n

subgroup of the centre S^1. The Lie algebra of any $\widetilde{\mathcal{L}G}_n$, $n > 0$, is isomorphic to the unique nontrivial central extension of the loop algebra $\mathcal{L}\bar{\mathfrak{g}}$.

We're interested in continuous projective representations of $\mathcal{L}G$ by bounded operators in a Hilbert space $\mathcal{H}$. We want these as usual to be $\mathbb{Z}$-graded. But an S^1 action is the same as a $\mathbb{Z}$-grading. More precisely, consider the group S^1 of rigid rotations R_θ in $\mathcal{L}G$ – that is, a loop $\gamma(t) \in \mathcal{L}G$ gets sent to the loop $(R_\theta \gamma)(t) = \gamma(t - \theta)$ for some fixed $0 \le \theta < 2\pi$. We can decompose this S^1 action on $\mathcal{H}$ Fourier-like into (the completion of) a direct sum

$$\oplus_{\ell=-\infty}^{\infty} \mathcal{H}(\ell)$$

of subspaces $\mathcal{H}(\ell)$ on which R_θ acts like $e^{-i\ell\theta}$. In other words, $e^{-iL_0\theta}$ represents R_θ.

We require $\mathcal{H}(\ell)$ to vanish for all ℓ sufficiently close to $-\infty$. Because of the conformal field theory interpretation given next chapter, these eigenvalues ℓ are thought of as energy, and these representations are called *positive energy representations*. Any such projective representation of $\mathcal{L}G$ lifts to one of the semi-direct product of this S^1 with any central extension $\widetilde{\mathcal{L}G}_n$. This double S^1-extension of $\mathcal{L}G$ corresponds to the double $\mathbb{C}$-extension of the (polynomial) loop algebra performed in Section 3.2.2.

Let G be semi-simple. Any projective representation $\mathcal{H}$ of $\mathcal{L}G$ of positive energy is unitary and hence is completely reducible into a discrete direct sum of irreducible representations. The above action of S^1 (through the operators $e^{-iL_0\theta}$) extends to a projective action of $\mathrm{Diff}^+(S^1)$. The L_0-eigenspaces $\mathcal{H}(\ell)$ of any irreducible representation $\mathcal{H}$ are all finite-dimensional. We can refine these eigenspaces by choosing a maximal torus T of G (it will be isomorphic to $S^1 \times \cdots \times S^1$ (r times), where r is the rank of G). We can diagonalise this action of $S^1 \times T \times S^1$, where the first S^1 is from the rigid rotations, and the second from the central extension; then

$$\mathcal{H}(n) = \oplus_{\mu \in P_+(\bar{\mathfrak{g}})} \mathcal{H}(n, \mu, k)$$

is the corresponding diagonalisation into weight spaces. Of course we are rediscovering the weight-space decomposition in, for example, (3.2.9a). The 'rigid rotation' S^1 corresponds to the extension of the loop algebra $\mathcal{L}_{poly}\bar{\mathfrak{g}}$ by the derivation $-\ell_0$, and the projective $\mathrm{Diff}^+(S^1)$ action corresponds to the Virasoro action (3.2.15). The maximal torus $S^1 \times T \times S^1$ of the double extension of $\mathcal{L}G$ corresponds to the (real) Cartan subalgebra $\mathfrak{h}$ of $\bar{\mathfrak{g}}^{(1)}$. Given any irreducible projective representation of $\mathcal{L}G$ of positive energy, then the derived projective representation of $\mathcal{L}\bar{\mathfrak{g}}$, restricted to $\mathcal{L}_{poly}\bar{\mathfrak{g}}$, is an integrable highest-weight representation $L(\lambda)$ of $\bar{\mathfrak{g}}^{(1)}$. Conversely, any such representation of $\bar{\mathfrak{g}}^{(1)}$ lifts ('integrates') to a projective representation of positive energy of $\mathcal{L}G$.

Any irreducible projective representation of $\mathcal{L}G$ lifts to a true representation of the simply-connected $\widetilde{\mathcal{L}G}_1$. It lifts to a true representation of $\widetilde{\mathcal{L}G}_n$ iff n divides the level k.

The analogue of Borel–Weil applies here much as in Section 1.5.5; the role of the symmetric space G/T is played here by the infinite Grassmannian $\mathcal{L}G/T$ (see chapter 11 of [**465**]). The irreducible representations also fit in well with Kirillov's orbit method [**198**].

It is tempting to hope more generally that the group $\mathrm{Map}(M, G)$, for any manifold M and compact G, should be a relatively accessible class of infinite-dimensional Lie groups. However, the theory is much more difficult than $\mathcal{L}G = \mathrm{Map}(S^1, G)$ and little is known about their representations (see chapter 3 and section 9.1 in [**465**]).

Question 3.2.1. Define A to be the space of all differential operators of the form $\sum_{m,n \in \mathbb{Z}} a_{m,n} x^m \mathrm{d}^n / \mathrm{d}x^n$, where all $a_{m,n} \in \mathbb{C}$ and all but finitely many $a_{m,n}$ equal 0. Define a Lie algebra structure on A in the obvious way. Prove that A is a simple $\mathbb{Z}$-graded Lie algebra of polynomial growth.

Question 3.2.2. For a manifold X and Lie algebra L, when is $\mathrm{Map}(X, L)$ a simple Lie algebra?

Question 3.2.3. Show that the Witt algebra acts on the affine algebra $\overline{\mathfrak{g}}^{(1)}$ as derivations.

Question 3.2.4. Show that a highest-weight representation of a nontwisted affine Lie algebra $\mathfrak{g} = X_r^{(1)}$ with highest weight $(\overline{\lambda}, k, u)$ is isomorphic as a $\mathfrak{g}$-module to one with highest weight $(\overline{\lambda}, k, 0)$, when $k \neq 0$.

Question 3.2.5. Classify all invariant symmetric bilinear forms for $A_1^{(1)}$.

Question 3.2.6. Compute the cardinality $\|P_+^k\|$ for all series $A_r^{(1)}, B_r^{(1)}, C_r^{(1)}, D_r^{(1)}$. (*Hint*: this can always be done using one or two binomial coefficients.)

Question 3.2.7. The affine Weyl group of $A_1^{(1)}$ has two generators, which we call here ω and t. These act on $\mathbb{Z}^2$ as follows:

$$\omega(a, b) = (-a, b + 2a), \qquad t(a, b) = (3a + 2b, -2a - b).$$

(a) Find a formula for the action of t^n on (a, b). Find the orders of ω and t, and the determinants $\det(\omega)$ and $\det(t)$.
(b) Let $\beta = (a, b) \in \mathbb{Z}^2$ obey $k := a + b > 0$. Write $\rho = (1, 1)$. Show that the affine Weyl orbit of $\beta + \rho$ intersects

$$P_{++}^{k+2} := \{(1, k + 1), (2, k), \ldots, (k, 2), (k + 1, 1)\}$$

in at most one point, and that the orbit fails to intersect P_{++}^{k+2} iff $\beta + \rho$ is fixed by some nontrivial element of the affine Weyl group.

3.3 Generalisations of the affine algebras

Affine algebras are fascinating because they draw together so many different areas of mathematics and physics. Like anything else, they embed into assorted families in plenty of ways, each embedding preserving some properties and losing others. But do they embed into a much larger family of algebras that are also of interest outside Lie theory?

Generalisation is not the point of mathematics, and in fact, one must be honest, is usually rather dry. The challenge is to generalise in a rich and revealing direction. One of the more reliable ways of doing this is *closure*. Suppose we like to perform a certain activity, which unfortunately sometimes results in our toys being flung from our sandbox.

Then we build a bigger sandbox. When we divide integers, we don't always get integers, so we construct the rationals. When we take limits of rationals, we don't always get rationals, so we construct the reals. When we take square-roots of reals, we don't always get reals, so we construct the complex numbers.

Another appealing strategy for generalisation – *analogy* – was followed by Moody at the birth of Kac–Moody algebras (Section 3.2.1). However this strategy, even in the hands of a master, will not always be successful. This section reviews various generalisations of affine algebras, all obtained through analogy. Most important for our story are the Borcherds–Kac–Moody algebras, which have played a key role for instance in the proof of the Monstrous Moonshine conjectures.

3.3.1 Kac–Moody algebras

Recall the presentation R1, R2 of simple Lie algebras given in Definition 1.4.5, defined in terms of a Cartan matrix C1–C4. From the point of view of generators and relations, the step from 'finite-dimensional simple' to 'Kac–Moody' is rather easy: the only difference is that we drop the 'positive-definite' condition C4 (which was responsible for finite-dimensionality). That is:

Definition 3.3.1 (a) *A* Cartan$_{KM}$ *matrix* A *is any* $\ell \times \ell$ *integral matrix* A *obeying* C1, C2, C3 *(see Definition 1.4.5(a)), together with*

C4′ *there exists a positive diagonal matrix* D *such that the product* AD *is symmetric (i.e.* $(AD)^t = AD$).

(b) *Given any* Cartan$_{KM}$ *matrix* A, *the* Kac–Moody *algebra* $\mathfrak{g} = \mathfrak{g}(A)$ *is the Lie algebra with generators* e_i, f_i, h_i, *subject as before to the relations* R1 *and* R2 *(see Definition 1.4.5(b)).*

What we call Kac–Moody algebras are usually called *symmetrisable* Kac–Moody algebras in the literature. The adjective 'symmetrisable' emphasises the requirement C4′, which we shall always assume; dropping it means losing the invariant bilinear form, among other things. What we call 'Cartan$_{KM}$ matrix' here is usually called 'generalised symmetrisable Cartan matrix', but although that use of the word 'generalised' is traditional, it is now inappropriate (see Definition 3.3.4 below). More generally, appending 'generalised' to a term is an unimaginative empty cop-out that should be banned.

The theory of Kac–Moody algebras is quite parallel to that of the finite-dimensional simple Lie algebras. They are also generated by (finitely many) A_1 subalgebras. Most entries of A again are zero, so it is most convenient to graphically represent A using the Coxeter–Dynkin diagram (recall their definition in Section 1.4.3). As before, we may without loss of generality take the Cartan$_{KM}$ matrices to be indecomposable (i.e. consider connected diagrams).

Lemma 3.3.2 ([328], section 4.3) *Let* A *be an indecomposable* Cartan$_{KM}$ *matrix. Then exactly one of the following possibilities holds:*
(Fin) $\det(A) \neq 0$ – *there exists a column vector* $u > 0$ *such that* $Au > 0$;

(Aff) the nullspace (i.e. 0-eigenspace) of A is one-dimensional – there is a column vector
$u > 0$ *such that* $Au = 0$;
(Hyp) there is a column vector $u > 0$ *such that* $Au < 0$.

If the Cartan$_{KM}$ matrix A is of finite type, then the corresponding Lie algebra $\mathfrak{g}(A)$
is finite-dimensional and simple. If the matrix A is of affine type, then the algebra
$\mathfrak{g}(A)$ is infinite-dimensional, but has a $\mathbb{Z}$-grading $\mathfrak{g}(A) = \sum_j \mathfrak{g}_j$ into finite-dimensional
subspaces $\mathfrak{g}_j$ where dimensions $\dim(\mathfrak{g}_j)$ grow at most polynomially with j (see Section 3.2.1). The affine algebras come in two flavours – nontwisted and twisted –
and are listed in Figures 3.2 and 3.3. For A of hyperbolic type, again $\mathfrak{g}(A)$ has a
$\mathbb{Z}$-grading into finite-dimensional subspaces $\mathfrak{g}_j$, but their dimensions $\dim(\mathfrak{g}_j)$ grow
exponentially with j. We are mostly interested in the nontwisted affine algebras
(Section 3.2). Relatively little is known about the hyperbolic ones (but see Section 3.4.3).

The relation between the realisation in Section 3.2.2 of an affine algebra as a loop
algebra and the presentation of Definition 3.3.1(b) is as follows. Consider for simplicity
$A_1^{(1)}$. The relevant Cartan$_{KM}$ matrix is $A = \begin{pmatrix} 2 & -2 \\ -2 & 2 \end{pmatrix}$; then $\mathfrak{g}(A) \cong \mathcal{L}_{poly}(A_1) \oplus$
$\mathbb{C}C$, with the isomorphism identifying

$$e_1 \mapsto \begin{pmatrix} 0 & 1 \\ 0 & 0 \end{pmatrix}, \quad f_1 \mapsto \begin{pmatrix} 0 & 0 \\ 1 & 0 \end{pmatrix}, \quad h_1 \mapsto \begin{pmatrix} 1 & 0 \\ 0 & -1 \end{pmatrix},$$

$$e_0 \mapsto \begin{pmatrix} 0 & t \\ 0 & 0 \end{pmatrix}, \quad f_0 \mapsto \begin{pmatrix} 0 & 0 \\ t^{-1} & 0 \end{pmatrix}, \quad h_0 \mapsto C - \begin{pmatrix} 1 & 0 \\ 0 & -1 \end{pmatrix}.$$

More generally, the central term C of the affine algebra is given by $C = \sum_i a_i^\vee h_i$. Note
though that we are missing the derivation ℓ_0; we will return to that shortly.

For indecomposable A, $\mathfrak{g}(A)$ is simple iff the determinant $\det(A) \neq 0$. When $\det(A) =$
0, $\mathfrak{g}(A)$ has a centre of dimension $\ell - m$ where m is the rank of the matrix A.

The basic structure theorem for Kac–Moody algebras is:

Theorem 3.3.3   *Let* $\mathfrak{g} = \mathfrak{g}(A)$ *be a symmetrisable Kac–Moody algebra (over* $\mathbb{R}$*). Then:*
(a) $\mathfrak{g}$ *has triangular decomposition* $\mathfrak{g} = \mathfrak{g}_+ \oplus \mathfrak{h} \oplus \mathfrak{g}_-$ *where* $\mathfrak{g}_+$ *is the subalgebra gener-*
    ated by the e_i*,* $\mathfrak{g}_-$ *is generated by the* f_i *and* $\mathfrak{h} = \text{span}\{h_i\}$ *is the Cartan subalgebra;*
(b) $\mathfrak{g}$ *has a root space decomposition – formally calling* e_i *degree* α_i *and* f_i *degree* $-\alpha_i$
    and defining $\mathfrak{g}_\alpha$ *to be the subspace of degree* $\alpha \in \mathbb{Z}\alpha_1 + \mathbb{Z}\alpha_2 + \cdots$*, we get* $\mathfrak{h} = \mathfrak{g}_0$
    and $\mathfrak{g}_\pm = \oplus_{\alpha \in \Delta_\pm} \mathfrak{g}_\alpha$*, where* $[\mathfrak{g}_\alpha, \mathfrak{g}_\beta] \subset \mathfrak{g}_{\alpha+\beta}$ *and* $\Delta_- = -\Delta_+$;
(c) *there is an involution* ω *on* $\mathfrak{g}$ *for which* $\omega e_i = f_i$*,* $\omega h_i = -h_i$ *and* $\omega \mathfrak{g}_\alpha = \mathfrak{g}_{-\alpha}$;
(d) $\dim \mathfrak{g}_\alpha < \infty$ *and* $\dim \mathfrak{g}_{\pm \alpha_i} = 1$;
(e) *there is an invariant symmetric bilinear form* $(\cdot|\cdot)$*, that is* $([ab]|c) = -(b|[ac])$*, such*
    that for each root $\alpha \neq 0$ *the restriction of* $(\cdot|\cdot)$ *to* $\mathfrak{g}_\alpha \times \mathfrak{g}_{-\alpha}$ *is nondegenerate and*
    $(\mathfrak{g}_\alpha|\mathfrak{g}_\beta) = 0$ *whenever* $\beta \neq -\alpha$;
(f) *there is a linear assignment* $\alpha \mapsto h_\alpha \in \mathfrak{h}$ *such that for all* $a \in \mathfrak{g}_\alpha, b \in \mathfrak{g}_{-\alpha}$ *we have*
    $[a, b] = (a|b) h_\alpha$.

These α are called *roots* and the α_i *simple roots*, as before. The roots α can be regarded
as linear functionals on $\mathfrak{h}$, in such a way that for any $x \in \mathfrak{g}_\alpha$ and $h \in \mathfrak{h}$, we have $[hx] =$

$\alpha(h) x$. The involution in (c) is the Cartan involution, and is needed in defining unitary representations. The bilinear form in (e) is the generalisation here of the Killing form. For simple roots α_i, h_{α_i} in (f) is h_i, and is sometimes denoted $\alpha_i^\vee$ and called a *co-root*. The field was taken to be $\mathbb{R}$ here for convenience (Question 3.3.1).

When $\det(A) = 0$, the bilinear form restricted to $\mathfrak{h}$ will be degenerate and the simple roots interpreted as linear functionals on $\mathfrak{h}$ will be linearly dependent. To get around this, extend the Cartan subalgebra by $\dim(\text{Null}(A)) = \ell - m$ more vectors. Call $\mathfrak{h}^e$ the resulting $(2\ell - m)$-dimensional space. Extend the bilinear form to $\mathfrak{h}^e$ so that it becomes nondegenerate, and the domain of the simple roots $\alpha_i \in \mathfrak{h}^*$ to all of $\mathfrak{h}^e$ so they become linearly independent. Up to equivalence, there is a unique way to do this. The space $\mathfrak{g}(A)^e := \mathfrak{g}(A) + \mathfrak{h}^e$ is given a Lie algebra structure by extending the relations of Definition 3.3.1(b) to include

$$[hh'] = 0, \qquad \forall h, h' \in \mathfrak{h}^e, \tag{3.3.1a}$$

$$[he_i] = \alpha_i(h), \qquad \forall h \in \mathfrak{h}^e, \tag{3.3.1b}$$

$$[hf_i] = -\alpha_i(h), \qquad \forall h \in \mathfrak{h}^e. \tag{3.3.1c}$$

For a Cartan$_{KM}$ matrix A of affine type, $\mathfrak{g}(A)^e$ is isomorphic to the corresponding algebra $\mathfrak{g} = \overline{\mathfrak{g}}^{(N)}$ we defined in Section 3.2.2: the extra vector is the derivation ℓ_0. Whenever $\det(A) = 0$, $\mathfrak{g}(A)^e$ and not $\mathfrak{g}(A)$ is the correct algebra to consider. Write $\mathfrak{g}(A)^e := \mathfrak{g}(A)$ when $\det(A) \ne 0$. Theorem 3.3.3 holds for $\mathfrak{g}(A)^e$, provided $\mathfrak{h}$ there is replaced with $\mathfrak{h}^e$.

Unlike the finite-dimensional case, some root multiplicities $\text{mult}(\alpha) := \dim \mathfrak{g}_\alpha$ may be > 1. The roots of $\mathfrak{g}(A)^e$ come in two flavours: *real* (with $(\alpha|\alpha) > 0$) and *imaginary* (with $(\alpha|\alpha) \le 0$). The simple roots are all real. Real roots behave exactly like the roots of finite-dimensional $\mathfrak{g}$: for example, $\text{mult}(\alpha) = 1$ and the only multiples of α that are also roots are $\pm\alpha$. Imaginary roots behave more like the nonroot $0 \in \mathfrak{h}^*$: for example, $\text{mult}(\alpha) \ge 1$ and any multiple $\mathbb{Z}\alpha$ is also a root.

The Weyl group W here is generated by the reflections through the simple roots α_i, or equivalently by reflections through all real roots. It has the usual properties: for example, root multiplicities are constant within the W-orbits.

A Kac–Moody algebra $\mathfrak{g}(A)^e$ has all the familiar representation-theoretic definitions and properties. For any weights $\lambda \in \mathfrak{h}^{e*}$, Verma modules $M(\lambda)$ and the irreducible highest-weight module $L(\lambda)$ are defined as usual. In particular, highest-weight modules are spanned by vectors of the form

$$f_{i_m} f_{i_{m-1}} \cdots f_{i_1} v, \tag{3.3.2}$$

where v is the highest-weight vector. Weight-space decompositions hold as before, and characters $\text{ch}_M(h)$ are defined as in (1.5.9a). The character of the Verma module $M(\lambda)$ again equals (3.2.6). Integrability is defined by the locally nilpotent condition (Section 3.2.3); again, $L(\lambda)$ is integrable iff all Dynkin labels $\lambda(h_i) \in \mathbb{N}$, iff $L(\lambda)$ is unitarisable. The character of an integrable $L(\lambda)$ is given by the *Weyl–Kac character formula*

$$\text{ch}_{L(\lambda)} = \frac{\sum_{w \in W} \det(w)\, e^{w(\lambda + \rho)}}{e^\rho \prod_{\alpha > 0} (1 - e^{-\alpha})^{\text{mult}(\alpha)}}. \tag{3.3.3}$$

This is identical to the Weyl character formula (1.5.11), except that the sum and product are infinite, and the multiplicities of (imaginary) root spaces can be > 1. For affine algebras, it reduces to (3.2.11c).

Apart from the affine and finite-dimensional simple algebras, the other Kac–Moody algebras have yet to make a real impact on other areas of mathematics and mathematical physics. However, [127] and [171] anticipate that the hyperbolic Kac–Moody algebras E_{10} and E_{11} will appear in *M-theory*, the still-hypothetical physics underlying strings.

3.3.2 Borcherds' algebras

In his efforts to prove the Monstrous Moonshine conjectures, Borcherds further generalised affine algebras. It is easy to associate a Lie algebra to a matrix A, but which class of matrices will yield a deep theory? Borcherds found such a class by holding in his hand a single algebra – the fake Monster Lie algebra (Section 7.2.2) – which acted much like a Kac–Moody algebra, even though it had imaginary simple roots.

Definition 3.3.4 (a) *A* Cartan$_{BKM}$ *matrix* A *is a (possibly infinite) matrix* $A = (a_{ij})$, $a_{ij} \in \mathbb{R}$, *obeying*

GC1. *either* $a_{ii} = 2$ *or* $a_{ii} \leq 0$;
GC2. $a_{ij} \leq 0$ *for* $i \neq j$, *and* $a_{ij} \in \mathbb{Z}$ *when* $a_{ii} = 2$; *and*
GC3. *there is a diagonal matrix* D *with each* $d_{ii} > 0$ *such that* DA *is symmetric*.

(b) *The* universal Borcherds–Kac–Moody algebra $\widehat{\mathfrak{g}} = \widehat{\mathfrak{g}}(A)$ *is the Lie algebra with generators* e_i, f_i, h_{ij}, *subject to the relations* [71]:

GR1. $[e_i f_j] = h_{ij}$, $[h_{ij} e_k] = \delta_{ij} a_{ik} e_k$ *and* $[h_{ij} f_k] = -\delta_{ij} a_{ik} f_k$, *for all* i, j;
GR2. $(\mathrm{ad}\, e_i)^{1-a_{ij}} e_j = (\mathrm{ad}\, f_i)^{1-a_{ij}} f_j = 0$, *whenever both* $a_{ii} = 2$ *and* $i \neq j$; *and*
GR3. $[e_i e_j] = [f_i f_j] = 0$ *whenever* $a_{ij} = 0$.

As before, the adjective 'symmetrisable' is usually appended in the literature. Unfortunately, the name 'Borcherds' is often replaced with the abomination 'generalised'. Note that for each i, span$\{e_i, f_i, h_{ii}\}$ is isomorphic to $\mathfrak{sl}_2(\mathbb{C})$ when $a_{ii} \neq 0$ and to $\mathfrak{Heis}$ (recall (1.4.3)) when $a_{ii} = 0$. Immediate consequences of the definition are that: (i) $[h_{ij} h_{mn}] = 0$; (ii) $h_{ij} = 0$ unless the ith and jth column of A are identical; (iii) the h_{ij} for $i \neq j$ lie in the centre of $\widehat{\mathfrak{g}}$. Setting all $h_{ij} = 0$ for $i \neq j$ gives the definition of the *Borcherds–Kac–Moody algebra* $\mathfrak{g} = \mathfrak{g}(A)$ [69]. This central extension $\widehat{\mathfrak{g}}$ of $\mathfrak{g}$ is introduced for its role in Theorem 3.3.6 below. If A has no zero columns, then $\widehat{\mathfrak{g}}$ equals its own universal central extension [71]. Because a Borcherds–Kac–Moody algebra can satisfy fewer relations, it typically contains a large free Lie subalgebra [323] (a free Lie algebra is analogous to a free group).

A universal Borcherds–Kac–Moody algebra differs from a Kac–Moody algebra in that it is built up from Heisenberg algebras as well as A_1, and these subalgebras intertwine in more complicated ways. Nevertheless, *much of the theory for finite-dimensional simple Lie algebras continues to find an analogue in this much more general setting* (e.g.

root-space decomposition, Weyl group, character formula, ...). This unexpected feature is the point of Borcherds–Kac–Moody algebras.

To get a feel for these algebras, let us prove a few simple results concerning the h_{ij}. Note first that, using the above relations together with anti-associativity, we obtain $[h_{ij}h_{k\ell}] = \delta_{ij}(a_{jk} - a_{j\ell})h_{k\ell}$. Comparing this with $[h_{k\ell}h_{ij}] = -[h_{ij}h_{k\ell}]$, we see that bracket must always equal 0. Hence all h's pairwise commute, and $h_{ij} = 0$ unless the ith and jth columns of A are identical.

The basic structure theorem is that of Kac–Moody algebras (Theorem 3.3.3):

Theorem 3.3.5 [69] *Let* $\mathfrak{g} = \mathfrak{g}(A)$ *be a Borcherds–Kac–Moody algebra (over* $\mathbb{R}$*). Then:*

(a) *$\mathfrak{g}$ has triangular decomposition $\mathfrak{g} = \mathfrak{g}_+ \oplus \mathfrak{h} \oplus \mathfrak{g}_-$ where $\mathfrak{g}_+$ is the subalgebra generated by the e_i, $\mathfrak{g}_-$ is generated by the f_i and $\mathfrak{h} = \mathrm{span}\{h_i\}$ is the Cartan subalgebra;*

(b) *$\mathfrak{g}$ has a root space decomposition – formally calling e_i degree α_i and f_i degree $-\alpha_i$, and defining $\mathfrak{g}_\alpha$ to be the subspace of degree $\alpha \in \mathbb{Z}\alpha_1 + \mathbb{Z}\alpha_2 + \cdots$, we get $\mathfrak{h} = \mathfrak{g}_0$ and $\mathfrak{g}_\pm = \oplus_{\alpha \in \Delta_\pm} \mathfrak{g}_\alpha$, where $[\mathfrak{g}_\alpha, \mathfrak{g}_\beta] \subset \mathfrak{g}_{\alpha+\beta}$ and $\Delta_- = -\Delta_+$;*

(c) *there is an involution ω on $\mathfrak{g}$ for which $\omega e_i = f_i$, $\omega h_i = -h_i$ and $\omega \mathfrak{g}_\alpha = \mathfrak{g}_{-\alpha}$;*

(d) *$\dim \mathfrak{g}_\alpha < \infty$ and $\dim \mathfrak{g}_{\pm \alpha_i} = 1$;*

(e) *there is an invariant symmetric bilinear form $(\cdot|\cdot)$ such that for each root $\alpha \ne 0$, the restriction of $(\cdot|\cdot)$ to $\mathfrak{g}_\alpha \times \mathfrak{g}_{-\alpha}$ is nondegenerate and $(\mathfrak{g}_\alpha|\mathfrak{g}_\beta) = 0$ whenever $\beta \ne -\alpha$;*

(f) *there is a linear assignment $\alpha \mapsto h_\alpha \in \mathfrak{h}$ such that for all $a \in \mathfrak{g}_\alpha, b \in \mathfrak{g}_{-\alpha}$, we have $[a, b] = (a|b) h_\alpha$.*

The condition that $\mathfrak{g}$ be symmetrisable (i.e. condition GC3) is necessary for the existence of the bilinear form in Theorem 3.3.5(e). As in Section 3.3.1, it is common to add derivations. In particular, define $D_i(a) = n_i a$ for any $a \in \mathfrak{g}_{n_1\alpha_1 + \cdots}$; then each linear map D_i is a derivation, and adjoining these to $\mathfrak{h}$ defines an abelian algebra $\mathfrak{h}^e$. The simple root α_i can be interpreted as the element of $\mathfrak{h}^{e*}$ obeying $\alpha_j(h_i) = a_{ij}$ and $\alpha_j(D_i) = \delta_{ij}$. The role of the derivations is to make these simple roots linearly independent. Construct the induced bilinear form $(\cdot|\cdot)$ on $\mathfrak{h}^{e*}$, obeying $(\alpha_i \,|\, \alpha_j) = d_i a_{ij}$ (see [322] for details).

The properties in Theorem 3.3.5 characterise Borcherds–Kac–Moody algebras (see e.g. [72] for a proof):

Theorem 3.3.6 *Let L be a Lie algebra (over $\mathbb{R}$) satisfying the following conditions:*

(i) *L has a $\mathbb{Z}$-grading $\oplus_i L_i$, and $\dim L_i < \infty$ for all $i \ne 0$;*

(ii) *L has an involution ω sending L_i to L_{-i} and acting as -1 on L_0;*

(iii) *L has a contravariant bilinear form $(\cdot|\cdot)$ such that $(L_i|L_j) = 0$ if $i \ne -j$, and such that $-(a \,|\, \omega(a)) > 0$ if $0 \ne a \in L_i$ for $i \ne 0$.*

Then there is a homomorphism π from some $\widehat{\mathfrak{g}}(A)$ to L whose kernel is contained in the centre of $\widehat{\mathfrak{g}}(A)$, and L is the semi-direct product of the image of π with a subalgebra of the abelian subalgebra L_0. That is, L is obtained from $\widehat{\mathfrak{g}}$ by modding out some of the centre and adding some commuting derivations.

Conversely, any (real) Borcherds–Kac–Moody algebra obeys conditions (i), (ii) and (iii). For example, let $L = \mathfrak{sl}_2(\mathbb{R})$ and recall (1.4.2b). Then L has $\mathbb{Z}$-grading $L_{-1} \oplus L_0 \oplus L_1 = \mathbb{C}e \oplus \mathbb{C}h \oplus \mathbb{C}f$, $\omega(x) = -x^t$ and $(x|y) = \text{tr}(xy)$. Theorem 3.3.6 tells us that *Borcherds–Kac–Moody algebras are the ultimate generalisation of simple Lie algebras*, in the sense that any further generalisation will lose some basic structural ingredient.

Let Π^{re} be the set of all *real simple roots*, i.e. all α_i with $a_{ii} = 2$; the remainder are the *imaginary simple roots* $\alpha \in \Pi^{im}$. The *Weyl group* W of $\mathfrak{g}^e$ is generated by the reflections $r_{\alpha_i} : \mathfrak{h}^{e*} \to \mathfrak{h}^{e*}$ for each $\alpha_i \in \Pi^{re}$: $r_i(\lambda) = \lambda - \lambda(h_i)\alpha_i$. It is a crystallographic Coxeter group (Section 3.2.1). The *real roots* of $\mathfrak{g}^e$ are defined to be those in $W(\Pi^{re})$; all other roots are called *imaginary*. For all real roots, $\dim(\mathfrak{g}^e)^\alpha = 1$ and $(\alpha \mid \alpha) > 0$.

Integrable highest-weight modules are defined as before: namely, each e_α, f_α must act locally nilpotently for all real roots α. More precisely, $V = \oplus_{\mu \in \mathfrak{h}^{e*}} V_\mu$ where the weight-space $V_\mu := \{v \in V \mid h.v = \mu(h)v\}$, with $\dim V_\mu < \infty$, and whenever $a_{ii} = 2$, $(e_i)^k.v = 0 = (f_i)^k.v$ for all $v \in V$ and all sufficiently large k. By the *character* we mean the formal sum $\text{ch}_V := \sum_{\mu \in \mathfrak{h}^{e*}} (\dim V_\mu) e^\mu$. Let P_+ be the set of all weights $\lambda \in \mathfrak{h}^{e*}$ obeying $\lambda(h_i) \in \mathbb{N}$ whenever $a_{ii} = 2$, and $\lambda(h_i) \geq 0$ for all other i. Define the highest-weight $\mathfrak{g}^e$-module $L(\lambda)$ in the usual way as the quotient of the Verma module by the largest proper graded submodule. Choose $\rho \in \mathfrak{h}^{e*}$ to satisfy $(\rho \mid \alpha_i) = \frac{1}{2}(\alpha_i \mid \alpha_i)$ for all i, and define $S_\lambda = e^{\lambda+\rho} \sum_s \epsilon(s) e^s$ where s runs over all sums of imaginary simple roots and $\epsilon(s) = (-1)^m$ if s is the sum of m distinct mutually orthogonal imaginary simple roots, each of which is orthogonal to λ, otherwise $\epsilon(s) = 0$. Then we get the *Weyl–Kac–Borcherds character formula*:

$$\text{ch}_{L(\lambda)} = \frac{\sum_{w \in W} \epsilon(w) \, w(S_\lambda)}{e^\rho \prod_{\alpha \in \Delta_+} (1 - e^{-\alpha})^{\text{mult}\,\alpha}} \tag{3.3.4}$$

(compare (3.3.3)). S_λ is the correction factor due to imaginary simple roots.

Thus Borcherds' algebras strongly resemble Kac–Moody ones and constitute a natural and nontrivial generalisation. The main differences are that they can be generated by copies of the Heisenberg algebra as well as $\mathfrak{sl}_2(\mathbb{R})$, and that there can be imaginary simple roots. For more on their theory, see, for example, [328] chapter 11.13, [272], [322], [469]. Interesting examples are the Monster Lie algebra (Section 7.2.2), whose (twisted) denominator identity supplied the relations needed to complete the proof of the Monstrous Moonshine conjectures, and the fake Monster [70]. A Borcherds–Kac–Moody algebra can be associated with any even Lorentzian lattice, and also with any Calabi–Yau manifold [275]. Of course it is a broad enough class that almost all of them will be uninteresting; an intriguing approach to identifying the interesting ones is sketched at the end of Section 3.4.3.

We know simple Lie algebras arise in both classical and quantum physics, and the affine Kac–Moody algebras are important in conformal field theory, as we see next chapter. Borcherds–Kac–Moody algebras have appeared in the physics literature in the context of BPS states in string theory (see [275]), and as a possible symmetry of M-theory [285].

3.3.3 Toroidal algebras

As mentioned in Section 3.2.6, replacing the loop algebra $S^1 \to \overline{\mathfrak{g}}$ with more general spaces $M \to \overline{\mathfrak{g}}$ has a very different theory and seems much more complicated. The most obvious generalisation of affine algebras, which has a chance of retaining some of their special properties, is to replace the loop algebra $S^1 \to \overline{\mathfrak{g}}$ with a space of maps $S^1 \times \cdots \times S^1 \to \overline{\mathfrak{g}}$. As $S^1 \times \cdots \times S^1$ (n times) is topologically the n-dimensional torus, these are called *toroidal algebras*. We will try to mimic the theory of loop algebras as far as we can. If nothing else, we will identify some features responsible for making the earlier theory so special.

Let $\overline{\mathfrak{g}}$ be a simple finite-dimensional Lie algebra. Choose any $n \geq 1$, and let $\widetilde{\mathfrak{g}}$ be the *multi-loop algebra*, i.e. tensor product $\overline{\mathfrak{g}} \otimes \mathbb{C}[t_0^{\pm 1}, \ldots, t_n^{\pm 1}]$ of $\overline{\mathfrak{g}}$ with Laurent polynomials in formal variables t_i. Then $\widetilde{\mathfrak{g}}$ is a Lie algebra with $\mathbb{Z}^{n+1}$-grading into finite-dimensional subspaces. The following theory treats as distinguished one of these $n + 1$ variables, namely t_0. To complete the construction of the toroidal algebra, we take the universal central extension $0 \to \mathcal{K} \to \widetilde{\mathfrak{g}} \otimes \mathcal{K} \to \widetilde{\mathfrak{g}} \to 0$ of the multi-loop algebra $\widetilde{\mathfrak{g}}$, and then adjoin sufficiently many derivations (as we've done throughout this chapter). However, both of these extensions are infinite-dimensional. More precisely, write $d_i = t_i \mathrm{d}/\mathrm{d}t_i$ for the degree-derivation for variable t_i. Let $\mathcal{D}^*$ denote the algebra of derivations $\oplus_{i=1}^n \mathbb{C}[t_0^{\pm 1}, \ldots, t_n^{\pm 1}]d_i \oplus \mathbb{C}d_0$. The resulting Lie algebra structure on the space $\widetilde{\mathfrak{g}} \oplus \mathcal{K} \oplus \mathcal{D}^*$ is uniquely determined up to a 2-cocycle $\tau : \mathcal{D}^* \times \mathcal{D}^* \to \mathcal{K}$, which defines how the bracket of derivations contributes a central term. There is a two-dimensional space of these τ; choosing any of them defines a *toroidal Lie algebra* $\mathfrak{g}_\tau$. Adding $\mathcal{D}^*$ reduces the centre from the infinite-dimensional $\mathcal{K}$ to an $(n + 1)$-dimensional space. See [**53**] for more details of the construction of $\mathfrak{g}_\tau$.

The role of the Virasoro algebra (which as we know is a central extension of $\mathrm{Der}(\mathbb{C}[t^{\pm 1}]) = \mathrm{Vect}(S^1) \otimes \mathbb{C}$) is here replaced by an abelian extension [**173**] of the complex vector fields on a torus or equivalently of $\mathrm{Der}(\mathbb{C}[t_0^{\pm 1}, \ldots, t_n^{\pm 1}])$. It is a Lie algebra $\mathfrak{V}_\tau$ parametrised by the 2-cocycle τ, defined on the space $\mathcal{K} \oplus \mathrm{Der}(\mathbb{C}[t_0^{\pm 1}, \ldots, t_n^{\pm 1}])$. $\mathfrak{V}_\tau$ acts for instance on the Verma modules of $\mathfrak{g}_\tau$. We will be more interested in the Lie subalgebra $\mathfrak{v}_\tau = \mathcal{K} \oplus \mathcal{D}^*$ of $\mathfrak{g}_\tau$. The modules constructed below carry a projective action of the Witt algebra $\mathbb{C}[t_0^{\pm 1}]d_0$, as in the affine setting.

Affine algebras exist for their (integrable) modules and in particular their characters, so we need to find an interesting class of modules for the toroidal algebras. This isn't easy to do, but major progress was made in [**53**]. Let L_τ be an irreducible highest-weight module of level $k \neq 0$, for the affine algebra $\overline{\mathfrak{g}}^{(1)}$, and let W be any finite-dimensional module for $\mathfrak{gl}_N$. Then [**53**] constructs an irreducible $\mathfrak{g}_\tau$-module $M_{\lambda,W}$ that has finite-dimensional homogeneous spaces with respect to the natural $\mathbb{Z}^{n+1}$-grading, and thus has a character. More precisely, they first obtain a $\mathfrak{v}_\tau$-module by applying a Verma-like construction to $W \otimes \mathbb{C}[t_1^{\pm 1}, \ldots, t_n^{\pm 1}]$, and then they take the irreducible quotient M_W as usual; finally, they define a $\mathfrak{g}_\tau$-module structure on the tensor product $M_{\lambda,W} := L_\lambda \otimes M_W$. In [**54**] they show that these are modules of a 'near-vertex operator algebra' (see Definition 5.1.3(c)) closely related to affine algebra vertex operator algebras at generic level. From this, their characters can be computed, and familiar modular forms arise. This is promising because interesting Lie algebra modules seem to be the ones that arise as modules of related structures (e.g. Lie groups or vertex operator algebras). It is too easy to be a Lie algebra module. On the other hand, these are surely not the best $\mathfrak{g}_\tau$-modules – they have only found the analogue of generic $L(\lambda)$, but not yet the analogue of the 'integrable' modules. Their characters are like (3.1.7a), but we would like to identify modules with characters analogous to the discrete series. By analogy with better-understood algebras, we should look for modules with maximal numbers of 'null vectors' quotiented out.

It may seem artificial to choose a distinguished direction (namely the 0th), but to some extent this is inevitable. It is an elementary consequence of Schur's Lemma (recall Lemma 1.1.3) that in these irreducible $\mathfrak{g}_\tau$-modules, the centre span$\{K_0, \ldots, K_n\}$ should act as scalars, and thus an n-dimensional subspace must act trivially. These representations are designed so that K_0 is nontrivial but the other K_i act trivially.

What is natural to pursue from, for example, an algebraic point of view, and what is a successful theory from that point of view, is not necessarily of more general interest. It is from this broader, multidisciplinary standpoint that we (unfairly) judge the value of these generalisations. There is a large class of $\mathfrak{g}_\tau$-modules (namely those described above) whose characters have (fairly weak) modularity properties, but this seems to arise solely from the well-milled Heisenberg algebra combinatorics and it isn't clear yet that they have independent value. Possible physical relevance in Wess–Zumino–Witten models in more than two space-time dimensions is explored in, for example, [**306**]. The jury is still out on the greater relevance of toroidal algebras to, for example, Moonshine or physics, and certainly more work is needed.

3.3.4 Lie algebras and Riemann surfaces

The previous subsection emphasises the difficulties of higher-dimensional analogues of loop algebras. Perhaps the best generalisation of the affine algebras, particularly in the sense of retaining and enriching automorphic properties of the characters, associates infinite-dimensional Lie algebras to each Riemann surface with marked points. This theory has been developed in a series of papers by Krichever–Novikov, Bremner, Schlichenmaier, Sheinman and others – see [**491**] for a list of references. The starting point is a reinterpretation of the Laurent polynomials $\sum a_n t^n \in \mathcal{L}_{poly}\overline{\mathfrak{g}}$. Before, we interpreted the formal variable t as a point on the unit circle $S^1 \subset \mathbb{C}$, but now we regard t as lying in the punctured plane $\mathbb{C}\backslash\{0\}$, or equivalently the twice-punctured Riemann sphere $\mathbb{P}'(\mathbb{C})$. Similarly, the Witt algebra $\text{Vect}(S^1)$ can be interpreted as the Lie algebra of meromorphic vector fields on $\mathbb{P}'(\mathbb{C})$ with possible poles only at 0 and ∞.

Let Σ be any Riemann surface of genus g, and choose $p > 1$ distinct ordered points $P = (z_1, \ldots, z_p)$, $z_i \in \Sigma$. In the language of string theory described next chapter, we can think of Σ as being a world-sheet corresponding to p asymptotic incoming or outgoing strings (Section 4.3.1). Let $\mathcal{A}_{\Sigma,P}$ be the space of functions meromorphic on Σ, with possible poles only at P, and let $\mathcal{L}_{\Sigma,P}$ be the space of meromorphic vector fields on Σ, again with possible poles only at P. The bracket of $\mathcal{L}_{\Sigma,P}$ comes from the Lie derivative, as usual with vector fields, while the bracket of $\mathcal{A}_{\Sigma,P}$ is taken to be trivial. Let $\bar{\mathfrak{g}}$ be any simple finite-dimensional Lie algebra. The loop algebra $\mathcal{L}_{poly}\bar{\mathfrak{g}}$ is replaced with $\mathfrak{g}_{\Sigma,P} := \mathcal{A}_{\Sigma,P} \otimes \bar{\mathfrak{g}}$, with bracket $[\sum_i f_i \otimes x_i, \sum_j g_j \otimes y_j] = \sum_{i,j} f_i g_j [x_i y_j]$. The Laurent polynomials $\mathbb{C}[t^{\pm 1}]$ are replaced with $\mathcal{A}_{\Sigma,P}$. The Witt algebra is replaced with $\mathcal{L}_{\Sigma,P}$. Just as $\mathfrak{Witt}$ acts on $\mathcal{L}_{poly}\bar{\mathfrak{g}}$ by derivations, so does $\mathcal{L}_{\Sigma,P}$ act on $\mathfrak{g}_{\Sigma,P}$.

There are some subtle differences with the more familiar loop algebras. The loop algebras have an important $\mathbb{Z}$-grading. These higher-genus algebras $\mathcal{L}_{\Sigma,P}$ and $\mathfrak{g}_{\Sigma,P}$ have instead an *almost-grading* by $\mathbb{Z}$, in the sense that $\mathcal{L}_{\Sigma,P}$ (say) can be decomposed $\mathcal{L}_{\Sigma,P} = \oplus(\mathcal{L}_{\Sigma,P})_n$ as a vector space into finite-dimensional subspaces $(\mathcal{L}_{\Sigma,P})_n$, such that

$$[(\mathcal{L}_{\Sigma,P})_m, (\mathcal{L}_{\Sigma,P})_n] \subseteq \oplus_{\ell=m+n+L}^{m+n+M}(\mathcal{L}_{\Sigma,P})_\ell$$

for some fixed integers $L, M \in \mathbb{Z}$. This would be a true grading if $M = L = 0$. The algebra $\mathfrak{g}_{\Sigma,P}$ behaves similarly. The subspaces $(\mathcal{L}_{\Sigma,P})_n$ and $(\mathfrak{g}_{\Sigma,P})_n$ are defined by considering orders of poles (and splitting P into incoming and outgoing points).

In the loop algebra situation, for $\bar{\mathfrak{g}}$ simple, there is a unique nontrivial central extension. On the other hand, $\mathfrak{g}_{\Sigma,P}$ typically has several. However, only one will be compatible with the almost-grading, and so that is the one we choose. Call it $\widehat{\mathfrak{g}_{\Sigma,P}}$. Similarly, we get a unique central extension $\widehat{\mathcal{L}_{\Sigma,P}}$ of $\mathcal{L}_{\Sigma,P}$, which in the special case of a sphere with one incoming and one outgoing puncture is $\mathfrak{Vir}$.

Verma modules, etc. for $\widehat{\mathfrak{g}_{\Sigma,P}}$ can be defined as before using the universal enveloping algebra, and are parametrised by $p = \|P\|$ highest weights $\lambda^{(1)}, \ldots, \lambda^{(p)} \in \mathfrak{h}^*$ and a complex number k (the level). For these modules $W_{(\lambda,k)}$, $\lambda = (\lambda^{(1)}, \ldots, \lambda^{(p)})$, there is an analogue of the Sugawara construction (3.2.15), which shows that each of these $\widehat{\mathfrak{g}_{\Sigma,P}}$-modules $W_{(\lambda,k)}$ is simultaneously a $\widehat{\mathcal{L}_{\Sigma,P}}$-module, in perfect analogy with the affine situation.

Physically, these algebras $\widehat{\mathfrak{g}_{\Sigma,P}}$ and $\widehat{\mathcal{L}_{\Sigma,P}}$ should be regarded as higher-genus global symmetries for, for example, the Wess–Zumino–Witten models discussed next chapter. Locally, that is in terms of local coordinates at each marked point z_i, we get a copy of the affine algebra $\bar{\mathfrak{g}}^{(1)}$ and Virasoro algebra $\mathfrak{Vir}$. A module for, for example, $\widehat{\mathfrak{g}_{\Sigma,P}}$ similarly specialises to the $\bar{\mathfrak{g}}^{(1)}$-module $L(\lambda^{(i)})$ at each point $z_i \in P$.

The theory is still a work in progress – see, for example, [**491**], [**492**] and references therein. But it can be expected that for each positive level k and choice of Σ, and p highest weights $\lambda^{(i)} \in P_+^k(\mathfrak{g})$, a number of level-$k$ representations of $\widehat{\mathfrak{g}_{\Sigma,P}}$ will be singled out (the exact number being given by Verlinde's formula (6.1.2)), and these will 'transform covariantly' with respect to the mapping class group of $\Sigma \setminus P$. Obviously this is an exciting direction that should be pursued, with direct relevance to higher-genus Moonshine (Section 6.3.1).

Question 3.3.1. (a) Define $D = \prod_{\alpha \in \Delta_+}(1 - e^{-\alpha})^{\text{mult}(\alpha)}$. Verify $r_i(D) = e^{-\alpha_i} D$.
(a) Find a vector $r \in \mathfrak{h}$ such that $w(e^r D) = \epsilon(w) e^r D$.

Question 3.3.2. Let A be a Cartan$_{BKM}$ matrix, and $\mathfrak{g}$ the corresponding universal Borcherds–Kac–Moody algebra.
(a) Prove h_{ij} lies in the centre.

(b) Suppose the ith and jth rows of A are identical. Then show that $h_{ii} - h_{jj}$ is in the centre of $\mathfrak{g}$.

Question 3.3.3. In what ways (if any) do Theorems 3.3.3, 3.3.5, 3.3.6 change if the field is $\mathbb{C}$ and not $\mathbb{R}$?

Question 3.3.4. Prove that for any Lie algebra L obeying conditions (i), (ii), (iii) of Theorem 3.3.6, L_0 will be an abelian subalgebra.

3.4 Variations on a theme of character

3.4.1 Twisted #3: twisted representations

In this subsection we complete the introduction of the twisted character which we began in Section 1.5.4. These are to the usual character what the McKay–Thompson series are to the j-function. In Section 5.3.6 we generalise this construction, but as always the special case of affine algebras is particularly pretty and significant. The reader is encouraged to reread Section 1.5.4 for background.

Let's start with a twisted affine algebra $\overline{\mathfrak{g}}^{(N)}$, obtained as in (3.2.4) from the nontwisted algebra $\mathfrak{g} = \overline{\mathfrak{g}}^{(1)}$ and an order-N symmetry α of the Coxeter–Dynkin diagram of $\overline{\mathfrak{g}}$. Consider any integrable highest-weight $\overline{\mathfrak{g}}^{(N)}$-module $L(\lambda)$, $\lambda \in P_+^k(\overline{\mathfrak{g}}^{(N)})$. Think of this as a representation ρ. We can extend ρ linearly to $\mathfrak{g}$, by defining

$$\rho(xt^n) = \xi_N^{i-n} \rho(xt^n), \tag{3.4.1a}$$

for x in the α-eigenspace $(\overline{\mathfrak{g}})_i$ (Section 1.5.4). This isn't a true representation of $\mathfrak{g}$ – it's called a *twisted representation* of $\mathfrak{g}$, as it obeys

$$[\rho(xt^n), \rho(yt^m)] = \xi_N^{i+j-n-m} \rho([xt^n, yt^m]), \tag{3.4.1b}$$

when $x \in (\overline{\mathfrak{g}})_i$ and $y \in (\overline{\mathfrak{g}})_j$. Thus a true representation of the twisted affine algebra $\overline{\mathfrak{g}}^{(N)}$ corresponds to a twisted representation of the nontwisted algebra $\overline{\mathfrak{g}}^{(1)}$. In Section 5.4.6 we extend this notion of twisted representation to vertex operator algebras.

Twisted representations are vaguely reminiscent of projective representations. But a projective representation becomes a true representation when the algebra is extended, while a twisted representation becomes a true representation when the algebra is shrunk. Groups most naturally have projective representations, vertex operator algebras most naturally have twisted ones, and affine algebras have both.

Consider more generally any symmetry α of the Coxeter–Dynkin diagram of $\mathfrak{g}$. As in Section 3.2.2, α extends to an automorphism of $\mathfrak{g}$ (e.g. $\alpha(e_i) = e_{\alpha i}$, and α fixes the centre and derivation). Because of this, α permutes the $\mathfrak{g}$-modules as in Section 1.5.4. In particular, α takes the highest-weight module $L(\lambda)$ to $L(\lambda^\alpha)$, where $(\lambda^\alpha)_i = \lambda_{\alpha i}$, and moreover takes weight-space $L(\lambda)_\mu$ to weight-space $L(\lambda^\alpha)_{\mu^\alpha}$. All of this generalises to any Borcherds–Kac–Moody algebra.

Now suppose $\lambda^\alpha = \lambda$, that is λ is a fixed point of α. Then $L(\lambda)$ and $L(\lambda)^\alpha$ are isomorphic as $\mathfrak{g}$-modules, so let τ_α be a linear isomorphism of the space $L(\lambda)$ that intertwines their $\mathfrak{g}$-actions: that is, $\alpha(x).v = x.\tau_\alpha(v)$ in terms of the $\mathfrak{g}$-action of $L(\lambda)$. Because $L(\lambda)$ is irreducible, τ_α is uniquely determined up to a scalar multiple; scaled appropriately, it will permute all vectors of the form (3.3.2). By the α-*twisted character* or *twining character* χ_λ^α we mean

$$
\begin{aligned}
\chi_\lambda^\alpha(h) &= \exp\left[\left(-h_\lambda + \frac{c_\lambda}{24} + \frac{k(d - d^{orb})}{24h^\vee}\right)\delta\right] \mathrm{tr}_{L(\lambda)}\tau_\alpha e^h \\
&= \exp\left[\left(-h_\lambda + \frac{c_\lambda}{24} + \frac{k(d - d^{orb})}{24h^\vee}\right)\delta\right] \sum_{\mu=\alpha\mu} \mathrm{tr}(\tau_\alpha)e^{\mu(h)}, \quad \forall h \in \mathfrak{h} \quad (3.4.2a)
\end{aligned}
$$

where d and d^{orb} are the dimensions of the semi-simple Lie algebras $\overline{\mathfrak{g}}$ and $\overline{\mathfrak{g}^{orb}}$ (the algebra $\mathfrak{g}^{orb}$ is defined in Theorem 3.4.1) and h_λ, c_λ are in (3.2.9). As in (3.2.9a), the normalisation here is chosen to make modularity simplest – see (3.4.2b) below. As we see from (3.2.5d), the vector $\delta \in \mathfrak{h}^*$ in (3.4.2a) isolates the coefficient $2\pi i \tau$ of the derivation ℓ_0.

Theorem 3.4.1 [213]   *Let $\mathfrak{g} = X_r^{(1)}$ be a nontwisted affine algebra, and let α be a symmetry of the Coxeter–Dynkin diagram of $\mathfrak{g}$. Then for any integrable highest-weight λ of $\mathfrak{g}$, with $\alpha\lambda = \lambda$, the α-twisted character $\chi_\lambda^\alpha(h)$, restricted to any $h \in \mathfrak{h}$ fixed by α, equals some true character $\chi_{\widetilde{\lambda}}(h)$ of the 'orbit Lie algebra' $\mathfrak{g}^{orb} = ((\mathfrak{g}^{op})_0)^{op}$.*

'$\mathfrak{g}^{op}$' is the affine Kac–Moody algebra whose Coxeter–Dynkin diagram is that of $\mathfrak{g}$ except with all arrows reversed. Note that $\mathfrak{g}^{orb}$ is not a subalgebra of $\mathfrak{g}$, although its Cartan subalgebra $\mathfrak{h}^{orb}$ can be identified with that $\mathfrak{h}_0$ of the fixed-point subalgebra $\mathfrak{g}_0$. What is special about $\mathfrak{g}^{orb}$ is that there is a natural map P_α (see Section 3.3 of [213] for its precise construction) sending $\mathfrak{g}$-weights fixed by α to the weights of $\mathfrak{g}^{orb}$, and preserving all inner-products. The weight $\widetilde{\lambda}$ in Theorem 3.4.1 is $P_\alpha(\lambda)$. The normalisation in (3.4.2a) is exactly what one would expect for a character of $\mathfrak{g}^{orb}$:

$$
h_{\widetilde{\lambda}}^{orb} - \frac{c^{orb}}{24} = h_\lambda - \frac{c}{24} + \frac{k(d - d^{orb})}{24h^\vee}. \quad (3.4.2b)
$$

For example, consider $\mathfrak{g} = A_{2n-1}^{(1)}$ and $\mathfrak{g} = A_{2n}^{(1)}$, respectively, with α being the left–right reflection symmetry ('charge-conjugation') 'C' fixing the 0th node. Then the orbit Lie algebra $\mathfrak{g}^{orb}$ is the twisted affine algebras $D_{n+1}^{(2)}$ and $A_{2n}^{(2)}$, respectively. For $\mathfrak{g} = \mathfrak{sl}_n^{(1)}$ with a cyclic symmetry ('simple-current') '$J^{n/d}$' of order d (so d divides n), $\mathfrak{g}^{orb} = \mathfrak{sl}_{n/d}^{(1)}$. The map P_α in these examples is

$$
P_C : \lambda_0\omega_0 + \sum_{i=1}^{n-1} \lambda_i(\omega_i + \omega_{2n-i}) + \lambda_n\omega_n \mapsto \sum_{j=0}^{n} \lambda_i\omega_i^{orb},
$$

$$
P_C : \lambda_0\omega_0 + \sum_{i=1}^{n} \lambda_i(\omega_i + \omega_{2n-i}) \mapsto \sum_{j=0}^{n} \lambda_i\omega_i^{orb},
$$

$$
P_{J^{n/d}} : \sum_{i=0}^{n/d-1} \lambda_i(\omega_i + \omega_{i+n/d} + \cdots + \omega_{i+n-n/d}) \mapsto \sum_{j=0}^{n/d-1} \lambda_i\omega_i^{orb}.
$$

The map P_α is not mysterious. For example, for $\mathfrak{g} = \mathfrak{sl}_{2n}{}^{(1)}$ and $\alpha = C$, the fundamental weights ω_i^{orb} of $\mathfrak{g}^{orb}$ are the obvious basis for the C-invariant weights of $\mathfrak{g}$, namely $\omega_i^{orb} = \omega_i + \omega_{2n-i}$ (for $1 \le i < n$) together with $\omega_0^{orb} = \omega_0$ and $\omega_n^{orb} = \omega_n$.

The most important case in Theorem 3.4.1 is the degenerate one. The Coxeter–Dynkin diagram of $\mathfrak{sl}_n{}^{(1)}$ has an order-n cyclic symmetry J. In this case, an α-fixed point looks like $\lambda = (\lambda_0, \lambda_0, \ldots, \lambda_0)$ for $\lambda_0 = k/n$, and the α-twisted character $\chi_\lambda^\alpha(h)$, restricted to h fixed by α, equals the τ-independent function $\exp[2\pi i\,(\overline{\lambda}(\overline{h}) + ku)]$ – that is, only the top weight-space survives.

A good question in Lie theory is always rewarded with a beautiful answer. Theorem 3.4.1 holds more generally for any Borcherds–Kac–Moody algebra. The proof follows that of the Weyl–Kac–Borcherds character formula.

We get from Theorems 3.4.1 and 3.2.4 that the twisted characters are modular functions, and obey an analogue of Theorem 3.2.3. As an isolated example, this is rather surprising, but it fits into a much larger context (Section 5.3.6). We also find there how modular transformations relate the twisted characters to twisted representations – it is quite analogous to (2.3.10b). From this greater context of vertex operator algebra modules and characters twisted by automorphisms, the modularity of these twisted characters is not so surprising. What is more surprising is positivity, that is, the q-expansion has positive integer coefficients. This is true, for instance, for only two-thirds of the McKay–Thompson series T_g. See Section 7.3.5, especially Conjecture 7.3.3, for an analogous result for the Moonshine module $V^\natural$.

3.4.2 Denominator identities

A very useful formula for the characters of simple finite-dimensional Lie algebras $\overline{\mathfrak{g}}$ is the Weyl character formula (1.5.11). It is rare indeed when the trivial special case of a theorem or formula is interesting. But that happens here. Consider the trivial representation: i.e. $x \mapsto 0$ for all $x \in \overline{\mathfrak{g}}$. Then the character (1.5.9a) is identically 1: $\mathrm{ch}_0 \equiv 1$. Thus the character formula tells us that a certain alternating sum over the Weyl group W equals a certain product over positive roots $\alpha \in \Delta_+$:

$$\prod_{\alpha \in \Delta_+} \left(1 - e^{-\alpha(z)}\right) = e^{-\rho(z)} \sum_{w \in W} \epsilon(w)\, e^{w(\rho)(z)}. \qquad (3.4.3)$$

Here, z lies in the Cartan subalgebra $\mathfrak{h}$, and the Weyl vector ρ is $\omega_1 + \cdots + \omega_r$. Equation (3.4.3) is called a *denominator identity*. For the smallest simple algebra A_1, (3.4.3) is trivial: $1 - e^{-z} = e^{-z/2}(e^{z/2} - e^{-z/2})$. For A_2 we get a sum of six terms equalling a product of three terms, and the complexity continues to rise from there.

In particular, look at $\overline{\mathfrak{g}} = \mathfrak{sl}_n(\mathbb{C})$. We can realise the roots, etc. of $\overline{\mathfrak{g}}$ in terms of an orthonormal basis $\{e_i\}$ of $\mathbb{C}^n$ as follows: the positive roots are $e_i - e_j$ for $1 \le i < j \le n$; the Cartan subalgebra $\mathfrak{h}$ is the hyperplane orthogonal to $\sum_i e_i$; the Weyl group is the symmetric group $\mathcal{S}_n$, acting on $\mathbb{C}^n$ and hence $\mathfrak{h}$ by permuting the e_i; the Weyl vector $\rho = \frac{1}{2}\sum_i (n + 1 - 2i)e_i$. Write $z = \sum_i z_i e_i \in \mathfrak{h}$ and $x_i = e^{-z_i}$ (so $\prod_i x_i = 1$). Then the left side of (3.4.3) becomes

$$\prod_{1\le i<j\le n} (1 - e^{-z_i+z_j}) = x_2^{-1}x_3^{-2}\cdots x_n^{1-n} \prod_{1\le i<j\le n} (x_j - x_i).$$

The right side of (3.4.3) becomes

$$\prod_j x_j^{(n+1-2j)/2} \sum_{\pi\in S_n} \epsilon(\pi) \prod_i x_{\pi i}^{-(n+1-2i)/2} = x_2^{-1}x_3^{-2}\cdots x_n^{1-n} \sum_{\pi\in S_n} \epsilon(\pi) \prod_i x_{\pi i}^i.$$

Thus the denominator identity for $\mathfrak{sl}_n(\mathbb{C})$ is simply the formula for the determinant of the Vandermonde matrix

$$\det \begin{pmatrix} x_1 & x_2 & \cdots & x_n \\ x_1^2 & x_2^2 & \cdots & x_n^2 \\ \vdots & \vdots & & \vdots \\ x_1^n & x_2^n & \cdots & x_n^n \end{pmatrix} = \prod_{1\le i<j\le n} (x_j - x_i). \tag{3.4.4}$$

In the early 1970s Macdonald [**396**] generalised these finite denominator identities to infinite identities, corresponding to the extended Coxeter–Dynkin diagrams. The simplest of his was known classically as the *Jacobi triple product identity*:

$$\prod_{m=1}^{\infty}(1 - x^{2m})(1 - x^{2m-1}y)(1 - x^{2m-1}y^{-1}) = \sum_{n=-\infty}^{\infty} (-1)^n x^{n^2} y^n. \tag{3.4.5a}$$

To Macdonald these were purely combinatorial, but soon Kac, Moody and others reinterpreted his formulae as denominator identities for nontwisted affine algebras, that is substituting $\lambda = 0$ into the Weyl–Kac character formula (3.3.3).

For example, parametrise the Cartan subalgebra of $A_1^{(1)}$ by $z\alpha_1 + z\ell_0 + uC$; then (3.2.5d) says $(m\alpha_1 + n\delta)(z\alpha_1 + \tau\ell_0 + uC) = 2mz - n\tau$. The positive roots of $A_1^{(1)}$ are $\alpha_1 + n\delta$ ($n \ge 0$), $-\alpha_1 + n\delta$ ($n \ge 1$) and $n\delta$ ($n \ge 1$). The Weyl group acts on the Weyl vector ρ by $t_{n\alpha_1}\rho = \rho + 2n\alpha_1 - (2n^2 + n)\delta$ and $t_{n\alpha_1}r_{\alpha_1}\rho = \rho + (2n - 1)\alpha_1 - (2n^2 - n)\delta$. Thus the $A_1^{(1)}$ denominator identity is

$$\prod_{n=0}^{\infty}(1 - rq^n)\prod_{n=1}^{\infty}(1 - r^{-1}q^n)\prod_{n=1}^{\infty}(1 - q^n) = \sum_{m=-\infty}^{\infty} (-1)^m r^{-m} q^{(m^2+m)/2}, \tag{3.4.5b}$$

where $q = e^{-\tau}$ and $r = e^{-2z}$. Equation (3.4.5a) is recovered from (3.4.5b) by setting $x = \sqrt{q}$ and $y = qr^{-1}$.

Freeman Dyson is a famous quantum physicist, but started his academic life in number theory and still enjoys it as a hobby. Dyson [**166**] found a curious formula for the Ramanujan τ-function, defined by $\sum_{n=1}^{\infty} \tau(n)q^n = \eta(q)^{24} := q\prod_{m=1}^{\infty}(1 - q^m)^{24}$:

$$\tau(n) = \sum \frac{\prod_{1\le i<j\le5}(a_i - a_j)}{1!\,2!\,3!\,4!}, \tag{3.4.6}$$

where the sum is over all 5-tuples a_i with $a_i \equiv i$ (mod 5) obeying $\sum_i a_i = 0$ and $\sum_i a_i^2 = 10n$. Using this, an analogous formula can be found for η^{24}. Dyson knew that similar formulae were also known for η^d for the values $d = 3, 8, 10, 14, 15, 21, 24, 26, 28, 35, 36, \ldots$

What was ironic was that Dyson found (3.4.6) at the same time that Macdonald was finding his own identities. Both were at Princeton then, and would often chat a little when they bumped into each other after dropping off their daughters at school. But they never discussed work. Dyson didn't realise that his strange list of numbers has a simple interpretation: they are precisely the dimensions of the simple Lie algebras! $3 = \dim(A_1)$, $8 = \dim(A_2)$, $10 = \dim(C_2)$, $14 = \dim(G_2)$, etc. In fact these formulae for η^d are none other than (specialisations of) the Macdonald identities. For example, Dyson's formula is the denominator formula for $A_4^{(1)}$ ($24 = \dim(A_4)$). If they had spoken, they would surely have anticipated the affine algebra denominator identity interpretation.

Incidentally, no simple Lie algebra has dimension 26, so the formula for η^{26} can't correspond to any of Macdonald's identities. Its algebraic meaning is still uncertain.

Macdonald certainly didn't close the book on denominator identities. Any algebra with a character formula analogous to (1.5.11) (e.g. Borcherds–Kac–Moody algebras (3.3.4)) will have one. Kac and Wakimoto [336] use denominator identities for Lie superalgebras to obtain nice formulae for various generating functions involving sums of squares, sums of triangular numbers (triangular numbers are numbers of the form $\frac{1}{2}k(k+1)$), etc. For instance, the number of ways n can be written as a sum of 16 triangular numbers is

$$\frac{1}{3 \cdot 4^3} \sum ab\,(a^2 - b^2)^2,$$

where the sum is over all odd positive integers a, b, r, s obeying $ar + bs = 2n + 4$ and $a > b$.

The most important application of denominator identities from our perspective is Borcherds' use of them (Section 7.2.2) in proving the Monstrous Moonshine conjectures. Indeed, this possibility was what motivated his introduction of the Borcherds–Kac–Moody algebras. Other applications are discussed next subsection.

Explicitly writing down denominator identities for Borcherds–Kac–Moody algebras tends to be quite difficult, because their root multiplicities are hard to find. The denominator identity of the Monster Lie algebra $\mathfrak{m}$ is a remarkable identity originally due to Zagier, but discovered independently by Borcherds and others:

$$p^{-1} \prod_{\substack{m>0 \\ n\in\mathbb{Z}}}(1 - p^m q^n)^{a_{mn}} = J(z) - J(\tau), \tag{3.4.7a}$$

with $p = e^{2\pi i z}$, where the powers 'a_i' are the coefficients of the q-expansion of the modular function $J(\tau) = \sum_i a_i q^i$. This yields infinitely many nontrivial polynomial identities in the coefficients a_n – for example, comparing third-degree terms on both sides gives

$$a_4 = \binom{a_1}{2} + a_3. \tag{3.4.7b}$$

In fact, (3.4.7a) is older than $\mathfrak{m}$ and is proved independently (Hecke operators permit a quick proof); turning the logic around, it is used to tell us the root multiplicities of $\mathfrak{m}$. This is its direct use in the proof of the Monstrous Moonshine conjectures.

Unfortunately, the numerator of the Weyl character formula for $L(\lambda)$ rarely has a product formula. However, certain specialisations of the numerator can manifestly equal certain (λ-dependent) specialisations of the denominator, and thus inherit the product expansion of the latter. Consider a simple example: any finite-dimensional A_n-module $L(\lambda)$ has a character satisfying

$$\mathrm{ch}_{L(\lambda)}(t\rho) = x^{(n+1)t(\lambda)/2} \prod_{1 \le i < j \le n+1} \frac{x^j y_j - x^i y_i}{y^j - y^i}, \qquad (3.4.8a)$$

for any $t \in \mathbb{C}$, where $x = e^t$ and $y_i = \exp[(i - \sum_{j=i}^{n} \lambda_j)t]$. Similar formulae hold for all Kac–Moody algebras [374]. In particular, from these we obtain instantly Weyl's dimension formula for finite-dimensional semi-simple Lie algebras:

$$\dim L(\lambda) = \prod_{\alpha > 0} \frac{(\alpha|\lambda + \rho)}{(\alpha|\rho)}. \qquad (3.4.8b)$$

3.4.3 Automorphic products

In Section 2.4.1 we explain the important notion of *lifting* a modular form $f : \mathbb{H} \to \mathbb{C}$ for a discrete subgroup Γ of $G = \mathrm{SL}_2(\mathbb{R})$. The result is an automorphic function $\phi : G \to \mathbb{C}$ obeying the transformation (2.4.2b).

Borcherds discovered an unexpected way to lift (meromorphic) modular forms for discrete Γ in $\mathrm{SL}_2(\mathbb{R})$ to much larger Lie groups. His starting point was (3.4.7a), where the coefficients of a modular function appear in the exponents of a product expansion. In hindsight, another example of this phenomenon is the product formula(2.2.6b) for η:

$$\eta(\tau) = q^{1/24} \prod_{n=1}^{\infty} (1 - q^n)^1, \qquad (3.4.9)$$

where the powers '1' are the coefficients of the q-expansion of the modular form $\theta_3(\tau)/2$. Moreover, both (3.4.7a) and (3.4.9) are the denominators of the Monster algebra $\mathfrak{m}$ and the affine algebra $\mathfrak{u}_1^{(1)}$ (recall (3.2.12c)). Are these hints of a much more general phenomenon?

Indeed. Borcherds found a far-reaching generalisation of (3.4.7a):

Theorem 3.4.2 [76] *Suppose $f(\tau) = \sum_n a_n q^n$ is a meromorphic modular form for $SL_2(\mathbb{Z})$ of weight $-s/2$, holomorphic in $\mathbb{H}$ (so its only possible pole is at the cusp), and with integer coefficients a_n. We require $s = 0, 8, 16, \ldots$; if $s = 0$ we also require that* 24 *divides a_0. Let $v_0 \in \mathbb{R}^{s+1,1}$ be a generic vector of negative norm. Then there is a unique lattice vector $\rho \in II_{s+1,1} \subset \mathbb{R}^{s+1,1}$ such that*

$$F(v) = e^{-2\pi i \rho \cdot v} \prod_{r \in II_{s+1,1},\ r \cdot v_0 > 0} (1 - e^{-2\pi i r \cdot v})^{a_{-r \cdot r/2}} \qquad (3.4.10)$$

can be analytically extended to a meromorphic modular form on $\mathbb{H}_{s+1,1}$ of weight $a_0/2$ for the group $O_{s+2,2}(\mathbb{Z})^+$.

Since f in Theorem 3.4.2 has nonpositive weight and is holomorphic in $\mathbb{H}$, it will necessarily have poles at the cusps $\mathbb{Q} \cup \{i\infty\}$ (unless it is constant). The set $II_{s+1,1}$ is the unique even self-dual lattice of signature $(s + 1, 1)$ (Section 1.2.1). $O_{s+2,2}(\mathbb{R})$ is the group of $(s + 4) \times (s + 4)$ matrices A with real entries, which obey $ADA^t = D$ for $D = \text{diag}(1, \ldots, 1, -1, -1)$. By a modular form for $O_{s+2,2}(\mathbb{Z})^+$, we mean the following. First, the imaginary norm vectors in $\mathbb{R}^{s+1,1}$ lie in two disjoint cones; denote by C the cone containing $-v_0$. The analogue of the upper half-plane $\mathbb{H}$ is here the set $\mathbb{H}_{s+1,1} \subset \mathbb{C}^{s+1,1}$ consisting of all vectors v with imaginary part $\text{Im}(v) \in C$. Then

$$F(v + \lambda) = F(v), \quad \forall \lambda \in II_{s+1,1}, \tag{3.4.11a}$$

$$F(w(v)) = \pm F(v), \qquad \forall w \in \text{Aut}(II_{s+1,1})^+, \tag{3.4.11b}$$

$$F\left(\frac{2v}{v \cdot v}\right) = \pm \left(\frac{v \cdot v}{2}\right)^{a_0/2} F(v), \tag{3.4.11c}$$

for appropriate choice of signs, where $\text{Aut}(II_{s+1,1})^+$ are the automorphisms of the lattice $II_{s+1,1}$ that send the cone C to itself. The transformations on $\mathbb{H}_{s+1,1}$ given in (3.4.11) generate a subgroup of $O_{s+2,2}(\mathbb{Z})$, denoted $O_{s+2,2}(\mathbb{Z})^+$. Now F can be lifted to the Lie group $O_{s+2,2}(\mathbb{R})^+$ in the usual way. This lifting of a modular form for a subgroup Γ of $SL_2(\mathbb{R})$ to automorphic forms for $O_{s+2,2}(\mathbb{R})^+$ is called a *Borcherds lift*.

Of course (3.4.7a) is recovered from taking $f(\tau) = j(\tau) - 744$; then $s = 0$, and the real Lie group $O_{2,2}(\mathbb{R})$ is essentially $SL_2(\mathbb{R}) \times SL_2(\mathbb{R})$ – that is, they share the same Lie algebra (recall Theorem 1.4.3) – with each $SL_2(\mathbb{R})$ contributing a copy of $\mathbb{H}$ and $SL_2(\mathbb{Z})$.

We can recover from F more familiar modular forms by restricting the domain of F to multiples τv of imaginary norm vectors v in $II_{s+1,1}$. For example, we get:

Theorem 3.4.3 [76] *Let $f(\tau) = \sum_{n=-\infty}^{\infty} a_n q^n$ be any meromorphic modular form for $\Gamma(4)$, holomorphic in $\mathbb{H}$ but possibly with poles at the cusps, and with integer coefficients a_n. We require $a_n = 0$ unless $n \equiv 0, 1 \pmod{4}$. Then for some choice of $h \in \mathbb{Z}/12$,*

$$F(\tau) = q^h \prod_{n=1}^{\infty} (1 - q^n)^{a_{n^2}}$$

is a meromorphic modular form of weight a_0, with all poles and zeros at cusps.

For example, (3.4.9) (or rather its square) is recovered by taking $f(\tau) = \theta_3(2\tau)$. Modular forms for SL_2 arise here because $O_{1,2}(\mathbb{R})$ is essentially $SL_2(\mathbb{R})$.

In this section we find several examples of product expansions of modular forms, Jacobi forms, etc. coming from the denominators of characters. An exciting development is provided by Gritsenko and Nikulin [**264**], [**265**]. Given any hyperbolic Kac–Moody algebra of rank $n \geq 3$ with certain properties (making them close in spirit to semi-simple Lie algebras), there exists a Borcherds–Kac–Moody algebra of the same rank with identical real roots (hence Weyl group, which will be a subgroup of $O_{n-1,1}(\mathbb{R})$), but with precisely the imaginary simple roots needed so that its denominator is an automorphic form for $O_{n,2}(\mathbb{R})$. It is reminiscent of Macdonald's identities: he found he needed to introduce extra factors to get modularity (namely the third product in (3.4.7b)), and we now interpret those as due to the imaginary roots of the corresponding affine algebra.

Most Borcherds–Kac–Moody algebras are of course not interesting; those that are (e.g. the Monster and fake Monster Lie algebras) have automorphic denominator identities. Thus this provides a systematic construction of what should be interesting Borcherds–Kac–Moody algebras. It is known that there are only finitely many such hyperbolic Kac–Moody algebras, and so this is a finite family of Borcherds–Kac–Moody algebras. Clearly, we should study their representation theory, and compute the characters of their 'interesting' (presumably integrable) modules. In analogy with affine algebras, we may hope that the numerators of those characters will also be automorphic.

Relations of these automorphic forms with mirror symmetry and string theory are beyond this book, but see, for example, [266], [342], [275], [276], [434]. The review article [358] is a good treatment of many of the topics of this subsection.

Question 3.4.1. Let $f(q) = \sum_{n=0} a_n q^n$, with $a_0 = 1$. Verify that, at least formally (i.e. without any regard to convergence), this can be written as $f(q) = \prod_{n=1}^{\infty}(1 - q^n)^{b_n}$ for some unique numbers b_n. If all a_n are integers, then so are all b_n.

Question 3.4.2. Prove (3.4.8a) and the Weyl dimension formula (3.4.8b) for $\mathfrak{sl}_n$.

Question 3.4.3. Express the character χ_λ of any integrable representation λ of $A_1^{(1)}$, specialised appropriately, as an infinite product.

4

Conformal field theory: the physics of Moonshine

This chapter presents the physical context for Moonshine. Rather than diving into a conventional discourse of conformal field theory (CFT), it might be more helpful to take several steps back and begin with Galileo. Physics even more than mathematics is interwoven with history. Our treatment of CFT is sketchy but should supply the reader with all that is necessary to appreciate the absolutely profound role physics has played in Moonshine and other aspects of 'pure' mathematics in recent years. It is hoped that this chapter will make it easier for the interested reader to pursue more standard treatments of CFT and string theory. It is written primarily with the mathematician in mind.

The third section explores the physics of CFT, and the fourth describes some mathematical formulations. CFT is to a generic quantum field theory what finite-dimensional semi-simple Lie algebras are to generic Lie algebras. Background for both sections is provided by the review of classical and quantum physics sketched in the first two sections.

For a mathematician studying physics, important to keep in mind is that physics has been driven historically more by its predictive power than by conceptual concerns (with a few remarkable exceptions, such as Einstein's general relativity). Given enough time, however, the theory becomes polished to a state of pristine mathematical elegance, as classical mechanics amply demonstrates. In particular, one has the sense that quantum theory is *ad hoc* and rather unsound – and it is both – but these features are due to the historical accident that we were born too close to its inception. Much more important is what it can teach mathematics, which is considerable. *The essence of quantum field theory is completely accessible to mathematicians and,* as mathematics of the late twentieth century shows, *should at least in its broad strokes be part of their standard repertoire.*

A special feature of classical physics is that the behaviour of a system – for example, its trajectory in phase space – becomes much simpler when looked at infinitesimally. The simple universal regularities are captured by differential equations; the complicated incidental features of a specific situation are relegated to the initial conditions. Among mathematicians, this central role of partial differential equations in classical physics was responsible for what had been a near-identification of their study with the subject they call mathematical physics. It was largely with the arrival of string theory that a much richer range of mathematics became relevant to physics, and it is this happy development that made this book possible.

Almost every facet of Moonshine fits comfortably into CFT, where it often was discovered first. Some have questioned though the necessity of involving such a complicated beast, or the closely related 'vertex operator algebras' of the next chapter, in our mathematical explanation of Moonshine. Although CFT has been an invaluable guide so far, they would argue, perhaps we are a little too steeped in its lore. Undoubtedly there is truth in this, but CFT still has new insights to share. It is an integral part of Moonshine's future as much as its past. Sections 4.3 and 4.4 are central to the whole book.

4.1 Classical physics

4.1.1 Nonrelativistic classical mechanics

Temporarily forget what you know of physics. One of the most blatant empirical facts must be that anything in motion on Earth eventually slows to a stop. On the other hand, stars and planets clearly behave otherwise, therefore earthly laws can't apply directly to the Heavens. Those observations are fundamental to Aristotelian physics. The starting point, however, for classical physics is *Newton's First Law*: the remarkable thought (due to Galileo, 1632) that anything anywhere will continue to move in a straight line and at constant speed, unless something (by definition a *force*) acts on it. Although in isolation it has no real content, it presents a powerful strategy for analysing Nature. For example, to first approximation the Moon travels in a circle about the Earth; rather than trying to conceive of some strange mechanism responsible for pushing or dragging the Moon in its nonlinear orbit, the First Law instead leads us to imagine some 'force' that always pulls the Moon towards the Earth. This second possibility is much more promising of course, and led Newton to his theory of gravitation.

Classical mechanics describes systems with finitely many degrees of freedom. The *configuration* (snapshot, instantaneous state) of a classical system at an instant t of time can be identified with the precise values of all degrees of freedom (e.g. position coordinates) at that time. The basic challenge is to predict the configuration at later times. This amounts to setting up and solving a system of differential equations, called the equations of motion of the system.

Consider a system of N particles, with positions $\mathbf{x}_i = (x_{i1}, x_{i2}, x_{i3})$. The $3N$ degrees of freedom are the position coordinates x_{ij}. The equations of motion, which determine the trajectories of the N particles by giving their response to the stimulus, are

$$m_i \frac{\mathrm{d}^2}{\mathrm{d}t^2}\mathbf{x}_i = \mathbf{F}_i, \tag{4.1.1}$$

where $\mathbf{F}_i$ is the net force experienced by the ith particle and the proportionality constant m_i is called its mass. Dots are used to denote time derivatives: for example, velocity is $\dot{\mathbf{x}}$ and acceleration is $\ddot{\mathbf{x}}$. Note that (4.1.1) is compatible with Newton's First Law.

In general the force $\mathbf{F}_i$ can be a function of all positions $\mathbf{x}_j$, velocities $\mathbf{v}_j$ and time t – for example, air resistance is approximately proportional to $\mathbf{v}_i^2$. We will restrict attention to the typical ones (from which can be derived all others), which are of the form

$$(\mathbf{F}_i)_j = -\frac{\partial}{\partial x_{ij}} V(\mathbf{x}_1, \ldots, \mathbf{x}_N)$$

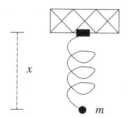

Fig. 4.1 The harmonic oscillator.

Fig. 4.2 Singular motion of five gravitationally interacting particles.

for some real-valued function V called the *potential*. These are called *conservative* forces because they conserve (keep constant) energy. The potential has units of energy, and the sign is introduced so that V contributes positively to total energy. In quantum mechanics the potential V is more fundamental than the force $\mathbf{F}$.

For example, Newton's gravitational potential is $V = -\sum_{i<j} G \frac{m_i m_j}{|\mathbf{x}_i - \mathbf{x}_j|}$, where G is a positive constant. Einstein found it profoundly significant that the gravitational 'charge' m_i here is numerically (though certainly not conceptually) identical to the 'inertial' mass m_i in (4.1.1) (see Section 4.1.2).

For a one-dimensional example, consider a *harmonic oscillator* – for example, the spring in Figure 4.1. Hooke's Law says that the force $F = -k\,(x - x_0)$, where k is a positive constant and x_0 is the resting length of the spring. Hence $-k\,(x - x_0) = m\ddot{x}$, so

$$x = x_0 + a \cos\left(\sqrt{\frac{k}{m}}\,t\right) + b \sin\left(\sqrt{\frac{k}{m}}\,t\right) = x_0 + A \cos\left(\sqrt{\frac{k}{m}}\,t + B\right). \qquad (4.1.2)$$

This force is conservative, with potential $V = \frac{1}{2}k\,(x - x_0)^2$. This elementary system is fundamental to theoretical physics, as it describes small oscillations about stable equilibrium states (i.e. points at which all forces $\mathbf{F}_i$ vanish). Indeed, if $\mathrm{d}V/\mathrm{d}x$ vanishes at $x = x_0$, for some potential V, then the Taylor expansion of $V(x)$ would begin like $a_0 + a_2(x - x_0)^2$, and so it would behave like a harmonic oscillator. We encounter the harmonic oscillator repeatedly in the following pages; in classical field theory these humble oscillations describe, for example, sound waves, and in quantum field theory they are the particles.

The mathematical difficulties faced by quantum field theory are notorious, but remarkably singular behaviour occurs in classical mechanics as well. For one example, consider five point particles interacting gravitationally, positioned as in Figure 4.2. Particle 5 moves horizontally between the orbiting pairs 1 and 2, and 3 and 4. It is possible [**485**] to arrange for particle 5 to zip back-and-forth between those pairs, picking up speed, until in a finite time it reaches infinite speed without ever colliding with the other particles.

Many other examples of singular behaviour in classical mechanics are possible [**485**]; it is not known yet how typical they are among all possible motions.

Later in this section and the next, we touch on other mathematical difficulties plaguing our physical theories. Generally speaking, these difficulties of classical and quantum physics have to do with probing space to arbitrarily high precision. Whenever we push scientific theories far beyond their established realm of reliability, our arrogance inevitably gets us punished.[1] The infinitesimal structure of space and time is surely such an unjustified speculative extrapolation. Unfortunately, all our physics is built on it. It is tempting to guess that when we understand how the illusion of a macroscopic four-dimensional space-time continuum arises from more fundamental concepts, these mathematical difficulties should become more tractable.

We know from our childhood that global properties can arise from second-order differential equations ('The shortest distance between two points is a straight line'). *Hamilton's principle* says that the solution to the equation of motion $m\ddot{x} = -\frac{d}{dx}V$, subject to the boundary conditions $x(t_1) = x_1, x(t_2) = x_2$, is the path $t \mapsto x(t)$ obeying the given boundary conditions, for which the *action*

$$S := \int_{t_1}^{t_2} \left(\frac{1}{2} m\, \dot{x}(t)^2 - V(x(t)) \right) \, dt \qquad (4.1.3)$$

is stationary (minimal if $|x_1 - x_2|$ and $|t_1 - t_2|$ are both small). The integrand is called the *Lagrangian* $L = T - V$, where $T = \frac{1}{2}m\dot{x}^2$ is the *kinetic energy*. The combination $T + V$ for the stationary path $x(t)$ will be independent of the time t, and is called the *energy*. Historically, a hard lesson to learn (even for men like Gauss and Hertz) was that energy is an abstract mathematical notion and not a measure of some physical quantity (see the excellent discussion in chapter 4, vol. I of [**188**]).

This observation leads to a formulation of classical physics called Lagrangian mechanics, which will be central to our discussion of quantum field theory in Section 4.2 (in quantum theory concepts like force, velocity and acceleration cease to play fundamental roles). The possible configurations of our physical system can be regarded as forming a manifold, called the *configuration space* $\mathcal{M}$. For example, for a rigid body such as a potato, the configuration space is $\mathbb{R}^3 \times SO_3(\mathbb{R}) \cong \mathbb{R}^3 \times \mathbb{P}^3(\mathbb{R})$:$\mathbb{R}^3$ gives its centre-of-mass, and $\mathbb{P}^3(\mathbb{R})$ its orientation. The behaviour of a system is regarded geometrically as a parametrised path $t \mapsto q(t)$ on $\mathcal{M}$, called the trajectory. Let q_i be a complete set of local coordinates on $\mathcal{M}$, obtained by restricting to some open set $U_\alpha \subset \mathcal{M}$ (recall Definition 1.2.3). The q_i represent the degrees of freedom of the system. The Lagrangian $L = T - V$ is a function of q_i and $\dot{q}_j$ – that is, a function on the tangent bundle $T\mathcal{M}$. In particular, in order to capture the kinetic energy T, which usually will be quadratic in the $\dot{q}_i$, we typically want $\mathcal{M}$ to be Riemannian, with T proportional to the norm-squared $\dot{q} \cdot \dot{q}$. The potential V will be a differentiable function on $\mathcal{M}$. The equations of motion

---

[1] Examples abound. There is, for instance, the famous remark of Lord Kelvin in 1899 that all of physics has been finished. Socrates' theory near the end of *Phaedo* as to the nature of the Earth makes a merry read. In mathematics recall the humbling experiences of Russell's Paradox and Gödel's Incompleteness Theorem.

in the coordinate patch U_α are the *Euler–Lagrange equations*

$$\frac{\mathrm{d}}{\mathrm{d}t}\left(\frac{\partial L}{\partial \dot{q}_i}\right) = \frac{\partial L}{\partial q_i}, \tag{4.1.4}$$

which say that the action (4.1.3) is stationary for the physical solutions $q_i(t)$. Equation (4.1.4) is obtained from the calculus of variations by varying q_i.

To solve a physical system in Lagrangian mechanics, the first task would be to choose good local coordinates q_i on the configuration space $\mathcal{M}$, then to express the kinetic and potential energies in terms of q_i and $\dot{q}_i$, and finally to write down and solve the corresponding partial differential equations (4.1.4). Lagrangian mechanics (and Hamiltonian mechanics, to be discussed shortly) are essentially equivalent to Newtonian mechanics (4.1.1). Their appeal though should be clear to any mathematician: by freeing the formulation from adherence to a specific choice of coordinates, the formal structure of classical mechanics becomes more evident. This is especially valuable when extensions of the theory are needed – for example, when handling enormous numbers of particles in statistical mechanics, or when we were struggling to obtain the laws of quantum mechanics.

Returning to the harmonic oscillator, take $q = x - x_0$. Then $L = T - V = \frac{1}{2}m\dot{q}^2 - \frac{1}{2}kq^2$ and the Euler–Lagrange equation (4.1.4) yields the differential equation $m\ddot{q} = -kq$. The configuration space is $\mathbb{R}$, and trajectories consist of segments $[-A, A]$ traversed periodically. Energy $T + V = \frac{1}{2}kA^2$ is constant on each trajectory.

The pervasive habit of writing physical quantities with 'units' (metres, seconds, ...) leads us into thinking of those mysterious entities as real and indispensable. In fact, many would regard as profound, or at least meaningful, the following question: What is the number of fundamental units in physics? However, Lagrangian mechanics should have led us to a somewhat more sophisticated understanding of units. Units themselves have no fundamental significance; choosing units is a special case of selecting a coordinate patch on the configuration space (together with a choice of time parameter). The common and useful practise of rejecting or anticipating formulae based on unit considerations ('dimensional analysis') merely captures some homogeneity information stored in the Lagrangian, and is the analogue here of the conservation laws of the following paragraphs. In particular, suppose we've selected a coordinate patch $\varphi : U \to \mathbb{R}^n, q \mapsto (q_i)$, and we want to change the scales (i.e. units) on each coordinate axis (which as expressions of nationalistic pride is fairly common). That is, we choose nonzero constants λ_i and consider the rescaling $q_i \mapsto q_i' = \lambda_i q_i$ of local coordinates, as well as $t \mapsto t' = \lambda_0 t$. This has two consequences. Firstly, we can write locally $L(q_i', \dot{q}_j', t') = L'(q_i, \dot{q}_j, t)$, that is, we can continuously deform the Lagrangian. Inevitably, some choices of units will simplify L and hence ease the resulting arithmetic. Secondly and more importantly, it typically will be possible to absorb the rescalings λ_i into the various 'physical constants', that is, the parameters in L, which will tell us invariance properties of L and hence of the equations of motion (4.1.4). This is how to obtain the convenient and well-known meta-theorem that says the units of each term of any physical expression should agree.

For example, note that the harmonic oscillator Lagrangian is invariant under the rescalings $q \mapsto \lambda_1 q, t \mapsto \lambda_0 t, k \mapsto \lambda_1^{-2} k$ and $m \mapsto \lambda_0^2 \lambda_1^{-2} m$; we see that each term of the solution (4.1.2) has a well-defined and consistent scaling behaviour (as they must). Also, for a preferred choice of λ_i, the Lagrangian simplifies to $\dot{q}^2 - q^2$. For another example, note that the gravitational Lagrangian $L = \frac{1}{2} m (\dot{x}_1^2 + \dot{x}_2^2 + \dot{x}_3^2) + G \frac{mM}{r}$ is invariant under the rescaling $x_i \mapsto \lambda x_i, t \mapsto \lambda_0 t$, provided m rescales like $\lambda^{-2} \lambda_0^2 m$ and GM rescales like $\lambda^3 \lambda_0^{-2}$. In all cases, this scaling behaviour can be taken as defining the 'units' of the corresponding quantity – our definition here that the units of L be trivial differs from the usual one (where L has units of 'energy'), but this is merely a matter of convention.

This discussion should lead us to suspect that other invariance properties of L may yield other 'meta-theorems', generalising in a way the dimensional analysis. Indeed that is beautifully the case. By a *symmetry* of our system, we mean a diffeomorphism α of the configuration space $\mathcal{M}$ respected by the physics:

$$L(\alpha(q), \hat{\alpha}(\dot{q})) = L(q, \dot{q}),$$

where $\hat{\alpha}(\dot{q})$ is the induced map (derivative) on the tangent space with ith component $\sum_j \frac{\partial \alpha(q)_i}{\partial q_j} \dot{q}_j$. Note that, unlike the rescalings considered in the previous paragraph, here we're requiring that L and hence all the physical constants be unchanged by α. Then $q(t)$ is a possible trajectory (i.e. a solution of (4.1.4)) iff $\alpha(q(t))$ is.

Now, suppose we have a *continuous* family α_s of symmetries, that is a one-parameter subgroup $s \mapsto \alpha_s$ in the Lie group of symmetries. This symmetry can be used to vary the coordinates $q_i, \dot{q}_j$ – and hence the action S (4.1.3) – infinitesimally. What does Hamilton's principle ($\delta S = 0$) tell us here? The answer (*Noether's Theorem*[2]) is remarkable: continuous symmetries yield conservation laws! Define the quantity ('charge')

$$Q := \frac{\partial L}{\partial \dot{q}} \left(\frac{\partial \alpha_s(q)}{\partial s} \right) \in \mathbb{R}.$$

This expression is meaningful because the 'generalised momentum' $p := \frac{\partial L}{\partial \dot{q}}$ is a section of the cotangent bundle $T^* \mathcal{M}$, while the derivative $\frac{\partial \alpha_s(q)}{\partial s}$ of the path $\alpha_s(q)$ (q fixed) defines a section of the tangent bundle $T\mathcal{M}$. Less formally, suppose α_s sends q_i to $q_i + s f_i(q, \dot{q}, t)$, keeping only first order in the parameter s; then $\hat{\alpha}_s$ sends $\dot{q}_i$ to $\dot{q}_i + s \frac{df_i}{dt}$, to first order, and $Q = \sum_i p_i f_i$. In either case, an easy calculation from (4.1.4) shows that Q is constant along each trajectory, that is Q is 'conserved'. (A deeper reason for this is that the Poisson bracket (4.1.6a) gives the space of solutions to (4.1.4) a symplectic structure.)

For example, the gravitational potential $V = -G \frac{m_1 m_2}{|\mathbf{x}_1 - \mathbf{x}_2|}$ is invariant with respect to translations $\alpha_s(\mathbf{x}) = \mathbf{x} + s\mathbf{a}$ for any fixed vector $\mathbf{a} \in \mathbb{R}^3$. The charge Q here is $\mathbf{a} \cdot \mathbf{p}$ where $\mathbf{p}$ is the 'total momentum' $m_1 \frac{d\mathbf{x}_1}{dt} + m_2 \frac{d\mathbf{x}_2}{dt}$. Varying $\mathbf{a}$, we find that momentum is conserved. We could say that the independence of the physics on absolute position

---

[2] As is typical, this designation is a little unfair: Noether published this in 1918, but Jacobi already knew in 1842 the connection between translation symmetry and momentum conservation, and rotational symmetry and angular momentum.

implies conservation of momentum. Likewise, independence of the physics on absolute time implies conservation of energy. In classical mechanics, Poincaré showed that all conservation laws are due to an underlying symmetry: if Q is conserved, then the Poisson bracket $\{Q, q\}_P$ of (4.1.6a) generates the corresponding symmetry. What is fundamental here isn't the Lie group action on $T\mathcal{M}$, but rather the infinitesimal generators (Lie algebra action), which need not be derived from a Lie group symmetry.

Another formulation of classical physics, useful for extensions to statistical and quantum mechanics, is Hamiltonian mechanics. Recall the generalised momenta $p_i = \frac{\partial L}{\partial \dot{q}_i}$. Together, the variables q_i, p_j parametrise a $2n$-dimensional manifold, the cotangent bundle $T^*\mathcal{M}$, called *phase space*. The *Hamiltonian* $H(q_i, p_j)$ is the quantity $\sum_i p_i \dot{q}_i - L$, expressed in variables q_i, p_i. Typically, it equals the total energy. The equations of motion here, obtained by varying both q_i and p_j, are *Hamilton's equations*:

$$\dot{q}_i = \frac{\partial H}{\partial p_i}, \qquad \dot{p}_i = -\frac{\partial H}{\partial q_i}, \tag{4.1.5}$$

that is $2n$ first-order differential equations, rather than the n second-order differential equations of Lagrangian mechanics (4.1.4). Although Hamiltonian mechanics is not always equivalent to Lagrangian mechanics, it is for typical systems. Because Hamilton's equations (4.1.5) are first-order, the configuration of the physical system at any time t is uniquely determined by the point in phase space it occupies at a given instant t_0. Thus phase space serves as a moduli space for physics. A more careful treatment of Hamiltonian mechanics requires the language of symplectic geometry – see, for example, [15] for details.

In classical mechanics the *observables*, that is the physically measurable quantities such as position, momentum or energy, are by definition real-valued smooth functions $A(q, p)$ on phase space. It is through the observables that a physical theory is compared to experiment. The observables $C^\infty(T^*\mathcal{M})$ form an infinite-dimensional Lie algebra, with bracket (in local coordinates) given by the *Poisson bracket*

$$\{A, B\}_P := \sum_i \left(\frac{\partial A}{\partial q_i} \frac{\partial B}{\partial p_i} - \frac{\partial A}{\partial p_i} \frac{\partial B}{\partial q_i} \right) \tag{4.1.6a}$$

(see Question 4.1.2). Then Hamilton's equations (4.1.5) imply

$$\frac{\mathrm{d}A}{\mathrm{d}t} = \{A, H\}_P, \tag{4.1.6b}$$

where on the left A is evaluated on a trajectory $(q(t), p(t))$. The term 'first integral' refers to any observable that is constant along each trajectory; the first integrals form a Lie subalgebra of dimension $< 2\mathrm{dim}\mathcal{M}$ in the observables $C^\infty(T^*\mathcal{M})$. Equation (4.1.6a) may seem obscure, but it is essentially equivalent to the natural bracket $[X, Y]$ of vector fields on a manifold – see corollary 5, page 217 of [15] for details. As we see in the next section, algebra arises in quantum field theory through the analogue there of Poisson bracket.

For example, recall the harmonic oscillator. The generalised momentum $p = m\dot{q}$ is the usual momentum. The Hamiltonian $H = \frac{1}{2m}p^2 + \frac{1}{2}kq^2$ is the energy. Hamilton's equations tell us $\dot{q} = p/m$ and $\dot{p} = kq$. Phase space is the plane $\mathbb{R}^2$, with ellipses as trajectories. The basic Poisson bracket $\{q, p\}_P = 1$ says the observables $q, p, 1$ span $\mathfrak{Heis}$ (recall (1.4.3)).

4.1.2 Special relativity

The fundamental theoretical advance of the nineteenth century was Maxwell's electromagnetism (Section 4.1.3), which unified light, electricity and magnetism. Although both Newtonian mechanics and Maxwell's theory were enormously successful, they were in some conflict. For instance, in Maxwell's theory is obtained the formula

$$c := \text{speed of light} = \frac{1}{\sqrt{\epsilon_0 \mu_0}},$$

where ϵ_0, μ_0 are numerical constants associated with the vacuum. This seems to suggest that the speed of light is itself a constant, independent of the observer. However Newton – and common sense – would have us believe that the speed at which light, or anything else, travels is variable. If light is emitted from a headlight with speed c, and a bug approaches the oncoming car with speed v, then surely to it that light travels with speed $v + c$.

The standard resolution in the nineteenth century was to regard Maxwell's equations as valid only with respect to a substance called the aether. The aether would be the stuff in which light-waves wave (propagate) – it would be to light what air is to sound. This aether concept was getting increasingly awkward as the century turned. Einstein's act of genius here was to flip the logic and trust Maxwell's message. Thus, the speed of light is the same for all observers: the light from that approaching car strikes the bug with the same speed c it left the headlights. Special relativity consists of the modifications this message implies for Newtonian physics. Indeed what we call magnetism can be thought of as a relativistic correction to the electrostatic force; Maxwell's electromagnetism was the first relativistic theory, created years before Einstein's birth.

The word 'special' in 'special relativity' arises because the equations are simplest and fundamental only for a certain class of privileged observers called 'inertial' – uniformly moving observers for which Newton's First Law holds. A car rounding a corner is certainly not inertial, but a coasting isolated spaceship could be treated as one to good approximation. Special relativity also applies to accelerating observers, provided one works infinitesimally. Physically speaking, *general* relativity (Section 4.1.3), which removes this preferential treatment of inertial observers, is a mathematically elegant global integration of the equivalence principle and locally applied special relativity.

An inertial observer is simply a choice of fixed basis in $\mathbb{R}^4$; the coordinates $(\mathbf{x}, t)$ with respect to this basis, of a point ('event') x in $\mathbb{R}^4$ ('space-time'), have the physical interpretation to that observer as space and time coordinates. Not every choice of basis is permitted: we require them to be orthonormal in the sense that the straight-line trajectory ('world-line') $(\mathbf{x}(t), t)$ traced in space-time $\mathbb{R}^4$ by a beam of light is required to satisfy $(\mathbf{x}(t) - \mathbf{x}(0)) \cdot (\mathbf{x}(t) - \mathbf{x}(0)) = c^2 t^2$ – this is what we mean by the speed of light being constant. Thus we are led to endow space-time $\mathbb{R}^4$ with the indefinite Minkowski metric $\eta = (\eta_{\mu\nu}) = \mathrm{diag}(1, 1, 1, -c^2)$. We write x^2 for $x \cdot x = \sum_{\mu,\nu=1}^{4} x_\mu x_\nu \eta_{\mu\nu}$ and $\mathbf{x}^2 = \sum_{\mu,\nu=1}^{3} \mathbf{x}_\mu \mathbf{x}_\nu$. Basis transformations between inertial observers belong to the Lie group $O_{3,1}(\mathbb{R})$. As mentioned in Section 1.4.2, it has four connected components; the component containing the identity is the *Lorentz group* $SO_{3,1}^{+}(\mathbb{R})$. Its universal cover $SL_2(\mathbb{C})$ and their semi-direct products with translations $\mathbb{R}^4$ (the *Poincaré group* and its double-cover) also arise in physics. Thus in special relativity space and time are coupled, just as in Euclidean geometry the x, y, z coordinates are coupled (i.e. their independent objective significance is denied). The disturbing dissimilarity between our qualitative experiences of time and space is ignored by Einstein's theory. Discovering what relation this dissimilarity has to the different signs in the metric, or to the apparent magnitude of c, clearly should be a fundamental task. By contrast, in Newtonian mechanics space-time $\mathbb{R}^4$ factorises globally as $\mathbb{R}^3 \times \mathbb{R}$, and the basis transformations are taken from $O_3(\mathbb{R}) \times \{\pm 1\}$.

That Maxwell's equations are invariant under the Lorentz group was known before Einstein. Einstein's contribution was to interpret the Lorentz group as giving the transformation of physical space and time. For example, the space-time transformation Λ between two observers with parallel spatial coordinate axes but travelling with uniform relative velocity $\mathbf{v} = (v, 0, 0)$, according to Einstein and Newton, is

$$\Lambda = \begin{pmatrix} \frac{1}{\sqrt{1-v^2/c^2}} & 0 & 0 & \frac{v}{\sqrt{1-v^2/c^2}} \\ 0 & 1 & 0 & 0 \\ 0 & 0 & 1 & 0 \\ \frac{v/c^2}{\sqrt{1-v^2/c^2}} & 0 & 0 & \frac{1}{\sqrt{1-v^2/c^2}} \end{pmatrix}, \tag{4.1.7a}$$

$$\Lambda = \begin{pmatrix} 1 & 0 & 0 & v \\ 0 & 1 & 0 & 0 \\ 0 & 0 & 1 & 0 \\ 0 & 0 & 0 & 1 \end{pmatrix}, \tag{4.1.7b}$$

respectively. Note that in the limit $c \to \infty$, (4.1.7a) tends to (4.1.7b). Physically, matrix (4.1.7a) says that the lengths of moving objects shrink, and their clocks run more slowly. This is *not* some illusion, optical or otherwise. For example, the muon is an unstable elementary particle with an average lifespan of 2×10^{-6} seconds when at rest. When travelling at speed v, it will last on average $2 \times 10^{-6}/\sqrt{1 - v^2/c^2}$ seconds. It will travel further than it would have if (4.1.7b) had been the correct transformation, and because of that will be able to participate in interactions that would have been too distant for a muon behaving nonrelativistically. Other physical quantities transform similarly – for example, the parameter m playing the role of relativistic mass equals $m_0/\sqrt{1 - v^2/c^2}$, for some constant m_0 called *rest-mass*. Now, expand this out using the binomial series:

$$m = m_0 + \frac{1}{2}m_0\frac{v^2}{c^2} + \frac{3}{8}m_0\frac{v^4}{c^4} + \cdots$$

Multiplying by c^2, we recognise the second term as kinetic energy and we are led to suspect that mc^2 is the relativistic analogue of kinetic energy – that is, $E = mc^2$ for a free particle.[3]

In order to compare observations, we need to understand how the physical quantities change when we switch inertial observers, that is, how they transform with respect to the Lorentz group $SO_{3,1}^+(\mathbb{R})$. Typically, they transform like matrix entries of $SO_{3,1}^+$-representations. For example, the 4-vector $(\mathbf{x}, t)$ transforms with respect to the defining representation of the Lorentz group, as does the energy-momentum 4-vector $(\mathbf{p}, E/c^2)$, and thus its Minkowski norm-squared $\mathbf{p}^2 - E^2 c^{-2}$ is an observer-independent quantity (a *Lorentz scalar*) and equals $-m_0^2 c^2$. It is conventional to denote with superscripts the components of any such 4-vector: for example, $(\mathbf{x}, t) = (x^1, x^2, x^3, x^4)$.

Writing equations of motion presents us with a challenge: in Newtonian physics we always want to differentiate or integrate with respect to time; however, relativity teaches that we shouldn't treat time distinctly from the spatial coordinates. Moreover, '$dt = dx^4$' transforms like a component of a 4-vector, which isn't necessarily what we want. The solution is that the infinitesimal norm-squared $d\mathbf{x}^2 - c^2 dt^2 =: -c^2 d\tau^2$ is $O_{3,1}$-invariant, defining the 'proper time' τ, and so we should differentiate/integrate with respect to τ. Physically, τ is the time coordinate in the (usually only infinitesimally inertial) reference frame in which the particle is at rest. The Lagrangian L is a Lorentz scalar, and the action (4.1.3) becomes $\int L \, d\tau$. For example, the Lagrangian for a free particle ('free' means no forces act on it, so the potential V is 0) can be taken to be

$$L = \frac{1}{2} m_0 \left(\left(\frac{d\mathbf{x}}{d\tau} \right)^2 - c^2 \left(\frac{dx^4}{d\tau} \right)^2 \right).$$

The Hamiltonian, being energy, transforms like time.

But what if there are several particles: which proper times τ_i do we use? The τ for the centre-of-mass, perhaps? In fact, this is a serious problem. The 'No-Interaction Theorem' (beginning with [**124**]) says that there can be no direct Lorentz-invariant interaction between particles, except through forces localised at a point causing an instantaneous change of velocity that don't change the number of particles. As there do seem to be unstable elementary particles (e.g. the muon) and gravity for instance isn't localised to a point, we have a problem. The obvious solution is to copy the first relativistic interaction theory, namely Maxwell's, and use *fields* (Section 4.1.3).

Special relativity says that the speed of *light* is fundamental to space-time. Modern physics helps us to accept this seeming glorification of light, by saying that there is a special speed c, and any particle with zero rest-mass m_0 (such as the photon, which mediates light) will always travel at that speed. But perhaps more can be said. Surely space-time is not a fundamental physical quantity; eventually it will be recognised as a fairly macroscopic epiphenomenon, and it will be understood how it arises operationally.

---

[3] The equivalence of matter and energy was proposed 50 years before Einstein, by Mendeleev, the father of the periodic table. Although his reasons were correct, his proposal was ignored and forgotten.

For instance, we can measure distance using rigid bodies called metersticks and time using quartz watches, but both this rigidity and periodicity are electromagnetic phenomena. Perhaps the constancy of the speed of, for example, light will be understood ultimately as a reflection of this circularity.

Einstein found the special treatment of inertial observers quite artificial. But it seems that accelerating observers can experience interesting phenomena. For instance, consider an observer S standing at the North Pole and an inertial observer T hovering above her, so T watches S uniformly spinning at the rate of one cycle every 24 hours. Let's assume for simplicity that the Earth's equator is a perfect circle; to T, the ratio of its circumference to the diameter of the Earth at the equator should be π. However, if S was to measure precisely the circumference and the diameter, she would find their ratio for this 'circle' to be (very slightly) greater than π. The reason for this is because S's observations must be consistent with T's: (4.1.7a) tells us that lengths parallel to the motion (such as S's metersticks along the equator as seen by T) will dilate by some factor $\sqrt{1 - v^2/c^2}$, while lengths perpendicular to the motion (e.g. the diameter) will remain unchanged. Likewise, S will find that her wristwatch will tick more quickly than a clock placed on the equator, even though both are at rest relative to her. Thus both geometry and physics change for non-inertial observers! (For a fairly convincing argument that gravity requires *curved* space-time, see section 7.3 of [**422**].)

In fact relaxing the inertial observer restriction provided Einstein with the key to his remarkable explanation of gravity. As mentioned earlier, the gravitational 'charge' numerically equals the mass m seen in formulae such as $\mathbf{F} = m\mathbf{a}$ or $T = \frac{1}{2}mv^2$ – this is precisely what Galileo's Pisa experiment was designed to verify. There are other 'forces' with this same property, for example the pull we feel when riding a merry-go-round. This got Einstein thinking: perhaps gravity is as fictitious as a centrifugal force? When we are in free-fall – whether in an orbiting spaceship or in an elevator suddenly decoupled from its cable – it is as if we are free of gravity, much as we are suddenly free of the centrifugal force when we step off the merry-go-round. This is the *equivalence principle*, which constitutes the only new physical content of general relativity. We are led to the thought that the gravitational 'force' experienced while sitting in a chair isn't due to the matter in the Earth pulling us towards it, but rather merely a consequence of the chair interfering with our natural inertial motion, just as does a car rounding a corner. All observers are physically valid, but awkward choices (such as me in a chair or in a turning car) introduce fictitious forces such as gravity. Everything tries to move in as straight a line, and with as constant a speed, as possible (at least if it's not under the influence of a true force like magnetism); that astronomical effect we call 'gravity' is merely a consequence of the fact that 'straight' has only a *local* significance. Space-time is not the vector space $\mathbb{R}^4$, but rather a nontrivial (curved) four-dimensional pseudo-Riemannian manifold. Gravity is the convergence or twisting of nearby geodesics; what we perceive as the elliptical revolution of the Earth about the Sun is merely the gentle entwining of the Earth's geodesic with the Sun's (Figure 4.3). General relativity, which we discuss briefly at the end of the next subsection, makes these thoughts mathematically precise.

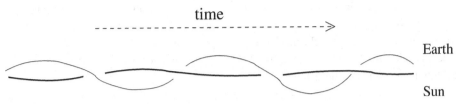

Fig. 4.3 The revolution of the Earth about the Sun.

4.1.3 Classical field theory

In physics, a 'field' is as in 'vector field' rather than 'number field'. It means a function of (usually) space-time, or more precisely a section of some vector bundle whose base is space-time. The most familiar example is Newton's gravitational field, namely the gravitational potential $V(\mathbf{x}, t)$. Another example is Maxwell's electromagnetic field $F(\mathbf{x}, t)$, which is matrix-valued.

Until now, we've been interested in particle dynamics, and the fields were auxiliary. To analyse how object A gravitationally influences object B, we first calculate how A influences the gravitational field, and then how the gravitational field influences B. In classical field theory, the field is a mechanical system in its own right – for example, it carries energy much like a fluid. It allows us to avoid the No-Interaction Theorem of relativistic dynamics. In quantum field theory discussed in the next section, the field is primary and the particle becomes an auxiliary phenomenon called a quantum, apparent only asymptotically.

A cherished physical principle, going back at least to Faraday, is called *locality*. The idea is that the only way we can *directly* affect something, is by nudging it. In order to influence something not touching us, we must propagate a disturbance from us to it, such as a sound-wave in air or a ripple in water. Special relativity sharpened locality into the requirement that no disturbance or influence can travel faster than light, so that space-time points $(\mathbf{x}, t)$, $(\mathbf{x}', t')$ that are *space-like separated* (i.e. obey $(\mathbf{x} - \mathbf{x}')^2 > c^2 (t - t')^2$) are causally independent.[4] As Faraday himself noted, locality leads to the concept of *field*. This is the main purpose for both classical and quantum fields – they provide a natural vehicle for realising locality.

Before, configuration space was finite-dimensional, with coordinates $(q_1, \ldots, q_N)$. Now our coordinates have a continuous index, $q_x = q(x)$, and configuration space is a space of functions. The Lagrangian in particle dynamics looks like $\sum_i T_i - \sum_{i,j} V_{ij}$. Now the sums are replaced by integrals and the Lagrangian becomes

---

[4] Strictly speaking this isn't a consequence of relativity, and in fact some physicists have entertained the possible existence of particles ('tachyons') that travel faster than light. These would behave curiously (e.g. they slow down the more energised they become), but like us they would require infinite energy to reach the speed of light – sadly, once a tachyon, always a tachyon. The difficulties facing the existence of tachyons are causality paradoxes. If P and Q are two space-like separated events, then there are reference frames in which P occurs before Q, and others in which Q occurs before P (why?). Hence if we had a gun that shot tachyonic bullets, then to some observers our victim would die before we pulled the trigger. Though not a logical contradiction, it is distinctly odd. Almost all physicists dismiss tachyons and faster-than-light influences as science fiction.

$L = \int \int \int \mathcal{L} \, dx \, dy \, dz$ for some function $\mathcal{L}$ called the Lagrangian density. $\mathcal{L}$ is a function of the fields $\phi(x, y, z, t)$ and their partial derivatives $\partial_x \phi$, etc. (together with contributions from particles). In field theory, $\mathcal{L}$ is more elementary and fundamental than L. Locality takes the form here of requiring that $\mathcal{L}$ only involves one space-time point. For each field ϕ^a there is a field equation

$$\frac{\partial}{\partial t} \frac{\partial \mathcal{L}}{\partial (\partial_t \phi^a)} + \sum_i \frac{\partial}{\partial x_i} \frac{\partial \mathcal{L}}{\partial (\partial_i \phi^a)} = \frac{\partial \mathcal{L}}{\partial \phi^a}, \qquad (4.1.8)$$

which describes the behaviour of the field. Additional equations (4.1.4) exist for each particle degree-of-freedom q_i present. The easiest example is the one-dimensional continuous Hooke's Law (e.g. vibrations in a rod). Our field here will be the amplitude $\phi(x, t)$ of the vibration at a point x on the rod. The Lagrangian density is

$$\mathcal{L}(x, t) = \frac{1}{2} \left\{ \mu \left(\frac{\partial \phi}{\partial t}(x, t) \right)^2 - y \left(\frac{\partial \phi}{\partial x}(x, t) \right)^2 \right\},$$

where μ is a constant called the mass density and y is a constant playing the role here of k. The first term is the kinetic energy density and the second (up to a sign) is the strain, or potential energy, in the rod. The field equation (4.1.8) gives us $\mu \frac{\partial^2 \phi}{\partial t^2} - y \frac{\partial^2 \phi}{\partial x^2} = 0$. This is easy to solve; physically it corresponds to a wave propagating with speed $v = \sqrt{y/\mu}$.

Define the momentum $\pi(\mathbf{x}, t) = \frac{\partial \mathcal{L}}{\partial (\partial_t \varphi)}$ conjugate to each field $\varphi(\mathbf{x}, t)$. Then the field equations (4.1.8) can be written as Poisson brackets involving Dirac deltas:

$$\{\varphi(\mathbf{x}, t), \pi(\mathbf{x}', t)\}_P = \delta(\mathbf{x} - \mathbf{x}'), \qquad (4.1.9a)$$

$$\{\varphi(\mathbf{x}, t), \varphi(\mathbf{x}', t)\}_P = \{\pi(\mathbf{x}, t), \pi(\mathbf{x}', t)\}_P = 0. \qquad (4.1.9b)$$

In special relativity, the Lagrangian density $\mathcal{L}$ transforms trivially (i.e. is a 'scalar') under the Lorentz group, and the fields ϕ^a span various representations R of the Lorentz group: that is, $\phi'^a(x') = \sum_b R(\Lambda)_{ab} \phi^b(x)$ where primes denote quantities in the reference frame (or $\mathbb{R}^4$-basis) obtained from the unprimed one using Lorentz transformation Λ.

An example important to physics (but not to us) is electromagnetism. The electromagnetic field has components $F_{\mu\nu} := \frac{\partial A_\nu}{\partial x^\mu} - \frac{\partial A_\mu}{\partial x^\nu}$, where A_4 is the electric potential and $\mathbf{A} = (A_1, A_2, A_3)$ is the magnetic potential. This field F transforms in a six-dimensional representation of the Lorentz group. The Lagrangian density is

$$\mathcal{L} = \frac{-1}{4} \sum_{\mu,\nu,\alpha,\beta} F_{\mu\nu} F_{\alpha\beta} (\eta^{-1})_{\mu\alpha} (\eta^{-1})_{\nu\beta} - \frac{1}{c} \sum_\mu j_\mu (\eta^{-1})_{\mu\nu} A_\nu =: \frac{-1}{4} F_{\mu\nu} F^{\mu\nu} - \frac{1}{c} j_\mu A^\mu,$$

where j is the electric current 4-vector describing the distribution and motion of charged particles. The matrix η^{-1} arises here in its Riemannian role defining an inner-product. The second expression is much more transparent, and uses $\eta^{\pm 1}$ to lower/raise indices, and summing over repeated indices. Of course to the Lagrangian must be added the (relativistic) kinetic energy of the particles or fields. The resulting field equations, called

Maxwell's equations, tell us for instance how charged particles create an electromagnetic field.

We see in Section 4.1.1 that even the simplest classical systems can have singular solutions, so the situation for classical field theory can only be worse. Most famous is the self-energy of charged particles in electromagnetism, discussed beautifully in chapter 28, vol. II of [**188**]: a charged particle localised to a point has infinite mass coming from the electromagnetic field. To see this, imagine that we hold half an electron in our left hand and the other half in our right; to make the electron whole we would have to connect these two repulsive halves, and an easy calculation (namely the integral $-\int_1^0 r^{-1} \mathrm{d}r = \infty$) says this requires infinite energy. This problem persists in its quantisation.

A remarkable classical field theory is Einstein's general relativity, in which space-time is a pseudo-Riemannian manifold with metric tensor $g(x)$, locally (but not globally) equivalent to the Minkowski metric η. Ignoring for convenience other forces, the Lagrangian density for a single particle is

$$\mathcal{L}(x) = \frac{1}{2} m_0 \sum_{\mu,\nu} g_{\mu\nu}(x) \frac{\mathrm{d}x^\mu}{\mathrm{d}\tau} \frac{\mathrm{d}x^\nu}{\mathrm{d}\tau} \delta^4(x - x(\tau)) + \frac{c^3}{16\pi G} \sqrt{-\det g} R, \qquad (4.1.10)$$

where G is Newton's gravitational constant and R is a geometric quantity (a measure of the radius of curvature of space-time at x). δ^4 is the highly singular Dirac delta. The numerical constant $c^3/16\pi G$, establishing the coupling strength between space-time and matter, is chosen so that Einstein's theory agrees with Newton's in the appropriate limit. Varying the particle's coordinates x^μ yields the geodesic equation

$$\frac{\mathrm{d}^2 x^\mu}{\mathrm{d}\tau^2} + \sum_{\nu,\kappa} \Gamma^\mu_{\nu\kappa} \frac{\mathrm{d}x^\nu}{\mathrm{d}\tau} \frac{\mathrm{d}x^\kappa}{\mathrm{d}\tau} = 0,$$

describing the straightest possible curves in the manifold ($\Gamma^\mu_{\nu\kappa}$ are the Christoffel symbols). Varying the metric g yields Einstein's field equations

$$R_{\mu\nu} - \frac{1}{2} R g_{\mu\nu} = \frac{8\pi G}{c^4} T_{\mu\nu}. \qquad (4.1.11)$$

$R_{\mu\nu}$ are components of the Ricci tensor and $T_{\mu\nu}$ are those of the stress-energy tensor defined below. The left side is geometrical, depending on first and second partial derivatives of $g_{\mu\nu}$, while the right side is physical, depending on the matter fields. Einstein's field equations (4.1.11), which tell us how matter and energy curve space-time, consist of 10 coupled nonlinear second-order partial differential equations for the components $g_{\mu\nu}$.

The relation between symmetries and conserved quantities in field theory takes the following form (generalised in Question 4.1.1). Suppose the Lagrangian density $\mathcal{L}$ is invariant under a continuous symmetry α_s. Associate with α_s the 4-vector

$$j^\mu(x) = \frac{\partial \mathcal{L}}{\partial(\partial\phi/\partial x^\mu)} \left(\frac{\partial \alpha_s(\phi)}{\partial s} \right), \qquad (4.1.12a)$$

for $\mu = 1, 2, 3, 4$, called the 'current'. Then $j(x)$ is *conserved*, that is it obeys

$$\partial_\mu j^\mu := \sum_{\mu=1}^{4} \frac{\partial j^\mu}{\partial x^\mu} = 0. \tag{4.1.12b}$$

This equation tells us to think of $j^4(x)$ as the density of some abstract fluid, and $\mathbf{j}(x) = (j^1(x), j^2(x), j^3(x))$ as its velocity at each space-time point x. Equation (4.1.12b) tells us that this 'fluid' is neither created nor destroyed, so that the total quantity ('charge') $Q(t) = \int j^4(x) \, dx^1 \, dx^2 \, dx^3$ (if the integral exists) is constant: $\frac{dQ}{dt} = 0$.

For example, the invariance of the Lagrangian density $\mathcal{L}$ with respect to time and space translations $x^\nu \mapsto x^\nu + a^\nu$ gives us the 'current' $T^{\mu\nu}(x)$ (one for each ν) called the stress-energy tensor. The 'charges' Q^ν here are the total momentum and energy. Or consider the full Lagrangian density for the coupling of the electromagnetic field F to a complex scalar field ϕ with mass m, charge e and potential V:

$$\mathcal{L} = \frac{-1}{4} \sum F_{\mu\nu} F^{\mu\nu} + \sum \left(\frac{\partial}{\partial x^\mu} - ieA_\mu \right) \phi \left(\frac{\partial}{\partial x_\mu} + ieA^\mu \right) \phi - m^2 \phi^* \phi - V(\phi^* \phi). \tag{4.1.13}$$

The terms only involving ϕ and ϕ^* form the Lagrangian for the field ϕ alone, while the terms involving both ϕ and A define the interaction. Note that there is a U_1 group symmetry of $\mathcal{L}$, which acts trivially on F and A but acts on ϕ by $\alpha_s(\phi) = e^{ie\alpha}\phi$. Then Q is indeed proportional to e. We return to this example next section.

Question 4.1.1. (a) Prove the following generalisation of Noether's Theorem. Suppose we have a continuous family α_s of diffeomorphisms of configuration space such that

$$L(\alpha_s(q), \hat{\alpha}_s(\dot{q})) = L(q, \dot{q}) + \frac{d}{dt} \Delta(q, \dot{q}),$$

for some function Δ. First, verify that $q(t)$ is a possible trajectory iff $\alpha_s(q(t))$ is. Next, verify that the quantity

$$Q = \frac{\partial L}{\partial \dot{q}} \left(\frac{\partial \alpha_s(q)}{\partial s} \right) - \Delta$$

is constant along any trajectory.

(b) The Lagrangian for a free Newtonian particle is $L = \frac{1}{2} m \dot{\mathbf{x}}^2$. Take $\alpha_s(\mathbf{x}) = \mathbf{x} + s \mathbf{a}$ for some constant vector $\mathbf{a} \in \mathbb{R}^3$. Find Δ here, and verify that the 'charge' Q is $m \mathbf{x}(0)$.

Question 4.1.2. Verify that the space $C^\infty(T^*\mathcal{M})$ of observables, with bracket given by (4.1.6a), defines a Lie algebra, and that the first integrals form a Lie subalgebra.

4.2 Quantum physics

We tend to have a naive view of progress in science, namely that the old theory gets superseded by a new theory that is better in every meaningful respect: any phenomenon the older theory could explain, and any question the older theory could answer, the new theory would explain and answer at least as accurately; moreover, there would be phenomena and questions that the older theory avoids but the newer, better theory handles adroitly. In reality, progress in science (in contrast to progress in technology) has much in common with progress in popular music or in, say, America's ability to elect great presidents. Copernicus' circular orbits match observation worse than Ptolemy's epicycles. More significantly, Copernicus required the Earth to move at incredible speeds, which mysteriously no experiment could ever detect (e.g. when we jump straight up, we come straight down). Ptolemy himself rejected the heliocentric hypothesis for these and several other good reasons. It was only after Galileo explained the role of inertia, *after* Copernicus' time, that Copernicus' unoriginal idea became scientifically reasonable. Of course to us today all motion is relative and the proceedings of that Great Debate belong in the voluminous Library-of-Dead-Religions. For another example, Aristotelian physics regarded friction as fundamental and the pendulum as complicated derived motion, whereas Newtonian physics regarded the pendulum as simple and friction as compound. In fact, classical physics never successfully explained friction – our present explanation requires quantum mechanics to correctly handle the relevant molecular forces (namely the van der Waals forces, which are residuals of the underlying electromagnetic forces). At least in part, 'progress' in science is a sociological phenomenon, a mantra bubbling on the lips of scientists as they pursue questions they are willing and able to address.

In any case, the conceptually and mathematically elegant classical mechanics has been superseded by the fairly incoherent quantum physics. A century has passed since the birth of the quantum, and although almost all physicists today regard quantum theory as having successfully transcended classical physics, it is dangerous to conclude much from this. But one thing is certain: mathematics has been a great beneficiary of this 'transcendence'.

4.2.1 Nonrelativistic quantum mechanics

For fixed time t, the state of a single particle in quantum mechanics can be captured by a complex-valued *wave-function* $\mathbf{x} \mapsto \psi(\mathbf{x}, t)$. Its interpretation is rather different from 'state' in classical physics: the quantity $|\psi(\mathbf{x}, t)|^2$ is the probability density that the particle is at position $\mathbf{x}$ at time t. Probability arises here *not* because of uncertainty of our knowledge, *nor* because of unavoidable disturbances caused by our heavy-handed measuring processes. Rather, it is a fundamental ingredient of quantum reality. God's analysis too would stop at this probability.

Recall the discussion of Hilbert spaces in Section 1.3.1, in particular the rigged Hilbert space $\mathcal{S}(\mathbb{R}^n) \subset L^2(\mathbb{R}^n) \subset \mathcal{S}(\mathbb{R}^n)^*$, where the Schwartz space $\mathcal{S}(\mathbb{R}^n)$ consists of all smooth functions falling off with their derivatives to 0 quickly as $|x| \to \infty$ and where the Hilbert space $L^2(\mathbb{R}^n)$ consists of the square-integrable functions with inner-product

$$\langle \phi, \psi \rangle := \int_{\mathbb{R}^n} \overline{\phi(\mathbf{x})}\, \psi(\mathbf{x})\, d^n\mathbf{x}.$$

For each time t, the span of the possible time-slices (states) $\psi(\star, t)$ form the Schwartz space $\mathcal{S} = \mathcal{S}(\mathbb{R}^3)$, while their topological span forms the Hilbert space $\mathcal{H} = L^2(\mathbb{R}^3)$.

We require the wave-function ψ to be normalised: $\langle \psi, \psi \rangle(t) = 1 \; \forall t$. Observables here correspond to self-adjoint operators $\widehat{A} : \mathcal{S} \to \mathcal{S}$. For example, the operator associated with measuring the ith coordinate of position takes $\psi \mapsto x_i \psi$, while energy is associated with the operator $i\hbar \frac{\partial}{\partial t}$ (we can use (4.2.1) below to express it using spatial derivatives) and the ith component of momentum with the operator $-i\hbar \frac{\partial}{\partial x_i}$.

The role of phase space is (loosely) played here by the projectification $\mathcal{S}/\mathbb{C}$, since the physical states corresponding to nonzero multiples $c\psi$ are the same. This is significant because it tells us that groups can act on $\mathcal{S}$ via *projective* representations, and still be well-defined. This persists in all quantum theories and has many consequences. Not all $\psi \in \mathcal{S}$ though are actually physical states – for example, it appears that every physical state must have a definite electric charge, that is be an eigenvector of some charge operator, and of course most $\psi \in \mathcal{S}$ aren't.

There are two independent ways the wave-function evolves in time. The first way is through Schrödinger's equation, which is the linear partial differential equation

$$i\hbar \frac{\partial \psi}{\partial t} = -\frac{\hbar^2}{2m} \nabla^2 \psi + V(\mathbf{x})\, \psi, \tag{4.2.1}$$

where V is the potential energy (which acts multiplicatively on ψ), $\hbar$ is Planck's constant and ∇^2 is the Laplacian $\frac{\partial^2}{\partial x_1^2} + \frac{\partial^2}{\partial x_2^2} + \frac{\partial^2}{\partial x_3^2}$. Schrödinger's equation governs the deterministic, unitary evolution of ψ occurring between measurements. It is standard to choose units so that Planck's constant $\hbar$ equals 1 (recall the discussion in Section 4.1.1); however, in units natural to our familiar macroscopic world (e.g. metres, kilograms and seconds) its magnitude (about 10^{-34}) emphasises just how invisible quantum effects are to us.

Schrödinger's equation can be formally integrated, and we obtain

$$\psi(\mathbf{x}, t) = U(t)\, \psi(\mathbf{x}, 0), \tag{4.2.2}$$

where $U(t) = \exp[-i\widehat{H}t/\hbar]$ is a unitary operator on $\mathcal{S}$ (hence $\mathcal{H}$) for the Hamiltonian operator $\widehat{H}$ given by the right side of (4.2.1). Conversely, we could have anticipated (4.2.1) by the following reasoning. The time evolution (4.2.2) should be given by a linear operator $U(t)$ independent of ψ (so $U(s)\,U(t) = U(s+t)$), which preserves the normalisation: $\|U(t)\,\psi(\mathbf{x}, 0)\| = \|\psi(\mathbf{x}, t)\| = 1$. This *implies* that $U(t) = \exp[iH't]$, that is $\partial \psi / \partial t = iH'\psi$, for some self-adjoint operator H'. For physical reasons we would expect H' to have something to do with energy, that is the classical Hamiltonian H, since energy is the conjugate observable to time just as momentum is to position. Indeed, Schrödinger's equation (4.2.1) comes from the nonrelativistic formula for energy ($E = \frac{1}{2m}\mathbf{p}^2 + V$), together with the quantum mechanical substitutions $E \mapsto i\hbar \frac{\partial}{\partial t}$ and $\mathbf{p} \mapsto -i\hbar\nabla$.

The second type of wave-function evolution is indeterministic and discontinuous, and occurs at the instant t_0 when a measurement is made. Let $\widehat{A}$ be the self-adjoint operator corresponding to the observable being measured. Assume for simplicity that its spectrum (i.e. its set of eigenvalues) is discrete and nondegenerate. Then there is an orthonormal set $\{\psi_a(\mathbf{x})\} \subset \mathcal{S}$ of eigenvectors spanning $\mathcal{H}$ (topologically). So $\widehat{A}\psi_a = a\psi_a$ and $\langle \psi_a, \psi_b \rangle = \delta_{ab}$. If ψ is the wave-function of the particle being observed, write $\psi(\mathbf{x}, t_0) = \sum_a c_a \psi_a(\mathbf{x})$. The result of the observation will be one of the eigenvalues a,

a_0 say, but which one cannot be predicted in advance. All that can be said is that $|c_a|^2$ is the probability that a will be the one observed. Nothing was responsible for the given eigenvalue a_0 arising – two completely identical quantum systems can (and usually will) yield different observed values. At time t_0 the wave-function ψ suffers a spontaneous and discontinuous change $\psi \mapsto \psi_{a_0}$ (or more generally the orthogonal projection of ψ into the a_0-eigenspace). For times immediately after t_0, the wave-function then proceeds to evolve by (4.2.1). This second type of evolution is necessary for the experimental consistency of the theory: experimental results can be reproduced! It is a truly physical evolution, and not merely book-keeping reflecting a change in our knowledge of the system.

For example, the simultaneous eigenvalues $\mathbf{p} = (p_1, p_2, p_3) \in \mathbb{R}^3$ of the three momentum operators correspond to eigenfunction $\psi_{\mathbf{p}}(\mathbf{x}, t) = e^{i\mathbf{p}\cdot\mathbf{x}/\hbar}$, while the simultaneous eigenvalues $\mathbf{a} = (a_1, a_2, a_3) \in \mathbb{R}^3$ of the position operators have eigenfunctions given by the three-dimensional Dirac delta $\delta^3(\mathbf{x} - \mathbf{a})$. These spectra aren't discrete and (generalised) eigenfunctions aren't square-integrable (rather they are tempered distributions – Section 1.3.1), because exact position and momentum observations in quantum theory are nonphysical idealisations (e.g. probing infinitesimal distances requires infinite energy). Moreover, since the position and momentum operators don't share any eigenvectors, it is meaningless to speak simultaneously of the (numerical) position and momentum of a particle: in quantum mechanics a particle cannot have a well-defined trajectory.

This framework generalises in the obvious ways. For n particles, the wave-function ψ looks like $\psi(\mathbf{x}_1, \ldots, \mathbf{x}_n, t)$ and on the right side of (4.2.1) the Laplacian ∇^2 get replaced by the sum of n Laplacians ∇_i^2, one for each $\mathbf{x}_i$.

This treatment of many particles indicates a weak point of quantum mechanics. Experiment tells us that the number of elementary particles can change, for example, a muon can decay into an electron and two neutrinos. It is rather difficult to believe that the fundamental equation of motion in physics changes *discontinuously* with time, but that is how quantum mechanics would model the decay of, for example, the muon: at some time t_0 the wave-function would acquire six more variables and Schrödinger's equation six more terms. The way out (Section 4.2.2) simultaneously handles all numbers of particles.

The fascinating *measurement problem* of quantum physics, present in any quantum theory, is the struggle to understand this dichotomy of wave-function evolutions. What is so special about measurement, that it should obey special laws? After all, surely a measurement is merely a certain kind of physical process. Many remarkable elaborations have been proposed by respected physicists, for example, that the universe splits into different 'parallel universes' after each measurement, or that a measurement involves the imposition of mind on matter. Precisely what constitutes a measurement? Any quantum measurement involves the amplification of a microscopic quantum property or effect to a macroscopic one. What does quantum physics tell us about the macroscopic (classical) world? The linearity of Schrödinger's equation implies that linear combinations ('superpositions') of solutions will again be solutions. Now, *microscopic* superpositions are well-observed and fundamental to the theory; during a quantum measurement (if not at other times) macroscopic superpositions should be unavoidable. However, what would

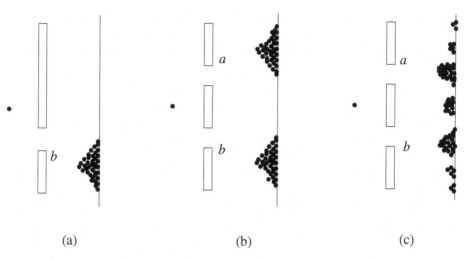

(a)   (b)   (c)

Fig. 4.4 The golf ball experiment.

a macroscopic superposition look like? Why have we never observed the superposition of, for example, a live and dead cat? We are led to the suspicion that quantum physics is incompatible with our most elementary qualitative observations of (macroscopic) physical reality.

To make this more precise, consider the situation depicted in Figure 4.4, where a machine randomly putts golf balls towards two barriers, one behind the other. When a hole is cut into the first barrier, as in Figure 4.4(a), the balls that reach the second barrier (i.e. pass through the hole) will impact it at roughly the same spot – the trajectories of golf balls over short distances are approximately linear. And if we cut two holes into the first barrier, we will get the result depicted in Figure 4.4(b). (We ignore all balls that get stopped by the first barrier.) Now suppose that whenever we avert our eyes for a few minutes, the golf balls make instead the impact pattern of Figure 4.4(c). That unbelievable phenomenon would suggest that changing the nature of our observation can dramatically affect golf ball trajectories. Classically, there is no evidence of this.

Of course that is precisely what occurs in the remarkable two-slit experiment, where electrons are fired at a screen. The electron wave-function ψ is the normalised *superposition* $\frac{1}{\sqrt{2}}(\psi_a + \psi_b)$ of wave-functions corresponding to travel through the a-slit or the b-slit. Individually, the wave-functions $\psi_a(\mathbf{x}, t)$ and $\psi_b(\mathbf{x}, t)$ both give rise to the probability density (for the arrival spot on the screen behind the two slits) we would expect from the golf balls of Figure 4.4(a). However, their superposition ψ gives rise to probabilities $\frac{1}{2}|\psi_a + \psi_b|^2 \neq \frac{1}{2}|\psi_a|^2 + \frac{1}{2}|\psi_b|^2$ – the two possible paths of the electron *interfere* with each other, much as they would if an electron were, for example, a water ripple. If we were to try to detect which slit the electron goes through, say by setting up a detector at each slit (as in Figure 4.4(b)), this additional measurement would first 'collapse' ψ into either ψ_a or ψ_b (with equal probabilities). The resulting probability density for the arrival spot would be the particle-like $|\psi_a|^2$ or $|\psi_b|^2$, respectively (or $\frac{1}{2}|\psi_a|^2 + \frac{1}{2}|\psi_b|^2$ if we don't keep track of which slit the electron passed through).[5]

So why can't *macroscopic* states interfere? The special feature ('decoherence') of a macroscopic system seems to be that it is under unavoidable continuous interaction with the environment, through gravity if nothing else. Macroscopically distinct states (e.g. different pointer positions on an instrument, or golf balls rolling through different holes) couple differently to the environment, and so the macroscopic system becomes thoroughly and irreversibly entangled with the environment. This entanglement is essentially irreversible because any interaction that succeeded in untangling the coupling of the state with the environment would require enormous numbers (10^{27} or so) of degrees of freedom to conspire appropriately. This has the effect of making the macroscopic states essentially 'decohere' from each other, that is, the interference terms $\frac{1}{2}\psi_A\overline{\psi_B} + \frac{1}{2}\psi_B\overline{\psi_A}$, when expanded into the disordered microscopic degrees of freedom, get averaged away to zero. To get the flavour of decoherence, consider the wave-functions $\psi_{A,B}$ describing classical objects A, B. They are actually functions of 10^{27} or so space variables x_{ij}, but because they are macroscopic we would expect them effectively to be functions of our familiar three-dimensional space. Moreover, they would be essentially localised in this space, so $|\psi_A(\mathbf{x}, t) + \psi_B(\mathbf{x}, t)|^2 = |\psi_A(\mathbf{x}, t)|^2 + |\psi_B(\mathbf{x}, t)|^2$, provided A and B are situated a macroscopic distance apart (i.e. provided the supports of the effective functions ψ_A and ψ_B are disjoint). This is decoherence.

Of course alone this doesn't resolve the measurement problem. At best decoherence can only explain why macroscopically distinct states in superpositions don't 'see' each other. A (perhaps overly zealous) application of quantum mechanics insists that macroscopic superpositions must occur; from this, the 'Many-Worlds' interpretation is inevitable. The explanation for the mysterious wave-function collapse then would be that measurement entangles the quantum system $\psi^q = \sum c_i \psi_i^q$ with a macroscopic system ψ^c – that is, via Schrödinger's equation, the decoupled wave-function $\psi^q \psi^c$ relevant just prior to measurement would be replaced with the coupled wave-function $\sum c_i \psi_i^q \psi_i^c$ just after. Each coupled state ('world') $\psi_i^q \psi_i^c$ in this superposition would decohere from the others, and so the various quantum states ψ_i^q could no longer 'see' each other. It would be as if at the moment of measurement, the universe split into parallel universes, one for each possible experimental outcome. The 'Many-Worlds' interpretation is quantum mechanics in its purest form; in this framework measurement is a physical process subject only to Schrödinger's equation, and neither wave-function collapse nor the splitting of universes actually occurs. The price of this demystification of measurement is a reality in which almost everything is hidden from us, including infinitely many near-copies of ourselves.[6] A derivation of sorts of the probability rule is also possible within this framework.

---

[5] We shouldn't over-emphasise this 'wave–particle duality'. 'Waves' and 'particles' are classical metaphors; an electron is neither. Even the name 'wave-function' for ψ is an anachronism going back to de Broglie's hypothesis that an electron behaves like a wave with wavelength h/p.

[6] In defence of this uncomfortable aspect of Many-Worlds, Nature – unlike us – clearly loves enormous numbers of nearly identical copies. Consider blades of grass in a field, or water molecules in a lake (or

We've only sketched one possible interpretation. There are many others. For instance, the presence of probability in quantum mechanics strongly suggests that we are ignoring certain degrees of freedom – after all, this is what probability signifies in classical mechanics. It is possible to formulate quantum mechanics as a deterministic classical theory, by introducing 'hidden variables'. In the case of one particle, these hidden degrees of freedom would be the position coordinates $\mathbf{x}(t)$ of the particle. The coordinates $\mathbf{x}(t)$ obey a differential equation involving the wave-function ψ, which in turn obeys Schrödinger's equation. A similar formulation can be made for any number of particles. However, 'Bell's Theorem' says that any multi-particle hidden variables theory must possess the notorious feature called 'nonlocality'. This means that an influence (e.g. measurement) done on one particle can *instantaneously* affect the state of a distant particle. Nonlocality in a theory warns of possible difficulties in making the theory relativistic.

The approaches to the quantum measurement problem illustrate the desperate imagination that squirts from our pores when we're backed into a corner. See the book [**556**] for more details, examples and references to the literature. Like any other metaphysical doctrine, an interpretation is chosen not for its approximation to Truth, but because we find intriguing (and publishable!) the avenues of study it suggests.

For a one-dimensional example of a quantum system, consider once again the harmonic oscillator. The potential is $V = -\frac{k}{2}x^2$, so Schrödinger's equation here reads

$$i\hbar \frac{\partial \psi}{\partial t} = -\frac{\hbar^2}{2m}\frac{\partial^2 \psi}{\partial x^2} + \frac{k}{2}x^2\,\psi. \tag{4.2.3a}$$

Because the potential V is independent of time, this is separable into energy eigenstates: write $\psi(x, t) = e^{-iEt/\hbar}\psi_E(x)$, where

$$-\frac{\hbar^2}{2m}\frac{d^2\psi_E}{dx^2} + \left(\frac{k}{2}x^2 - E\right)\psi_E = 0. \tag{4.2.3b}$$

In order for ψ to be normalisable, we require the boundary conditions $\psi(x, t) \to 0$ as $|x| \to \infty$; this implies (with a little work) that $E = (n + \frac{1}{2})\hbar\sqrt{\frac{k}{m}}$ for $n \in \mathbb{N}$, that is energy is quantised and bounded from below.

A useful idealisation is the step-function potential $V(x) = \begin{cases} 0 & \text{if } x < 0 \\ V_0 & \text{otherwise} \end{cases}$, where V_0 is constant. Solving the corresponding one-dimensional Schrödinger's equation with the requirement that both ψ and its derivative $\frac{\partial \psi}{\partial x}$ be continuous at $x = 0$, we obtain

$$\psi(x, t) = e^{-iEt/\hbar} \begin{cases} A\exp(ip_+x/\hbar) & \text{for } x > 0 \\ \exp(ip_-x/\hbar) + B\exp(-ip_-x/\hbar) & \text{for } x < 0 \end{cases},$$

where $p_+ = \sqrt{2m(E - V_0)}$ and $p_- = \sqrt{2mE}$ are the classical momenta (at least for $E > V_0$), and $A = 2\frac{p_-}{p_+ + p_-}$ and $B = \frac{p_- - p_+}{p_+ + p_-}$. Physically, this describes a wave (energy

---

perhaps research publications?). Or more to the point, consider the uncountably many moments making up each life.

eigenstate) travelling to the right from $x = -\infty$, with energy $E > 0$; it hits the wall at $x = 0$, part of it continuing to positive x and some of it reflecting back to negative x. If we were to measure whether or not reflection happened, we would find that reflection happened with probability $|B|^2 = 1 - |A|^2$. Note that we get some very nonclassical behaviour: classically, when $E > V_0$ the whole wave would be transmitted to positive x, but here some of the wave is reflected, even when $V_0 < 0$! It is as if we are about to tumble over Niagara Falls in a barrel, only to bounce back the instant we reach the precipice. Related to this is quantum tunnelling (Question 4.2.2).

Quantum mechanics was born around 1926 when Schrödinger obtained (4.2.1) and, simultaneously, when Heisenberg and others developed an equivalent formulation. Unlike Schrödinger's picture, in Heisenberg's the state Ψ of the system is regarded as constant in time, and the time-evolution is carried by the observables $\widehat{A}$. It is completely analogous to the two attitudes towards observables carried in classical mechanics: we can view an observable $A(q, p)$ as a time-independent C^∞-function on phase space, or we can regard it as a function $A(q(t), p(t))$ of time. The equivalence between these two pictures of quantum mechanics is straightforward: the Heisenberg state $\Psi \in S$ can be taken to be the wave-function $\psi(\star, 0)$ at time $t = 0$, while the Heisenberg operator $\widehat{A}(t)$ corresponds to Schrödinger's operator $\widehat{A}$ via the relation $\widehat{A}(t) = U(t)^{-1}\widehat{A}U(t)$, where $U(t) = \exp[-i\widehat{H}t/\hbar]$ as before. Differentiating, we find that the equation of motion in Heisenberg's picture is given by commutation with $\widehat{H}$:

$$\frac{\mathrm{d}}{\mathrm{d}t}\widehat{A}(t) = -\frac{\mathrm{i}}{\hbar}\left[\widehat{A}(t), \widehat{H}\right]. \tag{4.2.4}$$

In relativistic quantum theory, Heisenberg's picture is more convenient because time doesn't play as privileged a role. In particular, just as $U(t)$ describes translations in time, a unitary operator $V(\mathbf{x})$ describes translations in space, and so we can regard the state Ψ as independent also of space. More generally, we have a unitary (projective) representation $(a, \Lambda) \mapsto U_{(a,\Lambda)}$ of the Poincaré group, acting on the infinite-dimensional space of states.

Equation (4.2.4) should look familiar: it is formally identical to the classical evolution (4.1.6b) of observables, provided we replace the Poisson bracket of classical observables there with the commutator of the quantum observables (up to the factor $i\hbar$). Other examples of this are the calculations $\{x, p\}_P = 1$ and $[\widehat{x}, \widehat{p}] = i\hbar I$. In other words, the process ('quantisation') of going from classical mechanics to the corresponding quantum mechanics defines a representation of the Lie algebra $C^\infty(T^*\mathcal{M})$ (with Poisson bracket) into the Hilbert space $\mathcal{H}$. However, this quantisation is clouded somewhat by the observation that the classical space $C^\infty(T^*\mathcal{M})$ is also an associative commutative algebra using pointwise product $(fg)(y) = f(y)g(y)$ of the functions, and that this product is also important as it is how we can build up general observables from the elementary ones x_i, p_j. Unfortunately, there is no direct analogue of this second product for the space of self-adjoint operators on $\mathcal{H}$ (or S). The closest would be the operation $A * B = \frac{1}{2}(AB + BA)$, which makes the space of quantum operators into a (non-associative) *Jordan* algebra, originally named after the quantum physicist Pascual Jordan but now part of standard algebraic repertoire.

An alternate, rather intriguing approach to quantisation seeks to formulate quantum mechanics in terms of a one-parameter deformation of the pointwise product algebra $\mathcal{A} = C^\infty(T^*\mathcal{M})$ (see [141] for a review). In particular, let $\mathcal{A}[[\lambda]]$ denote the space of all formal power series in λ with coefficients in $\mathcal{A}$. We add these power series term by term in the obvious way, but the product in $\mathcal{A}[[\lambda]]$ is more complicated (though necessarily associative). Expand out the product: $f \star g = \sum_{k=0}^\infty C_k(f, g) \lambda^k$, where for each $f, g, C_k(f, g) \in \mathcal{A}$. Because it is a deformation we require $C_0(f, g)$ to equal the usual pointwise product fg. In order to relate this to quantum mechanics, we also require that the coefficient $C_1(f, g) - C_1(g, f)$ of the leading term in the commutator $f \star g - g \star f$ be the Poisson bracket $2\{f, g\}_P$. We think of the deformation parameter λ as equalling $i\hbar/2$. The main appeal of this approach to quantum mechanics is that classical and quantum mechanics are placed on the same page, so rigorous sense can be made of the statement that we recover classical physics from the $\hbar \to 0$ limit. However, it can be criticised for making classical mechanics logically prior to quantum mechanics, when the reverse would seem more natural. Also there are some quantum mechanical systems that don't seem to have a classical analogue. Kontsevich was awarded his Fields medal in 1998 in part for his proof that such a deformation exists not only for any phase space $X = T^*\mathcal{M}$ (this was known before), but more generally for any differentiable manifold X on which can be defined a Poisson bracket (a Lie algebra structure for $C^\infty(X)$).

Consider the harmonic oscillator in Heisenberg's picture. The possible states span a space $\mathcal{S}$, dense in a Hilbert space $\mathcal{H}$. Define the operators

$$\widehat{a} = \frac{(km)^{1/4}}{\sqrt{2\hbar}} \left[\widehat{x} + \frac{1}{\sqrt{km}} i\widehat{p} \right], \qquad \widehat{a}^\dagger = \frac{(km)^{1/4}}{\sqrt{2\hbar}} \left[\widehat{x} - \frac{1}{\sqrt{km}} i\widehat{p} \right] \qquad (4.2.5)$$

acting on $\mathcal{S}$. These are called annihilation and creation operators, respectively. Note that $[\widehat{a}, \widehat{a}^\dagger] = I$, the identity operator. Hence $I, \widehat{a}, \widehat{a}^\dagger$ define a representation of $\mathfrak{Heis}$ (1.4.3) on the infinite-dimensional space $\mathcal{S}$. Let's find a more explicit realisation of this representation. This requires identifying the *vacuum state* $|0\rangle \in \mathcal{S}$, that is an eigenvector of the Hamiltonian $\widehat{H}$ with minimal eigenvalue (i.e. a state with lowest energy), normalised so that $\||0\rangle\| = 1$. Physically, the vacuum denotes the ground state, containing no particles. The energy operator, that is the Hamiltonian, becomes $\widehat{H} = \frac{\widehat{p}^2}{2m} + \frac{\widehat{x}^2}{2} = (\widehat{a}^\dagger \widehat{a} + \frac{1}{2}) \hbar \sqrt{\frac{k}{m}}$ (as usual it is time-independent). The vacuum obeys $\widehat{a}|0\rangle = 0$ (why?) and has energy $E_0 = \frac{1}{2}\hbar\sqrt{\frac{k}{m}}$ (i.e. that is its $\widehat{H}$-eigenvalue). Assume that the vacuum is nondegenerate, that is the eigenspace associated with energy E_0 has dimension 1 – a degenerate vacuum would correspond to a number of non-interacting equivalent oscillators working in parallel. This assumption implies that the vacuum vector will be unique up to a phase $e^{i\alpha}|0\rangle$ (choose one), and that the vacuum state is well-defined. Define vectors $|n\rangle := (n!)^{-\frac{1}{2}}(\widehat{a}^\dagger)^n|0\rangle$. This curious notation is due to Dirac: the functional $\langle\star| \in \mathcal{S}^*$ is called a *bra*, the vector $|\star\rangle \in \mathcal{S}$ a *ket*, and the evaluation $\langle\star|\star\rangle \in \mathbb{C}$ a *bra(c)ket*. This bracket also captures inner-products, using the adjoint $|\star\rangle^\dagger = \langle\star|$. Note that $|n\rangle$ has norm 1, and it is an eigenvector of $\widehat{H}$ with eigenvalue $E_n := (2n + 1)E_0$. Construct the operator $\widehat{N} = \widehat{a}^\dagger\widehat{a}$, then $\widehat{N}|n\rangle = n|n\rangle$. We are to think of $\widehat{N}$ as a number operator, as

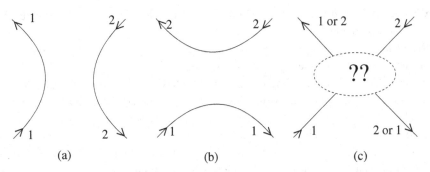

Fig. 4.5 A collision of two identical particles.

it counts the number of *quanta* (or excitations or quantum particles) in the given state. We say that the operator $\widehat{a}^\dagger$ *creates* a quanta, and $\widehat{a}$ *annihilates* a quanta. The vectors $|n\rangle$ $(n = 0, 1, 2, \ldots)$ form an orthonormal set; the state space $\mathcal{S}$ here consists of all $\sum_{n=0}^{\infty} c_n |n\rangle$ with $\sum n^m |c_n| < \infty$ for all m, while the Hilbert space $\mathcal{H}$ here consists of all $\sum c_n |n\rangle$ with $\sum |c_n|^2 < \infty$. In this algebraic way we can recover all of the physics.

When our system consists of a number of subsystems (e.g. different particles), the collective Hilbert space $\mathcal{H}$ will be given by the tensor product $\mathcal{H}_1 \otimes \cdots \otimes \mathcal{H}_n$ of the individual Hilbert spaces (this was implicit in our treatment of measurement, where the two subsystems were the observed and the observer). Given vectors $v_i \in \mathcal{H}_i$, we are to think of the 'diagonal' vector $v_1 \otimes \cdots \otimes v_n =: |v_1, \ldots, v_n\rangle$ as describing the situation where subsystem i is in state v_i. However, as we know, a typical vector u in the tensor product $\mathcal{H}$ won't be of this diagonal form. Only for such states $|v_1, \ldots, v_n\rangle$ do the subsystems themselves possess well-defined states. Even if the system begins in diagonal form (e.g. we start with two distant particles), it will lose this as soon as the subsystems interact. In this way, interacting systems lose their independent existence. This entangling of quantum subsystems doesn't occur in classical mechanics.

Something special, and also nonclassical, happens when the subsystems are *identical* (i.e. the subsystems obey identical laws, and differ only in incidental characteristics such as position). The collective Hilbert space $\mathcal{H}$ now is smaller than the full tensor product: it will be the symmetric product of n copies of the subsystem $\mathcal{H}_1$. More precisely, $\mathcal{H}$ is spanned by 'symmetric' vectors of the form $|v_1, \ldots, v_n\rangle := \frac{1}{\sqrt{n!}} \sum_{\sigma \in \mathcal{S}_n} v_{\sigma 1} \otimes \cdots \otimes v_{\sigma n}$. The physical reason for this is given in Figure 4.5. The first two diagrams represent classically distinct scatterings, but in quantum mechanics trajectories don't exist and we can't tell whether it is particle 1 or rather particle 2 moving northwest after the collision – Figure 4.5(c) applies. The labels '1' and '2' have no physical significance here: the vectors $|v_1, v_2\rangle$ and $|v_2, v_1\rangle$ now correspond to the same state – namely, the one where one of the particles (we cannot ask which) is in state v_1 and the other is in state v_2 – and should be identified. Perhaps we can say that here is the precise pen with which This August Personage signed That Important Document, but we cannot say (pointing) that this electron here was part of the pen at that Propitious Moment. An easy combinatorial consequence of this is that the identical particles here (but not those in the next paragraph!) tend to clump into similar states. This is responsible, for instance, for the existence of the laser.

Recall however that proportional vectors in the space S correspond to physically equivalent states. Thus it merely suffices to identify, for example, $|v_1, v_2\rangle$ and $|v_2, v_1\rangle$ *up to a scalar factor*. The preceding paragraph describes the *bosons* like photons of light (named after S. N. Bose, who with Einstein first considered their statistical mechanics). The next simplest possibility, describing the *fermions* such as electrons, obeys $|v_1, v_2\rangle = -|v_2, v_1\rangle$. Their Hilbert space is spanned by antisymmetric vectors of the form $|v_1, \ldots, v_n\rangle := \frac{1}{\sqrt{n!}} \sum_{\sigma \in S_n} (-1)^\sigma v_{\sigma 1} \otimes \cdots \otimes v_{\sigma n}$, where '$(-1)^\sigma$' equals ± 1 for an even/odd permutation σ, respectively. Note that antisymmetry forbids two fermions from sharing the same state. This simple fact is directly responsible for the remarkable diversity of chemical compounds, for if electrons obeyed instead the bosonic possibility $|v_1, v_2\rangle = +|v_2, v_1\rangle$, then there wouldn't be a chemical difference between the elements hydrogen, helium, lithium, . . . It is also responsible for large-scale structure, for example, why we don't fall through the floor.

These bosonic and fermionic 'statistics' correspond to the two one-dimensional representations of the symmetric group S_n, but there are other possibilities (e.g. parastatistics, which involves higher-dimensional representations of S_n, and braid statistics, which can occur when space-time is two-dimensional – both are discussed in, for example, chapter IV of [**269**]). However, only bosons and fermions seem to arise in Nature (except perhaps for some compound systems). Assuming this, a deep result of quantum field theory (Fierz and Pauli's Spin-Statistics Theorem – for a proof see section 4-4 of [**518**]) relates statistics to the Poincaré group. In particular, particles in relativistic quantum mechanics carry a representation of the universal cover of the Poincaré group. When that representation reduces to a representation of the Poincaré group itself, that is when spatial rotations through 2π correspond to the identity (we say the 'spin' is an integer), then the particle is a boson. Otherwise, that is when rotations through 2π correspond to $-I$ (so the spin is a half-integer), the particle will be a fermion. A connection between spin and statistics can be anticipated by the observation that the simple exchange of locations of two objects involves an implicit rotation by 2π of one relative to the other. We discuss this further in Section 4.3.5 below.

An important formulation of quantum physics is due to Feynman, and starts from an observation of Dirac: the infinitesimal quantum mechanical amplitude is governed by the value of the classical action (4.1.3). Suppose we know the wave-function $\mathbf{x} \mapsto \psi(\mathbf{x}, t_i)$ at some fixed initial time t_i. Then ψ at some other time t_f is given by

$$\psi(\mathbf{x}'', t_f) = \int K(\mathbf{x}'', \mathbf{x}'; t_f - t_i)\, \psi(\mathbf{x}', t_i)\, \mathrm{d}^3\mathbf{x}', \qquad (4.2.6a)$$

where K, called the 'propagation kernel', is the amplitude for a particle to go from position $\mathbf{x}'$ at time t_i to position $\mathbf{x}''$ at time t_f. The point is that K is given by the 'path integral' $\int \exp(\mathrm{i}\, S(\mathbf{x})/\hbar)\, \mathcal{D}\mathbf{x}$ over all paths $\mathbf{x} : t \mapsto \mathbf{x}(t)$ with endpoints $\mathbf{x}(t_i) = \mathbf{x}'$, $\mathbf{x}(t_f) = \mathbf{x}''$. For each choice of path $\mathbf{x}(t)$, $S(\mathbf{x})$ here is the classical action $\int_{t_i}^{t_f} L(\mathbf{x}, \dot{\mathbf{x}})\, \mathrm{d}t$. Integrals over spaces of paths arise here for much the same reason that the entries of

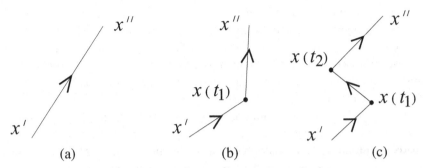

Fig. 4.6 Feynman diagrams in quantum mechanics.

powers A^n of a matrix could be described as sums over length-n walks through the entries of A. The path integral formulation intuits that the particle takes every conceivable trajectory from $(\mathbf{x}', t_i)$ to $(\mathbf{x}'', t_f)$, and each of these (appropriately weighted) contributes to the amplitude K and hence probability $|K|^2$. The precise mathematical meaning of Feynman's path integral is a little elusive, but attempts to define it in terms of, for example, Wiener integrals have been made. It is probably simplest though to regard it heuristically, as is done in Section 4.4.1.

Consider the classical limit $\hbar \to 0$ of (4.2.6a): using the stationary phase approximation, the dominant path $\mathbf{x}'(t)$ in the Feynman integral is one that satisfies the Euler–Lagrange equation (4.1.4). This provides an explanation for the mysteriously teleological Hamilton's principle of classical mechanics, discussed in Section 4.1.1.

The perturbative approach to quantum theories is particularly transparent in the path integral formalism. Write the Lagrangian as the sum $L = L_0 + \lambda L_{int}$ of the free part L_0 and the interaction part $\lambda L_{int} = -\lambda V$; the 'coupling constant' λ is a numerical constant (hopefully small), and we aim to expand the kernel K (and hence the wave-function ψ) in a Taylor expansion in λ. Explicitly, we have

$$
\begin{aligned}
K(\mathbf{x}'', \mathbf{x}'; t_f - t_i) &= \int \exp\left[\frac{i}{\hbar} \int_{t_i}^{t_f} (L_0 - \lambda V)\,dt\right] \mathcal{D}\mathbf{x} \\
&= \int \exp\left[\frac{i}{\hbar} \int_{t_i}^{t_f} L_0\,dt\right] \sum_{n=0}^{\infty} \frac{(-i\lambda/\hbar)^n}{n!} \left(\int_{t_i}^{t_f} V(\mathbf{x}(t))\,dt\right)^n \mathcal{D}\mathbf{x}.
\end{aligned}
$$

(4.2.6b)

We can represent this pictorially. The $n = 0$ term describes a particle propagating freely from $(\mathbf{x}', t_i)$ to $(\mathbf{x}'', t_f)$; the Feynman diagram for this term is given in Figure 4.6(a). The $n = 1$ term describes a particle propagating freely from $(\mathbf{x}', t_i)$ to some intermediate point $(\mathbf{x}, t_1)$, at which instant the potential V acts multiplicatively, and then the particle resumes free propagation to the final position $(\mathbf{x}'', t_f)$; we then integrate over all intermediate times (and finally over all paths $\mathbf{x}(t)$). The Feynman diagram is given in Figure 4.6(b), where the integration over t_1 is implicit. The kink there is called a 'vertex' – this is the same word as in *vertex* operator algebra. Likewise, the λ^n term corresponds to a Feynman diagram with n vertices, corresponding to the n integrals $\int V\,dt_j$ in (4.2.6b). The factor $n!$ in (4.2.6b) is removed by taking these intermediate times in the order $t_i < t_1 < \cdots < t_n < t_f$, as the diagrams suggest. In this way, we have replaced the actual physical situation, where of course the interaction V is always present, with a situation where the interaction is only present at discrete moments of time. It is as if the particle only interacts with V at the vertices. These are called *virtual* interactions, as they are mathematical artifacts and don't correspond directly to actual events in Nature.

We'll say more about perturbations and Feynman diagrams later. Typically, the sum (4.2.6b) won't converge, but the first few terms (when interpreted correctly) give good comparison with experiment. Conformal field theory – the physics of Moonshine – arises from the perturbative expansion of the quantum field theory called string theory.

Its treatment of measurement demonstrates that quantum mechanics is heuristic and idealised, and not at all in its finished form. But just as classical physics achieved a profound understanding of the concept of 'rest', and relativity provided a deep reanalysis of space and time, so is quantum mechanics forcing us to reconsider the seemingly harmless notion of observation. After all, we never observe an object, but rather the interaction between objects. Also profound, quantum mechanics teaches us that interacting subsystems become entangled, and physically this means that the whole is indeed much more than the disjoint union of its parts.

4.2.2 Informal quantum field theory

It is surprising that the next three natural tasks – namely, to bring in special relativity, to handle the experimental fact that the number of elementary particles can change, and to quantise classical field theories – are all accommodated by quantum field theories, the quantum theories of systems with infinitely many degrees of freedom. The sketch we provide here won't seem very satisfactory, but this is roughly the treatment to be found in physics textbooks. We avoid as too tangential most calculational issues and many technicalities (e.g. the quirks of fermions). Section 4.2.4 provides a more careful axiomatic treatment of quantum field theory, but knowing the informal physics background, at least in its broader strokes, is essential. A dated though otherwise excellent treatment of quantum field theory, somewhat in our style, is [479]; modern and masterful is [555].

To the working physicist, quantum field theory is the following conceptual hierarchy.

(i) *Experiment.* The experimenter measures half-lives of particles and scattering cross-sections. How well does experiment compare to theory?

(ii) *Amplitudes.* These observable quantities depend on the magnitude-squared of the appropriate transition amplitude $|\text{in}\rangle \to |\text{out}\rangle$. Unfortunately, transition amplitudes are too hard to calculate from the theory, except in infinite time ($t \to \pm\infty$) limits, which by definition are the entries of the S-matrix. Those limits, though mathematically dubious, are physically intuitive. So the theoretician needs to compute the S-matrix.

(iii) *Correlation functions.* The typical way to compute S-matrix entries is using correlation functions, via the so-called reduction formulae. So the theoretician wants to compute correlation functions.

(iv) *Feynman diagrams*. Typically, correlation functions are calculated 'perturbatively' by Taylor-expanding in some coupling constant. Each term in this (usually divergent) infinite series is computed separately using Feynman diagrams.

Moonshine is interested in the correlation functions of a class of extremely symmetrical and well-behaved quantum field theories called rational conformal field theories – these theories are so special that their correlation functions can be computed exactly and perturbation is not required. But before we turn to them, let's flesh out some of this hierarchy.

It would seem trivial to make quantum mechanics consistent with special relativity. Consider, for simplicity, a free particle of mass m. Recall that Schrödinger's equation (4.2.1) corresponds to the nonrelativistic energy $E = \frac{1}{2m}\mathbf{p}^2$. Since relativistic energy satisfies $E^2 - \mathbf{p}^2 c^2 = m^2 c^4$, the natural guess for the relativistic Schrödinger equation would be

$$\left(\hbar^2 \frac{\partial^2}{\partial t^2} - \hbar^2 c^2 \nabla^2 + m^2 c^4\right) \phi(\mathbf{x}, t) = 0. \qquad (4.2.7)$$

This is called the Klein–Gordon equation, and was proposed independently by Schrödinger, Klein and Gordon shortly after (4.2.1) was written down.[7] They expected it to describe the relativistic wave-function ϕ of a free 'scalar' particle (i.e. $\phi(x)$ is invariant under the action of the Lorentz group $\mathrm{SO}_{3,1}^+(\mathbb{R})$ on x), but such a theory is sick (see Question 4.2.4): for example, it suffers from negative probabilities and the energy eigenvalues have no lower bound (this means that we won't have a vacuum state $|0\rangle$, which is bad). The way to make (4.2.7) into a sensible physical theory is to interpret it as a quantum field theory.

Quantum field theory is far deeper than quantum mechanics, both physically and mathematically. Witten predicts [566] that one of the major themes of twenty-first century mathematics will involve coming to grips with quantum field theory.

Let $\Omega \subset \mathcal{H} \subset \Omega^*$ be a rigged Hilbert space; Ω is the span of the states in the theory, and is constructed below, while $\mathcal{H}$ is their topological span. We obtained nonrelativistic quantum mechanics by replacing classical observables by operators, so we would expect that the fields $\varphi(x)$ in quantum field theories are operator-valued functions of space-time. Unfortunately this is too optimistic, even in the simplest free theories. Rather, the correct statement is that quantum fields φ are operator-valued *distributions* of space-time: for any states $u, v \in \Omega$, the matrix entries $\langle u, \varphi v \rangle$ of φ are tempered distributions of space-time. In other words, the Schwartz space $\mathcal{S} = \mathcal{S}(\mathbb{R}^4)$ is a space of test functions of space-time that 'smear' the fields; the values $\varphi(f)$, for each $f \in \mathcal{S}$, are (unbounded) linear operators $\Omega \to \Omega$. Nevertheless, it is traditional to write $\varphi(x)$, as if the fields were functions of space-time, and informally think of $\varphi(f)$ as the integral $\int_{\mathbb{R}^4} f(x)\varphi(x)\,d^4x$. Unlike the wave-functions of quantum mechanics, a quantum field is not directly a probability

---

[7] Apparently, Schrödinger first derived the relativistic equation, noticed that it didn't work but that its nonrelativistic approximation (4.2.1) looked good, and so first published the approximation! See the historical discussion on page 4, vol. I of [555].

amplitude; rather, it is a linear combination of operators that increase or decrease by one the numbers of particles in any state.

Let $\varphi_1, \ldots, \varphi_n$ be the complete list of quantum fields in the theory. All operators (e.g. observables) occurring in the theory are constructed from these fields. More precisely, locality says that any operator at a given space-time point x is a function of fields and their derivatives, all evaluated at that point.

The mathematical meaning of a theory being (special-)relativistic is that its quantities transform nicely with respect to (i.e. in projective representations of) the Lorentz and Poincaré groups $SO_{3,1}^+$ and $\mathbb{R}^4 \rtimes SO_{3,1}^+$. As in Theorem 3.1.1, those projective representations are true representations of the universal covers $SL_2(\mathbb{C})$ and $\mathbb{R}^4 \rtimes SL_2(\mathbb{C})$, respectively. Firstly, the state space $\mathcal{H}$ carries a unitary representation $(a, \Lambda) \mapsto U_{(a,\Lambda)}$ of the universal cover of the Poincaré group. These operators $U_{(a,\Lambda)}$ send the state space Ω onto itself; on Ω, we can write $U_{(a,I)} =: \exp[-i \sum_\mu a_\mu P^\mu / \hbar]$, where the self-adjoint operators P^μ are the observables for momentum and (up to a constant) energy. In particular, $c^2 P^4$ is the Hamiltonian density. The absence of tachyons (footnote 4 in this chapter) says that the simultaneous eigenvalues $(\mathbf{p}, p^4)$ of the energy-momentum operators P^1, P^2, P^3, P^4 all have nonpositive Minkowski norm-squared $\sum_\mu p_\mu p^\mu = \mathbf{p}^2 - c^2 (p^4)^2 =: -m^2 c^2$. This parameter m is constant in any irreducible representation of $\mathbb{R}^4 \rtimes SL_2(\mathbb{C})$, and is called the *(rest-)mass*.

The span of the fields φ_i carries a projective representation of all symmetries of the theory. In particular, there is an n-dimensional representation V of $SL_2(\mathbb{C})$, governing how the n fields transform relativistically: that is,

$$U_{(a,\Lambda)} \varphi_i(f) U_{(a,\Lambda)}^{-1} = \sum_{i=1}^n V(\Lambda^{-1})_{ij} \, \varphi_j((a, \Lambda)^{-1}.f) \qquad (4.2.8a)$$

holds in Ω, where the Poincaré transformation $(a, \Lambda) \in \mathbb{R}^4 \rtimes SL_2(\mathbb{C})$ acts on test functions by $((a, \Lambda).f)(x) = f(\Lambda x + a)$. The inverses on the right side are needed in order for (4.2.8a) to be consistent with $U_{(a',\Lambda')} \circ U_{(a,\Lambda)} = U_{(a',\Lambda')\circ(a,\Lambda)}$. Restricting to translations $\mathbb{R}^4$, the derived representation of (4.2.8a) becomes the important equation of motion

$$\partial_\mu \varphi(x) = \frac{i}{\hbar} [P_\mu, \varphi(x)]. \qquad (4.2.8b)$$

Since the finite-dimensional representations of $SL_2(\mathbb{C})$ are completely reducible, we can collect the fields together that form irreducible representations, parametrised by Dynkin label $\lambda_1 = \mathbb{N}$. Mysteriously, physicists prefer to use *spin $s = \lambda_1/2$*.

In classical field theory, the particles and fields are phenomenologically independent even though they mutually influence each other. In quantum field theory, particles are secondary, arising from fields, as we see shortly. A great definition, due to Wigner, is:

Definition 4.2.1 *A particle is an irreducible projective representation of the Poincaré group, with real mass m and energy $c^2 p^4 \geq 0$, in the space $\mathcal{H}$ of states of the theory.*

More precisely, the spectra $(\mathbf{p}, p^4)$ of the energy-momentum operators P^μ in an irreducible representation are required to obey $\mathbf{p}^2 \leq c^2 (p^4)^2$; the mass $m \geq 0$ is the constant

$\sqrt{c^2(p^4)^2 - \mathbf{p}^2}$. Only the vacuum has 0 energy. Unlike the mass, the energy varies within the irreducible representation, and for a particle of mass m is never less than mc^2.

Subatomic experiments suggest that there are elementary (i.e. noncomposite) particles, for instance electrons. Each species of elementary particle in the theory arises from an irreducible $SL_2(\mathbb{C})$-module in the span of the fields φ_i. In particular, a particle with spin $s \in \frac{1}{2}\mathbb{N}$ requires $2s + 1$ fields $\varphi_{i_1}, \ldots, \varphi_{i_{2s+1}}$, called its components. Other symmetries of the theory combine with $SL_2(\mathbb{C})$ to form higher-dimensional representations. For example, in *quantum electrodynamics*,[8] 'parity' (i.e. the space-reflection $\mathbf{x} \mapsto -\mathbf{x}$) collects the two-component 'left-' and 'right-handed' electrons into an irreducible four-dimensional representation, while in the Standard Model parity is no longer a symmetry, but the left-handed electron and neutrino transform together as components in a four-dimensional representation of the symmetry group $SU_3 \times SU_2 \times U_1$, while the right-handed electron forms a two-dimensional representation by itself.

A Lagrangian density $\mathcal{L}(x)$ here is a self-adjoint operator, invariant under $SL_2(\mathbb{C})$, built up polynomially from the various φ_i and $\partial_\mu \varphi_i$, all evaluated at the same space-time point x. Each field φ_i obeys the corresponding Euler–Lagrange equation (4.1.8). As in classical field theory, define the 'canonical momentum field' $\pi_i(x) = \partial\mathcal{L}/\partial(\partial_4\varphi_i)$ (not to be confused with the momentum operators P^μ). The *equal-time commutation relations*

$$[\varphi_i(\mathbf{x}, t), \pi_j(\mathbf{x}', t)] = i\hbar \, \delta_{ij}\delta(\mathbf{x} - \mathbf{x}'), \qquad (4.2.9a)$$

$$[\varphi_i(\mathbf{x}, t), \varphi_j(\mathbf{x}', t)] = [\pi_i(\mathbf{x}, t), \pi_j(\mathbf{x}', t)] = 0 \qquad (4.2.9b)$$

are obtained from the classical Poisson brackets (4.1.9) via standard ('canonical') quantisation. When both φ_i, φ_j are fermionic (i.e. have fractional spin), then (4.2.9) should be replaced with anti-commutation relations. For simplicity, we consider only bosonic fields.

Because disturbances shouldn't travel faster than light, measurements occurring at space-time points x, x' that are space-like separated (i.e. $(x - x')^2 > 0$) should be independent. Quantum theory translates this into the statement that the corresponding observables $\mathcal{O}(x), \mathcal{O}'(x')$ should commute: $[\mathcal{O}(x), \mathcal{O}'(x')] = 0$ when $(x - x')^2 > 0$. Since the observables are built out of the fields φ_i, this is closely related to the commutation relations (4.2.9). Nevertheless, the relations (4.2.9) are controversial, as we'll see.

To see how to use the field equations and (4.2.9), consider for example the density

$$\mathcal{L}(x) = \frac{-1}{2}\left(m^2c^4\hbar^{-2}\phi(x)^2 + c^2\partial_\mu\phi(x)\,\partial^\mu\phi(x)\right), \qquad (4.2.10a)$$

where $\phi = \phi^\dagger$ is self-adjoint. (We will see shortly that this $\mathcal{L}$ has to be modified slightly to be physically sensible.) The field equation here is the Klein–Gordon equation (4.2.7). It can be solved by a trick: the Fourier transform of ϕ from 'position-space' into

---

[8] Quantum electrodynamics ('QED' for short) is the quantum theory of Maxwell's electromagnetism applied to electrons, positrons (the anti-particle of the electron) and photons (the particle of light). QED is subsumed by the *Standard Model*, the quantum field theory describing all known physics except for gravity.

'momentum-space' converts the Klein–Gordon equation into decoupled classical simple harmonic oscillator equations, so the field ϕ can be formally written

$$
\phi(\mathbf{x}, t) = \int \sqrt{\frac{\hbar}{(2\pi)^3 2\omega_{\mathbf{p}}}} \left[\widehat{a}(\mathbf{p}) \exp\left[\frac{\mathrm{i}}{\hbar} \mathbf{p} \cdot \mathbf{x} - \mathrm{i}\omega_{\mathbf{p}} t \right] \right.
$$
$$
\left. + \widehat{a}(\mathbf{p})^\dagger \exp\left[-\frac{\mathrm{i}}{\hbar} \mathbf{p} \cdot \mathbf{x} + \mathrm{i}\omega_{\mathbf{p}} t \right] \right] \mathrm{d}^3\mathbf{p}, \qquad (4.2.10b)
$$

where $\omega_{\mathbf{p}} = \hbar^{-1} p^4 = c\hbar^{-1}\sqrt{\mathbf{p}^2 + m^2 c^2}$. If ϕ were a real-valued function, (4.2.10b) would give the general solution, for arbitrary coefficients obeying $\overline{a(\mathbf{p})} = \widehat{a}(\mathbf{p})^\dagger \in \mathbb{C}$. Here the coefficients are operators, with $\widehat{a}(\mathbf{p})^\dagger$ the adjoint of $\widehat{a}(\mathbf{p})$ (hence the notation). The canonical momentum is $\pi = \partial_4 \phi$. Solving (4.2.10b) for $\widehat{a}(\mathbf{p})$ and $\widehat{a}(\mathbf{p})^\dagger$ in terms of ϕ, equations (4.2.9) become

$$
\left[\widehat{a}(\mathbf{p}), \widehat{a}(\mathbf{p}')^\dagger \right] = \delta^3(\mathbf{p} - \mathbf{p}'), \qquad \left[\widehat{a}(\mathbf{p}), \widehat{a}(\mathbf{p}') \right] = \left[\widehat{a}(\mathbf{p})^\dagger, \widehat{a}(\mathbf{p}')^\dagger \right] = 0. \qquad (4.2.10c)
$$

This trick of switching from position variables to momentum variables is common in field theory, and it isn't surprising that it should simplify the mathematics: the momentum degrees of freedom are uncoupled because the theory is translation-invariant (Noether's Theorem!). If instead ϕ is not self-adjoint, then we should expand ϕ into independent coefficients $a(\mathbf{p})$, $b(\mathbf{p})^\dagger$.

How do we accommodate particles in quantum field theory? First note that the particle interpretation pertains directly to state vectors $v \in \Omega$, and not the fields – for example, our universe corresponds to some vector $|universe\rangle \in \Omega$. There are, for example, only four electron fields (i.e. one component for each internal degree of freedom); all of the nearly infinitely many electrons in the universe are *created* by those fields in a way we'll describe shortly. The *number* of electrons is an observable quantity, and hence an eigenvector of the 'electron-number' operator $\widehat{N}_e$. Thus a typical vector $v \in \Omega$ will *not* have a well-defined number of (say) electrons.

The most important vector in Ω is the vacuum state $|0\rangle \in \Omega$, which contains zero particles of each type. It is fixed by the representation of the universal cover of the Poincaré group, i.e. $U_{(a,\Lambda)}|0\rangle = |0\rangle$, so in particular the state $|0\rangle$ has total momentum $\mathbf{0}$ and energy 0. As before, it is unique up to scalar multiplication, nondegenerate and has norm 1: $\langle 0|0\rangle := \||0\rangle\|^2 = 1$. (Actually, in quantum field theories with spontaneous symmetry breaking, such as the Standard Model, the vacuum will be degenerate, but we will ignore this possibility here.)

The particle interpretation is simplest in the free scalar field theory (4.2.10). Equations (4.2.10b) and (4.2.10c) tells us to think of the free field ϕ as infinitely many independent quantum harmonic oscillators (4.2.5), one for each possible momentum. The analogue of the one-particle state $|1\rangle$ there should be the one-particle state $|\mathbf{p}\rangle$ with momentum $\mathbf{p}$ and energy $\omega_{\mathbf{p}}\hbar$, defined by $|\mathbf{p}\rangle := \widehat{a}(\mathbf{p})^\dagger |0\rangle$. The problem is that its normalisation

$$
\||\mathbf{p}\rangle\|^2 = \langle 0| \widehat{a}(\mathbf{p})\widehat{a}(\mathbf{p})^\dagger |0\rangle = \delta(0),
$$

obtained using (4.2.10c), is infinite. This is why a quantum field ϕ is an operator-valued *distribution*. The one-particle states can't have well-defined momenta, but rather are 'wave-packets', linear combinations ('superpositions') of those momentum states $|\mathbf{p}\rangle$ constructed using test functions f. In particular, let f be in the Schwartz space $\mathcal{S}(\mathbb{R}^{3k})$. The k-particle states in Ω are of the form

$$|f\rangle := \int \cdots \int f(\mathbf{p}_1, \ldots, \mathbf{p}_k) \widehat{a}(\mathbf{p}_1)^\dagger \cdots \widehat{a}(\mathbf{p}_k)^\dagger |0\rangle \, \mathrm{d}^3 \mathbf{p}_k \cdots \mathrm{d}^3 \mathbf{p}_1.$$

The state $|f\rangle$ is an eigenvector of the number operator $\widehat{N}_\phi = \int \widehat{a}(\mathbf{p})^\dagger \widehat{a}(\mathbf{p}) \, \mathrm{d}^3 \mathbf{p}$, with eigenvalue k. The operators $\widehat{a}(\mathbf{p})$ again are annihilation operators and take a k-particle state to a $(k-1)$-particle state. Together, all these k-particle states, for $k = 0, 1, \ldots$, span the space Ω. The commutation relation $[a^\dagger, a^\dagger] = 0$ means that the particles obey bosonic statistics, that is both $f \in \mathcal{S}(\mathbb{R}^{3k})$ and its symmetrisation $\frac{1}{k!} \sum_{\sigma \in \mathcal{S}_k} f(\mathbf{p}_{\sigma 1}, \ldots, \mathbf{p}_{\sigma k})$ define physically identical states.

Just as a pendulum in classical mechanics undergoes small oscillations about its (vertical) stationary equilibrium position, so does the vacuum in quantum field theory. The oscillations of the quantum vacuum are the electrons, photons, etc. observed in Nature. This particle concept is the kinematics of quantum field theory.

In these free theories, the k particles in $|f\rangle$ move independently and freely. The notion of wave-packets explains the tracks of particles in the cloud chambers of high-energy experiments: such tracks seem to indicate that the particle has, to a good approximation, both a well-defined position and momentum. By contrast, the (nonphysical) momentum eigenstates $|\mathbf{p}\rangle$ are diffused throughout the universe.

Similarly, particles in any *free* quantum field theory arise by interpreting the Fourier coefficients of the fields as creation and annihilation operators (theories with interactions are considered shortly). Now, any operator can be expressed as an integral of sums and products of these creation and annihilation operators (see section 4.2 of [**555**] for a proof). For example, the free scalar theory (4.2.10) has energy–momentum operators

$$P^\mu = \frac{1}{2} \int p^\mu \left(\widehat{a}(\mathbf{p})^\dagger \widehat{a}(\mathbf{p}) + \widehat{a}(\mathbf{p}) \widehat{a}(\mathbf{p})^\dagger \right) \mathrm{d}^3 \mathbf{p}.$$

Since $[\widehat{N}_\phi, P^4] = 0$, we see from (4.2.4) that in this *free* theory the number of particles won't change. It can change only when we include interactions.

Note that in the free scalar theory $P^\mu |0\rangle = 0$ for $\mu = 1, 2, 3$, as it should, but $P^4 |0\rangle$, which gives the energy of the vacuum, is

$$P^4 |0\rangle = \int \hbar \omega_\mathbf{p} \left(\widehat{a}(\mathbf{p})^\dagger \widehat{a}(\mathbf{p}) + \frac{1}{2} \right) |0\rangle \, \mathrm{d}^3 \mathbf{p} = 0 + \frac{\hbar}{2} \int \omega_\mathbf{p} \mathrm{d}^3 \mathbf{p} |0\rangle,$$

so is divergent. This is a typical infinity in quantum field theory, but is easy to remedy, as it tells us that the Hamiltonian density $\mathcal{H}(\mathbf{p})$ (hence our original Lagrangian density $\mathcal{L}(x)$) is off by an additive (infinite) constant. It isn't surprising in hindsight that the naive guess (4.2.10a) for $\mathcal{L}(x)$ runs into problems: for one thing, classical energy is only defined up to an additive constant; for another, the order in which the numerical coefficients $a, a^\dagger$ appear in classical expressions for energy doesn't matter, while the order of the operators

$\widehat{a}, \widehat{a}^\dagger$ in quantum mechanics certainly does. Replacing $\mathcal{L}(x)$ and $\mathcal{H}(\mathbf{p})$ with their 'normal orders' $:\mathcal{L}:$ and $:\mathcal{H}:$, respectively, gives the vacuum zero energy and doesn't otherwise change the physics. The normal order $:\mathcal{O}:$ of an operator $\mathcal{O}$ given by an integral over $\mathbf{p}$'s of a product of $\widehat{a}(\mathbf{p})$'s and $\widehat{a}(\mathbf{p})^\dagger$'s is obtained by moving all annihilation operators $\widehat{a}(\mathbf{p})$ to the right of all creation operators $\widehat{a}(\mathbf{p})^\dagger$. This has the effect of making the evaluation of operators on states as simple as possible. For example, the Hamiltonian density becomes

$$: P^\mu := \int p^\mu \, \widehat{a}(\mathbf{p})^\dagger \widehat{a}(\mathbf{p}) \, \mathrm{d}^3\mathbf{p}.$$

The same procedure works in any quantum field theory to give the vacuum zero energy, with a minor change when there are fermions. We also used normal-ordering in, for example, (3.2.14a) to remove an analogous infinity in Lie theory.

The existence of negative energy states, which we recall was a serious sickness for relativistic quantum mechanics, is handled naturally in quantum field theory. Return for simplicity to the scalar theory, but now with $\phi \neq \phi^\dagger$. The positive energy coefficients $a(\mathbf{p})$ of ϕ annihilate a positive energy particle; the negative energy coefficients $b(\mathbf{p})^\dagger$ create a positive energy particle. The particle annihilated by the field ϕ is not quite the same as the particle created by ϕ: The various parameters describing particles will either be the same (e.g. mass) or opposite (e.g. electric charge), for these two kinds of particles. That is, the pair $\phi, \phi^\dagger$ of fields is associated with *pairs* of particles; one of these we arbitrarily call the *anti-particle*. Physically, an anti-particle can be interpreted as the corresponding particle 'travelling backwards in time with negative energy', and that is how it is depicted in Feynman diagrams. When $\phi = \phi^\dagger$, the particle is its own anti-particle.

This is how particles arise in *free* quantum field theories. The physically interesting quantum field theories have interactions, that is additional terms in $\mathcal{L}(x)$ corresponding to potential energy. Experiments (e.g. the cloud chambers) tell us that a particle interpretation is still appropriate there. A typical experiment begins and ends with several particles separated by macroscopic distances; interactions occur only at intermediate times when some particles are microscopically separated. What we observe are the initial ('incoming') and final ('outgoing') states, and the transition probabilities $|\langle \text{out}|\text{in}\rangle|^2$. Now, macroscopically separated particles should behave independently to good accuracy. Thus these initial and final states are described by the corresponding *free* theory, at least in the limits $t \to \mp\infty$. A particle interpretation applies directly only to these asymptotic states.

In particular, to each field φ_i in a quantum field theory[9] there are fields φ_i^{in} and φ_i^{out}. The field equations (4.1.8) for the φ_i of course include interaction effects, whereas the asymptotic fields $\varphi_i^{\text{in}}, \varphi_i^{\text{out}}$ obey the free field equations, such as the Klein–Gordon equation (4.2.7). Because $P^4|0\rangle = 0$, the vacuum is constant in time ('stable') and is its

---

[9] Many of the following comments assume the associated particle is stable and can exist in isolation of the other particles, at least asymptotically. This is the case, for example, for an electron, but not the muon or quark, which are also elementary and have their own fields. See the literature for the necessary modifications.

own incoming and outgoing asymptotic state. All other incoming states are built up from the vacuum $|0\rangle$ and φ^{in} by the process described earlier. The collection of all incoming states spans the space Ω. Similarly, $|0\rangle$ and φ_i^{out} create all outgoing states, and these also span Ω. Thus the 'in-fields' φ_j^{in} describe the (hypothetical) physics that would occur if the initial particles never interacted; the field φ_j interpolates between these free initial and final asymptotic situations (up to a multiplicative constant, as we'll see), and embodies the true physics by carrying the dynamical information of the system.

As mentioned earlier, experiments obtain information on the transition amplitudes $\langle\text{out}|\text{in}\rangle$ between (prepared) initial states and the (observed) final states, and the complicated machinery of quantum field theory is designed to compute these. These inner products can be thought of as matrix entries of an operator S, the *S(cattering)-matrix*, which defines the equivalence $\varphi^{\text{out}} = S^{-1}\varphi^{\text{in}}S$ between the algebras of in-fields and of out-fields, and the equivalence $|\text{in}\rangle = S|\text{out}\rangle$ between the corresponding incoming and outgoing states. Without going into the technical details, the so-called 'Lehmann–Symanzik–Zimmermann reduction formulae' (see e.g. section 7.2 of [**479**], or section 5-1-3 of [**310**]) express the transition amplitudes in terms of an n-fold integral $\int d^4x_1 \cdots d^4x_n$ over space-time, of 'n-point (correlation) functions', or 'Green's functions', or 'vacuum-to-vacuum expectation values' of 'time-ordered products' of the physical fields:

$$\langle\varphi_{j_1}(x_1)\cdots\varphi_{j_n}(x_n)\rangle := \langle0|T(\varphi_{j_1}(x_1)\cdots\varphi_{j_n}(x_n))|0\rangle. \qquad (4.2.11)$$

We will usually use the statistical term 'correlation function', standard in conformal field theory. The symbol 'T' here reorders the fields $\varphi_{j_i}(x_i)$ in increasing order of the time x_i^4, and is needed to guarantee convergence. The number n here is the total number of particles in $|\text{in}\rangle$ and $|\text{out}\rangle$ together.

In classical physics, Noether's Theorem associates with a continuous symmetry a conserved current $j^\mu(x)$ and a conserved charge Q. Now, a symmetry of a classical system may become broken in quantisation – this is called an *anomaly* (see e.g. section 11-5 of [**310**]). Usually an anomaly is bad news, but a harmless anomaly important to us is the soft breaking of the conformal symmetry in CFT. It is measured by a parameter called the *central charge* or *conformal anomaly* c (Section 4.3.1).

When a symmetry survives quantisation, the analogue of Noether's Theorem here is the *Ward identities* (see e.g. section 10.4 of [**555**]), which are differential equations satisfied by the correlation functions. They take the form

$$\frac{\partial}{\partial x^\mu}\langle j^\mu(x)\,\varphi_{j_1}(x_1)\cdots\varphi_{j_n}(x_n)\rangle = -i\sum_i \delta(x - x_i)\,\langle\varphi_{j_1}(x_1)\cdots G_i\varphi_{j_i}(x_i)\cdots\varphi_{j_n}(x_n)\rangle,$$
$$(4.2.12)$$

where G_i is the associated representation of the symmetry on the field φ_{j_i}.

The typical, and only general, way to compute correlation functions is perturbation theory. The correlation functions (4.2.11) play the role here of the propagation kernel K in (4.2.6a); their path integral expression looks like

$$\langle\varphi_{j_1}(x_1)\cdots\varphi_{j_n}(x_n)\rangle = \frac{1}{\mathcal{Z}}\int \phi_{j_1}(x_1)\cdots\phi_{j_n}(x_n)\,\exp[iS(\phi)/\hbar]\,\mathcal{D}\phi, \qquad (4.2.13a)$$

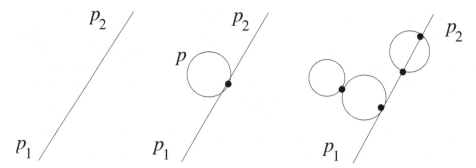

Fig. 4.7 Some two-point Feynman diagrams in the ϕ^4 model.

where S is the classical action (4.1.3) and the integral $\int \mathcal{D}\phi$ is over the space of complex-valued functions $\mathbb{R}^3 \to \mathbb{C}$ (one such 'wave-function' for each field φ_i in the theory). The normalisation factor $1/\mathcal{Z}$ in (4.2.13a) is

$$\mathcal{Z} = \int \exp[iS(\phi)/\hbar] \, \mathcal{D}\phi, \qquad (4.2.13b)$$

called a *partition function* for statistical reasons. We're glossing over technicalities, but the technicalities are (too) easily found in the literature. Once again the mathematical meaning (such as it is) of (4.1.13a) is best ignored; more important are the heuristics it suggests for perturbation.

For that purpose consider a toy model: a single self-adjoint scalar field $\phi = \phi^\dagger$, with ϕ^4 interaction term: $\mathcal{L} = -\frac{1}{2} \sum_\mu \partial_\mu \phi \partial^\mu \phi - \frac{1}{2} m^2 - \frac{\lambda}{4!} \phi^4$ (for typographical clarity we adopt here the usual conventions $c = \hbar = 1$). As always, the equations are simpler if we Fourier-transform to momentum space. The two-point function yields

$$\langle \phi(p_1)\phi(p_2) \rangle = (2\pi)^4 \delta^4(p_1 + p_2) \left\{ \frac{i}{p_1^2 - m^2} \right.$$
$$\left. - \lim_{\epsilon \to 0} \frac{\lambda}{(p_1^2 - m^2)^2} \int \frac{1}{(2\pi)^4} \frac{d^4 p}{p^2 - m^2 + i\epsilon} + O(\lambda^2) \right\}. \quad (4.2.13c)$$

The Dirac delta factor expresses momentum conservation. The integral in (4.2.13c) doesn't converge – this infinity is analogous to the infinite self-energy of the electron in classical electromagnetism (Section 4.1.3), and provides the first example of renormalisation, as we will see shortly. The first two terms within the braces of (4.2.13c) correspond to the first two diagrams in Figure 4.7. The second diagram can be interpreted as a particle emitting a pair of virtual particles, which then annihilate themselves. The four-point function $\langle \phi(p_1)\phi(p_2)\phi(p_3)\phi(p_4) \rangle$, computed to λ^1 accuracy, includes the diagrams of Figure 4.8.

The *Feynman rules* describe how to go from the finitely many Feynman diagrams at each perturbation order λ^k, to the corresponding integral expressions. Any book on quantum field theory (e.g. [310] or [555]) describes them in detail, as they are how the theory makes practical contact with experiment. We will make only general remarks.

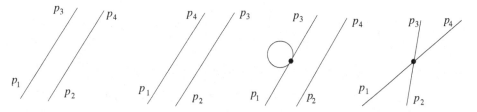

Fig. 4.8 Some four-point Feynman diagrams in the ϕ^4 model.

Fig. 4.9 A typical fourth-order term in the scattering of two electrons.

We can write (4.2.13a) symbolically as

$$\int \phi_{j_1}(x_1) \cdots \phi_{j_n}(x_n) \, \exp[iS(\phi)/\hbar] \, \mathcal{D}\phi = \sum_{\mathcal{G}} c(\mathcal{G}) \int \prod_e \mathrm{d}p_e \prod_v \vartheta_v \delta, \quad (4.2.13\mathrm{d})$$

where the sum is over all Feynman diagrams $\mathcal{G}$ with the external lines (i.e. edges with a free endpoint) corresponding to the fields ϕ_{j_i} in the n-point function. The numerical quantity $c(\mathcal{G})$ is combinatorial. For each internal edge e there is a 'propagator', a momentum p_e and an integral over p_e. At each vertex v there is an operator ϑ_v, which is proportional to the coupling constant, as well as a Dirac delta δ, which expresses momentum conservation at that vertex. Thus each vertex contributes a factor of the coupling constant (which is assumed to be small). The vertices in Figures 4.7 and 4.8 are all of valence 4, because the only interaction term in the Lagrangian density $\mathcal{L}$ here is ϕ^4. More interesting (and physically relevant) quantum field theories involve several types of particles, with several different interaction terms in the Lagrangian, and so the corresponding Feynman diagrams have several types of edges (one for each kind of particle) and several kinds of vertices (one for each term in the interaction Lagrangian). For example, in QED (footnote 8 in this chapter) the interaction term is $-e\overline{\psi} A\psi$, where e is the coupling constant (proportional to the charge of the electron) and where ψ is the (multi-component) field of the electron, $\overline{\psi}$ (essentially the adjoint of ψ) can be thought of as the positron field and A can be identified with the photon field. A vertex here must consist of three particles: a single incoming or outgoing photon, with an incoming and outgoing electron or positron. A typical Feynman diagram involved in the calculation of the four-point function $\langle \psi(p_1) \, \psi(p_2) \, \overline{\psi}(p_1') \, \overline{\psi}(p_2') \rangle$ is shown in Figure 4.9. It describes the virtual event where the incoming electrons (the bottom two solid lines) exchange a virtual photon (the horizontal wavy line), which in transit spontaneously breaks into an electron–positron pair, which then annihilate, returning the photon. All vertices in

Fig. 4.10 Feynman diagrams contributing to the mass shift.

Figure 4.9 are consistent with the interaction term; as there are four of them there, that diagram contributes to the e^4 term.

In order for an expansion in λ^n (or e^n) to make sense, the individual terms should tend to 0 with n. Embarrassingly, in a typical quantum field theory most individual terms are infinite! A simple example is the two-point function (4.2.13c) at one loop – the problem there is that the integrand doesn't go to 0 fast enough for large p. A different infinity provides a clue how to make sense of these perturbative expansions.

We know from free field theory that the term $-\frac{1}{2}m^2\phi^2$ in the ϕ^4 Lagrangian is a kinetic energy term, and so it is tempting to identify m there with the mass of the ϕ particle. However, that parameter m is not directly observable. The (squares of the) true masses of the particles are defined to be the corresponding eigenvalues of the operator $\sum_\mu P^\mu P_\mu$ (again ignoring $\hbar$'s and c's). The easiest way to compute these eigenvalues is through the two-point function $\langle \phi(p_1)\phi(p_2) \rangle$ (called the *propagator* of ϕ): by nonperturbative arguments (see e.g. section 10.2 of [**555**]), the propagator of ϕ should equal the Dirac delta $(2\pi)^4\delta^4(p' + p'')$ times a meromorphic function with a simple pole at $p'^2 = m_\phi^2$ (the physical mass-squared of the particle)[10] with residue i. In the ϕ^4 theory, the propagator to zeroth order (corresponding to the free theory) is $i/(p'^2 - m^2)$, ignoring the Dirac delta factor. However, the perturbative expansion contains geometric series that change the pole. In particular the sequence of diagrams in Figure 4.10 contributes to shifting the denominator, and hence the pole, of the propagator. We call the nonphysical parameter m appearing in the Lagrangian the 'bare mass', in contrast to the true observed mass $m_\phi = m - \delta m$ that is 'dressed' with the cloud of virtual particles arising by virtue of the interaction terms.

The actual values of m and δm can be ignored, since in any physically relevant expression they appear only in the combination $m - \delta m$, which can be replaced by the measured

---

[10] There is some evidence (by studying the 'running coupling constant') that the propagator of the photon in QED has, in addition to the pole at mass zero (corresponding to the massless photon), a pole at *imaginary mass*. This would correspond to a tachyon (footnote 4 in this chapter) called the Landau ghost, which presumably shouldn't exist. This calculation could indicate a fundamental inconsistency with QED at high energies, but more conservatively may merely indicate a collapse of the perturbative approximation at high energies. Even if each term in the perturbative expansion of QED can be made finite and well-defined (which at present requires *ad hoc* constructions like 'infrared cut-offs'), the full sum over all perturbative orders probably won't converge in any sense. Indeed, the perturbative expansion is a power series in the coupling constant e; if it converged for some small (positive) value of e, then it should also converge for some negative values of e, which for physical reasons is impossible. More generally, many suspect that a consistent quantum field theory must be 'asymptotically free' (i.e. the particles act as if they are free of interactions when the momenta are large). QED is not asymptotically free, but the Standard Model is. However, the Standard Model has other problems (due to the Higgs scalar field) and many suspect that it too is inconsistent.

value m_ϕ of the physical mass. This is an example of *renormalisation*, and in itself is a standard and uncontroversial ingredient in any physical theory.

However, the mass shift δm can be calculated perturbatively, and in a typical quantum field theory is infinite. Thus in order to account for the observed masses of the particles, the mass parameters in the Lagrangian would also be infinite, which is silly. Nevertheless, the renormalisation scheme given in the previous paragraph works to give sensible and accurate answers.

Likewise, the fields ϕ and coupling constants λ – in short, everything! – appearing in the Lagrangian are also unobservable. The coupling constants λ are renormalised analogously to mass, using the observed strengths of the corresponding interaction, and as usual the rescaling is by an infinite factor. The physical 'renormalised' fields, properly interpolating between the incoming and outgoing free fields, are scalar multiples $Z_\phi^{1/2}\phi$ of the Lagrangian 'bare' fields. This follows, for example, by the residue (call it Z_ϕi) of the propagator: it must equal i, but in a theory with interactions we'll have $Z_\phi \neq 1$ (in fact typically Z_ϕ is infinite). In short, the equal-time commutation relation (4.2.9a) (obeyed by the bare fields) and the residue i of the propagator (necessarily satisfied by the physical fields) are incompatible, and so the bare fields aren't physical. Once again it is not surprising that we must renormalise; what is disturbing is that the renormalisation is infinite.

Quantum field theory makes sense of (i.e. systematically removes) the infinities arising in perturbation theory by a combination of two procedures. The first, called regularisation (Section 4.2.3), introduces some new parameter, call it Λ, and replaces the divergent quantity by a limit as Λ goes to ∞, say, of finite quantities. This nonphysical parameter Λ may be a large momentum cutoff (which corresponds to a small distance cutoff), although more sophisticated cutoffs are common. As long as Λ is finite, the calculation will also be finite, but it will depend on Λ (as well as the various parameters $m, \lambda, \ldots$ in the Lagrangian). However, if we choose ('renormalise') those parameters $m, \lambda, \ldots$ so as to depend on Λ in such a way that the physically relevant quantities are independent of Λ (or at least have a finite limit), we can then take the limit $\Lambda \to \infty$ and get a sensible answer (even though the bare parameters $m, \lambda, \ldots$ will diverge in that limit). We then take those 'sensible answers' to be the predictions of the theory.

In order to remove all infinities, it may be necessary to introduce new bare parameters by adding new terms to $\mathcal{L}$. A quantum field theory is called *renormalisable* if this procedure terminates, that is if all Feynman diagrams will be finite after introducing only finitely many regularisors Λ_i and renormalising the finitely many Lagrangian parameters appropriately. The ϕ^4 model, QED and the Standard Model are all renormalisable. On the other hand, a quantum field theory for gravity in four dimensions, in the spirit of general relativity, is doomed to be nonrenormalisable. Renormalisability is a strong constraint on a theory – for example, it forbids fields with high spin and interaction terms involving many derivatives or products of many fields. For example, the only interaction terms allowed in the Lagrangian of a renormalisable four-dimensional quantum field theory of a single self-adjoint scalar ϕ are ϕ^i for $1 \leq i \leq 4$ and $\sum \partial_\mu \phi \, \partial^\mu \phi$.

A nonrenormalisable theory can always be renormalised (i.e. its divergences all removed) by adding infinitely many new terms to the Lagrangian (along with infinitely many regularisors Λ_i). The problem is that to fix the renormalised values of all those new coupling constants, we would need to perform infinitely many experiments. It would thus appear (and is often argued) that renormalisability would be a necessary condition for a physically relevant, predictive quantum field theory. Such a nonrenormalisable theory would display behaviour that is sensitive to the detailed structure at a much more microscopic level. This behaviour would appear random at the scale on which we are trying to focus. For a macroscopic example, consider the propagation of cracks in glass.

On the other hand, it is possible that all but finitely many of those new parameters will arise in perturbation terms that will be insignificant until the energies of the particles are sufficiently large (e.g. they could involve new particles with very large masses). That is, the contributions from all but finitely many of those parameters could be exponentially suppressed and thus be ignored. Such a theory would be essentially predictive as long as we kept the energies of the collisions far less than the masses of these new and irrelevant particles. Such a nonrenormalisable theory would describe the low energy limit of a more fundamental theory – its nonrenormalisability arises because there is pertinent physics that is not yet accounted for, which occurs at a smaller, deeper scale. For example, quantum gravity could be the low-energy limit of string theory.

In other words, nonrenormalisability could be the norm, as presumably all of our theories are merely limits of deeper ones. A renormalisable theory is merely one in which the deeper physics involves a much higher energy scale (equivalently, a smaller distance scale) than the ones attained in our present experiments. It is a happy accident that the Standard Model is renormalisable. For example, QED applied to a hydrogen atom (an electron moving about a proton) is renormalisable, but is nonrenormalisable when applied instead to a deuteron (an electron moving about a proton–neutron nucleus). The difference is that the physics describing the single proton concerns much smaller distances (approximately 10^{-13} cm) and higher energies than that describing the electron's motion in hydrogen (which involves distances on the order of 10^{-8} cm), while the physics describing the deuteron nucleus also occurs at roughly the same 10^{-8} cm scale.

On a conceptual level, this renormalisation scheme is clearly unsatisfactory. The infinities appearing throughout renormalisation tell us that the fields and parameters appearing in $\mathcal{L}$ are not only nonphysical, but are also nonmathematical. The former is not surprising; the latter gives powerful evidence that the Lagrangian approach to quantum field theory should be avoided. Nevertheless, it works: not only does it permit unambiguous numerical predictions from the Standard Model, but those predictions match up admirably with experiment.

It is easy to get the impression that, whatever its value may be to the pragmatic working physicist, renormalisation should best be avoided by the much more delicately disposed mathematician. Indeed much effort, though with comparatively little success, has been directed at nonperturbative quantum field theory. However, there are many situations where the mathematics arising in perturbation is fascinating. For example, the modular forms arising in string theory, and the Riemann surfaces of conformal field theory, arise directly in the perturbation expansion of string theory. Kreimer, Broadhurst and Connes (see [105], [361] and references therein) are studying the knot theoretic, Hopf algebraic and number theoretic structure arising in perturbative quantum field theory. Perturbative Chern–Simons theories give both Vassiliev link invariants [**38**] and Gromov–Witten invariants (see e.g. the review [**403**]), depending on how it is perturbatively expanded. We know that what we call perturbative quantum field theory has direct relevance to both mathematics and physics; what hasn't been worked out yet in a conceptually satisfying manner is its precise relationship with 'true' quantum field theory (whatever that is).

This relationship is still mysterious after half a century of work. But recall that Newton's calculus took well over a century to make *mathematical* sense, even though it gave good physics from the beginning. Dirac's use of his delta functions was a much humbler example, but still took several years before Schwartz mathematically legitimised them as distributions. Attempts to make direct sense of quantum field theory are discussed in Section 4.2.4. We are not merely discussing here the rigorous proof of physical conjectures that are almost certainly true – the importance of that activity is easy to overestimate. Rather, we are speaking of making coherent, of finding the meaning of, quantum field theory. There have also been several proposals for a new mathematics underlying quantum field theory. For example, we have the Barrett and Crane interpretation of Feynman diagrams as morphisms in a tensor category (dynamics here comes from representations of the Poincaré group thought of as a 2-category), or Connes' noncommutative geometry (where the geometry of space-time is replaced with an algebra of functions). Some of these approaches are discussed in [**28**].

Of course quantum field theory cannot be identified with perturbative quantum field theory. There are important nonperturbative effects, which cannot be seen in the perturbative expansion. Typical examples are quantum effects due to topologically nontrivial extended solutions to the classical field theory, such as magnetic monopoles (particles carrying magnetic charge) and instantons (solutions concentrated near a point in space-time rather than along a world-line as happens for particles).

There are other challenges to the coherence of quantum field theory as it is practised today. A famous example is *Haag's Theorem* (1955), which is rigorously proved in the context of the Wightman axioms (see e.g. [**518**]). It says that, given the assumptions built into the picture of quantum field theory sketched above, the S-matrix is very ill-defined unless the theory is free (which isn't physically interesting). We know (Theorem 2.4.2) that there is a unique irreducible unitary representation of the finite-dimensional Heisenberg algebras, but this breaks down for infinite-dimensional ones (Question 2.4.2). Thanks to the equal-time commutation relations (4.2.9), the space-smeared fields $\varphi_j(f)$ of a quantum field theory define at each time t a unitary representation of an infinite-dimensional Heisenberg algebra (just use countably many test functions f with disjoint support). For a fixed quantum field theory, the representations at different times t are unitarily equivalent via the time-evolution operator $U(t) := e^{-iHt/\hbar}$, so each theory defines a unique fixed representation. Haag's Theorem tells us that the representations for different values of the coupling constant will be equivalent only if the theories are equivalent. So if our theory is nontrivial, its Heisenberg representation will be different from that of the free theory, that is from that of our so-called asymptotic $t \to \pm\infty$ theories. Thus the limits $U(\pm\infty)$ can't be well defined, and *the justification* *for quantum field theory as interpolating between incoming and outgoing states must be dropped* (or at least seriously weakened).

One escape is to throw away the equal-time commutation relations (Section 4.2.4). After all, we know that the renormalised (physical) fields won't satisfy them. Also, it seems highly dubious to claim that (4.2.9) are physically relevant, if (4.2.9) permits us to smear fields only in the space direction. We should also smear in the time direction, which means we can no longer speak of *equal-time* relations and the simplicity of (4.2.9) will be lost. On the other hand, (4.2.9) are important, for example, for the usual interpretation of the number operator, and hence are central to the particle interpretation.

The attitude taken by most practitioners of quantum field theory towards these various mathematical difficulties is much like that taken by the author of this book towards most of Life's Little Crises: avoidance. 'Tomorrow they may just go away.' After all, this strategy worked fine with those monsters haunting the night-time shadows of our childhood.

There are formal similarities between quantum field theory and (classical) statistical mechanics. More precisely, path integral expressions in quantum field theory in d-dimensional space-time are the same as, or at least analogous to, thermal averages in statistical mechanics in $d + 1$ space-time dimensions, when the time t is replaced by $-ik/T$ where k is Boltzmann's constant and T is the temperature. The weak coupling limit in quantum field theory corresponds to the high-temperature limit. Quantum fluctuations about a classical solution correspond to statistical fluctuations about a thermodynamic equilibrium. We won't have much more to say about this connection, though it has been extremely fruitful. For example, spontaneous symmetry breaking in the Standard Model, needed to give masses to particles like the electron, is a phase transition. The Klein–Gordon equation, governing as we know scalar fields, also describes excitations of a dense plasma, or of vortex motions in liquid helium. Conformal field theories, as we shall see next section, can arise both from quantum field theories (string theories) and from statistical mechanics. Incidentally, the transition to imaginary time has an important place in quantum field theory, where it is called 'Wick rotation', and is related to the holomorphicity of the Wightman functions discussed in Section 4.2.4.

The operators in both classical and quantum mechanics form an algebra. This cannot be directly true in quantum field theory, because the product of distributions is not usually a distribution. *It does not make mathematical sense to multiply fields $\varphi_1(x)$, $\varphi_2(y)$ at the same space-time point $x = y$.* Nevertheless, the Lagrangian density, as well as the equal-time commutation relations and many other familiar expressions in quantum field theory, do precisely that. Kenneth Wilson proposed the *operator product expansion* (OPE) as a way to make sense of this. As it is a standard tool of conformal field theory, we defer its treatment to Section 4.3.2. Wilson intended this OPE to be an alternative to the problematic (4.2.9), but as too often happens, his attempt at reformation was absorbed into The System and has become one of its standard tools. The other way to make the operators into an algebra is to smear them, and that is the approach taken by Wightman.

Modern quantum field theory is based on the notion of a *gauge symmetry*. To help understand this important concept, consider the following toy model: a two-dimensional classical particle $(x(t), y(t))$, with equations of motion

$$\frac{d^2}{dt^2}x(t) + u(t)x(t) = 0 = \frac{d^2}{dt^2}y(t) + v(t)y(t), \qquad (4.2.14a)$$

for some fixed functions u, v. Writing $z = x + iy$ and $w = u + iv$, this becomes the simpler

$$\frac{d^2}{dt^2}z(t) + w(t)z(t) = 0. \qquad (4.2.14b)$$

Of course, this system has a $U_1(\mathbb{C})$ symmetry, corresponding to a rotation of the z-plane: for any fixed $e^{i\theta} \in U_1(\mathbb{C})$, $z(t)$ is a solution of (4.2.14b) iff $e^{i\theta}z(t)$ is a solution. We call this a *global* (as opposed to *local*) symmetry, because $e^{i\theta}$ must be constant if it is to define a symmetry of (4.2.14b). However, we can rewrite our system so that $U_1(\mathbb{C})$ becomes a *local* (time-dependent) symmetry. Introduce a function $A(t)$ (which will serve as a book-keeping or compensating device) and replace each derivative d/dt in (4.2.14b) with the differential operator $d/dt - iA(t)$, so (4.2.14b) becomes

$$\left(\frac{d}{dt} - iA(t)\right)\left(\frac{d}{dt} - iA(t)\right)z(t) + w(t)z(t) = 0. \qquad (4.2.14c)$$

This system (4.2.14c) has a *local* $U_1(\mathbb{C})$ symmetry: for any smooth function $\theta : \mathbb{R} \to U_1(\mathbb{C})$, $(z(t), A(t))$ is a solution to (4.2.14c) iff $(e^{i\theta(t)}z(t), A(t) + \frac{d}{dt}\theta(t))$ is a solution to (4.2.14c). Physically, this local symmetry corresponds to the freedom of rotating the system (or the observer) differently at each moment of time. We know from elementary physics that doing this requires introducing the centrifugal forces intimate to all amusement park aficionados. Indeed, we can think of (4.2.14c) as being the equation of motion of a particle z under the influence of a new external force described by A, in addition to the original force described by w. *This is the origin of the 'new external force' A.*

For historical reasons, local symmetries such as the $U_1(\mathbb{C})$ of (4.2.14c) are called 'gauge symmetries' (gauge here means calibration or scaling). What is significant here is that 'gauging' a global symmetry associates with it a new force; changing the gauge (e.g. rotating the z-plane) is indistinguishable from the action of an apparent force (e.g. a centrifugal one). In the trivial example given above, the force is globally 'fictitious' and the gauging process (4.2.14b) $\to$ (4.2.14c) involves no new physics, since we can always solve $A(t) + \dot{\theta}(t) = 0$ for θ and thus 'gauge away' the force A.

Remarkably, all fundamental forces in Nature (namely, gravity, electromagnetism, and the strong and weak nuclear forces) can be obtained by gauging a global symmetry. Consider first special relativity (Section 4.1.2) and for simplicity a single free particle $x(t)$. There, the Poincaré group acts as a global symmetry. It says that the laws of physics shouldn't depend on the choice of origin and inertial observer (coordinate axes). It is a global symmetry, in the sense that once those two choices are made, all observers (regardless of the space-time point x they animate) must agree to use that same origin and coordinate axes in comparing their observations, in order to have a symmetry. This rigidity, this global collaboration, seems physically artificial. What happens if we gauge this symmetry? That is, permit each observer (i.e. each space-time point) to independently choose an origin and coordinate axes. What does that anarchy mean for our description of the relativistic particle? Simply that its coordinates will have changed: $x(t) \mapsto x'(t) = \alpha(x(t))$ where $\alpha : \mathbb{R}^{3,1} \to \mathbb{R}^{3,1}$ encapsulates our new gauge. We require this global change of variables to be invertible, that is to be a diffeomorphism of Minkowski space. So our choice of gauge reduces to a choice of diffeomorphism α. Making the equation of motion independent of that choice α requires introducing book-keeping functions, $A^\lambda_{\mu\nu}$, so that the original equation of motion $d^2 x^\lambda / dt^2 = 0$ becomes

$$\frac{d^2 x'^\lambda}{dt^2} - \sum_{\mu,\nu} A^\lambda_{\mu\nu} \frac{dx'^\mu}{dt} \frac{dx'^\nu}{dt} = 0.$$

Requiring this equation to be equivalent to the original one, we recognise that the components $A^\lambda_{\mu\nu}$ are (up to a sign) none other than the Christoffel symbols $\Gamma^\lambda_{\mu\nu}$, and that the equation of motion is simply the geodesic equation. The new force corresponding to these A's is identified by Einstein's equivalence principle with gravity. The question of whether gravity can be 'gauged away', that is whether it is globally fictitious and our calculations have been merely a formal mathematical game, reduces to the question of whether space-time is globally flat. It is here – allowing for the suddenly natural possibility that space-time is not flat – that new physics enters. *The real purpose of gauging the symmetry of Minkowski space-time* (Einstein's requirement of 'general covariance') *was to lead us to the idea of curved space-time and the associated force* (which by independent reasoning we identify with gravity). More generally, gauging is a guide for introducing a new force into a theory with a global symmetry: the so-called *principle of minimal interactions*.

Gauging works similarly in quantum field theory. QED results from gauging the $U_1(\mathbb{C})$ symmetry of free theories. The global U_1 symmetry, $\psi(x) \mapsto e^{i\theta} \psi(x)$, corresponds to the ambiguity of defining the phase of, for example, the electron field ψ. Once we make the choice at one space-time point, then we must be consistent at all other points. Incidentally, that global symmetry leads to the conservation of global electric charge, by Noether's Theorem. Gauging it means the phase can be changed arbitrarily at each point, that is θ can depend on x. The associated book-keeping field $A_\mu(x)$ corresponds to the force we call electromagnetism, and the gauge symmetry implies *local* conservation of charge. For example, in the case of a charged scalar particle, the Klein–Gordon equation (4.2.7) gauges to

$$\sum_{\mu,\nu} \eta^{\mu\nu} (\partial_\mu - i A_\mu)(\partial_\nu - i A_\nu)\phi - m^2 \phi = 0.$$

It is straightforward to construct a Lagrangian from the original (free) one, which yields the new equations of motion: for example, the free Lagrangian $\sum (\partial_\mu \phi^\dagger)(\partial^\mu \phi) + m^2 \phi^\dagger \phi$ for a scalar field with charge e yields

$$\sum_\mu (\partial_\mu + i e A_\mu)\phi^\dagger (\partial^\mu + i e A^\mu)\phi + m^2 \phi^\dagger \phi.$$

But how should we think of A_μ? As another elementary field in the theory. But that means we should add a new term to the gauged Lagrangian, containing partial derivatives of A (otherwise the Euler–Lagrange equations (4.1.8) would be trivial). The simplest gauge-invariant, Lorentz-invariant way to do this is (4.1.13) (with $V = 0$),where $F_{\mu\nu} = \partial_\mu A_\nu - \partial_\nu A_\mu$ is called the field strength. This is the correct Lagrangian describing the QED of a charged scalar particle. Changing the gauge is indistinguishable from the matter field moving through an electromagnetic field. The associated perturbation theory involves, in Feynman's language, the exchange of virtual particles associated with this new A_μ field – those new particles are called photons.

General relativity tells us to expect a geometric picture here, and indeed that is the case. We think of the matter fields as being sections of a fibre bundle with base $\mathbb{R}^{3,1}$ and fibre $U_1(\mathbb{C})$; the electromagnetic field A_μ defines a connection for this bundle and $F_{\mu\nu}$ is the curvature tensor.

Similarly, the Standard Model is a gauge theory associated with the gauge group $SU_3(\mathbb{C}) \times SU_2(\mathbb{C}) \times U_1(\mathbb{C})$. SU_3 here corresponds to the strong nuclear force, responsible, for example, for the binding of quarks together to form protons and neutrons, and the binding of protons and neutrons together to form nuclei. $SU_2 \times U_1$ describes a unification of electromagnetism with the weak nuclear force (which describes, for example, the decay of the neutron). What this symmetry group $SU_3 \times SU_2 \times U_1$ means physically is less clear than it was for general relativity (or QED), and so the Standard Model lacks the conceptual clarity of Einstein's masterpiece. For example, many believe a deeper quantum field theory will involve a larger gauge group, such as E_6.

Describing other important ingredients of the Standard Model – the fundamental fields and how they transform under $SU_3 \times SU_2 \times U_1$ – would drag us even further from the main thread of this book. For detailed treatments of the Standard Model see, for example, [310], [555]. Although its comparison with experiment has been fabulous, it is surely not the 'final theory'. For one thing, it suffers from all the conceptual and mathematical flaws mentioned in this subsection. Also, it has 18 free parameters – for example, the electron mass – which must be experimentally determined and (depending on how one counts) there are 61 'elementary' particles in the theory. The Standard Model is an effective theory, valid only for a relatively narrow range of physics. The question is, how different from it will the theory superseding it look?

Quantum field theory challenges our concept of matter. In Newtonian physics reality obtained its solid objective structure from an inert unanalysable 'stuff', from which all substance came; though it could change form (e.g. ice to water), it was the clay on which the Laws of Physics acted. As we moved into the twentieth century we learned that this clay could be transformed into energy ('$E = mc^2$'), and that it is composed of atoms that are mostly empty space. Quantum field theory goes a step beyond: the particles composing atoms are to empty space like sound waves are to air. Bertrand Russell was more accurate than he thought when, in 1956, he compared matter to Lewis Carroll's Cheshire Cat which gradually faded until nothing was left but the grin – matter's grin, Russel speculated, was caused by amusement at those who still think it's there.

Likewise, our notion of force has changed from Newton's definition $\mathbf{F} = m\mathbf{a}$, to something that more generally changes the state of a particle, and that is due not to an active agent but to an indirect effect like a well-hidden symmetry – a further movement of physics away from the prerelativistic infatuation with intuitive space and time.

4.2.3 *The meaning of regularisation*

The mathematics of classical physics (symplectic geometry) is well understood, while that of quantum field theory isn't. But it's already clear that, mathematically speaking, quantum field theory is by far the more profound. Much as mechanics helped develop calculus, our standard tool for studying finite-dimensional systems, we can expect quantum field theory to supply us one day with sophisticated new tools for studying infinite dimensions. We are already seeing hints of this.

To a theoretical physicist, quantum field theory is a recipe book, an infinite sequence of finite calculations. To a mathematician, these recipes seem *ad hoc*, and surprisingly classical and finite-dimensional for something that is emphatically neither. A hundred years from now we'll look back at that recipe book much as a modern doctor reflects on medieval medicine: this herb is antiseptic, that incantation is mostly harmless, but leeches and blood-letting were simply bad ideas.

Of all these recipes, those connected with renormalisation and regularisation generate the most ire. For example, even mathematical stoics cannot be unmoved by the substitution (2.3.1). Yet it is in these places where most of the magic lives, as for example the derivation of the Atiyah–Singer Index Theorem from anomaly cancellation indicates.

It isn't difficult for a mathematician to appreciate the inevitability of some form of renormalisation. Consider, for example, the two-body Lagrangian

$$L = \frac{1}{2}m_1\dot{\mathbf{x}}_1^2 + \frac{1}{2}m_2\dot{\mathbf{x}}_2^2 + G\frac{m_1m_2}{|\mathbf{x}_1 - \mathbf{x}_2|}. \tag{4.2.15a}$$

We can integrate out one of the particles, since the centre-of-mass $m_1\mathbf{x}_1 + m_2\mathbf{x}_2$ is constant (without loss of generality, say it equals $\mathbf{0}$). The resulting one-particle system is

$$L = \frac{1}{2}m\dot{\mathbf{x}}^2 + \frac{k}{|\mathbf{x}|}, \tag{4.2.15b}$$

where $m = m_1(m_1 + m_2)/m_2$ and $k = Gm_1m_2^2/|m_2 + m_1|$. We say that the mass and coupling constants – the 'bare' parameters in (4.2.15a) – have been 'renormalised'.

Something similar happens whenever we integrate away degrees-of-freedom, or account for some effect (e.g. the unavoidable geometric series in Figure 4.10): the new parameters will be readjusted or *renormalised* compared to the old ones. This is completely noncontroversial. What is disturbing about renormalisation in quantum field theory is that you are asked to add/subtract/multiply/divide *infinite* quantities. Regularisation is the procedure of obtaining precise numbers from such an ill-defined operation.

In some sense, regularisation also arises in mathematics. We see it in our Dedekind eta calculation in (2.2.9), or the Virasoro action on affine algebra modules in (3.2.13). Sometimes analytic concerns become significant (e.g. the natural integrals or series one would naively write down turn out to diverge). If those concerns are ignored, we obtain incorrect answers (such as $\eta(-1/\tau) = \eta(\tau)$, or an action of the Witt algebra on affine algebra modules). Of course what we must do is go back and do the analysis properly. Regularisation is merely a symptom of sloppy analysis. It isn't supposed to be the place where the magic appears. The magic was there all along. But the penalty of pretending that (semi-)classical calculations can capture quantum field theory is the introduction of regularisation schemes. The classical calculations fail to pick up that magic, which is then forced to arise in that final step. It's like trying to straighten a Möbius band: as you move your hand around the strip, trying to keep the paper vertical, the twist is relegated to a smaller and smaller portion of paper until eventually the paper tears. That tear is called regularisation. The problem isn't inherent to quantum field theory, the problem is with the fantasy that we can treat quantum field theory semi-classically.

Feynman once asked why the same tricks work over and over in physics. Regularisation is Nature's way of telling us that they don't quite. Unfortunately, we don't yet know how to go back and do the quantum field theory calculations properly. But regularisation must supply some deep hints. For instance, the presence of infinite renormalisation seems to suggest that quantum field theory should be formulated without Lagrangians. Perhaps another hint is that the point ∞ is the difference between the (Riemann) sphere and the (complex) plane, suggesting that regularisation can be interpreted as a (global) topological effect. In [105], [106], a projective limit of certain Lie groups, corresponding to the Hopf algebra of Feynman graphs, acts on the coupling constants of renormalisable quantum field theories, and contains the renormalisation group as a one-parameter subgroup; dimensional regularisation can in some theories be interpreted as the index theorem in noncommutative geometry.

4.2.4 Mathematical formulations of quantum field theory

Making rigorous sense of quantum field theory is very difficult, as several comments made earlier should indicate. Even the free theories are very subtle; theories with interactions are filled with unresolved problems (Section 4.2.2). One thing is clear: quantum field theory as it is typically practised today (i.e. the informal theory) is mathematically incoherent.

However, quantum field theory *is* a part of mathematics in the sense that important aspects of it have been encoded axiomatically and several examples (mathematically if not physically interesting) have been rigorously constructed. Mathematicians under-appreciate just how accessible quantum field theory is. The purpose of this subsection is to briefly describe two of the most influential of these mathematical treatments. These lead to two different formulations of conformal field theories, which we study in later chapters. The fundamental difficulty in the subject lies in rigorously constructing nontrivial examples of quantum field theories within these formulations. Only the very simplest theories (e.g. the free ones) have been rigorously constructed.

The simplest and best-known mathematical treatment of quantum field theory, the Wightman axioms [518], was first formulated in the 1950s by Gårding and Wightman. Lagrangians and the equal-time commutation relations (4.2.9) are avoided, and instead attention is focused on the interpolating renormalised 'physical' fields. This makes rigour much easier to attain, but contact with the particle interpretation is more difficult. One unexpected gain is the holomorphicity of the vacuum-to-vacuum expectation values.

According to Wightman, a quantum field theory consists of the data collected in the following seven axioms w.I–w.VII. For convenience, put $c = \hbar = 1$. Naturally, there is much overlap with the preceding material – the main clarification provided here is what from Section 4.2.2 can (and should?) be avoided.

w.I. (*relativistic state space*) Let $\mathcal{H}$ be a separable Hilbert space, carrying a continuous unitary representation $U_{(a,\Lambda)}$ of the universal cover $\mathbb{R}^4 \rtimes \mathrm{SL}_2(\mathbb{C})$ of the Poincaré group. Define the self-adjoint operators P^μ by $U_{(a,I)} = \exp[i \sum_\mu P^\mu a_\mu]$; they mutually commute so we can speak of simultaneous eigenstates. All the (simultaneous) eigenvalues p^μ of P^μ are required to satisfy the conditions $p^4 \geq 0$ and $\sum_\mu p_\mu p^\mu \leq 0$.

w.II. (*vacuum*) There is a state $|0\rangle \in \mathcal{H}$, unique up to scalar multiple, invariant under all $U_{(a,\Lambda)}$.

w.III. (*fields*) There is a space $\mathcal{D} \subset \mathcal{H}$, dense in $\mathcal{H}$ and containing $|0\rangle$. There are a finite number $\varphi_1, \ldots, \varphi_M$ of operator-valued tempered distributions over space-time $\mathbb{R}^4$, such that for any 'test function' $f \in \mathcal{S}(\mathbb{R}^4)$, each $\varphi_i(f)$ is an operator from $\mathcal{D}$ to $\mathcal{D}$. The set of fields φ_i is closed under adjoint (i.e. $\varphi_i^\dagger$ equals some φ_j).

w.IV. (*covariance of fields*) For all $(a, \Lambda) \in \mathbb{R}^4 \rtimes \mathrm{SL}_2$, $U_{(a,\Lambda)}(\mathcal{D}) = \mathcal{D}$. Equation (4.2.8a) holds in $\mathcal{D}$, and so the matrices $V(\Lambda)$ define an M-dimensional $\mathrm{SL}_2(\mathbb{C})$-representation.

Physically, the vectors in $\mathcal{H}$ (or rather the rays) are interpreted as the possible states of the theory, and the φ_i are the (renormalised interpolating) quantum fields. We discuss tempered distributions and the Schwartz space $\mathcal{S}$ in Section 1.3.1, and the Poincaré and Lorentz groups and their doubles in Section 4.1.2. If there are any other symmetries of the theory, then $\mathcal{H}$ will also carry a unitary projective representation of those groups. The energy–momentum operators P^μ, generating space-time translations, exist because of the assumed unitarity of the U's. They mutually commute because their exponentiations $U_{(a,I)}$ do. Up to a factor of c^2, the eigenvalue p^4 is the energy of the state and $\sqrt{-\sum p_\mu p^\mu}$ its mass m. We call the vector $|0\rangle \in \mathcal{D}$ in w.II the *vacuum*, and normalise it so that $\langle 0|0\rangle = 1$.

Postulating a common domain $\mathcal{D}$ is necessary because (Section 1.3.1) unbounded operators on a Hilbert space aren't defined everywhere (think of differentiation on the space of square-integrable functions $L^2(\mathbb{R})$). We see from w.III that $\mathcal{D}$ certainly contains the vectors obtained from the vacuum $|0\rangle$ by applying all polynomials in the smeared fields $\varphi_i(f)$, and we learn in w.VI below that those vectors $p(\varphi(f))|0\rangle$ are indeed dense in $\mathcal{H}$. To some approximation, $\mathcal{D}$ can be identified with that subspace (see page 98 of [**518**]).

w.V. (*local commutativity*) For any pair of test functions $f, g \in \mathcal{S}(\mathbb{R}^4)$ satisfying $f(x) g(x) = 0$ whenever $(x - y)^2 \geq 0$ (in other words, the supports of f and g are space-like separated), then for any fields φ_i, φ_j, a sign $\pm$ (depending on i, j) can be chosen so that on $\mathcal{D}$

$$[\varphi_i(f), \varphi_j(g)]_\pm := \varphi_i(f) \varphi_j(g) \pm \varphi_j(g) \varphi_i(f) = 0.$$

w.vi. (*completeness*)   The vacuum is cyclic for the smeared fields. That is, polynomials in the smeared fields $\varphi_i(f)$, applied to the vacuum $|0\rangle$, form a subspace dense in $\mathcal{H}$.

Completeness w.vi implies irreducibility of the smeared field operators, in the following sense (inspired by Schur's Lemma): if $B : \mathcal{D} \to \mathcal{D}$ is a bounded operator satisfying

$$\langle u, B\varphi_i(f)v \rangle = \langle \varphi_i(f)^* u, Bv \rangle, \qquad \forall u, v \in \mathcal{D}, \ \forall f \in \mathcal{S}(\mathbb{R}^4), \ \forall i = 1, \ldots, M$$

(so in this weak sense B commutes with all φ_i), then B is a constant multiple of the identity. Completeness corresponds here to the remark in Section 4.2.2 that any operator in the theory can be expressed as a function of the smeared fields.

Physically, local commutativity w.v concerns the quantum mechanical fact that measurements localised at space-time points x and y should commute (i.e. be simultaneously measurable without mutual interference) when x and y are space-like separated. It is a consequence of the axioms which sign to take, as is discussed below.

A final axiom is needed to make content with particles (that is to say, with experiment). As it is more technical, it is often avoided in treatments of Wightman's axioms, and we too will be sketchy. The basic idea is that any single particle state $|\lambda\rangle \in \mathcal{H}$ (as usual, $\lambda = \lambda(p)$ describes the decomposition of the state into momentum eigenstates $|p\rangle$) will be an eigenvector for the operator $\sum_\mu P^\mu P_\mu$, with eigenvalue $-m^2 c^2$ independent of λ (m is the mass of the particle). On the other hand, eigenstates $|\lambda_1, \ldots, \lambda_n\rangle$ of $\sum P^\mu P_\mu$ corresponding to $n > 1$ particles will have eigenvalue varying continuously with the λ_i. In other words, considering the spectral decomposition of the self-adjoint operator $\sum P^\mu P_\mu$ in $\mathcal{H}$, the single particle states $|\lambda\rangle$ correspond to the discrete part of the spectrum. Call $\mathcal{H}^{(1)}$ the Hilbert space they span – it is a proper subspace of $\mathcal{H}$. There need be no direct relation between the number of elementary fields φ_i and the types of single particles. For example, in the Standard Model quarks correspond to elementary fields but not particles, and protons are particles without a corresponding elementary field. We can now construct incoming $|\lambda_1, \ldots, \lambda_n\rangle^{\text{in}}$ and outgoing $|\lambda_1, \ldots, \lambda_n\rangle^{\text{out}}$ n particle states, corresponding in the $t \to \mp\infty$ limits to tensor products $|\lambda_1\rangle \otimes \cdots \otimes |\lambda_n\rangle$ – see section II.V of [269] for the detailed construction. Then the final axiom is:

w.vii. (*asymptotic completeness*)   The incoming particle states $|\lambda_1, \ldots, \lambda_n\rangle^{\text{in}}$ topologically span $\mathcal{H}$, as do the outgoing particle states $|\lambda_1, \ldots, \lambda_n\rangle^{\text{out}}$.

Unfortunately, this treatment requires all particles in the theory to have nonzero mass, and so isn't realistic. For example, in quantum electrodynamics the photon is massless and the electron is always surrounded by a cloud of photons, so the single electron states don't belong to a discrete eigenspace of the operator $\sum P^\mu P_\mu$, but rather the eigenvalue varies continuously with upper bound $-m^2 c^2$ corresponding to the mass of the electron. For a more sophisticated treatment of the particle concept within quantum field theory, see chapter VI in [269].

The role of the n-point functions (4.2.11) are played here by the *Wightman functions*, which are also vacuum-to-vacuum expectation values but aren't time-ordered. Let

$\varphi_{i_1}, \ldots, \varphi_{i_n}$ be n fields, not necessarily distinct. Define W_n to be the inner-product

$$W_n(x_1, \ldots, x_n) := W_{\varphi_{i_1}, \ldots, \varphi_{i_n}}(x_1, \ldots, x_n) := \langle 0|\varphi_{i_1}(x_1) \cdots \varphi_{i_n}(x_n)|0\rangle.$$

Of course, to make sense of this expression we must smear the points x_i, that is, replace them with test functions f_i. Thus W_n is a complex-valued function of $\mathcal{S}(\mathbb{R}^4) \times \cdots \times \mathcal{S}(\mathbb{R}^4)$. Thanks to Schwartz's Nuclear Theorem, W_n has a unique extension to a tempered distribution on $\mathcal{S}(\mathbb{R}^{4n})$, and it is this extension that is studied. Nevertheless, the inaccurate and occasionally misleading notation $W_n(x_1, \ldots, x_n)$ is too standard to change.

It is possible to convert the data and properties in w.I–w.VII into constraints on the Wightman functions. For example, the relativistic invariance of the vacuum leads to the expression, valid for any (Λ, a),

$$\sum_{j_1, \ldots, j_n=1}^{M} V_{i_1 j_1}(\Lambda) \cdots V_{i_n j_n}(\Lambda) \, W_{\varphi_{j_1}, \ldots, \varphi_{j_n}}(x_1, \ldots, x_n)$$
$$= W_{\varphi_{i_1}, \ldots, \varphi_{i_n}}(\Lambda x_1 + a, \ldots, \Lambda x_n + a). \qquad (4.2.16)$$

As always of course, everything should be smeared, that is evaluated at $f_i \in \mathcal{S}(\mathbb{R}^4)$ (or $f \in \mathcal{S}(\mathbb{R}^{4n})$). In its unsmeared form, (4.2.16) suggests that W_n is actually a 'generalised function' $w_n(\xi_1, \ldots, \xi_{n-1})$ of the differences $\xi_i = x_i - x_{i+1}$; the precise statement and proof for smeared W_n is given in pages 39–40 of [**518**].

A central result (due to Wightman) is the Reconstruction Theorem: these vacuum-to-vacuum functions W_n uniquely determine the quantum field theory. More precisely, if a collection of tempered distributions W_n satisfies all of the 'obvious' properties (such as the covariance (4.2.16)) that the set of all Wightman functions *should* obey, then the Hilbert space $\mathcal{H}$ and the various fields φ_i obeying axioms w.I–w.VI can be constructed, and moreover any quantum field theory realising the given Wightman functions will be equivalent to the one constructed. The general proof is notationally laborious though fairly straightforward (it is closely related to the Gel'fand–Naimark–Segal construction of a Hilbert space $\mathcal{H}_\rho$ and a representation π_ρ of a C^*-algebra $\mathcal{A}$, associated with a functional $\rho : \mathcal{A} \to \mathbb{C}$). See section 3-4 of [**518**] for the explicit statement and proof for the theory of a single free boson. The Reconstruction Theorem does not tell us when w.VII (i.e. the particle interpretation) holds.

Wightman also proved another remarkable property of his functions: each 'generalised function' $w_n(\xi_1, \ldots, \xi_{n-1})$ is the limit as $z_i \to \xi_i$ of a *holomorphic function* $w_n(z_1, \ldots, z_{n-1})$ of complex variables $z_i \in \mathbb{C}^4$. The domain of holomorphicity contains the following points: $\mathrm{Re}(z_i)$ can be arbitrary but $y_i := \mathrm{Im}(z_i)$ lies in the forward light-cone (i.e. $y_i^4 > 0$ and $y_i \cdot y_i < 0$). So the *distributions* $W_n(x_1, \ldots, x_n)$ are boundary values of the holomorphic *functions* $w_n(z_1, \ldots, z_{n-1})$. The proof of this is not difficult, and involves writing $w_n(z_1, \ldots, z_{n-1})$ as the Laplace transform of the Fourier transform of $w_n(\xi_1, \ldots, \xi_{n-1})$. Physically, this amounts to holomorphically extending from real time (i.e. the Minkowski space-time of physics) to imaginary time (i.e. Euclidean space-time, with better analytic properties).

As mentioned earlier, the choice of sign in w.v is fixed. In particular, if φ_1 and φ_2 have spins s_1 and s_2, then we take the sign $-(-1)^{(2s_1)(2s_2)}$. In small space-time dimensions alternatives to bosons and fermions are possible – see section 4.3.5 below – but these exotic possibilities are precluded here by the local commutativity axiom.

Apart from free theories, very few quantum field theories obeying the Wightman axioms have been constructed. In 1953, Thirring rigorously constructed the first inter-acting theories, but these live in two-dimensional space-time. In the 1960s and 1970s several nontrivial theories with interactions (e.g. a single scalar with ϕ^4 interaction term) were constructed in three and especially two space-time dimensions. One of the \$1 mil-lion Clay Institute problems (see http://www.claymath.org/) is to rigorously construct four-dimensional gauge quantum field theories. Quite probably there are easier ways of becoming a millionaire.

In the 1960s Haag and Kastler proposed a different axiomatic approach to quantum field theory, which although more abstract and complicated, appears to be more flexible. We will only sketch it here – see the excellent book [**269**] for a complete treatment, as well as several insights into general quantum field theory. This approach avoids fields, focusing instead on the algebra of observables – as the existence of very different-looking but equivalent field theories emphasises, it is the observables and not the fields that have a direct physical meaning. Remarkably, the entire physical content of the theory can be recovered from these algebras of observables.

Their starting point is to associate with each bounded open set $\mathcal{O}$ in space-time $\mathbb{R}^{3,1}$, a von Neumann algebra $\mathcal{A}(\mathcal{O})$ of bounded operators on a fixed Hilbert space $\mathcal{H}$. This is the same state space $\mathcal{H}$ as in the Wightman axioms, but its role here is much more minor. The self-adjoint elements in $\mathcal{A}(\mathcal{O})$ correspond to the measurements performable within the region $\mathcal{O}$, and so $\mathcal{O}_1 \subset \mathcal{O}_2$ implies $\mathcal{A}(\mathcal{O}_1) \subset \mathcal{A}(\mathcal{O}_2)$. If fields φ were present, $\mathcal{A}(\mathcal{O})$ would be obtained from polynomials in the smeared fields $\varphi(f)$, for test functions with support in $\mathcal{O}$. Conversely, one may hope to define fields $\varphi(x)$ by sending $\mathcal{O} \to \{x\}$. Thus this approach is related to that of Wightman, and it shares with the latter the near-absence of nontrivial examples.

Question 4.2.1. The nonrelativistic analogue of the Poincaré group is the Galilei group, generated by all translations $(\Delta\mathbf{x}, \Delta t)$, all rotations $R \in SO_3$ and all 'boosts' in velocity $\Delta\mathbf{v} \in \mathbb{R}^3$, as in (4.1.7b). Galilean invariance for nonrelativistic quantum mechanics says that, for any element $\alpha = (R, \Delta\mathbf{v}, \Delta\mathbf{x}, \Delta t)$ of the Galilei group, a wave-function $\psi(x)$ satisfies Schrödinger's equation (4.2.1) iff the corresponding transformed wave-function $\psi'(x')$ (whatever that is) satisfies

$$i\hbar\frac{\partial\psi'(x')}{\partial t'} = -\frac{\hbar^2}{2m}\nabla'^2\psi'(x') + V(\mathbf{x}')\,\psi'(x'),$$

where $x' = \alpha.x = (t\Delta v + R\mathbf{x} + \Delta\mathbf{x}, t + \Delta t)$ as usual. Show that the obvious trans-formation formula $\psi'(x') = \psi(x)$ (corresponding to a nonrelativistic scalar) fails here. Rather than transforming in a representation of the Galilei group, ψ must transform in a *projective* representation. Show that the transformation law $\psi'(x') = \exp[i\Delta_\alpha(x)/\hbar]\psi(x)$ works, where $\Delta_\alpha(x) = m\,(\Delta\mathbf{v})\cdot\mathbf{x} + \frac{m}{2}(\Delta\mathbf{v})^2\,t$.

Question 4.2.2. Let V_0 be a constant. Solve the one-dimensional Schrödinger equation (4.2.1) for the potential $V(x) = \begin{cases} V_0 & \text{for } -1 < x < 1 \\ 0 & \text{otherwise} \end{cases}$, with the condition that both ψ and $\partial \psi$ be continuous at $x = \pm 1$.

Question 4.2.3. (a) The vacuum $|0\rangle$ for the harmonic oscillator is the state with minimum possible energy. Find its normalised wave-function $\phi(x, t)$. (See equations (4.2.3).)
(b) Use your answer in (a) to find the average value (expectation value) $\int \psi^* \hat{x}^4 \psi$ of the observable $\hat{x}^4$ in the vacuum.
(c) Now do the same calculation using the Heisenberg picture (4.2.5): calculate the expectation value $\langle 0 | \hat{x}^4 | 0 \rangle$ using creation/annihilation operators.

Question 4.2.4. (a) In nonrelativistic quantum physics, the current density is $\mathbf{j}(x) = \frac{i}{2m} (\overline{\psi} \nabla \psi - (\nabla \overline{\psi}) \psi)$ and the probability density is $\rho(x) = |\psi(x)|^2$. Verify that they obey the equation of continuity $\partial \rho / \partial t + \nabla \cdot \mathbf{j} = 0$. (The equation of continuity says that the spatial integrals $\int \rho(\mathbf{x}, t) \, d^3 \mathbf{x}$ are independent of t.)
(b) Suppose ϕ was a wave-function obeying the Klein–Gordon equation (4.2.7). The relativistic version of $(\mathbf{j}, \rho)$ is $j^\mu(x) = \frac{i}{2m} (\overline{\phi} \, \partial^\mu \phi - (\partial^\mu \overline{\phi}) \phi)$. Verify that this obeys the relativistic equation of continuity $\sum_\mu \partial_\mu j^\mu = 0$, but that the corresponding probability density j^4 is not positive. (This is the first sickness of relativistic quantum physics based on the Klein–Gordon equation. The reason for these negative probabilities is that j^4 involves a time derivative, due to the Klein–Gordon equation being second order in time.)
(c) Verify that $\phi_k(x) = \exp[-i \sum k_\mu x^\mu]$ satisfies the Klein–Gordon equation and is also an eigenfunction of energy and momentum, provided k and m are related in a certain way. Verify that negative energy solutions to the Klein–Gordon equation do exist. (This is the second, related sickness.)

Question 4.2.5. Mathematically speaking, bounded operators are much nicer than unbounded ones. Explain why, physically speaking, we don't lose any generality restricting to bounded self-adjoint observables.

4.3 From strings to conformal field theory

In this section we introduce rational conformal field theory (RCFT), as it is known in physics. Standard references for this material are the book [131] and the review articles [239], [209], [224]. We also touch on one of its motivations: string theory. A more mathematical treatment of RCFT is provided in the following section.

We essentially identify conformal field theory (CFT) and perturbative string theory, but this is an oversimplification. For instance, a string theory exists simultaneously on several Riemann surfaces, and the corresponding amplitudes are added together. These surfaces correspond to the various terms in a perturbative expansion (a Taylor series in the string tension parameter T) of the true physical amplitudes. In string theory, the quantities for each surface are of no direct significance by themselves, any more than the term '196 884q' by itself means anything special to $SL_2(\mathbb{Z})$. In CFT, on the other hand, the Riemann surface is fixed – for example, the theory on the torus could be realised by a statistical mechanical model on the plane where the fields obey doubly-periodic boundary conditions. In fact, it is the deep connection to string theory that gave conformal field theorists the compulsion to explore their theories in arbitrary genus.

Conformal field theory and string theory have impacted remarkably on mathematics. For instance, five of the twelve Fields medals awarded in the 1990s were to men (Drinfel'd, Jones, Witten; Borcherds, Kontsevich) whose work directly concerned aspects of CFT. Probably no other structure has affected so many areas of mathematics in so short a time. Moonshine (and this book) have been deeply influenced by CFT.

The impact so far on physics has been less profound. String theory is still our best hope for a unified theory of everything, and in particular a consistent theory of quantum gravity. It goes through periods of boom and periods of bust, not unlike the breathing of a snoring drunk, and it is still too early to draw any definite conclusions.

However, recall Dirac's quote in Section 1.2.2 about the deep relation between mathematics and physics. For example, the inverse-square law ('force is proportional to $|x - y|^{-2}$') is so mathematically elegant that it must play a role in physics, at least in certain limiting situations. We see it in Newton's gravitation, and the Coulomb force between electric charges, and we now understand it to be the effective macroscopic theory associated with a massless boson in an abelian gauge theory. The same, it can be argued, should be true with string theory.[11]

4.3.1 String theory

The Standard Model describes the quantum theory of the electromagnetic, weak and strong forces. It ignores the force that to us plodding behemoths is the most blatant: *gravity*. The direct approach to quantising gravity fails: the resulting quantum field theory is easy to write down but it is nonrenormalisable and computationally useless. This strongly suggests that new physics should be entering in at high energies (= small distances). Indeed, naive calculations involving general relativity (which relates energy densities to the space-time metric) suggest that as we zoom in on space-time at distances of around 10^{-33} cm (the so-called *Planck length*), the virtual quantum oscillations will change the *topology* of space-time. Far from being a continuum (manifold), space-time at small scales would seem to be some sort of quantum foam.

Because this issue is so fundamental, there are several approaches to resolving it. One of these is string theory, which was created by accident in 1968, where it was applied to the wrong problem, and gave, it was soon realised, the wrong answers. The explosion of interest in it as a theory of quantum gravity, and everything else, began in 1984.

The electron is a *particle*, that is, it can be localised to a point. The Standard Model, say, contains several other equally fundamental particles, each distinguished by different abstract assignments (e.g. representations) attached to that point. In string theory, the

---

[11] I owe this thought to Peter Goddard.

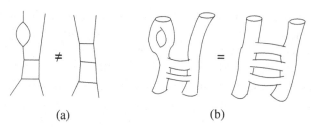

Fig. 4.11 Some two-loop Feynman diagrams of (a) particles and (b) strings.

fundamental object is a string (i.e. a finite curve of length approximately 10^{-33} cm). Depending on the particular theory, this string can be open or closed, oriented or unoriented.

There are several advantages to having extended objects. One is that the particle zoo is simplified, as those abstract assignments can be modelled geometrically using the changing shape of the string. For example, the difference between a string realising an electron, and a string realising a photon, is in how it oscillates. In place of the several dozen 'elementary particles' of the Standard Model, we have only one string, whose precise physical properties at a given time depend not only on its momentum but also its vibrational mode. Likewise, the possible interactions are simplified. Recall that to each term in a particle Lagrangian $\mathcal{L}$, we have a possible vertex for the Feynman diagrams of perturbation theory. On the other hand the interactions of strings are purely topological: for example, a single string can split into two, or two join into one. Most importantly, a theory of quantum gravity seems to arise naturally and seems far better behaved than other quantum theories of gravity.

The weary reader may wonder whether future physicists could initiate new 'revolutions' by replacing strings with membranes or other higher-dimensional manifolds. Such a reader may find some solace in the No-Go theorem described in chapter 2.1.1 of [**261**]. Nevertheless, modern string theory interprets *D-branes* (membranes where the endpoints of open strings reside) as dynamical objects in their own right, corresponding to higher-energy semi-classical solutions. Just as for low-energy approximations we study perturbations about a vacuum, for higher-energy approximations we need to study perturbations about D-branes. It is hoped (though with little justification) that together those perturbative patches cover all of parameter space.

The Lagrangian of a free particle says that the classical particle travels in such a way that its arc-length is minimised. The natural analogue for a string says that the classical string tries to minimise the area of the surface ('world-sheet') it traces out. This *Nambu–Goto action* describes what we now call the bosonic string. An equivalent formulation, called the *Polyakov action*, expresses it as an integral over moduli space.

We are interested in perturbative string theory. Recall (4.2.13d). Figure 4.11 gives some two-loop Feynman diagrams arising in the scattering of two particles/strings. As usual, we take the incoming and outgoing states to be asymptotic (this simplifies things considerably). For simplicity, make the particle theory ϕ^3 (so the diagrams are trivalent)

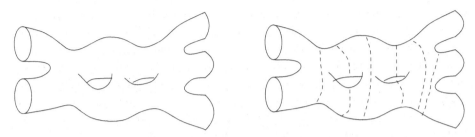

Fig. 4.12 Dissecting a surface into pairs-of-pants.

Fig. 4.13 The punctured surface corresponding to Figure 4.11(b).

and the string closed. For the particle, both diagrams in (a) would contribute a term. For the string, the equality in (b) reflects the fact that in Polyakov's formulation, conformally equivalent world-sheets correspond to the same term in the perturbative expansion, and should only be counted once. This is why the Feynman sum reduces to an integral over moduli space (in this case $\mathfrak{M}_{2,4}$).

In any quantum field theory, each vertex v contributes some operator ϑ_v to that perturbation summand. To what does this correspond in (b)? We obtain our 'vertices' by dissecting our world-sheet into spheres with three legs ('*pairs-of-pants*'), as in Figure 4.12. The operator in string theory is called a *vertex (intertwining) operator*. It is a local operator describing the absorption or emission of a string state by another. Surprisingly, these vertex operators are central to the rest of our story.

Because we're really interested in asymptotic $t \to \pm\infty$ initial/final states, the external tubes of the world-sheets are semi-infinite. We can conformally shrink those tubes into punctures (one for each incoming/outgoing string), so Figure 4.11(b) becomes Figure 4.13. The easiest example of this map is also the most important: send a cylindrical world-sheet, with local coordinates $-\infty < t < 0$ and $0 \le \theta < 2\pi$, to the complex plane using $(t, \theta) \mapsto z = e^{t-i\theta}$; then the cylinder goes to the unit disc and $t = -\infty$ corresponds to the puncture at $z = 0$. It thus suffices to consider world-sheets that are compact surfaces, with marked points indicating the external lines. The data of those external string states are stored in the appropriate vertex operator attached to that point. This is one of the remarkable features of string theory: that space-time string amplitudes (in, for example, 26 dimensions) can be expressed as correlation functions (4.2.11) in a point-particle quantum field theory in two dimensions, where the fields are vertex operators.

String theory is important to Moonshine because modular functions arise there. That amplitudes in string theory could be modular functions was known almost from the very beginning, and by 1971 we even knew the modern geometric explanation: one-loop vacuum-to-vacuum amplitudes in string theory are path integrals $\int \mathcal{Z}(\text{torus})\, d[\text{torus}]$ over conformal equivalence classes of tori; because the moduli space of tori is $\mathbb{H}/SL_2(\mathbb{Z})$ (Section 2.1.4), this makes the modularity of $\mathcal{Z}(\tau) := \mathcal{Z}(\mathbb{C}/(\mathbb{Z} + \tau\mathbb{Z}))$ manifest. The meromorphicity of the amplitudes at the cusps follows from the good behaviour ('factorisation') of the amplitudes when the surface is deformed into one with nodes (Section 2.1.4). In short, modular forms and functions appear very naturally in perturbative string theory. Elsewhere, especially Section 7.2.4, we study why this is in more depth.

The modularity (Theorem 3.2.3) of the affine algebra characters χ_λ arises from strings living on the corresponding compact simply-connected Lie group G (this is the so-called Wess–Zumino–Witten model). Likewise, quadratic moonshine (i.e. the modularity of theta functions) arises from the theory of strings living on the torus $\mathbb{R}^n/L$. There is also a string theory responsible for the modularity of the j-function (0.1.8). Much of the remainder of this book tries to explain this.

It is often argued that string theory makes no experimental predictions, other than the dimension of space, which it over-estimates by a factor of 3. This is perhaps a little unfair. String theory predicts a world qualitatively much like that we observe: a world with quantum gravity governed by Einstein's equations at the low-energy, long-distance limit, and gauge groups large enough to include the Standard Model with its zoo of particles. String theory also seems more finite than usual quantum field theories. Unlike the 18 adjustable parameters of the Standard Model, and the fairly arbitrary choices of gauge groups and particles possible in quantum field theories, there is a unique (M-)theory!

But that too is a little dishonest. There are enormous numbers of classical solutions, and each of these serves as a possible vacuum to perturb about. Each choice of vacuum corresponds to a different effective dimension of space-time, gauge group, etc. – different physics. So the problem for the perturbative approach is which vacuum to choose. This isn't so strange: the dynamic role of the vacuum is also important in the Standard Model, where the vacuum is less symmetric than the Lagrangian, and this gives rise to the masses of particles, etc. Also, we know that perturbation theory is only an approximation (probably ill-defined) to the full quantum theory, where for instance we have quantum tunnelling between different vacua. To really understand the effective physics and thus make precise experimental predictions would require a truly nonperturbative treatment of string theory, and this is difficult (D-branes are our most reliable probe for this). In fact, when we have large numbers of strongly interacting strings, the string picture probably ceases as a good way of capturing the physics. But these issues, though important for physics, don't concern Moonshine.

Whether a believer, sceptic or agnostic, one must concede that string theory is truly remarkable. To Witten, physics without strings is like mathematics without complex numbers: just as the particle traces out a real curve (its world-line), the string traces out a complex curve (its world-sheet). Standard string theory books are [261], [463].

4.3.2 Informal conformal field theory

A conformal field theory is a quantum field theory, usually on a two-dimensional space-time, whose symmetries include the conformal transformations. The first two-dimensional CFT (the $c = 1/2$ free fermion) was constructed by Thirring in 1953. CFT really took off in the 1980s, starting with [**50**]. It arises in string theory, as well as the statistical mechanics describing certain phase transitions. Higher-dimensional CFT appears in the so-called AdS/CFT correspondence (see e.g. [**5**]).

The relation between CFT and string theory is that CFT lives on the world-sheet Σ traced by the strings as they evolve (colliding and separating) through time. Of course, a quantum string only collides and separates in the virtual sense of a Feynman diagram, and so CFT arises in perturbative string theory. More precisely, each term in the Feynman perturbation expansion of S-matrix entries in closed string theory will be a correlation function in a CFT living on the world-sheet. The world-sheets of these scattering strings have a boundary component for every incoming and outgoing string, as in Figure 4.11(b). Any such surface is conformally equivalent to a compact Riemann surface Σ with marked points $p_1, \ldots, p_n$ (one for every incoming and outgoing string), as in Figure 4.13. For reasons we will explain shortly, we also require a choice of local coordinate z_i for each p_i – that is, an explicit identification of a neighbourhood of $p_i \in \Sigma$ with one of $0 \in \mathbb{C}$, so that $z_i = 0$ is the coordinate for p_i. We discuss the moduli space $\widehat{\mathfrak{M}}_{g,n}$ of these 'enhanced surfaces' in Section 2.1.4.

This space-time Σ can be any conformal surface, and we identify conformally equivalent Σ. We restrict to compact orientable Σ, although we don't fix an orientation on it. Because of the string theory interpretation, it is tempting but incorrect to give each such Σ a Lorentzian metric (i.e. locally $dt^2 - dx^2$), but for compact Σ such a metric exists only for the torus. Instead, we give each Σ the usual Euclidean signature (i.e. locally $dx^2 + dy^2 = dz\,d\bar{z}$) of Riemann surfaces. We think of the same CFT as living simultaneously on all such Σ. This leads inevitably to a moduli space formulation.

The simplest indication why two dimensions are so special for CFT is that the space of local conformal transformations, which forms a Lie algebra isomorphic to $\mathfrak{so}_{n+1,1}(\mathbb{R})$ in $\mathbb{R}^n$ for $n > 2$, becomes infinite-dimensional in two dimensions. More precisely, if $f(z)$ is any holomorphic map with nonzero derivative $f'(z_0)$ at some point $z_0 \in \mathbb{C}$, then f is conformal in a neighbourhood of z_0 (the converse is also true – see, for example, theorem 14.2 in [**481**]). Similarly, anti-holomorphic maps preserve the absolute value of angles but reverse the sign. This is essentially the statement that the Lie algebra of conformal Killing vector fields in $\mathbb{R}^n$ is infinite-dimensional iff $n = 2$ (see chapter 1 of [**495**] for a definition and proof); when $n = 2$ it contains two commuting copies of the Witt algebra $\mathfrak{Witt}$ (1.4.9) (one copy for the holomorphic maps and one for the anti-holomorphic ones), arising as dense polynomial subalgebras in this conformal algebra. In our approach, this is how the Virasoro algebra arises. As mentioned in Section 3.1.2, n copies of $\mathfrak{Witt}$ act on the enhanced moduli space $\widehat{\mathfrak{M}}_{g,n}$, either by changing the local coordinate z_i, moving the insertion point p_i or changing the complex structure of Σ.

The CFT literature is very sloppy when discussing the conformal *group* in two dimensions. In spite of numerous published claims to the contrary, it is not the conformal group of $\mathbb{R}^2$ versus that of $\mathbb{R}^n$ ($n > 2$) that singles out two dimensions. The conformal group is isomorphic to the finite-dimensional $SO_{n+1,1}(\mathbb{R})$ *in any* $\mathbb{R}^n$. Although we can identify $\mathbb{R}^2$ with $\mathbb{C}$, and although holomorphic functions f are locally conformal (provided we avoid the zeros of f'), these f don't form a group. Although the conformal group of $\mathbb{R}^2 \cong \mathbb{C}$ (or its compactification S^2, if we permit poles) is finite-dimensional, the conformal group of 'Minkowski space' $\mathbb{R}^{1,1}$ (or better, its compactification $S^1 \times S^1$ – one S^1 for each null-direction $x^1 \pm x^2$) is infinite-dimensional, and for $S^1 \times S^1$ consists of two copies of $\mathrm{Diff}^+(S^1) \times \mathrm{Diff}^+(S^1)$, where $\mathrm{Diff}^+(S^1)$ is the oriented diffeomorphism-group of the circle (Section 3.1.2). Thus its Lie algebra is $\mathfrak{Witt} \oplus \mathfrak{Witt}$. If one wants an infinite-dimensional conformal group in CFT, one must put a Minkowski metric on the cylinder or plane.

The subtle and poorly understood role of two dimensions for the conformal group is carefully discussed in [**495**]. Also interesting is how it arises in Segal's picture (Section 4.4.1). For the interplay and representation theories of $\mathfrak{Witt}$, its central extension $\mathfrak{Vir}$ and the real Lie group $\mathrm{Diff}^+(S^1)$, see Section 3.1.2.

On the cylindrical world-sheet in string theory, given a Minkowski metric, the standard light-cone coordinates would be $t \pm x$, where t is time and x is a periodic angle parameter. The solutions to the classical equations of motion on the cylinder would be functions of $t \pm x$ (i.e. left- and right-moving disturbances travelling at the speed of light). As always, the Hamiltonian is proportional to the generator $\partial/\partial t$ of time translations. The Euclidean version (which is what we use) is $w, \overline{w} = t \mp ix$, and so the left- and right-movers become holomorphic/anti-holomorphic functions of the cylindrical coordinate $w = t - ix$. As is traditional but slightly disturbing, w and $\overline{w}$ are usually to be treated as independent complex variables; we will return to this subtle point shortly. By a formal application of the chain rule, the Hamiltonian in the Euclidean picture will be

$$\frac{\partial}{\partial t} = \frac{\partial w}{\partial t}\frac{\partial}{\partial w} + \frac{\partial \overline{w}}{\partial t}\frac{\partial}{\partial \overline{w}} = \frac{\partial}{\partial w} + \frac{\partial}{\partial \overline{w}}.$$

In CFT, we prefer to use compact surfaces with marked points, so we should conformally map the semi-infinite tubes of the world-sheets to punctures on a compact surface. Locally, such a map looks like $z = \exp(w)$. This conformally maps our Euclidean cylinder to the punctured plane $\mathbb{C} \setminus 0$. Likewise, $\overline{z} = \exp(\overline{w})$ becomes to the right-moving coordinate. We can now write the Hamiltonian $\mathfrak{Witt}$ generators $\ell_n = \overline{z}^{n+1}\partial_z$:

$$\frac{\partial}{\partial t} = z\frac{\partial}{\partial z} + \overline{z}\frac{\partial}{\partial \overline{z}} = -\ell_0 - \overline{\ell}_0.$$

Basic data in the CFT are the quantum fields $\varphi(z, \overline{z})$ – the vertex operators of last subsection – centred at $z = \overline{z} = 0$ on the Riemann sphere $\Sigma = \mathbb{P}^1(\mathbb{C})$. The notation $\varphi(z, \overline{z})$ emphasises that these fields may depend neither holomorphically nor anti-holomorphically on z. These φ are 'operator-valued distributions' on Σ, acting on the space $\mathcal{H}$ of states for the punctured plane (i.e. corresponding to a propagating string); as usual in quantum field theory, they create the various states by acting on the vacuum $|0\rangle \in \mathcal{H}$. As usual, $\mathcal{H}$ comes with a Hermitian product, which allows us to compare $|\mathrm{in}\rangle$ with $|\mathrm{out}\rangle$; in a physical theory it should be positive-definite (a theory without this positive-definiteness is called *non-unitary*). When we say $\varphi(z, \overline{z})$ is 'centred at 0', we mean that the matrix entry $\langle u, \varphi(z, \overline{z})v \rangle$ will be a Laurent polynomial in the local coordinates z and $\bar{z}$, for any u, $v \in \mathcal{H}$, with a singularity only at 0 (unless the outgoing state u isn't the vacuum, in which case infinity can also be singular).

In a CFT, anything that looks like a quantum field is called a quantum field. In the quantum field theories of Section 4.2, only the finitely many generating fields (e.g. the ones appearing in the Lagrangian) are usually called quantum fields.

Any quantum field theory has a *state-field correspondence*: to a field φ is associated its incoming state, that is the $t \to -\infty$ limit of $\varphi|0\rangle$. Typically, different fields can correspond to the same state. In CFT though, this correspondence becomes a bijection: to a given field $\varphi(z, \bar{z})$ on $\mathbb{P}^1(\mathbb{C})$, we associate the state $\varphi(0, 0)|0\rangle = v \in \mathcal{H}$ (recall that $z = e^{t-\mathrm{i}x}$). Let φ_v denote the unique field corresponding to state v.

As for any quantum field theory, solving a CFT requires calculating all n-point correlation functions (4.2.11):

$$\langle \varphi_{v_1}(z_1, \bar{z}_1)\, \varphi_{v_2}(z_2, \bar{z}_2) \cdots \varphi_{v_n}(z_n, \bar{z}_n) \rangle_{\Sigma; p_1, \ldots, p_n}, \tag{4.3.1a}$$

for any choice of enhanced surface (Σ, p_i, z_i) and states $v_i \in \mathcal{H}$. We think of $\varphi_{v_i}(z_i, \bar{z}_i)$ as being centred at p_i; the local coordinates $z_i, \bar{z}_i$ describe it as an 'operator-valued distribution' on Σ about p_i. Simplest is the sphere $\Sigma = \mathbb{P}^1(\mathbb{C})$, because then we can fix a global variable w, and choose $z_i = w - p_i$. In this case the time-ordering of (4.2.11), necessary for convergence, becomes the radial-ordering

$$|p_1| < |p_2| < \cdots < |p_n|, \tag{4.3.1b}$$

because of our map $e^{t-\mathrm{i}x}$. The interpretation of n-point functions for other surfaces is more subtle and will be discussed shortly.

The partition functions $\mathcal{Z}_\Sigma$ (4.2.13b) correspond to vacuum-to-vacuum string amplitudes, and are functions on the moduli space of Σ. For example, a sphere is the world-sheet traced by a closed string spontaneously created from and then reabsorbed into the vacuum. As usual in quantum field theory, we can organise these amplitudes by how many internal 'loops' are involved (i.e. the *genus* of the surface): topologically, 0-loop (i.e. 'tree-level') world-sheets are spheres, 1-loop world-sheets are tori, etc. The 0-loop contribution isn't very interesting (all spheres are conformally equivalent), but we'll see shortly that the 1-loop partition function contains considerable information.

Next we describe two general tools introduced by Kenneth Wilson in the 1960s (see e.g. [**558**]). The first is the operator product expansion (OPE). The idea is to replace the ill-defined product $\varphi_1(x)\varphi_2(x)$ of quantum fields by

$$\varphi_1(x)\,\varphi_2(x') = \sum_{n=0}^{\infty} C_n(x - x')\, O_n(x), \tag{4.3.2}$$

so the singularity structure as $x' \to x$ becomes manifest. The singular terms of (4.3.2) are physically the relevant ones. Here, the O_n are fields in the theory, and are expressible as polynomials in the fields φ_i and their various derivatives. The coefficients C_n are complex-valued functions with singularities of the form $|x|^{-p}$ (for $p > 0$) or $\log|x|$, with the more singular coefficients C_n corresponding to simpler fields O_n. Equation (4.3.2) is meant to hold for x' close to x, in the weak sense of matrix entries, that is correlation functions (4.3.1a). The significance of (4.3.2) to (4.3.1a) should be clear. A derivation and clarification of this fundamental concept (4.3.2) is made in (5.1.6), in the context of vertex operator algebras. The scalar quantum field theory in four dimensions, with ϕ^4 interaction term, is worked out in detail in section 13-5-1 of [**310**], where we find for example that the only singular coefficient in the OPE of $\phi(x)\phi(y)$ is proportional to $\log(x^2)$. The reader may find helpful the discussion of OPE given in lecture 3 of [**567**].

The OPE can be made more explicit here because CFT (unlike most theories) is scale-invariant, and this is Wilson's second tool. We apply it separately to z and $\bar{z}$. Scale-invariance means we have a unitary representation $s \mapsto U(s)$ of the multiplicative group $\mathbb{R}_{>}^{\times}$ of positive real numbers, which is a symmetry of the Lagrangian; an eigenfield φ transforms by $U(s)^{-1}\varphi(z, \bar{z})\, U(s) = s^h \varphi(sz, \bar{z})$ for some real number h (the 'scaling dimension' or *conformal weight* of φ). Similarly, scaling $\bar{z}$ yields an independent conformal weight $\bar{h}$. Scale-invariance requires that the coefficient C_n in (4.3.2) scales like

$$C_n(sz, \overline{sz}) = s^{-h_1-h_2+h(n)}\bar{s}^{-\bar{h}_1-\bar{h}_2+\bar{h}(n)}C_n(z, \bar{z}),$$

where $h(n)$ is the conformal weight of O_n. Since

$$U(s)^{-1}\partial_z\varphi\, U(s) = \frac{\partial}{\partial z}s^h\varphi(sz, \bar{z}) = s^h s\frac{\partial}{\partial(sz)}\varphi(sz, \bar{z}) = s^{h+1}(\partial_z\varphi)(sz, \bar{z}),$$

the field $\partial_z\varphi$ has conformal weight $h + 1$. Thus the possible conformal weights of the fields O_n lie in $\mathbb{N}h_1 + \mathbb{N}h_2$. This means that (4.3.2) involves only finitely many singular coefficients C_n. We see this more explicitly in (5.1.6).

Recall that, classically, a continuous symmetry implies by Noether's Theorem the existence of a conserved current and conserved charges. In the case of the conformal symmetry of CFT, the conserved current is the *stress–energy tensor*, which has nonzero components $T(z) := T_{zz}(z)$ and $\overline{T}(\bar{z}) := T_{\bar{z}\bar{z}}(\bar{z})$. The conserved charges $L_n := \frac{1}{2\pi i}\oint T(z)z^{n-1}dz$ satisfy

$$T(z) = \sum_{n\in\mathbb{Z}} L_n z^{-n-2} \tag{4.3.3}$$

(and similarly for $\overline{L}_n$). In a quantum field theory, these arise in the Ward identities (4.2.12). Here these say, roughly, that taking a derivative of a correlation function $\langle\cdots\rangle_\Sigma$ with respect to a component of the metric on Σ is equivalent to inserting some component of $T(z)$ into that correlation function. The OPE of the field $T(z)$ with itself can be computed:

$$T(z)T(z') = \frac{c}{2}(z - z')^{-4}\, id + 2(z - z')^{-2}\, T(z) + \cdots, \tag{4.3.4}$$

where we display only the singular terms. The number c is called the (holomorphic) *central charge* of the CFT. From this we obtain (see (5.1.6c)) the commutation relations for the modes L_n, and we recover (3.1.5a). In other words, the modes L_n define a representation of the Virasoro algebra on $\mathcal{H}$. Likewise, the modes $\overline{L_m}$ also define a representation of the Virasoro algebra (say with central charge $\bar{c}$). These two copies of

$\mathfrak{Vir}$ commute: $[L_n, \overline{L_m}] = 0$. From the Hermitian product we get that $c, \overline{c}$ and all the conformal weights h are nonnegative real numbers.

Thus, just as the usual quantum field theories (e.g. the Standard Model) carry projective representations of the Poincaré algebra, a CFT carries a projective representation of its conformal algebra, that is, of two commuting copies of the Witt algebra. Hence we get the true representation of $\mathfrak{Vir} \oplus \mathfrak{Vir}$ on $\mathcal{H}$ defined above. A nonzero central charge c (which is typical) amounts physically to a soft breaking of the conformal symmetry – an anomaly – caused by considering CFT on a surface with curvature. More precisely, the correlation functions (4.3.1a) of a CFT will always be invariant under complex diffeomorphisms of the surface Σ, but in genus > 1 when $c \neq 0$ the correlation functions change under local rescalings of the metric. The central charge can be interpreted physically [3] as a Casimir (vacuum) energy, something which depends on space-time topology.

As we have seen, everything in CFT comes in a combination of strictly holomorphic (left-moving) and strictly anti-holomorphic (right-moving) quantities. Here, 'holomorphic' is in terms of the two-dimensional space-time Σ (which locally looks like $\mathbb{C}$), or the local parameters on the appropriate moduli space (which usually locally looks like $\mathbb{C}^\infty$). These holomorphic and anti-holomorphic building blocks are called *chiral*. A CFT is studied by first analysing its chiral parts, and then determining explicitly how they piece together to form the physical quantities. For the applications of CFT to Moonshine, the chiral parts and not the full CFT are what's important. More generally, almost all attention in CFT by mathematicians has focused on the chiral data.

Let $\mathcal{V}$ consist of all the holomorphic fields $\varphi(z)$, and $\overline{\mathcal{V}}$ the anti-holomorphic ones. For example, $\mathcal{V}$ contains $T(z)$. Both $\mathcal{V}$ and $\overline{\mathcal{V}}$ are closed under the OPE (4.3.2), and so form algebras called the *chiral algebras* of the theory. In the next chapter these algebras are axiomatised. $\mathcal{V}$ and $\overline{\mathcal{V}}$ mutually commute and the symmetry algebra of the CFT is often identified with $\mathcal{V} \oplus \overline{\mathcal{V}}$. However, the vacuum is not invariant under most of $\mathcal{V} \oplus \overline{\mathcal{V}}$; we say this symmetry is 'spontaneously broken'. Under the state-field correspondence, $\mathcal{V}$ and $\overline{\mathcal{V}}$ correspond to subspaces V and $\overline{V}$ of the state space $\mathcal{H}$. We call the quantum fields $\varphi(z) \in \mathcal{V}$ *(chiral) vertex operators*.

Since L_0 acts like $-z\partial_z$, the scaling operator $U(s)$ defined earlier is s^{-L_0}. The Virasoro operators $L_0, L_{\pm 1}$ are special in that they generate the three-dimensional conformal group $SL_2(\mathbb{C})$ of the (Riemann) sphere. We have

$$s^{L_0} \varphi_v(z) s^{-L_0} = s^h \varphi_v(sz), \tag{4.3.5a}$$

$$e^{xL_{-1}} \varphi_v(z) e^{-xL_{-1}} = \varphi_v(z + x), \tag{4.3.5b}$$

$$e^{xL_1} \varphi_v(z) e^{-xL_1} = (1 - xz)^{-2h} \varphi_v\left(\frac{z}{1 - xv}\right), \tag{4.3.5c}$$

for any $v \in V$, provided $L_0 v = hv$ (we say v has conformal weight h) and $L_1 v = 0$. Such states v are called *conformal quasi-primaries*. If in addition v satisfies $L_n v = 0$ for all $n > 0$, then v is called a *conformal primary* state. They are precisely the lowest-weight states (Section 3.1.2) for the irreducible $\mathfrak{Vir}$-submodules of state-space $\mathcal{H}$; $\mathcal{H}$

will be the direct integral (Section 1.3.1) over all conformal primaries of the associated lowest-weight $\mathfrak{Vir}$-modules. Equations (4.3.5) are generalised in (5.3.15).

More generally, the state-space $\mathcal{H}$ carries a representation of the symmetry algebra $\mathcal{V} \oplus \overline{\mathcal{V}}$, and decomposes into a direct integral of irreducible $\mathcal{V} \oplus \overline{\mathcal{V}}$-modules (proposition 3.1 of [187]). A *rational conformal field theory* (RCFT) is one whose state-space $\mathcal{H}$ decomposes into a *finite* sum

$$\mathcal{H} = \oplus M \otimes \overline{N}, \tag{4.3.6a}$$

where M and $\overline{N}$ are irreducible modules of the chiral algebras $\mathcal{V}$ and $\overline{\mathcal{V}}$, respectively. One of the summands in (4.3.6a) is $V \otimes \overline{V}$. The rational ones are the CFTs we are interested in; the name 'rational' was chosen because for them the central charge c and all conformal weights h are rational numbers. The chiral algebras of an RCFT will have only finitely many irreducible modules M; for later convenience let $\Phi = \Phi(\mathcal{V})$ denote the set of these. The $M \in \Phi$ are called *chiral primaries* even though they don't necessarily correspond to a unique vector in $\mathcal{H}$. It is more convenient to write (4.3.6a) in the equivalent form

$$\mathcal{H} = \oplus_{M \in \Phi, \overline{N} \in \overline{\Phi}} \mathcal{Z}_{M,\overline{N}} \, M \otimes \overline{N}, \tag{4.3.6b}$$

where $\mathcal{Z}_{M,N}$ are multiplicities (many of which may be 0). It turns out (because $\mathcal{V}$ is maximal) that $\mathcal{Z}$ will be a permutation matrix. This decomposition (4.3.6b) is reminiscent of the decomposition of a group algebra into irreducible modules. A beautiful interpretation in terms of Frobenius algebras in category theory is given in [211].

An important class of RCFT are the *Wess–Zumino–Witten* (WZW) models. These correspond to strings living on a compact Lie group G. Their mathematics is especially pretty, and any natural question seems to have an elegant Lie-theoretic answer. The chiral algebra $\mathcal{V}$ is closely related to the affine Kac–Moody algebra $\overline{\mathfrak{g}}^{(1)}$ associated with G (Section 5.2.2); its modules $M \in \Phi$ can be identified with the integrable highest-weight modules $L(\lambda)$ at a level k determined by c and (3.2.9c).

As with everything else in CFT, the correlation functions (4.3.1a) can be expressed in terms of purely chiral quantities called *conformal* or *chiral blocks*

$$\mathcal{F} = \langle \mathcal{I}_1(v_1, z_1) \, \mathcal{I}_2(v_2, z_2) \cdots \mathcal{I}_n(v_n, z_n) \rangle_{(\Sigma; p_1, \ldots, p_n; M^1, \ldots, M^n)}. \tag{4.3.7}$$

Once again, Σ is a compact Riemann surface with marked points p_i; to each point p_i we assign a local coordinate z_i as before, and also a choice of irreducible module $M^i \in \Phi$. The state v_i is taken from M^i, and the fields $\mathcal{I}_i(v_i, z_i)$, centred at p_i, are called *intertwining operators* and generalise the vertex operators $\varphi_v \in \mathcal{V}$. See Definition 6.1.9 (roughly, each $\mathcal{I}_i(v_i, z_i)$ is an operator-valued distribution sending vectors in some module to another). In the case of higher genus Σ, (4.3.7) cannot be taken too literally, and the study of higher-genus chiral blocks is more difficult [573], [296]; roughly, the points p_i are first taken in the same coordinate patch of Σ; the function is then extended holomorphically. It will need branch-cuts in Σ to be well-defined.

To solve a given RCFT, it suffices to:

(a) construct all possible chiral blocks (4.3.7); and
(b) reconstruct the correlation functions (4.3.1a) from those chiral blocks.

In its broad strokes, part (a) was explained in work of Moore–Seiberg [**436**] (and more carefully in [**32**]) – see Section 6.1.4. In deep work, Huang is pursuing the explicit solution to (a) for all sufficiently nice chiral algebras $\mathcal{V}$ (see e.g. [**295**] for the genus-0 story and [**296**] for genus-1). Likewise, in a series of papers written by Fuchs, Schweigert and collaborators, topological field theories (Section 4.4.3) are used to find a solution to (b) (see the reviews [**211**], [**496**]).

In CFT the Ward identities (4.2.12) are especially useful, since the symmetries are so considerable. For example, they imply that it suffices to evaluate the chiral blocks (4.3.7) when all v_i are conformal primaries. Recall that $\mathfrak{Witt}$ acts on moduli spaces (Section 3.1.2); this lifts to one of $\mathfrak{Vir}$ on chiral blocks, and the resulting partial differential equations are the KZ equations of Section 3.2.4. Their monodromy is what makes the chiral blocks so interesting, especially to Moonshine.

The most important example of chiral block is for the torus $\mathbb{C}/(\mathbb{Z} + \tau\mathbb{Z})$ with one marked point (it doesn't matter where), assigned $\mathcal{V}$-module $M^1 = V$ and state $v_1 = |0\rangle$. Taking any operator $\mathcal{I}_1$ intertwining some $M \in \Phi(\mathcal{V})$ with itself, the corresponding chiral block (up to a constant multiple) will be the *graded dimension*

$$\chi_M(\tau) := \mathrm{tr}_M e^{2\pi i\tau\,(L_0 - c/24)}, \tag{4.3.8a}$$

where c is the central charge and L_0 is the Virasoro generator corresponding to energy. We explain in Section 5.3.4 how this arises. Using (4.3.6), the 0-point correlation function for the torus – the 1-loop partition function $\mathcal{Z}$ – becomes

$$\mathcal{Z}(\tau, \overline{\tau}) := \mathrm{tr}_{\mathcal{H}} e^{2\pi i\,[\tau\,(L_0 - c/24) - \overline{\tau}\,(\overline{L}_0 - \overline{c}/24)]} = \sum_{M \in \Phi, \overline{N} \in \Phi} \mathcal{Z}_{M,\overline{N}}\,\chi_M(\tau)\,\chi_{\overline{N}}(\overline{\tau}). \tag{4.3.8b}$$

This is a very typical decomposition of a physical correlation function into chiral blocks.

The reviews [**496**], [**216**] provide careful explanations of why sometimes we treat z and $\overline{z}$ as independent, and other times we must treat one as the complex conjugate of the other. In short, from the point of view of chiral data, the single space-time Σ of the full CFT is really two disjoint copies with opposite orientation (the Schottky double). For example, the torus with modular parameter $\tau \in \mathbb{H}$ is paired with the one with parameter $-\overline{\tau} \in \mathbb{H}$. As in (4.3.8b), the correlation functions of the full CFT involve both modular parameters, but at the chiral level the two tori don't see each other.

In particular, for a given choice $(\Sigma; \{p_i\}; \{M^i\})$, an RCFT assigns a finite-dimensional space $\mathfrak{B}^{(g,n)}_{\{p_i\},\{M^i\}}$ of chiral blocks. Each chiral block depends multi-linearly on the $v_i \in M^i$, and meromorphically on the z_i, though branch-cuts in Σ between p_i will be needed. The dimension of this space $\mathfrak{B}^{(g,n)}_{\{p_i\},\{M^i\}}$ is called the *Verlinde dimension*, and is given by Verlinde's formula (6.1.2) below.

For example, consider a WZW model associated with an affine algebra $\mathfrak{g} = \overline{\mathfrak{g}}^{(1)}$ and level $k \in \mathbb{N}$. Fix an extended surface (Σ, p_i, z_i). We have a copy of $\mathfrak{g}$ at each p_i, built in the usual way (Section 3.2.2) from the loop algebra $\overline{\mathfrak{g}} \otimes \mathbb{C}[z_i^{\pm 1}]$. The chiral primaries $M \in \Phi$ are the integrable highest weights $\lambda \in P_+^k(\mathfrak{g})$; to each point p_i choose some $\lambda^{(i)} \in P_+^k(\mathfrak{g})$. The associated space $\mathfrak{B}$ of chiral blocks is constructed in [**530**], and these have an important geometric interpretation as spaces of generalised theta functions (see chapter 10 of [**495**]).

The affine algebra characters χ_λ of (3.2.9a), as well as the j-function (0.1.8) are examples of chiral blocks. As we see next subsection, the spaces $\mathfrak{B}^{(g,n)}_{\{p_i\},\{M^i\}}$ naturally carry a representation of the mapping class group $\widehat{\Gamma}_{g,n}$, and this is the source of the relation of the braid group to subfactors, as well as the modularity of Moonshine. In particular, the RCFT characters (4.3.8a) transform nicely under $\mathrm{SL}_2(\mathbb{Z})$: for example,

$$\chi_M(-1/\tau) = \sum_{N \in \Phi} S_{M,N}\,\chi_N(\tau), \qquad (4.3.9a)$$

$$\chi_M(\tau + 1) = \sum_{N \in \Phi} T_{M,N}\,\chi_N(\tau), \qquad (4.3.9b)$$

where S, T are finite complex matrices. This T matrix is given by

$$T_{M,N} = e^{2\pi \mathrm{i}(h_M - c/24)}\delta_{M,N}, \qquad (4.3.10)$$

where h_M is a real number (called the *conformal weight*) associated with the chiral primary $M \in \Phi$. The matrix S is, however, more complicated (Section 6.1.2). For example, the matrix T for the WZW models involves the quadratic Casimir of $\bar{\mathfrak{g}}$, while the matrix S involves characters of G evaluated at elements of finite order.

The simplest class of RCFT are the *minimal models*, which have the smallest possible chiral algebra (generated only by the identity field and the stress–energy field $T(z)$) and nevertheless still have a finite decomposition (4.3.6a). They are well understood (see e.g. [**131**]). They are the RCFT with central charge $0 < c < 1$, and correspond to the discrete series (3.1.6) of $\mathfrak{Vir}$.

The smallest nontrivial minimal model is the Ising model. It has central charge $c = 0.5$. The associated chiral algebra has three irreducible modules, which we label $\Phi = \{0, \epsilon, \sigma\}$ as in [**131**]. Their graded dimensions (4.3.8a) are

$$\chi_0(\tau) = q^{-1/48}\,(1 + q^2 + q^3 + 2q^4 + 2q^5 + 3q^6 + 3q^7 + \cdots), \qquad (4.3.11a)$$

$$\chi_\epsilon(\tau) = q^{23/48}\,(1 + q + q^2 + q^3 + 2q^4 + 2q^5 + 3q^6 + 3q^7 + \cdots), \qquad (4.3.11b)$$

$$\chi_\sigma(\tau) = q^{1/24}\,(1 + q + q^2 + 2q^3 + 2q^4 + 3q^5 + 4q^6 + 5q^7 + \cdots), \qquad (4.3.11c)$$

where as always $q = e^{2\pi \mathrm{i}\tau}$. From this we can read off the conformal weights $h_0 = 0, h_\epsilon = 1/2, h_\sigma = 1/16$, and hence the T matrix of (4.3.9b):

$$T = \begin{pmatrix} e^{-\pi \mathrm{i}/24} & 0 & 0 \\ 0 & e^{23\pi \mathrm{i}/24} & 0 \\ 0 & 0 & e^{\pi \mathrm{i}/12} \end{pmatrix}. \qquad (4.3.11d)$$

The matrix S is more difficult to find, but it equals

$$S = \frac{1}{2}\begin{pmatrix} 1 & 1 & \sqrt{2} \\ 1 & 1 & -\sqrt{2} \\ \sqrt{2} & -\sqrt{2} & 0 \end{pmatrix}. \qquad (4.3.11e)$$

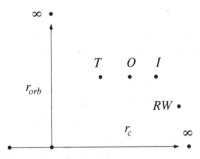

Fig. 4.14 The moduli space of conformal field theories with central charge $c = 1$.

The 1-loop partition function $\mathcal{Z}(\tau)$ of (4.3.8b) is

$$\mathcal{Z}(\tau) = |\chi_0(\tau)|^2 + |\chi_\epsilon(\tau)|^2 + |\chi_\sigma(\tau)|^2.$$

The CFT corresponding to open string perturbation – *boundary CFT* – is also interesting (see e.g. the review [**461**]). In this direction, see the proposals in [**215**], [**458**] (building on the α-induction of subfactors [**65**]). For instance, the 1-loop partition function corresponds to a Frobenius algebra (Section 4.4.3) in the modular category of modules of the associated chiral algebra, and the boundary CFT data arise as a 'category module'. However, boundary CFT isn't so relevant for Moonshine and will mostly be ignored in this book.

The space of CFTs can be probed using 'marginal operators' – fields φ_v with conformal weight $(h, \overline{h}) = (1, 1)$ obeying certain other properties (see e.g. [**137**] and [**246**] section 8.6). A given CFT can be deformed (changing its spectrum but not central charge c), provided it contains such a field. If the given CFT has n marginal operators, then the space of CFTs in its neighbourhood is expected (typically) to look like an n-dimensional real manifold. When the given CFT has more marginal operators than the neighbouring ones, the space of CFTs at that point may look like two manifolds intersecting transversely, or it can mean an orbifold singularity where you get different realisations for the same CFTs. The RCFTs are special points in this space. The space of known $c = 1$ CFTs is drawn in Figure 4.14. Points on the horizontal and vertical lines are parametrised by a radius $\sqrt{2}^{-1} \leq r_{orb}, r_c \leq \infty$; these two half-lines intersect at $r_{orb} = 1/\sqrt{2}, r_c = \sqrt{2}$. The known *rational* $c = 1$ CFT consists of the three isolated theories *T(etrahedral)*, *O(ctahedral)* and *I(cosahedral)*, together with those theories with $r_{orb}^2 \in \mathbb{Q}$ or $r_c^2 \in \mathbb{Q}$. The fourth isolated point, *RW*, is irrational and described in [**483**]. Theories with radii r_c and $r_c' = 1/(2r_c)$ are equivalent, as are those with radii r_{orb} and $r_{orb}' = 1/(2r_{orb})$ (this is an example of 'T-duality', and arises from the extra marginal operator possessed by the $r_c = \sqrt{2}^{-1}$ and $r_{orb} = \sqrt{2}^{-1}$ theories). The intersection point also has two, while the isolated points have no marginal operators, and the remainder have one (which permits r to be continuously varied). The moduli space for CFT with central charge $c < 1$ consists of countably many isolated points [**91**]. Very little is known about the moduli space for $c > 1$.

4.3.3 Monodromy in CFT

One way to make conformal symmetry manifest is to make the relevant physical quantities be holomorphic functions of (or more precisely, sections of bundles over) the appropriate moduli spaces. Let $\mathcal{V}$ be the chiral algebra of an RCFT and let Φ label its (finitely many) irreducible modules, that is the chiral primaries. Let 0 denote the one corresponding to the subspace V of $\mathcal{H}$. Let's investigate more closely what chiral blocks (4.3.7) are.

In any RCFT, there are differential equations that the chiral blocks must satisfy. The most well known of these are the *Knizhnik–Zamolodchikov* (or KZ) equations. We studied these for WZW models at genus 0 in Section 3.2.4. Good expositions of this material are given in [355], [207], [186]. Differential equations can also be found using null vectors [50], and using the Ward identities.

Return to the Ising model, introduced last subsection. We know its chiral blocks in genus-0 with two or three marked points (Question 4.3.5). Consider now four marked points on the Riemann sphere, at positions $w_i \in \mathbb{C} \cup \{\infty\}$. The chiral block will be the product of the quantity

$$\prod_{1 \le i < j \le 4} (w_i - w_j)^{-h_i - h_j + \frac{1}{3} \sum_k h_k} \tag{4.3.12a}$$

with some function of the cross-ratio

$$w := \frac{(w_1 - w_2)(w_3 - w_4)}{(w_1 - w_3)(w_2 - w_4)}. \tag{4.3.12b}$$

We can simplify this using the Möbius symmetry of the Riemann sphere to move w_i to $0, w, 1, \infty$, respectively. If we label all four marked points with the primary field $\sigma \in \Phi$, then the space of chiral blocks is two-dimensional, spanned by

$$\mathcal{F}_1(w) = \frac{\sqrt{1 + \sqrt{1 - w}}}{\sqrt{2}\,(w(1 - w))^{1/8}}, \tag{4.3.13a}$$

$$\mathcal{F}_2(w) = \frac{\sqrt{1 - \sqrt{1 - w}}}{\sqrt{2}\,(w(1 - w))^{1/8}}. \tag{4.3.13b}$$

The fractional powers tell us these chiral blocks have branch-point singularities – that is, to get a holomorphic function on the w-plane, we need to make semi-infinite cuts. Nevertheless, we can analytically continue these functions along any curve. Take a point w_0 so that $0 < |w_0| < 1$, and consider the circle $w(t) = w_0\, e^{2\pi i t}$ for $0 \le t \le 1$. Nothing special happens to the numerator of the $\mathcal{F}_i(w)$: its values at $t = 0$ and $t = 1$ are equal. The denominator however picks up a factor $e^{2\pi i/8}$, and thus both blocks $\mathcal{F}_i(w)$ pick up a net factor of $e^{-2\pi i/8}$. We call this the *monodromy* about $w = 0$ (Section 3.2.4).

Consider next their monodromy about $w = 1$. Here our circle will be $w(t) = 1 + w_0 e^{2\pi i t}$, again for w_0 small. Note that the numerators of $\mathcal{F}_1$ and $\mathcal{F}_2$ switch, and the denominators again pick up a factor of $e^{2\pi i/8}$. Thus this monodromy can be written

$$\begin{pmatrix} \mathcal{F}_1(w) \\ \mathcal{F}_2(w) \end{pmatrix} \mapsto \begin{pmatrix} 0 & e^{-2\pi i/8} \\ e^{-2\pi i/8} & 0 \end{pmatrix} \begin{pmatrix} \mathcal{F}_1(w) \\ \mathcal{F}_2(w) \end{pmatrix}.$$

In Section 3.2.4 we explain how to think of this. Reintroducing the four coordinates w_i, the chiral blocks $\mathcal{F}_i$ will be holomorphic on the universal cover $\widetilde{\mathfrak{C}}_4$ of the configuration space $\mathfrak{C}_4$ of (1.2.6). Analytically continuing along any closed path γ in $\mathfrak{C}_4$ (across any of those branch cuts) defines an action of the fundamental group $\pi_1(\mathfrak{C}_4)$ on the space $\mathfrak{B}^{(0,4)}$ of chiral blocks. This group $\pi_1(\mathfrak{C}_4)$ is the pure braid group of the sphere with four strands. An element β of the full braid group of the sphere maps the space $\mathfrak{B}^{(0,4)}_{m^1,m^2,m^3,m^4}$ to $\mathfrak{B}^{(0,4)}_{m^{\beta 1},m^{\beta 2},m^{\beta 3},m^{\beta 4}}$, where βi is the associated permutation, so in our example (4.3.13) the full braid group acts. We can recover the usual planar braid groups $\mathcal{P}_3$ and $\mathcal{B}_3$ here by fixing one of the four points at say ∞, and letting the others wander around.

Equivalently, as a 'function' on the configuration space, the chiral blocks form (multivalued) holomorphic sections of a projective flat vector bundle. What this means is that each chiral block satisfies a system of partial differential equations (the KZ equations) describing how to parallel-transport it around the configuration space, and flatness says it will *locally* depend only on the moduli space parameters (and not on the path chosen). *Globally*, however, there will be monodromy [**437**], [**32**], [**355**].

More generally, a chiral block $\mathcal{F}$ on an enhanced surface Σ is a multi-valued function on the corresponding moduli space. To make it well defined, $\mathcal{F}$ can be lifted to the corresponding Teichmüller space. There will be an action of the corresponding mapping class group $\widehat{\Gamma}_{g,n}$ coming from monodromy (a *projective* action, if as usual the central charge c is nonzero). How to centrally extend these $\widehat{\Gamma}_{g,n}$ so that the projective representation becomes a true one is discussed, for example, in [**404**]. This picture, which is explained quite clearly in [**32**] and is developed further in, for example, Section 7.2.4, encompasses not only the braid group monodromy of the KZ equation (Section 3.2.4) but also the modular group action (4.3.9) on the graded dimensions (4.3.8a). It is the source of the modularity in Moonshine.

Although the chiral blocks themselves are multi-valued functions on the moduli spaces $\widehat{\mathfrak{M}}_{g,n}$, conformal invariance requires that the n-point correlation functions (4.3.1) themselves be well-defined functions on $\widehat{\mathfrak{M}}_{g,n}$. For example, even though the graded dimensions χ_M transform as in (4.3.9), the 1-loop partition function in (4.3.8b) is $SL_2(\mathbb{Z})$-invariant. See also Question 4.3.7.

As we know from Section 2.2.1, there is more to being a modular form or function than transforming nicely with respect to $SL_2(\mathbb{Z})$. The behaviour at the cusps of $\mathbb{H}$ is also crucial, as it says our function lives on a compact space. Something similar also holds in RCFT. The analogue of cusps for the other moduli spaces – that is, the surfaces corresponding to the extra points needed for compactification – are surfaces with nodes (Section 2.1.4). What we need is nice behaviour of chiral blocks as we move in moduli space towards surfaces with nodes, that is, as we shrink a closed curve about a handle on our surface down to zero radius. This is given by (4.4.3) and is called *factorisation* [**203**], [**539**]. It connects the moduli spaces of different topologies, and tells us CFT is defined on a 'universal tower' of moduli spaces (Sections 3.1.2 and 6.3.3).

Incidentally, it is tempting to try to extend this formalism to the 'surfaces of infinite genus' given by projective limits $\varprojlim \Gamma \backslash \mathbb{H}$ (see Section 2.4.1). The discrete groups Γ appearing in each such limit must all be commensurable (i.e. intersections of any two of them should have finite index in both), in order for the limit to be defined. In Section 2.4.1 we describe the most famous piece of such a limit: the modular tower $\lim_{\leftarrow} \Gamma(N)\backslash\mathbb{H}$, so important to number theory. The assignment of, for example, chiral blocks to such 'surfaces' may be built up from those of each $\Gamma\backslash\mathbb{H}$, in a relatively straightforward way; because of this, perhaps we could interpret the string-theoretic data for $\lim_{\leftarrow}\Gamma\backslash\mathbb{H}$ as the (nonperturbative?) contribution ('sum') associated collectively with all world-sheets $\Gamma\backslash\mathbb{H}$ appearing in that limit. In any case, we are led to speculate from (2.4.3) that both CFT and the theory of vertex operator algebras (and indeed Moonshine itself) may extend quite nicely to the p-adics $\widehat{\mathbb{Q}}_p$. Some moves in this direction are [562], [520]. To a number theorist, the usual perturbation about a vacuum would correspond to the infinite prime, but would mysteriously ignore the contributions from all the finite primes. It would be interesting to see if nonperturbative phenomena like D-branes can be sensed by these projective limits.

As discussed at the end of Section 2.2.1, the analogue of q-expansions, for chiral blocks and partition functions in higher genus, are expansions about surfaces with nodes. A natural projectively flat connection on these spaces $\mathfrak{B}^{(g,n)}$ of chiral blocks is given by the stress–energy tensor $T(z)$ [203], [530]; this connection is responsible for the KZ equations, and is the analogue here of the $\mathfrak{Witt}$ action on moduli spaces, and the meaning of $T(z)$ insertions into correlation functions discussed in Section 4.3.2.

4.3.4 Twisted #4: the orbifold construction

To particles, a space-time singularity is a problem; to strings, it is merely a region where stringy effects are large. The most tractable way to introduce such singularities is by quotienting ('gauging') by a finite group. This construction plays a fundamental role for CFTs and vertex operator algebras; it is the physics underlying what Norton calls *generalised Moonshine* (Section 7.3.2). This is where finite group theory touches CFT.

Let M be a manifold and G a finite group of symmetries of M. The set M/G of G-orbits inherits a topology from M, and forms a manifold-like space called an *orbifold*. Fixed points become conical singularities. For example, $\{\pm1\}$ acts on $M = \mathbb{R}$ by multiplication. The orbifold $\mathbb{R}/\{\pm1\}$ can be identified with the interval $x \geq 0$. The fixed point at $x = 0$ becomes a singular point on the orbifold, that is, a point where locally the orbifold does not look like some open n-ball (open interval in this one-dimensional case). For other examples, see Question 4.3.8.

Orbifolds were introduced into geometry in the 1950s as spaces with mild singularities; recalling Definition 1.2.3, they are V_α/G_α patched together, where $V_\alpha \subset \mathbb{R}^n$ is open and G_α is a finite group. They were introduced into string theory in [143], which greatly increased the class of background space-times in which the string could live and still be amenable to calculation. This subsection briefly sketches the corresponding construction for CFT; our purpose is to motivate Section 5.3.6.

For concreteness think of a closed string whose world-sheet $\Sigma \subset M$ is a torus, since the 1-loop partition function (4.3.8b) is the easiest way to obtain the spectrum (4.3.6) of the theory. Think of Σ being parametrised by $z \in \mathbb{C}/(\tau\mathbb{Z} + 2\pi\mathbb{Z})$, with τ being the time-period of the 1-loop and 2π being the space-period of the closed string. Here, G is a finite group of symmetries of the theory – it acts not only on space-time M, but also on the internal states of the string (i.e. the state-space $\mathcal{H}$ carries a representation of G). Assume for now that G is abelian and that $\mathcal{H} = V \otimes \overline{V}$. For example, this is satisfied by the WZW theory for $E_8^{(1)}$ at level 1, or strings living on the torus $\mathbb{R}^n/L$ for an even self-dual n-dimensional lattice L. Consider first the chiral data. The orbifold chiral algebra $\mathcal{V}^{orb}$ is the subalgebra $\mathcal{V}^G$ of $\mathcal{V}$ consisting of all G-invariant fields. More difficult to answer is what the orbifold state-space $\mathcal{H}^{orb}$ looks like.

In the case of a point particle, a 1-loop world-line $\mathbf{x}(t) \in M/G$ would be a circle, the motion $\mathbf{x}(t)$ would be periodic (say with period T); lifting $\mathbf{x}(t)$ to M, we would require that $\mathbf{x}(T) = g.\mathbf{x}(0)$ for some $g \in G$. The closed string also requires this twisted periodicity in the time direction, but being closed it will similarly have a twisted periodicity in the space direction. Thus we are led to consider string processes satisfying the boundary conditions

$$\mathbf{x}(z + \tau) = g.\mathbf{x}(z), \qquad \mathbf{x}(z + 2\pi) = h.\mathbf{x}(z). \tag{4.3.14a}$$

The strings satisfying $\mathbf{x}(2\pi) = h.\mathbf{x}(0)$ form the *h-twisted sector* $\mathcal{V}^h$ – these twisted sectors are the special feature of strings living on orbifolds. They don't live in the original chiral space V, and are hard to construct; in particular, there isn't a systematic twisted analogue of the vertex operator construction (i.e. exponentials of free fields) of untwisted sectors.

The contribution of the processes (4.3.14a) to the 1-loop path integral will be

$$\mathcal{Z}_{(g,h)}(\tau) := \text{tr}_{\mathcal{V}^h}\, g\, e^{2\pi i \tau\, (L_0 - c/24)}, \tag{4.3.14b}$$

for reasons that will become clearer next section (the trace comes from obtaining the torus by sewing together the inner and outer boundaries of an annulus). Each (finite-dimensional) L_0-eigenspace in $\mathcal{V}^h$ carries a representation of the group $\langle g \rangle$, so that is the matrix to substitute into the trace (4.3.14b). The modular group $SL_2(\mathbb{Z})$ acts on the cycles (homology H_1) of the torus in the usual way, which gives the behaviour of $\mathcal{Z}_{(g,h)}$ under modular transformations:

$$\mathcal{Z}_{(g,h)}\left(\frac{a\tau + b}{c\tau + d}\right) = \mathcal{Z}_{(g^a h^c, g^b h^d)}(\tau). \tag{4.3.14c}$$

Actually, we will find shortly that in general this transformation has to be modified slightly.

The twisted sector $\mathcal{V}^h$ is an irreducible (twisted) module for the original chiral algebra $\mathcal{V}$ (Section 5.3.6). In terms of the orbifold chiral algebra $\mathcal{V}^G$, $\mathcal{V}^h$ will be a true module, though not an irreducible one. Its decomposition ('branching rules') into irreducible $\mathcal{V}^G$-modules is

$$\mathcal{V}^h = \oplus_\rho \mathcal{V}^h_\rho \otimes \rho, \tag{4.3.15a}$$

where the sum is over all irreducible G-representations ρ (when G is non-abelian, this will be modified slightly). Plugging this into (4.3.14b) gives the equivalent expressions

$$\mathcal{Z}_{(g,h)}(\tau) = \sum_\rho \mathrm{ch}_\rho(g)\, \chi_{(h,\rho)}(\tau), \qquad (4.3.15b)$$

$$\chi_{(h,\rho)}(\tau) := \mathrm{tr}_{\mathcal{V}_\rho^h} e^{2\pi i \tau\,(L_0 - c/24)} = \frac{1}{\|C_G(h)\|} \sum_{g \in G} \overline{\mathrm{ch}_\rho(g)}\, \mathcal{Z}_{(g,h)}(\tau). \quad (4.3.15c)$$

The graded dimension $\chi_{(h,\rho)}$, unlike $\mathcal{Z}_{(g,h)}$, has a q-expansion with coefficients in $\mathbb{N}$, but $\mathcal{Z}_{(g,h)}$ has the simpler modular behaviour, in perfect analogy to Θ_{t+L} versus $\Theta_{L;r,s}$ (compare (2.2.11) and (2.3.10)).

An important example of this orbifold construction is the Moonshine module $V^\natural$ (Sections 5.3.6 and 7.2.1). Its starting point is the chiral algebra $\mathcal{V}(\Lambda)$ for the torus $\mathbb{R}^{24}/\Lambda$, where Λ is the Leech lattice. The symmetry group G corresponds to the centre $\{\pm 1\}$ of $\mathrm{Aut}(\Lambda)$. The graded dimension of the untwisted sector $\mathcal{V}(\Lambda)$ is $\mathcal{Z}_{(1,1)}(\tau) = J(\tau) + 24$, and has -1-twisted graded dimension

$$\mathcal{Z}_{(-1,1)}(\tau) = q^{-1} \prod_{n=0}^{\infty} (1 - q^{2n+1})^{24} = q^{-1} - 24 + 276q - 2048q^2 + \cdots$$

The -1-twisted sector $\mathcal{V}(\Lambda)^{-1}$ has untwisted/twisted graded dimension

$$\mathcal{Z}_{(\pm 1,-1)} = 2^{12} q^{1/2} \prod_{n=0}^{\infty} \left(1 \mp q^{(2n+1)/2}\right)^{-24}$$

$$= q^{1/2} \pm 98304q + 1228800q^{3/2} \pm 10747904q^2 + \cdots$$

The Moonshine module $V^\natural$ consists of the sectors $\mathcal{V}(\Lambda)^1_+ \oplus \mathcal{V}(\Lambda)^{-1}_+$ and so has graded dimension

$$\chi_{V^\natural}(\tau) = \chi_{(1,+)}(\tau) + \chi_{(-1,+)}(\tau)$$

$$= \frac{1}{2}\left(\mathcal{Z}_{(1,1)}(\tau) + \mathcal{Z}_{(-1,1)}(\tau) + \mathcal{Z}_{(1,-1)}(\tau) - \mathcal{Z}_{(-1,-1)}(\tau)\right) = J(\tau). \qquad (4.3.16)$$

So far we have discussed only the *chiral* orbifold CFT – our main interest. The state-space (4.3.6) of the full orbifold CFT can look like

$$\mathcal{H}^{orb} = \oplus \mathcal{V}_\rho^g \otimes \overline{\mathcal{V}}_\rho^g. \qquad (4.3.17)$$

There are other possibilities for $\mathcal{H}^{orb}$; a systematic but far from exhaustive source is provided by discrete torsion [**136**]. The lattice construction $L\{T\}$ of Section 2.3.3 (applied to indefinite lattices L) is this orbifold construction of $\mathcal{H}^{orb}$, coming largely from discrete torsion. The construction of $V^\natural$ is a heterotic version (i.e. with trivial 'anti-holomorphic' chiral algebra $\overline{\mathcal{V}}$). In any case, the full orbifold theory will typically involve most sectors $\mathcal{V}_\rho^g$. Modular invariance (4.3.14c) is one way to see the necessity of this; another is string dynamics (see figure 8.1 in [**463**], vol. I).

There are three significant generalisations of this orbifold construction as outlined above. *Non-abelian* orbifold groups G are at least as interesting to us (e.g. Maxi-Moonshine concerns $V^\natural/\mathbb{M}$), and introduce new subtleties. For example, using (4.3.14a) to evaluate $\mathbf{x}((z+\tau)+2\pi) = \mathbf{x}((z+2\pi)+\tau)$ requires $hg.\mathbf{x}(z) = gh.\mathbf{x}(z)$. That is, we should limit ourselves to boundary conditions (4.3.14a) whose pairs (g, h) commute. Moreover, consider the h-twisted sector $\mathbf{x}(2\pi) = h.\mathbf{x}(0)$; hitting both sides with $g \in G$ yields $(g\mathbf{x})(2\pi) = (ghg^{-1}).(g\mathbf{x})(0)$, that is, the twisted sectors $\mathcal{V}^h$ and $\mathcal{V}^{ghg^{-1}}$ are naturally isomorphic. In fact, $\mathcal{Z}_{(g,h)} = \mathcal{Z}_{(kgk^{-1}, khk^{-1})}$ for any $k \in G$, so we should identify each boundary condition (g, h) with all simultaneous conjugations (kgk^{-1}, khk^{-1}). This will be clearer in Sections 5.3.6 and 6.2.4. The sums in (4.2.15) are over all $g \in C_G(h)$ and all irreducible $C_G(h)$-representations ρ, where $C_G(h)$ is the centraliser of h in G.

For the second generalisation, note that $g \in C_G(h)$ takes the sector $\mathcal{V}^h$ to $\mathcal{V}^{ghg^{-1}} = \mathcal{V}^h$ so (as in Section 1.5.4) we get a linear map $\phi_g^{(h)} : \mathcal{V}^h \to \mathcal{V}^h$. So far we have implicitly assumed that these assignments $g \mapsto \phi_g^{(h)}$ define a representation of $C_G(h)$. But $\mathcal{V}^h$ are chiral data and so group actions, etc. may be projective. That is, we only know that $g \mapsto \phi_g^{(h)}$ defines a *projective* representation of $C_G(h)$. In this case, (4.3.14c) must be replaced by

$$\mathcal{Z}_{(g,h)} \left(\frac{a\tau + b}{c\tau + d} \right) = \gamma \, \mathcal{Z}_{(g^a h^c, g^b h^d)}(\tau), \tag{4.3.18}$$

for some root of unity γ. See [138], and Section 5.3.6 below, for details. For example, the Maxi-Moonshine orbifold $V^\natural/\mathbb{M}$ will necessarily be of that projective type [408].

For the final generalisation, we have discussed orbifolding the CFTs with one chiral primary (i.e. with $\|\Phi\| = 1$) only because they are simpler. The behaviour of more typical multi-primary orbifolds is analogous (Section 5.3.6). For example, the horizontal line of $c = 1$ CFTs in Figure 4.14 corresponds to bosons compactified on a circle of radius r, while the vertical line there corresponds to bosons on the orbifold $S^1/\mathbb{Z}_2$ (see the treatment in [246]); most of these theories have infinitely many chiral primaries (i.e. aren't rational). The WZW theory for $A_1^{(1)}$ at level 1 is a $c = 1$ theory with two chiral primaries corresponding to a string living on S^3; we can orbifold this rational theory by any of the finite subgroups of $\mathrm{SU}_2(\mathbb{C})$. These subgroups fall into an A–D–E pattern (Section 2.5.2). Orbifolding by the (cyclic) A-series of subgroups gives the $c = 1$ theories $r_c = n/\sqrt{2}$, and by the (dihedral) D-series gives the $c = 1$ theories $r_{orb} = n/\sqrt{2}$. The (tetrahedral) E_6-, (octahedral) E_7- and (icosahedral) E_8-subgroups give us the isolated theories T, O, I of Figure 4.14.

Choose any CFT $\mathcal{H}$ and tensor it with itself n times to get a new CFT $\mathcal{H}^{\otimes n}$. The orbifold $\mathcal{H}^{\otimes n}/\mathcal{S}_n$ is called a *permutation orbifold*. Requiring that $\mathcal{H}^{\otimes n}/\mathcal{S}_n$ possesses the standard CFT properties imposes highly nontrivial conditions on the chiral data of $\mathcal{H}$. See, for example, [37] for applications of this powerful theoretical tool.

4.3.5 Braided #4: the braid group in quantum field theory

Much of Moonshine is implicit in two-dimensional CFT. What is the most distinctive physical feature of two-dimensional quantum field theory?

In three or more dimensions, the rotation group $\mathrm{SO}_n(\mathbb{R})$ is non-abelian. We know everything about the finite-dimensional unitary projective representations of this simple Lie group: there are countably many, namely the highest-weight representations of its universal cover $\text{Spin}_n(\mathbb{R})$. Physically, we know these fall into two families ('superselection sectors'), depending on what happens after a rotation by 2π: the true representations of $SO_2(\mathbb{R})$ (the 'integer-spin' bosons) and those that are merely projective (the 'half-integer spin' fermions).

In two dimensions, this familiar picture collapses, as the rotation group $SO_2(\mathbb{R})$ is isomorphic to S^1 and has universal cover $\mathbb{R}$. The unitary representations are parametrised by the 'unitary duals' $\widehat{S^1} \cong \mathbb{Z}$ and $\widehat{\mathbb{R}} \cong \mathbb{R}$, respectively. In particular, the element $x \in \mathbb{R}$ is sent to the 1×1 matrix $e^{2\pi i \alpha x}$ for 'spin' $\alpha \in \widehat{\mathbb{R}} = \mathbb{R}$. The behaviour (monodromy) of these representations under rotations by 2π again determines the physics, and instead of the boson/fermion alternative, we get superselection sectors parametrised by $\widehat{\mathbb{R}}/\widehat{S^1} \cong S^1$.

The different physics of bosons and fermions is revealed by the spin–statistics relation. Define as in (1.2.6) the configuration space $\mathfrak{C}_n(\mathbb{R}^d)$ of n distinct points $x^{(i)}$ in $\mathbb{R}^d$, consisting of n copies of $\mathbb{R}^d$ with all diagonals $x^{(i)} = x^{(j)}$ deleted. We are interested in these describing the positions of n identical particles, so for each permutation $\sigma \in \mathcal{S}_n$ identify $(x^{(1)}, \ldots, x^{(n)}) \in \mathfrak{C}_n(\mathbb{R}^d)$ with $(x^{(\sigma 1)}, \ldots, x^{(\sigma n)})$. A closed loop in $\mathfrak{C}_n(\mathbb{R}^d)/\mathcal{S}_n$ corresponds to an explicit rearrangement of the n particles. It is important to note that, for any n, d, the space of trajectories will be disconnected. In Feynman's formalism, this means we have the freedom to introduce relative factors between the corresponding disjoint path integrals. By unitarity these factors should be phases (complex numbers of modulus 1), and consistency requires them to define a representation of the fundamental group $\pi_1(\mathfrak{C}_n(\mathbb{R}^d))$. For $d > 2$ this fundamental group is the symmetric group $\mathcal{S}_n$, and so there are only two possible choices for these relative phases, corresponding to the two one-dimensional representations of $\mathcal{S}_n$: all $+1$'s, or $\det(\sigma)$. The spin–statistics theorem [518] tells us that $+1$ corresponds to bosons and $\det(\sigma)$ to fermions.

In two dimensions, the fundamental group is the braid group $\mathcal{B}_n$, and its one-dimensional unitary representations are parametrised by $t \in \mathbb{R}/\mathbb{Z}$ and defined by $\sigma_i \mapsto e^{2\pi i t}$. This t parametrises the different consistent assignments of phases to the disjoint integrals in the Feynman expressions. Again, the spin–statistics theorem relates this phase assignment to spin: this t is the same as the spin α (mod 1). This is called *braid statistics* for obvious reasons. Such particles are called *plektons* (after the Greek word for 'braid') or *anyons* (since they can have *any* spin).

One-dimensional representations of $\mathcal{S}_n$ or $\mathcal{B}_n$ are the simplest. Higher-dimensional representations would indicate an internal structure and are considered in, for example, parastatistics. In Section 4.3.3 we see how higher-dimensional representations arise in a similar way in CFT. See, for example, [204], [191] for some general treatments of braid statistics in CFT. Possible physical realisations of braid statistics are reviewed in [557], [345]. In particular, subjecting certain semiconductors to large magnetic fields and cold temperatures yields the so-called fractional quantum Hall effect, and its quasi-particles provide an actual realisation of anyons. Since braid statistics is a topological effect, it is intimately related to the Aharanov–Bohm effect (a notorious topological effect in quantum theories).

So two dimensions are special for quantum field theory. We know four dimensions are special in differential geometry [195]. For example, in any $\mathbb{R}^n$ all differential structures are equivalent, except $n = 4$ where there are uncountably many inequivalent ones (Section 1.2.2). Are those two dimensions related to these four dimensions, and are they related to the apparent four-dimensionality of macroscopic space-time? This isn't clear to this author.

The possibility of braid statistics arises in two dimensions because the space-like vectors in two-dimensional space-time are disconnected. The other special features of two dimensions are all related to this. As we discuss in Section 4.3.2, the space of local conformal transformations is finite-dimensional in n dimensions, except for $n = 2$ where it is infinite. The light-cone minus the origin is also disconnected in two dimensions, and this implies the existence of infinitely many conserved currents.

What makes four dimensions special in differential geometry is the behaviour of embedded 2-discs (many proofs in n dimensions are based on understanding that behaviour). A generic map of a disc into an n-manifold has self-intersections that are one-dimensional if $n = 3$, which consist of isolated points if $n = 4$, and are non-existent if $n \geq 5$. Also, the Seiberg–Witten equations (so useful for studying 4-manifolds) exploit the fact that the rotation algebra $\mathfrak{so}_4 \cong \mathfrak{so}_3 \oplus \mathfrak{so}_3$ (corresponding to a group $SO_4(\mathbb{R})$ homeomorphic to $S^3 \times \mathbb{P}^3(\mathbb{R})$) is nonsimple, while in all other dimensions $n > 2$ $\mathfrak{so}_n$ is simple.

Question 4.3.1. (a) Consider the free scalar theory in d dimensions, given by Lagrangian $\mathcal{L} = -\frac{1}{2} \sum_\mu \partial_\mu \phi \, \partial^\mu \phi$. Assuming scale-invariance of $\mathcal{L}$, deduce the scaling dimension of ϕ.
(b) This theory is massless. What happens when the mass term is introduced?

Question 4.3.2. Prove that when $p + q \neq 2$, the infinitesimal conformal symmetries of $\mathbb{R}^{p,q}$ form a finite-dimensional Lie algebra, but that it is infinite-dimensional when $p + q = 2$. (That is, write $x^\mu \mapsto x^\mu + \epsilon^\mu(x)$; we're interested in those infinitesimal ϵ^μ for which the metric ds^2 goes to a multiple of itself.)

Question 4.3.3. Let Σ be a Riemann surface of genus g with n discs removed. Suppose it is dissected into N 'pairs-of-pants' (i.e. spheres with three discs removed). Prove that this dissection is possible only if $n + 2g > 2$, in which case $N = n + 2g - 2$.

Question 4.3.4. Assuming (4.3.5) and the state-field correspondence, prove $L_1 v = 0$ and $L_0 v = h v$.

Question 4.3.5. Suppose $L_1 v_i = 0$ and $L_0 v_i = h_i v_i$. Compute the chiral blocks

$$\langle \varphi_{v_1}(z_1) \varphi_{v_2}(z_1) \rangle = \begin{cases} C_{12} |z_1 - z_2|^{-2h_1} & \text{if } h_1 = h_2 \\ 0 & \text{otherwise} \end{cases},$$

$$\langle \varphi_{v_1}(z_1) \varphi_{v_2}(z_2) \varphi_{v_3}(z_3) \rangle = \frac{C_{123}}{|z_1 - z_2|^{h_1 + h_2 - h_3} |z_2 - z_3|^{h_2 + h_3 - h_1} |z_1 - z_3|^{h_1 + h_3 - h_2}}$$

for constants C_{12}, C_{123}, using (4.3.5).

Question 4.3.6. Describe the monodromy (if any) about $w = \infty$ of the chiral blocks in (4.3.13).

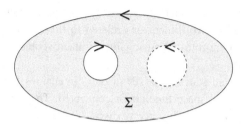

Fig. 4.15 A morphism $\Sigma : C_1 \to C_2$.

Question 4.3.7. Find the sesquilinear combinations $\sum_{i=1,2} c_{ij} \mathcal{F}_i(w) \overline{\mathcal{F}_j(w)}$ of the chiral blocks in (4.3.13), which are invariant under the various monodromies. (The physical correlation functions will be of that form.)

Question 4.3.8. Describe the following orbifolds: (a) $(\mathbb{R}/\mathbb{Z})/\{\pm 1\}$; (b) $(\mathbb{C}/(\mathbb{Z}\tau + \mathbb{Z}))/\{\pm 1\}$; (c) $(\mathbb{C}/(\mathbb{Z} + i\mathbb{Z}) \setminus \Delta)/\mathbb{Z}_2$, where Δ is the diagonal $x + ix$, and $\mathbb{Z}_2$ acts by identifying (x, y) and (y, x).

4.4 Mathematical formulations of conformal field theory

In Sections 4.3.2 and 4.3.3 we gave a quick standard sketch of the basics of CFT, introducing the reader to the main notions. In this section, as well as Chapter 5 and Section 6.1, we explore certain aspects of CFT more carefully, clarifying them considerably. Surprisingly, many of these aspects are fundamental to Moonshine.

4.4.1 Categories

A deeply influential formulation of CFT is due to Graeme Segal [**500**], [**502**], [**498**]; see also [**241**]. It is motivated by string theory (Section 4.3.1) and is phrased using category theory (Section 1.6.1). According to Segal, a CFT is a functor S from a category **C** of Riemann surfaces (the world-sheets) to the category **Hilb** of Hilbert spaces (the state-spaces).

The objects of category **C** are finite disjoint unions C_n of n circles, for all $n \geq 0$. We fix a parametrisation on these circles – that is, a smooth identification t of each circle C with $\mathbb{R}/\mathbb{Z}$; this induces an orientation on C. A morphism $C_m \to C_n$ is a (not necessarily connected) Riemann surface Σ with boundary $\partial\Sigma$ consisting of $m + n$ parametrised circles; exactly n of those boundary circles come with parametrisations consistent with the orientation of Σ induced from its complex structure. We think of these n as 'outgoing' strings and the remaining m as 'incoming' ones. For example, in Figure 4.15 the solid circles are outgoing and the dashed one is incoming. We identify two such morphisms $\Sigma : C_m \to C_n$, $\Sigma' : C_m \to C_n$ if there is a conformal map $f : \Sigma \to \Sigma'$ such that the parametrisations t_i and $t_i' \circ f$ of the boundaries $\partial\Sigma$ and $\partial\Sigma'$ agree.

The space $\mathrm{Hom}(C_m, C_n)$ is topological, with a connected component $\mathbf{C}_\Sigma$ for each homeomorphism class $[\Sigma]$ of (not necessarily connected) surfaces with boundary having

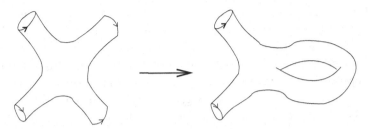

Fig. 4.16 An example of sewing.

$m + n$ components. For example, $\text{Hom}(C_0, C_0)$ has one component for every choice of n_0 spheres, n_1 tori, ..., n_g compact genus-g surfaces, ..., provided $\sum_g n_g < \infty$.

Finally, the composition $\Sigma' \circ \Sigma$ of morphisms $\Sigma : C_m \to C_n$, $\Sigma' : C_n \to C_p$ is obtained by sewing together the surfaces Σ and Σ' along the circles in C_n by using the parametrisation to identify corresponding points on the boundaries. In fact, this sewing construction is the main reason we require these boundary circles to be parametrised.

Recalling Definition 2.1.6, this space $\mathbf{C}_\Sigma$ can be regarded as the quotient of the space of complex structures on Σ, by the group of all diffeomorphisms of Σ that are the identity on the boundary $\partial\Sigma$. Thus, $\mathbf{C}_\Sigma$ is an infinite-dimensional moduli space. Write $\mathbf{C}_{g,k}$ for the component of $\text{Hom}(C_m, C_n)$ corresponding to connected genus-g surfaces (with $k = m + n$ punctures) – this is the most interesting part of $\mathbf{C}_\Sigma$. Recall the enhanced moduli space $\widehat{\mathfrak{M}}_{g,k}$ defined in Section 2.1.4; provided only that $k > 0$, $\mathbf{C}_{g,k}$ is a finite-dimensional complex *manifold*, unlike $\widehat{\mathfrak{M}}_{g,k}$, and can be expressed as a bundle over $\widehat{\mathfrak{M}}_{g,k}$ with infinite-dimensional fibre (page 453 of [**502**]). The mapping class group for $\mathbf{C}_{g,k}$ is the $\widehat{\Gamma}_{g,k}$ of Section 2.1.4, that is an extension of $\Gamma_{g,k}$ by k copies of $\mathbb{Z}$.

The most important space is that $\mathbf{C}_{0,2}$ of annuli. We get the easy homeomorphism

$$\mathbf{C}_{0,2} \cong (0, 1) \times (\text{Diff}^+(S^1) \times \text{Diff}^+(S^1))/S^1. \qquad (4.4.1)$$

The interval $(0,1)$ arises because any annulus is diffeomorphic to $r \le |z| \le 1$ for some $0 < r < 1$. The two copies of $\text{Diff}^+(S^1)$ correspond to reparametrisations of the two boundary circles – this is where the two copies $L_n, \overline{L}_m$ of $\mathfrak{Vir}$ arise. We factor out by S^1 since rotations are the only holomorphic automorphisms of $r \le |z| \le 1$.

A CFT is (among other things) a projective representation of category $\mathbf{C}$: to each object C_n we assign a vector space $\mathcal{S}(C_n)$, and to each morphism $\Sigma : C_m \to C_n$ a linear map $\mathcal{S} : \mathcal{S}(C_m) \to \mathcal{S}(C_n)$, such that for any objects C_m, C_n, C_p and morphisms $\Sigma' : C_m \to C_n$, $\Sigma : C_n \to C_p$, we obtain the functorial *sewing axiom*

$$\mathcal{S}(\Sigma \circ \Sigma') = c(\Sigma, \Sigma') \mathcal{S}(\Sigma) \circ \mathcal{S}(\Sigma') \qquad (4.4.2)$$

for some nonzero $c(\Sigma, \Sigma') \in \mathbb{C}$. More precisely, $\mathcal{S}(C_n)$ is the tensor product $\mathcal{H} \otimes \cdots \otimes \mathcal{H} =: \mathcal{H}^{\otimes n}$ of the state-space $\mathcal{H}$ of our CFT, and $\mathcal{H}^{\otimes 0} := \mathbb{C}$. Here, $\mathcal{H}$ is something like the space $L^2(\mathcal{L}M)$ of wave-functions on the loop-space $\mathcal{L}M := \{f : s' \to M\}$, where M is the space-time in which the string lives. Convergence in the Figure 4.16 sewing operation described below requires the operator $\mathcal{S}(\Sigma)$ to be trace class.

The idea is for $\mathcal{S}(\Sigma)$ to mimic the Feynman path integral (4.2.13a), while avoiding the latter's analytic challenges. In string theory, the incoming state |in⟩ consists of a choice of string state for each of the m circles, so |in⟩ $\in \mathcal{H}^{\otimes m}$; similarly |out⟩ $\in \mathcal{H}^{\otimes n}$. Segal's operator $\mathcal{S}(\Sigma)$ is none other than the (finite) scattering matrix, or the time-evolution operator e^{iHt} (holomorphically extended to imaginary time): the desired string amplitude is⟨out|$\mathcal{S}(\Sigma)$|in⟩. This is what Segal is trying to capture formally.

If Σ is the disjoint union of surfaces Σ_1 and Σ_2, then $\mathcal{S}(\Sigma) = \mathcal{S}(\Sigma_1) \otimes \mathcal{S}(\Sigma_2)$. That the fundamental identity (4.4.2) should hold can be seen by cutting open a Feynman path integral: an integral over all paths starting from α at time 0 to ω at time 1 can be expressed as the integral over all possible μ of all paths starting from α at $t = 0$ to μ at $t = 0.5$, and all paths from μ at $t = 0.5$ to ω at $t = 1$. This is just matrix multiplication, as (4.4.2) suggests. A physical description of sewing can be found, for instance, in section 9.3 of [**253**]. To construct the projective factor in (4.4.2), Segal uses the 'determinant line bundle' [**192**] (see e.g. [**498**], [**502**] for details). An alternate approach to central charge $c \neq 0$ within the Segal formalism is given in lecture 2 of [**241**].

Another kind of sewing occurs when two oppositely oriented boundary components of Σ are sewn together, increasing the genus by 1, as is illustrated in Figure 4.16. Algebraically, this corresponds to taking a trace or a sum using the Hermitian form. (To see why this is compatible with (4.4.2), interpret matrix multiplication as a trace of the tensor product of the matrices.)

Segal's use of surfaces with boundary differs from that of Section 4.3.2. Usually, quantum field theory restricts to the (easier to calculate) limiting case where the incoming and outgoing states are at $t = \mp\infty$. This is the strategy followed in Section 4.3. Segal is instead trying to capture the string amplitudes for finite times, because it makes the $\mathfrak{Vir}$ action manifest, as we'll see shortly. The relation of Segal's picture with that of enhanced compact surfaces is made in pages 6–7 of [**295**].

The multiplication $\mathbf{C}_{0,2} \times \mathbf{C}_{0,2} \to \mathbf{C}_{0,2}$ makes the annuli space $\mathbf{C}_{0,2}$ into an infinite-dimensional complex Lie semi-group (it has no identity and inverses). Its multiplication is described explicitly in section 9 of [**448**], but to get a taste for it, forget temporarily the parametrisations on the boundary circles: then the sewing of annuli $r < |z| < 1$ and $r' < |z| < 1$ obviously yields the annulus $rr' < |z| < 1$, and so this annulus semi-group is isomorphic to that of the interval $(0, 1)$ under multiplication. Recall from Section 3.1.2 that the complex Lie algebra $\mathfrak{Witt}$ has no Lie group, or equivalently that the real Lie group $\mathrm{Diff}^+(S^1)$ has no complexification. The semi-group $\mathbf{C}_{0,2}$ should be regarded as the complexification of $\mathrm{Diff}^+(S^1)$; it plays the same role for $\mathrm{Diff}^+(S^1)$ that the punctured disc $0 < |z| < 1$ plays for S^1. One hint of this is (4.4.1). Another (proposition 3.1 of [**502**]) is that there is a one-to-one correspondence between positive energy projective representations of $\mathrm{Diff}^+(S^1)$ (recall their definition in Section 3.1.2) and holomorphic projective representations of $\mathbf{C}_{0,2}$. The positive energy representations of $\mathrm{Diff}^+(S^1)$ are the only ones with a hope to extend to $\mathbf{C}_{0,2}$, and all of them are necessarily projective. By a conjecture of Kac, these are all highest-weight modules.

In applications to string theory (namely in the presence of 'ghosts'), the positive-definiteness of the Hermitian product in the Hilbert space $\mathcal{H}$ should be weakened. Also, one may wish to supersymmetrise the state-spaces, that is, give them a $\mathbb{Z}_2$-grading (in order to include fermions). See [**502**] for some comments along these lines.

Note that there is an action of $\mathbf{C}_{0,2}$ on each $\mathbf{C}_\Sigma$ – in fact, one for each boundary circle. This semi-group action amounts to lengthening the arms of each end (equivalently, shrinking the boundary circle); physically, this corresponds to time evolution $t \to \infty$ of outgoing states, or time devolution $t \to -\infty$ of incoming states. We are used to time evolution being a unitary (hence invertible) process, but here time is imaginary, that is, space-time is Euclidean, so time evolution is a contraction. As mentioned in Section 4.2.4, Euclidean space-time is better behaved mathematically than the more physical Minkowski space-time, though in a healthy quantum field theory they should be equivalent.

This semi-group action is the integration of the action of $\mathfrak{Witt}$ on the moduli spaces (Section 3.1.2). By (4.4.2), this action means that each space $\mathcal{S}(\mathbf{C}_\Sigma)$ carries a projective $\mathbf{C}_{0,2}$-representation. In particular, we get an action of $\mathbf{C}_{0,2}$ on the state-space $\mathcal{H}$, projective if $c \neq 0$. This is how we recover the representation of $\mathfrak{Vir} \oplus \mathfrak{Vir}$ on $\mathcal{H}$ that is so important in Section 4.3.2.

The higher-genus behaviour of an RCFT is determined from the lower-genus behaviour, by composition of 'arrows' (i.e. the sewing together of surfaces) in category C, as we see in Figures 4.12 and 4.16. Note that several different sewings can yield the same surface. That they must each give the same answer turns out to be a powerful constraint on CFT, called *duality* (Section 6.1.4).

Thanks to sewing, a CFT is uniquely determined by the chiral algebras $\mathcal{V}, \overline{\mathcal{V}}$; the 1-loop partition function (which gives the spectrum of the theory, i.e. the structure of $\mathcal{H}$ as a $\mathcal{V} \oplus \overline{\mathcal{V}}$-module); and the OPE (4.3.2) (see e.g. section 4 of [**502**]).

The simplest interesting example here is the 'tree-level creation of a string from the vacuum', i.e. $\Sigma : C_0 \to C_1$. In this case the world-sheet looks like a bowl, that is homeomorphic with a disc D, and so is associated with a linear map $\mathcal{S}(D) : \mathbb{C} \to \mathcal{H}$. Equivalently, $\mathcal{S}(D)$ is the assignment of the vector $\mathcal{S}(D)(1)$ in $\mathcal{H}$ to D. In the case of the standard unit disc (i.e. where $D = \{z \in \mathbb{C} \,|\, |z| \leq 1\}$ and the parametrisation of the boundary S^1 is simply $\theta \mapsto e^{2\pi i \theta}$), this vector is called the *vacuum state* $|0\rangle$. In section 9 of [**502**] it is explained how to recover the stress–energy tensors $T(z), \overline{T}(\overline{z})$, by deforming the complex structure on the disc; this idea is borrowed from CFT.

For another important example, a surface $\Sigma : C_2 \to C_1$, that is a pair-of-pants, corresponds to a bilinear map $\mathcal{H} \otimes \mathcal{H} \to \mathcal{H}$, and makes $\mathcal{H}$ into an algebra. Choosing Σ appropriately, this gives the OPE (4.3.2). A different choice defines the physical vertex operators (this is explicitly given on page 770 of [**241**]).

Finally, suppose the initial and final objects here are both C_0, so the world-sheets Σ are closed Riemann surfaces. Segal's functor $\mathcal{S}(\Sigma)$ is a linear map $\mathbb{C} \to \mathbb{C}$, so is completely determined by its value at $1 \in \mathbb{C}$. This value $\mathcal{S}(\Sigma)(1) =: \mathcal{Z}(\Sigma) \in \mathbb{C}$ is the partition function. Consider now Σ a torus. Up to conformal equivalence, Σ can be written as the quotient $\Sigma_\tau := \mathbb{C}/(\mathbb{Z} + \mathbb{Z}\tau)$, and so the 1-loop partition function $\mathcal{Z}(\Sigma_\tau)$ becomes a function on $\mathbb{H}$. As we know, Σ_τ and $\Sigma_{\alpha.\tau}$ are conformally equivalent when $\alpha \in \mathrm{SL}_2(\mathbb{Z})$, and so $\mathcal{Z}$ must be modular invariant. We can construct a torus by sewing together the two ends of a cylinder, or equivalently an annulus $A_q = \{z \in \mathbb{C} \mid |q| \leq |z| \leq 1\}$ for $q \in \mathbb{C}$ where the boundaries are parametrised by $qe^{2\pi i\theta}$ and $e^{2\pi i\theta}$. We know that this recovers Σ_τ up to conformal equivalence, if $q = e^{2\pi i\tau}$. Then $\mathcal{S}(A_q) = q^{L_0}\overline{q}^{\overline{L_0}}$ and so by the sewing axiom (with $c = 0$ for convenience) the torus partition function becomes

$$\mathcal{Z}(\tau) = \mathrm{tr}_{\mathcal{H}} q^{L_0}\overline{q}^{\overline{L_0}}.$$

It must be invariant under the usual action of $\mathrm{SL}_2(\mathbb{Z})$. Of course, if the central charge is nonzero, then the sewing axiom picks up a multiplicative factor that recovers (4.3.8b). See page 768 of [**241**] for details.

So far Segal is addressing general CFT. He defines an RCFT – our main interest – as a *modular functor* $\mathfrak{B}$. It assigns to each surface Σ its space of chiral blocks (4.3.7). Let Φ be a finite set of labels – this parametrises the irreducible modules of chiral algebra $\mathcal{V}$. One of these labels, call it 0, is distinguished (it corresponds to the vacuum, and was called V in Section 4.3.2). We require that Φ has an involution $i \mapsto i^*$, called *charge conjugation* and related to complex conjugation. By a labelled Riemann surface with boundary (Σ, α) we mean to assign a label $\alpha_i \in \Phi$ to each (parametrised) boundary circle of Σ. These are the objects in a category **Riem**$_\Phi$. The morphisms are 'holomorphic collapsing maps' (see section 5 of [**502**]), which sew together pairs of boundary circles in the usual way. The target is the category **Vect**$_f$ of finite-dimensional vector spaces, since the spaces of chiral blocks live there; morphisms are linear transformations.

Definition 4.4.1 [502]   *A* modular functor *is a functor $\mathfrak{B}$ from* **Riem**$_\Phi$ *to* **Vect**$_f$*, such that:*

(i) *$\mathfrak{B}$ takes the disjoint union $\Sigma \cup \Sigma'$ to $\mathfrak{B}(\Sigma) \otimes \mathfrak{B}(\Sigma')$.*

(ii) *$\mathfrak{B}(\Sigma) = \mathfrak{B}(-\Sigma)$, where '$-\Sigma$' means that we reverse the orientation of all boundary circles of Σ (i.e. interchange incoming with outgoing circles), and also replace each label α_i with its conjugate α_i^*.*

(iii) *Suppose surface Σ is obtained from surface Σ' by cutting along a closed curve. For each label $i \in \Phi$, let Σ_i be the surface Σ labelled the same as Σ', except its two additional circles are both given the label i. Then*

$$\oplus_{i \in \Phi} \mathfrak{B}(\Sigma_i) \cong \mathfrak{B}(\Sigma'). \tag{4.4.3}$$

(iv) *If D is the standard disc then $\mathfrak{B}(D)$ is $\mathbb{C}$ if the boundary is labelled 0, and $\{0\}$ otherwise.*

(v) *Finally, if Σ_w is a family of surfaces varying holomorphically with a parameter w, then the spaces $\mathfrak{B}(\Sigma_w)$ fit together to form a holomorphic vector bundle.*

We won't spell out precisely what condition (v) means (roughly, it says that the chiral blocks are holomorphic functions on the moduli space), but certainly it implies that the dimension of $\mathfrak{B}(\Sigma)$ only depends on the orientations of the boundary circles and the labels, and not on the complex structure of Σ. We discuss chiral blocks in Section 4.3.3. Their most important property is that they carry a projective representation of the mapping class group of Σ. The definition of modular functor using closed surfaces with marked points, as well as an alternate approach to $c \neq 0$, is given in chapter 5 of [**32**].

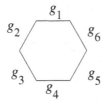

Fig. 4.17 A natural depiction of an identity $g_1 g_2 g_3 g_4 g_5 g_6 = e$.

There are still no known examples of modular functors, though it is expected that any sufficiently nice vertex operator algebra will yield one. Nevertheless, this picture of RCFT is incomplete, as it only captures some elements of the chiral halves of an RCFT. For instance, the modular functor corresponding to Monstrous Moonshine is trivial. The 1-loop partition function (4.3.8b) is important data for the RCFT, but its presence here is obscure (to this author at least), as more generally is the explicit relation between the full CFT and the two chiral halves.

4.4.2 Groups are decorated surfaces

This short subsection motivates topological field theory and can be skipped on first reading.

Fix a group G. We can think of G as a set of identities $g_1 g_2 \cdots g_k = e$. Conjugating by g_1, we observe

$$g_1 g_2 \cdots g_k = e \text{ iff } g_2 g_3 \cdots g_k g_1 = e. \tag{4.4.4}$$

Thus, an identity '$g_1 \cdots g_k = e$' in G really should be written circularly, as in Figure 4.17. In other words, we can think of G as a way to assign to each polygon, whose sides are labelled consecutively by elements g_i of G, a number $\mathcal{P}(g_1, g_2, \ldots, g_k) \in \{0, 1\}$. We assign '1' to a given labelled polygon if, starting anywhere on the circumference and reading counterclockwise, the product of the labels equals e; otherwise assign '0' to it. We get a dihedral symmetry,

$$\mathcal{P}(g_1, g_2, \ldots, g_k) = \mathcal{P}(g_2, \ldots, g_k, g_1), \tag{4.4.5a}$$

$$\mathcal{P}(g_1, g_2, \ldots, g_k) = \mathcal{P}(g_k^{-1}, \ldots, g_2^{-1}, g_1^{-1}), \tag{4.4.5b}$$

corresponding to the symmetries of the k-gon.

Of course not every assignment of 0's and 1's to labelled polygons will come from groups. Most importantly, we have

$$\mathcal{P}(g_1, \ldots, g_m, h_1, \ldots, h_n) = \sum_{g \in G} \mathcal{P}(g_1, \ldots, g_m, g) \, \mathcal{P}(g^{-1}, h_1, \ldots, h_n). \tag{4.4.5c}$$

This can be depicted pictorially as the dissection rule of Figure 4.18. We also get the normalisation rule

$$\sum_{g \in G} \mathcal{P}(g_1, \ldots, g_k, g) = 1. \tag{4.4.5d}$$

This polygonal definition is completely equivalent to the usual one of a group:

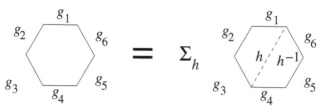

Fig. 4.18 The dissection rule: $g_1g_2g_3g_4g_5g_6 = e$ iff $g_1g_2g_3 = (g_4g_5g_6)^{-1}$.

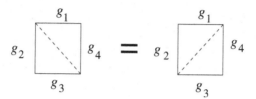

Fig. 4.19 Associativity in a group.

Proposition 4.4.2 *Let S be a set and let $\Pi(S)$ be the set of all polygons labelled with elements of S. Suppose $\mathcal{P} : \Pi(S) \to \{0, 1\}$ obeys all equations (4.4.5), where for $g \in S, \text{'}g^{-1}\text{'}$ denotes the unique element of S satisfying $\mathcal{P}(g, g^{-1}) = 1$. Define $e \in S$ by $\mathcal{P}(e) = 1$ and the multiplication 'gh' by $\mathcal{P}(g, h, (gh)^{-1}) = 1$. Then this defines a group structure on S compatible with the values $\mathcal{P}$ of the polygons in $\Pi(S)$.*

Thus, knowing the values of 1-gons, 2-gons and triangles fixes all other values. Associativity is equivalent to Figure 4.19, and all other generalised associativity relations can be derived from it. The entire group structure is encoded in a few polygons – the rest are redundant – and indeed that is how a group is usually defined. But there is an aesthetic appeal to considering this global (albeit highly redundant) structure provided by all identities in G, and this charm is lost if we focus only on the banal building blocks. It is reminiscent of interpreting the presentation (1.1.9) as a group of braids.

Nevertheless, this rephrasing of the definition of a group is unsatisfactory for several reasons. It seems artificial that the values $\mathcal{P}$ are always either 0's or 1's. Why should we limit the right side of (4.4.4) to being e – for example, any central element will work equally well. Can we consistently sew together two sides of the same polygon, and get more interesting topologies? What does the normalisation condition really mean group-theoretically? These thoughts lead to the following construction.

Fix a group G and irreducible character ch (Section 1.1.3). A polygon whose sides are labelled with elements g_i of G is assigned the complex number $\mathcal{P}(g_1, \ldots, g_k) = \frac{\text{ch}(e)}{\|G\|}\text{ch}(g_1 \cdots g_k)$ (recall that ch(e) is the dimension of ch). Equation (4.4.5a) continues to hold, while (4.4.5b) becomes $\overline{\mathcal{P}(g_1, \ldots, g_k)} = \mathcal{P}(g_k^{-1}, \ldots, g_1^{-1})$. Equation (4.4.5c) follows from the generalised orthogonality relation (theorem 2.13 in [**308**])

$$\frac{1}{\|G\|} \sum_{g \in G} \text{ch}_i(gh) \, \text{ch}_j(g^{-1}) = \delta_{ij} \frac{\text{ch}_i(h)}{\text{ch}_i(e)},$$

valid for irreducible ch_i, ch_j. The 'normalisation condition' (4.4.5d) should be replaced by

$$\sum_{g \in G} |\mathcal{P}(g_1, \dots, g_k, g)|^2 = \frac{ch(e)^2}{\|G\|},$$

where $ch = \sum m_i ch_i$ expresses ch as a sum of irreducible characters. We see that, as before, two consecutive arcs, labelled g, h, can always be replaced by a single arc labelled gh; so a polygon can always be replaced with a disc. Moreover, the label on a disc depends only on the conjugacy class.

More generally, we can use any character of the form $ch = \sum ch_i(e) ch_i$, where we sum over any subset of the irreducible characters; then $\mathcal{P} = ch/\|G\|$ works. For instance, the original assignment (with values in $\{0, 1\}$) corresponds to the character ch of the regular representation of G. The normalisation condition (4.4.5d) is thus seen to be a consequence of orthogonality of characters.

There is no need to stop here. The dissection rule applied to an annulus labelled with conjugacy classes K_g, K_h (h inner, g outer) implies it is assigned $ch(g)\overline{ch(h)}$; more generally, a disc with n smaller discs removed will have value $ch(g)\overline{ch(h_1)} \cdots \overline{ch(h_n)}$. In these more general settings, the orientation of the boundary circle should be made explicit (here they're all taken to be counter-clockwise). A torus with a disc removed, and the boundary circle labelled K_g, has value $\frac{\|G\|}{ch(e)} ch(g)$.

Likewise, any surface with (oriented) punctures labelled by conjugacy classes can be assigned a well-defined complex number. This is, in fact, a slightly enhanced topological field theory (Question 4.4.4).

4.4.3 Topological field theory

The essence of mathematics involves seeing that two different-looking things are actually (from the appropriate perspective) the same. What are different ways of going from point a to point b? In algebra these are functions, the simplest being linear; in geometry, these are cobordisms; in physics, this is time evolution. A topological field theory is their identification.

This subsection strays a little from the main thread of this book, and so we will only sketch the basic idea. The following definition, that topological field theory is a monoidal functor from the cobordism category to $\mathbf{Vect}_f$, is due to Atiyah and was heavily influenced by Segal's definition of CFT (Section 4.4.1). Topological field theory is a beautiful language that has elegantly formulated several deep mathematical ideas (e.g. Morse theory, the Jones polynomial, Donaldson invariants) – see the reviews [25], [564], [62], [534], [32]. The first topological field theories were constructed in physics by Schwarz (1978) and Witten (1982) (see [62] for references). Physically, a topological field theory should arise from the large-distance limit of any quantum field theory with mass gap.

Definition 4.4.3 [25] *A topological field theory in $d + 1$ dimensions assigns to each compact oriented smooth d-dimensional manifold Σ a finite-dimensional complex vector*

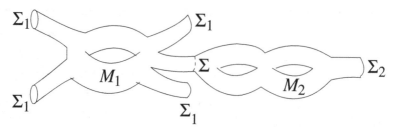

Fig. 4.20 Sewing.

space $T(\Sigma)$, *and to each compact oriented* $(d+1)$-*dimensional manifold* M *with bound-ary* Σ, *a vector* $T(M) \in T(\Sigma)$, *such that:*

(i) $T(\Sigma^*) = T(\Sigma)^*$, *where* Σ^* *denotes* Σ *with opposite orientation, and* $T(\Sigma)^*$ *is the dual space.*

(ii) T *takes the disjoint union* $\Sigma_1 \cup \Sigma_2$ *to* $T(\Sigma_1) \otimes T(\Sigma_2)$.

(iii) *If* $\partial M_i = \Sigma \cup \Sigma_i$ *(disjoint union) and* M *is obtained from* M_1 *and* M_2 *by sewing along a common boundary component* Σ, *as in Figure 4.20, then*
$$T(M) = T(M_2) \circ T(M_1).$$

(iv) T *takes the empty* d-*manifold* $\emptyset$ *to* $\mathbb{C}$.

(v) $T(\Sigma \times I)$ *is the identity endomorphism of* $T(\Sigma)$, *where* I *is the unit interval.*

(vi) *If* $f : \Sigma \to \Sigma'$ *is a homeomorphism, then there is a vector space isomorphism* $T_f : T(\Sigma) \to T(\Sigma')$; *if* $F : M \to M'$ *is a homeomorphism, then*

$$T_{F|_{\partial M}}(T(M)) = T(M').$$

Some technicalities are implicit here; see section 4.2 of [**32**] for any needed clarifications. The book [**534**] is also helpful. If the boundary of M is Σ, and we write Σ as the disjoint union $\Sigma_1 \cup \Sigma_2$, then $T(\Sigma) = T(\Sigma_1) \otimes T(\Sigma_2^*)^*$ and thus the 'vector' $T(M)$ can be regarded as a linear map $T(\Sigma_2^*) \to T(\Sigma_1)$. This functional interpretation is implied in (iii) and (v).

M plays the role here of space-time, and Σ that of space (i.e. a space-like time-slice of M). $T(\Sigma)$ is the space of all states at the given instant, while the map $\mathcal{Z}(M)$ is the time-evolution operator e^{iHt}. Condition (iv) can be interpreted as saying that the Hamiltonian H is 0, so the only evolution is topological.

Question 4.4.6 asks for a proof of the homotopy invariance of T. This means that the mapping class group of Σ, that is the group of components of the group $\mathrm{Diff}^+(\Sigma)$ of orientation-preserving diffeomorphisms, acts on the space $T(\Sigma)$. This is obviously important to us.

Condition (iv) is needed to eliminate the trivial theory. If M is a closed manifold (i.e. it has no boundary), then $T(M) \in T(\emptyset) = \mathbb{C}$. Thus a topological field theory assigns a numerical invariant to closed $(d+1)$-dimensional manifolds.

Let Σ be any d-manifold and put $M_1 = \Sigma \times I$, $M_2 = \Sigma^* \times I$. Sewing these together along corresponding copies of Σ, we get $M = \Sigma \times S^1$. From (v) we get that $T(M_i)$ are the identity maps $T(\Sigma) \to T(\Sigma)$ and $T(\Sigma)^* \to T(\Sigma)^*$, respectively. But we can also think of them as vectors in $\mathcal{T}(\Sigma) \otimes \mathcal{T}(\Sigma)^*$ and $\mathcal{T}(\Sigma)^* \otimes \mathcal{T}(\Sigma)$, so these vectors must be $\sum_i e_i \otimes e_i^*$ and $\sum_j e_j^* \otimes e_j$, respectively, where e_i is any basis of $\mathcal{T}(\Sigma)$ and e_i^* is the dual basis. Thus

$$\mathcal{T}(\Sigma \times S^1) = \left\langle \sum_i e_i \otimes e_i^*, \sum_j e_j^* \otimes e_j \right\rangle = \dim(\mathcal{T}(\Sigma)). \qquad (4.4.6a)$$

Now, we know that 'dimension' can be twisted into 'character' whenever a group is present. So let γ lie in the mapping class group and define $\Sigma \times_\gamma S^1$ to be the $(d+1)$-dimensional manifold obtained by sewing $\Sigma \times I$ to $\Sigma^* \times I$ by identifying the boundary $\Sigma^* \times 0$ with $\Sigma \times 0$ and $\gamma(\Sigma) \times 1$ with $\Sigma \times 1$. Repeating the earlier calculation yields

$$\mathcal{T}(\Sigma \times_\gamma S^1) = \left\langle \sum_i \mathcal{T}_\gamma(e_i) \otimes e_i^*, \sum_j e_j^* \otimes e_j \right\rangle = \mathrm{tr}(\mathcal{T}_\gamma). \qquad (4.4.6b)$$

Theorem 4.4.4   *A topological field theory in* $1+1$ *dimensions is equivalent to a finite-dimensional commutative associative algebra A over $\mathbb{C}$ with unit 1, together with a linear map* $\mathrm{tr}\colon A \to \mathbb{C}$ *such that the bilinear form* $(a, b) \mapsto \mathrm{tr}(ab)$ *is nondegenerate.*

Nondegenerate here means that the only $a \in A$ with $\mathrm{tr}(ab) = 0$, $\forall b \in A$, is $a = 0$. Such an algebra A is called a *Frobenius algebra* – see, for example, chapter 2 of [**353**]. Frobenius algebras were introduced by Frobenius in 1903. The association of a Frobenius algebra to a (1+1)-dimensional topological field theory is straightforward. The vector space A is given by $\mathcal{T}(S^1)$. The boundary of the disc D can be thought of as $\partial D = S^1$ or $\partial D = \emptyset \cup (S^1)^*$, the former interpretation defines a vector $1 := \mathcal{T}(D) \in A$, while the latter defines the map $\mathrm{tr} := \mathcal{T}(D)\colon A \to \mathbb{C}$. The product structure on A comes from $\mathcal{T}$ applied to a pair-of-pants, with boundary $S^1 \cup (S^1 \cup S^1)^*$. The various properties obeyed by multiplication, 1 and tr follow inductively from the various pictures – it's a good idea for the reader to work these out. The proof that a Frobenius algebra defines a unique and well-defined topological field theory is based on the fact that any surface can be obtained by sewing together discs, cylinders and pairs-of-pants; the only difficulty is verifying well-definedness: as we know, the same surface can be decomposed this way in many different ways. The details of this proof are given in section 4.3 of [**32**]; see also section 3.3 of [**353**] for a more pedagogical treatment. This proof is practise for Section 6.1.4, where we do the same for RCFT.

Our symbol 'Σ' in Definition 4.4.3 is due to the special importance of $d = 2$. In his analysis of the Jones polynomial, Witten discovered the explicit relation between topological field theory in $2+1$ dimensions and CFT (in the usual two dimensions): the spaces $\mathcal{T}(\Sigma)$ are the spaces $\mathcal{B}(\Sigma)$ of chiral blocks. The relation between CFT and $(2+1)$-dimensional topological field theory is carefully explained in chapter 5 of [**32**]. In particular, the association of a modular functor with a topological field theory is easy, but (according to [**32**]) the association of a topological field theory with each modular functor is only conjectural at present. $(2+1)$-dimensional topological field theory has been used recently in a series of papers (see [**211**] for a review) for constructing (boundary) RCFT correlation functions.

4.4.4 From amplitudes to algebra

The final rigorous approach to CFT we sketch reconstructs the chiral theory directly from the vacuum-to-vacuum amplitudes. The physical appeal of this approach is that it starts with 'observational' data. For us, it's excellent motivation for the material of the next chapter. We focus on the chiral halves of RCFT – the parts of CFT of greatest interest to mathematics.

A chiral half of a CFT on a sphere consists of a state-space $\mathcal{H}$ and a collection

$$\langle Y(\psi_1, a_1) Y(\psi_2, z_2) \cdots Y(\psi_n, z_n) \rangle \tag{4.4.7}$$

of correlation functions, where z_i lie in the Riemann sphere $\mathbb{P}^1(\mathbb{C}) = \mathbb{C} \cup \{\infty\}$. To avoid circularity, restrict (4.4.7) to states ψ_i in some (typically finite-dimensional) subspace $\mathcal{H}_{gen}$ that generates $\mathcal{H}$. For now, all we need to know about these correlation functions (4.4.7) is that they are multi-linear in the states ψ_i, symmetric under permutation of ψ_i and analytic in the points z_i, except possibly for poles when $z_i = z_j$. At this point the notation in (4.4.7) is purely formal, so for example '$Y(\psi_i, z_i)$' has no meaning. Our first task is to associate with the states $\psi \in \mathcal{H}_{gen}$, vertex operators $Y(\psi, z)$.

Let O be any open set in $\mathbb{P}^1(\mathbb{C})$, with the property that its complement is path-connected and contains a disc. A counterintuitive result of axiomatic quantum field theory (the Reeh–Schlieder Theorem [518], [269]) says that the states $\sum \varphi_1(f_1) \cdots \varphi_n(f_n) |0\rangle$ generated from the vacuum $|0\rangle$ by fields φ_i smeared by test-functions f_i localised to O, will be dense in $\mathcal{H}$. This observation motivates the following construction.

Define the space $\mathcal{V}_O$ formally spanned by all words $Y(\psi_1, z_1) \cdots Y(\psi_n, z_n) |0\rangle$, where $\psi \in \mathcal{H}_{gen}$ and $z_i \in O$, z_i pairwise distinct, and we require any word to be bilinear and symmetric in the ψ_i. We want to complete these infinite-dimensional spaces (i.e. include the limits of certain sequences), topologise them (i.e. decide when vectors are 'close') and identify vectors that are physically indistinguishable (i.e. quotient by null-vectors). We can do all three, using the amplitudes (4.4.7) to define a bilinear pairing $\mathcal{V}_O \times \mathcal{V}_{O'} \to \mathbb{C}$, for any open set O' in the complement of O:

$$\left(\sum_i Y\left(\psi_1^{(i)}, z_1^{(i)}\right) \cdots Y\left(\psi_{m^{(i)}}^{(i)}, z_{m^{(i)}}^{(i)}\right) |0\rangle, \sum_j Y\left(\phi_1^{(j)}, w_1^{(j)}\right) \cdots Y\left(\phi_{n^{(j)}}^{(j)}, w_{n^{(j)}}^{(j)}\right) |0\rangle \right)$$

$$\mapsto \sum_{i,j} \left\langle Y\left(\psi_1^{(i)}, z_1^{(i)}\right) \cdots Y\left(\phi_{n^{(j)}}^{(j)}, w_{n^{(j)}}^{(j)}\right) \right\rangle \tag{4.4.8}$$

for all $\psi_k^{(i)}, \phi_\ell^{(j)} \in \mathcal{H}_{gen}$, $z_k^{(i)} \in O$, $w_\ell^{(j)} \in O'$. A topology on say $\mathcal{V}_{O'}$ is obtained by defining this pairing to be continuous. We identify vectors in $\mathcal{V}_{O'}$ by quotienting by those vectors in $\mathcal{V}_{O'}$ that are orthogonal to all of $\mathcal{V}_O$. This pairing (4.4.8) also allows us to complete the space $\mathcal{V}_{O'}$. The resulting space turns out to be independent of O' – call it $\mathcal{V}^O$. See [227] for details.

If $O_1 \subset O_2$, then we get a natural continuous embedding of $\mathcal{V}^{O_2}$ into $\mathcal{V}^{O_1}$. The role here of the space $\mathcal{H}$ of states is played by this collection $\mathcal{V}$ of topological vector spaces, just as the role of the algebra $\mathcal{A}$ of observables in quantum field theory is played in *algebraic* quantum field theory by the net $\mathcal{A}(\mathcal{O})$ (Section 4.2.4). However, if $O \subset \mathbb{P}^1(\mathbb{C})$ contains ∞ but not 0, then we can define the modes $\psi_{(n)}$, for $\psi \in \mathcal{H}_{gen}$, in the usual way and from this get a Fock space $\mathcal{H}^O \subset \mathcal{V}^O$ spanned by all $(\psi_1)_{(n_1)} \cdots (\psi_k)_{(n_k)} |0\rangle$. It is easy to see that it is dense in $\mathcal{V}^O$ and independent of the choice of O. This Fock space will be the VOA (Definition 5.1.3) of the CFT.

It is now easy to define the vertex operators $Y(\psi, z)$. Choose any $\psi \in \mathcal{H}_{gen}$ and $z \in O$, and any subset $O' \subset O$ with $z \notin O'$. Then the operator $Y(\psi, z) : \mathcal{V}^O \to \mathcal{V}^{O'}$ is defined by

$$\sum_i Y\left(\psi_1^{(i)}, z_1^{(i)}\right) \cdots Y\left(\psi_{m^{(i)}}^{(i)}, z_{m^{(i)}}^{(i)}\right) |0\rangle$$

$$\mapsto \sum_i Y(\psi, z) Y\left(\psi_1^{(i)}, z_1^{(i)}\right) \cdots Y\left(\psi_{m^{(i)}}^{(i)}, z_{m^{(i)}}^{(i)}\right) |0\rangle$$

(there is a little work to see that this operator lifts from $\mathcal{V}_C$ to $\mathcal{V}^O$ – again see [227]). Note that we automatically obtain commutativity: the identity

$$Y(\psi, z) Y(\phi, w) = Y(\phi, w) Y(\psi, z)$$

holds in $\mathcal{V}^O$ provided $z, w \in O$, $z \neq w$, $\psi, \phi \in \mathcal{H}_{gen}$ (compare VA4 in Definition 5.1.3).

So far we have assumed only the most basic properties of the amplitudes (4.4.7). The full splendour of CFT begins to reveal itself once we impose Möbius invariance, which says that it shouldn't matter how we identify the sphere $\mathbb{P}^1(\mathbb{C})$ with the complex coordinates $\mathbb{C} \cup \{\infty\}$. This invariance implies the usual Möbius covariance of the amplitudes and vertex operators. It allows us to extend the definition of vertex operators to, for example, $\mathcal{V}^O$, and to establish Jacobi's identity (5.1.7a). Although this is where things start getting interesting, this is where we leave off.

We know the state-space $\mathcal{H}$ of the CFT is a module for the chiral algebra. This is recovered in this formalism through the two-point functions, which are of the form

$$\langle Y''(\varphi_2, w_2) Y(\psi_1, z_1) \cdots Y(\psi_m, z_m) Y'(\varphi_1, w_1) \rangle, \tag{4.4.9}$$

where the states ψ_i lie in $\mathcal{H}_{gen}$ as before, and φ_j lie in spaces $\mathcal{W}_j$ (which we can take to be dual to each other, although this isn't necessary). We can construct spaces $\mathcal{W}^O$ much as before, generated by

$$\sum_i Y(\psi_1, z_1) \cdots Y(\psi_m, z_m) Y'(\varphi_1, w_1) |0\rangle,$$

and interpret the symbol $Y'(\varphi_1, w_1)$ as a vertex operator sending $\mathcal{W}^O \to \mathcal{W}^{O'}$, much as before. This leads quite naturally to the notion of a VOA-module (Definition 5.3.1).

An observation that will be helpful in Section 5.3.2 in motivating Zhu's algebra is that each representation corresponds to a linear functional on the chiral algebra:

Proposition 4.4.5 [429], [227] *The amplitudes (4.4.9) define a representation of the chiral algebra $\mathcal{V}$, provided that for each open O with path-connected complement, and each states $\varphi_j \in \mathcal{W}_j$ and points $w_i \notin O$, there is a state $v = v(\varphi_1, \varphi_2, w_1, w_2) \in \mathcal{V}^O$ satisfying*

$$\langle Y''(\varphi_2, w_2) Y(\psi_1, z_1) \cdots Y(\psi_m, z_m) Y'(\varphi_1, w_1) \rangle = \langle Y(\psi_1, z_1) \cdots Y(\psi_m, z_m) v \rangle$$

for all choices of $z_i \in O$, $\psi_i \in \mathcal{H}_{gen}$.

The proof of the proposition isn't difficult (see theorem 6 in [**227**]). This proposition permits us to characterise the representations of a chiral algebra by states v. It turns out that these v, which can be interpreted as linear functionals on the Fock space $\mathcal{H}^{O'}$ using the pairing (4.4.8), vanish on a certain large subspace $0_{w_1, w_2}^{O'}$ of $\mathcal{H}^{O'}$, and so define linear functionals on the quotient $\mathcal{H}^{O'} / O_{w_1, w_2}^{O'}$. In the case of a *rational* CFT, this quotient space will be finite-dimensional and is called *Zhu's algebra* (Section 5.3.2).

Question 4.4.1. What is the value $\mathfrak{B}(S^2)$ that Segal's functor associates with the sphere?

Question 4.4.2. Suppose labelled surfaces Σ and Σ' are sewed end-to-end (so the corresponding labels match, and the corresponding circle orientations are opposite), to produce a new labelled surface Σ''. Construct a canonical map $\mathfrak{B}(\Sigma) \otimes \mathfrak{B}(\Sigma') \to \mathfrak{B}(\Sigma'')$. If $\mathfrak{B}(\Sigma), \mathfrak{B}(\Sigma'), \mathfrak{B}(\Sigma'')$ are all nonzero, can that map be identically 0?

Question 4.4.3. (a) Let A be any annulus with oppositely oriented boundary circles. Prove $\mathfrak{B}(A) = \{0\}$, unless both circles are given the same label $i \in \Phi$, in which case $\mathfrak{B}(A) = \mathbb{C}$.
(b) If T is any torus, prove that $\mathfrak{B}(T)$ has dimension equal to the cardinality of Φ.

Question 4.4.4. Find a relation between the assignments $\mathcal{P}^\chi$ to surfaces with punctures labelled with conjugacy classes of G, and two-dimensional topological field theory.

Question 4.4.5. (a) If M is the disjoint union of M_1 and M_2, what is $\mathcal{T}(M)$ in terms of $\mathcal{T}(M_i)$?
(b) What does $\mathcal{T}$ send the empty $(d+1)$-manifold $\emptyset$ to?

Question 4.4.6. Prove that if $f : \Sigma \to \Sigma'$ is a homeomorphism homotopic to the identity, then the linear map $\mathcal{T}_f$ of (vi) is the identity.

Question 4.4.7. Classify all topological field theories of dimension $d = 0$.

<h1 style="text-align:center">5</h1>

<h1 style="text-align:center">Vertex operator algebras</h1>

Vertex operator algebras (VOAs) are a mathematically precise formulation of the notion of *chiral algebra* (Section 4.3.2), the symmetry algebra of conformal field theory. They constitute the simplest expression we have of the machine that associates the Monster $\mathbb{M}$ with the Hauptmoduls. VOAs were first defined by Borcherds, and their theory has since been developed by a number of people. We begin with the rather complicated definition, before turning to our greatest interest: their representation theory. The final section sketches some relations of vertex algebras to geometry. See, for example, [201], [330], [197], [376] for more complete treatments; a more physically minded introduction is provided in [242].

Vertex operator algebras are not a type of operator algebra; rather, they are an algebra of *vertex operators*. Vertex operators arose first in string theory back in the early 1970s as a device for computing string amplitudes. They appeared independently in the mathematical literature (starting with [377]) in order to realise affine Kac–Moody algebras and their modules as algebras of differential operators. Today, just as we define a 'vector' to be an element of a vector space, we define a 'vertex operator' to be a formal power series $Y(u, z)$ appearing in a vertex algebra.

5.1 The definition and motivation

5.1.1 Vertex operators

In bosonic string theory, the vertex operator (Section 4.3.1) corresponding to the absorption of a tachyon with momentum $k = (k^\mu)$ at world-sheet position z and space-time position $X(z) = (X^\mu(z))$ is the normal-ordered expression $V(k, z) = {:}e^{\mathrm{i} k \cdot X(z)}{:}$. Write

$$X^\mu(z) = x^\mu - \mathrm{i} p^\mu \log(z) + \mathrm{i} \sum_{n \neq 0} \frac{1}{n} \alpha_n^\mu z^{-n},$$

where x^μ and p^μ are classically the position and momentum of the string's centre-of-mass, and α_n^μ its oscillation coordinates. Then the vertex operator is (chapter 2.2 of [261])

$$V(k, z) = \exp\left(k \cdot \sum_{n \geq 1} \frac{\alpha_{-n}}{n} z^n\right) z^{k \cdot p - 1} e^{\mathrm{i} k \cdot x} \exp\left(-k \cdot \sum_{n \geq 1} \frac{\alpha_n}{n} z^{-n}\right). \qquad (5.1.1a)$$

Independently, Lepowsky and Wilson realised the affine algebra $A_1^{(1)}$ using differential operators (they tried to do this because finite-dimensional Lie algebras often act as differential operators, for example, on the space of functions on an associated Lie group):

Theorem 5.1.1 [**377**]   *A basis for the affine algebra $A_1{}^{(1)}$ consists of the operators*

$$1, \; y_n, \; \frac{\partial}{\partial y_n}, \; Y_k \qquad \forall n \in \left\{ \frac{1}{2}, \frac{3}{2}, \frac{5}{2}, \dots \right\}, \; k \in \frac{1}{2}\mathbb{Z},$$

thought of as operators on the space $\mathbb{C}[y_{1/2}, y_{3/2}, y_{5/2}, \dots]$ of polynomials in the y_n (we are ignoring the derivation in $A_1{}^{(1)}$). The differential operators Y_k are the homogeneous components of the formal generating function

$$Y(z) = \sum_{k \in \frac{1}{2}\mathbb{Z}} Y_k \, z^k = \exp\left(\sum_n \frac{y_n}{n} z^n \right) \exp\left(-2 \sum_n \frac{\partial}{\partial y_n} z^{-n} \right). \qquad (5.1.1b)$$

In particular, (ignoring the derivation ℓ_0) $A_1{}^{(1)}$ is spanned by a central term C, as well as $e \otimes t^m$, $f \otimes t^m$, $h \otimes t^m$ for each $m \in \mathbb{Z}$ (Section 3.2.2). In Theorem 5.1.1, 1 corresponds to C. For each $n \in \mathbb{N} + 1/2$, the operators y_n and $\partial / \partial y_n$ correspond (up to numerical proportionality factors) respectively to $e \otimes t^{\mp n - 1/2} + f \otimes t^{\mp n + 1/2}$, and $Y_{\pm n}$ corresponds to $-e \otimes t^{\pm n - 1/2} + f \otimes t^{\pm n + 1/2}$. For $k \in \mathbb{Z}$, the operator Y_k corresponds to $h \otimes t^k$ (for $k \neq 0$) and $h \otimes 1 - C/2$ for $k = 0$.

It was Garland who first recognised the formal resemblance between these transcendental expressions (5.1.1a) and (5.1.1b). Note that when expanded out they both involve a sum over powers of z, unbounded in both the positive and negative directions. Doubly-infinite series scream of convergence difficulties. The fractional indices n, k in Theorem 5.1.1 are a signature of what we today call twisted vertex operators.

The geometric meaning of the vertex operator is perhaps best explained in the context of the loop group (Section 3.2.6). Suppose the loop group $\mathcal{L}S^1$ acts on some space $\mathcal{H}$. For each $0 \leq s \leq 2\pi$ and $\epsilon > 0$, consider the loop $\gamma_s^\epsilon \in \mathcal{L}S^1$ defined by

$$\gamma_s^\epsilon(t) = \begin{cases} 1 \in S^1 & \text{for } |s - t| \geq \epsilon \\ \exp\left(\pi i \frac{s-t}{2\epsilon} \right) \in S^1 & \text{for } s - \epsilon < t < s + \epsilon \end{cases},$$

for all $0 \leq t < 2\pi$. In words, γ_s^ϵ stays at the identity $1 \in S^1$ for all time t, except for a small interval around $t \approx s$ where the loop rapidly winds around S^1 once. This loop corresponds to some operator on $\mathcal{H}$; the limit (appropriately taken) as $\epsilon \to 0$ is an operator-valued distribution on $\mathcal{H}$ called a vertex operator (see chapter 13 of [**465**] for details).

5.1.2 Formal power series

As we saw last chapter, the basic object of quantum field theory is the quantum field. It is tempting to think of it as a choice of operator $\hat{A}(x)$ at each space-time point x, but 'function' (or 'section of a vector bundle' for that matter) is too narrow a concept even in free theories.

The *analytic* way to make sense of 'functions' like quantum fields is through distributions, and this was the approach taken in Section 4.2.4. We will describe now the *algebraic* alternative. These two approaches are not equivalent: you can do some things in one approach that you can't do in the other, at least not without difficulty (Section 5.4.1). But as always the algebraic approach is considerably simpler technically – there are no convergence concerns to address – and it is remarkable how much can still be captured. It was first created around 1980 by Garland and Date–Kashiwara–Miwa to make sense of doubly-infinite series like (5.1.1), and is now the language of VOAs. Good introductions to the material in this subsection are [**201**], [**330**], [**376**].

Keep in mind that in CFT we are trying to capture operator-valued 'functions' on two-dimensional Euclidean space-time (Section 4.3.2). Locally space-time looks like $\mathbb{C}$; as explained in Section 4.3, we like to compactify the external legs – for example, for an incoming string tracing a cylindrical world-sheet, the space-time point (x, t) is associated with the complex number $z = e^{t+ix}$, so time $t = -\infty$ corresponds to $z = 0$.

Let $\mathcal{W}$ be any vector space. Define $\mathcal{W}[[z^{\pm 1}]]$ to be the set of all formal series $\sum_{n=-\infty}^{\infty} w_n z^n$, where the coefficients w_n lie in our space $\mathcal{W}$. We don't ask here whether a given series converges or diverges; z is merely a formal place-keeping variable. We will also be interested in the space $\mathcal{W}[z^{\pm 1}]$ of Laurent polynomials, that is, expressions of the form $\sum_{n=-M}^{N} w_n z^n$. $\mathcal{W}[[z^{\pm 1}]]$ itself forms a vector space, using the obvious addition and scalar multiplication.

Our aim here is to describe quantum fields, so we want our formal series to be operator-valued. To do this, choose $\mathcal{W}$ to be a vector space of operators (matrices if you prefer): $\mathcal{W} = \mathrm{End}(V)$, for some space V. We are actually interested in V being the infinite-dimensional state-space of the theory, but in the following examples we take $V = \mathbb{C}$, that is power series with numerical coefficients.

We can now multiply our formal series in the obvious way. For example, consider $V = \mathbb{C}$, and take $c(z) = z^{21} - 5z^{100}$ and $d(z) = \sum_{n=-\infty}^{\infty} z^n$. Then

$$c(z)\,d(z) = \sum_{n\in\mathbb{Z}} z^{n+21} - 5 \sum_{n\in\mathbb{Z}} z^{n+100} = \sum_{n\in\mathbb{Z}} z^n - 5 \sum_{n\in\mathbb{Z}} z^n = -4d(z).$$

This simple calculation tells us many things.

(i) *We can't always divide*: $c(z)\,d(z) = -4d(z)$ shows that the cancellation law fails and that $\mathbb{C}[[z^{\pm 1}]]$ isn't even an integral domain.

(ii) Try to compute the square $d(z)^2$: we get infinity. That is, you can't always multiply in $\mathcal{W}[[z^{\pm 1}]]$.

(iii) Working out a few more multiplications of this kind, we find that $f(z)\,d(z) = f(1) \times d(z)$ for any f for which $f(1)$ exists (e.g. any Laurent polynomial $f \in \mathcal{W}[z^{\pm 1}]$). Thus $d(z)$ is what we have called the Dirac delta $\delta(z - 1)$ centred at $z = 1$. (You can think of it as the Fourier expansion of the Dirac delta, followed by a change of variables.) So of course it makes perfect sense that we couldn't work out $d(z)^2$ – we were trying to square the Dirac delta, which we know is impossible!

There is a certain divergence of notations here: should δ be written additively (i.e. $\delta(z-1)$), in the familiar way, or should it be written multiplicatively (i.e. $\delta(z)$), in the more honest way? Throughout this chapter *we use the multiplicative notation*. So we get

$$\delta(z) := \sum_{n=-\infty}^{\infty} z^n. \tag{5.1.2}$$

In fact, the best notation of all would be the awkward $\delta(z)\,dz$, since the Dirac delta centred at $z = a$ is $\sum_n z^n a^{-n-1} = a^{-1}\delta(z/a)$.

Making contact with Section 1.3.1, the Laurent polynomials $(\text{End}\,\mathcal{V})[z^{\pm 1}]$ play the role here of the smooth functions C^{∞}_{cs} with compact support, and the formal power series $(\text{End}\,\mathcal{V})[[z^{\pm 1}]]$ play the role of its dual. So these power series $f \in (\text{End}\,\mathcal{V})[[z^{\pm 1}]]$ are formal distributions – this is why $f(z)$ usually diverges. The evaluation $f(p)$ of a distribution $f \in (\text{End}\,\mathcal{V})[[z^{\pm 1}]]$ on the test function $p \in (\text{End}\,\mathcal{V})[z^{\pm 1}]$ is given by the 'formal residue' $\text{Res}_z(f(z)\,p(z)) \in \text{End}\,\mathcal{V}$, where

$$\text{Res}_z\left(\sum_{n\in\mathbb{Z}} b_n z^n\right) = b_{-1}. \tag{5.1.3a}$$

The idea is that, up to a factor of $(2\pi\mathrm{i})^{-1}$, this would equal the contour integral of $g(z) = \sum b_n z^n$ around a small circle about $z = 0$, at least for meromorphic g. Hence Res_z obeys many of the familiar properties of integrals, such as integration by parts:

$$\text{Res}_z(g\,\partial_z f) = -\text{Res}_z(f\,\partial_z g), \tag{5.1.3b}$$

where $\partial_z f$ is the formal (term-by-term) derivative of $f(z)$. For example, the formal distribution $a^{-k-1}(\partial_z^k \delta)(z/a)$ takes the test function $f(z)$ to the value $(-1)^k(\partial_z^k f)(a)$. Because of the usefulness of the notion of residue, we write

$$f(z) = \sum_{n\in\mathbb{Z}} a_n z^n =: \sum_{m\in\mathbb{Z}} a_{(m)} z^{-m-1}, \tag{5.1.3c}$$

where $a_{(m)} = \text{Res}_z(z^m f(z)) = a_{-m-1}$ is called a *mode*.

Similar remarks hold for several variables z_i. The distributions are the formal series

$$f(z_1,\ldots,z_k) = \sum_{n_i\in\mathbb{Z}} a_{n_1,\ldots,n_k} z_1^{n_1}\cdots z_k^{n_k} = \sum_{m_i\in\mathbb{Z}} a_{(m_1,\ldots,m_k)} z_1^{-m_1-1}\cdots z_k^{-m_k-1}$$

in $\mathcal{W}[[z_1^{\pm 1},\ldots,z_k^{\pm 1}]]$, and the test functions $f(z_1,\ldots,z_k) \in \mathcal{W}[z_1^{\pm 1},\ldots,z_k^{\pm 1}]$ consist of those power series with only finitely many nonzero terms. The Dirac delta centred at $z_1 = z_2$ is given by $z_2^{-1}\delta(z_1/z_2) = z_1^{-1}\delta(z_2/z_1)$.

But we must not get overconfident:

Paradox 5.1   *Consider the following product:*

$$\delta(z) = \left[\left(\sum_{n\geq 0} z^n\right)(1-z)\right]\delta(z) = \left(\sum_{n\geq 0} z^n\right)[(1-z)\,\delta(z)] = \left(\sum_{n\geq 0} z^n\right)[0\,\delta(z)] = 0.$$

When physicists are confronted with 'paradoxes' such as this, they respond by treading with care when they are involved in a calculation reminiscent of the paradoxes, and otherwise trusting their instincts. Mathematicians typically over-react: after kicking themselves for walking head first into a 'paradox', they devise a rule absolutely guaranteeing that the paradox will always be safely avoided in the future. We will follow the mathematicians' approach, and in the next few paragraphs describe how to avoid Paradox 5.1 by forbidding certain innocent-looking products.

Recall that we are actually interested in the space $\mathcal{W} = \text{End}(\mathcal{V})$. We call infinitely many linear maps $w^{(i)} \in \text{End}(\mathcal{V})$ *algebraically summable* if for every vector $v \in \mathcal{V}$, only finitely many values $w^{(i)}v \in \mathcal{V}$ are different from 0. In other words, fixing a basis for $\mathcal{V}$, only finitely many of the matrices $w^{(i)}$ have a nonzero first column, only finitely many have a nonzero second column, etc. The usual notation '$\sum_i w^{(i)}$' will denote the well-defined endomorphism sending each $v \in \mathcal{V}$ to that effectively finite sum $\sum_i w^{(i)}v$.

Consider a family (possibly infinite) of formal series $w^{(i)}(z) \in \mathcal{W}[[z^{\pm 1}]]$. We certainly have a well-defined sum $\sum_i w^{(i)}(z)$ if for each fixed n, the set $\{w_n^{(i)}\}$ (as i varies) of maps is algebraically summable. We shall call such a sum *algebraically defined*, and write

$$\sum_i w^{(i)}(z) = \sum_{n \in \mathbb{Z}} \left(\sum_i w_n^{(i)} \right) z^n.$$

All other sums are forbidden. Likewise, we certainly have a well-defined product $\prod_{i=1}^m w^{(i)}(z)$ of finitely many formal power series if for each n, the set

$$\left\{ w_{n_1}^{(1)} \, w_{n_2}^{(2)} \cdots w_{n_m}^{(m)} \right\}_{\sum n_i = n}$$

(vary the n_i subject to the constraint $\sum_i n_i = n$) is algebraically summable. Again, call such a product *algebraically defined* and set it equal to

$$\prod_{i=1}^m w^{(i)}(z) = \sum_{n \in \mathbb{Z}} \left(\sum_{n_1 + \cdots + n_m = n} w_{n_1}^{(1)} \, w_{n_2}^{(2)} \cdots w_{n_m}^{(m)} \right) z^n,$$

where the second sum is over all m-tuples (n_i) obeying $\sum_i n_i = n$. All other products (e.g. all infinite ones) are forbidden. An algebraically defined product is necessarily associative.

There are certainly more general ways to have a well-defined product or sum. For example, according to our rule, the series $\sum_n 2^{-n}$ would be forbidden. In this way we avoid the more complicated realm of convergence issues. In short, we are doing *algebra* here, and don't want to be distracted by the dust clouds kicked up by mere *analytic* concerns. Such restrictions are common in infinite-dimensional algebra (recall footnote 14 in chapter 1). The product of a distribution $f \in \mathcal{W}[[z_1^{\pm 1}, \ldots, z_k^{\pm 1}]]$ with a test function $p \in \mathcal{W}[z_1^{\pm 1}, \ldots, z_k^{\pm 1}]$ is always defined, and will be a distribution. The explanation of Paradox 5.1 is that although $(\sum z^n)(1 - z)$ exists and equals 1, and $(1 - z)\delta(z)$ exists and equals 0, the triple product $(\sum z^n)(1 - z)\delta(z)$ is forbidden.

A consequence of our algebraic approach is that the product $z^{\frac{1}{2}}\delta(z)$ does not equal $1^{\frac{1}{2}}\delta(z) = \delta(z)$ – their formal power series are very different. In hindsight this 'failing' is understandable: it is artificial here to prefer the positive root of 1 over the negative root.

Proposition 5.1.2   *Let $\mathcal{W}$ be any vector space, and $f \in \mathcal{W}[[z_1^{\pm 1}, z_2^{\pm 1}]]$. Then $(z_1 - z_2)^N f(z_1, z_2) = 0$ for some integer $N \geq 1$, iff*

$$f(z_1, z_2) = \sum_{j=0}^{N-1} c_j(z_2)\, \partial_{z_2}^j \delta(z_1/z_2),$$

where $c_j(z_2) = \mathrm{Res}_{z_1}((z_1 - z_2)^j f(z_1, z_2)) \in \mathcal{W}[[z_2^{\pm 1}]]$.

Proof: First, $(z_1 - z_2) f(z_1, z_2) = 0$ iff $a_{m-1,n} = a_{m,n-1}\ \forall m, n$, iff $a_{m,n} = a_{0,m+n}$ $\forall m, n$, iff

$$f(z_1, z_2) = \left(\sum_{n \in \mathbb{Z}} a_{0,n} z_2^{n+1} \right) \delta(z_1/z_2).$$

Also, for any $j \geq 1$,

$$(z_1 - z_2)\, \partial_{z_2}^j \left(z_2^{-1} \delta(z_1/z_2) \right) = (z_1 - z_2) \sum_{n \in \mathbb{Z}} n\,(n-1) \cdots (n - j + 1) z_1^{-n-1} z_2^{n-j}$$

$$= j\, \partial_{z_2}^{j-1} \left(z_2^{-1} \delta(z_1/z_2) \right).$$

Hence

$$(z_1 - z_2)\, f(z_1, z_2) = \sum_{j=0}^{M} b_j(z_2)\, \partial_{z_2}^j \left(z_2^{-1} \delta(z_1/z_2) \right)$$

has general solution

$$f(z_1, z_2) = \sum_{j=0}^{M} \frac{1}{j+1} b_j(z_2)\, \partial_{z_2}^{j+1} \left(z_2^{-1} \delta(z_1/z_2) \right).$$

∎

For reasons given next subsection, we call any formal distributions $a(z)$, $b(z)$ *mutually local* if $f(z_1, z_2) := [a(z_1), b(z_2)]$ satisfies the condition in Proposition 5.1.2. In a vertex algebra or VOA (Definition 5.1.3), all fields are mutually local.

We need ways to make new formal power series from old ones. First, for any $n \in \mathbb{Z}$, we define the binomial formula to hold:

$$(z_1 + z_2)^n := \sum_{k \in \mathbb{N}} \binom{n}{k} z_1^{n-k} z_2^k, \tag{5.1.4a}$$

where we define $\binom{n}{k} = n(n-1) \cdots (n-k+1)/k!$ for any n. Equation (5.1.4a) lets us define, for any formal power series $f(z) = \sum_n a_n z^n \in \mathcal{W}[[z^{\pm 1}]]$,

$$f(z_1 + z_2) := \sum_{n \in \mathbb{Z}} \sum_{k \geq 0} a_n \binom{n}{k} z_1^{n-k} z_2^k \in \mathcal{W}\left[[z_1^{\pm 1}, z_2]\right]. \tag{5.1.4b}$$

Paradox 5.2   *Expand $(1 - z)^{-1}$ in a formal series in z to get $\sum_{n \geq 0} z^n$, and $(1 - z)^{-1} = -z^{-1}(1 - z^{-1})^{-1}$ in a formal series in z^{-1} to get $-\sum_{n < 0} z^n$. Subtract these equal expressions; we presumably should get 0, but we actually get $\delta(z)$. Similarly, applying (5.1.4a) to $(1 + z)^{-1} = (z + 1)^{-1}$ again gives us the contradiction $0 = \delta(z)$.*

The analytic explanation is that the left expansions in Paradox 5.2 converge only for $|z| < 1$, while the second converges for $|z| > 1$, so it would be naive to expect their formal difference to be 0. We see from this that *it really matters in which variable we expand rational functions*. The seemingly harmless (5.1.4a) is actually a *convention* saying that we'll expand in positive powers of the second variable. For instance, at first glance

$$z_0^{-1}\delta\left(\frac{z_1 - z_2}{z_0}\right) - z_0^{-1}\delta\left(\frac{z_2 - z_1}{-z_0}\right) = z_2^{-1}\delta\left(\frac{z_1 - z_0}{z_2}\right) \qquad (5.1.5)$$

is nonsense; it only holds if you expand the terms in positive powers of z_2, z_1 and z_0 respectively. A rational function by itself does not define a unique formal power series. When we need to be explicit, we write $\iota_z(f)$ to expand a rational function f in positive powers of z (i.e. for expanding it about $z = 0$). For example,

$$\iota_z\left(\frac{1}{w-z}\right) - \iota_{z^{-1}}\left(\frac{1}{w-z}\right) = \delta(z/w).$$

Recall the operator product expansion (OPE) of quantum fields (4.3.2), introduced to interpret pointwise products. Here we can study this more explicitly. For most pairs $a(z), b(z) \in (\text{End } \mathcal{V})[[z^{\pm 1}]]$, the naive product $a(z) b(z)$ will not be algebraically defined. It is easy to prove directly from Proposition 5.1.2 (see theorem 2.3 of [330]) that if $(z_1 - z_2)^N [a(z_1), b(z_2)] = 0$, then

$$a(z_1) b(z_2) = \sum_{j=0}^{N-1} \frac{c^j(z_2)}{(z_1 - z_2)^{j+1}} + :a(z_1) b(z_2): \qquad (5.1.6a)$$

separates $a(z_1)b(z_2)$ into its singular and regular parts, where

$$:a(z_1) b(z_2): := \left(\sum_{n\geq 0} a_n z_1^n\right) b(z_2) + b(z_2)\left(\sum_{n<0} a_n z_1^n\right), \qquad (5.1.6b)$$

$$c_{(k)}^j(z_2) = \sum_{\ell=0}^{N-1} \frac{j!}{\ell!\,(j-\ell)!} a_{(j-\ell-k)} b_{(\ell)}. \qquad (5.1.6c)$$

By $1/(z_1 - z_2)^{j+1}$ in (5.1.6a) we mean to expand z_2 in powers from $-j$ to ∞. The point of (5.1.6a) is that the *normal-ordered product* (5.1.6b) is algebraically defined even at $z_1 = z_2$ (Question 5.1.6) so any singular behaviour of $a(z_1) b(z_2)$ as $z_1 \to z_2$ is captured by the finitely many series c^j. Equations (5.1.6) are the desired relation in CFT between the singular part of the OPE of quantum fields and the commutators of modes, mentioned in Section 4.3.2. The clarity that vertex algebras bring to quantum field theory (especially CFT) alone makes its definition worth all the pain.

5.1.3 Axioms

We are now prepared to introduce the important new structure called vertex operator algebras (VOAs). Although VOAs are natural from the CFT perspective and appear to be an important and rapidly developing area in mathematics, their definition is difficult and nontrivial examples are not easy to find.

A VOA is an infinite-dimensional graded vector space $V = \oplus_{n \geq 0} V_n$ with infinitely many bilinear products $u *_n v$ respecting the grading (in particular $V_k *_n V_\ell \subseteq V_{k+\ell-n-1}$), obeying infinitely many constraints. We can collect all these products into one generating function: to each $u \in V$ associate the formal power series (a *vertex operator*)

$$Y(u, z) := \sum_{n \in \mathbb{Z}} u_{(n)} z^{-n-1} \in (\text{End } V)[[z^{\pm 1}]].$$

For each $u \in V$, the coefficients $u_{(n)}$ (called modes (5.1.3c)) are maps from V to V. The product $u *_n v$ is now written $u_{(n)}v := u_{(n)}(v)$. The bilinearity of $*_n$ translates into two things: that $u \mapsto Y(u, z)$ is linear, and that each function $v \mapsto u_{(n)}v$ is itself linear (i.e. $u_{(n)}$ is an endomorphism of V).

Definition 5.1.3   (a) *Let V be a graded vector space $V = \oplus_{n=-\infty}^{\infty} V_n$ such that each subspace V_n is finite-dimensional. Suppose we have a linear assignment $u \mapsto Y(u, z) = \sum_{n \in \mathbb{Z}} u_{(n)} z^{-n-1}$ from V into $(\text{End } V)[[z^{\pm 1}]]$ and a distinguished vector $\mathbf{1} \in V$ in V_0, obeying the following properties $\forall u, v \in V$:*

VA1. (grading) *For $u \in V_k$, $u_{(n)}$ is a linear map from V_ℓ into $V_{k+\ell-n-1}$;*
VA2. (vacuum) *$Y(\mathbf{1}, z)$ is the identity (i.e. $\mathbf{1}_{(n)}v = \delta_{n,-1}v$);*
VA3. (state-field correspondence) *$Y(u, 0)\mathbf{1}$ exists and equals u;*
VA4. (locality) *$(z_1 - z_2)^M [Y(u, z_1), Y(v, z_2)] = 0$ for some integer $M = M(u, v)$;*
VA5. (regularity) *there is an $N = N(u, v)$ such that $u_{(n)}v = 0$ for all $n \geq N$.*

Any such triple $(V, Y, \mathbf{1})$ is called a vertex algebra. *The distributions $Y(u, z)$ are called* vertex operators, *and the vector $\mathbf{1}$ is called the* vacuum.
(b) *A vertex algebra $(V, Y, \mathbf{1})$ is called a* vertex operator algebra (VOA) *if there is a distinguished vector $\omega \in V_2$ such that*

VOA1. (conformal symmetry) *$L_n := \omega_{(n+1)}$ forms a $\mathfrak{Vir}$-module, whose central term C in (3.1.5) acts as $c \, \text{id}_V$ for some $c \in \mathbb{C}$;*
VOA2. (conformal weight) *$L_0 v = nv$ whenever $v \in V_n$;*
VOA3. (translation generator) *$Y(L_{-1}v, z) = \partial_z Y(v, z)$;*
VOA4. (CFT type) *$V_0 = \mathbb{C}\mathbf{1}$ and $V_n = \{0\}$ for all $n < 0$.*

The vector ω is called the conformal vector, *and c is called the* central charge, *conformal anomaly or rank. The grading n of $u \in V_n$ is called its* conformal weight.
(c) *A quadruple $(V, Y, \mathbf{1}, \omega)$ is called a* near-VOA *if all axioms of a VOA are satisfied, except for VOA4, and in addition the homogeneous subspaces V_n are allowed to be infinite-dimensional.*

We prefer the more descriptive name 'conformal vertex algebra' to the historical 'vertex operator algebra', although it is probably too late to dislodge the latter name. We study the Virasoro algebra in Section 3.1.2, where we discuss its relation to conformal transformations. We are more interested in VOAs than vertex algebras, since the Virasoro algebra is essential for the relation of V to higher genus and in particular to modular functions. The central charge c is an important invariant of V. The original axioms

[**68**] by Borcherds didn't involve $\mathfrak{Vir}$ nor require $\dim(\mathcal{V}_n) < \infty$. The conformal axioms VOA1–VOA3 were introduced in [**201**] along with the name 'vertex operator algebra'. Although VOA4 holds for most important VOAs and yields the richest theory, it is not standard and is included here for simplicity. Note though that with it, VA5 becomes redundant and can be dropped. The name 'near-VOA' is not standard; we need the notion in Section 7.2.2.

In the physics literature, the vacuum **1** is often denoted $|0\rangle$, and in place of the expansion $Y(u, z) = \sum_n u_{(n)} z^{-n-1}$ for $u \in \mathcal{V}_k$ appears the expression $\sum_n u_{\{n\}} z^{-n-k}$ (so $L_n = \omega_{\text{ins}}$). This new expansion cleans up some formulae a little; it has the disadvantage though of artificially favouring the 'homogeneous' vectors $u \in \mathcal{V}_k$.

By Proposition 5.1.2, the peculiar-looking VA4 simply says that the commutator $[Y(u, z_1), Y(v, z_2)]$ of two vertex operators is a finite linear combination of derivatives of various orders of the Dirac delta centred at $z_1 = z_2$; this powerful locality axiom is at the heart of a vertex algebra. A recommended exercise is to show that in a VOA, $M = 4$ works in VA4 for $u = v = \omega$; more generally see Question 5.1.4.

By $\mathcal{V} = \oplus \mathcal{V}_n$ here, we mean that any vector $u \in \mathcal{V}$ can be expressed as a *finite* sum $\sum_n u(n)$ of homogeneous vectors $u(n) \in \mathcal{V}_n$. To emphasise this finiteness, the notation

$$\mathcal{V} = \coprod_{n \in \mathbb{N}} \mathcal{V}_n$$

is often used. Note that in a vertex algebra, any series $Y(u, z)v$ will be a finite sum – that is, the infinite sum $Y(u, z)$ is algebraically defined (Section 5.1.2).

An immediate consequence of VA1, VA2 and VOA2 is that $\mathbf{1} \in \mathcal{V}_0$ and $\omega \in \mathcal{V}_2$ – we needn't assume these.

Let $\mathcal{V}$ be a vector space with a linear map $Y : \mathcal{V} \to \text{End}\,\mathcal{V}$, such that $Y(u)Y(v) = Y(v)Y(u)$. Also, assume that there exists a distinguished vector $\mathbf{1} \in \mathcal{V}$ such that $Y(\mathbf{1})$ is the identity, and such that $Y(u)\mathbf{1} = u$ for all $u \in \mathcal{V}$. It isn't hard to identify such a structure. Given any $u, v \in \mathcal{V}$, define the 'product' $u * v$ to be the value $Y(u)v$. The linearity of $Y : \mathcal{V} \to \text{End}\,\mathcal{V}$, as well as the linearity of each map $Y(u)$, yields the distributivity laws. Also, $\mathbf{1} * u = Y(\mathbf{1})u = Iu = u$ and $u * \mathbf{1} = Y(u)\mathbf{1} = u$, so $\mathbf{1}$ is a unit. Evaluating $Y(u)Y(v) = Y(v)Y(u)$ on the right by w, gives

$$u * (v * w) = Y(u)(Y(v)w) = Y(v)(Y(u)w) = v * (u * w).$$

Substituting $w = \mathbf{1}$ gives $u * v = v * u$, that is the product is commutative. Likewise, $u * (v * w) = u * (w * v) = w * (u * v) = (u * v) * w$, so the product is associative. Thus *a vertex algebra is an analogue of a commutative associative algebra with unit, where there is a product $u *_z v = Y(u, z)v$ at each point z in a punctured disc.* A vertex algebra isn't as obscure as it may first look.

Theorem 5.1.4 *The following are equivalent:*

(i) $\mathcal{V}$ *is a commutative vertex algebra, i.e. $Y(u, z_1)Y(v, z_2) = Y(v, z_2)Y(u, z_1)$ for all $u, v \in \mathcal{V}$;*

(ii) $\mathcal{V} = \oplus_{n=0}^{\infty} \mathcal{V}_n$ *is a $\mathbb{Z}$-graded commutative associative algebra with unit and derivation, with each* $\dim(\mathcal{V}_n) < \infty$;

(iii) $\mathcal{V}$ *is a vertex algebra where each vertex operator $Y(u, z)$ involves only nonnegative powers of z, i.e. $u_{(n)} = 0$ for all $n \geq 0$.*

Proof: The equivalence (i) $\Leftrightarrow$ (ii) was essentially established in the previous paragraph. (i) $\Rightarrow$ (iii): Consider the equality

$$\sum_{n\in\mathbb{Z}} u_{(-n-1)}v\, z_1^n = Y(u, z_1)v = Y(u, z_1)Y(v, z_2)\mathbf{1}|_{z_2=0}$$

$$= Y(v, z_2)Y(u, z_1)\mathbf{1}|_{z_2=0} = \sum_{n\geq 0} v_{(-1)}u_{(-n-1)}\,\mathbf{1}\, z_1^n.$$

Since the expression on the right side involves nonnegative powers of z_1 only, the same must hold for the left side.

(i) $\Leftarrow$ (iii): For any power series $f(z_1, z_2) = \sum_{m,n=0}^{\infty} a_{mn}z_1^m z_2^n \in \mathcal{W}[[z_1, z_2]]$, Proposition 5.1.2 implies that $(z_1 - z_2)^M f(z_1, z_2) = 0 \Rightarrow f(z_1, z_2) = 0$, since each residue of $f(z_1, z_2)$ will be 0. Applying this to $f(z_1, z_2) = [Y(u, z_1), Y(v, z_2)]$ gives the desired result. ∎

Locality VA4 can be rewritten in the form (see Section 3.2 of [**376**])

$$z_0^{-1}\delta\left(\frac{z_1 - z_2}{z_0}\right) Y(u, z_1)Y(v, z_2) - z_0^{-1}\delta\left(\frac{z_2 - z_1}{-z_0}\right) Y(v, z_2)Y(u, z_1)$$

$$= z_2^{-1}\delta\left(\frac{z_1 - z_0}{z_2}\right) Y(Y(u, z_0)v, z_2), \qquad (5.1.7a)$$

where the formal series are expanded appropriately. This embodiment of commutativity and associativity in the vertex algebra is called the *Jacobi identity* since it plays an analogous role in VOAs as the Jacobi identity plays in Lie algebras. It corresponds directly to the duality of the sphere with four points removed (namely Figure 6.3(a)). Expanding it out, the coefficient in front of $z_0^\ell z_1^m z_2^n$ gives *Borcherds' identity*:

$$\sum_{i\geq 0}(-1)^i \binom{\ell}{i} \left(u_{(\ell+m-i)} \circ v_{(n+i)} - (-1)^\ell v_{(\ell+n-i)} \circ u_{(m+i)}\right)$$

$$= \sum_{i\geq 0}\binom{m}{i} (u_{(\ell+i)}v)_{(m+n-i)}. \qquad (5.1.7b)$$

Specialising (5.1.7b) to $\ell = 0$ and $m = 0$, respectively, gives us

$$[u_{(m)}, v_{(n)}] = \sum_{i\geq 0}\binom{m}{i} (u_{(i)}v)_{(m+n-i)}, \qquad (5.1.7c)$$

$$(u_{(\ell)}v)_{(n)} = \sum_{i\geq 0}(-1)^i \binom{\ell}{i} \left(u_{(\ell-i)} \circ v_{(n+i)} - (-1)^\ell v_{(\ell+n-i)} \circ u_{(i)}\right). \qquad (5.1.7d)$$

In any vertex algebra, define an endomorphism $T : \mathcal{V} \to \mathcal{V}$ by

$$Tu = u_{(-2)}\mathbf{1}. \qquad (5.1.8a)$$

This is the derivation of Theorem 5.1.4(ii). Indeed, applying (5.1.7d) to it and using VA2, we get $Y(Tu, z) = \partial_z Y(u, z)$. Thus in any VOA, VOA3 says

$$L_{-1}u = u_{(-2)}\mathbf{1}. \tag{5.1.8b}$$

Moreover, (5.1.7c) tells us that any $u \in V_2$ automatically obeys $[u_{(0)}, Y(v, z)] = Y(u_{(0)}v, z)$. Thus in any VOA

$$[L_{-1}, Y(u, z)] = \partial_z Y(u, z).$$

More generally, a more subtle argument (see e.g. proposition 3.1.19 of [376]) shows that in any vertex algebra, we have

$$Y(u, z)v = e^{zT} Y(v, -z)u.$$

These equations also allow us to compute explicitly the grading of $u_{(n)}v$ in a VOA, recovering VA1: let $u \in V_k$, $v \in V_\ell$, then

$$L_0(u_{(n)}v) = \omega_{(1)}(u_{(n)}(v)) = u_{(n)}(\omega_{(1)}v) + (\omega_{(1)}u)_{(n)}v + (\omega_{(0)}u)_{(n)}v = (k + \ell - n - 1)u_{(n)}v.$$

Duality (5.1.7a) also implies (see section 3.8 of [376])

$$Y\left(u_{(-m)}v, z\right) = \frac{1}{(m-1)!} : \left(\partial_z^{m-1} Y(u, z)\right) Y(v, z):, \tag{5.1.9a}$$

$$Y\left(u_{(n)}v, z\right) = \mathrm{Res}_{z_1}(z_1 - z)^n [Y(u, z_1), Y(v, z)], \tag{5.1.9b}$$

where $m \geq 1$ and $n \geq 0$. As we see next section, this is quite useful as a way of obtaining the full VOA from a small number of generators.

Unexpectedly, modular functions arise in VOA theory through the generating functions of the dimensions of the homogeneous spaces:

$$\mathrm{tr}_V q^{L_0} = \sum_{n=0}^{\infty} \dim V_n \, q^n. \tag{5.1.10a}$$

We also see this important theme in, for example, Section 3.1.2. As in (3.1.10), a small refinement should be made: by the *graded dimension* $\chi_V(\tau)$ of V we mean

$$\chi_V(\tau) := \mathrm{tr}_V e^{2\pi i\tau (L_0 - c/24)} = q^{-c/24} \sum_{n=0}^{\infty} \dim V_n \, q^n, \tag{5.1.10b}$$

where as always $q = e^{2\pi i\tau}$. The reason for the $q \mapsto \tau$ change-of-variables here will turn out to be the same as why Gauss and Jacobi introduced it into Euler's generating function $1 + 2x + 2x^4 + 2x^9 + \cdots$: both the graded dimension of V and Euler's generating function are naturally associated with tori. Explanations for the now-familiar $-c/24$ shift are given in Sections 3.2.3 and 5.3.4. Incidentally, the term *character* is also used in the literature for $\chi_V(\tau)$, but Section 5.3.3 contains our diatribe against this misnomer.

Section 1.5 illustrates the usefulness of the Killing form in Lie theory. Similarly, our VOAs all have a nondegenerate invariant bilinear form [199] – a bilinear pairing

$(u|v) \in \mathbb{C}$ for $u, v \in \mathcal{V}$, such that

$$(Y(u, z)v|w) = \left(v|Y(e^{zL_1}(-z^{-2})^{L_0}u, z^{-1})w\right), \qquad \forall u, v, w \in \mathcal{V}. \qquad (5.1.11a)$$

That this complicated definition is right is explained in remark 5.3.3 of [**199**] and equation (54) of [**242**]. For such a form, the homogeneous spaces $\mathcal{V}_m$ are mutually orthogonal, symmetry $(u|v) = (v|u)$ holds, and we recover familiar RCFT formulae such as $(L_n u|v) = (u|L_{-n}v)$. It is known (section 3 of [**380**]) that there is a unique invariant bilinear form (up to a scalar factor), provided that $\mathcal{V}$ is *simple* (defined in Section 5.3.1) and

$$L_1\mathcal{V}_1 = 0 \qquad (5.1.11b)$$

– both conditions are always satisfied by our VOAs. In this case the bilinear form restricted to each space $\mathcal{V}_n$ will be nondegenerate. The most convenient normalisation is

$$(\mathbf{1}|\mathbf{1}) = -1, \qquad (5.1.12a)$$

because for this choice the bilinear form on the homogeneous space $\mathcal{V}_1$ becomes

$$(u|v) = u_1 v, \qquad \forall u, v \in \mathcal{V}_1. \qquad (5.1.12b)$$

The invariant bilinear form plays an important role in CFT as well as Moonshine.

By a *vertex operator superalgebra* we mean there is a $\mathbb{Z}_2$-grading of $\mathcal{V} = \mathcal{V}_{\bar{0}} \oplus \mathcal{V}_{\bar{1}}$ into even and odd parity subspaces, and for u, v both odd the commutator in, for example, Axiom VA4 is replaced by an anti-commutator. Their basic theory is very similar to that of VOAs (see e.g. [**330**]). For instance, we write

$$\chi_\mathcal{V}(\tau) := \chi_{\mathcal{V}_{\bar{0}}}(\tau) - \chi_{\mathcal{V}_{\bar{1}}}(\tau).$$

Although we occasionally allude to vertex operator superalgebras (e.g. in Sections 5.4.2 and 7.3.5), we won't develop their theory.

In RCFT, $\mathcal{V}$ would be the 'Hilbert space of states' (more carefully, $\mathcal{V}$ is a dense subspace of it), and $z = e^{t+\mathrm{i}x}$ would be a local complex coordinate on a Riemann surface. L_0 generates time translations, and so its eigenvalues (the conformal weights) are energy. For each state u, the vertex operator $Y(u, z)$ is a meromorphic (chiral) quantum field. $Y(\omega, z)$ is the stress–energy tensor T. Physically, the requirement that $\mathcal{V}_n = 0$ for $n < 0$ says that the vacuum $\mathbf{1} = |0\rangle$ is the state with minimal energy. Also, $z = 0$ in VA3 corresponds to the time limit $t \to -\infty$. The most important axiom, VA4, says that quantum fields commute away from $z_1 = z_2$, and so are local. It is equivalent to the duality of chiral blocks in CFT, discussed in Sections 4.3.2, 4.4.1, 6.1.4.

In Segal's language (Section 4.4.1), $Y(u, z)$ appears quite naturally. Consider the virtual event of two strings combining to form a third. To first order (i.e. the tree-level Feynman diagram), this would correspond in Segal's language to a 'pair-of-pants', or a sphere with three punctures, two of which are negatively oriented (corresponding to incoming strings) and the other positively oriented. We can think of this as the Riemann sphere $\mathbb{C} \cup \{\infty\}$; put the punctures at ∞ (outgoing) and z and 0 (incoming). Segal's functor $\mathcal{T}$ associates with this a z-dependent homomorphism $\varphi_z : \mathcal{V} \times \mathcal{V} \to \mathcal{V}$, where

$\varphi_z(u, v) = Y(u, z)v \in \mathcal{V}$. Incidentally, the symbol 'Y' *could* have been chosen because of this 'pair-of-pants' picture (time flows from the top of the 'Y' to the bottom).[1]

By VOA1, any VOA is a $\mathfrak{Vir}$-module. For most VOAs, this module is highly reducible. By a *conformal primary* v of conformal weight k we mean $L_n v = 0$ for all $n > 0$ and $L_0 v = kv$ for some k. These states are especially well behaved. Any such primary generates a highest-weight module for $\mathfrak{Vir}$, on the space spanned by the elements $L_{-n_1} \cdots L_{-n_m} v$. The VOAs we are interested in are generated by the conformal primaries together with the operators L_n, in the sense that $\mathcal{V}$ can be decomposed into a direct sum (usually infinite) of highest-weight $\mathfrak{Vir}$-modules.

Question 5.1.1. Theorem 5.1.1 actually provides a realisation for a highest-weight representation of $A_1^{(1)}$. Identify that representation.

Question 5.1.2. Using the notion of algebraic summability, write down an algebraic definition of $\lim_{z_1 \to z_2} F(z_1, z_2)$ valid for formal power series $F(z_1, z_2) \in \mathcal{W}[[z_1^{\pm 1}, z_2^{\pm 1}]]$ realising the intuition of substituting in $z_1 = z_2$. Prove that $\lim_{z_1 \to z_2} F(z_1, z_2)$ 'algebraically exists' iff the product $F(z_1, z_2) \delta(z_1/z_2)$ does, in which case $F(z_1, z_2) \delta(z_1/z_2) = F(z_2, z_2) \delta(z_1/z_2)$.

Question 5.1.3. (a) Given any formal power series $F(z) \in \mathcal{W}[[z^{\pm 1}]]$, prove that

$$e^{w \frac{d}{dz}} F(z) = F(z + w).$$

(b) Prove (5.1.5).

Question 5.1.4. (a) Let $\mathcal{V}$ be any VOA, and $u, v \in \mathcal{V}$. Then for any $k \in \mathbb{Z}$, prove that

$$(z_1 - z_2)^k \left[Y(u, z_1), Y(v, z_2) \right] = \sum_{\ell \geq 0} \frac{1}{\ell!} \left(\partial_{z_2}^\ell z_1^{-1} \delta(z_2/z_1) \right) Y \left(u_{(k+\ell)} v, z_2 \right).$$

(b) Let $u \in \mathcal{V}_m$, $v \in \mathcal{V}_n$ be homogeneous vectors in any vertex algebra $\mathcal{V}$. Prove that $M(u, v) = m + n$ works in VA4.

Question 5.1.5. (a) Prove that in any vertex algebra, the vacuum $\mathbf{1}$ is translation-invariant, i.e. $T\mathbf{1} = 0$.
(b) In any VOA, verify that the span of L_{-1}, L_0, L_1 is the Lie algebra $\mathfrak{sl}_2(\mathbb{C})$. Verify that the vacuum is invariant under it.

Question 5.1.6. Prove that for any a, b, c in a vertex algebra $\mathcal{V}$, every coefficient z^n of $: a(z) b(z): u$ involves a finite sum, and for all but finitely many negative n this sum is 0.

5.2 Basic theory

A VOA is a remarkably rich algebraic structure, with infinitely many heavily constrained products. In this section we continue to work out the easy consequences of the axioms.

---

[1] But it wasn't. Remarkably, the actual historical reason is that Y comes after X, and X was the name arbitrarily chosen in [**201**] for a pre-vertex operator. The symbol Y first appeared in their chapter 8; Borcherds used the symbol Q.

The deep role of the Virasoro algebra remains hidden in this section. We also associate VOAs with lattices and affine algebras.

5.2.1 Basic definitions and properties

For any $u \in \mathcal{V}_n$, define $o(u) = u_{(n-1)}$. Then VA1 tells us $o(u)$ preserves each grade, that is it maps each homogeneous space $\mathcal{V}_m$ to itself. In particular, every space $\mathcal{V}_n$ has an algebraic structure defined by $u \times v = o(u)\,v$. In the CFT literature, these are called the *zero-mode algebras* (because $u_{(n-1)} = u_{\{0\}}$).

Typically, the zero-mode algebras $\mathcal{V}_n$ are quite complicated. However, consider $\mathcal{V}_1$. Put $\ell = m = n = 0$ in (5.1.7b) and hit it with any $w \in \mathcal{V}$: we get $u_{(0)}(v_{(0)}w) - v_{(0)}(u_{(0)}w) = (u_{(0)}v)_{(0)}w$. If we now formally write $[xy] := x_{(0)}y$, then this becomes $[u[vw]] - [v[uw]] = [[uv]w]$, which is one of the forms of the Lie algebra Jacobi identity (1.4.1b). Thus our bracket will be anti-associative if it is anti-commutative, in which case $\mathcal{V}_1$ will be a Lie algebra. But is it anti-commutative? From (5.1.9) we get

$$u_{(n)}v = \sum_{i \geq 0} \frac{1}{i!}(-1)^{i+n+1}(L_{-1})^i (v_{(n+i)}u) \qquad (5.2.1)$$

so $u_{(0)}v \equiv -v_{(0)}u \pmod{L_{-1}\mathcal{V}}$. However, from VA1, VOA4 and Question 5.1.5, we get

$$(L_{-1}\mathcal{V})_1 = L_{-1}(\mathcal{V}_0) = L_{-1}(\mathbb{C}\mathbf{1}) = \{0\}.$$

Thus, in any VOA, $\mathcal{V}_1$ is a finite-dimensional Lie algebra. Each homogeneous space $\mathcal{V}_n$ is a module for $\mathcal{V}_1$.

Given any $u, v \in \mathcal{V}_1$, $u_{(1)}v \in \mathcal{V}_0 = \mathbb{C}\mathbf{1}$, and so define $(u|v) \in \mathbb{C}$ by $(u|v)\mathbf{1} = u_{(1)}v$. From (5.2.1), $(u|v) = (v|u)$, so $(\star|\star)$ defines a symmetric bilinear form on $\mathcal{V}_1$. We would like $(\star|\star)$ to respect the Lie algebra structure, that is be $[\star\star]$-invariant. We compute from (5.1.7) and VA2

$$([uv]|t)\mathbf{1} = -v_{(0)}((u|t)\mathbf{1}) + (u|[vt])\mathbf{1} = (u|[vt])\mathbf{1}, \qquad (5.2.2)$$

that is $([uv]|t) = (u|[vt])$ and $(\star|\star)$ is indeed $[\star\star]$-invariant. Of course, this bilinear form is identical with that of (5.1.12b), and so provided (5.1.11b) is satisfied, it will be nondegenerate.

The existence of this bilinear form severely restricts the possibilities for the Lie algebra $\mathcal{V}_1$. Such Lie algebras are called *self-dual* and are precisely those for which the Sugawara construction (3.2.15) works. They are studied, for instance, in [415], [189], [384] – see also example 2.1 in [156]. If we also demand that the VOA be *weakly rational* (Definition 5.3.2), then $\mathcal{V}_1$ will be *reductive* (i.e. a direct sum of simple and trivial Lie algebras) [156].

The affinisation $\mathcal{V}_1^{(1)}$ of the Lie algebra $\mathcal{V}_1$ also appears naturally in the VOA $\mathcal{V}$. In particular, the modes $u_{(n)}$, for all $u \in \mathcal{V}_1$ and $n \in \mathbb{Z}$, have the commutators

$$u_{(m)} \circ v_{(n)} - v_{(n)} \circ u_{(m)} = ([u, v])_{(m+n)} + m\,(u|v)\delta_{m+n,0}\mathbf{1}_{(-1)}.$$

Thus these $u_{(n)}$, together with centre $\mathbb{C}\mathbf{1}_{(-1)}$ and derivation L_{-1}, span a $\mathcal{V}_1^{(1)}$-module.

More generally, in Section 7.2.2 we need to obtain a Lie algebra from a near-VOA $\mathcal{V}$. As before, we obtain a Lie algebra structure on $\mathcal{V}/L_{-1}\mathcal{V}$, and it has an invariant bilinear form if we restrict to $\mathcal{V}_1/L_{-1}\mathcal{V}_0$. In the situations we will be interested in, this algebra is too large, but it can be reduced as follows. Define

$$\mathcal{P}V_n := \{u \in \mathcal{V}_n \mid L_m u = 0 \text{ for all } m > 0\}, \qquad (5.2.3)$$

i.e. the conformal primaries with conformal weight n. Then a straightforward calculation verifies that $\mathcal{P}V_1/(L_{-1}\mathcal{V}_0 \cap \mathcal{P}V_1)$ is itself a Lie algebra, with the usual bracket. Through the map $u \mapsto u_{(0)}$, this Lie algebra acts on $\mathcal{V}$ and this action commutes with that of L_m. These associations of Lie algebras to (near-)VOAs are due to Borcherds [68].

By an *automorphism* (or symmetry) α of a VOA $\mathcal{V}$ we mean an invertible linear map $\alpha : \mathcal{V} \to \mathcal{V}$ obeying

$$\alpha(Y(u,z)v) = Y(\alpha(u),z)\alpha(v),$$

together with $\alpha(\mathbf{1}) = \mathbf{1}$ and $\alpha(\omega) = \omega$. This is how group theory arises in VOAs. The automorphism group can be finite (e.g. $\mathrm{Aut}(V^\natural) = \mathbb{M}$) or infinite (e.g. $\mathrm{Aut}(\mathcal{V}(\Lambda)) \cong (\mathbb{R}^\times)^{24} \rtimes Co_0$), but it can be finite only if $\mathcal{V}_1 = 0$ (Question 5.2.2). Conjecturally, at least when $\mathcal{V}$ is sufficiently nice, $\mathrm{Aut}(\mathcal{V})$ will be finite if (and only if) $\mathcal{V}_1 = 0$.

Similar arguments (Question 5.2.3) show that when $\mathcal{V}_1 = 0$, $\mathcal{V}_2$ is a commutative non-associative algebra with product $u \times v := u_{(1)}v \in \mathcal{V}_2$ and identity element $\frac{1}{2}\omega$. Moreover, an 'associative' bilinear form can be defined on $\mathcal{V}_2$ (Question 5.2.3). For example, the Moonshine module $V^\natural$ satisfies $V_1^\natural = 0$ (Section 7.2.1), and $V_2^\natural$ is none other than the Griess algebra [263] extended by an identity element.

The operators $u_{(0)}$, $u \in \mathcal{V}$, are derivations (i.e. infinitesimal automorphisms) of $\mathcal{V}$, that is

$$[u_{(0)}, Y(v,z)] = Y(u_{(0)}(v),z), \qquad (5.2.4)$$

and so $\exp(u_{(0)})$ is an automorphism of $\mathcal{V}$ if it is defined. This is important to the BRST cohomology construction (Question 5.2.4), borrowed from string theory.

5.2.2 Examples

Unlike more classical algebraic structures, VOAs are notorious for having no easy examples. In this section we construct families of them, in the most direct way possible. This explicitness has the drawback of making the constructions seem *ad hoc*. The reader interested in seeing the naturality of these constructions should consult the more sophisticated treatments in, for example, [330], [376].

Recall from (3.2.12a) the oscillator algebra $\mathfrak{g} = \mathfrak{u}_1{}^{(1)}$, with basis consisting of a_n, $n \in \mathbb{Z}$, together with the central term C. For any nonzero level $k \in \mathbb{C}$, we get a 'vacuum module' $\mathcal{V}(\mathfrak{g}, k)$ defined to have basis consisting of the formal combinations

$$a_{-m_1} \cdots a_{-m_r} \mathbf{1} \qquad (5.2.5a)$$

for $r \geq 0$, where $m_1 \geq m_2 \geq \cdots \geq m_r \geq 1$. Using the actions $C\mathbf{1} = k\mathbf{1}$, $a_n\mathbf{1} = 0$ for $n \geq 0$, we see $\mathcal{V}(\mathfrak{g}, k)$ has a $\mathfrak{u}_1^{(1)}$-module structure. Of course $\mathfrak{u}_1$ embeds into $\mathcal{V}(\mathfrak{g}, k)$ by $x \in \mathfrak{u}_1$ goes to $xa_{-1}\mathbf{1}$.

We claim that $\mathcal{V}(\mathfrak{u}_1^{(1)}, k)$ has a VOA structure, for $k \neq 0$. For the assignment of vertex operators, it suffices by (5.1.9) to define $Y(x, z)$ for $x \in \mathfrak{u}_1$: we get the 'current'

$$Y(xa_{-1}\mathbf{1}, z) := x \sum_{n \in \mathbb{Z}} a_n z^{-n-1}. \tag{5.2.5b}$$

All other vertex operators follow from (5.1.9). For example, for $m \geq 1$,

$$Y(a_{-m}\mathbf{1}, z) = \frac{1}{(m-1)!} \partial_z^{m-1} \sum_n a_n z^{-n-1}.$$

The unique singular term in the OPE (5.1.6) of the basic current with itself is

$$Y(a_{-1}\mathbf{1}, z_1) Y(a_{-1}\mathbf{1}, z_2) = \frac{C}{(z_1 - z_2)^2} + \cdots \tag{5.2.5c}$$

The Sugawara construction (3.2.14a) says here that the conformal vector is

$$\omega = \frac{1}{2k} a_{-1} a_{-1}\mathbf{1}, \tag{5.2.5d}$$

which makes $\mathcal{V}(\mathfrak{g}, k)$ into a (highly reducible) $\mathfrak{Vir}$-module with central charge $c = 1$. We also get the commutation relations

$$[L_m, a_n] = -na_{m+n}. \tag{5.2.5e}$$

In particular, the grading, given as we know by L_0, assigns the basis vector (5.2.5a) conformal weight $m_1 + \cdots + m_r$, so the current (5.2.5b) has conformal weight 1.

There is an obvious generalisation to any abelian Lie algebra $\overline{\mathfrak{h}} = \mathbb{C}^d$ with a choice of nondegenerate inner product on the space $\overline{\mathfrak{h}}$ (this defines the central term of the affine bracket (3.2.12a)). Namely, replace a with an orthonormal basis $a^1, \ldots, a^d$ of $\mathbb{C}^d$; the basis of the VOA is built up from all the operators a_{-n}^i as in (5.2.5a). These VOAs $\mathcal{V}(\overline{\mathfrak{h}}^{(1)}, k)$ are often called *Heisenberg VOAs*, because $\overline{\mathfrak{h}}^{(1)}$ is a Heisenberg algebra (i.e. a Lie algebra $\mathfrak{h}$ with $[\mathfrak{h}, \mathfrak{h}]$ equal to the centre of $\mathfrak{h}$). It turns out (Question 5.2.6) that the VOA $\mathcal{V}(\overline{\mathfrak{h}}^{(1)}, k)$ is independent of the choice of level k, provided $k \neq 0$, and also the choice of inner product, provided it is nondegenerate. We will let $\mathcal{V}(\mathbb{C}^n)$ denote the Heisenberg VOA with level $k = 1$ and standard inner product on the abelian Lie algebra $\overline{\mathfrak{h}} = \mathbb{C}^n$.

The generalisation to any affine algebra $\mathfrak{g} = \overline{\mathfrak{g}}^{(1)}$ [68], [201], [202], [384] is also straightforward. To any level $k \in \mathbb{C}$, $k \neq -h^\vee$ ($h^\vee$ the dual Coxeter number of $\overline{\mathfrak{g}}$), we get a natural VOA structure $\mathcal{V}(\mathfrak{g}, k)$ on the Verma module $M(k\omega_0)$ associated with highest weight $k\omega_0$, with central charge (3.2.9c). For example, from the Sugawara construction (3.2.15), the conformal vector is

$$\omega = \frac{1}{2(k + h^\vee)} \sum_i a_{(-1)}^i b_{(-1)}^i \mathbf{1}, \tag{5.2.6}$$

where $a^i, b^j \in \overline{\mathfrak{g}}$ are bases for $\overline{\mathfrak{g}}$, dual with respect to the Killing form on $\overline{\mathfrak{g}}$: $(a^i | b^j) = \delta_{ij}$. Any pair of dual bases give the same ω – the element $\frac{1}{2} \sum_i a^i b^i$ in the universal enveloping algebra $U(\overline{\mathfrak{g}})$ is simply the Casimir operator, and lies in the centre of $U(\overline{\mathfrak{g}})$. The only important difference here from the Heisenberg VOA is that sometimes there are 'null vectors', that is the Verma module $M(\lambda)$ may not be irreducible. In fact, maximal numbers of null vectors is the signature of the most interesting levels, namely $k \in \mathbb{N}$. We should quotient out all null vectors: by $\mathcal{V}(\mathfrak{g}, k)$ we mean the VOA structure (5.2.6) on the irreducible $\mathfrak{g}$-module $L(k\omega_0)$ defined in Section 3.2.3. Most interesting (because of its representation theory – Section 5.3) is $\mathcal{V}(\mathfrak{g}, k)$ when $k \in \mathbb{N}$, what we will call *integrable affine VOAs*.

The Lie algebra $\mathcal{V}_1$ associated with these affine algebra VOAs $\mathcal{V} = \mathcal{V}(\mathfrak{g}, k)$ is isomorphic to the reductive Lie algebra $\overline{\mathfrak{g}}$. Its affinisation, defined last subsection, equals $\mathfrak{g}$.

The forbidden level $k = -h^\vee$ is called the *critical level* and is very interesting in its own way. The conformal structure is lost (the conformal vector (5.2.6) won't exist), but the Möbius symmetry remains. The affine algebra vertex algebras at critical level have a highly nontrivial centre, and through it are related to geometric Langlands (see e.g. the discussion in section 17.4 of [197]). For this reason, it should be interesting to study it from the context of CFT.

Another relatively simple class of VOAs are associated with lattices [68], [201]. The simplest possibility is an n-dimensional positive-definite lattice L (Section 1.2.1), all of whose inner products $a \cdot b$ are *even* integers. By $\mathbb{C}\{L\}$ we mean the (infinite-dimensional) group algebra of the additive group L, written using formal exponentials: for each vector $v \in L$, we have a basis vector e^v of $\mathbb{C}\{L\}$, which multiply by $e^u e^v = e^{u+v}$. Let $\overline{\mathfrak{h}} = \mathbb{C} \otimes L \cong \mathbb{C}^n$ be the underlying complex vector space of L, interpreted as an abelian Lie algebra. It inherits the inner product of L. The underlying vector space of the VOA $\mathcal{V}(L)$ is $\mathcal{V}(\overline{\mathfrak{h}}) \otimes \mathbb{C}\{L\}$, where $\mathcal{V}(\overline{\mathfrak{h}})$ is the Heisenberg VOA constructed earlier. The vertex operator $Y(h \otimes 1, z)$, for $h \in \mathcal{V}(\overline{\mathfrak{h}})$, equals the vertex operator $Y(h, z)$ in $\mathcal{V}(\overline{\mathfrak{h}})$. Less clear is how to define the vertex operators $Y(1 \otimes e^\alpha, z)$, but once we know how the affine algebra $\overline{\mathfrak{h}}^{(1)}$ acts on the group algebra $\mathbb{C}\{L\}$, they will be heavily constrained by the OPEs (5.1.6a). Define $ht^m.e^\alpha = (h|\alpha)\delta_{m,0}e^\alpha$, for any $h \in \overline{\mathfrak{h}}$ and $\alpha \in L$, where we identify $\alpha \in L$ with the corresponding vector in $\overline{\mathfrak{h}} = \mathbb{C} \otimes L$. Then the OPE (5.1.6a) tells us (as usual displaying only the singular terms)

$$h(z_1) Y(1 \otimes e^\alpha, z_2) = \frac{(h|\alpha)}{z_1 - z_2} Y(1 \otimes e^\alpha, z_2) + \cdots$$

From this, and the pairwise locality of these vertex operators, we derive the formula

$$Y(1 \otimes e^\alpha, z) = e^\alpha \exp\left(-\sum_{j<0} \frac{z^{-j}}{j} \alpha_j\right) \exp\left(-\sum_{j>0} \frac{z^{-j}}{j} \alpha_j\right) z^{\alpha_0}.$$

In the usual way, this determines all vertex operators $Y(h \otimes e^\alpha, z)$. The vacuum is $\mathbf{1} \times 1$ and conformal vector ω is $\omega \otimes 1$; the central charge c though now equals the dimension n of L. The vectors $h \otimes 1$ for $h \in \overline{\mathfrak{h}}$ have conformal weight 1, while $\mathbf{1} \otimes e^\alpha$ have conformal weight $(\alpha|\alpha)/2$.

The construction is the same for any even positive-definite lattice L (i.e. all norm-squareds are even), except that the group algebra $\mathbb{C}\{L\}$ should be 'twisted' so that $e^\alpha e^\beta = (-1)^{(\alpha|\beta)} e^\beta e^\alpha$. If instead L is an *odd* positive-definite lattice (i.e. an integral lattice with some vectors of odd norm-square), the same construction yields vertex operator superalgebras (i.e. VOAs except the locality axiom VA4 can involve anti-commutators). For example, $L = \mathbb{Z}$ describes two fermions.

Repeating this construction for an *indefinite* even lattice L will yield a near-VOA. To see this, note that the conformal weight of $\mathbf{1} \otimes e^\alpha$ is $(\alpha|\alpha)/2$. If we regard $\mathcal{V}(L)$ as graded by L rather than by $\mathbb{Z}$, we obtain a grading into finite-dimensional subspaces.

There are several ways to construct new VOAs from old ones. For example, one can take the direct sum of VOAs with equal central charge (this doesn't change the central charge), or tensor products of arbitrary VOAs (the central charge adds) – see section 3.12 of [376]. The *orbifold construction* mods out by discrete symmetries: for a finite group G of symmetries of a VOA $\mathcal{V}$, let $\mathcal{V}^G$ denote the subspace of $\mathcal{V}$ fixed by G; then $\mathcal{V}^G$ is a vertex operator subalgebra of $\mathcal{V}$ – see Sections 4.3.4 and 5.3.6.

Finally, *Goddard–Kent–Olive (GKO) coset construction* [250] mods out by continuous symmetries. In particular, let $(\mathcal{V}, Y, \mathbf{1}, \omega)$ and $(\mathcal{V}', Y, \mathbf{1}, \omega')$ be VOAs with $\mathcal{V}' \subset \mathcal{V}$. So $\mathcal{V}'$ would be a vertex operator subalgebra of $\mathcal{V}$ except the conformal vectors need not be equal. Assume, however, that $\omega' \in \mathcal{V}_2$ and $L_1 \omega' = 0$. The coset construction finds a VOA structure on the centraliser

$$C_{\mathcal{V}}(\mathcal{V}') := \{v \in \mathcal{V} \mid [Y(v, z_1), Y(u, z_2)] = 0 \; \forall u \in \mathcal{V}'\}$$
$$= \{v \in \mathcal{V} \mid v_n u = 0 \; \forall u \in \mathcal{V}', n \in \mathbb{Z}\}. \tag{5.2.7}$$

The equality in (5.2.7) follows from Question 5.2.5. Then $(C_{\mathcal{V}}(\mathcal{V}'), Y, \mathbf{1}, \omega - \omega')$ is a VOA with central charge $c - c'$. In the VOA language, this was developed in [202]; see also the lucid treatment in section 3.11 of [376].

A conjecture of Moore and Seiberg [436], [437] states that every RCFT arises from orbifold and coset constructions applied to lattice and affine algebra theories (generously enough interpreted). They speculate that this would be the analogue here of Tannaka–Krein duality (Section 1.6.2). We seem a long way from proving this optimistic guess, even in a more limited context of sufficiently nice VOAs.

The most famous VOA is the *Moonshine module* $V^\natural$, constructed in 1984 in a *tour de force* by Frenkel–Lepowsky–Meurman [200]. It has central charge $c = 24$, with $V^\natural = V_0^\natural \oplus V_1^\natural \oplus V_2^\natural \oplus \cdots$, where $V_0^\natural = \mathbb{C}\mathbf{1}$ is one-dimensional, $V_1^\natural = \{0\}$ is trivial and $V_2^\natural = (\mathbb{C}\omega) \oplus$ (Griess algebra) is $(1 + 196883)$-dimensional. Its automorphism group is precisely the Monster $\mathbb{M}$. Thus each graded piece $V_n^\natural$ is a finite-dimensional $\mathbb{M}$-module. It has graded dimension J, and is the space (0.3.1) lying in the heart of Conway and Norton's Monstrous Moonshine (see Sections 4.3.4 and 7.2.1).

A formal parallel exists between integral lattices L and VOAs $\mathcal{V}$ [201], [248]. The dimension n of L corresponds to the central charge c of $\mathcal{V}$. An even lattice corresponds to a VOA while an odd lattice corresponds to a vertex operator superalgebra. As we see in the next section, the determinant $|L|$ relates to a measure of how many irreducible modules the VOA has. The norm-$\sqrt{2}$ vectors in L correspond to the vectors in

$\mathcal{V}_1$ – indeed, the norm-$\sqrt{2}$ vectors in a lattice L are special because they generate a Coxeter subgroup in Aut(L); the vectors in $\mathcal{V}_1$ are special because they generate a continuous subgroup (a Lie group) of Aut($\mathcal{V}$). In particular, the Leech lattice Λ and the Moonshine module $V^{\natural}$ play analogous roles (Section 7.2.1). Analogies of these kinds are always useful in their easy role as squirrels. The battle-cry 'Why invent when one can profitably copy?' is heard not only in Hollywood.

Question 5.2.1. Let $\mathcal{V}$ be a VOA, and let a finite group G act as automorphisms on $\mathcal{V}$, so each space $\mathcal{V}_n$ is a (finite-dimensional) G-module. Prove that for each n, $\mathcal{V}_n$ is a G-submodule of $\mathcal{V}_{n+1}$. (*Hint*: Consider the map L_{-1}.)

Question 5.2.2. In any VOA, define a map $e^{o(v)} : \mathcal{V} \to \mathcal{V}$ for each $v \in \mathcal{V}_1$, and show that for $v \neq 0$ it defines a nontrivial automorphism of $\mathcal{V}$. Verify that $e^{\mathcal{V}_1}$ generates a normal subgroup of Aut($\mathcal{V}$), and hence that Aut($\mathcal{V}$) will be uncountable if $\mathcal{V}_1 \neq 0$.

Question 5.2.3. Suppose a VOA $\mathcal{V}$ has $\mathcal{V}_1 = 0$. For $u, v \in \mathcal{V}_2$, define $u \times v = u_1 v$. Verify that $\mathcal{V}_2$ is commutative with this product, with identity $\omega/2$. Define a $\mathbb{C}$-valued bilinear form on $\mathcal{V}_2$ and discover how it is compatible with $\times$.

Question 5.2.4. Let $\mathcal{V}$ be a vertex algebra, and suppose $u \in \mathcal{V}_k$ satisfies $(u_{(0)})^2 = 0$. Prove that $\mathcal{V}^{(u)} = \ker u_{(0)}/\operatorname{im} u_{(0)}$ is itself a vertex algebra.

Question 5.2.5. Prove that $[Y(u, z_1), Y(v, z_2)] = 0$ iff $u_n v = 0$ for all $n \geq 0$.

Question 5.2.6. (a) Suppose both V, V' are complex n-dimensional vector spaces together with choices of nondegenerate inner-products. Verify that the Heisenberg VOAs $\mathcal{V}(V^{(1)}, k)$ and $\mathcal{V}(V'^{(1)}, k')$ are isomorphic as VOAs, provided only that k, k' are both nonzero.
(b) Let $\mathfrak{g} = \bar{\mathfrak{g}}^{(1)}$ be the nontwisted affine algebra associated with a simple finite-dimensional Lie algebra $\bar{\mathfrak{g}}$, and let $k \neq k'$ be two complex numbers, both distinct from the critical level $-h^{\vee}$. When are the affine algebra VOAs $\mathcal{V}(\mathfrak{g}, k)$ and $\mathcal{V}(\mathfrak{g}, k')$ isomorphic as VOAs?
(c) Let L, L' be two positive-definite lattices, all of whose inner-products $u \cdot v$ are even integers. When are the lattice VOAs $\mathcal{V}(L)$ and $\mathcal{V}(L')$ isomorphic as VOAs?

Question 5.2.7. Find an even indefinite lattice L such that the near-VOA $\mathcal{V}(L)$ has finite-dimensional homogeneous spaces $\mathcal{V}(L)_n$ for all $n \in \mathbb{Z}$.

5.3 Representation theory: the algebraic meaning of Moonshine

We know affine algebras have modules (namely the integrable ones) with interesting characters. However they have many other modules that are far less interesting, even if we restrict to highest weight ones with positive integer level. What general principle distinguishes the interesting ones from the generic? Of the uncountably many level $k \in \mathbb{N}$ highest-weight $X_r^{(1)}$-modules, the integrable ones are precisely those that are unitary. It is tempting then to guess that unitarity is the key principle. However, the reason to doubt its fundamental role is that there are RCFTs (e.g. the Yang–Lee model with central charge $c = -22/5$, see section 7.4.1 of [131]) whose graded dimensions obey all of the properties the affine characters do, but whose modules are not unitary.

The key feature possessed by the integrable affine modules is that they are unexpectedly small – that is, the null vectors in the associated Verma module, all of which are quotiented away, are maximally numerous. In other words, they are also modules of a sufficiently nice ('rational') VOA. The appearance of an affine algebra here is not directly significant, rather it is the appearance of that rational VOA. Modules of those VOAs may or may not be unitary. *VOAs serve as the unifying mathematics underlying the modules singled out by Moonshine.*[2]

The *raison d'être* of VOAs are their modules, and in Moonshine we are primarily interested in their graded dimensions and characters. It is to this important topic – *the algebraic meaning of Moonshine* – that we finally turn. See also [199], [376].

5.3.1 Fundamentals

A module of a VOA $\mathcal{V}$ is a vector space on which $\mathcal{V}$ acts, in such a way that this action preserves all possible structure. More precisely:

Definition 5.3.1 [199]   *Let $\mathcal{V}$ be a VOA. A weak $\mathcal{V}$-module (M, Y_M) is an $\mathbb{N}$-graded vector space $M = \oplus_{n \in \mathbb{N}} M_{[n]}$, and a linear map $Y_M : \mathcal{V} \to \text{End } M[[z^{\pm 1}]]$, written $Y_M(u, z) = \sum_{n \in \mathbb{Z}} u_{(n)} z^{-n-1}$, such that for any $u \in \mathcal{V}_k$, the mode $u_{(n)}$ is a linear map from $M_{[\ell]}$ into $M_{[k+\ell-n-1]}$,*

$$Y_M(\mathbf{1}, z) = id_M, \tag{5.3.1a}$$

$$z_0^{-1} \delta \left(\frac{z_1 - z_2}{z_0} \right) Y_M(u, z_1) Y_M(v, z_2) - z_0^{-1} \delta \left(\frac{z_2 - z_1}{-z_0} \right) Y_M(v, z_2) Y_M(u, z_1)$$

$$= z_2^{-1} \delta \left(\frac{z_1 - z_0}{z_2} \right) Y_M(Y(u, z_0)v, z_2), \tag{5.3.1b}$$

where each mode $u_{(n)}$ operates on M. The $Y_M(u, z)$ are also called vertex operators. *A weak $\mathcal{V}$-module (M, Y_M) is called a $\mathcal{V}$-module if in addition it comes with a grading $M = \oplus_{\alpha \in \mathbb{C}} M_\alpha$, with $M_\alpha = 0$ for $\text{Re}(\alpha)$ sufficiently negative, obeying*

$$M_\alpha = \{ x \in M \mid L_0 x = \alpha x \} \tag{5.3.1c}$$

(the eigenvalue α is again called the conformal weight *of $y \in M_\alpha$), and all homogeneous spaces M_α are finite-dimensional.*

We are interested in $\mathcal{V}$-modules. For the VOAs of interest to us (see Definition 5.3.2), the conformal weights are always rational (hence the name). Definition 5.3.1 uses the Jacobi identity (5.1.7a) rather than the simpler locality VA4 because, although locality and the Jacobi identity are equivalent for VOAs, for modules the Jacobi identity is stronger (see chapter 4 of [376]).

---

[2] Victor Kac expresses a related position by isolating locality as the key principle [329].

As before, the modes $L_n = \omega_{(n+1)}$ of the conformal vector $\omega \in \mathcal{V}$ yield on M a representation of the Virasoro algebra $\mathfrak{Vir}$, with the same central charge c as $\mathcal{V}$. In analogy with (5.1.10b), the *graded dimension* of a $\mathcal{V}$-module M is defined to be

$$\chi_M(\tau) := \mathrm{tr}_M e^{2\pi i \tau (L_0 - c/24)} = q^{-c/24} \sum_{\alpha \in \mathbb{C}} \dim M_\alpha \, q^\alpha. \qquad (5.3.2)$$

It is fundamental to the whole theory that these χ_M are modular, at least for 'nice' $\mathcal{V}$ and M (see Theorem 5.3.8 below). The automorphism group of $\mathcal{V}$ acts on each homogeneous space M_α – that is, each M_α carries a representation of $\mathrm{Aut}(\mathcal{V})$, and so the q-coefficients of $\chi_M(\tau)$ are dimensions of $\mathrm{Aut}(\mathcal{V})$-representations (famous examples being (0.2.1)).

It is straightforward [199], [376] to write down the definitions of $\mathcal{V}$-module homomorphism, direct sum of $\mathcal{V}$-modules, submodule, irreducible module (no nontrivial submodule), completely reducible module (i.e. M can be written as a direct sum of irreducible $\mathcal{V}$-modules), etc. Invariant bilinear forms can be defined for modules as in (5.1.11a), and have analogous properties [199], [380].

The easiest example of a $\mathcal{V}$-module, of course, is $\mathcal{V}$ itself, called the *adjoint* module. If $\mathcal{V}$ is irreducible as a $\mathcal{V}$-module, it is called *simple* (see Definition 6.2.3). All VOAs of interest in this book are simple. An example of a nonsimple vertex algebra is the affine algebra vertex algebra at critical level $k = -h^\vee$.

The notion of tensor product – called *fusion* $M \boxtimes N$ – for VOA modules is unexpectedly subtle. For example, the infinite-dimensional adjoint module $\mathcal{V}$ should have trivial fusions, just like the one-dimensional Lie algebra module $\mathbb{C}$ has trivial tensor products. See, for example, [298], [222], [382] for various approaches. Fusion products in a weakly rational VOA can be decomposed into irreducible modules as usual:

$$M \boxtimes N \cong \oplus_{P \in \Phi(\mathcal{V})} \mathcal{N}_{MN}^P \, P, \qquad (5.3.3)$$

where the multiplicities $\mathcal{N}_{MN}^P$ are called *fusion coefficients*. These numbers are most easily defined (via Schur's Lemma) as the dimension of the space of intertwiners [199] (Definition 6.1.9). For semi-simple Lie algebras, the tensor product of modules defines a symmetric monoidal category (Section 1.6.2); for nice VOAs, the fusion of modules defines a braided monoidal category and the structure constants $\mathcal{N}_{MN}^P$ a fusion ring (Section 6.2.2).

Definition 5.3.2 [574] *A VOA $\mathcal{V}$ is called* weakly rational *if every $\mathcal{V}$-module is completely reducible, $\mathcal{V}$ has only a finite number of irreducible modules, and every irreducible weak $\mathcal{V}$-module is a $\mathcal{V}$-module.*

Let $\Phi(\mathcal{V})$ denote the set of irreducible $\mathcal{V}$-modules. Most of our VOAs will be weakly rational. The term 'weakly rational' is not standard; *rational* is sometimes used. However, a rational VOA should enjoy all properties of the chiral algebra of a RCFT, which is why we reserve the term 'rational' for the stronger notion presented in Definition 6.2.3.

Lemma 5.3.3 [574] *Let $\mathcal{V}$ be a weakly rational VOA, and let M be any irreducible $\mathcal{V}$-module. Then there is a number $h \in \mathbb{Q}$ such that the homogeneous subspace M_h is nonzero, and such that if $M_\alpha \neq 0$ for some $\alpha \in \mathbb{C}$, then $\alpha - h \in \mathbb{N}$.*

The proof isn't difficult – see page 244 of [**574**] for a more general argument. We call $h = h(M)$ the *conformal weight* of M, and the space $M_h = M_{[0]}$ the *lowest-weight space* of M. For example, the conformal weight $h(\mathcal{V})$ of the adjoint module is 0. The lowest-weight space M_h generates the whole module, in the sense (5.1.9a) that M is spanned by vectors of the form $(u_1)_{(n_1)} \cdots (u_k)_{(n_k)} y$ for $u_i \in \mathcal{V}$ and $y \in M_h$. The lemma implies that for such a module M, we have $\chi_M(\tau + 1) = e^{2\pi i h(M)} \chi_M(\tau)$ as formal power series.

In both finite group theory and Lie theory, given any module M, a module structure can also be found on the vector space dual M^* of M in a straightforward way. This module is called the *dual* or *contragredient* of M. Something similar happens for VOAs. However, the naive dual of an infinite-dimensional space tends to be too large (recall that in infinite dimensions, the double-dual $(V^*)^*$ properly contains V), so here we take instead the *restricted dual* $M^\star$ of M, defined by

$$M^\star = \oplus_\alpha (M_\alpha)^*. \tag{5.3.4a}$$

The explicit $\mathcal{V}$-module structure on $M^\star$ (see section 5.2 of [**199**]) is quite complicated and closely related to the definition of invariant bilinear form in (5.1.11a). Note that

$$\chi_{M^\star}(\tau) = \chi_M(\tau) \tag{5.3.4b}$$

even though $M^\star$ and M are usually non-isomorphic as $\mathcal{V}$-modules. Thus our graded dimensions (5.3.2) won't always distinguish modules, something that was independently observed in the context of Monstrous Moonshine, as we'll see. We return to this bothersome but not unexpected fact in Section 5.3.3. The more obscure term 'contragredient' is usually used for $M^\star$, as 'dual' has too many unfortunately independent meanings. The notion of contragredient module plays a large role in RCFT: roughly, $M^\star$ is the anti-particle of M, and they are related by charge-conjugation C.

All VOAs $\mathcal{V}$ of interest to us have an anti-linear involution $u \mapsto u^*$ such that the invariant bilinear form $(u|v)$ of (5.1.11a) satisfies

$$(u|v^*) = \overline{(v|u^*)}, \qquad \forall u, v \in \mathcal{V}. \tag{5.3.5a}$$

The notion of *unitary module* M is important in physics: it is a $\mathcal{V}$-module in which the bilinear form on M satisfies

$$(ux|y)_M = (x|u^*y)_M, \qquad \forall u \in \mathcal{V}, x, y \in M. \tag{5.3.5b}$$

Consider first the lattice VOA $\mathcal{V}(L)$ constructed in Section 5.2.2, where L is a positive-definite even lattice (recall the definitions in Section 1.2.1). It is weakly rational, and its irreducible modules are parametrised naturally by the cosets L^*/L, where $L^* \supseteq L$ is the dual lattice to L [**144**]. The explicit construction of these modules $M[t]$, for $[t] \in L^*/L$, is very similar to that of the VOA $\mathcal{V}_L$ itself – see section 6.5 of [**376**]. Thus the number $\|\Phi(\mathcal{V}(L))\|$ of its irreducible modules is given by the determinant $|L|$ of the lattice. The adjoint module is $M[0]$. The module $M[t]$ has contragredient $M[-t]$ and graded dimension

$$\chi_{M[t]}(\tau) = \frac{\Theta_{t+L}(\tau)}{\eta(\tau)^n}, \tag{5.3.6}$$

where n is the dimension of L, η is the Dedekind eta function (2.2.6b) and Θ_{t+L} is the theta series of (2.2.11a). The fusion product here is $M[t] \boxtimes M[t'] = M[t + t']$.

The Heisenberg VOAs are not weakly rational. For example, $\mathcal{V}(\mathbb{C})$ has a distinct irreducible module $M(\lambda)$ (namely the Verma module $V(\lambda)$ of (3.2.12b)) for every $\lambda \in \mathbb{C}$. The adjoint module is $M(0)$, and the contragredient of $M(\lambda)$ is $M(-\lambda)$. Only the modules with $\lambda \in \mathbb{R}$ are unitary. The graded dimension of $M(\lambda)$ is given in (3.2.12c).

However, if $\overline{\mathfrak{g}}$ is a simple Lie algebra and $\mathfrak{g} = \overline{\mathfrak{g}}^{(1)}$ is the associated nontwisted affine algebra, then the VOA $\mathcal{V}(\mathfrak{g}, k)$ will be weakly rational iff the level k lies in $\mathbb{N}$. Just as the VOA $\mathcal{V}(\mathfrak{g}, k)$ is the $\mathfrak{g}$-module $L(k\omega_0)$ with additional structure, the irreducible $\mathcal{V}(\mathfrak{g}, k)$-modules can be identified with the $\mathfrak{g}$-modules $L(\lambda)$, for level-k integrable highest weights $\lambda \in P_+^k(\mathfrak{g})$ [202]. In particular, the VOA graded dimension will equal the corresponding *specialised* affine algebra characters $\chi_\lambda(2\pi i \tau \ell_0) = \chi_\lambda(\tau, 0, 0)$ of (3.2.11c). The usual tensor product $L(\lambda) \otimes L(\mu)$ of affine algebra modules is less interesting than the fusion product $L(\lambda) \boxtimes L(\mu)$ – in the former, levels add and the tensor product coefficients $T_{\lambda\mu}^\nu$ can be infinite, while the latter is studied in Section 6.2.1.

A weakly rational VOA is called *holomorphic* if it has a unique irreducible module. As usual this terminology comes from RCFT: a holomorphic VOA can be the left-moving chiral algebra of a CFT with trivial right-moving chiral algebra, so the physical correlation functions (4.3.1a) of such a CFT would be holomorphic (at least locally, when all insertion points z_i are distinct). Thus the lattice VOA $\mathcal{V}(L)$ is holomorphic iff the lattice L is self-dual. The most famous example of a holomorphic VOA though is the Moonshine module $V^\natural$ [145]. In fact, its holomorphicity is one of the keys to Monstrous Moonshine (see Question 5.3.4).

5.3.2 Zhu's algebra

In many ways a VOA resembles a Lie algebra, and this analogy has often been exploited to flesh out the theory of VOAs. However, the representation theory of the *weakly rational* VOAs resembles that of a finite group.

Consider for concreteness the symmetric group $G = \mathcal{S}_3$. Its representation theory is captured by its group algebra $\mathbb{C}G$ (Section 1.1.3), that is the formal span of the elements $\sigma \in G = \{(1), (12), (23), (13), (123), (132)\}$, where G acts by left multiplication. The associative algebra $\mathbb{C}G$ is semi-simple, and so is a direct sum of matrix algebras: here,

$$\mathbb{C}G \cong M_{1\times1} \oplus M_{1\times1} \oplus M_{2\times2}, \tag{5.3.7a}$$

where the first summand $M_{1\times1}$ contains one copy of the trivial one-dimensional irreducible representation $\rho_1(\sigma) = 1$, the second summand $M_{1\times1}$ contains one copy of the 'sign' one-dimensional irreducible representation $\rho_s(\sigma) = (-1)^\sigma$, and the four-dimensional algebra $M_{2\times2}$ contains a continuum of copies of the two-dimensional irreducible representation ρ_2. More precisely, the three subspaces of the group algebra $\mathbb{C}G$

specified by (5.3.7a) are

$$V_1 = \mathbb{C}\{(1) + (12) + (23) + (13) + (123) + (132)\} \cong \rho_1, \tag{5.3.7b}$$

$$V_s = \mathbb{C}\{(1) - (12) - (23) - (13) + (123) + (132)\} \cong \rho_s, \tag{5.3.7c}$$

$$V_2 = \mathbb{C}\{(1) - (123), (1) - (132), (12) - (23), (12) - (13)\} \cong \rho_2 \oplus \rho_2. \tag{5.3.7d}$$

Incidentally, the different copies of the irreducible module ρ_2 in the subspace V_2 are parametrised by the projective line $\mathbb{P}^1(\mathbb{R}) \cong S^1$: choosing a nonzero point x in

$$\mathbb{C}\{(1) - (12) + (23) - (132), (23) - (13) + (123) - (132)\}, \tag{5.3.7e}$$

and hitting with arbitrary $\sigma \in G$, spans a copy $V_2(x)$ of the two-dimensional module ρ_2, and $V_2(x) \cap V_2(x') = \{0\}$ unless x and x' are complex multiples of each other, in which case $V_2(x)$ and $V_2(x')$ are equal as sets. On the other hand, choosing a generic element of V_2 (respectively $\mathbb{C}G$) will span all of V_2 (respectively $\mathbb{C}G$).

The representation theory of a finite group G is equivalent to that of the associative algebra $\mathbb{C}G$. Likewise, for semi-simple Lie algebras $\mathfrak{g}$ there is also an associative algebra, generated by $\mathfrak{g}$, which classifies all irreducible $\mathfrak{g}$-modules: the universal enveloping algebra $U(\mathfrak{g})$ (Section 1.5.3). However, it is infinite-dimensional, reflecting the fact that $\mathfrak{g}$ has infinitely many inequivalent irreducible modules.

Remarkably, weakly rational VOAs V have (like finite G), a finite-dimensional associative semi-simple algebra, denoted $A(V)$, which classifies the finitely many irreducible V-modules. As we know, the full module M can be generated from its lowest-weight space M_h, by repeatedly acting by modes of V, and so it suffices to study M_h. Now, the zero-modes $o(u)$, defined at the beginning of Section 5.2.1, act on each homogeneous space M_α; Zhu's algebra $A(V)$ is the algebra of zero-modes, as seen by the lowest-weight spaces M_h. A more formal construction, which will begin next paragraph, is due to Zhu [574], although it was anticipated in physics [429], [87]. Similar to the above, each irreducible V-module M corresponds to a linear functional f_M on V (Section 4.4.4); a certain large subspace $O(V)$ of V lies in the kernel of all functionals $f_M \circ o(v) \ \forall v \in V$, so each of these defines a well-defined functional on the quotient $A(V) := V/O(V)$. The quotient $A(V)$ has a product $u * v$ making it into an associative algebra; the space of functionals $f_M \circ o(v)$ carries a module action of $A(V)$, and as such can be identified with the dual M_h^* of the lowest-weight space of M. Conversely, any (irreducible) right-module for $A(V)$ is the lowest-weight space of an (irreducible) V-module M. This physically motivated treatment of Zhu's algebra is fleshed out in [227].

Zhu's treatment is similar. For $u, v \in V$, where $u \in V_k$, define a product

$$u * v = \mathrm{Res}_z \left(Y(u, z) v \frac{(z + 1)^k}{z} \right), \tag{5.3.8a}$$

or equivalently, in terms of the modes,

$$(u * v)_{(n)} = \sum_{m \geq k} u_{(-1-m)} \circ v_{(m+n)} + \sum_{m \leq k-1} v_{(m+n)} \circ u_{(-1-m)}. \tag{5.3.8b}$$

Extend $*$ linearly to all $u \in \mathcal{V}$. Let $O(\mathcal{V})$ be the subspace of $\mathcal{V}$ spanned by elements

$$(L_{-1}u + L_0u) * v, \qquad \forall u, v \in \mathcal{V}. \tag{5.3.8c}$$

By *Zhu's algebra* $A(\mathcal{V})$ we mean the quotient $\mathcal{V}/O(\mathcal{V})$.

The point of these definitions is that, on the lowest-weight space M_h of any irreducible $\mathcal{V}$-module M, a straightforward calculation (see page 250 of [**574**]) verifies that

$$o(u * v) = o(u) \circ o(v). \tag{5.3.9a}$$

Using (5.1.8b), (5.1.7d) and VA2, we see that

$$o(L_{-1}u + L_0u) = 0 \tag{5.3.9b}$$

identically on $\mathcal{V}$. Together, (5.3.9) tell us $o(u) = 0$ on each lowest-weight space M_h, for any $u \in O(\mathcal{V})$. Thus for any class $[u] \in A(\mathcal{V})$, the zero-mode $o(u)$ is a well-defined operator on each M_h.

Theorem 5.3.4 [**574**] *Let $\mathcal{V}$ be a weakly rational VOA (recall Definition 5.3.2) and let $A(\mathcal{V}) = \mathcal{V}/O(\mathcal{V})$ be Zhu's algebra. Then $A(\mathcal{V})$ is a finite-dimensional, associative and semi-simple algebra, isomorphic as an algebra to the matrix algebra*

$$A(\mathcal{V}) \cong \oplus_{M \in \Phi(\mathcal{V})} M_{n(M) \times n(M)},$$

where $\Phi(\mathcal{V})$ is the set of all irreducible $\mathcal{V}$-modules, and $n(M)$ is the dimension of the lowest-weight space M_h.

In other words, there is a one-to-one correspondence between the irreducible modules of $A(\mathcal{V})$ and $\mathcal{V}$; the irreducible $A(\mathcal{V})$-modules can in fact be naturally identified with the lowest-weight spaces M_h of the irreducible $\mathcal{V}$-modules. It is almost identical to what happens with the group algebra of a finite group. Note that the dimension $n(M)$ is the coefficient of the first nontrivial term $n(M)q^{h-c/24}$ of the graded dimension χ_M. The hard part of the proof of Theorem 5.3.4 is establishing that an irreducible $A(\mathcal{V})$-module lifts to an irreducible $\mathcal{V}$-module (the basic idea is sketched above). Incidentally, there are non-weakly rational VOAs (coming from 'logarithmic' CFTs) with Zhu's algebra $A(\mathcal{V})$ finite-dimensional but not semi-simple.

For example, Zhu's algebra $A(V^{\natural})$ for the Moonshine module $V^{\natural}$ is one-dimensional, while the integrable affine VOA $\mathcal{V}(\mathfrak{g}, k)$ at level $k \in \mathbb{N}$ has Zhu's algebra

$$A(\mathcal{V}(\mathfrak{g}, k)) \cong \oplus_{\lambda \in P_+^k(\mathfrak{g})} M_{\dim L(\bar{\lambda}) \times \dim L(\bar{\lambda})},$$

where $L(\bar{\lambda})$ is a highest-weight $\bar{\mathfrak{g}}$-module (to get $\bar{\lambda}$, drop λ_0 from λ). In general though, it is hard to compute $A(\mathcal{V})$ (unless the $\mathcal{V}$-modules are already known!) because we lose the grading – expressions like $L_{-1}u + L_0u$ are not homogeneous.

The definition (5.3.8a) of the product '$*$' in Zhu's algebra can be modified to give the more familiar 'normal-ordered product' (recall (5.1.6))

$$u \cdot v = \mathrm{Res}_z(Y(u, z) v z^{k-1}) = u_{(-1)}v \tag{5.3.10a}$$

for $u \in \mathcal{V}_k$, or equivalently in terms of modes

$$(u \cdot v)_{(n)} = \sum_{m \geq 0} u_{(-1-m)} \circ v_{(m+n)} + \sum_{m \leq -1} v_{(m+n)} \circ u_{(-1-m)}. \qquad (5.3.10b)$$

Let $O_2(\mathcal{V})$ be the span of all elements of the form $u_{(-2)}v$, and $A_2(\mathcal{V})$ the quotient $\mathcal{V}/O_2(\mathcal{V})$. Then $A_2(\mathcal{V})$ is a graded commutative associative algebra with product '$\cdot$'. It also has a Lie algebra structure, with bracket given by $[uv] = u_{(0)}v$; together, the Lie and associative products define a commutative Poisson algebra. Its main role in VOA theory is in a finiteness condition:

Definition 5.3.5 [574]   *A VOA $\mathcal{V}$ is said to be C_2-cofinite if the $A_2(\mathcal{V}) = \mathcal{V}/O_2(\mathcal{V})$ is finite-dimensional.*

Most of the important weakly rational VOAs (e.g. the Moonshine module, the lattice VOAs, the affine algebra VOAs at positive integer level) satisfy this condition. The term 'C_2-cofinite' comes from Zhu's name for what we call $O_2(\mathcal{V})$. It has several consequences. Most importantly, the graded dimensions $\chi_M(\tau)$ of a C_2-cofinite VOA converge to functions holomorphic in the upper half-plane $\mathbb{H}$ (theorem 4.4.2 of [574]). A C_2-cofinite VOA will have well-defined finite fusion coefficients (5.3.3) (see theorem 13 in [229]).

It is conjectured that a VOA is weakly rational if and only if it is C_2-cofinite, but although this would significantly simplify the definition of weakly rational, it seems difficult to prove. Weakly rational VOAs satisfy $\dim A_2(\mathcal{V}) \geq \dim A(\mathcal{V})$ (generalised in lemma 3 of [229]), but inequality can occur – for example, the integrable affine algebra VOA $\mathcal{V}(E_8^{(1)}, 1)$ has a one-dimensional Zhu's algebra but $A_2(\mathcal{V})$ is at least 249-dimensional [224].

A C_2-cofinite VOA is finitely generated in the sense that there will be finitely many vectors $u^1, \ldots, u^n \in \mathcal{V}$ (namely, choose u^i to be the lifts to $\mathcal{V}$ of a basis of $A_2(\mathcal{V})$) such that $\mathcal{V}$ is spanned by all vectors of the form

$$u^{i_1}_{(-m_1)} \cdots u^{i_k}_{(-m_k)} \mathbf{1}, \qquad (5.3.11a)$$

where $m_1 > \cdots > m_k > 0$ [229]. Something similar (but weaker) holds for $\mathcal{V}$-modules. Using this we quickly obtain a growth estimate: given any C_2-cofinite VOA $\mathcal{V}$, there is a constant $C > 0$ such that, for any irreducible $\mathcal{V}$-module M, the dimension of the homogeneous space M_α is bounded above by

$$\dim M_\alpha < C_M e^{C\sqrt{\alpha - h}}, \qquad (5.3.11b)$$

for some constant C_M, where as always $h = h(M)$ is the conformal weight of M. The constant C depends only on $\dim A_2(\mathcal{V})$, while C_M is essentially $\dim M_h$, adjusted slightly to ensure (5.3.11b) also holds for small α.

Various interesting generalisations of Zhu's algebras have appeared in the literature [149], [150], [229], [410]. From our point of view, these algebras play a crucial technical role in the statement and proof of the modularity of VOA characters.

5.3.3 The characters of VOAs

The next four subsections mark a climax for the book, as we discuss the modularity of the graded dimensions (5.1.10b), (5.3.2). We also explain why this was anticipated by physicists. But first let's reflect on the notion of character.

Calling the quantities $\chi_V(\tau)$ and $\chi_M(\tau)$ 'characters', as is common in the literature, is a misnomer – they are merely graded dimensions. Defining characters for an algebraic object is as much art as science. The beautiful success of the character theory of semi-simple and Borcherds–Kac–Moody Lie algebras hides the nontrivial intuition that went into the original definitions. Presumably the starting point was that the characters of finite groups are given by the trace. Also, exponentiation associates a Lie group with a Lie algebra. Putting this together leads to the character of (1.5.9a). The characters of (Borcherds–)Kac–Moody algebras then follow by analogy. Unfortunately, the situation for VOAs isn't nearly as clear.

The main properties we may hope a character χ_M to obey are: it specialises to dimension (or graded dimension); it distinguishes inequivalent modules; and it respects direct sum and tensor product (fusion for us), in the sense that $\chi_{M \oplus N} = \chi_M + \chi_N$ and $\chi_{M \boxtimes N} = \chi_M \chi_N$. We would also expect the VOA characters in the special case of the integrable affine VOA $V(\mathfrak{g}, k)$ to equal the corresponding affine algebra characters χ_λ in (3.2.9a) (recall that the $V(\mathfrak{g}, k)$-modules can be identified with the integrable $\mathfrak{g}$-modules).

This wish-list is hopelessly optimistic for even the nicest VOAs. The graded dimensions $\chi_{M(\lambda)}(\tau)$ for the integrable affine VOA $V(\mathfrak{g}, k)$ will not respect the fusion product:

$$\chi_{M(\lambda) \boxtimes M(\mu)}(\tau) \neq \chi_{L(\lambda) \otimes L(\mu)}(\tau) = \chi_{L(\lambda)}(\tau)\, \chi_{L(\mu)}(\tau) = \chi_{M(\lambda)}(\tau)\, \chi_{M(\mu)}(\tau),$$

where $L(\lambda) \otimes L(\mu)$ denotes the tensor product of $\mathfrak{g}$-modules. On the other hand, fusion respects the asymptotic dimensions: for all sufficiently nice VOAs V, the limit

$$\mathcal{D}(M) = \lim_{\tau \to 0} \frac{\chi_M(\tau)}{\chi_V(\tau)}, \tag{5.3.12}$$

called the *quantum dimension* of $M \in \Phi(V)$, satisfies $\mathcal{D}(M \boxtimes N) = \mathcal{D}(M)\,\mathcal{D}(N)$. 'Sufficiently nice' here means any C_2-cofinite weakly rational VOA V obeying the additional very common property that of all irreducible V-modules $M \in \Phi(V)$, a unique one realises the smallest conformal weight $\min_{M \in \Phi(V)} h(M)$ (in the most familiar examples the unique minimal conformal weight belongs to the adjoint module $M = V$).

Recall from (5.3.4b) that the graded dimensions $\chi_M(\tau)$ of inequivalent V-modules can be equal. A further example occurs whenever an even positive-definite lattice L has an automorphism α; then any pair $M[t]$, $M[\alpha t]$ of $V(L)$-modules will have identical graded dimension. However, such equalities need not always have an easy algebraic explanation: for example, in Monstrous Moonshine two McKay–Thompson series (namely, $T_{27A}(\tau) = T_{27B}(\tau)$, corresponding to unrelated elements of order 27) accidentally coincide for no obvious reason. None of this is surprising, since dimensions certainly don't uniquely specify Lie algebra or finite group modules.

We certainly would like VOA characters to distinguish inequivalent V-modules, and in fact be linearly independent. How to do this is clear from the study of lattice theta functions or affine algebra characters: in order to retain more information of the homogeneous spaces M_α than merely their dimensions, we must include more variables in χ_M.

Definition 5.3.6   *The* character *of a $\mathcal{V}$-module M is the* one-point function $\chi_M(\tau, v)$

$$\chi_M(\tau, v) := \mathrm{tr}_M o(v) \, q^{L_0 - c/24} = q^{-c/24} \sum_{n=0}^{\infty} \mathrm{tr}_{M_{h+n}} o(v) q^{h+n}. \tag{5.3.13}$$

$h = h(M)$ is the conformal weight of M, and $o(v)$ is the zero-mode (Section 5.2.1) of $v \in \mathcal{V}$, which is an endomorphism on each homogeneous space M_{h+n} (so its trace can be computed by choosing bases and writing $o(v)$ as a matrix for each n). This function χ_M arises naturally in CFT, as the one-point chiral block (Section 4.3.2) on the torus. We explain shortly why it is associated with a torus – this is the source of its modularity.

Note that $\chi_M(\tau, \mathbf{1})$ equals the graded dimension $\chi_M(\tau)$. By definition, the dependence of $\chi_M(\tau, v)$ on $v \in \mathcal{V}$ is linear. Provided $\mathcal{V}$ is C_2-cofinite, theorem 4.4.1 of [574] tells us that, for each $v \in \mathcal{V}$, $\chi_M(\tau, v)$ is holomorphic for $\tau \in \mathbb{H}$. This is proved by finding and studying a differential equation satisfied by $\chi_M(\tau, v)$. Their modularity is established in Section 5.3.5.

When $\mathcal{V}$ is weakly rational and C_2-cofinite, the one-point functions are linearly independent and thus distinguish inequivalent $\mathcal{V}$-modules. In fact, we see from the proof of theorem 5.3.1 in [574] that if $\mathcal{V}_A$ is any lift from Zhu's algebra $A(\mathcal{V})$ to $\mathcal{V}$, then the one-point functions $\chi_M(\tau, v)$ will remain linearly independent even if v is restricted to the finite-dimensional subspace $\mathcal{V}_A$. For example, the graded dimensions $\chi_M(\tau)$ and $\chi_{M^*}(\tau)$ are equal, but for $v \in \mathcal{V}_n$ the one-point functions obey

$$\chi_{M^*}(\tau, v) = (-1)^n \chi_M(\tau, v).$$

Although one-point functions (5.3.13) don't directly respect the fusion product (but recall (5.3.12)), they deserve the title 'character' as they are the simplest linearly independent extension of graded dimension. However, since they depend *linearly* and not *exponentially* on v, how can we reconcile them with the Jacobi theta functions (2.3.7) and the affine algebra characters (3.2.9a)? Mindlessly defining a function

$$\exp[2\pi i w] \, \mathrm{tr}_M \exp[2\pi i \, o(v)] \, q^{L_0 - c/24} \tag{5.3.14}$$

for $v \in \mathcal{V}$ and $w \in \mathbb{C}$ will lose modularity.

The key is to realise that, although the exponential $q = e^{2\pi i \tau}$ is *topological* in origin, the exponential $e^{2\pi i z}$ in (2.3.7) and (3.2.9a) is *Lie theoretic* in origin. In particular:

Definition 5.3.7   *Let $\mathcal{V}$ be a weakly rational C_2-cofinite VOA. For any $\mathcal{V}$-module $M \in \Phi(\mathcal{V})$, define the* Jacobi character *to be the quantity $\chi_M^J(\tau, v, w)$ given by (5.3.14), except we restrict v to the Lie algebra $\mathcal{V}_1$.*

Of course $v = 0$ and $w = 0$ recovers the graded dimensions. As we know, $e^{o(v)}$ is an automorphism of M for $v \in \mathcal{V}_1$, and as we recall from the McKay–Thompson series the graded trace of automorphisms is worthy of study. Question 5.3.5 asks the reader to verify that χ_M^J recovers affine algebra characters. Of course the complex variable 'w' is merely included for book-keeping. We return to Jacobi characters in Theorem 5.3.9.

If we hadn't restricted v in Definition 5.3.7 to $\mathcal{V}_1$, then linear independence would have been assured by that of the one-point functions $\chi_M(\tau, v)$ (why?). In the familiar examples (e.g. lattice or affine algebra VOAs) we still have linear independence of the Jacobi characters, but it won't hold for all other VOAs.

5.3.4 Braided #5: the physics of modularity

Let's turn next to one of the central questions in the book: why should the VOA characters χ_M have anything to do with modularity? In short, it is because they are toroidal chiral blocks of RCFT, and the mapping class group $\Gamma_{1,1}$ (which must act on those chiral blocks) is $SL_2(\mathbb{Z})$. While filling in this explanation we'll finally explain the shift '$c/24$' appearing in the definition of the affine algebra characters and more generally the VOA characters χ_M.

Lurking in the background of the following argument is the closed string, with period-1 arc-parameter σ and time-parameter t (recall Section 4.3.2). For the left-moving (holomorphic) sector it is convenient to introduce complex parameters $\sigma - it$ and $e^{2\pi i(\sigma - it)}$, which we now call z and w, respectively.

From the perspective of VOAs and CFT, the easiest way to realise the torus $\mathbb{C}/(\mathbb{Z} + \mathbb{Z}\tau)$ for $\tau \in \mathbb{H}$, starting with the space $\mathbb{C}$, is by first considering the map $z \mapsto e^{2\pi iz}$ (the '$2\pi i$' is merely a convenient normalisation). This is a holomorphic map sending neighbourhoods of 0 to neighbourhoods of 1. It changes the global topology, however, sending the plane $\mathbb{C}$ to the annulus $\mathbb{C}\backslash\{0\}$. Now it is simple to obtain our torus: we simply identify z and qz, where as always $q = e^{2\pi i\tau}$. This is equivalent to taking the finite annulus $\{z \in \mathbb{C} \mid |q| < |z| < 1\}$ and sewing together its two boundary circles by identifying z on the outer circle with qz on the inner. The resulting torus is conformally equivalent to $\mathbb{C}/(\mathbb{Z} + \mathbb{Z}\tau)$ (why?). The point is that the chiral blocks on the torus can be obtained from those of the plane, through this construction of the torus from $\mathbb{C}$. Let us now give the details.

Let $\mathcal{V}$ be any VOA. For any coordinate transformation $z \mapsto w = f(z)$ sending 0 to 0, and holomorphic in a neighbourhood of 0, the Virasoro algebra lets us calculate its effect on any vertex operator: we can write

$$Y(v, z) \mapsto T_f \circ Y(v, z) \circ T_f^{-1} \tag{5.3.15a}$$

for some invertible linear map $T_f : \mathcal{V} \to \mathcal{V}$ (see [223], [295] for the explicit and general calculation). More precisely, there are $a_i \in \mathbb{C}$ such that (see proposition 2.1.1 in [295])

$$f(z) = \exp\left[\sum_{n=0}^{\infty} a_n z^{n+1} \frac{\mathrm{d}}{\mathrm{d}z}\right] z \tag{5.3.15b}$$

as formal power series, where 'exp' is defined by its Taylor series. Then we obtain

$$T_f v = \exp\left[\sum_{n=0}^{\infty} a_n L_n\right] v \tag{5.3.15c}$$

(regularity VA5 implies this map $T_f : V \to V$ is always defined). When v is a conformal primary of conformal weight k (recall (5.2.3)), the transformation is particularly nice:

$$T_f \circ Y(v, z) \circ T_f^{-1} = Y(v, w)(f'(z))^k. \qquad (5.3.15d)$$

The other important special case is the stress–energy tensor $T(z) = Y(\omega, z)$:

$$T_f \circ Y(\omega, z) \circ T_f^{-1} = Y(\omega, w)(f'(z))^2 + \frac{c}{12}\{f(z), z\}, \qquad (5.3.15e)$$

where $\{f, z\}$ is the *Schwarzian derivative*

$$\{f, z\} := \frac{f'''(z)}{f'(z)} - \frac{3}{2}\left(\frac{f''(z)}{f'(z)}\right)^2. \qquad (5.3.15f)$$

The factor '$c/12$' in (5.3.15e) is the same as in (3.1.5a). The Schwarzian derivative vanishes if and only if f is a Möbius transformation (i.e. if and only if f conformally maps the Riemann sphere to itself), and so is a measure of how f changes the global topology.

Provided $f(z)$ is holomorphic near 0 and obeys $f(0) = 0$, a second VOA structure can be defined on the vector space V as follows. The vertex operators are $Y_f(v, z) = Y(T_f v, f(z))$, the vacuum is $\mathbf{1}_f = T_f(\mathbf{1}) = \mathbf{1}$, and conformal vector is $\omega_f = T_f(\omega)$. Let V_f denote this second VOA. Then V and V_f are isomorphic. (See [**293**] for a generalisation dropping the $f(0) = 0$ condition.)

We are interested in the transformation $w = f(z) = e^{2\pi i z} - 1$. Then everything simplifies and we get

$$\omega_f = 4\pi^2(\omega - c/24), \qquad (5.3.16a)$$

$$Y_f(v, z) = Y(v, w)e^{2\pi i z k}, \qquad \forall v \in V_k. \qquad (5.3.16b)$$

Although V_f is a VOA isomorphic to V sharing the same underlying space, modes and conformal weights are quite different. We will use square brackets to indicate the modes of V_f, and denote its Virasoro generators by $L[n] = (\omega_f)_{[n+1]}$. We find for instance that

$$L[-1] = 2\pi i (L_{-1} + L_0), \qquad (5.3.16c)$$

$$L[0] = L_0 + \sum_{i \geq 1} \frac{(-1)^{n-1}}{n(n+1)} L_n. \qquad (5.3.16d)$$

Although by the isomorphism of V and V_f the homogeneous spaces V_n and $V_{[n]}$ must be equal dimension, and in fact carry isomorphic representations of Aut V, we only have $V_n = V_{[n]}$ for $n = 0$ or if dim $V_n = 0$. On the other hand, if $v \in V$ is a conformal primary of conformal weight k with respect to the operators L_n, then it will be one with respect to the operators $L[n]$ as well (see Question 5.3.2).

For a technical reason, we are also interested in the simple relation between the usual power series modes $L[n]$ of V_f, and the Fourier modes L'_n of V, defined by

$$T(z) = Y(\omega, z) = -\sum_{m=-\infty}^{\infty} L'_m e^{2\pi i m w}.$$

We get (recall Question 3.1.8)

$$L'_n = L[n] - \delta_{n,0}\frac{c}{24}.$$

The occurrences of '$-c/24$' in, for example, the characters of affine algebras and VOAs can be traced back to its occurrence in (5.3.16a). Mathematically, it is a symptom of the change of global topology, from the plane to an annulus. Physically this is interpreted as the Casimir energy of the cylinder [3]; see also the discussion in section 5.4 of [131].

Our map f mapped the plane to the annulus $\mathbb{C}\backslash\{-1\}$. To get the torus, we need to identify z on the outer circle $e^{i\theta} - 1$, with the point $q(z+1) - 1 = qe^{i\theta} - 1$ on the inner circle. By the axioms of CFT (e.g. Section 4.4.1), this identification ('sewing') corresponds to taking a trace. For simplicity consider first the vacuum-to-vacuum amplitude ('partition function') on this torus, and write $\tau = s + it$. The desired trace will be over the full space of states $\mathcal{H}$, and will be of the 'propagator' for the cylinder, which takes the string and evolves it $2\pi t$ ahead in time and twists it $2\pi s$ arcwise. The infinitesimal generator of twists is the corresponding momentum operator, call it P, and the infinitesimal generator for time evolution is the Hamiltonian H, both in the z-coordinate frame. Thus the partition function will be

$$\mathcal{Z}(\tau) = \mathrm{tr}_{\mathcal{H}} \exp[2\pi i s P - 2\pi t H].$$

To find, for example, the Hamiltonian, note that changing time by δt changes the w-coordinate by the factor $e^{-2\pi\delta t}$, so the Hamiltonian generates dilations in w (recall the calculation in Section 4.3.2); similarly, the momentum operator generates rotations in w. We obtain

$$P = L'_0 - \overline{L}_0{}' = L[0] - \overline{L}[0] - \frac{c}{24} + \frac{\overline{c}}{24},$$
$$H = L'_0 + \overline{L}'_0 = L[0] + \overline{L}[0] - \frac{c}{24} - \frac{\overline{c}}{24},$$

where we use bars to denote the anti-holomorphic quantities. Thus we obtain the familiar expression for the partition function:

$$\mathcal{Z}(\tau) = \mathrm{tr}_{\mathcal{H}} q^{L[0]-c/24} \overline{q}^{\overline{L}[0]-\overline{c}/24} = \mathrm{tr}_{\mathcal{H}} q^{L_0-c/24} \overline{q}^{\overline{L}_0-\overline{c}/24},$$

where the final equality follows from the isomorphism of VOAs $\mathcal{V}$ and $\mathcal{V}_f$. CFT or string theory requires that $\mathcal{Z}(\tau)$ be a function only of the conformal equivalence class of the torus $\mathbb{C}/(\mathbb{Z} + \mathbb{Z}\tau)$ – in other words, $\mathcal{Z}(\tau)$ must be invariant under the action of the modular group $\mathrm{SL}_2(\mathbb{Z})$.

We are more interested here in the associated chiral quantities, since a VOA is the chiral algebra of the theory. From the previous paragraph, together with the decomposition (4.3.6) of $\mathcal{H}$ into modules of $\mathcal{V} \otimes \mathcal{V}'$, we can now read off the decomposition of $\mathcal{Z}(\tau)$ into chiral blocks (see (4.3.8b)) in a RCFT. Hence the chiral blocks for the torus are

$$\mathrm{tr}_M q^{L_0-c/24}$$

– that is, they are simply the graded dimensions of the irreducible $\mathcal{V}$-modules, including the strange shift by $c/24$. RCFT requires that this space must carry a projective representation of the mapping class group of the torus $SL_2(\mathbb{Z})$.

By the same reasoning, we can calculate the n-point chiral blocks on the torus. For $L[0]$-homogeneous vectors $u^i \in \mathcal{V}_{[k_i]}$, they are simply

$$e^{2\pi i z_1 k_1} \cdots e^{2\pi i z_n k_n} \mathrm{tr}_M Y_M(u^1, e^{2\pi i z_1}) \cdots Y_M(u^n, e^{2\pi i z_n}) q^{L_0 - c/24}, \tag{5.3.17a}$$

where $u^i \in \mathcal{V}$ are the inserted states and $z_i \in \mathbb{C}$ are the points of insertion. As usual, the definition for nonhomogeneous vectors follows by linearity. By construction these functions automatically have period 1 in each z_i, and it is an easy calculation to verify that they also have period τ in each z_i, and thus the insertion points z_i lie on the torus $\mathbb{C}/(\mathbb{Z} + \mathbb{Z}\tau)$, as they should. In particular, the reader can verify that the one-point chiral blocks are indeed what we call the one-point functions: for $u \in \mathcal{V}_{[k]}$,

$$e^{2\pi i z k} \mathrm{tr}_M Y_M(u, e^{2\pi i z}) q^{L_0 - c/24} = \chi_M(\tau, u), \tag{5.3.17b}$$

hence the name of the latter. By the general principles of RCFT, the space of say one-point chiral blocks should carry a projective representation of the mapping class group of the once-punctured torus, i.e. $SL_2(\mathbb{Z})$ (recall (4.3.9)), called *modular data* (Section 6.1.2). In Section 7.2.4 we find that a much larger group acts naturally on these one-point functions.

In (5.3.17) we inserted states u^i from only the vacuum sector. More generally, however, the states u^i can come from any sector, that is be vectors in any module $M \in \Phi(\mathcal{V})$. In that case the vertex operators Y_M should be replaced by intertwiners $\mathcal{Y}$ (Definition 6.1.9). Although this generalisation is fundamental to VOAs and RCFT, it is less so for Monstrous Moonshine (since $V^{\natural}$ is holomorphic).

The point of this subsection is to see in some detail how physics (RCFT) anticipates the statement and proof of Zhu's Theorem, to which we now turn.

5.3.5 *The modularity of VOA characters*

The most important property of the one-point functions is their modularity:

Theorem 5.3.8 (Zhu [574]) *Suppose $\mathcal{V}$ is a C_2-cofinite weakly rational VOA (see Definitions 5.3.2 and 5.3.5), and let $\Phi(\mathcal{V})$ be the finite set of irreducible $\mathcal{V}$-modules. Then there is a representation ρ of $SL_2(\mathbb{Z})$ by complex matrices $\rho(A)$ indexed by $\mathcal{V}$-modules $M, N \in \Phi(\mathcal{V})$, such that the one-point functions (5.3.13) obey*

$$\chi_M \left(\frac{a\tau + b}{c\tau + d}, v \right) = (c\tau + d)^n \sum_{N \in \Phi(\mathcal{V})} \rho \begin{pmatrix} a & b \\ c & d \end{pmatrix}_{MN} \chi_N(\tau, v) \tag{5.3.18a}$$

for any $v \in \mathcal{V}$ obeying $L[0] v = nv$ for some $n \in \mathbb{N}$ (see (5.3.16d)).

In particular, the graded dimensions (5.3.2) obey

$$\chi_M \left(\frac{a\tau + b}{c\tau + d} \right) = \sum_{N \in \Phi(\mathcal{V})} \rho \begin{pmatrix} a & b \\ c & d \end{pmatrix}_{MN} \chi_N(\tau), \qquad \forall \begin{pmatrix} a & b \\ c & d \end{pmatrix} \in SL_2(\mathbb{Z}). \tag{5.3.18b}$$

In (5.3.18), the quantity 'c' is an entry of a matrix in $\mathrm{SL}_2(\mathbb{Z})$ and should not be confused with the central charge. As we saw last subsection, $L[0]$ plays the role of L_0 in a Virasoro representation obtained from L_n by a change-of-variables z: $\mathcal{V} = \oplus_n \mathcal{V}_{[n]}$, where $n \in \mathbb{N}$ and $\mathcal{V}_{[n]}$ is the eigenspace of $L[0]$ with eigenvalue n. We can summarise (5.3.18a) by saying that $\chi_M(\tau, v)$ is a vector-valued modular form of weight n and multiplier ρ (recall Definition 2.2.2). We will summarise the proof of Theorem 5.3.8 shortly; see [**442**] for an independent argument.

One-point functions for the Moonshine module $\mathcal{V} = \mathcal{V}^\natural$ are studied in [**155**], where we find that all meromorphic modular forms for $\mathrm{SL}_2(\mathbb{Z})$ appear as some $\chi_{V^\natural}(\tau, v)$, provided the obvious constraints (namely that they be holomorphic in $\mathbb{H}$, have zero constant term in their q-expansion and have at worst a simple pole at $q = 0$) are satisfied – clearly, if the coefficient of q^α in $\chi_M(\tau)$ is zero then it must vanish in all other $\chi_M(\tau, v)$. Thus although we see the Monster in the graded dimension of $V^\natural$, we won't see it in most one-point functions of $V^\natural$.

However, if $v \in \mathcal{V}$ is fixed by some subgroup G_v of the automorphism group of $\mathcal{V}$, then the q^α coefficient of $\chi_M(\tau, v)$ relates to the representations of G_v and the eigenvalues of $o(v)|_{M_\alpha}$ (see Question 5.3.3). Note that in each homogeneous space $\mathcal{V}_n \neq 0$ there will be nonzero vectors invariant under the full automorphism group of $\mathcal{V}$ (why?). For example, we read off from Table 7.3 that in the homogeneous spaces $(V^\natural)_n$ of the Moonshine module for $0 \leq n \leq 7$, the $\mathbb{M}$-invariant subspace has dimension $1, 0, 1, 1, 2, 2, 4, 4, 7$, respectively.

The representation ρ in Zhu's Theorem is called modular data (Section 6.1.2). The diagonal matrix $\rho \begin{pmatrix} 1 & 1 \\ 0 & 1 \end{pmatrix}$ is given in (4.3.10). The matrix $S = \rho \begin{pmatrix} 0 & -1 \\ 1 & 0 \end{pmatrix}$ relates to the fusion multiplicities $\mathcal{N}_{MN}^P$ via Verlinde's formula (6.1.1b) (at least for nice VOAs – see Section 6.2.2). It is conjectured that, for sufficiently nice VOAs, the representation ρ should be trivial on a congruence subgroup $\Gamma(N)$ (see the Congruence Property 6.1.7). When this is true, each graded dimension $\chi_M(\tau)$ will be a modular function for that $\Gamma(N)$.

If we weaken the hypothesis of weak rationality or C_2-cofiniteness (recall that these are conjectured to be equivalent) in Zhu's Theorem, then we can still recover some kind of modularity. In particular, physicists speak of *quasi-rational* CFTs, which are CFTs with finite fusions; in examples it seems that they still obey some weakened form of Zhu's Theorem (see Section 6.2.2).

Note that Zhu's Theorem is already strong enough to imply that the Moonshine module $V^\natural$ must have graded dimension $J(\tau)$. To see this, note that holomorphicity implies that $\rho(A)$ is a one-dimensional representation of $\mathrm{SL}_2(\mathbb{Z})$. However, $\rho \begin{pmatrix} 1 & 1 \\ 0 & 1 \end{pmatrix}$ must be trivial and thus

$$\chi_{V^\natural}(A.\tau) = \chi_{V^\natural}(\tau), \qquad \forall A \in \mathrm{SL}_2(\mathbb{Z}).$$

We know $\chi_{V^\natural}(\tau)$ must be holomorphic in $\mathbb{H}$ (all graded dimensions are), has constant term 0 and a simple pole at the cusp. Therefore it equals $J(\tau)$. See also Question 5.3.4.

The proof of the Hauptmodul property for the other McKay–Thompson series T_g is much more subtle, unfortunately.

Zhu's Theorem rigorously generalises RCFT modularity to that of any sufficiently nice VOA. Its proof is long and complicated, but follows closely the intuition of CFT.

Zhu first defines abstractly a space of sequences $(S_1, S_2, \ldots)$ of functions, where each S_n maps n-tuples $(a_1, \ldots, a_n) \in \mathcal{V}^{\otimes n}$ to meromorphic functions of $(z_1, \ldots, z_n, \tau) \in \mathbb{C}^n \times \mathbb{H}$. They obey several conditions, for example they are doubly-periodic in each variable z_i, with periods 1 and τ. Each function S_n is what we would call a chiral block on the torus $\mathbb{C}/(\mathbb{Z} + \mathbb{Z}\tau)$ with n marked points at z_i; it lies in the space $\mathfrak{B}^{(1,n)}_{\mathcal{V},\ldots,\mathcal{V}}$. Zhu's definition abstracts out the manifest properties of this space. It is immediate from his definition that $SL_2(\mathbb{Z})$ acts on this space, in exactly the way we would expect from CFT. Verlinde's formula (6.1.2) tells us that the dimensions of these spaces should be independent of the number n of punctures, and in fact CFT tells us that a canonical basis for $\mathfrak{B}^{(1,n)}_{\mathcal{V},\ldots,\mathcal{V}}$ should be

$$(a_1, \ldots, a_n) \mapsto \mathrm{tr}_M\left(Y_M(a_1, e^{2\pi i z_1}) \cdots Y_M(a_n, e^{2\pi i z_n}) q^{L_0}\right) \tag{5.3.19}$$

(appropriately normalised), for each irreducible $\mathcal{V}$-module M. However, showing rigorously that these functions (5.3.19) in fact satisfy his definition, and that they do indeed span his space, are both more difficult. But we see that the modularity in Zhu's Theorem arises through that $SL_2(\mathbb{Z})$ action on the space of chiral blocks.

The modularity of the Jacobi characters $\chi_M^J(\tau, v, w)$ of Definition 5.3.7 is now easy.

Theorem 5.3.9   *Let $\mathcal{V}$ be a weakly rational C_2-cofinite VOA. Then the Jacobi characters $\chi_M^J(\tau, v, w)$ are holomorphic in $\mathbb{H}$ for any fixed v, w, and obey*

$$\chi_M^J\left(\frac{a\tau + b}{c\tau + d}, \frac{v}{c\tau + d}, w - c\frac{(v|v)}{2(c\tau + d)}\right) = \sum_{N \in \Phi(\mathcal{V})} \rho\begin{pmatrix} a & b \\ c & d \end{pmatrix}_{MN} \chi_N^J(\tau, v, w),$$

$$\tag{5.3.20}$$

for all $\begin{pmatrix} a & b \\ c & d \end{pmatrix} \in SL_2(\mathbb{Z})$, $v \in \mathcal{V}_1$, *and* $w \in \mathbb{C}$, *where* ρ *is as in Theorem 5.3.8 and where the inner-product $(v|v)$ is given by* $v_{(1)}v = -(v|v)\mathbf{1}$.

Again, 'c' in (5.3.20) refers to a matrix entry and not the central charge. The transformation on the left side of (5.3.20) is exactly that of, for example, Jacobi theta functions. Theorem 5.3.9 is an easy corollary of the main theorem of [**426**] (which in turn is a corollary of the proof of Theorem 5.3.8 as given in [**574**]). In particular, define

$$Z_M(\tau, u, v) = \mathrm{tr}_M e^{2\pi i (o(v) - (v|u)/2)} q^{L_0 + o(u) - (c + 12(u|u))/24}, \tag{5.3.21a}$$

for any $u, v \in \mathcal{V}_1$, so $\chi_M^J(\tau, v, w) = \exp[2\pi i w] Z_M(\tau, 0, v)$. Then provided $o(v)u = 0$ (i.e. u and v commute in the Lie algebra $\mathcal{V}_1$), [**426**] obtained the transformation law

$$Z_M\left(\frac{a\tau + b}{c\tau + d}, u, v\right) = \sum_{N \in \Phi(\mathcal{V})} \rho\begin{pmatrix} a & b \\ c & d \end{pmatrix}_{MN} Z_N(\tau, cv + du, av + bu), \tag{5.3.21b}$$

for any $\begin{pmatrix} a & b \\ c & d \end{pmatrix} \in \mathrm{SL}_2(\mathbb{Z})$. To prove (5.3.20), it suffices to prove it for the two generators $\begin{pmatrix} 1 & 1 \\ 0 & 1 \end{pmatrix}$ and $\begin{pmatrix} 0 & -1 \\ 1 & 0 \end{pmatrix}$, and this follows directly from (5.3.21b). Holomorphicity of Z_M follows from Proposition 1.8 of [**151**].

5.3.6 Twisted #5: twisted modules and orbifolds

Last subsection we saw how the modularity of VOA modules permits a one-paragraph proof that the graded dimension of the Moonshine module $V^\natural$ must equal $J(\tau)$. How about the other McKay–Thompson series? In this subsection we find that the notion of $\mathcal{V}$-module must be generalised to the equally fundamental notion of *twisted* $\mathcal{V}$-modules. Twisted modules are vaguely reminiscent of projective representations of groups, but while a projective representation of G is a true representation of some central extension of G, a twisted $\mathcal{V}$-module is a true module of a vertex operator subalgebra of $\mathcal{V}$. Most groups don't have twisted modules, and VOAs don't seem to have a natural notion of a projective module, but Lie algebras have a foot in each camp and as we see in Chapter 3 have both kinds of modules.

Far from being an esoteric development, twisted modules are crucial to Monstrous Moonshine and absolutely central to the whole theory. In CFT and string theory, they arise in the important orbifold construction (Section 4.3.4). Twisted modules of Lie algebras – a baby example of twisted modules of VOAs – are discussed in Sections 1.5.4 and 3.4.1. Moonshine is the relation of VOAs to modular functions; the modular function analogue of this twisting has long been understood and also plays a central role (Section 2.3.3).

Fix a VOA $\mathcal{V}$ and any automorphism $g \in \mathrm{Aut}(\mathcal{V})$ of order N. We can define g-twisted modules [**185**], by blending together the definitions in Sections 3.4.1 and 5.3.1. In particular, decompose $\mathcal{V}$ into eigenspaces of g: $\mathcal{V} = \oplus_{j=0}^{N-1} \mathcal{V}^j$ where $\mathcal{V}^j = \{v \in \mathcal{V} \mid g.v = \xi_N^{-j} v\}$. A g-twisted $\mathcal{V}$-module (M, Y_M) has a $\mathbb{C}$-grading $M = \oplus_{\alpha \in \mathbb{C}} M_\alpha$, with $\dim M_\alpha < \infty$, as in Definition 5.3.1, as well as a linear map $\mathcal{V} \to \mathrm{End}[[z^{\pm 1/N}]]$, written $Y_M(u, z) = \sum_{r \in \mathbb{Z}/N} u_{(r)} z^{-r-1}$, such that (5.3.1a), (5.3.1c) hold,

$$Y(u, z) = \sum_{r \in -j/N + \mathbb{Z}} u_{(r)} z^{-r-1}, \qquad \forall u \in \mathcal{V}^j, \tag{5.3.22a}$$

and (5.3.1b) becomes

$$z_0^{-1} \delta\left(\frac{z_1 - z_2}{z_0}\right) Y_M(u, z_1) Y_M(v, z_2) - z_0^{-1} \delta\left(\frac{z_2 - z_1}{-z_0}\right) Y_M(v, z_2) Y_M(u, z_1)$$

$$= z_2^{-1} \left(\frac{z_1 - z_0}{z_2}\right)^{-j/N} \delta\left(\frac{z_1 - z_0}{z_2}\right) Y_M(Y(u, z_0)v, z_2), \tag{5.3.22b}$$

where $u \in \mathcal{V}^j$. We say two g-twisted $\mathcal{V}$-modules M, N are *isomorphic* if there is an isomorphism $\varphi : M \to N$ satisfying $Y_M(\varphi v, z) = Y_N(v, z)\varphi$ for all $v \in N$. Note that an e-twisted $\mathcal{V}$-module (e being the identity of G) is an ordinary $\mathcal{V}$-module.

Any $h \in \text{Aut}(\mathcal{V})$ permutes the twisted $\mathcal{V}$-modules as follows. Let M be g-twisted, and for each $v \in V$ define

$$_h Y_M(v, z) := Y_M(h.v, z).$$

Then $(M, {}_h Y_M)$ is an $h^{-1} g h$-twisted $\mathcal{V}$-module. When h and g commute, we say the module (M, Y_M) is *h-stable* if (M, Y_M) and $(M, {}_h Y_M)$ are isomorphic. We call $h \in \text{Aut}(\mathcal{V})$ an *inner automorphism* of $\mathcal{V}$, and write $h \in \text{Inn}(\mathcal{V})$, if every untwisted $\mathcal{V}$-module is h-stable.

Now let M be an irreducible g-twisted $\mathcal{V}$-module, and G any group of automorphisms $h \in \text{Aut}(\mathcal{V})$ commuting with g such that M is h-stable for all $h \in G$. Then for each $h \in G$, we get an automorphism $\varphi(h) : M \to M$ of M, satisfying $\varphi(h) Y_M(v, z) \varphi(h)^{-1} = Y_M(h.v, z)$. Hence we can perform Thompson's trick (0.3.3) and write

$$\mathcal{Z}(M, h; \tau) := q^{-c/24} \, \text{tr}_M \varphi(h) q^{L_0}. \tag{5.3.23}$$

These $\mathcal{Z}(M, h)$'s are the building blocks of the graded dimensions of various eigenspaces of h in M: for example, if h has order m, then the subspace of M fixed by the automorphism $\varphi(h)$ will have graded dimension $m^{-1} \sum_{i=1}^m \mathcal{Z}(M, h^i)$.

This assignment φ does not necessarily define a representation of G in $\text{End}(M)$. However, $\varphi(h_2)^{-1} \varphi(h_1)^{-1} \varphi(h_1 h_2)$ clearly commutes with all vertex operators $Y_M(v, z)$ and so by irreducibility of M is a scalar multiple $c_g(h_1, h_2) I$ of the identity. Equivalently, φ is a projective representation of G:

$$\varphi(h_1 h_2) = c_g(h_1, h_2) \varphi(h_1) \varphi(h_2). \tag{5.3.24}$$

For any $h, k \in C_G(g)$ (i.e. commuting with g), $\varphi(khk^{-1}) = \alpha_{k,h} \varphi(k) \varphi(h) \varphi(k)^{-1}$ for some scalar $\alpha_{k,h}$, and thus $\mathcal{Z}(M, khk^{-1}; \tau) = \alpha_{k,h} \mathcal{Z}(M, h; \tau)$ by the cyclic property of trace. This means that, for fixed g, it suffices to restrict to one h from each $C_G(g)$-conjugacy class. By a similar argument (Question 5.3.6), we get that $\mathcal{Z}(M, h; \tau)$ vanishes identically, unless for all $k \in C_G(g)$ commuting with h, the 2-cocycle of (5.3.24) satisfies $c_g(h, k) = c_g(k, h)$. Thus we can further restrict to those h.

Conjecture 5.3.10 [136], [138], [152] *Suppose $\mathcal{V}$ is a weakly rational VOA, with exactly n irreducible $\mathcal{V}$-modules $M_1, \ldots, M_n$. Fix any finite subgroup G of $\text{Inn}(\mathcal{V})$. Then:*

(a) *For any $g \in \text{Inn}(\mathcal{V})$, there will be exactly n irreducible g-twisted $\mathcal{V}$-modules $M_1^g, \ldots, M_n^g$. Moreover, each M_i^g has a conformal weight $h_i^g \in \mathbb{Q}$ as in Lemma 5.3.3, and any g-twisted $\mathcal{V}$-module is completely reducible into a direct sum of the M_i^g. Labelling the modules appropriately, we get $(M_i^g, {}_h Y_{M_i^g}) \cong (M_i^{h^{-1} g h}, Y_{M_i^{h^{-1} g h}})$. This defines a projective representation $\varphi(h)$ of the centraliser $C_G(g)$ as in (5.3.24).*

(b) *For each commuting pair $g, h \in G$, define $\mathcal{Z}^i_{(g,h)}(\tau) := \mathcal{Z}(M_i^g, h; \tau)$. Then each $\mathcal{Z}^i_{(g,h)}(\tau)$ is holomorphic in $\mathbb{H}$, and is a modular function for (i.e. is fixed by) some congruence subgroup. For any $A = \begin{pmatrix} a & b \\ c & d \end{pmatrix} \in \text{SL}_2(\mathbb{Z})$, there exist scalars $a(A, g, h)_{ij}$*

such that

$$\mathcal{Z}^i_{(g,h)} \left(\frac{a\tau + b}{c\tau + d} \right) = \sum_{j=1}^n a(A, g, h)_{ij} \, \mathcal{Z}^j_{(g^a h^c, g^b h^d)}(\tau). \tag{5.3.25}$$

(c) *Let $\mathcal{V}^G$ be the vertex operator subalgebra consisting of all $v \in \mathcal{V}$ fixed by all elements of G. Then the $\mathbb{C}$-span of the graded dimensions of all nontwisted $\mathcal{V}^G$-modules will equal that of all $\mathcal{Z}^i_{(g,h)}(\tau)$ for commuting $g, h \in G$, and the total number of irreducible $\mathcal{V}^G$-modules will equal n times the sum, over representatives g of all conjugacy classes in G, of the number of inequivalent irreducible projective representations of $C_G(g)$ with 2-cocycle c_g as in (5.3.24).*

(d) *In the special case that $\mathcal{V}$ is holomorphic (i.e. $n = 1$), $\mathrm{Inn}(\mathcal{V}) = \mathrm{Aut}(\mathcal{V})$ and the coefficients a_{ij} in (5.3.25) are roots of unity. There is a 3-cocycle $\alpha \in H^3(G, U_1(\mathbb{C}))$ such that the 2-cocycle c_g of (5.3.24) is given by*

$$c_g(h_1, h_2) = \alpha(g, h_1, h_2)\alpha(h_1, h_2, g)\alpha(h_1, g, h_2)^*.$$

Some progress towards this important conjecture is provided by, for example, [150]. Monstrous Moonshine is interested in the holomorphic case (i.e. $n = 1$), which is by far the best understood; we return to it in Section 6.2.4. The number of irreducible projective representations in (c) is described in Section 3.1.1. We find in (d) that the cohomology group $H^3(G, U_1(\mathbb{C})) \cong H^4(G, \mathbb{Z})$ (trivial action of G on the coefficients) classifies all the possibilities for the orbifold; the analogous result for nonholomorphic VOAs is much more subtle, being more sensitive to the structure of $\mathcal{V}$, and is still poorly understood.

Part (c) leads us to a Galois theory for $\mathcal{V}^G$. But considering the depth of Jones' Galois theory for subfactors, and the 'Galois theory' for lattices sketched in Section 2.3.5, it is clear that a far more interesting theory is possible for VOAs. It would certainly be interesting to develop this.

The easiest examples of the orbifold construction are of a self-dual lattice VOA $\mathcal{V}(L)$ by a subgroup G of the automorphism group of L (see e.g. [150]). We learn in Section 5.2.2 that there is a deep analogy between lattices and VOAs. This orbifold construction of VOAs corresponds directly to the shift construction of lattices outlined in Section 2.3.3.

The most famous VOA, the Moonshine module $V^\natural$, was the original orbifold. Frenkel–Lepowsky–Meurman [201] obtained it as the orbifold of the Leech lattice VOA $\mathcal{V}(\Lambda)$ by the ±1-symmetry of Λ. Since Λ is self-dual, $\mathcal{V}(\Lambda)$ is holomorphic. As predicted by Conjecture 5.3.10, there is a unique -1-twisted $\mathcal{V}(\Lambda)$-module. We discuss this orbifold more in Sections 4.3.4 and 7.2.1; see also [201] for details.

Question 5.3.1. Let $\mathcal{V}$ be any VOA, and let $\mathcal{W}$ be a vector space and $T : \mathcal{V} \to \mathcal{W}$ be any isomorphism of vector spaces. Use this linear map T to carry the VOA structure on $\mathcal{V}$ to one on $\mathcal{W}$.

Question 5.3.2. Let $\mathcal{V}$ be any VOA and let $\mathcal{V}_{[n]}$ be the grading induced by $L[0]$ in (5.3.16d). Prove for any $N \geq 0$, that

$$\oplus_{n=0}^N \mathcal{V}_n = \oplus_{n=0}^N \mathcal{V}_{[n]}.$$

Question 5.3.3. Find an expression for the coefficient of the q^α term in the one-point function $\chi_M(\tau, v)$, using the representation theory of the stabiliser $G_v < \mathrm{Aut}\, V$ and the eigenvalues of the zero-mode $o(v)$ restricted to the homogeneous space M_α.

Question 5.3.4. Let V be any holomorphic weakly rational C_2-cofinite VOA with central charge $c = 24$. Prove that its graded dimension $\chi_V(\tau)$ must equal $J(\tau) + c$, where the constant c is dim V_1.

Question 5.3.5. (a) Relate the Jacobi character $\chi^J_{V(L)}$ of a lattice VOA $V(L)$, for L positive-definite and with even integer inner-products, and the theta series Θ_L of (2.3.7). (b) Relate the Jacobi character $\chi^J_{M(\lambda)}$ of an irreducible module of an integrable affine VOA $V(\mathfrak{g}, k)$, for $\bar{\mathfrak{g}}$ simple, with the affine algebra character χ_λ of (3.2.9a).

Question 5.3.6. Let M be g-twisted. Show that the series $\mathcal{Z}(M, h)$ of (5.3.23) is identically 0, unless $h \in C_G(g)$ has the property that, for all $k \in C_G(g)$ commuting with h, $c_g(h, k) = c_g(k, h)$. (*Hint*: first show that $\mathcal{Z}(M, hk)$ is identically 0 if $c_g(h, k) \neq c_g(k, h)$; then use the 2-cocycle condition (3.1.1b).)

5.4 Geometric incarnations

Vertex (operator) algebras are a deep construct and, in spite of their complexity, are here to stay. In this section we describe some connections with geometry. Section 5.4.1 describes the programme to rigorously construct CFTs in Segal's sense (Section 4.4.1), from VOAs. Section 5.4.2 reviews the geometric side of vertex operator superalgebras.

5.4.1 Vertex operator algebras and Riemann surfaces

The introductory chapter stated that the physics of Moonshine exploits the duality between Hamilton's and Feynman's pictures of CFT. Manin put it this way back in 1985:

> The quantum theory of (super)strings exists at present in two entirely different mathematical fields. Under canonical quantization it appears to a mathematician as the representation theory of algebras of Heisenberg, Virasoro, and Kac–Moody and their superextensions. Quantization with the help of the Polyakov path integration leads to the analytic theory of algebraic (super)curves and their moduli spaces, to invariants of the type of the analytic curvature, *etc.* Establishment of direct mathematical connections between these two forms of a single theory is a big and important problem. [**402**]

Our best answer to Manin is the theory of geometric vertex operator algebras.

Note that any time we have an algebraic structure with a binary operation (e.g. 'product') ab, we can express multiple products using *binary trees*, which keep track of the brackets. For example, the binary trees in Figure 5.1 correspond to the products XY and $A((BC)D)$, respectively. The external (i.e. valance 1) vertices are assigned vectors,

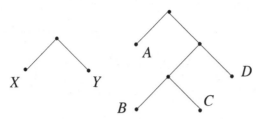

Fig. 5.1 Some binary trees.

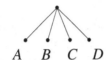

Fig. 5.2 Associativity.

while each internal vertex corresponds to a single product. Different algebraic structures can be axiomatised from this 'geometric' point of view. For instance, if the product is associative (e.g. we have a group), then it doesn't matter where we place the brackets – for example, the above *ABCD*-binary tree can be replaced with the tree in Figure 5.2.

More interesting for us are the geometrical axioms for Lie algebras [**294**]. Let V be any Lie algebra. Then to any binary tree with n legs corresponds a linear map φ from n copies $V \otimes \cdots \otimes V$ of the vector space V, to V. The map corresponding to the *ABCD*-binary tree of Figure 5.1 takes the Lie algebra vectors A, B, C, D to the nested Lie bracket $[A[[BC]D]]$. It is then fairly straightforward to encode all properties of the Lie algebra in the language of trees. For example, anti-commutativity says that if we flip the two descendents of an inner vertex of the tree – for example, in Figure 5.1 flipping D with the 3-vertex tree containing B and C – then the corresponding maps φ differ by a factor of -1. Gluing the root (uppermost vertex) of one tree to an external vertex of another corresponds in the Lie algebra to inserting one nested bracket into the middle of another. The only nontrivial property is anti-associativity (see Question 5.4.2). The result is a formulation of Lie algebra that is completely equivalent to the usual algebraic one [**294**].

Now, if we 'two-dimensionalise' that definition of 'geometric Lie algebra', we get something called a *geometric VOA* [**295**] that is equivalent to the 'algebraic' VOA of Definition 5.1.3. In place of binary trees (Figure 5.1), we have spheres with tubes (Figure 5.3). Equivalently, a sphere with n tubes is the Riemann sphere with n marked points and a choice of local coordinate at each point – an enhanced surface of type $(0,n)$ (Section 2.1.4). The moduli space of binary trees with n legs is a finite set, but the moduli space of spheres with n tubes is an infinite-dimensional complex space. To each such sphere with tubes we get a linear map φ from n copies of our vector space $\mathcal{V}$ (which is our VOA) to $\mathcal{V}$ (or rather a certain completion of $\mathcal{V}$ – a complication caused by the infinite-dimensionality of $\mathcal{V}$). A geometric VOA satisfies meromorphicity requirements, and most importantly the sewing axiom. In fact this map φ is Segal's functor $\mathcal{S}$ described in Section 4.4.1.

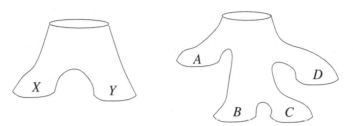

Fig. 5.3 The surfaces corresponding to Figure 5.1.

The point is that the resulting notion of geometric VOA is equivalent to that of algebraic VOA [295], though it takes considerable effort to show this. Thus a VOA is an 'algebra' with a two-dimensional analogue of a binary operation. In particular, let P_w be the simplest pair-of-pants, namely the Riemann sphere $\mathbb{PC}^1$ with marked points 0, ∞ and w and local coordinates given by z, $1/z$ and $z - w$ (z being the global coordinate on $\mathbb{C}$). Then the formal series $Y(u, w)v$ corresponds to $\mathcal{S}(P_w)(u \otimes v)$. On the other hand, consider the annulus, that is the Riemann sphere with marked points 0 and ∞, with local coordinates z and $\exp[-\epsilon z^{-1} d/dz\, z^{-1}]$. Recalling the realisation $-z^{-1}d/dz = \ell_{-2} \in \mathfrak{Witt}$ and the formula $\omega = L_{-2}\mathbf{1}$, we can recover the conformal vector ω by differentiating with respect to ϵ the map obtained from $\mathcal{S}$. The Virasoro algebra is fundamental here, capturing the effect of changing local coordinates (recall (5.3.15c)), and is responsible for the meromorphicity in the geometric VOA. The Jacobi identity (5.1.7a) is obtained from the sewing axiom. This equivalence relates formal power series (algebra) to distribution theory (analysis). It proves that the chiral blocks $\langle v, Y(u_1, z_1) \cdots Y(u_n, z_n)v' \rangle$ will be meromorphic, except for poles at $z_i = z_j$.

As mentioned before, a group corresponds to trees such as Figure 5.2. We can also two-dimensionalise that, and obtain what Huang calls a *vertex group* [293]. The easiest examples are $\mathbb{C}^{\times}$ and the enhanced moduli space $\widehat{\mathfrak{M}}_{0,1}$. Vertex groups should be to VOAs what Lie groups are to Lie algebras.

The motivation for this deep work is to construct examples satisfying Segal's definitions of CFT and modular functor. We know at present no nontrivial examples, although the general belief is that any sufficiently nice VOA will provide one. Huang's work [295] establishes this for genus 0, and more recently he has pushed it to genus 1 [297].

We end this subsection on a more speculative note [560], [295]. According to Witten, to understand string theory conceptually, we need a new analogue of Riemannian geometry. In contrast to the more classical 'particle-math', there is a more modern 'string-math'. We have the real numbers (particle physics) versus the complex numbers (string theory); binary trees versus spheres with tubes; Lie algebras versus VOAs; the representation theory of Lie algebras versus RCFT, etc. What are the stringy analogues of calculus, ordinary differential equations, Riemannian manifolds, the Atiyah–Singer Index theorem, . . . ? Huang suggests that just as we could imagine Moonshine as a mystery that is explained in some way by RCFT, perhaps the stringy version of calculus would similarly explain the mystery of two-dimensional gravity, stringy ODEs would explain the mystery of infinite-dimensional integrable systems, stringy Riemannian manifolds would help explain the mystery of mirror symmetry, and the stringy index theorem would help explain the elliptic genus (for this latter possibility, consider the work of Tamanoi reviewed next subsection).

What makes this more subtle is that complexification is not unique. To give a simple example, S^1 can be thought of as the real projective space $\mathbb{P}^1(\mathbb{R})$ and as the Lie group $SO_2(\mathbb{R})$. The obvious complexification of $\mathbb{P}^n(\mathbb{R})$ is $\mathbb{P}^n(\mathbb{C})$. An obvious complexification of $SO_n(\mathbb{R})$ is $SO_n(\mathbb{C})$. But if we think of $O_n(\mathbb{R})$ more geometrically as the real matrices that preserve the quadratic form $x_1^2 + \cdots + x_n^2$, then its complexification should be those complex matrices that preserve the Hermitian form $|x_1|^2 + \cdots + |x_n|^2$, i.e. $U_n(\mathbb{C})$. Thus the complexifications of S^1 in these cases would be the 2-sphere $\mathbb{P}^1(\mathbb{C})$, the cylinder $SO_2(\mathbb{C})$ (i.e. the multiplicative group $\mathbb{C}/\{0\}$) and the 3-sphere $SU_2(\mathbb{C})$ (as a real Lie group). So the specific complexification obtained depends on the context. In all cases, the way to proceed is to convert the defining relations of the given object into symbols that make sense over $\mathbb{C}$.

What sense can we make of the statement that the complexification of a binary tree is a sphere with punctures? Consider the simplest case: the segment $0 \leq x \leq 1$. This can be thought of geometrically as the locus $(a, b, c) \in \mathbb{R}^3$ satisfying $a + b^2 - 1 = a - c^2 = 0$. Over the complex numbers the parameter 'a' is redundant, and this locus has the obvious complexification $w^2 + z^2 = 1$. We know this is a sphere with two punctures, that is, a cylinder as we would like it to be.

Incidentally, Arnol'd speculates that there is in fact a triality: the reals, the complex numbers and the quaternions. He discusses several examples in [**18**], as well as some applications of this thought. This suggests that there is a third structure, generalising vertex algebras much as vertex algebras generalise Lie algebras.

5.4.2 Vertex operator superalgebras and manifolds

Through the work of Witten and others, we have discovered that much can be learned about a space X, by studying a string theory living in X. Much of this is reviewed in [**291**]. For example, to a Calabi–Yau manifold X [**299**], [**571**] and an element of its complexified Kähler cone, string theory associates two $N = 2$ superconformal field theories, called the A and B models (which focus on respectively the Kähler and complex structures of X). To clarify (and rigorise) these ideas, Malikov–Schechtman–Vaintrob [**401**] suggested how one may construct, given X, the vertex algebra of the $N = 2$ superconformal field theory (the A model) associated with X. This work is clearly fundamental. We can only sketch it here.

To any smooth complex variety X, reference [**401**] associates a sheaf of ($N = 1$) vertex operator superalgebras, called the *chiral de Rham complex* $\mathcal{MSV}_X$. In other words, to every open set $U \subset X$, there is a near-vertex operator superalgebra $\mathcal{MSV}_X(U)$ (the 'space of sections of $\mathcal{MSV}_X$ over U'). Whenever sets $U \subset V$ are open, there is a surjective restriction map $r_U^V : \mathcal{MSV}_X(V) \to \mathcal{MSV}_X(U)$, which is a homomorphism of near-vertex operator superalgebras. We briefly discuss vertex operator superalgebras in Section 5.1.3. These near-vertex operator superalgebras are bi-graded, by commuting operators L_0 (the Hamiltonian) and J_0 (the fermionic charge) with eigenvalues $\mathbb{N}$ and $\mathbb{Z}$, respectively. They form a complex in the sense that there is a differential Q_{BRST} obeying $Q_{BRST}^2 = 0$ and increasing fermionic charge by 1. When the open set U is homeomorphic to an open ball in $\mathbb{C}^n$, then $\mathcal{MSV}_X(U)$ is essentially the tensor product of n copies of what string theory calls a bosonic ($\beta\gamma$) ghost system (similar to the Heisenberg VOA), with n copies of a (bc) fermionic ghost system. The physics of these ghost systems is described in [**463**].

The prototypical example of 'sheaf' is the structure sheaf $\mathcal{O}_X$, which associates with each open set U the space of functions $f : U \to \mathbb{C}$. The prototypical example of 'complex' is the de Rham complex, given by the space of differential forms on X, with a differential d obeying $\mathrm{d}^2 = 0$ and taking p-forms to $p + 1$-forms. Of course the point of a complex is to take the cohomology $H^* = \ker \mathrm{d}/\mathrm{im}\,\mathrm{d}$. The books [**537**] are a readable introduction to algebraic geometry; in particular section 2.2 provides elementary examples of sheaves, and section 6.1 treats sheaf cohomology. For a sheaf $\mathcal{F}$ over X, $H^0(X, \mathcal{F})$ is always the global section $\mathcal{F}(X)$, and it is common for the other $H^i(X, \mathcal{F})$ to all vanish. The name 'chiral de Rham complex' was chosen because the $L_0 = 0$ subspace can be identified with the familiar space of differential forms ('chiral' refers to the chiral algebra of Section 4.3.2 or the chiral ring discussed in [**291**]).

In the case of $\mathcal{MSV}_X$, the sheaf cohomology $H^*(X, \mathcal{MSV}_X)$ yields the global section $\mathcal{MSV}_X(X)$, which is a near-vertex operator superalgebra. The case where X is Calabi–Yau is the most interesting, as $\mathcal{MSV}_X(X)$ has $N = 2$ (rather than merely $N = 1$) supersymmetry, which makes it much richer. $\mathcal{MSV}_X(X)$ is a fundamental invariant associated with X, and much information of X can be recovered from it. For example, the usual de Rham cohomology $H_{DR}^*(X)$ of X is $H^*(\mathcal{MSV}_X(X); Q_{BRST})$. For another example, the elliptic genus (discussed shortly) of X equals $\mathrm{tr}_{\mathcal{MSV}_X(X)} q^{L_0} y^{J_0}$ [**81**].

Elliptic genus appeared in the mid-1980s in both string theory and topology. For details see, for example, [**287**], [**499**], [**523**]. In Thom's cobordism ring Ω, elements are equivalence classes of cobordant manifolds, addition is connected sum and multiplication is Cartesian product. The *universal elliptic genus* $\phi(M)$ is a ring homomorphism from $\mathbb{Q} \otimes \Omega$ to the ring of power series in q, which sends n-dimensional manifolds with spin connections (see [**369**] for the relevant geometry) to a weight $n/2$ modular form of $\Gamma_0(2)$ with integer coefficients. Several variations and generalisations have been introduced, for example, the Witten genus assigns spin manifolds with vanishing first Pontrjagin class a weight $n/2$ modular form of $SL_2(\mathbb{Z})$ with integer coefficients. On a finite-dimensional manifold M, the index of the Dirac operator (in the heat kernel interpretation) is a path integral in supersymmetric quantum mechanics, that is an integral over the loop space $\mathcal{L}M = \{\gamma : S^1 \to M\}$; the string theory version of this is that the index of the Dirac operator on $\mathcal{L}M$ should be an integral over $\mathcal{L}(\mathcal{L}M)$, that is over smooth maps of tori into M, and this (heuristically) is just the elliptic genus, and explains why it should be modular.

The important rigidity property of the Witten genus with respect to any compact Lie group action on the manifold is a consequence of the modularity of the characters of affine algebras (our Theorem 3.2.3) [**388**]. In physics, elliptic genera arise as partition functions of $N = 2$ superconformal field theories [**561**]. The Witten genus (normalised by η^8) of the Milnor–Kervaire manifold M_0^8, an eight-dimensional manifold built from the E_8 diagram, equals $j^{\frac{1}{3}}$ [**287**]. Also, the elliptic genus of even-dimensional projective spaces $\mathbb{P}^{2n}(\mathbb{C})$ unexpectedly has only nonnegative coefficients and in fact equals the graded dimension of a certain vertex algebra [**400**]; this suggests interesting representation-theoretic questions in the spirit of Monstrous Moonshine. Exciting developments are described in [**517**], including relations with von Neumann (sub)factors.

Related to $\mathcal{MSV}_X$ must be the work of Tamanoi [**521**]. The index of an operator d is $\ker d - \operatorname{coker} d$; we can interpret this geometrically as the superdimension associated with the 'superpair' $(\ker d; \operatorname{coker} d)$ of vector spaces. This is what Tamanoi does with the elliptic genus. In particular, to each closed Riemannian manifold X he associates a vertex operator superalgebra $T(X)$, determined from its geometry. It has a nonnegative half-integer grading and central charge $N = \dim X/2$. The Riemannian metric of X yields the conformal vector ω. In the special case of a Kähler manifold, the Kähler forms (i.e. the closed real differential forms of type $(1,1)$) form a level 1 representation of the affine algebra $D_N{}^{(1)}$. Again, the elliptic genus is recovered as the graded dimension of $T(X)$. It is obviously desirable to relate these invariants $T(X)$ and $\mathcal{MSV}_X(X)$. We return to elliptic genus in Section 7.3.7.

Question 5.4.1. Find a complexification for the Möbius band.

Question 5.4.2. In a non-associative algebra, the ambiguous product $v_1 \cdots v_n$ can only be evaluated when the $n - 1$ pairs of brackets are placed. Let L be any Lie algebra. Prove that for any $n \geq 3$, L has an identity of the form

$$v_1 \cdots v_n = v_1 v_n v_2 v_3 \cdots v_{n-1} + \cdots + v_1 \cdots v_i v_n v_{i+1} \cdots v_{n-1} + \cdots + v_1 v_2 \cdots v_{n-1} v_n.$$

More precisely, for any choice of bracketing on the left, prove that there is a choice of bracketing for each of the $n - 1$ terms on the right such that the resulting formula holds for any $v_i \in L$. For example, $[[v_1 v_2]v_3]$ equals $[[v_1 v_3]v_2] + [v_1[v_2 v_3]]$ and $[[v_1[v_2 v_3]]v_4]$ equals $[[v_1 v_4][v_2 v_3]] + [v_1[[v_2 v_4]v_3]] + [v_1[v_2[v_3 v_4]]]$.

6

Modular group representations throughout the realm

There are two aspects to Moonshine. The more general one is the unexpected presence of modular group actions over a wide range of algebraic settings, and is now fairly well understood. We have seen instances of this already with, for example, the characters of affine algebras and VOAs. This chapter completes our treatment of these modular actions. The more specific aspect – the association of Hauptmoduls to the Monster – is still poorly understood and is the subject of the following chapter.

Much of this chapter is orthogonal to Monstrous Moonshine. For example, we discuss here fusion rings and modular data; both the fusion ring and modular data of the Moonshine module $V^\natural$ are maximally trivial. Nevertheless, this chapter helps to paint the general context of Monstrous Moonshine. In Section 7.2.4 we build on some of the lessons from this chapter to speculate on a possible second proof of Monstrous Moonshine.

6.1 Combinatorial rational conformal field theory

Recall the semi-simple Lie algebras: we study their structure and obtain their classification by abstracting out combinatorial features (e.g. roots, Coxeter–Dynkin diagrams). Of course this is easy to do with a finite-dimensional linear structure. RCFTs are infinite-dimensional, but by definition their infinite-dimensional symmetry and implicit rigidity again effectively reduces them to certain discrete structures. As we see next section, those discrete structures are remarkable for their ubiquity in modern mathematics. See [208], [207], [33], [131], [437], [236] for further background. As with all other chapters except Chapter 7, we've tended to avoid giving original references, as these are voluminous and can be recovered from the numerous review articles and books.

6.1.1 Fusion rings

Recall that the eigenvalues of a self-adjoint (equivalently, Hermitian) matrix are all real. Consider the following scenario. Let A, B and C be $n \times n$ Hermitian matrices with eigenvalues $\alpha_1 \geq \alpha_2 \geq \cdots \geq \alpha_n, \beta_1 \geq \cdots \geq \beta_n, \gamma_1 \geq \cdots \geq \gamma_n$. What are the conditions on these eigenvalues so that $C = A + B$? The answer consists of a number of inequalities involving the numbers $\alpha_i, \beta_j, \gamma_k$. Now discretise this problem:

Theorem 6.1.1   *Let* $\alpha_1 \geq \alpha_2 \geq \cdots \geq \alpha_n \geq 0, \beta_1 \geq \cdots \geq \beta_n \geq 0, \gamma_1 \geq \cdots \geq \gamma_n \geq 0$, *all be integers. Then the following are equivalent:*

(i) *Hermitian matrices A, B and $C = A + B$ exist with eigenvalues α, β, γ,*
   respectively;
(b) *the $\mathfrak{gl}_n(\mathbb{C})$ tensor product multiplicity $T_{\alpha\beta}^{\gamma}$ is nonzero.*

Recall from Section 1.5.1 that the finite-dimensional unitary irreducible modules of the Lie algebra $\mathfrak{gl}_n(\mathbb{C}) \cong \mathbb{C} \oplus \mathfrak{sl}_n(\mathbb{C})$ are naturally labelled by pairs $(a, \lambda) \in \mathbb{R} \times \mathbb{N}^{n-1}$, where $z \mapsto iaz$ is a representation of the abelian Lie algebra $\mathbb{C}$, and $\lambda = (\lambda_1, \ldots, \lambda_{n-1})$ is a highest weight for the simple Lie algebra $\mathfrak{sl}_n$. The eigenvalues α correspond to labels $a = \alpha_n$ and $\lambda_i = \alpha_i - \alpha_{i+1}$. The number $T_{\alpha\beta}^{\gamma}$ is the number of times the $\mathfrak{gl}_n$-module $L(\gamma)$ appears in the tensor product $L(\alpha) \otimes L(\beta)$ of modules. This remarkable theorem and related results are discussed in the review article [**218**].

Now consider instead $n \times n$ unitary matrices with determinant 1. Any such matrix $D \in \mathrm{SU}_n(\mathbb{C})$ can be assigned a unique n-tuple $\delta = (\delta_1, \ldots, \delta_n)$ as follows. Write its eigenvalues as $e^{2\pi i \delta_i}$, where $\delta_1 \geq \cdots \geq \delta_n$, $\sum_{i=1}^{n} \delta_i = 0$ and $\delta_1 - \delta_n \leq 1$. Let Δ_n be the set of all such n-tuples δ, as D runs through $\mathrm{SU}_n(\mathbb{C})$. Note that D will have finite order iff all $\delta_i \in \mathbb{Q}$, and that D will be a scalar matrix dI iff all differences $\delta_i - \delta_j \in \mathbb{Z}$. Of course, a sum of Hermitian matrices corresponds here to a product of unitary matrices.

Theorem 6.1.2 [**4**]   *Choose any rational n-tuples $\alpha, \beta, \gamma \in \Delta_n \cap \mathbb{Q}^n$. Then the following are equivalent:*
 (i) *there exist matrices $A, B, C \in \mathrm{SU}_n(\mathbb{C})$, with $C = AB$, with n-tuples α, β, γ;*
 (ii) *there is a positive integer k such that all differences $k\alpha_i - k\alpha_j, k\beta_i - k\beta_j$,*
   $k\gamma_i - k\gamma_j$ are integers, and the fusion multiplicity $\mathcal{N}_{k\alpha, k\beta}^{(k)\ k\gamma}$ of $\mathfrak{sl}_n^{(1)}$ at level k is
   nonzero.

We met the affine algebra $\mathfrak{sl}_n^{(1)} = A_{n-1}^{(1)}$ and its modules in Section 3.2. Here, $k\alpha$ corresponds to the level-k integrable highest weight $\lambda \in P_+^k(A_{n-1}^{(1)})$ with Dynkin labels $\lambda_i = k\alpha_i - k\alpha_{i+1}$. The $\mathfrak{sl}_n^{(1)}$ fusion multiplicities are studied in Section 6.2.1. Theorems 6.1.1 and 6.1.2 provide one instance of a general principle:

> *A result or construction valid for $\mathfrak{gl}_n$ or $\mathfrak{sl}_n$ tensor products should have an interesting analogue for the $\mathfrak{sl}_n^{(1)}$ fusion product.*

The $\mathfrak{gl}_n$ tensor product multiplicities are classical quantities, appearing in numerous and varied contexts. The $\mathfrak{sl}_n^{(1)}$ fusion multiplicities are equally fundamental, equally ubiquitous, but less well understood.

Just as the tensor product multiplicities are structure constants of the character ring of the Lie algebra, so do fusion multiplicities define a *fusion ring*, an aspect of Moonshine complementary to Monstrous Moonshine.

Definition 6.1.3   *A fusion ring $R = R(\beta, \mathcal{N})$ is a commutative ring R with unity 1, together with a finite basis $\beta = \{x_a \mid a \in \Phi\}$ (over $\mathbb{Z}$) containing $1 = X_0$, such that:*
 F1. *The structure constants $\mathcal{N}_{ab}^c$, defined by $x_a x_b = \sum_{c \in \Phi} \mathcal{N}_{ab}^c x_c$, are all nonnegative integers.*
 F2. *There is a ring homomorphism $x \mapsto x^*$ stabilising the basis Φ (we write $(x_a)^* = x_{a^*}$).*

F3. $\mathcal{N}^1_{ab} = \delta_{b,a^*}$.

F4. '$S = S^t$' *(we'll explain this shortly, but it says R is self-dual in a strong sense).*
The numbers $\mathcal{N}^c_{ab}$ are called fusion multiplicities, *the labels $a \in \Phi$ are called* primaries,
$0 \in \Phi$ *is called the* vacuum *and '$*$' is called* charge-conjugation.

The only reason for distinguishing the basis β from the labels Φ is that for fusion rings the multiplicative notation (e.g. unit 1) is natural, but in the traditional examples of modular data additive notation is used. The terminology here comes from RCFT.

An important ingredient of fusion rings, as with character rings, is their preferred basis β. Abstract rings don't come with a basis. Forgetting the basis β, fusion rings aren't interesting: for example, the algebra $R \otimes_{\mathbb{Z}} \mathbb{C}$ over $\mathbb{C}$ (i.e. the span over $\mathbb{C}$ of β, retaining the same multiplication and addition) is isomorphic as a $\mathbb{C}$-algebra to $\mathbb{C}^{\|\Phi\|}$ with operations defined component-wise (see Lemma 6.1.4 below). This is reminiscent of the character ring of the Lie algebra X_r, which is isomorphic (as a $\mathbb{C}$-algebra) to a polynomial algebra in r variables.

For each $a \in \Phi$, define the *fusion matrix* $\mathcal{N}_a$ by

$$(\mathcal{N}_a)_{b,c} = \mathcal{N}^c_{ab}.$$

Note that the fusion matrix $\mathcal{N}_0$ equals the identity matrix I, and $\mathcal{N}_{a^*} = (\mathcal{N}_a)^t$ (Question 6.1.1). The fusion matrices can be simultaneously diagonalised:

Lemma 6.1.4   *(a) Given any fusion ring $R = R(\Phi, \mathcal{N})$, there is a unique (up to ordering of the columns) unitary matrix S, with rows parametrised by Φ and columns by say Φ', obeying both*

$$S_{0i} > 0, \qquad\qquad\qquad\qquad\qquad (6.1.1a)$$

$$\mathcal{N}^c_{ab} = \sum_i \frac{S_{ai} \, S_{bi} \, \overline{S_{ci}}}{S_{0i}}, \qquad\qquad\qquad (6.1.1b)$$

for all $a, b, c \in \Phi$ and $i \in \Phi'$.
(b) All simultaneous eigenspaces of all the fusion matrices are of dimension 1, and are spanned by each column $S_{\ddagger,b}$.

The proof of Lemma 6.1.4 only involves F1–F3. The condition F4 can now be expressed by requiring that the S of Lemma 6.1.4 (for some ordering of the columns) obey $S = S^t$ (so $\Phi' = \Phi$). The proof of Lemma 6.1.4 is elementary – the fusion matrices commute with each other and hence with their transposes, and so are simultaneously diagonalisable – and analogues hold in much greater generality. Equation (6.1.1b) says that the bth column $S_{\ddagger,b}$ of S is an eigenvector of each $\mathcal{N}_a$, with eigenvalue $\frac{S_{ab}}{S_{0b}}$. From the unitarity of S, we know that $\frac{S_{ab}}{S_{0b}} = \frac{S_{ac}}{S_{0c}}$ can hold for all $a \in \Phi$, only if $b = c$, which gives us part (b).

> *The matrix S acts a lot like the character table of a finite group; a general theorem valid for character tables has a fusion ring analogue.*

Note that *a priori* the rows (parametrising basis vectors) and columns (parametrising eigenvectors) of S in Lemma 6.1.4 play entirely different roles. In a natural sense [236], the dual ring to R has structure constants given by replacing S in (6.1.1b) with its transpose S^t. This is what underlies calling F4 a self-duality condition. In contrast, the character ring of a finite group is fusion-like, is diagonalised by the character table, but its dual involves multiplying conjugacy classes and is isomorphic to the character ring only for abelian groups. The appearance of self-duality here may seem somewhat mysterious, but some sort of self-duality is pervasive in the mathematics of this chapter. In particular, Drinfel'd's 'quantum double' construction (Section 6.2.3) generates algebraic structures possessing fusion rings and modular data, by combining a given (inadequate) algebraic structure with its dual in some way. An example is provided by Section 6.2.4, where the true (self-dual) fusion ring of a finite group is built up out of the character ring and its dual.

Fusion rings arise naturally in RCFT (Sections 4.3.2 and 6.1.4). The 'primaries' are the chiral primaries, parametrising the irreducible modules of the chiral algebra $\mathcal{V}$. The fusion multiplicities $\mathcal{N}_{ab}^c$ are the dimension of the space of chiral blocks $\mathfrak{B}_{a,b}^{(0,3)c}$ on a sphere with three punctures (two 'incoming' and 1 'outgoing'), where we label those punctures with the primaries a, b, c. Equation (6.1.1b) is called *Verlinde's formula* [542], and S has an interpretation in terms of modular transformations of the characters (4.3.9a). A similar formula gives the dimension of any space of chiral blocks:

$$\dim \mathfrak{B}_{a_1,...,a_n}^{(g,n+m)\,b_1,...,b_m} := \mathcal{N}_{a_1,...,a_n}^{(g,n+m)\,b_1,...,b_m}$$

$$= \sum_{c \in \Phi} (S_{0c})^{2(1-g)} \frac{S_{a_1 c}}{S_{0c}} \cdots \frac{S_{a_n c}}{S_{0c}} \frac{\overline{S_{b_1 c}}}{S_{0c}} \cdots \frac{\overline{S_{b_m c}}}{S_{0c}}. \qquad (6.1.2)$$

The depth of Verlinde's formula (6.1.1b), (6.1.2), which is considerable, lies in this modular interpretation given to S. The S matrix is called the *modular matrix* for this reason. Historically [50], the fusion ring arose directly by interpreting the chiral OPE symbolically in terms of products of $\mathcal{V}$-families of chiral fields (see e.g. section 7.3 of [131]).

Recall Perron–Frobenius theory from Section 2.5.2. The fusion matrices $\mathcal{N}_a$ are non-negative, and it is indeed natural to multiply them:

$$\mathcal{N}_a \mathcal{N}_b = \sum_{c \in \Phi} \mathcal{N}_{ab}^c \mathcal{N}_c.$$

So we can expect Perron–Frobenius to tell us something interesting. By (6.1.1a), the Perron–Frobenius eigenvalue of $\mathcal{N}_a$ is $\frac{S_{a0}}{S_{00}}$; hence we obtain the important inequality

$$S_{a0} S_{0b} \geq |S_{ab}| \, S_{00}. \qquad (6.1.3a)$$

Unitarity of S applied to (6.1.3a) forces

$$\min_{a \in \Phi} S_{a0} = S_{00}. \qquad (6.1.3b)$$

The *quantum-dimension* $\mathcal{D}_{(a)}$ of (5.3.12) equals $\frac{S_{a0}}{S_{00}}$, and so is bounded below by 1.

The borderline case of (6.1.3b) are those primaries $a \in \Phi$, called *simple-currents* in RCFT, obeying $S_{a0} = S_{00}$. To any such simple-current $j \in \Phi$, there is a phase

$\varphi_j : \Phi \to \mathbb{C}$ and a permutation J of Φ such that $j = J0$ and

$$S_{Ja,b} = \varphi_j(b) S_{a,b}, \tag{6.1.4a}$$

$$\mathcal{N}^b_{j,a} = \delta_{b,Ja}. \tag{6.1.4b}$$

For example, we see from (4.3.11e) that ϵ is a simple-current for the Ising model, with phases $\varphi_\epsilon(0) = \varphi_\epsilon(\epsilon) = 1$ and $\varphi_\epsilon(\sigma) = -1$.

It is clear what plays the role of the endomorphism '$*$' in the character ring of a finite group: complex conjugation. So take the complex conjugate of (6.1.1b). We find that $\overline{S}$ also simultaneously diagonalises the fusion matrices $\mathcal{N}_a$. Hence from Lemma 6.1.4(b) there is a permutation of Φ, which we denote by C, and some $\alpha_b \in \mathbb{C}$, such that

$$\overline{S_{ab}} = \alpha_b S_{a,Cb}.$$

Unitarity of S forces each $|\alpha_b| = 1$. Looking at $a = 0$ and applying (6.1.1a), we see that the α_b must be positive. Hence

$$\overline{S_{ab}} = S_{Ca,b} = S_{a,Cb}, \tag{6.1.5}$$

so as a permutation matrix, $C = S^2$. Comparing F3 to Verlinde's formula (6.1.1b), we find that C is charge-conjugation: $Ca = a^*$. Note that C, like complex conjugation, is an involution, and that $C_{00} = 1$.

More generally, recall our discussion of cyclotomic fields and their Galois automorphisms from Section 1.7. The character values $\mathrm{ch}(g)$ of a finite group G lie in the cyclotomic field $\mathbb{Q}[\xi]$, for the root of unity $\xi = \xi_{\|G\|}$. Write σ_ℓ for the automorphism of $\mathbb{Q}[\xi]$ defined by $\sigma_\ell(\xi) = \xi^\ell$, for some integer ℓ coprime to $\|G\|$. Then σ_ℓ acts on the character table by

$$\sigma_\ell(\mathrm{ch}(g)) = \mathrm{ch}(g^\ell) = \mathrm{ch}^{\sigma_\ell}(g), \tag{6.1.6}$$

for some character $\mathrm{ch}^{\sigma_\ell}$ of G (to see which one, use the fact [308] that every G-representation is equivalent to a matrix representation with all entries in $\mathbb{Q}[\xi_{\|G\|}]$).

Theorem 6.1.5 [**114**]   *Choose any fusion ring, and let S be the associated modular matrix. The entries S_{ab} of the matrix S lie in some cyclotomic field $\mathbb{Q}[\xi_N]$. Given any Galois automorphism $\sigma \in \mathrm{Gal}(\mathbb{Q}[\xi_N]/\mathbb{Q})$,*

$$\sigma(S_{ab}) = \epsilon_\sigma(a) S_{a^\sigma,b} = \epsilon_\sigma(b) S_{a,b^\sigma} \tag{6.1.7}$$

for some permutation $b \mapsto b^\sigma$ of Φ, and some signs (parities) $\epsilon_\sigma : \Phi \to \{\pm 1\}$.

This is a fundamental symmetry of fusion rings, or rather their modular matrices. For example, for σ equal to complex conjugation, (6.1.7) reduces to (6.1.5). Equation (6.1.7) is essentially the statement that the fusion multiplicities are rational numbers; the cyclotomicity follows from Theorem 1.7.1 and depends crucially on self-duality F4. Any property of charge-conjugation seems to have an analogue for any of these Galois symmetries, although it is usually more complicated.

What has a fusion ring to do with 'modular stuff'? That is explained next.

6.1.2 Modular data

Choose any even integer $n > 0$. The matrix

$$S = \left(\frac{1}{\sqrt{n}} e^{-2\pi i mm'/n} \right)_{0 \le m, m' < n} \tag{6.1.8}$$

is the finite Fourier transform. Define the diagonal matrix T by $T_{mm} = \exp(\pi i m^2/n - \pi i/12)$. The assignment

$$\begin{pmatrix} 0 & -1 \\ 1 & 0 \end{pmatrix} \mapsto S, \qquad \begin{pmatrix} 1 & 1 \\ 0 & 1 \end{pmatrix} \mapsto T \tag{6.1.9}$$

defines an n-dimensional representation ρ of $SL_2(\mathbb{Z})$, using (2.2.1a). This is the simplest (and least interesting) example of *modular data*. Verlinde's formula (6.1.1b) associates a fusion ring with (6.1.8). Here the labels are $\Phi = \{0, 1, \ldots, n-1\}$ and the fusion ring is the ring of integers $\mathbb{Z}[\xi_n]$ with preferred basis

$$\beta = \left\{ 1, \xi_n, \ldots, \xi_n^{n-1} \right\}.$$

The fusion multiplicities are given by addition mod n.

This $SL_2(\mathbb{Z})$-representation (6.1.9) is realised by modular functions in the following sense. For each $a \in \{0, 1, \ldots, n-1\}$, define the functions

$$\chi_a(\tau) = \frac{1}{\eta(\tau)} \sum_{k=-\infty}^{\infty} q^{n(k+a/n)^2/2},$$

where as always $q = e^{2\pi i \tau}$ and $\eta(\tau)$ is the Dedekind eta function (2.2.6b). Then (4.3.9) hold. Thus $\vec{\chi} = (\chi_a)_{a \in \Phi}^t$ is a vector-valued modular function with multiplier ρ for $SL_2(\mathbb{Z})$ (Definition 2.2.2).

Definition 6.1.6   *Let Φ be a finite set of labels, one of which – denote it 0 – is distinguished. Modular data are matrices $S = (S_{ab})_{a,b \in \Phi}$, $T = (T_{ab})_{a,b \in \Phi}$ of complex numbers such that:*

MD1. *S, T are unitary and symmetric, and T is diagonal and of finite order. That is, $T^N = I$ for some N.*
MD2. *$S_{0a} > 0$ for all $a \in \Phi$.*
MD3. *$S^2 = (ST)^3$.*
MD4. *The numbers $\mathcal{N}_{ab}^c$ defined by (6.1.1b) are nonnegative integers.*

From the presentation (2.2.1a) of the modular group $SL_2(\mathbb{Z})$, we see that modular data defines a representation of $SL_2(\mathbb{Z})$, as in (6.1.9). Modular data abstracts out the $SL_2(\mathbb{Z})$ action arising in unitary RCFT (for non-unitary RCFT, MD2 should be weakened). It is a significant refinement of fusion rings. In particular, most fusion rings are not realised by any modular data (Question 6.1.5), but those that are are always realised by at least three sets of modular data.

We can generalise (6.1.8) using lattices (recall Section 1.2.1). If we write L for the lattice $\sqrt{n}\mathbb{Z}$, then $L^* = \frac{1}{\sqrt{n}}\mathbb{Z}$ is the dual lattice, the labels $\{0, \ldots, n-1\}$ parametrise the cosets L^*/L, and the modular function χ_a is the theta series of the ath coset, normalised by η. More generally, any even lattice L defines modular data in this way. The vacuum '0' will be $[0] = L$. The fusion multiplicities $\mathcal{N}_{[a],[b]}^{[c]}$ equal the Kronecker delta $\delta_{[c],[a+b]}$, so the fusion product is given by addition in the finite group L^*/L. All primaries $[a] \in \Phi$ are simple-currents (6.1.4), corresponding to permutation $J_{[a]}([b]) = [a+b]$ and phase $\varphi_{[a]}([b]) = e^{2\pi i a \cdot b}$. Charge-conjugation (6.1.5) is given by $C[a] = [-a]$. The Galois action (6.1.7) here is also simple: there is a Galois automorphism σ_ℓ for any integer ℓ coprime to the determinant $|L|$; σ_ℓ takes $[a]$ to $[\ell a]$, and all parities $\epsilon_\ell([a])$ equal $+1$. From our point of view, however, this lattice example is a little too trivial.

In RCFT (Section 4.3.2), the labels $a \in \Phi$ are the chiral primaries and '0' is the vacuum state. The matrix T equals (4.3.10). Charge-conjugation C is a symmetry in quantum field theory that interchanges particles with their anti-particles (and so reverses charge, hence the name). The modular data S, T arise through (4.3.9), where χ_a are the one-point functions on a torus. The above lattice example corresponds to the string theory of m free bosons compactified on the torus $\mathbb{R}^m/L$, where $m = \dim L$.

Every property of fusion rings should have an analogue in modular data. For example, the analogue of (6.1.5) is

$$T_{Ca,Cb} = T_{ab}, \qquad (6.1.10a)$$

which says that T and $C = S^2 = (ST)^3$ commute. The analogue of (6.1.4) is

$$T_{Ja,Ja}\overline{T_{aa}} = \overline{\varphi_j(a)}\, T_{jj}\, \overline{T_{00}}. \qquad (6.1.10b)$$

In all known examples, including all those associated with RCFT [37], Galois is intimately connected with the existence of characters χ_a realising the modular data as in (4.3.9), which are modular functions for a congruence subgroup (recall (2.2.4)). In particular, for all these examples, we get the remarkable property:

Definition 6.1.7 (congruence property) *Let S, T be modular data, and let ρ be the associated $SL_2(\mathbb{Z})$-representation. Let N be the order of the matrix T, so $T^N = I$. Then we say S, T obey the congruence property if the following are all satisfied: ρ is trivial (i.e. with value I) on the congruence subgroup $\Gamma(N)$, and so defines a representation of the finite group $SL_2(\mathbb{Z}_N)$; we have characters χ_a realising the modular data in the sense of (4.3.9), and those characters are modular functions for $\Gamma(N)$; the entries S_{ab} all lie in the cyclotomic field $\mathbb{Q}[\xi_N]$; and finally, the Galois automorphism σ_ℓ corresponds to the modular transformation $\begin{pmatrix} \ell & 0 \\ 0 & \ell^{-1} \end{pmatrix} \in SL_2(\mathbb{Z}_N)$, and so we get*

$$\left(\rho\begin{pmatrix} \ell & 0 \\ 0 & \ell^{-1} \end{pmatrix}\right)_{ab} = \epsilon_\ell(a)\,\delta_{b,a^{\sigma_\ell}}, \qquad \forall a, b \in \Phi, \qquad (6.1.11a)$$

$$T_{a^{\sigma_\ell},a^{\sigma_\ell}} = (T_{aa})^{\ell^2}, \qquad \forall a \in \Phi. \qquad (6.1.11b)$$

The finite group $SL_2(\mathbb{Z}_N)$ arises as $SL_2(\mathbb{Z})/\Gamma(N)$. The quantity '$\ell^{-1}$' denotes the multiplicative inverse of ℓ (mod N), and exists because $\gcd(\ell, N) = 1$. We return to the congruence property in Section 6.3.3. Probably Definition 6.1.6 is so weak that some 'sick' S, T are examples. It is expected, however, that all reasonably healthy modular data, for example, modular data associated with nice CFTs, VOAs or modular categories, would obey the congruence property (or at least something close to it). It is known [**169**] that modular data obeying the congruence property will typically (always?) be realised by some vector-valued modular function as in (4.3.9).

6.1.3 Modular invariants

Modular data axiomatises the appearance of $SL_2(\mathbb{Z})$ in unitary RCFT. Two places modular data directly impacts on RCFT are Verlinde's formula (6.1.2) and the partition function (4.3.8b).

Definition 6.1.8 *Choose any modular data* S, T. *A modular invariant is a matrix* $\mathcal{Z}$, *with rows and columns labelled by* Φ, *obeying:*

MI1. $\mathcal{Z}S = S\mathcal{Z}$ *and* $\mathcal{Z}T = T\mathcal{Z}$;

MI2. $\mathcal{Z}_{ab} \in \mathbb{N}$ *for all* $a, b \in \Phi$; *and*

MI3. $\mathcal{Z}_{00} = 1$.

It will be convenient at times to rewrite $\mathcal{Z}S = S\mathcal{Z}$ as $S\mathcal{Z}\overline{S} = \mathcal{Z}$ (recall that S is unitary). The easiest modular invariants are the identity $\mathcal{Z} = I$ and charge-conjugation $\mathcal{Z} = C$. More generally, $\mathcal{Z}$ is a modular invariant iff $C\mathcal{Z}$ is.

Modular invariants axiomatise the 1-loop partition functions $\mathcal{Z}(\tau)$ (4.3.8b) of RCFT. More precisely, an RCFT consists of two VOAs, called chiral algebras. For convenience we will take them to be isomorphic, though this is not necessary (when they aren't isomorphic, the theory is called 'heterotic'). The modular invariant describes how these VOAs act on the state space $\mathcal{H}$, that is how $\mathcal{H}$ decomposes into modules of the chiral algebras:

$$\mathcal{H} = \oplus_{a,b \in \Phi} \mathcal{Z}_{ab} \mathcal{H}_a \otimes \overline{\mathcal{H}_b}.$$

MI2 holds because the $\mathcal{Z}_{ab}$ are multiplicities. The adjoint module $\mathcal{H}_0 \otimes \overline{\mathcal{H}_0}$ contains the vacuum $\mathbf{1} \otimes \overline{\mathbf{1}}$, and MI3 says there should be only one vacuum. Finally, the 1-loop partition function $\mathcal{Z}(\tau)$, being a physical correlation function defined on the torus, must be invariant with respect to the modular group $SL_2(\mathbb{Z})$ of the torus. Equivalently, $\mathcal{Z}(\tau) = \mathcal{Z}(-1/\tau) = \mathcal{Z}(\tau + 1)$. Applying (4.3.9) and the unitarity of S and T gives the modular invariance condition MI1.

Perhaps it is because of their basic importance to RCFT, but the lists of modular invariants associated with affine algebras (Section 6.2.1) are quite remarkable. They also play natural roles for subfactors and VOAs, as we'll see.

A second partition function, playing the same role for boundary CFT (the open string) that $\mathcal{Z}(\tau)$ plays for bulk CFT (the closed string), is that corresponding to a cylinder. Its coefficient matrices $\mathcal{M}_{ax}^y$ define a fusion ring representation (6.2.6), called a NIM-rep [**47**], [**236**]. Although they are a fascinating part of the bigger picture, we'll say little about them in this book.

Fix a choice of modular data. Commutation MI1 of $\mathcal{Z}$ with T is trivial to solve, since T is diagonal: it yields the selection rule

$$\mathcal{Z}_{ab} \neq 0 \;\Rightarrow\; T_{aa} = T_{bb}. \tag{6.1.12}$$

More subtle and valuable is commutation with S. In particular, each symmetry of S yields a symmetry of $\mathcal{Z}$, a selection rule telling us certain entries of $\mathcal{Z}$ must vanish, and a way to construct new modular invariants.

First consider simple-currents j, j'. Equation (6.1.4a) and positivity tell us

$$\mathcal{Z}_{j,j'} = \left| \sum_{c,d\subset\Phi} \varphi_j(c)\, S_{0c}\, \mathcal{Z}_{cd}\, \overline{S_{d0}}\, \overline{\varphi_{j'}(d)} \right| \le \sum_{c,d} S_{0c}\, \mathcal{Z}_{cd}\, S_{d0} = \mathcal{Z}_{00} = 1.$$

Thus $\mathcal{Z}_{j,j'} \neq 0$ implies $\mathcal{Z}_{j,j'} = 1$, as well as the selection rule

$$\mathcal{Z}_{cd} \neq 0 \;\Rightarrow\; \varphi_j(c) = \varphi_{j'}(d). \tag{6.1.13a}$$

A similar calculation yields the symmetry

$$\mathcal{Z}_{J0,J'0} \neq 0 \;\Rightarrow\; \mathcal{Z}_{Ja,J'b} = \mathcal{Z}_{ab}, \qquad \forall a, b \in \Phi. \tag{6.1.13b}$$

The most useful application of simple-currents to modular invariants is to their construction. In particular, let $j = J_0$ be a simple-current of order n. Then (by Question 6.1.7(b)) we can find integers r_j and $Q_j(a)$ such that

$$\varphi_j(a) = \exp\left[2\pi i\,\frac{Q_j(a)}{n}\right], \qquad T_{jj}\overline{T_{00}} = \exp\left[2\pi i\, r_j\,\frac{n-1}{2n}\right].$$

Now define the matrix $\mathcal{Z}[j]$ by [489]

$$\mathcal{Z}[j]_{ab} = \sum_{\ell=1}^{n} \delta_{J^\ell a,b}\, \delta\!\left(Q_j(a) + \frac{\ell}{2n} r_j\right), \tag{6.1.14}$$

where $\delta(x) = 1$ when $x \in \mathbb{Z}$ and is 0 otherwise. This matrix will be a modular invariant iff $T_{jj}\overline{T_{00}}$ is an nth root of 1. For instance, $\mathcal{Z}[0] = I$.

Now look at the consequences of Galois. Applying the Galois automorphism σ to $\mathcal{Z} = S\mathcal{Z}\overline{S}$ yields, from (6.1.7) and $\mathcal{Z}_{ab} \in \mathbb{Q}$, the equation

$$\mathcal{Z}_{ab} = \sum_{c,d\in\Phi} \epsilon_\sigma(a)\, S_{\sigma a,c}\, \mathcal{Z}_{cd}\, \overline{S_{d,\sigma b}}\, \epsilon_\sigma(b) = \epsilon_\sigma(a)\, \epsilon_\sigma(b)\, \mathcal{Z}_{\sigma a,\sigma b}.$$

(Why must σ commute with complex conjugation?) Because $\mathcal{Z}_{ab} \geq 0$, this implies the selection rule and symmetry

$$\mathcal{Z}_{ab} \neq 0 \;\Rightarrow\; \epsilon_\sigma(a) = \epsilon_\sigma(b), \tag{6.1.15a}$$
$$\mathcal{Z}_{\sigma a,\sigma b} = \mathcal{Z}_{ab}, \tag{6.1.15b}$$

valid for any σ. Of all the equations (6.1.13) and (6.1.15), (6.1.15a) is the most useful. The reader can try to construct modular invariants from certain special σ_ℓ.

6.1.4 The generators and relations of RCFT

In fundamental and influential work of the late 1980s, Moore and Seiberg [436], [437] isolated the data (finite-dimensional vector spaces and linear transformations) defining each chiral half of RCFT, and provided a complete set of relations they satisfy. Roughly,

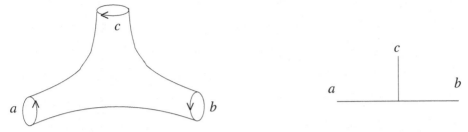

Fig. 6.1 A vertex.

they do for topological field theories in $2 + 1$ dimensions what Theorem 4.4.4 does in $1 + 1$ dimensions. Most of their work has been rigorously clarified in the important book [**32**]. This section sketches the basic ideas.

Their goal is to understand the spaces $\mathfrak{B}(\Sigma)$ of chiral blocks (Section 4.3.2). As in Section 4.4.1, incoming strings are those boundary circles oriented oppositely to the surface. We can change the orientation of a boundary circle provided we also replace its label (a module $M \in \Phi(\mathcal{V})$) with its charge-conjugate $M^\star$ (5.3.4a). Thus, for instance, the spaces $\mathfrak{B}^{(g,n+m)\, b_1,\dots,b_m}_{a_1,\dots,a_n}$ and $\mathfrak{B}^{(g,n+m)}_{a_1,\dots,a_n,b_1^\ast,\dots,b_m^\ast}$ are naturally isomorphic in this way.

We know from the proof of Theorem 4.4.4 that we can build up an arbitrary surface with boundary by sewing together discs, cylinders and pairs-of-pants. Hence the basic building block is the vertex in Figure 6.1. In the spirit of the diagrams of Section 1.6.2, it can be written as the graph on the right. This vertex represents an *intertwining operator* – the $\mathcal{I}_i$ in (4.3.7). They are a natural generalisation of vertex operators (in fact they are often called that), and they generate the chiral blocks $\mathcal{F}$ in exactly the same way that quantum fields generate correlation functions (4.3.1a).

Definition 6.1.9 [199], [436] *Let $\mathcal{V}$ be a VOA, and let (M^i, Y^i), for labels $i \in \Phi$, be its irreducible modules. For any $a, b, c \in \Phi$, an* intertwining operator *of type $\binom{c}{a\,b}$ is a linear map*

$$w \mapsto \mathcal{Y}(w, z) = \sum_{n \in \mathbb{Q}} w_{(n)} z^{-n-1} \qquad (6.1.16)$$

for each $w \in M^a$, where each mode $w_{(n)} \in \mathrm{Hom}(M^b, M^c)$ (hence the name 'inter-twiner'), such that for all $w^a \in M^a$, $w^b \in M^b$ and $v \in \mathcal{V}$, $w^a_{(n)}(w^b) = 0$ for all sufficiently large n (depending on both w^a, w^b), and we have both

$$z_0^{-1} \delta\left(\frac{z_1 - z_2}{z_0}\right) Y^c(v, z_1)\, \mathcal{Y}(w^a, z_2)w^b - z_0^{-1}\delta\left(\frac{z_2 - z_1}{-z_0}\right)\mathcal{Y}(w^a, z_2)\, Y^b(v, z_1)w^b$$

$$= z_2^{-1}\delta\left(\frac{z_1 - z_0}{z_2}\right)\mathcal{Y}(Y^b(v, z_0)w^a, z_2)w^b,$$

$$\frac{\mathrm{d}}{\mathrm{d}z}\mathcal{Y}(w^a, z) = \mathcal{Y}(L_{-1}w^a, z).$$

Let $\mathcal{V}\binom{c}{a\,b}$ denote the space of all $\mathcal{Y}$ of the given type.

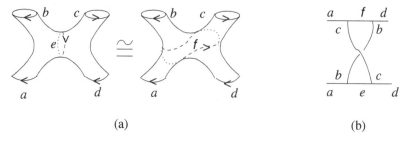

(a)   (b)

Fig. 6.2 The braiding operator $B_{ef}\begin{bmatrix} b & c \\ a & d \end{bmatrix}$.

In short, the intertwining operator obeys all the properties the vertex operator Y_M obeys in Definition 5.3.1. Of course the z-derivative in the definition completely specifies the z-dependence of an intertwining operator. Note that the defining vertex operator $Y(v, z)$ of a VOA is an intertwining operator of type $\binom{0}{0\,0}$, while the vertex operator Y_{M^a} of the module M^a is of type $\binom{a}{0\,a}$. Summing the formal power series in (6.1.16) over $\mathbb{Q}$ is a little lazy here: the sum really is over $n \in r + \mathbb{Z}$, where $r = \operatorname{wt} w^c - \operatorname{wt} w^a - \operatorname{wt} w^b \in \mathbb{Q}$. The analogue of the grading VA1 here is that $\operatorname{wt} w^a_{(n)} = \operatorname{wt} w^a - n - 1$.

The dimension of the space of intertwiners is just the fusion multiplicities:

$$\dim \left(\mathcal{V}\binom{c}{a\,b} \right) = \dim \mathfrak{B}\left(\Sigma^{(0,3)c}_{ab} \right) = \mathcal{N}^c_{ab} < \infty. \tag{6.1.17}$$

Given a surface Σ with $m + n$ boundary circles, finding a basis for the space $\mathfrak{B}(\Sigma^{b_1,...,b_n}_{a_1,...,a_m})$ is now trivial, at this formal level: simply perform the following Feynman rules.

 (i) Fix a basis for each space $\mathcal{V}\binom{c}{a\,b}$ of intertwining operators.
 (ii) Fix some dissection of Σ into pairs-of-pants, as in Figure 4.12 (it is more convenient but not necessary to draw the corresponding trivalent graph).
 (iii) Assign to each internal cut, or equivalently each internal edge of the trivalent graph, a dummy label.
 (iv) To each vertex in your dissection, bounded by labels $a, b, c \in \Phi$ (appropriately oriented), choose an intertwining operator from the basis of the appropriate space of intertwiners.
 (v) 'Evaluate' the corresponding chiral block in (4.3.7) – this is a desired basis vector.
 (vi) Repeat for each operator in your basis, and each possible value of all dummy labels.

For example, consider the left-most dissection in Figure 6.2(a) of a sphere with four boundary components. Let $\mathcal{Y}$ and $\mathcal{Y}'$ be any intertwining operators in $\mathcal{V}\binom{b}{a\,e}$ and $\mathcal{V}\binom{e}{d\,c}$, respectively. Then we get a chiral block

$$\mathcal{F} = \langle w^b, \mathcal{Y}(w^a, z)\,\mathcal{Y}'(w^d, z')w^c \rangle, \tag{6.1.18}$$

where Möbius invariance was used to send the b- and c-marked points to 0 and ∞. Section 9.3 of [**253**] gives a more physical description of sewing. Incidentally, each dissection corresponds to moving towards a 'maximally degenerate' boundary point on

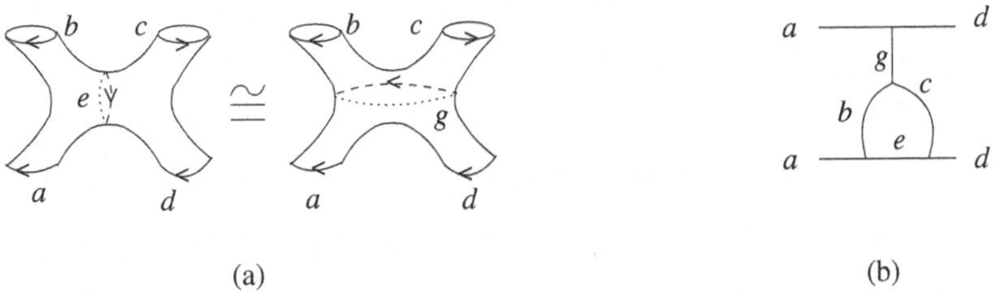

(a)   (b)

Fig. 6.3 The fusing operator $F_{eg}\begin{bmatrix} b & c \\ a & d \end{bmatrix}$.

$\overline{\mathfrak{M}}_{g,n}$ (recall Section 2.1.4), that is deforming the surface ever more closely to a trivalent graph.

For each dissection, the chiral blocks of (v) are linearly independent and form a basis for the desired space $\mathfrak{B}(\Sigma_{a_1,\ldots,a_m}^{b_1,\ldots,b_n})$. This linear independence implies a product formula for fusion multiplicities, for any pair of dissections of each labelled surface. For instance, the dissections in Figures 6.2(a) and 6.3(a) tell us the nontrivial fact that

$$\mathcal{N}_{acd}^{(0,4)\,b} = \dim \mathfrak{B}\left(\Sigma_{acd}^{(0,4)\,b}\right) = \sum_{e\in\Phi}\mathcal{N}_{ae}^{rb}\mathcal{N}_{cd}^{e} = \sum_{f\in\Phi}\mathcal{N}_{ac}^{f}\mathcal{N}_{df}^{rb} = \sum_{g\in\Phi}\mathcal{N}_{ad}^{g}\mathcal{N}_{cg}^{rb}. \qquad (6.1.19)$$

These identities imply that the fusion ring of an RCFT, defined here formally to have structure constants $\mathcal{N}_{ab}^{c}$, is both commutative and associative. All of these product formulae can be quickly deduced from Verlinde's formula (6.1.2).

As we've repeatedly mentioned, a given surface can be dissected in different ways. *Duality* here is the statement that although each dissection of Σ produces a different basis of chiral blocks, they must be bases for the same space $\mathfrak{B}(\Sigma)$, that is there must be invertible linear maps relating the chiral blocks of different dissections. Consider the easy examples in Figures 6.2 and 6.3. There we've given three dissections of the $(g,n) = (0,4)$ surface. The corresponding linear maps (actually matrices, given our explicit but noncanonical bases) are denoted $B\begin{bmatrix} b & c \\ a & d \end{bmatrix} = \oplus_{e,f\in\Phi}B_{ef}\begin{bmatrix} b & c \\ a & d \end{bmatrix}$ and $F\begin{bmatrix} b & c \\ a & d \end{bmatrix} = \oplus_{e,g\in\Phi}F_{eg}\begin{bmatrix} b & c \\ a & d \end{bmatrix}$. For the purposes of manipulating identities, it is convenient to represent these operators pictorially as in (b) (recall Section 1.6.2). Because of these pictures, they are usually called *braiding* and *fusing*. They play the same role here as the *Clebsch–Gordon* and *Racah* coefficients (or 3j- and 6j-symbols), respectively, play in the Lie theory of the quantum mechanics literature. See also the treatment in chapter 16 of [**214**].

The proposition at the end of [**278**] gives us four basic 'moves' from which any two dissections can be related. These occur for surfaces with

$$(g, n) = (0, 1), (0, 2), (0, 4), (1, 1) \qquad (6.1.20)$$

(namely, the surfaces that need at most one cut to unfold them into discs, cylinders or pairs-of-pants). The one for (1,1) is given in Figure 6.4. The corresponding operator is called $S(a)$ because it corresponds to the modular transformation $\tau \mapsto -1/\tau$. The result

Fig. 6.4 The S-operator $S(a)$.

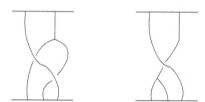

Fig. 6.5 A typical identity.

of [278] is the key to proving that a few dualities generate all others. In particular, all duality transformations can be written in terms of F, B, S, $e^{2\pi ic/24}$.

These duality operators obey several identities, coming from surfaces $(0, 5)$ and $(1, 2)$ (those requiring two cuts to decompose into pairs-of-pants). An example is Figure 6.5; another is the Yang–Baxter equation (Figure 1.29). The reader is encouraged to write these identities down explicitly. Figure 6.5 has the shape $FBB = BF$, while the Yang–Baxter equation looks like $BBB = BBB$. Other identities are given in section 3 of [437].

[436] argue, and [32] prove, that all mapping class group actions on the spaces $\mathfrak{B}(\Sigma)$ can be deduced from these relations. They also argue that Verlinde's formula (6.1.2) follows, by considering the space $\mathfrak{B}_{aa^*}^{(1,2)}$.

For example, consider the Ising model (Section 4.3.2). Here $\Phi = \{1, \epsilon, \sigma\}$. Its modular data S, T is given in (4.3.11), and a basis for the space of chiral blocks in $\mathfrak{B}_{\sigma\sigma\sigma}^{(0,4)\sigma}$ is given in (4.3.13). Its fusion ring is defined by $\epsilon \boxtimes \epsilon = 1$, $\epsilon \boxtimes \sigma = \sigma$ and $\sigma \boxtimes \sigma = 1 \oplus \epsilon$. Recall that these blocks assume that the four points $z_1, \ldots, z_4$ have been mapped to $0, w, 1, \infty$, respectively (so w goes to the cross-ratio). To find the fusing matrix, one way is to note that this duality interchanges the roles of $z_1 = 0$ and $z_3 = 1$, and therefore corresponds to the Möbius transformation $w \mapsto (1 - w)/(1 - 0) = 1 - w$. Likewise, braiding interchanges z_2 with z_3, and so corresponds to the Möbius transformation $w \mapsto (0 - 1)/(0 - w) = 1/w$. When applying Möbius transformations to chiral blocks, recall (4.3.5); equivalently, chiral blocks (of quasi-primaries) are often written as differential forms: here they are $\mathcal{F}_i \, dw^{-1}$. The braiding and fusing matrices here become

$$B \begin{bmatrix} \sigma & \sigma \\ \sigma & \sigma \end{bmatrix} = \frac{1}{\sqrt{2}} \begin{pmatrix} y & y^{-3} \\ y^{-3} & y \end{pmatrix}, \tag{6.1.21a}$$

$$F \begin{bmatrix} \sigma & \sigma \\ \sigma & \sigma \end{bmatrix} = \frac{y^2 + y^{-2}}{\sqrt{2}} \begin{pmatrix} 1 & 1 \\ 1 & -1 \end{pmatrix}, \tag{6.1.21b}$$

for some primitive 16th root y of 1. Also, $S(\sigma)$ is 0×0 (since $\mathcal{N}_{\sigma}^{(1,1)} = 0$ by (6.1.2)), and $S(\epsilon) = (y^{-2})$.

Question 6.1.1. (a) Directly from Definition 6.1.3, prove that the fusion ring homomorphism $*$ in F2 is an involution (i.e. $*^2 = id$).
(b) Again directly from the definition, prove that $\mathcal{N}_{a^*} = (\mathcal{N}_a)^t$ for any fusion matrix $\mathcal{N}_a$.
(c) Directly from the definition, prove that the numbers $\mathcal{N}_{abc} := \mathcal{N}_{ab}^{c^*}$ in any fusion ring are completely symmetric in a, b, c.

Question 6.1.2. Choose your favourite character table theorem in, for example, [308] and find and prove the fusion ring analogue.

Question 6.1.3. Prove that a fusion ring $R(\beta, \mathcal{N}) \otimes_{\mathbb{Z}} \mathbb{Q}$, considered as an algebra over $\mathbb{Q}$, is isomorphic to a direct sum of number fields. Construct these number fields explicitly, from the matrix S. (*Hint*: (6.1.7) may be helpful.)

Question 6.1.4. Prove Theorem 6.1.5.

Question 6.1.5. Classify all one- and two-dimensional fusion rings and modular data.

Question 6.1.6. What happens to the modular data of the lattice example when the lattice is integral but not even (i.e. it has odd norm-squared vectors).

Question 6.1.7. (a) Prove (6.1.13b).
(b) Prove that if $j = J_0$ is order n, then $\varphi_j(a)$ is an nth root of unity, and for n odd $T_{Ja,Ja}\overline{T_{aa}}$ is also an nth root of 1, while for n even it is a $2n$th root of 1.
(c) Prove that the set of all simple-currents forms an abelian group (with respect to composition of the permutations J).
(d) Prove that $\mathcal{N}_{Ja,J'b}^{JJ'c} = \mathcal{N}_{ab}^{c}$. Describe σj and $\epsilon_{\sigma}(j)$ of simple-currents, for any $\sigma \in \mathrm{Gal}(\mathbb{Q}[\xi_N]/\mathbb{Q})$.

Question 6.1.8. Suppose all $a \in \Phi$ are simple-currents. Prove that any modular invariant is of the form (6.1.14).

Question 6.1.9. Suppose we have four sets of functions, namely $a_i(z)$ and $b_i(z)$ (for $1 \le i \le n$), and $c_j(z)$ and $d_j(z)$ (for $1 \le j \le m$), and they are all holomorphic in some common domain (e.g. the unit disc). Suppose the equality

$$\sum_{i=1}^{n} a_i(z)\overline{b_i(z)} = \sum_{j=1}^{m} c_j(z)\overline{d_j(z)}$$

holds throughout that domain. Then $n = m$ and there is an invertible $n \times n$ matrix M such that both

$$a_i(z) = \sum_{j=1}^{n} M_{ij}c_j(z), \qquad b_i(z) = \sum_{j=1}^{n} \overline{(M^{-1})_{ij}}d_j(z).$$

6.2 Examples

6.2.1 Affine algebras

The mathematical riches of CFT go far beyond Lie theory, but CFT would have remained an esoteric part of mathematical physics, unknown to mathematics proper, if its deep connection to Lie theory hadn't been discovered.

The source of some of the most interesting modular data are the nontwisted affine Kac–Moody algebras $\mathfrak{g} = X_r{}^{(1)}$ (Section 3.2). We are interested in its integral highest weights $\lambda \in P_+^k(\mathfrak{g})$ with a given fixed level $k \in \mathbb{N}$.

Recall that the $\mathfrak{g}$-character $\chi_\lambda(\tau)$ (3.2.11c) is essentially a lattice theta function, and transforms nicely under the modular group $SL_2(\mathbb{Z})$. In fact, the $SL_2(\mathbb{Z})$-representation ρ of Theorem 3.2.3 defines modular data. The 'vacuum' is $0 = k\omega_0$, and the set of 'primaries' Φ are the highest weights $P_+^k(\mathfrak{g})$ given in (3.2.8). The matrix T is related to the eigenvalues of the second Casimir operator of $\bar{\mathfrak{g}} = X_r$, and S to elements of finite order in the Lie group of X_r [**333**]:

$$T_{\lambda\mu} = \exp\left[\frac{-\pi \mathrm{i}\,(\rho|\rho)}{h^\vee}\right] \exp\left[\frac{\pi \mathrm{i}\,(\bar{\lambda} + \rho|\bar{\lambda} + \rho)}{k + h^\vee}\right] \delta_{\lambda,\mu}, \tag{6.2.1a}$$

$$S_{\mu\nu} = \alpha \sum_{w \in \overline{W}} \det(w) \exp\left[-2\pi \mathrm{i}\,\frac{(w(\overline{\mu} + \rho)|\overline{\nu} + \rho)}{k + h^\vee}\right], \tag{6.2.1b}$$

$$\frac{S_{\lambda\mu}}{S_{0\mu}} = \mathrm{ch}_{L(\bar{\lambda})}\left(\exp\left[-2\pi \mathrm{i}\,\frac{(\bar{\lambda} \mid \overline{\mu} + \rho)}{k + h^\vee}\right]\right). \tag{6.2.1c}$$

The unimportant number α is given explicitly in theorem 13.8(a) of [**328**]. The inner-product is the usual Killing form of $\bar{\mathfrak{g}}$, $\overline{W}$ is the (finite) Weyl group of $\bar{\mathfrak{g}}$, ρ is the Weyl vector $\sum_{i=1}^r \omega_i$ and $h^\vee$ is the dual Coxeter number (the sum $\sum_{i=0}^r a_i^\vee$ of the colabels in Figure 3.2). Also, $\bar{\lambda}$ denotes the projection $\lambda_1\omega_1 + \cdots + \lambda_r\omega_r$, and '$\mathrm{ch}_{L(\bar{\lambda})}$' is the appropriate finite-dimensional Lie group character.

The combinatorics of Lie group characters at elements of finite order, that is the ratios (6.2.1c), are quite rich and have been studied by many people. For instance, [**431**] show that they lead to quick algorithms for computing, for example, tensor product multiplicities. Kac [**327**] used them in a Lie theoretic proof of quadratic reciprocity.

For example, for $A_1{}^{(1)}$ at level k, we may take $P_+^k = \{0, 1, \ldots, k\}$ (the value of λ_1), and then the S and T matrices and fusion multiplicities are given by

$$S_{ab} = \sqrt{\frac{2}{k + 2}} \sin\left(\pi \frac{(a + 1)(b + 1)}{k + 2}\right), \tag{6.2.2a}$$

$$T_{aa} = \exp\left[\frac{\pi \mathrm{i}(a + 1)^2}{2(k + 2)} - \frac{\pi \mathrm{i}}{4}\right], \tag{6.2.2b}$$

$$\mathcal{N}_{ab}^c = \begin{cases} 1 & \text{if } c \equiv a+b \ (\mathrm{mod}\ 2) \text{ and } |a-b| \le c \le \min\{a+b,\ 2k-a-b\} \\ 0 & \text{otherwise} \end{cases}. \tag{6.2.2c}$$

For $A_1{}^{(1)}$ the matrix S is real and so charge-conjugation $C = id$. More generally, for $X_r{}^{(1)}$ C corresponds to a symmetry of the Coxeter–Dynkin diagram of X_r. For $A_1{}^{(1)}$, there

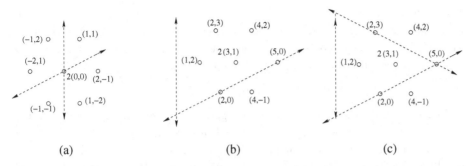

Fig. 6.6 Tensor and fusion products $L(\overline{2,0}) \otimes L(\overline{1,1})$ and $L(0,2,0) \boxtimes_2 L(0,1,1)$.

is precisely one nontrivial simple-current, namely $j = k$, corresponding to $Ja = k - a$ and $\varphi_j(a) = (-1)^a$. More generally, to any affine algebra (except for $E_8^{(1)}$ at $k = 2$), the simple-currents correspond to symmetries of the extended Coxeter–Dynkin diagram. For $A_1^{(1)}$ this symmetry interchanges the zeroth and first nodes, that is $J(\lambda_0\omega_0 + \lambda_1\omega_1) = \lambda_1\omega_0 + \lambda_0\omega_1$ (recall $a = \lambda_1$ and $k = \lambda_0 + \lambda_1$).

The fusion multiplicities $\mathcal{N}_{\lambda\mu}^\nu$, defined by (6.1.1b), are essentially the tensor product multiplicities $T_{\overline{\lambda\mu}}^{\overline{\nu}} := \text{mult}_{\overline{\lambda}\otimes\overline{\mu}}(\overline{\nu})$ for $\overline{\mathfrak{g}}$ (as opposed to the unrelated and less interesting tensor product multiplicities of $\mathfrak{g}$), except 'folded' in a way depending on the level k. This is seen explicitly by the Kac–Walton formula (see [**328**] page 288, [**552**], though there are other co-discoverers):

$$\mathcal{N}_{\lambda\mu}^\nu = \sum_{w \in W} \det(w)\, T_{\overline{\lambda\mu}}^{w.\overline{\nu}}, \tag{6.2.3a}$$

where $w.\gamma := w(\gamma + \rho) - \rho$ and W is the affine Weyl group of $X_r^{(1)}$ (the dependence on k arises through this action of W). The proof follows quickly from (6.2.1c). This practical formula is also described in Section 16.2 of [**131**] and Section 4.9 of [**553**].

Equation (6.2.3a) looks more natural when viewed as follows. The Racah–Speiser formula (there are other co-discoverers) for tensor product multiplicities says

$$T_{\overline{\lambda\mu}}^{\overline{\nu}} = \sum_{w \in \overline{W}} \det(w)\, \dim L(\overline{\mu})_{w.\overline{\nu}-\overline{\lambda}}. \tag{6.2.3b}$$

Combining (6.2.3) gives the 'affinisation' of Racah–Speiser:

$$\mathcal{N}_{\lambda\mu}^\nu = \sum_{w \in W} \det(w)\, \dim L(\overline{\mu})_{w.\overline{\nu}-\overline{\lambda}}. \tag{6.2.3c}$$

For example, the weights for the eight-dimensional A_2-module $L(\overline{1,1})$ are given in Figure 6.6(a). In Figure 6.6(b), we translate this weight space by $\rho + \overline{\lambda} = (3, 1)$. Equation (6.2.3b) now tells us to Weyl-reflect each dot not in the A_2 alcove $P_+^k + \rho$. Two of these dots are fixed by a Weyl reflection and so cancel themselves. Weight $(4, -1)$ gets Weyl reflected to $(3, 1)$ and so reduces the multiplicity there by 1. Shifting back by $\rho = (1, 1)$, we thus get the tensor product

$$L(\overline{2,0}) \otimes L(\overline{1,1}) = L(\overline{0,1}) \oplus L(\overline{2,0}) \oplus L(\overline{1,2}) \oplus L(\overline{3,1}).$$

The calculation of the $A_2^{(1)}$ fusion multiplicity at, for example, level 2 (Figure 6.6(c)) is identical, except we now have extra Weyl reflections and the alcove is much smaller. The weight $(4, 2)$ now lies outside the alcove, and reflects to $(3, 1)$ where it reduces that multiplicity to 0. Thus we obtain the fusion product (writing the level as subscript)

$$L(0, 2, 0) \boxtimes_2 L(0, 1, 1) = L(1, 0, 1).$$

Equation (6.2.3a) has the flaw that, although the $\mathcal{N}_{\lambda\mu}^{\nu}$ are manifestly integral, it is not clear why they are positive. An open problem in the theory is the discovery of a combinatorial rule, for example, in the spirit of the well-known Littlewood–Richardson rule [217], for the affine algebra fusions. Such a rule for $A_r^{(1)}$ is conjectured in [88], although it is quite complicated even for $A_1^{(1)}$.

Identical numbers $\mathcal{N}_{\lambda\mu}^{\nu}$ appear in several other contexts, many of which we'll see below. Because of these isomorphisms, we know that the $\mathcal{N}_{\lambda\mu}^{\nu}$ defined by (6.1.1b) and (6.2.1b) do indeed lie in $\mathbb{N}$, for any affine algebra, as predicted by RCFT.

As mentioned before, the fusion product here is *not* the usual tensor product of affine algebra modules. However, the fusion product has been interpreted algebraically (with much effort) as a new kind of tensor product of affine algebra modules, in a series of papers by Kazhdan and Lusztig; it was proved equivalent to fusions in [190].

Fusion multiplicities arise in the quantum cohomology or Gromov–Witten invariants of Grassmannians [565], [57], often called the 'quantum Schubert calculus'. Recall that 'points' in the projective plane consist of lines through the origin; more generally, the Grassmannian $\mathrm{Gr}(m, n)$ consists of m-dimensional subspaces in $\mathbb{R}^n$. The (classical) Schubert calculus (see e.g. [217]) uses the cohomology ring of $\mathrm{Gr}(m, n)$ to solve problems in enumerative geometry such as 'How many lines in projective 3-space $\mathbb{P}^3(\mathbb{R})$ meet four given lines?'. On the other hand, the Gromov–Witten invariants count surfaces lying in the Grassmannian, which satisfy certain conditions (see e.g. [359]). The quantum cohomology ring (which counts spheres) of $\mathrm{Gr}(m, n)$ is isomorphic to the fusion ring of $\mathfrak{gl}_m^{(1)} = (\mathfrak{u}_1 \oplus A_{m-1})^{(1)}$ at level $(nm, n - m)$, 'orbifolded' with a 'projection/field-identification' given by the order-m simple-current (J^{-n}, J); the Gromov–Witten invariants are the fusion multiplicities. Now, there is a classical isomorphism $\mathrm{Gr}(m, n) \cong \mathrm{Gr}(n - m, n)$ (why?); this implies that there is a close relation ('rank–level duality') between the fusion rings of $A_r^{(1)}$ level k and $A_{k-1}^{(1)}$ level $r + 1$. There are analogous rank–level dualities for the other classical algebras [428]. This is one of many symmetries of the $\mathfrak{g}$ fusion multiplicities that has no analogue for the $\bar{\mathfrak{g}}$ tensor product multiplicities. Another example is that any symmetry of the extended Coxeter–Dynkin diagram is a symmetry of fusion multiplicities. In short, *affine algebra fusion multiplicities are mathematically more interesting than their classical counterparts.*

We have long known that the representation theory of a Lie group G is related to K-theory. For example, the equivariant K-theory $K_G^{dim\,G}(p)$ of the (trivial) action of G on a point p is the representation ring (over $\mathbb{Z}$). The analogue of this for fusion rings is due to Freed–Hopkins–Teleman [193]: the fusion ring of $X_r^{(1)}$ at level k is the twisted equivariant K-theory $^h K_{LG}^{dim\,G}(p) := {}^{k+h^\vee} K_G^{dim\,G}(G)$, where G is the compact simply-connected Lie group corresponding to X_r, G acts on itself by conjugation and

$k + h^\vee \in \mathbb{Z} = H^3_G(G, \mathbb{Z})$ is the twist h. The strength of this important formulation is also its weakness: it pushes most technical difficulties under the carpet, but what remains is a clean conceptual characterisation of the fusion ring.

Fusion multiplicities also arise as dimensions of spaces of generalised theta functions [179] (see also the discussion in [565]), as tensor product multiplicities in Hecke algebras at roots of 1 [255] and modular representations for, for example, the Lie algebra $\bar{\mathfrak{g}}$ for fields $\mathbb{F}_p$ (see e.g. [392]). In Section 6.1.1 we give another appearance of the $A_{n-1}^{(1)}$ fusion multiplicities.

The Galois action for the affine algebras can be expressed geometrically using the action of the affine Weyl group on the weight lattice of X_r. The parity $\epsilon_\sigma(\lambda)$ is quite interesting (see e.g. [7] for cohomological and number-theoretic interpretations). For a concrete example, consider $A_1^{(1)}$: (6.2.2a) shows explicitly that S_{ab} lies in the cyclotomic field $\mathbb{Q}[\xi_{4(k+2)}]$. Write $\{x\}$ for the number congruent to $x \bmod 2(k + 2)$ satisfying $0 \leq \{x\} < 2(k + 2)$. Choose any Galois automorphism $\sigma = \sigma_\ell$. Then if $\{\ell(a + 1)\} < k + 2$, we will have $a^\sigma = \{\ell(a + 1)\} - 1$, while if $\{\ell(a + 1)\} > k + 2$, we will have $a^\sigma = 2(k + 2) - \{\ell(a + 1)\} - 1$. The parity $\epsilon_\sigma(a)$ depends on a contribution from $\sqrt{\frac{2}{k+2}}$ (which can usually be ignored), as well as the sign $+1$ or -1, respectively, depending on whether or not $\{\ell(a + 1)\} < k + 2$.

Affine algebra modular data corresponds to Wess–Zumino–Witten RCFT [245], where a closed string lives on a Lie group manifold G. The action is given by the sum of two terms: one is an integral over the world-sheet and corresponds to a so-called sigma model [343] of a bosonic field living on G; the other is a topological Wess–Zumino term, an integral over the volume bounded by the (compactified) world-sheet. Classically, the sigma model by itself would be conformally invariant, but quantisation breaks this. It was Witten who realised that conformal invariance would be retained if the Wess–Zumino term was added. For topological reasons the Wess–Zumino term comes with an integral prefactor (or coupling constant), which we call the level k.

Why is the level k always shifted by the dual Coxeter number $h^\vee$ in the formulae, and the weights by the Weyl vector ρ? The ρ-shift appears even for the simple finite-dimensional algebras (1.5.11), and arises from the combinatorics of geometric series. The algebraic explanation of the $h^\vee$-shift was given after (3.2.15). Physically, in the Wess–Zumino–Witten model, these ρ- and $h^\vee$-shifts also arise automatically: the former as a quantum effect, due to normal-ordering or regularisation, much like the $q^{1/24}$ shift in the Dedekind eta; the latter as an effect of latent supersymmetry caused by decoupling fermions (see e.g. section 8 of [248], or [206]).

The modular data (6.2.2) of $A_1^{(1)}$ level k is related to the dilogarithm by the remarkable formula

$$\frac{1}{L(1)} \sum_{b=1}^{k} L\left(\frac{S_{0a}^2}{S_{ba}^2}\right) = c_k - 24h_a + 6a \qquad (6.2.4a)$$

for each $a \in P_+^k$, where $c_k = 3k/(k + 2)$ is the central charge and $h_a = \frac{a(a+2)}{4(k+2)}$ the conformal weight (recall (3.2.9)). $L(x)$ here is *Roger's dilogarithm*, which for $0 < x < 1$ is given by

$$L(x) = \sum_{n=1}^{\infty} \frac{x^2}{n^2} + \frac{1}{2} \log x \log (1 - x). \tag{6.2.4b}$$

We put $L(1) := \lim_{x \to 1^-} L(x) = \pi/6$. $L(x)$ is strictly increasing, real-analytic, and obeys $L(x) + L(1 - x) = L(1)$ and

$$L(x) + L(y) = L(xy) + L\left(\frac{x - xy}{1 - xy}\right) + L\left(\frac{y - xy}{1 - xy}\right). \tag{6.2.4c}$$

As was discovered by Lobachevsky and Schläffli in the nineteenth century, the dilogarithm is related to volumes of tetrahedra, and several other appearances have been uncovered since. Equation (6.2.4a) is the tip of the iceberg; see [347] for several other identities and some history. (6.2.4a) can be proved by studying the $\tau \to 0$ asymptotics of certain character formulae. For a simple example, the two $k = 1$ $A_1^{(1)}$ characters can be written

$$\chi_{i\omega_1 + (1-i)\omega_0}(\tau) = \sum_{\substack{M+N+i \ even \\ M,N \in \mathbb{N}}} \frac{q^{(M+N)^2/2}}{(q)_M (q)_N}, \tag{6.2.5a}$$

where $(q)_N$ is the q-deformed factorial $\prod_{n=1}^{N}(1 - q^n)$. Similar expressions exist for all other affine algebras and conjecturally all RCFT – see [14] for the state-of-the-art, and below for a conjecture. Actually, (6.2.4a) is obtained from the asymptotics of these character identities for certain non-unitary RCFTs, which have essentially the same S matrix as (6.2.2a). An explanation of some of these identities (at least mod 1) has been made by [164], who use the dilogarithm to express a natural map from $H_3(\widetilde{SL}_2(\mathbb{R}), \mathbb{Z})$ to $\mathbb{R}/\mathbb{Z}$.

Choose any $r \times r$ rational positive-definite matrix $A = A^t$, $b \in \mathbb{Q}^r$ and $d \in \mathbb{Q}$. Define

$$f_{A,b,d}(\tau) := \sum_{n \in \mathbb{N}^n} \frac{\exp[2\pi i \tau (n^t An/2 + b^t n + d)]}{(q)_{n_1} \cdots (q)_{n_r}}. \tag{6.2.5b}$$

Conjecture 6.2.1 (Nahm [444])   *Let A be any $n \times n$ rational positive-definite matrix. Then there are finitely many vectors $b_1, \ldots, b_m \in \mathbb{Q}^n$ and numbers $d_1, \ldots, d_m \in \mathbb{Q}$ such that the functions $\chi_i(\tau) := f_{A,b_i,d_i}(\tau)$ are the entries of a vector-valued modular function for $SL_2(\mathbb{Z})$, iff these $\chi_i(\tau)$ are the graded-dimensions of the m primaries of some (not necessarily unitary) RCFT where $d_i = h_i - c/24$, iff there is a corresponding element of finite order in the Bloch group.*

The precise statement involving the Bloch group would take us too far afield, but see [444] for details. This beautiful conjecture has been verified only for $r = 1$ (which has three different A). A plausibility argument suggesting that RCFT characters should always be of that form involves considering their massive integrable perturbations [444]. Torsion in the Bloch group has known connections with modularity.

The affine algebra $\mathfrak{g}$ arises in the Wess–Zumino–Witten model, for the same reason the Virasoro does (recall the discussion around (4.3.4)): to each $g \in G$ we get a conserved current, and its conserved charges define the level-k representation of $\mathfrak{g}$. As before, we get two commuting actions of $\mathfrak{g}$ on the state-space $\mathcal{H}$, recovering the finite decomposition (4.3.6b).

For affine algebra modular data, the classification of modular invariants seems to be just barely possible, and the answer is that (generically) the only modular invariants are constructed in straightforward ways from symmetries of the Coxeter–Dynkin diagrams. For instance, consider $A_1{}^{(1)}$:

Theorem 6.2.2 [91] *Recall that $P_+^k = \{0, 1, \ldots, k\}$, and the simple-current is given by $Ja = k - a$. Then the complete list of $A_1{}^{(1)}$ modular invariants is*

$$\mathcal{A}_{k+1} = \sum_{a=0}^{k} |\chi_a|^2 \qquad \text{for all } k \geq 1,$$

$$\mathcal{D}_{\frac{k}{2}+2} = \sum_{a=0}^{k} \chi_a \, \chi_{J^a a}^* \qquad \text{when } \frac{k}{2} \text{ is odd,}$$

$$\mathcal{D}_{\frac{k}{2}+2} = |\chi_0 + \chi_{J0}|^2 + |\chi_2 + \chi_{J2}|^2 + \cdots + 2|\chi_{\frac{k}{2}}|^2 \qquad \text{when } \frac{k}{2} \text{ is even,}$$

$$\mathcal{E}_6 = |\chi_0 + \chi_6|^2 + |\chi_3 + \chi_7|^2 + |\chi_4 + \chi_{10}|^2 \qquad \text{for } k = 10,$$

$$\mathcal{E}_7 = |\chi_0 + \chi_{16}|^2 + |\chi_4 + \chi_{12}|^2 + |\chi_6 + \chi_{10}|^2$$
$$\qquad + \chi_8 (\chi_2 + \chi_{14})^* + (\chi_2 + \chi_{14}) \chi_8^* + |\chi_8|^2 \qquad \text{for } k = 16,$$

$$\mathcal{E}_8 = |\chi_0 + \chi_{10} + \chi_{18} + \chi_{28}|^2 + |\chi_6 + \chi_{12} + \chi_{16} + \chi_{22}|^2 \qquad \text{for } k = 28.$$

A simple proof is given in [234]. The modular invariants $\mathcal{A}_n$ and $\mathcal{D}_n$ are generic, given by (6.1.14), and correspond respectively to the order 1 (i.e. identity) and order 2 (i.e. simple-current J) Coxeter–Dynkin diagram symmetries. Physically, $\mathcal{A}_n$ and $\mathcal{D}_n$ are the partition functions (4.3.8b) of Wess–Zumino–Witten models on the $SU_2(\mathbb{C})$ and $SO_3(\mathbb{R})$ group manifolds, respectively. The exceptionals $\mathcal{E}_6$ and $\mathcal{E}_8$ correspond to strings living on Sp_4 and G_2 manifolds, at level 1. The $\mathcal{E}_7$ exceptional is harder to interpret, but is the first in an infinite series of exceptionals involving rank–level duality and D_4 triality.

Around Christmas 1985, Zuber wrote to Kac about the $A_1{}^{(1)}$ modular invariant problem, and mentioned the modular invariants they knew at that point (what we now call $\mathcal{A}_\star$ and $\mathcal{D}_{even}$). A few weeks later, Kac wrote back saying he found one more invariant, and jokingly pointed out that it must indeed be quite exceptional as the exponents of E_6 appeared in it. By summer 1986, Cappelli–Itzykson–Zuber found $\mathcal{E}_7$, $\mathcal{D}_{odd}$ and then $\mathcal{E}_8$, and at some point recalled by chance Kac's cryptic remark. They rushed to the library to find a list of the exponents of the other algebras, and were delighted to discover that they all matched. Thus the A–D–E pattern (Section 2.5.2) to their modular invariants was discovered!

The modular invariants for $A_1{}^{(1)}$ realise the A–D–E pattern, in the following sense [91]. The (dual) Coxeter number $h = h^\vee$ of the name $\mathcal{X}_n$ equals $k + 2$, and the exponents m_i of $\mathcal{X}_n$ equal 1 plus those $a \in P_+^k$ for which $\mathcal{Z}_{aa} \neq 0$ (for the algebras A_n, D_n, E_n, the integers m_i are defined by writing the eigenvalues of the corresponding Cartan matrix

(Definition 1.4.5) as $4\sin^2(\frac{\pi m_i}{2h})$). Probably what first led Kac to his observation about the E_6 exponents was that $k+2$ (this is how k enters most formulae), for his exceptional, equals the Coxeter number 12 for E_6. More recently, deeper connections between *A–D–E* and the $A_1^{(1)}$ modular invariants have been found, notably in subfactor theory (Section 6.2.6). This modular invariant classification, however, has never been directly reduced to the suggestion of Section 2.5.2.

The modular invariants have also been classified, for example, for $A_2^{(1)}$ [232], and they too seem quite interesting (Section 6.3.2). We are almost at the point where we can safely conjecture the complete list of modular invariants for $X_r^{(1)}$ at any k, for X_r a simple algebra (see e.g. [236]). The most surprising thing about these affine algebra modular invariant classifications is that there are so few surprises: almost every modular invariant is 'generic', that is constructable using a few simple uniform methods such as Coxeter–Dynkin diagram symmetries. Unfortunately, the classification for semi-simple algebras $X_{r_1} \oplus \cdots \oplus X_{r_s}$ does not reduce to that for simple ones, and will be hopeless.

Has *A–D–E* been discovered in the other modular invariant classifications? No, only in those classifications trivially reducible to Theorem 6.2.2. There is, however, a rather natural way to assign (multi-di)graphs to modular invariants, generalising the *A–D–E* pattern for $A_1^{(1)}$. It is called a NIM-rep, and is a *rep*resentation of the fusion ring by *non*negative *integer matrices. More precisely, for each weight $a \in P_+^k(A_1^{(1)})$ we want a nonnegative integer matrix $\mathcal{M}_a$ such that

$$\mathcal{M}_a \mathcal{M}_b = \sum_{c=0}^{k} \mathcal{N}_{ab}^c \mathcal{M}_c, \tag{6.2.6}$$

where $\mathcal{N}_{ab}^c$ are the fusion multiplicities of (6.2.2c). We also require $\mathcal{M}_0 = I$, and all these matrices to be symmetric: $\mathcal{M}_a = (\mathcal{M}_a)^t$. In Question 6.2.2 you are asked to find all such assignments $a \mapsto \mathcal{M}_a$. Surprisingly, there is a near-perfect correspondence between the $A_1^{(1)}$ modular invariants, and these NIM-reps. Physically, NIM-reps are associated with boundary conformal field theory or D-branes in string theory. See [47], [236] and references therein for the basic theory and examples of NIM-reps. They are an integral part of the combinatorial data of RCFTs. However, the simplicity of the correspondence for $A_1^{(1)}$ is an accident due to the small size of the relevant Perron–Frobenius eigenvalue here. In particular there appear to be far more NIM-reps for $A_2^{(1)}$ than modular invariants.

Hanany–He [271] suggest that the $A_1^{(1)}$ *A–D–E* pattern can be related to subgroups $G \subset \mathrm{SU}_2(\mathbb{C})$ by orbifolding four-dimensional $N=4$ supersymmetric gauge theory by G, resulting in an $N=2$ superconformal field theory whose 'matter matrix' can be read off from the Coxeter–Dynkin diagram corresponding to G. The same game can be played with finite subgroups of $\mathrm{SU}_3(\mathbb{C})$, resulting in $N=1$ superconformal field theories whose matter matrices resemble the NIM-reps of $A_2^{(1)}$. [271] use this to conjecture optimistically a McKay-type correspondence between singularities of type $\mathbb{C}^n/G$, for $G \subset \mathrm{SU}_n(\mathbb{C})$, and the modular invariants of $A_{n-1}^{(1)}$. This in their view would be the form *A–D–E* takes for higher-rank modular invariants. Their conjecture is still too vague to be probed.

So far we have considered only integrable modules, which are necessarily at level $k \in \mathbb{N}$. But their modular behaviour can be mimicked at certain fractional levels, by the so-called *admissible modules* [335]. It is tempting to guess that there should be natural CFT and VOA interpretations for these, analogous to the integrable ones. The matrix S there is symmetric, but has no column of constant phase and thus naively putting it into Verlinde's formula (6.1.1b) will necessarily produce some negative numbers (it appears that they'll always be integers though). A legitimate fusion ring has been obtained for $A_1^{(1)}$ at fractional level in other ways [26], [184], and initial steps for $A_2^{(1)}$ have been made in [221]. VOA interpretations for $A_1^{(1)}$ admissible modules are given in [2], [148]. Serious doubt, however, on the relevance of these efforts has been cast by [225], [378]. Sorting this out is a high priority.

Related roles for other Kac–Moody algebras are slowly being found. The *twisted* affine algebras also have modular-like data, and arise naturally in the data for NIM-reps [58], [226]. *Lorentzian* Kac–Moody algebras have been proposed [171], [285] as the symmetries of 'M-theory', the conjectural 11-dimensional theory underlying superstrings. Relations between strings and Borcherds–Kac–Moody algebras are discussed in [275], [276], [134].

6.2.2 Vertex operator algebras

Let $\mathcal{V}$ be a 'nice' VOA (more on this shortly). The primaries $a \in \Phi$ label the finitely many irreducible $\mathcal{V}$-modules M^a. The relation between VOAs and $\mathrm{SL}_2(\mathbb{Z})$ given in (4.3.9) was anticipated by RCFT, and proved by Zhu (Theorem 5.3.8). It gives (among other things) the modular matrices S and T. Do they define modular data? If so, does Verlinde's formula (6.1.1b) compute the dimensions of intertwiner spaces (6.1.17)?

Definition 6.2.3 *By a rational vertex operator algebra (RVOA) we mean a weakly rational vertex operator algebra $\mathcal{V}$ (Definition 5.3.2) obeying in addition*

  (i) *$\mathcal{V}$ is simple (that is is an irreducible module for itself) and the contragredient $\mathcal{V}^\star$ is isomorphic to $\mathcal{V}$ as a $\mathcal{V}$-module;*
 (ii) *$M_0 = \{0\}$ for all irreducible modules $M \neq \mathcal{V}$;*
(iii) *every $\mathbb{N}$-graded weak module is completely reducible;*
 (iv) *$\mathcal{V}$ is C_2-cofinite (Definition 5.3.5).*

C_2-cofiniteness is a technical condition with many consequences. As we know, every VOA is a module for itself; the contragredient of a module is discussed around (5.3.4a). In any unitary RCFT, all conformal weights $h_a, a \in \Phi$, are positive except for $a = 0$, so condition (ii) is then automatic. Condition (iii) is a little stronger than the usual complete reducibility requirement.

This use of the term 'rational' is not standard, and different definitions of 'RVOA' can be found in the literature (some of these are listed in appendix A of [224]). But the term 'rational VOA' should be limited to those VOAs that possess some variant of modular data. The justification for our use of the term is the following recent theorem:

Theorem 6.2.4 (Huang [297])   *Let $\mathcal{V}$ be a VOA, rational in the sense of Definition 6.2.3. Let Φ label its (finitely many) irreducible modules, let $\mathcal{N}_{ab}^c$ be the dimension of the space $\mathcal{V}\binom{c}{ab}$ of intertwiners, and let S be the matrix defined in Theorem 5.3.8, satisfying (4.3.9a). Then Verlinde's formula (6.1.1b) holds and S is symmetric. Also, the category Rep $\mathcal{V}$ of $\mathcal{V}$-modules has a natural structure as a modular category.*

The objects of the category Rep $\mathcal{V}$ are $\mathcal{V}$-modules, and the morphisms are $\mathcal{V}$-module homomorphisms. A modular category is described in Section 6.2.5 and is (among many other things) a braided monoidal category. Theorem 6.2.4 is a corollary to Huang's programme of constructing geometric VOAs (Section 5.4.1) in genus ≤ 1 from an algebraic VOA. It appears that additional minor conditions on the VOA $\mathcal{V}$ will be needed [**296**] in order that the higher-genus chiral blocks be constructed – once identified, these restrictions should be included in the definition of rationality for VOAs. Extending this work to genus > 1 would be the final step in associating a modular functor – that is, a chiral half of an RCFT, including all the Moore–Seiberg data – to a nice VOA.

Equation (6.1.1b) can be defined only if all $S_{M0} \neq 0$, so Theorem 6.2.4 certainly implies that. Some RVOAs (e.g. those associated with non-unitary RCFTs) won't possess modular data in the narrow sense of Definition 6.1.6. However, suppose in addition to being rational that $\mathcal{V}$ has the (common) property that any irreducible module $M \neq \mathcal{V}$ has positive conformal weight h_M (recall $h_M - c/24$ is the smallest power of q in the Fourier expansion of the graded dimension $\chi_M(\tau) = q^{-c/24} \sum_{n=0}^{\infty} a_n^M q^{n+h_M}$). This holds for instance in all VOAs associated with unitary RCFTs. Then consider the behaviour of $\chi_M(\tau)$ for $\tau \to 0$ along the positive imaginary axis: since each Fourier coefficient a_n^M is nonnegative, $\chi_M(\tau)$ will go to $+\infty$. But this is equivalent to considering the limit of $\sum_N S_{MN} \chi_N(\tau)$ as $\tau \to i\infty$ along the positive imaginary axis. By hypothesis, this latter limit is dominated by $S_{M0} a_0^0 q^{-c/24}$, at least when $S_{M0} \neq 0$. So what we find is that, under this hypothesis, the 0-column of S consists of nonnegative real numbers (and also that the central charge c is positive). But Verlinde's formula certainly requires that all numbers in the 0-column of S be nonzero. Thus we get:

Corollary 6.2.5   *Suppose $\mathcal{V}$ is a rational VOA and for all irreducible modules M, $M_n = 0$ for all $n < 0$. Then (4.3.9) (more precisely Theorem 5.3.8) define modular data.*

Of course the affine algebra modular data discussed in Section 6.2.1 is a special case of that considered here, corresponding to the integrable affine VOA $\mathcal{V}(\mathfrak{g}, k)$ constructed in Section 5.2.2.

Verlinde's formula (6.1.1b) is only a genus-0 special case of (6.1.2). What makes the proof of Theorem 6.2.4 difficult is the difficulty in constructing chiral blocks in genus > 0. At the time of writing, only special cases have been worked out in arbitrary genus (see, e.g., theorem 6.2 in [**573**]). Moore–Seiberg bypassed this difficulty by assuming the chiral blocks all exist and have all the required properties.

As mentioned in Section 5.3.5, one direction Huang's Theorem could possibly be extended is to 'quasi-rational' CFT [**436**]. These are VOAs with infinitely many irreducible modules, but with finite fusion products (5.3.3). They would correspond to a 'C_1-cofiniteness' condition and typically have infinite-dimensional Zhu's algebra. The easiest example is the Heisenberg VOA (5.2.5), associated with the oscillator algebra $\mathfrak{u}_1^{(1)}$ (3.2.12). We find directly from (3.2.12c) that the graded dimension of $V(\lambda)$ obeys

$$\chi_\lambda(\tau + 1) = e^{\pi i (\lambda^2 - \frac{1}{12})} \chi_\lambda(\tau), \tag{6.2.7a}$$

$$\chi_\lambda(-1/\tau) = \int_{-\infty}^{\infty} e^{2\pi i \lambda\mu} \chi_\mu(\tau) \, d\mu. \tag{6.2.7b}$$

In other words, on the Hilbert space $L^2(\mathbb{R})$ of square-integrable functions $f(\alpha)$, let $S(f)$ be the Fourier transform of f, and $T(f)$ the function given by

$$T(f)(\alpha) = e^{\pi i (\alpha^2 - \frac{1}{12})} f(\alpha)$$

Then S and T define a unitary representation of $SL_2(\mathbb{Z})$ on the space $L^2(\mathbb{R})$ spanned by the χ_λ (more precisely, they act on the space of functions $\chi_f(\tau) = \int_{-\infty}^{\infty} f(\alpha) \chi_\alpha(\tau) d\alpha$ for $f \in L^2(\mathbb{R})$). In Verlinde's formula (6.1.1b), the sum over Φ becomes an integral over $\mathbb{R}$, and yields the distribution

$$\mathcal{N}_{\lambda\mu}^\nu = \delta(\nu - \lambda - \mu),$$

in other words $L(\lambda) \boxtimes L(\mu) = L(\nu)$, so the 'fusion ring' $L^2(\mathbb{R})$ is given a convolution product.

It can be hoped that this modular behaviour would be typical for a wide class of other quasi-rational theories. The generalisation of Zhu's Theorem 5.3.8 and Huang's Theorem 6.2.4 to such quasi-rational theories would be wonderful to see.

Modular invariants have a VOA interpretation. Let M^a and M'^i be the irreducible modules of RVOAs $\mathcal{V} \subset \mathcal{V}'$ sharing the same conformal vector ω. Then each M'^i is a $\mathcal{V}$-module. An RVOA is completely reducible, so each M'^i should be expressible as a direct sum of M^a's – these are called the branching rules. The sum of $\sum_{i \in \Phi'} |\chi'_{M'^i}|^2$ is invariant under that $SL_2(\mathbb{Z})$-action; rewriting the $\chi'_{M'^i}$'s there in terms of the χ_{M^a}'s via the branching rules yields a nontrivial modular invariant for $\mathcal{V}$.

For instance, the VOA $L(\omega_0)'$ corresponding to the affine algebra $G_2^{(1)}$ at level 1 contains the VOA $L(28\omega_0) = L(0)$ for $A_1^{(1)}$ at level 28. We get the branching rules

$$L(\omega_0)' = L(0) \oplus L(10) \oplus L(18) \oplus L(28),$$
$$L(\omega_2)' = L(6) \oplus L(12) \oplus L(16) \oplus L(22).$$

Thus the $\mathcal{Z}' = I$ modular invariant for $G_2^{(1)}$ level 1 yields the $A_1^{(1)}$ modular invariant $\mathcal{E}_8$ in Theorem 6.2.2.

So knowing the modular invariants for an RVOA $\mathcal{V}$ gives considerable information concerning its possible 'nice' extensions $\mathcal{V}'$. For instance, we are learning from this that the only finite extensions of a generic integrable affine algebra VOA are those studied in [147] ('simple-current extensions'), and whose modular data is given in [212].

6.2.3 Quantum groups

The chiral data of affine algebras and Wess–Zumino–Witten models is also recovered by quantum groups (deformations of the universal enveloping algebra $U(\bar{\mathfrak{g}})$), though the reasons are still somewhat mysterious (i.e. indirect).

Over the years large numbers of two-dimensional models in statistical mechanics were found that are exactly solvable (completely integrable). Gradually it became clear that the underlying reason was the so-called *(quantum) Yang–Baxter equation* [**394**]:

$$R^{12} R^{13} R^{23} = R^{23} R^{13} R^{12}, \tag{6.2.8}$$

where $R : V \otimes V \to V \otimes V$ is linear and where, for example, $R^{13} : V \otimes V \otimes V \to V \otimes V \otimes V$ sends $v_1 \otimes v_2 \otimes v_3 \in V \otimes V \otimes V$ to $\sum_i a_i \otimes v_2 \otimes b_i$, where $R(v_1 \otimes v_3) = \sum_i a_i \otimes b_i$. (Generalisations of (6.2.8) exist but this is enough for us.) The Yang–Baxter equation should make us think of braids (recall Figure 1.29) and indeed an easy result is:

Proposition 6.2.6   *Given a solution R to (6.2.8), we obtain a representation of the braid group $\mathcal{B}_n$ on $V \otimes \cdots \otimes V$ (n times) by sending the braid generator σ_i to $(\tau R)^{i,i+1}$, defined by $(\tau R)^{i,i+1}(v_1 \otimes \cdots \otimes v_n) = v_1 \otimes \cdots v_{i-1} \otimes (\sum_j b_j \otimes a_j) \otimes v_{i+1} \otimes \cdots \otimes v_n$, where $R(v_i \otimes v_{i+1}) = \sum_j a_j \otimes b_j$.*

The 'transpose' τ in Proposition 6.2.6 is the flip of the two copies of V; we see it again in Definition 6.2.8. The reader should try to prove the proposition, but it's also proved in section 15.2A of [**98**].

We are interested in families $R = R(q)$ of solutions to (6.2.8), depending on a complex parameter q. Write $q = e^{i\hbar}$. If we Taylor expand $R(e^{i\hbar}) = \sum_{n=0}^{\infty} \hbar^n r_n$ and retain only the first-order terms in $\hbar$, we obtain the classical Yang–Baxter equation for $r := r_1$:

$$[r^{12}, r^{13}] + [r^{12}, r^{23}] + [r^{13}, r^{23}] = 0. \tag{6.2.9}$$

Being a sum of commutators, it's reminiscent of Lie algebras and indeed Lie theory provides classes of solutions [**98**], [**394**]. Roughly, *quantum groups* were proposed by Drinfel'd and Jimbo around 1985 as a Lie-like symmetry underlying (6.2.8), that is, as providing a way to solve the quantum Yang–Baxter equation using q-deformations of Lie theory.

The idea of deformations [**279**] is a beautiful one. For example, consider n-space $\mathbb{R}^n$ and fix a vector $q \in \mathbb{R}^n$ (the 'deformation parameter'). Define the new multiplication by scalars to be $k \cdot_q x := kx + (1 - k)q$ and vector addition to be $x +_q y := x + y - q$ (where the operations on the right sides are the usual $\mathbb{R}^n$ ones). The zero-vector here is $0_q := q$. This defines a new vector-space structure on the same underlying space. However, it is of course isomorphic (as a vector space) to the original one, since the dimension hasn't changed.

The finite-dimensional complex semi-simple Lie algebras $\bar{\mathfrak{g}}$ are also rigid in this sense (see Question 6.2.3(b)). However, nontrivial deformations of their universal enveloping algebras $U(\bar{\mathfrak{g}})$ (Section 1.5.3) do exist.

Consider for concreteness $\bar{\mathfrak{g}} = A_1$, with basis e, f, h of (1.4.2b). Define

$$[e, f] = \frac{q^h - q^{-h}}{q - q^{-1}}, \tag{6.2.10a}$$

$$q^h e = q^2 e q^h, \tag{6.2.10b}$$

$$q^h f = q^{-2} f q^h. \tag{6.2.10c}$$

Here by, for example, 'q^h' we mean the Taylor expansion in powers of h. These equations define the *quantum group* $U_q(A_1)$, a one-parameter deformation of $U(A_1)$. Given this, we get a solution $R(q)$ to (6.2.8):

$$R(q) = \sum_{n=0}^{\infty} q^{\frac{n(n+1)}{2}} \frac{(1 - q^{-2})^n}{\lfloor n \rfloor_q!} (q^{-h} e)^n \otimes (q^h f)^n e^{\frac{h \otimes h}{2}}, \tag{6.2.10d}$$

where $\lfloor n \rfloor_q! = \lfloor n \rfloor_q \lfloor n - 1 \rfloor_q \cdots \lfloor 1 \rfloor_q$ for $\lfloor k \rfloor_q = (q^k - q^{-k})/(q - q^{-1})$. Nevertheless, these equations look random and opaque (to this author at least). The next few paragraphs aim to make some sense out of them.

Definition 6.2.7 *Let k be a ring (take $k = \mathbb{C}$ if this generality makes you uncomfortable). A Hopf algebra A is:*

(i) *An associative algebra over k with unit $\mathbf{1}$ and multiplication μ.*

(ii) *A co-associative co-algebra over k, i.e. with co-multiplication $\Delta : A \to A \otimes A$ and co-unit $\epsilon : A \to k$.*

(iii) *The algebra and co-algebra structures are compatible, i.e. Δ and ϵ are algebra homomorphisms, and μ and $\mathbf{1}$ (regarded as a map $\iota : k \to A$ sending $x \mapsto x\mathbf{1}$) are co-algebra homomorphisms.*

(iv) *A has a map $S : A \to A$, called the antipode, which obeys*

$$\mu \circ (id \otimes S) \circ \Delta = \iota \circ \epsilon = \mu \circ (S \otimes id) \circ \Delta.$$

We've seen 'algebra' before. A Hopf algebra may or may not be commutative as an algebra. A 'co-algebra' is an 'algebra with the arrows reversed': just as an algebra has a bilinear map $A \otimes A \to A$ (multiplication), so a co-algebra has a linear map $A \to A \otimes A$ (co-multiplication), and similarly for unit and co-unit.

Perhaps [**51**] or the introduction to [**398**] can help make this definition seem more natural. Hopf algebras are algebras with a rich representation theory. If M, N are modules of a generic algebra A, then their usual vector-space tensor product $M \otimes N$ always has a natural structure as an $A \otimes A$-module, but generally not an A-module. But if A has a co-product, we get the A-module structure by the formula $a.(m \otimes n) := \Delta(a).(m \otimes n)$. The antipode converts left modules into right modules, and is used to define the representation M^* dual to a given representation M. It plays the role of inverse in the algebra. See also Question 6.2.4.

For example, a universal enveloping algebra $U(\bar{\mathfrak{g}})$ forms a Hopf algebra with co-product given by $\Delta(x) = x \otimes \mathbf{1} + \mathbf{1} \otimes x$ for $x \in \bar{\mathfrak{g}}$ and $\Delta(\mathbf{1}) = \mathbf{1} \otimes \mathbf{1}$; co-unit $\epsilon(x) = 0$ for $x \in \bar{\mathfrak{g}}$ and $\epsilon(\mathbf{1}) = \mathbf{1}$; and antipode $S(x) = -x$ for $x \in \bar{\mathfrak{g}}$ and $S(\mathbf{1}) = \mathbf{1}$. In a similar way, the space $F(G)$ of functions on a Lie group G is also a Hopf algebra (in fact a dual of $U(\bar{\mathfrak{g}})$). $U(\bar{\mathfrak{g}})$ is co-commutative, whereas $F(G)$ is commutative; in fact, these $U(\bar{\mathfrak{g}})$ are the only co-commutative, and $F(G)$ the only commutative, Hopf algebras (modulo certain technical assumptions). This is in fact why Drinfel'd [**160**] cooked up the name 'quantum group' for these q-deformations. $U_q(\bar{\mathfrak{g}})$ is a non-co-commutative deformation of $U(\bar{\mathfrak{g}})$, so we could imagine that just as the dual of $U(\bar{\mathfrak{g}})$ consists of the functions on a group G, the dual of $U_q(\bar{\mathfrak{g}})$, which will be a non-commutative Hopf algebra, should correspond to something like the functions on a group-like object G_q, which would be some sort of q-deformed version of G. This picture is in the same spirit as Connes' non-commutative geometry. In any case the term 'quantum group' has inappropriately slipped from G_q to apply directly to $U_q(\bar{\mathfrak{g}})$.

The co-product, etc. for these $U_q(\bar{\mathfrak{g}})$ are explicitly given in proposition 6.5.1 of [**98**] in full generality. Although $U_q(\bar{\mathfrak{g}})$ is not co-commutative, it is nearly so:

Definition 6.2.8   *A quasi-triangularisable Hopf algebra A is a Hopf algebra with invertible element $\mathcal{R} \in A \otimes A$ such that $\tau(\Delta(a)) = \mathcal{R}\,\Delta(a)\,\mathcal{R}^{-1}$ for all $a \in A$, as well as*

$$(\Delta \otimes id)(\mathcal{R}) = \mathcal{R}^{13}\mathcal{R}^{23} \in A \otimes A \otimes A,$$
$$(id \otimes \Delta)(\mathcal{R}) = \mathcal{R}^{13}\mathcal{R}^{12} \in A \otimes A \otimes A.$$

This element $\mathcal{R}$ is called the *universal R-matrix* (or braiding) of A. Of course if A is co-commutative, then $\mathcal{R} = \mathbf{1} \otimes \mathbf{1}$ works. The point: the element $\mathcal{R}$ satisfies the quantum Yang–Baxter equation (6.2.8). This is the origin of the word 'triangular' in Definition 6.2.8: an alternate name for the Yang–Baxter equation is the star–triangle relation. So given any representation of A, $\mathcal{R}$ maps to a matrix satisfying (6.2.8) – this representation-independent aspect of $\mathcal{R}$ justifies the word 'universal'. Any non-co-commutative quasi-triangularisable Hopf algebra is now called a quantum group.

Drinfel'd [**160**] found a remarkable way, independent of the Yang–Baxter equation, to construct quantum groups from any Hopf algebra A. The quasi-triangular Hopf structure is put on the space $A \otimes (A^*)^{op}$, where $(A^*)^{op}$ is the dual Hopf algebra A^* except that its co-multiplication is changed from Δ^* to its transpose $\tau \circ \Delta^*$. A nice discussion is in [**480**]; a general categorical interpretation is the 'centre construction' [**338**]. In particular, the quantum group $U_q(\bar{\mathfrak{g}})$ of (6.2.10) arises as a simple quotient of the quantum double of $U_q(B^+)$, where B^+ is the Borel subalgebra of $\bar{\mathfrak{g}}$, generated by h_i and e_i. See section 4.6 of [**207**], where this is discussed very explicitly. The point is that $U_q(B^+)$ is very easy to understand, so this gives an explicit way to compute $\mathcal{R}$ for $U_q(\bar{\mathfrak{g}})$.

As usual we're interested in representation theory. Recall that the modules of A_1 and $U(A_1)$ are identical. There is only one one-dimensional A_1-module: everything gets sent to 0. However, there are exactly two one-dimensional representations of the quantum group $U_q(A_1)$: $e.v = f.v = 0$ and $q^h.v = \pm v$. Call these $\psi_{\pm}$. ψ_+ is just the deformation of the trivial $U(A_1)$-representation, but ψ_- has no classical (i.e. $q \to 1$) analogue. The existence of ψ_- is the only difference between the representation theory of $U_q(A_1)$ and $U(A_1)$ (or A_1): every finite-dimensional irreducible $U_q(A_1)$ module is uniquely expressible as the tensor product of a one-dimensional representation $\psi_{\pm}$ with some highest-weight representation $L_q(m)$, for $m \in \mathbb{N}$, where $L_q(m)$ is a deformation of $L(m)$ with the same Weyl character. This generalises to any $U_q(\mathfrak{g})$.

We're more interested in $U_q(\mathfrak{g})$ 'at a root of unity'. The meaning of this is very subtle, but is explained very thoroughly in chapter 9 of [**98**] (we are interested in their second construction, the 'restricted integral form' $U_q^{res}(\mathfrak{g})$, which is a quotient of $U_q(\mathfrak{g})$); see also [**10**], [**392**]. The representation theory is also subtle, and most treatments (e.g. that of [**98**]) assume from the start that the order of the root of unity must be odd. See, for example, [**392**], [**10**] for their modules. There are now indecomposable modules that are not irreducible, a common situation in algebra (recall Question 1.1.6). The trick of how to proceed was discovered by physicists: throw the sick modules away! In particular, when we evaluate the Weyl characters at the root of unity q, the result is called the *quantum dimension* of the module. We keep those modules with nonzero quantum dimension, and discard the others. This prescription works because the direct product of any $U_q^{res}(\mathfrak{g})$-module with any sick one is a direct sum of sick ones. We can call this 'the reduced representation ring of the quantum group $U_q(\mathfrak{g})$ specialised to the root of unity q'. See section 4.5 of [**207**] for examples (though note that his q is the square of ours).

The result is somewhat surprising: this reduced representation ring, for $q = e^{\pi i/m(k+h^\vee)}$ (where m is defined below), is isomorphic to that of the fusion ring of $\overline{\mathfrak{g}}^{(1)}$ at level k [**190**]. Here, $m = 1$ for $\overline{\mathfrak{g}} = A_r, D_r, E_6, E_7, E_8$; $m = 2$ for $\overline{\mathfrak{g}} = B_r, C_r, F_4$; and $m = 3$ for $\overline{\mathfrak{g}} = G_2$.

More generally, much of the chiral data of the Wess–Zumino–Witten theories are recovered by the corresponding quantum group at a root of unity [**253**], [**207**]: along with the fusion multiplicities, also the braiding and fusing matrices of Section 6.1.4, and the associated knot invariants of Section 6.2.5. Explanations for these 'coincidences' are given in, for example, chapter 11 of [**253**], but they are all unsatisfying in that they are so indirect.

6.2.4 Twisted #6: finite group modular data

In many respects, a finite group G behaves much like a compact connected Lie group, and so we may hope that they possess an analogue of Section 6.2.1. Indeed that is beautifully the case.

For any finite group G (Section 1.1), let $K_1, \ldots, K_h$ be its conjugacy classes, and write k_i for $\sum_{g \in K_i} g \in \mathbb{C}G$. These k_i's form a basis for the centre of $\mathbb{C}G$. Write

$$k_i k_j = \sum_\ell c_{ij}^\ell k_\ell; \tag{6.2.11a}$$

then the structure constants c_{ij}^ℓ are nonnegative integers, and we obtain

$$c_{ij}^\ell = \frac{\|K_i\| \, \|K_j\|}{\|G\|} \sum_{\mathrm{ch} \in \mathrm{Irr}\,G} \frac{\mathrm{ch}(g_i) \, \mathrm{ch}(g_j) \, \overline{\mathrm{ch}(g_\ell)}}{\mathrm{ch}(e)}, \tag{6.2.11b}$$

where $g_i \in K_i$. This resembles (6.1.1b), with S_{ab} replaced by $S_{i,\mathrm{ch}} = \mathrm{ch}(g_i)$ and the vacuum 0 by the identity e. Unfortunately, the other axioms of modular data fail.

However, the group algebra $\mathbb{C}G$ is a Hopf algebra, with co-multiplication $\Delta(g) = g \otimes g$, co-unit $\epsilon(g) = 1$ and antipode $S(g) = g^{-1}$. The way to obtain true modular data is to take the quantum double of $\mathbb{C}G$. Its Hopf dual, the space $F[G]$ of functions $G \to \mathbb{C}$, is also a Hopf algebra, for example, with co-product $\Delta(f)(g_1, g_2) = f(g_1 g_2)$. The construction of the double $\mathcal{D}(G)$ is described nicely in [**406**]; we will simply describe its modular data.

Let Φ be the set of all pairs (a, ch), where the a are representatives of the conjugacy classes of G and ch is the character of an irreducible representation of the centraliser $C_G(a)$. (Recall that $C_G(a)$ is the set of all $g \in G$ commuting with a.) Φ parametrises the irreducible modules of the double $\mathcal{D}(G)$. Put [**393**], [**136**]

$$S_{(a,\mathrm{ch}),(a',\mathrm{ch}')} = \frac{1}{\|C_G(a)\|\,\|C_G(a')\|} \sum_{g \in G(a,a')} \overline{\mathrm{ch}'(g^{-1}ag)}\,\overline{\mathrm{ch}(ga'g^{-1})}, \quad (6.2.12a)$$

$$T_{(a,\mathrm{ch}),(a',\mathrm{ch}')} = \delta_{a,a'}\delta_{\mathrm{ch},\mathrm{ch}'}\frac{\mathrm{ch}(a)}{\mathrm{ch}(e)}, \quad (6.2.12b)$$

where $G(a, a') = \{g \in G \mid aga'g^{-1} = ga'g^{-1}a\}$ and $e \in G$ is the identity. For the 'vacuum' 0 take $(e, 1)$. Then (6.2.12) is modular data. Manifestly, $\mathbb{N}$-valued descriptions of the fusion multiplicity $\mathcal{N}^{(c,\mathrm{ch}'')}_{(a,\mathrm{ch}),(b,\mathrm{ch}')}$ exist (see section 2 of [**391**], who realises the fusion ring as the Grothendieck ring for G-equivariant vector bundles). For Lusztig, (6.2.12) arose in his determination of irreducible characters of Chevalley groups. The higher-genus fusion multiplicities in (6.1.2) also have interpretations as multiplicities of representations of $\mathcal{D}(G)$ in $\mathcal{D}(G) \otimes \cdots \otimes \mathcal{D}(G)$ [**35**].

For instance, the modular data associated with the finite group $\mathcal{S}_3$ is

$$S = \frac{1}{2}\begin{pmatrix} 1 & 1 & 2 & 2 & 2 & 2 & 3 & 3 \\ 1 & 1 & 2 & 2 & 2 & 2 & -3 & -3 \\ 2 & 2 & 4 & -2 & -2 & -2 & 0 & 0 \\ 2 & 2 & -2 & 4 & -2 & -2 & 0 & 0 \\ 2 & 2 & -2 & -2 & -2 & 4 & 0 & 0 \\ 2 & 2 & -2 & -2 & 4 & -2 & 0 & 0 \\ 3 & -3 & 0 & 0 & 0 & 0 & 3 & -3 \\ 3 & -3 & 0 & 0 & 0 & 0 & -3 & 3 \end{pmatrix}, \quad (6.2.13a)$$

$$T = \mathrm{diag}(1, 1, 1, 1, e^{2\pi \mathrm{i}/3}, e^{-2\pi \mathrm{i}/3}, 1, -1). \quad (6.2.13b)$$

See [**115**] for several more explicit examples.

This modular data can be twisted [**138**], [**135**], [**34**], [**115**] by a 3-cocycle $\alpha \in H^3(G, \mathbb{C}^\times)$. Indeed this twisted modular data is absolutely as fundamental as (6.2.12) – recall the discussion in Sections 4.3.4 and 5.3.6. This cocycle α plays the same role here that level does in affine algebra modular data, as $H^3(G, \mathbb{C}^\times) \cong \mathbb{Z}$ when G is simply-connected and simple. This sort of twist has a generalisation to arbitrary chiral data [**118**].

One of the remarkable features of affine algebra modular data – its ubiquity – is shared by finite group modular data. Most important for us, it arises in the orbifold of holomorphic VOAs (recall Section 5.3.6). Let G be a finite group of automorphisms

  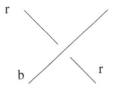

Fig. 6.7 Colourings at a crossing.

of a holomorphic VOA $\mathcal{V}$ – all finite groups arise in this way (Question 6.2.7). Let $\mathcal{V}^G$ be the space of fixed points of G; it inherits a VOA structure from $\mathcal{V}$. Then the modular data of $\mathcal{V}$ is trivial but that of $\mathcal{V}^G$ is expected to be (6.2.12) or some twisted version (see Conjecture 5.3.10). This modular data also appears in the crossed-product construction in von Neumann algebras (Section 6.2.6). In physics, it arises in $(2 + 1)$-dimensional Chern–Simons theory with finite gauge group G [138], [194], as well as $(2 + 1)$-dimensional quantum field theories where a continuous gauge group has been spontaneously broken to a finite group [31] (adding a Chern–Simons term here corresponds to the cohomological twist).

This modular data is quite interesting for nonabelian G, and deserves more study. It seems very effective at distinguishing groups – in fact, it is known to distinguish all groups of order < 128. Conversely, there are non-isomorphic groups of order $2^{15} \cdot 3^4 \cdot 5 \cdot 7$ with identical modular data up to reordering primaries [175]. Finite group modular data behaves very differently from the affine algebra data (see e.g. [115], [457], [178]). For instance, Eiichi Bannai has found that the alternating group $\mathcal{A}_5$, which has only 22 primaries, has a remarkably high number (8719) of modular invariants. By contrast, affine algebras have relatively few modular invariants.

6.2.5 Knots

The Jordan curve theorem states that all knots in $\mathbb{R}^2$ are trivial. Are there any nontrivial knots in $\mathbb{R}^3$?

In Figures 1.9 and 1.10 are some knots in $\mathbb{R}^3$, flattened into the plane of the paper. A moment's consideration will confirm that the second knot of Figure 1.9 is indeed trivial. What about the trefoil?

A knot diagram cuts the knotted S^1 into several connected components (*arcs*), whose endpoints lie at the various *crossings* (double-points of the projection). By a *3-colouring*, we mean to colour each arc in the knot diagram either red, blue or green, so that at each crossing either one or three distinct colours are used. For example, the first two colourings in Figure 6.7 are allowed, but the third isn't. By considering the 'Reidemeister moves' (Figure 1.12), which tell us how to move between equivalent knot diagrams, different diagrams for equivalent knots (such as the two in Figure 1.9) are seen to have equal numbers of distinct 3-colourings. Hence, the number of 3-colourings is a knot invariant.

For example, consider the diagrams in Figure 1.9 for the trivial knot: clearly, all arcs must be given the same colour, and thus there are precisely three distinct 3-colourings.

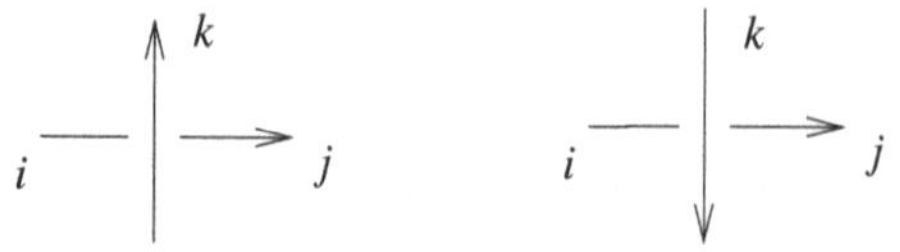

$$x_k x_i = x_j x_k \qquad\qquad x_i x_k = x_k x_j$$

Fig. 6.8 The Wirtinger presentation of the knot group.

On the other hand, the trefoil has nine distinct 3-colourings – the bottom two arcs of Figure 1.10 can be assigned arbitrary colour, and that choice fixes the colour of the top arc. Thus the trefoil is nontrivial!

Essentially what we are doing here is counting the number of homomorphisms φ from the *knot group* $\pi_1(\mathbb{R}^3 \setminus K)$ of knot K to the symmetric group $\mathcal{S}_3$. The reason is that any (oriented) knot diagram gives a presentation for $\pi_1(\mathbb{R}^3 \setminus K)$, where there is a generator x_i for each arc and a relation of the form $x_i x_j = x_k x_i$ for each crossing (Figure 6.8). See section 3.D of [**478**] for more details and a proof. For example, the knot group of the right knot of Figure 1.9 has presentation

$$\langle x_1, \ldots, x_7 \mid x_1 x_2 = x_4 x_1, x_5 x_1 = x_3 x_5, x_5 x_4 = x_3 x_5, x_2 x_1 = x_5 x_2,$$

$$x_2 x_7 = x_2 x_2, x_2 x_7 = x_6 x_2, x_2 x_5 = x_6 x_2 \rangle,$$

which is isomorphic to $\mathbb{Z}$. By contrast, the knot group of the trefoil is $\mathcal{B}_3$ (Question 6.2.8). Incidentally, the complement $\mathbb{R}^3 \setminus K$ of a knot determines the knot, and the extent to which the knot group determines the knot is also understood (see section 1 of [**61**]). Therefore, in this sense the trefoil and $\mathcal{B}_3$ are intimately connected (recall Section 2.4.3).

$\mathcal{S}_3$ is generated by the transpositions $(12), (23), (13)$. The homomorphism $\varphi : \pi_1(\mathbb{R}^3 \setminus K) \to \mathcal{S}_3$ is defined using, for example, the identification $r \leftrightarrow (12), b \leftrightarrow (23), g \leftrightarrow (13)$, and the above 3-colouring condition at each crossing is equivalent to requiring that φ obeys each relation in the Wirtinger presentation. Our homomorphism φ will be onto iff at least two different colours are used. By considering more general (non-abelian) colourings, the target ($\mathcal{S}_3$ here) can be made to be any other group G, resulting in a different knot invariant.

In the early 1980s, knot theory was dormant; by the late 1980s it was flourishing. But as a consequence, we suddenly had too many knot invariants. Reshitikhin and Turaev [**473**] brought order to this chaos, by proving that whenever we have a ribbon category **V**, we get invariants of (framed) knots and links, that is of knotted and linked ribbons. The reason for their result, as we explain in Section 1.6.2, is the universality of the topological category **Ribbon** of ribbons (Theorem 1.6.2). Given any knotted link, coloured with the objects of **V**, their functor associates the link with some morphism $\mathrm{Hom}(\emptyset, \emptyset)$ of **V**, and isotopic links get assigned the same morphism. This morphism is the desired link invariant. For example, the 3-colouring invariant comes from a ribbon category associated with the modular data (6.2.13).

We can express their result slightly differently. Suppose we have a representation of every braid group $\mathcal{B}_n$ (e.g. Proposition 6.2.6 says we get this from a solution to

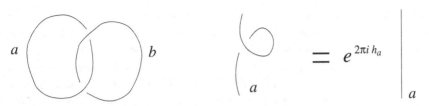

Fig. 6.9 The S_{ab} and T_{aa} matrix entries in modular categories.

the quantum Yang–Baxter equation). To every braid we get a link by closing it up, as in Figure 1.14. Unfortunately, different braids can get assigned the same link. As we explain in Section 1.2.3, the two Markov moves capture precisely this redundancy. Thus we get a link invariant from our braid representations if we can construct a quantity invariant with respect to these two moves. The first move $\beta' \leftrightarrow \beta\beta'\beta^{-1}$ suggests we assign to the braid the trace of its representing matrix; unfortunately, that usually won't respect the second move, $\beta \leftrightarrow \beta T_m^{\pm 1}$.

However, [**473**] explain how to enhance the braid representation coming from any quasi-triangularisable Hopf algebra (Definition 6.2.8), to get link invariants. See section XI.3.1 of [**534**] for details. Thus, combining their construction with the Drinfel'd double, which associates a quasi-triangularisable Hopf algebra with any Hopf algebra, we can construct (or recover) enormous numbers of link invariants.

So far we have discussed invariants of links embedded in $\mathbb{R}^3$ (equivalently, S^3). Much more difficult is to construct invariants of links in arbitrary 3-manifolds, but it is precisely this that is relevant to our story. There are (at least) two ways to do this: one uses 'Dehn surgery' to construct the manifold from S^3 [**474**], and the other uses triangulation by tetrahedra [**535**]. We allude to the *Turaev–Viro theory* [**535**] elsewhere. In the early 1960s Lickorish and Wallace established that any closed compact oriented 3-manifold M can be obtained by surgery on the 3-sphere S^3 along some framed link L (see section II.2.1 of [**534**] for details). The idea is to construct an invariant for M from the link invariant of L in S^3. For instance, the 3-manifold $S^1 \times S^2$ arises from S^3 by surgery along the trivial ribbon. The problem is that different links give rise to the same manifold. However, this redundancy is completely captured by the so-called 'Kirby moves' (see section II.3.1 of [**534**] for details). Once again, Reshitikhin and Turaev [**474**] find the necessary refinement to ribbon categories, as well as the precise expression for the 3-manifold invariant, which will make the quantity invariant under the Kirby moves. The result is called a *modular category* (see chapter 2 of [**534**] for complete details). Roughly speaking, it is a ribbon category with the additional property of direct sum, with a finite set of 'simple objects' (closed under ∗) and a complete reducibility property, whose Hopf link invariant (Figure 6.9) is nondegenerate. More generally, this procedure gives us link invariants in any 3-manifold. Again, the ultimate source of these topological invariants is a universality property of the appropriate topological category. All of these universalities have as their source the universality of **Braid** for braided monoidal categories (Theorem 1.6.1).

Any RCFT gives a modular category (in fact two of them, one for each chiral half). For an RCFT, the simple objects are the objects that are the chiral primaries, the monoidal structure is the fusion product and duality is charge-conjugation. Modular data is obtained directly from the Hopf link and twist, as in Figure 6.9. There are thus three different incarnations of the S-matrix in RCFT: the modular transformation (4.3.9a), Verlinde's formula (6.1.2), and the Hopf link. In fact, the notion of a modular category is equivalent to that of Segal's modular functor (Section 4.4.1) [**534**], [**32**]. For a sufficiently nice VOA $\mathcal{V}$, the simple objects are the irreducible $\mathcal{V}$-modules. The 3-colouring invariant of Figure 6.7 comes from a holomorphic orbifold VOA, and as such can be modified to yield a link invariant in any 3-manifold.

For instance, we get S^3 knot invariants from the quantum group $U_q(X_r)$ with generic parameter, but to get invariants for any closed 3-manifold requires specialising q to a root of unity. Modular categories are far less common than ribbon categories, but they can be obtained by an analogue of the Drinfel'd double.

6.2.6 Subfactors

The final general source of modular data that we discuss is from subfactor theory. The relations of subfactors to knots is reviewed in, for example, [**317**], [**318**], [**319**], while reviews of the relation between subfactors and CFT can be found in [**177**], [**66**].

Recall the definitions in Section 1.3.2. Let $N \subset M$ be an inclusion of type II_1 factors. We call N a *subfactor*, provided N includes the identity of M. Jones' motivation for looking at subfactors came from their formal similarity with Galois theory. After all, the very notation $\dim_M(\mathcal{H})$ for the 'coupling constant' of Section 1.3.2 suggests thinking of a type II_1 factor as a non-commutative analogue of 'field of scalars'.

In particular, let G be a finite group acting on some type II_1 factor N. Then the crossed-product $N \rtimes G$ is also a type II_1 factor, iff each $g \in G$, $g \neq e$, is 'outer'. By an outer automorphism g of N we mean that there are no unitary operators $u \in N$ such that $g.x = uxu^*$ for all $x \in N$. Any locally compact (e.g. finite) group G acts on, for example, the hyperfinite type II_1 factor by outer automorphisms, so this isn't a major restriction. This yields a Galois correspondence between subgroups H of G, and subalgebras of M containing the algebra M^G of fixed points, given by $H \leftrightarrow M^H$. This is analogous to the relation between subfields $\mathbb{K} \subset \mathbb{L}$ and Galois groups in Section 1.7.2. So what is the subfactor analogue of the index $[\mathbb{L} : \mathbb{K}]$?

Jones' answer is the *Jones index* of the subfactor $N \subseteq M$:

$$[M : N] := \dim_N(L^2(M)) \geq 1, \qquad (6.2.14)$$

where $L^2(M)$ is the Hilbert space of Question 1.3.6. For instance, for any $n \geq 1$, $[N \otimes M_n(\mathbb{C}) : N] = n^2$. If $H \leq G$ are finite groups of outer automorphisms, then $[M \rtimes G : M \rtimes H] = \|G\|/\|H\| = [M^H : M^G]$, where the crossed-product $M \rtimes H$ and fixed-point M^H factors are discussed in Section 1.3.2.

The following theorem was completely unexpected.

Theorem 6.2.9 [**316**]   *For any number*

$$d \in \{4\cos^2(\pi/n)\}_{n=3}^{\infty} \cup [4, \infty],$$

there is a subfactor $N \subseteq M$ of the unique hyperfinite type II_1 factor M, with index $[M : N] = d$. Conversely, the index of any subfactor of a (not necessarily hyperfinite) type II_1 factor will be in that set.

In fact, the following rigidity is true: if M is the hyperfinite type II_1 factor, then at most four inequivalent subfactors $N \subseteq M$ can possess the same index < 4. The reader, with Section 2.5.2 fresh in mind, may recognise the discrete sequence of indices in Theorem 6.2.9 as the square of the Perron–Frobenius eigenvalues of the A–D–E graphs – is this a coincidence?

The key to proving Theorem 6.2.9, as well as the further developments, is the so-called *basic construction*, which appears to have been found independently by a number of people in the late 1970s. Let $N \subseteq M$ be an inclusion of type II_1 factors. Even though M and N are isomorphic as factors, there is rich combinatorics surrounding how N is embedded in M. The Hilbert space $L^2(N)$ is naturally contained in $L^2(M)$. Let e_N be the orthogonal projection of $L^2(M)$ onto $L^2(N)$. Then M and e_N generate the von Neumann algebra $\langle M, e_N \rangle''$ acting on the space $L^2(M)$. If the index $[M : N]$ is finite, then $\langle M, e_N \rangle''$ will also be a type II_1 factor, with index $[\langle M, e_N \rangle'' : M] = [M : N]$. Moreover, since the trace (normalised so that $\mathrm{tr}(1) = 1$) on a type II_1 factor is unique, we can unambiguously speak of the trace $\mathrm{tr}(e_N)$, and we find it equals $1/[M : N]$. For later convenience define $\tau := 1/[M : N]$.

For example, taking N to be the fixed points M^G, for some finite group G of outer automorphisms, then $e_N = (1/\|G\|) \sum_g g$, $\mathrm{tr}(e_N) = 1/\|G\|$ and $\langle M, e_N \rangle'' = M \rtimes G$. This demonstrates the naturalness of this construction. What is the von Neumann algebra generated by M and e? The answer is the crossed-product $M \rtimes G$.

We can repeat the basic construction indefinitely. Put $M_0 := N$, $M_1 := M$ and define inductively

$$M_{i+1} := \langle M_i, e_{i-1} \rangle'',$$

where $e_i := e_{M_{i-1}}$ is the orthogonal projection from $L^2(M_i)$ onto $L^2(M_{i-1})$. We thus get a tower $M_0 \subset M_1 \subset \cdots$ of type II_1 factors, and a sequence $e_1, e_2, \ldots$ of projections. The limit $M_\infty := \cup_{n=0}^\infty M_n$ is also a type II_1 factor, with a unique (normalised) trace tr, which restricts to the unique trace on each M_n. Thus each $\mathrm{tr}(e_n) = \tau$. The algebra $\mathcal{A}_{\infty,\tau}$ spanned by the projections e_i obeys the relations

$$e_i^2 = e_i^* = e_i, \tag{6.2.15a}$$

$$e_i e_{i\pm 1} e_i = \tau e_i, \tag{6.2.15b}$$

$$e_i e_j = e_j e_i \quad \text{if } |i - j| \geq 2, \tag{6.2.15c}$$

$$\mathrm{tr}(x e_{n+1}) = \mathrm{tr}(x)\, \tau, \tag{6.2.15d}$$

where x is in the (finite-dimensional semi-simple) algebra $\mathcal{A}_{n,\tau}$ generated by $1, e_1, \ldots, e_{n-1}$. In fact these are the complete list of relations for $\mathcal{A}_{n,\tau}$, because the (normalised) trace tr on any type II_1 factor obeys $\mathrm{tr}(xx^*) \geq 0$ with equality only if $x = 0$. The (easy) proofs of all these statements are in [**319**]. The point is that the tower

$M_0 \subset M_1 \subset \cdots$ and the projections $e_1, e_2, \ldots$ depend only on the original subfactor. Positive-definiteness of the trace on $\mathcal{A}_{n,\tau}$ gives the discrete values of Theorem 6.2.9.

Of course we are now trained to recognise (6.2.15b) and (6.2.15c) as having to do with the braid groups. In particular, if we try to send the braid group generator σ_i to $ae_i + b$, we obtain the solution $a = t + 1$, $b = -1$, where t satisfies $t + t^{-1} + 2 = \tau^{-1}$. Thus to any finite index type II$_1$ subfactor, we get a representation of the braid group!

We know how to go from a braid group representation to a link invariant: we need to associate a number with each braid that is invariant under the two Markov moves (Section 1.2.3). For a braid $\beta \in \mathcal{B}_n$, the combination

$$J_\beta(t) = \left(-\left(\sqrt{t} + \frac{1}{\sqrt{t}} \right) \right)^{n-1} \sqrt{t}^{\deg \beta} \mathrm{tr}(\beta) \qquad (6.2.16)$$

works (verify this), where 'deg β' is defined in Section 1.1.4 and 'tr(β)' means the trace of the corresponding element in M_n. This function J_β is the famous Jones polynomial.

Witten showed that the Jones polynomial can be recovered from the topological field theory (or modular category) associated with affine algebra $A_1^{(1)}$ at level $k \in \mathbb{N}$, when the highest weight $\omega_1 + (k - 1)\omega_0$ is assigned to each strand of the link. Of course, there is no need to restrict to $A_1^{(1)}$ or that weight, and other choices yield other link invariants.

Can the subfactor approach also recover these other link polynomials, or is it inherently 'rank 1'? Is the full topological field theory (or if you prefer, the CFT or modular category) obtainable from the subfactor, or does the subfactor only see the link polynomials? The answer to both questions is yes; the construction was originally due to Ocneanu, and is explained carefully in [177] (see also [354] for a very accessible treatment of certain parts of the theory). The starting point is the realisation that the projections e_i are only a small part of the full tower $M_0 \subset M_1 \subset M_2 \subset \cdots$.

Subtleties in any representation theory arise through the interplay of addition with multiplication, and with contragredient (dual). Addition (direct sum) of modules comes for free here. Unfortunately, the modules of factors (which we briefly described at the end of Section 1.3.2) don't have an obvious tensor product, and in any case are rather colourless (e.g. there is a unique nontrivial module for type III factors).

The right objects to study here are *bimodules*. We call a Hilbert space $X = {}_M X_N$ an M–N bimodule if M acts on the left and N on the right. The point is that they have a natural multiplication: the relative tensor product ('Connes fusion') ${}_M X_N \otimes_N Y_P$ will be an M–P bimodule. The multiplicative identity (playing the role of the trivial one-dimensional module) is ${}_M L^2(M)_M$, usually abbreviated to ${}_M M_M$. Given any bimodule ${}_M X_N$, the conjugate Hilbert space $\overline{X}$ is naturally an N–M bimodule: $n\overline{x}m := m^* x n^*$. Moreover, the possibilities for bimodules are far richer than for modules.

Let $N \subset M$ be an inclusion of II$_1$ factors with finite Jones index $[M : N]$. Recall the tower $M_0 = N \subset M = M_1 \subset M_2 \subset \cdots$ arising from the basic construction. Let Φ_M denote the set of equivalence classes of irreducible M–M submodules of $\oplus_{n \geq 1} {}_M L^2(M_n)_M$, and Φ_N that for the irreducible N–N submodules of $\oplus_{n \geq 0} {}_N L^2(M_n)_N$. We require these sets to be finite ('finite depth'). Write $\mathcal{H}_{AB}^C$ for the

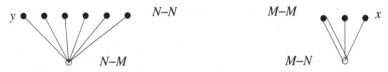

Fig. 6.10 The principal and dual principal graphs associated with $\mathcal{S}_3$.

(finite-dimensional) intertwiner space $\mathrm{Hom}_{M-M}(C, A \otimes_M B)$. For any $A, B \in \Phi_M$, the product $A \otimes_M B$ can be decomposed into a finite sum $\sum_{C \in \Phi_M} \mathcal{N}_{AB}^C C$, where $\mathcal{N}_{AB}^C = \dim \mathcal{H}_{AB}^C \in \mathbb{N}$ are the multiplicities. Indeed, all axioms of a fusion ring will be obeyed, except usually commutativity and self-duality.

Returning to the Galois theory analogy, the Jones index merely corresponds to the degree of the field extension. To what corresponds the Galois group? Ocneanu's answer is an intricate subfactor invariant called a *paragroup* [**453**] (see especially chapter 10 of [**177**]). It consists of two graphs (the *principal* and *dual principal*), whose vertices are bimodules for M and N; an order-2 involution of the vertices corresponding to the contragredient map $A \mapsto \overline{A}$; and a 'connection', that is an assignment of complex numbers to closed paths in the graphs, reminiscent of $6j$-symbols, describing the change between natural bases. The graphs are obtained from the fusion rings; their Perron–Frobenius eigenvalues equal the square-roots of the Jones index. For example, when the Jones index is < 4 (corresponding to eigenvalue < 2), those two graphs are equal, and are one of A_n, D_{even}, E_6 or E_8 (recall Figure 1.4) – it cannot be the tadpole T_n for elementary reasons, but D_{odd} and E_7 are excluded for their inability to support a connection. Two inequivalent connections are possible on the E_6 and E_8 graphs, corresponding to different subfactors. Thus Theorem 6.2.9 indeed constitutes another realisation of A–D–E, and for the ultimate reason suggested in Section 2.5.2.

A paragroup is a generalised ('quantised') sort of group. Figure 6.10 gives the graphs for $R \subset R \rtimes G$ (for R the hyperfinite II_1 factor and $G = \mathcal{S}_3$). The M–M bimodules are parametrised by the irreducible characters ch_i of G, with precisely $\mathrm{ch}_i(e)$ edges connecting the ith node to the root of the graph. The N–N bimodules are parametrised by elements of the group. The contragredient involution and fusion rings are the ones familiar to aficionados of character tables: complex-conjugate and the character ring, and $g \mapsto g^{-1}$ and the group ring $\mathbb{C}G$. The connection explicitly recovers the group structure, much as in the topological field theory of Section 4.4.2. On the otherhand, the graphs for $R^G \subset R$ are switched. More generally, given any subgroup $H < G$, we get subfactors $R^G \subset R^H$ and $R \rtimes H \subset R \rtimes G$, and their paragroups give a group-like interpretation to G/H even when H is not normal.

We say subfactors $N_i \subset M_i$ are equivalent if there is an isomorphism $\theta : M_1 \to M_2$ with $\theta(N_1) = N_2$. When M is hyperfinite type II_1, the paragroup identifies $N \subset M$ up to equivalence. Hence, when G is a finite abelian group, $R^G \subset R$ is equivalent to $R \subset R \rtimes G$ (when instead G is nonabelian, they are merely dual).

The paragroup yields a topological invariant for manifolds, generalising the Turaev–Viro one [**535**] (see [**354**] for a very readable treatment of this part of the theory).

However, it doesn't directly correspond to the data of an RCFT (e.g. the fusion rings of Figure 6.10 aren't self-dual). To get RCFT data, we must pass from $N \subset M$ to the 'asymptotic inclusion' $\langle M, M' \cap M_\infty \rangle \subset M_\infty$, where M_∞ is the (weak completion of the) union of all M_n. Asymptotic inclusion plays the role of Drinfel'd's quantum-double here, and corresponds physically to taking the continuum limit of the lattice model, yielding the CFT from the underlying statistical mechanical model (see section 12.6 of [**177**]). All chiral data of the VOA or RCFT, including the link invariants, are obtainable from the asymptotic inclusion. For instance, the Jones index $[M:N]$ equals $1/S_{00}^2$.

A very similar (but simpler) theory has been developed for type III factors. Bimodules now are equivalent to 'sectors', that is equivalence classes of endomorphisms $\lambda : N \to N$ (the corresponding subfactor is $\lambda(N) \subset N$). This use of endomorphisms is the key difference (and simplification) between the type II and type III fusion theories. Given $\lambda, \mu \in \mathrm{End}(N)$, we define $\langle \lambda, \mu \rangle$ to be the dimension of the vector space of intertwiners, that is all $t \in N$ such that $t\lambda(n) = \mu(n)t$ $\forall n \in N$. The endomorphism $\lambda \in \mathrm{End}(N)$ is irreducible if $\langle \lambda, \lambda \rangle = 1$. Let Φ be a finite set of irreducible sectors. The fusion product is given by composition $\lambda \circ \mu$; addition can also be defined, and the fusion multiplicity $\mathcal{N}_{\lambda\mu}^\nu$ is then the dimension $\langle \lambda \circ \mu, \nu \rangle$. The 'vacuum' 0 is the identity id_N. Restricting to a finite set Φ of irreducible sectors, closed under fusion, the result is again a (noncommutative non-self-dual) fusion ring (after all, why should the compositions $\lambda \circ \mu$ and $\mu \circ \lambda$ be related). The missing ingredients are nondegenerate braidings $\epsilon^\pm(\lambda, \mu) \in \mathrm{Hom}(\lambda \circ \mu, \mu \circ \lambda)$, which say roughly that λ and μ nearly commute (the $\epsilon^\pm$ must also obey some analogue of the Yang–Baxter equation (6.2.8)). Provided we have a nondegenerate braiding (which we can obtain from asymptotic inclusion as before), Rehren [**470**] proved that we will automatically have modular data. When we have a hyperfinite type III$_1$ subfactor $N \subset M$ with a braided system of endomorphisms, there is a simple expression (see [**65**] and references therein) for the corresponding modular invariant (Definition 6.1.8) using 'α-induction' (a process of inducing an endomorphism from N to M using the braiding $\epsilon^\pm$): we get $\mathcal{Z}_{\lambda\mu} = \langle \alpha_\lambda^+, \alpha_\mu^- \rangle$. The NIM-rep is defined similarly [**65**].

Wassermann and collaborators (see e.g. [**554**]) have explicitly constructed the affine algebra subfactors, recovering the affine algebra modular data, at least for $A_r^{(1)}$ and $B_r^{(1)}$. To any subgroup–group pair $G < H$, the subfactor $R \rtimes G \subset R \rtimes H$ of crossed-products has a (in general non-commutative) fusion-like ring. But sometimes it will have a braiding – for example, the diagonal embedding $G < G \times G$ recovers the finite group modular data of Section 6.2.4.

These approaches cannot reconstruct the full RCFT or VOA. To give a simple example, the VOA associated with any even self-dual lattice or the Moonshine module corresponds to the trivial subfactor $N = M$, where M is the unique hyperfinite type II$_1$ factor. The way to get more information uses nets of subfactors.

There are two standard axiomatisations of quantum field theory (Section 4.2.4). The Wightman axioms, applied to two-dimensional CFT, yield quite naturally a VOA (see chapter 1 of [**330**]). Algebraic quantum field theory [**269**], on the other hand, leads to subfactors. In particular, to any open set $\mathcal{O}$ in Minkowski space $\mathbb{R}^{1,1}$ we are to assign a von Neumann algebra $\mathcal{A}(\mathcal{O}) \subset \mathcal{L}(\mathcal{H})$ of observables localised to $\mathcal{O}$, obeying various properties (such as $\mathcal{O}_1 \subset \mathcal{O}_2$ implies $\mathcal{A}(\mathcal{O}_1) \subset \mathcal{A}(\mathcal{O}_2)$). The axioms imply these $\mathcal{A}(\mathcal{O})$ will all be type III$_1$ factors. In two dimensions, choosing 'light-cone' coordinates $x_0 \pm x_1$, we can take these $\mathcal{O}$ to be the product $\mathcal{I} \times \mathcal{J}$ of open intervals $\mathcal{I}, \mathcal{J} \subset \mathbb{R}$. This means that for most purposes the theory decomposes into a one-dimensional net $\mathcal{A}(\mathcal{I})$ – the chiral theory. The one-dimensional 'space-time' $\mathbb{R}$ is compactified to S^1, and requiring the theory to be covariant with respect to Diff(S^1), the result is called a *local conformal net*. The theory of these one-dimensional nets should be equivalent to that of VOAs, and that of the two-dimensional ones to the full RCFT, though most details of this equivalence are still to be established. Nevertheless, some aspects of the theory will likely remain much more accessible using, for example, subfactors than VOAs (in particular, orbifolds seem simpler in subfactor theory). For references and results, see, for example, [**341**], [**340**], [**568**], [**332**] and references therein.

Question 6.2.1. Prove equation (6.2.3a).

Question 6.2.2. Find all NIM-reps for $A_1^{(1)}$ at each level $k = 1, 2, 3, \ldots$ (*Hint*: Verify that the Perron–Frobenius eigenvalue of $\mathcal{M}_1$ is $S_{10}/S_{00} = 2\cos(\pi/(k+2)) < 2$.)

Question 6.2.3. (a) Find a continuous one-parameter deformation of the three-dimensional complex Lie algebra span$\{x, y, z\}$ with brackets $[xy] = x, [xz] = [yz] = 0$.
(b) Verify that any continuous deformation of A_1 is trivial.

Question 6.2.4. Let M, N be left A-modules, where A is a Hopf algebra. Prove that $\operatorname{Hom}_K(M, N)$ is a left A-module.

Question 6.2.5. (a) When does the character table of a finite group, with rows and columns appropriately normalised and ordered, equal the S-matrix of modular data?
(b) Let G be finite and abelian. Is the fusion ring for the quantum double $\mathcal{D}(G)$ (see Section 6.2.4) isomorphic to the group ring of $G \times G$?

Question 6.2.6. Let G be any finite group and consider the modular data of (6.2.12). Find the conjugation C, the simple-currents J and their action and monodromy φ_J, and identify the group of all simple-currents. Identify the Galois action and parities.

Question 6.2.7. Prove that any finite group can be realised as a subgroup of the group of automorphisms of a holomorphic VOA. (*Hint*: think of self-dual lattices.)

Question 6.2.8. Identify the knot group $\pi_1(\mathbb{R}^3 \setminus T)$ of the trefoil, using the Wirtinger presentation of Figure 6.8.

Question 6.2.9. Prove, using the Reidemeister moves, that the Wirtinger presentation yields the same group no matter which knot diagram is chosen for the given knot.

Question 6.2.10. Recall (6.2.15). Find all values a, b such that $\sigma_i \mapsto ae_i + b$, $i = 1, \ldots, n-1$, yields a representation of the braid group $\mathcal{B}_n$ in $\mathcal{A}_{n,\tau}$.

6.3 Hints of things to come

String theory has profoundly affected geometry (e.g. elliptic genus and mirror symmetry), algebra (e.g. VOAs) and topology (e.g. knot invariants), but so far it has had little impact on number theory. That may have something to do with the knowledge and interests of the individuals who have developed its mathematical side. There are in fact several indications of deep relations with number theory, waiting to be developed. In this section we sketch some of these.

6.3.1 Higher-genus considerations

String theory tells us that CFT can live on any surface Σ. The VOAs, including the geometric VOAs of Section 5.4.1, capture CFT in genus 0. The graded dimensions and traces considered above concern CFT quantities ('chiral blocks') at genus 1: $\tau \mapsto e^{2\pi i\tau}$ maps $\mathbb{H}$ onto a cylinder, and the trace identifies the two ends. But there are analogues of all this at higher genus [573] (though the formulae can rapidly become awkward). We have alluded to this throughout the book so will only add some quick remarks here. Our main point is that this is surely the direction for important future research, with direct implications to Moonshine.

For example, the graded dimension of the $V^\natural$ CFT in genus 2 is computed in [533], and involves, for example, Siegel theta functions. The higher-genus mapping class group representations coming from the $A_1^{(1)}$ RCFT are studied in [220]. A more radical suggestion, using projective limits, is given in Section 4.3.3.

The orbifold theory in Sections 5.3.6 and 7.3.2 is genus 1: each sector (g, h) corresponds to a homomorphism from the fundamental group $\mathbb{Z}^2$ of the torus into the orbifold group G (e.g. $G = \mathbb{M}$) – g and h are the targets of the two generators of $\mathbb{Z}^2$ and so must commute. More generally, the sectors correspond to homomorphisms $\varphi : \pi_1(\Sigma) \to G$, and for each we get a higher-genus trace $\mathcal{Z}(\varphi)$, which are functions on the Teichmüller space $\mathfrak{T}_g$ (generalising the upper half-plane $\mathbb{H}$ for genus 1). The action (7.3.3) of $SL_2(\mathbb{Z})$ on $N_{(g,h)}$ generalises to the action of the mapping class group on $\pi_1(\Sigma)$ and $\mathfrak{T}_g$.

For example, we can count the number of inequivalent homomorphisms $\pi_1(\Sigma) \to G$, for G a compact genus-g surface. This number is given by Verlinde's formula (6.1.2) together with the expression (6.2.12a) [194]:

$$\mathcal{N}^{(g,0)} = \sum_h \sum_{\mathrm{ch} \in \mathrm{Irr}(C_G(h))} \left(\frac{\|C_G(h)\|}{\mathrm{ch}(e)} \right)^{2(g-1)}, \tag{6.3.1}$$

where we sum over representatives h of the various conjugacy classes of G.

6.3.2 Complex multiplication and Fermat

A few years ago Philippe Ruelle was walking in a library in Dublin. He spotted a yellow book in the mathematics section, called *Complex Multiplication* [367]. A strange title for a book by Lang! Ruelle flipped it to a random page, which turned out to be 26. There he found what we would call the Galois selection rule (6.1.15a) for $A_2^{(1)}$, analysed and solved for the cases where $k + 3$ is coprime to 6. Lang, however, knew nothing of modular invariants; he was reviewing work by Koblitz–Rohrlich [351] on decomposing the Jacobians of the Fermat curve $x^n + y^n + z^n = 0$ into their prime pieces, called 'simple factors'.

Fix $n > 3$. Let F_n denote the nth Fermat curve, that is the projective complex curve $x^n + y^n + z^n = 0$. We will describe some similarities with the modular data of $A_2^{(1)}$ at level $k = n - 3$.

First, let's review some $A_2^{(1)}$ chiral data. Call a pair $(r, s) \in \mathbb{N} \times \mathbb{N}$ *admissible* if $1 \leq r, s$ and $r + s < n$. The integrable highest weights $\lambda \in P_+^k(A_2^{(1)})$ are in one-to-one correspondence with the admissible pairs, given by $\lambda_{(r,s)} := (n - r - s - 1)\omega_0 + (r - 1)\omega_1 + (s - 1)\omega_2$. For any admissible (r, s), define

$$H_{r,s} = \{\ell \in \mathbb{Z}_n^\times \mid \langle \ell r \rangle + \langle \ell s \rangle < n\},$$

where $\mathbb{Z}_N^\times$ is (as always) the multiplicative group (mod N) of integers coprime to N, and $\langle a \rangle$ is the unique integer $0 \leq \langle a \rangle < n$ congruent to a (mod n). Then $\mathbb{Z}_{3n}^\times$ is the Galois group over $\mathbb{Q}$ of the field generated by all entries $S_{\lambda\mu}$ of the $A_2^{(1)}$ level-k matrix S. The Galois selection rule (6.1.15a) says that if $\mathcal{Z}$ is a modular invariant, then

$$\mathcal{Z}_{\lambda_{(r,s)}, \lambda_{(r',s')}} \neq 0 \quad \Rightarrow \quad H_{r,s} = H_{r',s'}.$$

The hard part of the $A_2^{(1)}$ modular invariant classification involves solving this condition $H_{r,s} = H_{r',s'}$ [232].

Before we compare this to F_n, let's introduce some geometric terminology. An *abelian variety* is a torus of the form $\mathbb{C}^m/L$, where L is a $2m$-dimensional lattice in $\mathbb{C}^m$, which admits an embedding into projective space. This means there is a Hermitian form on $\mathbb{C}^m$ (defined in Section 1.1.3), whose imaginary part takes integer values when restricted to L. Most tori (when $m > 1$) don't satisfy this Hermitian form condition, though it is automatic when $m = 1$. We say two abelian varieties $\mathbb{C}^m/L$ and $\mathbb{C}^m/L'$ are *isogenous* if there exists a continuous group homomorphism from one to the other that is surjective; equivalently, if there is an invertible complex-linear endomorphism of $\mathbb{C}^m$ taking the lattice L onto a sublattice of L'. Isogeny is an equivalence relation preserving most things of interest.

Now suppose an abelian variety $\mathbb{C}^m/L$ contains another, $\mathbb{C}^n/L'$, of dimension $n < m$. Then the Hermitian form can be used to show that the original variety is isogenous to the product of $\mathbb{C}^n/L'$ with some $\mathbb{C}^{m-n}/L''$ (roughly, L'' is the orthogonal complement of L' in L). Continuing in this way, we get that any abelian variety is isogenous to the product of simple factors, where *simple factor* means an abelian variety containing no proper abelian subvariety.

A very special property that an abelian variety may possess is *complex multiplication*. The general definition is a little too complicated to get into here (see chapter 1.4 of [367]), so let's restrict to one-dimensional abelian varieties, that is the torus $A_\tau = \mathbb{C}/(\mathbb{Z} + \tau\mathbb{Z})$. We say A_τ has complex multiplication if its endomorphism ring $\mathrm{End}(A_\tau)$ is strictly greater than $\mathbb{Z}$; equivalently, if there is a non-integer $z \in \mathbb{C}$ such that

$z(\mathbb{Z} + \tau\mathbb{Z}) \subset \mathbb{Z} + \tau\mathcal{Z}$ (hence the name). It turns out that if A_τ has complex multiplication, then (among other things) $j(\tau)$ is an algebraic integer. This illustrates just how rare complex multiplication is: only countably many A_τ have it. It also illustrates its number-theoretic significance, which only becomes more profound as the dimension rises.

We get an abelian variety from any complex projective curve, by taking the Jacobian (Section 2.1.4), which is of complex dimension equal to the genus. In the case of the Fermat curve F_n, the genus is $\binom{n-1}{2}$, which equals the cardinality $\|P_+^k(A_2^{(1)})\|$. A bijection between $P_+^k(A_2^{(1)})$ and a basis of holomorphic 1-forms is

$$\lambda_{(r,s)} \leftrightarrow \omega_{(r,s)} := x^{r-1} y^{s-1} \frac{\mathrm{d}x}{y^{n-1}},$$

for any admissible (r, s). For each (r, s) let $[r, s]$ denote the $H_{r,s}$-orbit $\{(\langle \ell r \rangle, \langle \ell s \rangle)\}_{\ell \in H_{r,s}}$. Then the Jacobian $\mathrm{Jac}(F_n)$ is isogenous to the product, over all orbits $[r, s]$, of a $\|\mathbb{Z}_m^\times\|/2$-dimensional abelian variety $A_{[r,s]}$, for $m = n/\gcd(r, s, n-r-s)$. All $A_{[r,s]}$ have complex multiplication, which simplifies the following analysis.

We wish to decompose $\mathrm{Jac}(F_n)$ into a product of simple factors. Thus we need to know when the $A_{[r,s]}$ are isogenous to one another, and also when they are simple. Both questions reduce to knowing when $H_{r,s} = H_{r',s'}$, which as we mentioned earlier is also the key step in the $A_2^{(1)}$ modular invariant classification.

Similarly, Itzykson discovered traces of the $A_2^{(1)}$ exceptionals – these occur when $k + 3 = 8, 12, 24$ – in the Jacobian of F_{24}. See [46] for additional observations.

The point is that the combinatorial heart of two very different problems – the decomposition of the Jacobian of Fermat curves into simple factors, and the classification of RCFT associated with $A_2^{(1)}$ – are identical. Nevertheless, this must seem a little ad hoc. What is needed are other independent probes of this (still hypothetical) relationship. One possibility, suggested by the presence of complex multiplication, is the following.

Basic data associated with an algebraic variety V is its zeta-function $L(V, s)$, which counts its points over various finite fields. Isogenous varieties have equal zeta-functions. The Mellin transform of the zeta-function (Section 2.3.1) formally gives a q-series $f_V(\tau) = \sum_n a_n q^n$. For a typical variety V, f_V won't have any special properties, but when V has complex multiplication, the zeta-function decomposes into a product of Hecke L-functions, and their q-series do have modularity properties [505], [506].

Thus, associated with the abelian varieties $A_{[r,s]}$ – by virtue of complex multiplication – are various sorts of modular forms. And associated with the weights $\lambda_{(r,s)}$ – by virtue of being integrable highest weights of an affine algebra – are various sorts of modular forms.

Problem   *How are the modular forms associated with the zeta-functions of the factors $A_{[rs]}$ in the Jacobian of the Fermat curve F_n related to the modular forms associated with integrable highest-weight modules of $A_2^{(1)}$ at level $n - 3$?*

The easiest n to check will be $n = 4, 6, 8, 12$, since for them $\mathrm{Jac}(F_n)$ is isogenous to a product of elliptic curves. A somewhat related project, concerning $A_1^{(1)}$, is proposed in [490], though nothing definite has been achieved there yet.

In any case, these Fermat $\leftrightarrow A_2{}^{(1)}$ 'coincidences' are still not understood. It is tempting to guess that, more generally, the $A_r{}^{(1)}$ level-k modular invariant classification is somehow related to the hypersurface $x_1^n + \cdots + x_r^n = z^n$, for $n = k + r + 1$, but this is probably too naive. As with other meta-patterns, the most realistic hope wouldn't be to find a direct connection between Fermat curves and the RCFTs associated with $\mathfrak{sl}_3$. Rather, the idea is to identify the combinatorial nugget common to both. The real hope would be that this 'coincidence' lies in a series: A–D–E for $\mathfrak{sl}_2$, Fermat for $\mathfrak{sl}_3$, ..., and that this would lead to insights into $\mathfrak{sl}_4$ RCFT and beyond.

Complex multiplication in CFT has been the subject of other work – see [435] for several references. Let's mention two examples. Arithmetic varieties related to number fields seem to be naturally selected in the study of black holes in Calabi–Yau compactifications of string theory [435]. It has been conjectured [268] that superconformal field theory with target space given by a Calabi–Yau manifold M will be rational iff both M and its mirror have complex multiplication.

6.3.3 Braided # 6: the absolute Galois group

The absolute Galois group of the rationals is the group of symmetries of the field of algebraic numbers. It is the most important, and poorest understood, group in algebraic number theory. But it also has deep contacts with geometry (through the generalised Riemann existence theorem), and there have been several proposals conjecturing its relevance to RCFT (see e.g. [128], [435], [268] and references therein), and even quantum field theory [106], [93].

Recall the discussion of algebraic numbers and Galois groups in Section 1.7. The algebraic closure $\overline{\mathbb{Q}}$ of the rationals is the set of all algebraic numbers, or equivalently the union of all finite-dimensional field extensions of $\mathbb{Q}$. The absolute Galois group of $\mathbb{Q}$ is $\Gamma_{\mathbb{Q}} := \mathrm{Gal}(\overline{\mathbb{Q}}/\mathbb{Q})$. It's uncountably infinite, and extremely complicated. Only two of its elements have names: the identity and complex conjugation. If $\mathbb{K}$ is any finite Galois extension of $\mathbb{Q}$, then its Galois group $G = \mathrm{Gal}(\mathbb{K}/\mathbb{Q})$, which will be a finite group, is a homomorphic image of $\Gamma_{\mathbb{Q}}$ and so is a quotient $\Gamma_{\mathbb{Q}}/N$ of $\Gamma_{\mathbb{Q}}$. Much effort has been devoted to discovering which groups G can arise as Galois groups over $\mathbb{Q}$ (see [548] for a review of the so-called inverse Galois problem).

Conjecture 6.3.1 *Any finite group G is a quotient of $\Gamma_{\mathbb{Q}}$.*

This conjecture shows just how complicated $\Gamma_{\mathbb{Q}}$ is. Incidentally, there are many nontrivial points of contact between braid groups and inverse Galois theory (see e.g. [549]).

$\Gamma_{\mathbb{Q}}$ is an example of a *profinite* group, that is a projective limit of finite groups (here, of the Galois groups G). We define projective limit in Section 2.4.1 – the indexing set here are the fields $\mathbb{K}$, ordered by inclusion, to which is attached its Galois group G. This just means that $\sigma \in \Gamma_{\mathbb{Q}}$ consists of a choice of Galois automorphism $\sigma_{\mathbb{K}}$ for each finite extension $\mathbb{K} \supseteq \mathbb{Q}$, which obeys the obvious compatibility constraint (if $\mathbb{K} \subset \mathbb{L}$, then $\sigma_{\mathbb{L}}$ restricted to $\mathbb{K}$ must equal $\sigma_{\mathbb{K}}$). Thus, if the conjecture is true, $\Gamma_{\mathbb{Q}}$ would be the limit

$\lim_{\leftarrow} G$ of all finite groups, in this sense. Of course any finite group is also a quotient of some free group $\mathcal{F}_n$, and so we may wonder if $\Gamma_{\mathbb{Q}}$ and $\mathcal{F}_n$ are somehow related.

Thanks to their realisations as fundamental groups, the braid group $\mathcal{B}_n$ acts faithfully on $\mathcal{F}_n$ (Question 6.3.5) – in other words, $\mathcal{B}_n$ can be regarded as a subgroup of $\mathrm{Aut}(\mathcal{F}_n)$. This can be seen as follows. Recall the space $\mathfrak{C}_n$ of (1.2.6). We have the obvious projection $\pi : \mathfrak{C}_{n+1} \to \mathfrak{C}_n$, given by forgetting the $(n+1)$th point. Hence π induces an action of the fundamental group $\pi_1(\mathfrak{C}_n)$ of the base on the fundamental group of the fibre $\pi^{-1}(z_1, \ldots, z_n) = \mathbb{C} \setminus \{z_1, \ldots, z_n\}$, that is an action of the pure braid group $\mathcal{P}_n$ on $\mathcal{F}_n$. The action of $\mathcal{B}_n$ is obtained similarly. We will find that similar reasoning allows us to replace $\mathcal{B}_n$ by $\Gamma_{\mathbb{Q}}$, and $\mathcal{F}_n$ by its profinite completion.

Let X be an algebraic variety defined over $\mathbb{Q}$ – that is, X is defined as the set of solutions $(z_1, \ldots, z_n) \in \mathbb{C}^n$ to a collection of polynomials $p_i(z_1, \ldots, z_n) = 0$, and the polynomials have coefficients in $\mathbb{Q}$. Let $X(\mathbb{Q})$ be the set of points $(z_1, \ldots, z_n) \in X$ with all coordinates $z_i \in \mathbb{Q}$. Fix a base-point $p \in X(\mathbb{Q})$ (assuming one exists).

Let N be a finite-index normal subgroup of $\pi_1(X, p)$. Then by the geometric Galois correspondence (Section 1.7.2), N corresponds to a finite Galois cover $f_N : X_N \to X$ of X, with $\pi_1(X_N) \cong N$ and the quotient $\pi_1(X, p)/N$ can be identified with the set of homeomorphisms $\gamma : X_N \to X_N$ satisfying $f_N \circ \gamma = f_N$. Each γ, restricted to the finite set $f_N^{-1}(p)$, will be a permutation, and this permutation uniquely determines it.

By the *generalised Riemann existence theorem* (Grauert–Remmert, 1958), each finite cover X_N of X is an algebraic variety defined over $\overline{\mathbb{Q}}$. Thus each automorphism $\sigma \in \Gamma_{\mathbb{Q}}$ permutes the finite covers of X (or if you prefer, the normal subgroups N): it acts on X_N by acting simultaneously on the coefficients of all the defining polynomials of X_N.

Grothendieck [267] explained that $\Gamma_{\mathbb{Q}}$ acts on the profinite completion $\widehat{\pi}_1(X, p)$ of the fundamental group of X, called the *algebraic fundamental group* of X. This means the following. The profinite completion $\widehat{G}$ of a group G is the projective limit $\lim_{\leftarrow} G/N$ over all finite quotients G/N (i.e. N runs over all normal subgroups of finite index in G). An element $g \in \widehat{G}$ consists of a choice $g_N N$ of coset in G/N for each such N, such that whenever N_1 is a subgroup of N_2 then $g_{N_1} N_2 = g_{N_2} N_2$. This should remind us of the construction of the p-adic integers $\widehat{\mathbb{Z}}_p$ – indeed, $\widehat{\mathbb{Z}} = \prod_p \widehat{\mathbb{Z}}_p$ is the profinite completion of $\mathbb{Z}$. Profinite completion is the algebraic analogue of the topological completion of a space by Cauchy sequences (as in the construction of $\mathbb{R}$ from $\mathbb{Q}$). Its purpose is the same: just as $\mathbb{R}$ fills in the 'gaps' in $\mathbb{Q}$, so does $\widehat{G}$ supply the missing elements in G. For example, $\sqrt{2}$ exists in $\widehat{\mathbb{Z}}_7$ but not in $\mathbb{Z}$. Of course, being a projective limit, the profinite completion is also an 'integration' of all G/N, that is a way of treating them all simultaneously. A solution in $\widehat{\mathbb{Z}}$ to a polynomial equation gives us simultaneously a solution modulo any n.

For example, $\widehat{\ell} \in \widehat{\mathbb{Z}}$ corresponds, for each $n \in \mathbb{N}$, to an integer $\widehat{\ell}_n$ defined modulo n, subject to the obvious compatibility condition. Then an element $\widehat{\ell}$ is invertible, written $\widehat{\ell} \in \widehat{\mathbb{Z}}^\times$, iff for each $n > 1$, $\widehat{\ell}_n$ is invertible mod n. Hence any $\widehat{\ell} \in \widehat{\mathbb{Z}}^\times$ has a well-defined action on finite-order roots of unity: given any nth root of unity ξ, $\xi^{\widehat{\ell}}$ is defined to be $\xi^{\widehat{\ell}_n}$. In fact, consider the field $\overline{\mathbb{Q}}^{ab}$ obtained by taking the union of all cyclotomic fields (or equivalently, by Theorem 1.7.1, all abelian extensions of $\mathbb{Q}$). Its Galois group $\mathrm{Gal}(\overline{\mathbb{Q}}^{ab}/\mathbb{Q})$ can be naturally identified with the multiplicative group $\widehat{\mathbb{Z}}^\times$ in this way. This is just the action of $\Gamma_{\mathbb{Q}}$ restricted to cyclotomic fields – call this restriction the *cyclotomic character* $\chi^{cyclo} : \Gamma_{\mathbb{Q}} \to \widehat{\mathbb{Z}}^{\times}$ (this is a 'character' in the sense of a one-dimensional representation, not as a trace of a higher-dimensional character). This action has a large kernel – in fact, $\widehat{\mathbb{Z}}^{\times}$ is isomorphic to the abelianisation $\Gamma_{\mathbb{Q}}/[\Gamma_{\mathbb{Q}}\Gamma_{\mathbb{Q}}]$.

Let $\widehat{\gamma} \in \widehat{\pi}_1(X, p)$, that is for each finite-index normal subgroup N of $\pi_1(X, p)$, we have a coset representative $\widehat{\gamma}_N$ of some coset $\widehat{\gamma}_N N \in \pi_1(X, p)/N$ and these $\widehat{\gamma}_N$ – which we are to think of as permutations of finite sets $f_N^{-1}(p)$ – are compatible in the appropriate way. Then for any $\sigma \in \Gamma_{\mathbb{Q}}$ and $\widehat{\gamma} \in \pi_1(X, p)/N$, the action $\sigma.\widehat{\gamma}$ is defined by

$$(\sigma.\widehat{\gamma})_N = \sigma \circ \widehat{\gamma}_{\sigma^{-1}N} \circ \sigma^{-1}, \qquad (6.3.2)$$

where σ acts on the points in $f_N^{-1}(p) \subset \overline{\mathbb{Q}}^n$ component-wise, and acts on the normal subgroups N as above. As we will see, choosing the variety X appropriately, (6.3.2) includes the profinite analogue of the braid group action on $\mathcal{F}_n$ mentioned earlier: the image of $\Gamma_{\mathbb{Q}}$ in $\text{Aut}\,\widehat{\mathcal{F}}_n$ lies in this image of $\widehat{\mathcal{B}}_n$. Equation (6.3.2) generalises to an action of $\Gamma_{\mathbb{Q}}$ on the fundamental groupoids $\pi_1(X, p, q)$ of (homotopy equivalence classes of) paths in X with endpoints $p, q \in X(\mathbb{Q})$.

Now, generically $\pi_1(X, p)$ is isomorphic to the mapping class group $\Gamma_{g,n}$, when X is a surface of genus g with n punctures. By the *modular tower* we mean the collection of moduli spaces $\mathfrak{M}_{g,n}$, where the different spaces are related by the obvious topological actions such as forgetting marked points, or sewing surfaces together ('tower' means a family of objects linked by homomorphisms). In Section 2 of his *Esquisse d'un Programme*, Grothendieck conjectured that $\Gamma_{\mathbb{Q}}$ acts on the profinite completion of this tower (i.e. on the profinite completion of all $\Gamma_{g,n}$, and respecting those topological actions), and is in fact the full automorphism group of this completion, and that this provides an effective, almost combinatorial, way to study $\Gamma_{\mathbb{Q}}$, not directly related to its action on algebraic numbers. He conjectured that his profinite modular tower could be reconstructed from $\mathfrak{M}_{0,3}, \mathfrak{M}_{0,4}, \mathfrak{M}_{1,1}$, with all relations obtained from $\mathfrak{M}_{0,5}$ and $\mathfrak{M}_{1,2}$.

For example, the ordered moduli space $\mathfrak{M}_{0,4}$ is the thrice-punctured sphere $\mathbb{P}^1(\mathbb{C}) \setminus \{0, 1, \infty\}$ and can be defined over $\mathbb{Q}$ – indeed, it is just $\Gamma(2)\backslash\mathbb{H}$ and has defining equation $z_1 z_2^2(z_2 - 1)^2 = z_2^2 - z_2 + 1$. Its fundamental group is $\mathcal{F}_2$, the free group on two generators. Therefore, $\Gamma_{\mathbb{Q}}$ acts on $\widehat{\mathcal{F}}_2$. In fact, this action is known to be faithful (Belyi, 1987), so $\Gamma_{\mathbb{Q}}$ is a subgroup of $\text{Aut}\,\widehat{\mathcal{F}}_2$. Similarly, we get an action of $\Gamma_{\mathbb{Q}}$ on $\widehat{\mathcal{B}}_n$, which we will give shortly. This action yields one on $\widehat{\mathcal{B}}_n/Z(\widehat{\mathcal{B}}_n)$, and for $n = 3$ the latter equals the completion $\widehat{\text{PSL}_2(\mathbb{Z})}$ of the modular group (recall (1.1.10b)).

Does Moonshine (or if you prefer, RCFT or VOAs) see this same $\Gamma_{\mathbb{Q}}$-action? After all, modular data possesses a nice Galois action (6.1.7), as does the spectrum of the theory (6.1.15b). Also, Grothendieck's modular tower, with generators $(0, 3), (0, 4), (1, 1)$ and relations $(0, 5)$ and $(1, 2)$, reminds one of the Moore–Seiberg data of Section 6.1.4. There are a few difficulties with this hope. For instance, we should take profinite limits of these actions – for example, lift our action on $\text{SL}_2(\mathbb{Z})$ to one on $\widehat{\text{SL}_2(\mathbb{Z})}$. Can that have any natural meaning to RCFT? Also, and most disappointingly, the modular data always lies in cyclotomic fields, so the Galois action (6.1.7) in RCFT really only sees the rather uninteresting action of the abelianisation $\Gamma_{\mathbb{Q}}/[\Gamma_{\mathbb{Q}}\Gamma_{\mathbb{Q}}] \cong \widehat{\mathbb{Z}}^{\times}$, as explained earlier.

The first difficulty is easy to address. Subject to Conjecture 6.1.7, we obtain the following universal actions of $\Gamma_{\mathbb{Q}}$ on modular data: for any $\sigma \in \Gamma_{\mathbb{Q}}$,

$$\sigma.T = T^{\chi^{cyclo}(\sigma)}, \tag{6.3.3a}$$

$$\sigma.S = T^{\chi^{cyclo}(\sigma)} S T^{\chi^{cyclo}(\sigma^{-1})} S T^{\chi^{cyclo}(\sigma)} S^2. \tag{6.3.3b}$$

In order for (6.3.3) to make sense, these equations must live in the profinite completion of $SL_2(\mathbb{Z})$. This is the meaning of the profinite completions here: the 'integration' of the data of all RCFT (or VOAs) necessary for universal formulae. The generators S, T of $SL_2(\mathbb{Z})$ also generate $\widehat{SL_2(\mathbb{Z})}$, though in the topological sense (i.e. just as 1 topologically generates $\widehat{\mathbb{Z}}$). Since the action (6.3.3) is continuous, it defines a $\Gamma_{\mathbb{Q}}$-action on $\widehat{SL_2(\mathbb{Z})}$. It is very natural, in the sense that there is a map $\Gamma_{\mathbb{Q}} \to \widehat{SL_2(\mathbb{Z})}$ given by

$$\sigma \mapsto G_\sigma := T^{\chi^{cyclo}(\sigma)} S T^{\chi^{cyclo}(\sigma^{-1})} S T^{\chi^{cyclo}(\sigma)} S = \begin{pmatrix} \chi(\sigma) & 0 \\ 0 & \chi(\sigma^{-1}) \end{pmatrix} \in \widehat{SL_2(\mathbb{Z})}, \tag{6.3.4}$$

and $\sigma.S$ equals the matrix multiplication $G_\sigma S$. This map (6.3.4) is also what gives the Galois action (2.3.14) on modular functions for $\Gamma(N)$ or, in more suggestive language, the meromorphic functions on $\lim_{\leftarrow} \Gamma(N)\backslash\overline{\mathbb{H}}$ (see Section 2.4.1). Of course, in RCFT there is a preferred basis for this $\widehat{SL_2(\mathbb{Z})}$-representation (namely, that given by the VOA characters), and in that basis the matrices become signed permutation matrices $\epsilon_\sigma(a)\delta_{a^\sigma,b}$. It will be extremely interesting to find universal formulae for the Galois action on the remaining Moore–Seiberg data. The difficulty is that, in obtaining (6.3.3), we were guided by the presence of a preferred basis, and so (6.3.3) reduces to the usual Galois action on the corresponding matrices. For the braiding and fusing matrices, typically there isn't a preferred basis, and so other principles must be our guide.

Why do cyclotomic fields exhaust RCFT, hence demanding that the RCFT Galois action, unlike that on Grothendieck's modular tower, be far from faithful? Is it trying to tell us something? What other principles can guide us to a Galois action on the remaining Moore–Seiberg data?

Those questions lead us to Drinfel'd [161]. Recall from Section 1.6.2 that the pure braid group $\mathcal{P}_n$ acts on each set $\text{Hom}A_1 \oplus \cdots \oplus A_n, V)$ in any braided monoidal category. In particular, we can ask which subgroup of $\mathcal{P}_3 \times \mathcal{P}_2$ acts on the set of all braided monoidal categories, where $\beta \in \mathcal{P}_3$ and $y \in \mathcal{P}_2$ send the associativity constraint $a : (A \otimes B) \otimes C \to A \otimes (B \otimes C)$ and the commutativity constraint $c : A \otimes B \to B \otimes A$, respectively, of one such category to that of another. We require that $\beta.a$ and $\gamma.c$ satisfy the various axioms, most importantly the pentagon and hexagon equations.

Dualising this, Drinfel'd suggested to act with $\mathcal{P}_3 \times \mathcal{P}_2$ on the data of *quasi-triangular quasi-Hopf algebras* A (defined in e.g. [98]). These algebras are co-commutative up to conjugation by the R-matrix $\mathcal{R} \in A \otimes A$ (as in Definition 6.2.8), and co-associative up to conjugation by the *associator* $\Phi \in A \otimes A \otimes A$ (Φ measures how A fails to be Hopf). Φ and $\mathcal{R}$ are required to obey the triangle, pentagon and hexagon equations of Section 1.6.2. We met quasi-triangular Hopf algebras in Definition 6.2.8; it will be clear shortly why Drinfel'd prefers quasi-Hopf algebras. Identify $\mathcal{P}_2$ with $\mathbb{Z}$ and $\mathcal{P}_3$ with $\mathcal{F}_2 \times \mathbb{Z}$ (1.1.10c);

then $m \in \mathcal{P}_2$ acts on the R-matrix by $m.\mathcal{R} = \mathcal{R}.(\mathcal{R}_{21}\mathcal{R})^m$ and, for example, a word $f(x, y) \in \mathcal{F}_2 < \mathcal{P}_3$ acts on the associator by $f.\Phi = f(\mathcal{R}_{21}\mathcal{R}_{12}, \Phi\mathcal{R}_{32}\mathcal{R}_{23}\Phi^{-1})^{-1}\Phi$. The other quantities in the algebra A are left unchanged. Unfortunately, this nice idea fails: only the two elements $(\pm 1, 1) \in \mathcal{P}_2 \times \mathcal{P}_3$ satisfy the constraints and thus permute quasi-triangular quasi-Hopf algebras (the nontrivial one sending $\mathcal{R}$ to $\mathcal{R}_{21}$ and fixing everything else).

Drinfel'd then proposed that there would be more solutions if we take profinite completions (indeed, this is a *raison d'être* of completions), so in place of $\mathcal{P}_2 \cong \mathbb{Z}$ and $\mathcal{P}_3 \cong \mathbb{Z} \times \mathcal{F}_2$ we take $\widehat{\mathcal{P}_2} \cong \widehat{\mathbb{Z}}$ and $\widehat{\mathcal{P}_3} \cong \widehat{\mathbb{Z}} \times \widehat{\mathcal{F}_2}$. To get these profinite actions on the $\mathcal{R}$ and Φ, it suffices to take the scalars of the algebras A to be formal power series $\widehat{\mathbb{Q}}[[h]]$ rather than $\mathbb{C}$. The hope is that by completing the groups, there is more chance of nontrivial solutions to the triangle, pentagon and hexagon equations. The details would take us too far afield, but the result is that there are indeed several solutions.

Drinfel'd was interested in this because, in an earlier paper, he had found, for each choice of simple Lie algebra $\mathfrak{g}$, a universal formula for one solution $(\Phi, \mathcal{R})$ to those equations, using Kohno's monodromy theorem for the KZ connection. Unfortunately this formula for Φ is quite complicated. In [161] he investigates two commuting actions on the set of all solutions $(\Phi, \mathcal{R})$, which he uses to deduce the existence of a simpler solution (see [39]). One of these actions was this pure braid group action.

Let $\widehat{GT}$, the *Grothendieck–Teichmüller group*, be the group of all pairs $(\lambda, f) \in \widehat{\mathbb{Z}} \times \widehat{\mathcal{F}_2}$ (the $\widehat{\mathbb{Z}}$ of $\widehat{\mathcal{P}_3}$ can't contribute) satisfying those equations and thus permuting those quasi-triangular quasi-Hopf algebras. $\widehat{GT}$ is large, in fact as we will see $\Gamma_{\mathbb{Q}}$ embeds as a subgroup in it. Drinfel'd conjectured that $\widehat{GT}$ should act on the profinite completion of Grothendieck's tower. For example, on $\widehat{\mathcal{B}_n}$, topologically generated as we know by $\sigma_1, \ldots, \sigma_{n-1}$, we get the action by $(\lambda, f) \in \widehat{GT}$ given by

$$(\lambda, f).\sigma_i = f(y_i, \sigma_i^2)^{-1}\sigma_i^\lambda f(y_i, \sigma_i^2), \tag{6.3.5a}$$

$$(\lambda, f).Z = Z^\lambda, \tag{6.3.5b}$$

where $Z = (\sigma_{n-1}\cdots\sigma_1)^n$ topologically generates the centre of $\widehat{\mathcal{B}_n}$ (just as it does that of $\mathcal{B}_n$) and $y_i = \sigma_{i-1}\cdots\sigma_1^2\cdots\sigma_{i-1}$. This element y_i arises in presentations of the genus-0 mapping class groups $\Gamma_{0,n}$ or braid groups of the sphere [59]. The 'profinite word' $f(y_i, \sigma_i^2) \in \widehat{\mathcal{F}_2}$ means the value $\varphi(f)$ of the homomorphism $\varphi : \widehat{\mathcal{F}_2} \to \widehat{\mathcal{B}_n}$ defined by $\varphi(x) = y_i$ and $\varphi(y) = \sigma_i^2$.

Moreover, $\Gamma_{\mathbb{Q}}$ maps injectively into $\widehat{GT}$ and so can be identified with some subgroup of $\widehat{GT}$. Conjecturally, $\Gamma_{\mathbb{Q}}$ equals $\widehat{GT}$. For example, $(-1, 1)$ corresponds to complex-conjugation. See [305], [494], [493], [39] and section 16.4 of [98] for reviews of $\widehat{GT}$ and its action on, for example, the modular tower; [128] speculates on its relation to RCFT.

This is brought one step closer to RCFT by Kassel–Turaev [339]. It is relatively straightforward to extend Drinfel'd's action to certain braided monoidal categories. In [339] a 'pro-unipotent completion' $\widehat{\mathbf{R}}$ is defined for any ribbon category $\mathbf{R}$. $\widehat{\mathbf{R}}$ is itself a ribbon category, with the same objects as $\mathbf{R}$, but with each $\mathrm{Hom}(A, B)$ replaced by some projective limit of its linearisation over $\widehat{\mathbb{Q}} = \prod_p \widehat{\mathbb{Q}}_p$. For example, for the choice

R = **ribbon**, Hom($\emptyset$, $\emptyset$) can be identified with the space of formal finite linear combinations over $\widehat{\mathbb{Q}}$ of framed oriented links in $\mathbb{R}^3$. Drinfel'd's work yields an action of $\Gamma_{\mathbb{Q}}$ on the collection of these ribbon categories.

This category **ribbon** obeys a universality property as in Theorem 1.6.2. Now, any automorphism $\sigma \in \widehat{\Gamma_{\mathbb{Q}}}$ acts on the data of **ribbon** to produce a new ribbon category $\widehat{\textbf{ribbon}}^{\sigma}$. Its objects and Hom($\emptyset$, $\emptyset$) are unchanged. By universality, there is a functor from **ribbon** to **ribbon**$^{\sigma}$, sending Hom($\emptyset$, $\emptyset$) to itself. That is, we get an action of $\Gamma_{\mathbb{Q}}$ on the $\widehat{\mathbb{Q}}$-span of links: a (framed oriented) link L is taken to some linear combination (over $\widehat{\mathbb{Q}}$) of links. $\Gamma_{\mathbb{Q}}$ also acts on related spaces, such as $\widehat{\mathbb{Q}}$-valued Vassiliev invariants [**339**].

For example, complex-conjugation sends a link L to its mirror reflection (in general a link is not isotopic to its mirror reflection – see footnote 6 in chapter 1). However, [**339**] show that this $\Gamma_{\mathbb{Q}}$ action is trivial on the commutator $[\Gamma_{\mathbb{Q}}\Gamma_{\mathbb{Q}}]$, and thus really is an action of $\widehat{\mathbb{Z}}^{\times}$.

This action is clearly very similar to that of RCFT. As we know, RCFT attaches the matrix S to the Hopf link (Figure 6.9). Complex-conjugation ($\lambda = -1 \in \widehat{\mathbb{Z}}^{\times}$) sends the Hopf link to its mirror image; the mirror image corresponds to $\overline{S}$, which is what (6.3.3b) reduces to for $\lambda = -1$.

Problem *Identify the relation between [**339**] and the action (6.3.3) in RCFT. Can this be used somehow to identify the Galois action on arbitrary Moore–Seiberg data?*

We conjecture these actions are identical or at least very close. After all, they both factor through to $\widehat{\mathbb{Z}}^{\times}$ and agree with complex-conjugation applied to the Hopf link. Theorem 4 of [**40**] should make it possible to compute the [**339**] action on the Hopf link for any $\lambda \in \widehat{\mathbb{Z}}^{\times}$, thus allowing us to compare it directly to (6.3.3a). As we've learned, there are topological underpinnings of chiral RCFT data (e.g. the modular categories of [**534**], [**32**]) as well as full RCFT (see e.g. [**211**]); this seems the obvious way to attack this problem.

At least as interesting as this Galois action on the Moore–Seiberg data is that we can also hope that $\Gamma_{\mathbb{Q}}$ (or at least $\widehat{\mathbb{Z}}^{\times}$) will act on the spaces $\mathfrak{B}^{(g,n)}$ of chiral blocks, since they do on $\mathfrak{B}^{(1,1)}$, i.e. on the characters, which are modular functions (recall Section 2.3.3).

The Galois action (6.3.3) of RCFT is not directly related to Grothendieck's (6.3.5). The RCFT action would seem to be intimately related to Congruence Property 6.1.7, so more relevant to RCFT than $\widehat{SL_2(\mathbb{Z})}$ should be the much simpler $\lim_{\leftarrow} SL_2(\mathbb{Z})/\Gamma(N) = SL_2(\hat{\mathbb{Z}})$.

So far in this subsection we've only addressed CFT 'in the bulk'. What if anything does Galois do to, for example, D-branes? Indeed, an action persists in boundary RCFT, though it is more complicated [**235**]. In particular, *this Galois action will no longer be abelian* – the algebraic numbers involved belong to exponent-2 extensions of the cyclotomic field $\overline{\mathbb{Q}}^{ab}$. This complication opens the door to much more interesting mathematics.

It will be interesting to see if the $\hat{\mathbb{Z}}^{\times}$ action in [**106**] can be related to that of RCFT. We are to think of RCFT as being to generic quantum field theory what semi-simple finite-dimensional Lie algebras are to generic ones. In this spirit, this Galois action on RCFT, and its relation to $\Gamma_{\mathbb{Q}}$ and Grothendieck's *Esquisse*, can be regarded perhaps as a toy model for the much more ambitious Cosmic Galois Group of [**93**], which conjecturally underlies the multiple zeta values found by Kreimer and others in more physical quantum field theories.

As a final remark, it is quite possible that the Galois actions explored in this subsection are related to the Fermat remarks of last subsection (see in particular section II of [**304**]). The Fermat curve $F_N = \{x^N + y^N = 1\}$ $x^N + y^N = 1$ is an abelian cover of $\mathbb{P}^1(\mathbb{C}) \setminus \{0, 1, \infty\}$; in turn, its abelian covers are controlled by torsion points on its Jacobian Jac(F_N), and in [**304**] the action of $\Gamma_{\mathbb{Q}}$ on $\widehat{\mathcal{F}_2}$ is studied via those torsion points, with results somewhat reminiscent of Section 6.3.2.

Question 6.3.1. Use the fact that the S and T matrices of (6.1.8) define modular data to compute the sum $\sum_{m=1}^{n} e^{2\pi i m^2/n}$. (*Note*: This is called a Gauss sum. A similar calculation yields a generalisation of Gauss sums for any modular data.)

Question 6.3.2. Find all $\tau \in \mathbb{H}$ such that the torus $\mathbb{C}/(\mathbb{Z} + \mathbb{Z}\tau)$ is isogenous to $\mathbb{C}/(\mathbb{Z} + \mathbb{Z}i)$.

Question 6.3.3. Prove that the elliptic curves $y^2 = x^3 + ax$ and $y^2 = x^3 + b$ both have complex multiplication for any a, b.

Question 6.3.4. What is the profinite completion $\widehat{G}$ for finite groups G?

Question 6.3.5. (a) Define $\sigma_i.x_j = x_j$ if $j \neq i, i + 1$, and $\sigma_i.x_i = x_{i+1}$ and $\sigma_i.x_{i+1} = x_{i+1}^{-1}x_i x_{i+1}$. Verify that this is a well-defined action. (It turns out that this action is faithful.)
(b) Verify that for any $\beta \in \mathcal{B}_n$, β fixes $x_1 \cdots x_n$, and there is a permutation π_β and words $a_i \in \mathcal{F}_n$ such that $\beta.x_i = A_i x_{\pi_\beta} A_i^{-1}$. (It turns out that, conversely, any automorphism β obeying those two conditions must come from this braid group action. This gives a way to solve the word problem in $\mathcal{B}_n$.)

Question 6.3.6. Choosing X to be a sphere with two punctures, describe the associated $\Gamma_{\mathbb{Q}}$-action (6.3.2).

7

Monstrous Moonshine

Thomas Edison once said that to invent you need a good imagination and a pile of junk. Let's see what some imagination can do.

This book has been about *Moonshine*: a diverse collection of points-of-contact between algebra, number theory and mathematical physics, which nevertheless has a common theory. The most remarkable example of Moonshine is surely the association of Hauptmoduls with elements of the Monster $\mathbb{M}$. It is to this we finally turn.

The reader should reread the introductory chapter, which quickly sketches the basics of Monstrous Moonshine. In this chapter we explore this in more detail. The original article [**111**] is still very readable and contains a wealth of information not found in other sources. Other reviews are [**107**], [**410**], [**73**], [**154**], [**412**], [**249**], [**75**], [**469**], [**78**], [**237**] and the introductory chapter in [**201**], and each has its own emphasis.

7.1 The Monstrous Moonshine Conjectures

Recall from the introductory chapter the *McKay equation*

$$196\,884 = 196\,883 + 1. \tag{7.1.1}$$

The number on the left is the first nontrivial coefficient of the j-function, and the numbers on the right are the dimensions of the smallest irreducible representations of the Fischer–Griess Monster $\mathbb{M}$. On the one side, we have a modular function; on the other, a sporadic finite simple group. Monstrous Moonshine explores this completely unexpected connection between finite groups and modular functions.

The world is full of coincidences, and it isn't always clear how seriously they should be regarded. For instance, at the heart of Monstrous Moonshine is a holomorphic $c = 24$ VOA; the conjectured number of holomorphic $c = 24$ VOAs [**488**] is 71, and this is the largest prime dividing $\|\mathbb{M}\|$. There are 26 sporadics, 26 generators in a presentation of the Bimonster discussed shortly, and 26 conjugacy classes in the largest Mathieu group M_{24}. Are any of those numbers related to the 24 of Section 2.5.1, the k-group $\mathbb{Z}_{48}$ of the integers or the number (24) of 24-dimensional even self-dual lattices?[1]

Nor is physics immune to such thoughts. The great physicist Dirac noticed [**140**] that the ratio of the electrostatic to gravitational force between the proton and electron in a

---

[1] Perhaps this Mathieu group remark is related somehow to the fact that for subgroups G of $SL_3(\mathbb{C})$, the Euler number of a minimal resolution of the quotient singularity $\mathbb{C}^3/G$ equals the number of conjugacy classes of G [**143**], [**471**].

hydrogen atom is a number N of order 10^{40}. He computed that the ratio of the mass of the universe to the mass of a proton is roughly N^2, and that the ratio of the age of the universe with the time needed for light to travel across the classical radius of the electron is again roughly N. One can add that $\sqrt{N}$ is roughly Avagadro's number, so gives a measure of the minimum number of molecules needed in a macroscopic object. Dirac argued that the simple functional relation of these numbers indicates that they are all somehow physically related.

What distinguishes (7.1.1) from some of these other coincidences is that the more it was studied, the more the coincidences multiplied, and the more structure was revealed.

A noble goal for mathematics is surely to find interesting and fundamentally new theorems. Both history and common-sense suggest that to this end it is most profitable to look simultaneously at both exceptional structures and generic structures, to understand the special features of the former in the context of the latter, and to be led in this way to a new generation of exceptional and generic structures. That is the spirit in which Monstrous Moonshine should be studied.

7.1.1 The Monster revisited

Recall the finite simple group classification discussed in Section 1.1.2. The sporadics are summarised in Table 7.1 (its dates are only approximate and the list of investigators is taken from [**109**]). The Monster $\mathbb{M}$ is the largest of these 26 sporadic groups. Its existence was conjectured in 1973 by Fischer and Griess, and finally constructed (somewhat artificially) in 1980 by Griess [**263**]. Tits [**528**] showed that $\mathbb{M}$ is the automorphism group of a 196 883-dimensional commutative non-associative algebra also constructed by Griess and now called the *Griess algebra* (Griess showed only that $\mathbb{M}$ was a subgroup of that automorphism group). We now understand the Griess algebra as the first nontrivial tier (0-mode algebra) of a VOA, the *Moonshine module $V^\natural$*, lying at the heart of Monstrous Moonshine.

The Monster has 194 conjugacy classes, and so that number of irreducible representations. Its character table (and other useful information) is given in the Atlas [**109**], where we also find analogous data for the other simple groups of 'small' order. Table 7.2 gives the upper-left 0.25% or so of the character table of $\mathbb{M}$. The name '4C', for example, is given to the third smallest (hence 'C') conjugacy class of elements of order 4. Table 7.2 tells us that the dimensions of the smallest irreducible representations of $\mathbb{M}$ are 1, 196 883, 21 296 876 and 842 609 326.

The centralisers $C_G(g)$ of conjugate elements are isomorphic (why?). The centralisers for all classes of order up to 11 are given in table 2a of [**111**]. The first few are $C_\mathbb{M}(2A) \cong 2.\mathbb{B}$, $C_\mathbb{M}(2B) \cong 2^{25}.Co_1$, $C_\mathbb{M}(3A) \cong 3.Fi'_{24}$, $C_\mathbb{M}(3B) \cong 3^{13}.2.Suz$, $C_\mathbb{M}(3C) \cong 3 \times Th$. We follow the notation of [**109**]: by, for example, '2.$\mathbb{B}$' we mean a group with $\mathbb{Z}_2$ as a normal subgroup and $\mathbb{B}$ as the quotient, or equivalently an extension of $\mathbb{B}$ by $\mathbb{Z}_2$. Of course the centraliser $C_G(g)$ has $\langle g \rangle$ as a subgroup of its centre, hence $\langle g \rangle$ is normal in $C_G(g)$ – that is, for example, the '2' in 2.$\mathbb{B}$. Knowing the centraliser, the sizes of the conjugacy

Table 7.1. *The 26 sporadic groups*

Group	Exact order	Approximate order	Investigators
M_{11}	$2^4.3^2.5.11$	7.9×10^3	Mathieu (1861, 1873)
M_{12}	$2^6.3^3.5.11$	9.5×10^4	Mathieu (1861, 1873)
J_1	$2^3.3.5.7.11.19$	1.8×10^5	Janko (1965)
M_{22}	$2^7.3^2.5.7.11$	4.4×10^5	Mathieu (1861, 1873)
J_2	$2^7.3^3.5^2.7$	6.0×10^5	Hall, Janko (1960s)
M_{23}	$2^7.3^2.5.7.11.23$	1.0×10^7	Mathieu (1861, 1873)
HS	$2^9.3^2.5^3.7.11$	4.4×10^7	Higman, Sims (1968)
J_3	$2^7.3^5.5.17.19$	5.0×10^7	Janko, Higman, McKay (1960s)
M_{24}	$2^{10}.3^3.5.7.11.23$	2.4×10^8	Mathieu (1861, 1873)
McL	$2^7.3^6.5^3.7.11$	9.0×10^8	McLaughlin (1969)
He	$2^{10}.3^3.5^2.7^3.17$	4.0×10^9	Held, Higman, McKay (1960s)
Ru	$2^{14}.3^3.5^3.7.13.29$	1.5×10^{11}	Rudvalis, Conway, Wales (1973)
Suz	$2^{13}.3^7.5^2.7.11.13$	4.5×10^{11}	Suzuki (1969)
$O'N$	$2^9.3^4.5.7^3.11.19.31$	4.6×10^{11}	O'Nan, Sims (1970s)
Co_3	$2^{10}.3^7.5^3.7.11.23$	5.0×10^{11}	Conway (1968)
Co_2	$2^{18}.3^6.5^3.7.11.23$	4.2×10^{13}	Conway (1968)
Fi_{22}	$2^{17}.3^9.5^2.7.11.13$	6.5×10^{13}	Fischer (1970s)
HN	$2^{14}.3^6.5^6.7.11.19$	2.7×10^{14}	Harada, Norton, Smith (1975)
Ly	$2^8.3^7.5^6.7.11.31.37.67$	5.2×10^{16}	Lyons, Sims (1972)
Th	$2^{15}.3^{10}.5^3.7^2.13.19.31$	9.1×10^{16}	Thompson, Smith (1975)
Fi_{23}	$2^{18}.3^{13}.5^2.7.11.13.17.23$	4.1×10^{18}	Fischer (1970s)
Co_1	$2^{21}.3^9.5^4.7^2.11.13.23$	4.2×10^{18}	Conway, Leech (1968)
J_4	$2^{21}.3^3.5.7.11^3.23.29.31.37.43$	8.7×10^{19}	Janko, Norton, Parker, Benson, Conway, Thankray (1970s)
Fi'_{24}	$2^{21}.3^{16}.5^2.7^3.11.13.17.23.29$	1.3×10^{24}	Fischer (1970s)
$\mathbb{B}$	$2^{41}.3^{13}.5^6.7^2$ $.11.13.17.19.23.31.47$	4.2×10^{33}	Fischer, Sims, Leon (1970s)
$\mathbb{M}$	$2^{46}.3^{20}.5^9.7^6.11^2.13^3$ $.17.19.23.29.31.41.47.59.71$	8.1×10^{53}	Fischer, Griess (1973, 1982)

classes can be quickly determined through the formula $\|K_g\| = \|\mathbb{M}\|/\|C_\mathbb{M}(g)\|$. These centralisers play a large role in Section 7.3 below.

The Monster $\mathbb{M}$ has a remarkably simple presentation. As with any noncyclic finite simple group, it is generated by its involutions (i.e. elements of order 2) and so is a homomorphic image of a Coxeter group (Definition 3.2.1) – see Question 7.1.1.

Let $\mathcal{G}_{pqr}$, $p \geq q \geq r \geq 2$, be the graph consisting of three strands of lengths $p + 1, q + 1, r + 1$, sharing a common endpoint. Label the $p + q + r + 1$ nodes as in Figure 7.1 (this labelling is not standard). Given any graph $\mathcal{G}_{pqr}$, define Y_{pqr} to be the group consisting of a generator for each node, obeying the usual Coxeter group relations, together with an additional one (what Conway calls the 'spider relation'):

$$(ab_1b_2ac_1c_2ad_1d_2)^{10} = 1. \tag{7.1.2}$$

The relation (7.1.2) arises naturally in a generalisation of the Coxeter group due to Conway, called a *fabulous group*. Conway conjectured and, building on work by

Table 7.2. *The north-west corner of the Monster character table*

ch\K_g	1A	2A	2B	3A	3B	3C	4A	4B	4C	4D	5A	5B
ρ_0	1	1	1	1	1	1	1	1	1	1	1	1
ρ_1	196883	4371	275	782	53	−1	275	51	19	−13	133	8
ρ_2	21296876	91884	−2324	7889	−130	248	1772	−52	−20	12	626	1
ρ_3	842609326	1139374	12974	55912	−221	−248	8878	782	−82	78	2451	−49
ρ_4	18538750076	8507516	123004	249458	1598	248	28796	2652	380	156	6326	76
ρ_5	19360062527	9362495	−58305	297482	1508	−247	35903	−833	63	−65	8152	27
ρ_6	293553734298	53981850	98970	1055310	−3927	3876	94874	1274	−102	−454	17423	−77
ρ_7	3879214937598	337044990	−690690	4751823	−4173	−3876	345598	−3874	−258	286	54473	98
ρ_8	36173193327999	1354188159	2864511	12616074	18954	0	701823	20383	−897	351	91124	−126
ρ_9	1255107227015275	3215883115	1219435	24688454	−25375	248	1223531	19499	−661	−1365	145275	−350

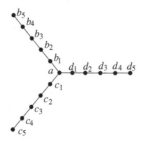

Fig. 7.1 The graph $\mathcal{G}_{555}$ presenting the Bimonster.

Ivanov [311], Norton proved [451] that $Y_{555} \cong Y_{444}$ is the *Bimonster*, the wreathed-square $\mathbb{M} \wr \mathbb{Z}_2 \cong (\mathbb{M} \times \mathbb{M}).2$ of the Monster (in fact it is a semi-direct product $(\mathbb{M} \times \mathbb{M}) \rtimes \mathbb{Z}_2$). We define the wreath product in Question 7.1.2; the wreathed-square $\mathbb{M} \wr \mathbb{Z}_2$ has $G = \mathbb{M}$ and $H = S = \mathbb{Z}_2$, where H acts on S by group multiplication. The group-theoretic significance of the wreath product is that any group G containing a normal subgroup N with quotient $G/N \cong H$ can be identified with a subgroup of $N \wr H$ with $S = H$. Thus any extension $\mathbb{M}.2$ of $\mathbb{Z}_2$ by $\mathbb{M}$ is a subgroup of the Bimonster. The Bimonster appears naturally in Section 7.3.9. A closely related presentation of the Bimonster has 26 involutions as generators and has relations given by the incidence graph of the projective plane of order 3; the Monster itself arises from 21 involutions and the affine plane of order 3. See [112] for details.

The groups Y_{pqr}, for $p \leq 5$, have now all been identified – see [312] for a unified treatment. The ones involving sporadic groups are

$$Y_{553} \cong Y_{443} \cong \mathbb{M} \times \mathbb{Z}_2,$$
$$Y_{533} \cong Y_{433} \cong \mathbb{Z}_2 \times (2.\mathbb{B}),$$
$$Y_{552} \cong Y_{442} \cong 3.(Fi'_{24}.2),$$
$$Y_{532} \cong Y_{432} \cong \mathbb{Z}_2 \times Fi_{23},$$
$$Y_{332} \cong \mathbb{Z}_2 \times (2.Fi_{22}).$$

The Coxeter groups of the graphs $\mathcal{G}_{555}$, $\mathcal{G}_{553}$, $\mathcal{G}_{533}$, $\mathcal{G}_{552}$ and $\mathcal{G}_{532}$ are all infinite groups of hyperbolic reflections in, for example, $\mathbb{R}^{17,1}$, and contain copies of groups such as the affine E_8 Weyl group, so there should be rich geometry here.

What role, if any, these remarkable presentations have in Monstrous Moonshine hasn't been established yet. As a first step though, [424] has found in the automorphism group of the Moonshine module $V^\natural$ the 21 involutions generating $\mathbb{M}$. Perhaps this can simplify the hardest part of [201] (see Section 7.2.1 below). Indeed, Miyamoto's simplified construction [427] of $V^\natural$ and proof that $\mathrm{Aut}(V^\natural) \cong \mathbb{M}$ uses Ivanov's characterisation [311] of $\mathbb{M}$. There is a correspondence [425] between certain involutions of a VOA $\mathcal{V}$ (e.g. class 2A in $\mathbb{M}$ for $V^\natural$) and certain vertex operator subalgebras of $\mathcal{V}$ isomorphic to the unique $c = 1/2$ rational VOA (the Ising model of Section 4.3.2); this technical tool has many applications, for example the association of various vertex operator superalgebras to $V^\natural$, and the VOA interpretation of McKay's $E_8^{(1)}$ observation in Section 7.3.6.

7.1.2 Conway and Norton's fundamental conjecture

As mentioned in the introductory chapter, the central structure in the attempt to understand equations (0.2.1) is an infinite-dimensional graded module for the Monster, $V = V_{-1} \oplus V_1 \oplus V_2 \oplus \cdots$, with graded dimension $J(\tau) = j(\tau) - 744$ (see (0.3.2)). If we let ρ_d denote the dth smallest irreducible $\mathbb{M}$-module, numbered as in Table 7.2, then the first few subspaces will be $V_0 = \rho_0$, $V_1 = \{0\}$, $V_2 = \rho_0 \oplus \rho_1$, $V_3 = \rho_0 \oplus \rho_1 \oplus \rho_2$ and $V_4 = \rho_0 \oplus \rho_0 \oplus \rho_1 \oplus \rho_1 \oplus \rho_2 \oplus \rho_3$. As we know from Section 1.1.3, a dimension can (and should) be twisted, by replacing it with the character. This gives us the graded traces

$$T_g(\tau) := \mathrm{ch}_{V_{-1}}(g) q^{-1} + \sum_{n=1}^{\infty} \mathrm{ch}_{V_n}(g) q^n, \qquad (7.1.3)$$

called the *McKay–Thompson series* for this module V. Of course, $T_e = J$.

Conjecture 7.1.1 (Conway–Norton [111]) *There exists a graded $\mathbb{M}$-module V such that, for each element g of the Monster $\mathbb{M}$, the McKay–Thompson series T_g is the Hauptmodul*

$$J_{\Gamma_g}(\tau) = q^{-1} + \sum_{n=1}^{\infty} a_n(g) q^n \qquad (7.1.4)$$

for a genus-0 group Γ_g of Moonshine-type. These groups each contain $\Gamma_0(N)$ as a normal subgroup, for some N dividing $o(g) \gcd(24, o(g))$, and the quotient group $\Gamma_g / \Gamma_0(N)$ has exponent ≤ 2.

So for each n the map $g \mapsto a_n(g)$ is a character $\mathrm{ch}_{V_n}(g)$ of $\mathbb{M}$. The quantity $o(g)$ is the order of g. We defined the groups of Moonshine-type in Definition 2.2.4 and $\Gamma_0(N)$ in (2.2.4b). By the exponent of a group we mean the smallest positive m such that $h^m = 1$ for all h in the group. [111] explicitly identify each of the groups Γ_g. The first 50 coefficients $a_n(g)$ of each T_g are given in [413]. Together with the recursions given in Section 7.1.4 below, this allows one to effectively compute arbitrarily many coefficients $a_n(g)$ of the Hauptmoduls. It is also this that uniquely defines V, up to equivalence, as a graded $\mathbb{M}$-module.

There are around 8×10^{53} elements in the Monster, so naively we may expect about 8×10^{53} different Hauptmoduls T_g. However, a character evaluated at g and at hgh^{-1} will always be equal, so $T_g = T_{hgh^{-1}}$. Hence there can be at most 194 distinct T_g (one for each conjugacy class). All coefficients $a_n(g)$ are integers (as are in fact most entries of the character table of $\mathbb{M}$). This implies that $T_g = T_h$ whenever the cyclic subgroups $\langle g \rangle$ and $\langle h \rangle$ are equal (why?). In fact, the total number of distinct McKay–Thompson series T_g arising in Monstrous Moonshine turns out to be only 171.

Of those many redundancies among the T_g, only one is unexpected (and unexplained): the McKay–Thompson series of two unrelated classes of order 27, namely 27A and 27B, are equal. It would be interesting to understand what general phenomenon (if any) is responsible for $T_{27A}(\tau) = T_{27B}(\tau)$. But as we know from Section 5.3.3, the McKay–Thompson series $T_g(\tau)$ are actually specialisations of 1-point functions and as such are functions of not only τ but of all $\mathbb{M}$-invariant vectors v in $V^{\natural}$. What we call $T_g(\tau)$

is really the specialisation $T_g(\tau, \mathbf{1})$ of this function $T_g(\tau, v)$. All 194 T_g (one for each conjugacy class) will be linearly independent, if we include this $v \in (V^\natural)^{\mathbb{M}}$ dependence. Thus the equality $T_{27A}(\tau) = T_{27B}(\tau)$ should be regarded as an accidental redundancy caused by specialisation, and is not of any deep significance. Plenty of other Norton's series $N_{(g,h)}(\tau)$ (Section 7.3.2) will likewise be accidentally equal. Modular aspects of the 1-point functions $T_g(\tau, v)$ are studied in [155].

Recall that there are two different conjugacy classes of order 2 elements: 2A and 2B. Class 2B corresponds to $\Gamma_0(2)$ and gives the Hauptmodul J_2 in (2.2.17a), while class 2A corresponds to $\Gamma_0(2)+$, where for any prime p we define

$$\Gamma_0(p)+ := \left\langle \Gamma_0(p), \frac{1}{\sqrt{p}} \begin{pmatrix} 0 & -1 \\ p & 0 \end{pmatrix} \right\rangle. \tag{7.1.5}$$

Similarly, (2.2.17b) corresponds to an order 13 element in $\mathbb{M}$, but J_{25} in (2.2.17c) doesn't equal any T_g. Recall that there are exactly 616 Hauptmoduls of Moonshine-type with integer coefficients [121], so most of these don't arise as T_g. Recently [110], a fairly simple characterisation has been found of the groups arising as Γ_g in Monstrous Moonshine:

Proposition 7.1.2 [110]   *A subgroup G of $SL_2(\mathbb{R})$ equals one of the modular groups Γ_g appearing in Conjecture 7.1.1, iff:*

(i) *G is genus 0;*
(ii) *G has the form '$\Gamma_0(n||h) + e, f, g, \ldots$';*
(iii) *the quotient of G by $\Gamma_0(nh)$ is a group of exponent ≤ 2; and*
(iv) *each cusp $\mathbb{Q} \cup i\infty$ can be mapped to $i\infty$ by an element of $SL_2(\mathbb{R})$ that conjugates the group to one containing $\Gamma_0(nh)$.*

The notation in (ii) is a little too technical to explain here, but it is given in [111] or [110]. We now understand the significance, in the VOA or CFT framework, of transformations in $SL_2(\mathbb{Z})$ (see especially Section 5.3.6), but (ii) emphasises that many modular transformations relevant to Moonshine are more general (called *Atkin–Lehner involutions*). Monstrous Moonshine will remain mysterious until we can understand its Atkin–Lehner symmetries. This isn't a hopeless task – for example, [433] provides an early attempt at studying string theories with Atkin–Lehner symmetries, as well as its possible physical significance. Some of these involutions appear naturally in Weil's Converse Theorem (see e.g. page 64 of [90]). Perhaps a topological interpretation for the groups Γ_g not contained in $SL_2(\mathbb{Z})$, in the spirit of Section 2.4.3, will help us understand their relevance in VOAs and the meaning of Atkin–Lehner involutions to CFT. This proposition is the answer to an important question, but unfortunately their proof of this characterisation is by exhaustion, and so by itself doesn't contribute anything conceptually.

7.1.3 E_8 and the Leech

There are other less important conjectures in [111]. We've already seen easy-to-understand relations of E_8 and the Leech lattice Λ to the J-function: (0.5.1) (explained in Section 3.2.3) and (0.5.2) (explained in Question 2.2.7). There is another way E_8 and Λ can be related to modular functions.

Lattices are related to groups through their automorphism groups, which are always finite for positive-definite lattices. The automorphism group $\text{Aut}(\Lambda) = Co_0$ of the Leech lattice has order about 8×10^{18}, and is a central extension by $\mathbb{Z}_2$ of Conway's simple group Co_1. Several other sporadic groups are also involved in Co_0, as we'll see in Section 7.3.1. To each automorphism $\alpha \in Co_0$, let θ_α denote the theta function of the sublattice of Λ fixed by α. Conway–Norton also associate with each automorphism α a certain function $\eta_\alpha(\tau)$ of the form $\prod_i \eta(a_i\tau)/\prod_j \eta(b_j\tau)$ built out of the Dedekind eta function (2.2.6b). Both θ_α and η_α are constant on each conjugacy class in Co_0, of which there are 167. [**111**] remarks that the ratio $\theta_\alpha/\eta_\alpha$ always seems to equal some McKay–Thompson series $T_{g(\alpha)}$.

It turns out that this observation isn't quite correct [**366**]. For each automorphism $\alpha \in Co_0$, the subgroup of $\text{SL}_2(\mathbb{R})$ that fixes $\theta_\alpha/\eta_\alpha$ is indeed always genus 0, but for exactly 15 conjugacy classes in Co_0, $\theta_\alpha/\eta_\alpha$ is not the Hauptmodul. Nevertheless, this construction proved useful for establishing Moonshine for M_{24} [**407**].

Similarly, one can ask this for the E_8 root lattice, whose automorphism group is the Weyl group of the Lie algebra E_8 (of order 696 729 600). The automorphisms of the lattice E_8 that yield a Hauptmodul were classified in [**95**]. On the other hand, Koike established a Moonshine of this kind for the groups $\text{PSL}_2(\mathbb{F}_7)$, $\text{PSL}_2(\mathbb{F}_5) \cong \mathcal{A}_5$ and $\text{PSL}_2(\mathbb{F}_3)$, of order 168, 60 and 12, respectively [**356**].

7.1.4 Replicable functions

A conjecture in [**111**] that played an important role in ultimately proving the main conjecture involves the *replication formulae*. Conway–Norton want to think of the Hauptmoduls T_g as being intimately connected with $\mathbb{M}$; if so, then the group structure of $\mathbb{M}$ should somehow directly relate different T_g. Considering the power map $g \mapsto g^n$ leads to the following.

It was well known classically that $J(\tau)$ (equivalently, $j(\tau)$) has the property that

$$s(\tau) := J(p\tau) + J\left(\frac{\tau}{p}\right) + J\left(\frac{\tau+1}{p}\right) + \cdots + J\left(\frac{\tau+p-1}{p}\right) \qquad (7.1.6a)$$

is a polynomial in $J(\tau)$, for any prime p. The proof is straightforward, and is based on the principle that the easiest way to construct a function invariant with respect to some group G is by averaging it over the group: $\sum_{g \in G} f(g.x)$. Here $f(x)$ is $J(p\tau)$ and G is $\text{SL}_2(\mathbb{Z})$, and we'll average over finitely many cosets rather than infinitely many elements. First, writing Γ for $\text{SL}_2(\mathbb{Z})$, note that

$$\Gamma \begin{pmatrix} p & 0 \\ 0 & 1 \end{pmatrix} \Gamma = \begin{pmatrix} p & 0 \\ 0 & 1 \end{pmatrix} \Gamma \cup \bigcup_{i=0}^{p-1} \begin{pmatrix} 1 & i \\ 0 & p \end{pmatrix} \Gamma = \{A \in M_{2\times 2}(\mathbb{Z}) \mid \det(A) = p\}.$$

$$\qquad (7.1.6b)$$

In Question 7.1.4 you show that this implies (7.1.6a) is a modular function for $\text{SL}_2(\mathbb{Z})$. Hence $s(\tau)$ equals a rational function $Q(J(\tau))/P(J(\tau))$ of $J(\tau)$, as in (0.1.7). Because the only poles of J are at the cusps, the same applies to $s(\tau)$. This implies that the denominator polynomial $P(z)$ must be trivial (recall that $J(\mathbb{H}) = \mathbb{C}$). QED

The map $J(\tau) \mapsto s(\tau)$ in (7.1.6a) is called a 'Hecke operator', and is an important ingredient of modular theory. More generally, the same argument says

$$\sum_{ad=n, 0 \leq b < d} J\left(\frac{a\tau + b}{d}\right) = Q_n(J(\tau)), \tag{7.1.7}$$

where Q_n is the unique polynomial for which $Q_n(J(\tau)) - q^{-n}$ has a q-expansion with only strictly positive powers of q. For example, $Q_2(x) = x^2 - 2a_1$ and $Q_3(x) = x^3 - 3a_1 x - 3a_2$, where we write $J(\tau) = \sum_n a_n q^n$. These equations (7.1.7) can be rewritten into recursions such as $a_4 = a_3 + (a_1^2 - a_1)/2$, or collected together into the remarkable expression (3.4.7a).

Conway and Norton conjectured that these formulae have an analogue for any McKay–Thompson series T_g. In particular, (7.1.7) becomes

$$\sum_{ad=n, 0 \leq b < d} T_{g^a}\left(\frac{a\tau + b}{d}\right) = Q_{n,g}(T_g(\tau)), \tag{7.1.8a}$$

where $Q_{n,g}$ plays the same role for T_g that Q_n plays for J. For example, we get

$$T_{g^2}(2\tau) + T_g\left(\frac{\tau}{2}\right) + T_g\left(\frac{\tau + 1}{2}\right) = T_g(\tau)^2 - 2a_1(g),$$

$$T_{g^3}(3\tau) + T_g\left(\frac{\tau}{3}\right) + T_g\left(\frac{\tau + 1}{3}\right) + T_g\left(\frac{\tau + 2}{3}\right) = T_g(\tau)^3 - 3a_1(g)\,T_g(\tau) - 3a_2(g).$$

These are called the replication formulae. Again, these yield recursions like $a_4(g) = a_2(g) + (a_1(g)^2 - a_1(g^2))/2$, or can be collected into the expression

$$p^{-1} \exp\left[-\sum_{k>0}\sum_{\substack{m>0 \\ n \in \mathbb{Z}}} a_{mn}(g^k)\frac{p^{mk} q^{nk}}{k}\right] = T_g(z) - T_g(\tau). \tag{7.1.8b}$$

This looks a lot more complicated than (3.4.7a), but you can glimpse the Taylor expansion of $\log(1 - p^m q^n)$ there and in fact for $g = e$, (7.1.8b) reduces to (3.4.7a).

Axiomatising (7.1.8a) leads to Conway and Norton's notion of *replicable function* [**449**], [**6**].

Definition 7.1.3 *Let f be any function of the form $f(\tau) = q^{-1} + \sum_{n=1}^{\infty} b_n q^n$, and write $f^{(1)} = f$ and $b_n^{(1)} = b_n$. Let $Q_{n,f}$ be the unique (degree n) polynomial such that the q-expansion of $Q_{n,f}(f(\tau)) - q^{-n}$ has only positive powers of q. Use*

$$\sum_{ad=n, 0 \leq b < d} f^{(a)}\left(\frac{a\tau + b}{d}\right) = Q_{n,f}\left(f^{(1)}(\tau)\right), \tag{7.1.9}$$

to recursively define each $f^{(n)}$. If each $f^{(n)}$ has a q-expansion of the form $f^{(n)}(\tau) = q^{-1} + \sum_{k=1}^{\infty} b_k^{(n)} q^k$ – that is, no fractional powers of q arise – then we call f replicable.

Proposition 7.1.4 [**6**] *Suppose f is of the form $f(\tau) = q^{-1} + \sum_{n=1}^{\infty} a_n q^n$, and define $Q_{n,f}$ as in Definition 7.1.3. Define $H_{m,n}$ by*

$$Q_{n,f}(f(\tau)) = q^{-n} + \sum_{n=1}^{\infty} n H_{n,m}\, q^m.$$

Then f is replicable iff $H_{n,m} = H_{r,s}$ holds whenever $mn = rs$ and $\gcd(n, m) = \gcd(r, s)$.

The proof isn't hard: if f is replicable, with replicates $f^{(n)} = q^{-1} + \sum_k a_k^{(n)} q^k$, then

$$H_{n,m} = \sum_{d | \gcd(n,m)} \frac{1}{d} a^{(d)}_{nm/d^2}$$

and the $H_{n,m} = H_{r,s}$ property is manifest. See Question 7.1.5 for the converse.

Equation (7.1.8a) conjectures that the McKay–Thompson series are replicable. In particular, we have $(T_g)^{(n)}(\tau) = T_{g^n}(\tau)$. [**123**] proved that the Hauptmodul of any genus-0 modular group of Moonshine-type is replicable, provided its coefficients are rational. Incidentally, if the coefficients $b_k^{(1)}$ are irrational, then Definition 7.1.3 should be modified to include Galois automorphisms (see section 8 of [**114**]). Replication in positive genus is discussed in [**510**].

Conversely, Norton has conjectured:

Conjecture 7.1.5 *Any replicable function with rational coefficients is either a Hauptmodul for a genus-0 modular group of Moonshine-type, or is one of the 'modular fictions' $f(\tau) = q^{-1} = \exp[-2\pi i \tau]$, $f(\tau) = q^{-1} + q = 2\cos[2\pi\tau]$, $f(\tau) = q^{-1} - q = -2i\sin[2\pi\tau]$.*

This conjecture seems difficult and is still open.

As is manifest in (7.1.8a), replication concerns the power map $g \mapsto g^n$ in $\mathbb{M}$. Can Moonshine see more of the group structure of $\mathbb{M}$? One step in this direction is explored in Section 7.3.6, where McKay models products of conjugacy classes using Coxeter–Dynkin diagrams. A different idea is given in Section 7.3.2. It would be very desirable to find other direct connections between the group operation in $\mathbb{M}$ and, for example, the McKay–Thompson series.

Question 7.1.1. Let G be a finite simple group, and let $K \neq \{e\}$ be any nontrivial conjugacy class. Prove that K generates G. Why is any noncyclic finite simple group a homomorphic image of a (possibly infinite) Coxeter group?

Question 7.1.2. Let G, H be any groups, and S any finite set on which H acts. By the wreath product $G \wr H$ we mean the set of all pairs (f, h), where f is any function from $S \to G$ and $h \in H$. Group multiplication is given by $(f, h)(f', h') = (f'', hh')$, where $f'' : S \to G$ is defined by $f''(s) = f(s)f'(h^{-1}.s)$.
(a) Verify that $G \wr H$ is a group. Compute its order.
(b) Find a normal subgroup in $G \wr H$, isomorphic to $G \times \cdots \times G$ ($\|S\|$ times). Identify the quotient of $G \wr H$ by this normal subgroup.
(c) Find a subgroup of $G \wr H$ isomorphic to H.

Question 7.1.3. Note that the dimensions 196 883 and 21 296 876 – see (0.2.1) – exactly divide the order of the Monster – see (0.2.2). Is this (i) merely a coincidence; (ii) a mysterious property of $\mathbb{M}$ perhaps relevant to Moonshine; or (iii) does it have a more mundane explanation?

Question 7.1.4. Prove (7.1.6b). Use that to prove that the sum $s(\tau)$ in (7.1.6a) is invariant under $\mathrm{SL}_2(\mathbb{Z})$.

Question 7.1.5. Complete the proof of Proposition 7.1.4.

Question 7.1.6. Suppose $f(\tau) = q^{-1} + \sum_{k=1}^{N} a_k q^k$ is a replicable Laurent polynomial. Prove that f is a modular fiction: $f(\tau) = q^{-1}$ or $f(\tau) = q^{-1} \pm q$.

Question 7.1.7. As we know from Section 3.2.3, $j^{\frac{1}{3}}$ is the graded dimension of the $E_8^{(1)}$-module $L(\omega_0)$. Thus j is the graded dimension of $L(\omega_0) \otimes L(\omega_0) \otimes L(\omega_0)$, on which the Lie group $(E_8(\mathbb{C}) \times E_8(\mathbb{C}) \times E_8(\mathbb{C})) \rtimes S_3$ acts. Explain why $L(\omega_0) \otimes L(\omega_0) \otimes L(\omega_0)$ cannot be the $\mathbb{M}$-module V whose graded characters (7.1.3) are the McKay–Thompson series (ignoring the irrelevant constant 744).

7.2 Proof of the Monstrous Moonshine conjectures

At first glance, any deep significance to the Moonshine conjectures seems very unlikely: they constitute after all a finite set of very specialised coincidences. The whole point though is to try to understand *why* such seemingly incomparable objects as the Monster and the Hauptmoduls can be so related, and to try to extend and apply this understanding to other contexts. Establishing the truth (or falsity) of the conjectures was merely meant as an aid to uncovering the meaning of Monstrous Moonshine. Indeed, in proving them, important new algebraic structures were formulated. We sketch this proof in this section.

The main Conway–Norton conjecture was attacked almost immediately. Thompson showed [524] (see also [476]) that if $g \mapsto a_n(g)$ is a character for all sufficiently small n (apparently $n \leq 1300$ is sufficient), then it will be for all n. He also showed that if certain congruence conditions hold for a certain number of $a_n(g)$ (all with $n \leq 100$), then all $g \mapsto a_n(g)$ will be *virtual* characters (i.e. differences of true characters of $\mathbb{M}$). Atkin, Fong and Smith (see [511] for details) used that and a computer to prove that indeed all $a_n(g)$ were virtual characters (they didn't quite get to $n = 1300$ though). But their work doesn't say anything more about the underlying (possibly virtual) representation V, other than its existence, and so adds no light to Moonshine. It plays no role in the following.

We want to prove Conjecture 7.1.1, that is, show that the McKay–Thompson series $T_g(\tau)$ of (7.1.3) equals the Hauptmodul $J_{\Gamma_g}(\tau)$ in (7.1.4). First, we need to construct the infinite-dimensional module V of $\mathbb{M}$. This we discuss in Section 7.2.1. Borcherds' strategy was to bring in Lie theory, by associating with the module V a 'Monster Lie algebra'. This example of a Borcherds–Kac–Moody algebra is described in Section 7.2.2. Next, we go from the Monster Lie algebra to the replication formula, and conclude the

Table 7.3. *The first few homogeneous spaces of
the Moonshine module $V^\natural$*

	M-module
$V_0^\natural$	ρ_0
$V_1^\natural$	0
$V_2^\natural$	$\rho_0 \oplus \rho_1$
$V_3^\natural$	$\rho_0 \oplus \rho_1 \oplus \rho_2$
$V_4^\natural$	$2\rho_0 \oplus 2\rho_1 \oplus \rho_2 \oplus \rho_3$
$V_5^\natural$	$2\rho_0 \oplus 3\rho_1 \oplus 2\rho_2 \oplus \rho_3 \oplus \rho_5$
$V_6^\natural$	$4\rho_0 \oplus 5\rho_1 \oplus 3\rho_2 \oplus 2\rho_3 \oplus \rho_4 \oplus \rho_5 \oplus \rho_6$
$V_7^\natural$	$4\rho_0 \oplus 7\rho_1 \oplus 5\rho_2 \oplus 3\rho_3 \oplus \rho_4 \oplus 3\rho_5 \oplus \rho_6 \oplus \rho_7$
$V_8^\natural$	$7\rho_0 \oplus 11\rho_1 \oplus 7\rho_2 \oplus 6\rho_3 \oplus 3\rho_4 \oplus 4\rho_5 \oplus 2\rho_6 \oplus 2\rho_7 \oplus \rho_8$

proof. In the final subsection, we explain the need for a second proof, and suggest what it may involve.

Thanks largely to Borcherds, the Monstrous Moonshine conjectures opened a door to mathematical riches far beyond what Conway and Norton could have originally hoped. For his work in Monstrous Moonshine and related topics, Richard Borcherds was awarded the Fields medal in 1998.

7.2.1 *The Moonshine module $V^\natural$*

The first essential step in the proof of the Monstrous Moonshine conjectures was the construction by Frenkel–Lepowsky–Meurman [200] of a graded infinite-dimensional representation $V^\natural$ of $\mathbb{M}$. They conjectured (correctly) that it is the representation V in (0.3.1). As we know, $V^\natural$ has a very rich algebraic structure: it is in fact a VOA. A somewhat simpler construction of $V^\natural$ is now available [427]; in particular, the fundamental fact that $\text{Aut}(V^\natural) \cong \mathbb{M}$ seems much clearer.

Each homogeneous space $V_n^\natural$ of $V^\natural$ is a finite-dimensional $\mathbb{M}$-module – see Table 7.3. Being a finite group, $\mathbb{M}$ only has finitely many (in fact exactly 194) irreducible representations, whereas $J(\tau)$ has infinitely many coefficients a_n, which grow polynomially with n. As can already be observed in the table, the decompositions of $V_n^\natural$ into irreducible $\mathbb{M}$-modules become increasingly complicated, with ever-increasing multiplicities. Thus the fact that 196 884 almost equals 196 883 is of no special significance, other than that it made it easier to anticipate that j and $\mathbb{M}$ are related.

Now, $V^\natural$ was constructed before VOAs had been defined. It was natural for Frenkel–Lepowsky–Meurman to use vertex operators to try to construct the $\mathbb{M}$-module V of (0.3.1), because there were already vertex operator constructions associated with lattices, affine algebra modules and string theory, and all of these have connections to modular functions. Borcherds' definition [68] of vertex algebras abstracted out algebraic properties of $V^\natural$ as well as those older vertex operator constructions.

As we discuss in Sections 4.3.4 and 5.3.6, the Moonshine module $V^\natural$ was constructed as the orbifold of the Leech lattice VOA $\mathcal{V}(\Lambda)$ by the ± 1-symmetry of Λ – more precisely, by an involution in $\mathrm{Aut}(\mathcal{V}(\Lambda))$ restricting to the automorphism -1 of Λ. This orbifold construction implies that $V^\natural$ is the direct sum of an invariant part $V^\natural_+ := \mathcal{V}(\Lambda)^1_+$ and a twisted part $V^\natural_- := \mathcal{V}(\Lambda)^{-1}_+$ (recall (4.3.16)). The underlying vector spaces can be (and usually are) chosen to be real, and in fact later we speculate that they can be taken to be $\mathbb{Z}$-modules (Conjecture 7.3.3).

The orbifold serves two purposes. First, it removes the constant term '24' from the graded dimension $J + 24$ of $\mathcal{V}(\Lambda)$. This means that the Lie algebra $V^\natural_1$ vanishes, giving $V^\natural$ a chance to have a finite automorphism group (Section 5.2.1). Second, this orbifold construction enhances the symmetry from the discrete part of $\mathrm{Aut}(\mathcal{V}(\Lambda))$, which is an extension of Co_0 by $(\mathbb{Z}_2)^{24}$, to all of $\mathbb{M}$. In particular, that discrete part of $\mathrm{Aut}(\mathcal{V}(\Lambda))$ preserves the decomposition $V^\natural = V^\natural_+ \oplus V^\natural_-$ and is isomorphic to the centraliser $C_\mathbb{M}(2B)$. An additional automorphism of $V^\natural$, an involution σ mixing $V^\natural_\pm$ and related to 'triality', was constructed by hand. A theorem of Griess [263] shows that together they generate $\mathbb{M}$. See [201] for more details. Establishing this symmetry enhancement is the most difficult part of [201].

A major claim of [201] is that $V^\natural$ is a 'natural' structure (hence their notation). This has been uncontested. We have $V^\natural_0 = \mathbb{C}\mathbf{1}$, as usual, and $V^\natural_1 = 0$. Hence the space $V^\natural_2$ will be a commutative non-associative algebra with product $u \times v := u_1 v$ and identity $\frac{1}{2}\omega$ (Question 5.2.3). In fact, it is the 196 883-dimensional Griess algebra [263] extended by an identity element, which is known to have automorphism group exactly $\mathbb{M}$ [528]. Using this, the automorphism group of $V^\natural$ can be seen to equal the Monster $\mathbb{M}$. The only irreducible module for $V^\natural$ is itself – such a VOA is called holomorphic (Section 5.3.1). Together with Zhu's Theorem 5.3.8, this implies that its graded dimension must be a modular function for $\mathrm{SL}_2(\mathbb{Z})$, and in fact $j(\tau) - 744$ (Question 5.3.4).

All arguments relating $V^\natural$ to $\mathbb{M}$ are complicated by the bipartite structure $V^\natural_\pm$ built into $V^\natural$. In particular, not all elements of $\mathrm{Aut}(V^\natural)$ are equally accessible. For example, [201] could prove Conjecture 7.1.1 when $g \in \mathbb{M}$ preserves $V^\natural_\pm$ – equivalently, for any $g \in \mathbb{M}$ commuting with some element in class 2B – but not for the other $g \in \mathbb{M}$. Perhaps the work of [424] will make the Monster's action on $V^\natural$ more uniformly accessible.

Conjecturally, there are 71 holomorphic VOAs with central charge $c = 24$ [488]. Recall that the Leech lattice Λ is the unique even self-dual positive-definite lattice of dimension 24 containing no norm-squared 2-vectors [113]. Under the lattice$\leftrightarrow$VOA correspondence mentioned at the end of Section 5.2.2, we are led to the following:

Conjecture 7.2.1 [201]   *The Moonshine module $V^\natural$ is the unique holomorphic VOA $\mathcal{V}$ with central charge $c = 24$ and with trivial $\mathcal{V}_1$.*

Proving Conjecture 7.2.1 is one of the most important and difficult challenges in the subject – the first small step towards this is [146]. If true, as is expected, it would tell us $V^\natural$ is a fundamental exceptional structure, on par with the Leech lattice or the E_8 Lie algebra or indeed the Monster $\mathbb{M}$. We return to this conjecture in Section 7.3.4; the analogue $A^{f\natural}$ for vertex operator superalgebras (holomorphic, $c = 12$ and $\mathcal{V}_{1/2} = 0$) is known and has automorphism group Co_1 [163]. Although the theta series Θ_L usually doesn't determine the lattice, Λ *is* the unique lattice with theta series Θ_Λ (this follows quickly from its above-mentioned uniqueness). It is thus tempting to also conjecture that the Moonshine module is the unique VOA with graded dimension J (see Question 7.2.7).

7.2.2 The Monster Lie algebra $\mathfrak{m}$

It was discovered early on that every Hauptmodul is replicable, and moreover that any replicable function is determined by its first few coefficients. An obvious approach to Conjecture 7.1.1 then is to show that the McKay–Thompson series T_g are also replicable. To get the necessary identities satisfied by their q-expansions, Borcherds used the denominator identity (Section 3.4.2) of a Lie algebra he associated with $V^\natural$.

We want to construct a Lie algebra $\mathfrak{m}$ from the Moonshine module $V^\natural = V_0^\natural \oplus V_1^\natural \oplus \cdots$. Of course, the direct choice $V_1^\natural$ is 0-dimensional, so we must modify $V^\natural$ first. Recall from Section 5.2.2 that a near-VOA $\mathcal{V}(L)$ is associated with any even *indefinite* lattice L. Let $\mathcal{V}_{1,1} := \mathcal{V}(II_{1,1})$ be the near-VOA associated with the two-dimensional even self-dual indefinite lattice $II_{1,1}$ defined in Section 1.2.1. We take both $V^\natural$ and $\mathcal{V}_{1,1}$ to be real. Define $\mathcal{V}$ to be the near-VOA $V^\natural \otimes \mathcal{V}_{1,1}$. As we know, the Monster $\mathbb{M}$ acts on $V^\natural$; extend this action to $\mathcal{V}$ by defining $\mathbb{M}$ to fix $\mathcal{V}_{1,1}$. An invariant positive-definite bilinear form on $V^\natural$ is constructed in [201]; extend it to $\mathcal{V}$ in the obvious way. Then the resulting form $(\star|\star)$ is $\mathbb{M}$-invariant.

The Monster Lie algebra $\mathfrak{m}$ is the quotient of $\mathcal{P}\mathcal{V}_1$ by the radical of the form $(\star|\star)$ on $\mathcal{V}$, where the spaces $\mathcal{P}\mathcal{V}_n$ are defined in (5.2.3). The radical contains $\mathcal{P}\mathcal{V}_0$, so $\mathfrak{m}$ has a natural (real) Lie algebra structure (see Question 7.2.4). From the $\mathcal{V}_{1,1}$ part of $\mathcal{V}$ we get the involution ω and $\mathbb{Z}$-grading (see e.g. section 6.2 of [323] for details). Then by Theorem 3.3.6, a certain central extension of $\mathfrak{m}$ is some universal Borcherds–Kac–Moody algebra $\widehat{\mathfrak{g}}(A)$ – see [72], [323] for details. More precisely, its Cartan$_{BKM}$ matrix

$$
A = \begin{pmatrix}
2 & 0 & \cdots & 0 & -1 & \cdots & -1 & \cdots \\
0 & -2 & \cdots & -2 & -3 & \cdots & -3 & \cdots \\
\vdots & \vdots & & \vdots & \vdots & & \vdots & \\
0 & -2 & \cdots & -2 & -3 & \cdots & -3 & \cdots \\
\vdots & \vdots & & & \vdots & & &
\end{pmatrix}
\tag{7.2.1}
$$

consists for each $i, j \in \{-1, 1, 2, 3, \ldots\}$ of a block in the (i, j) spot of size $a_i \times a_j$ and with entries $-(i + j)$, where a_i are the coefficients $J(\tau) = \sum_{i=-1}^\infty a_i q^i$.

Theorem 7.2.2 *The Monster Lie algebra is* $\mathfrak{m} = \widehat{\mathfrak{g}}(A)/\mathfrak{c}$, *where A is in (7.2.1) and $\mathfrak{c}$ is the (infinite-dimensional) centre of* $\widehat{\mathfrak{g}}(A)$. $\mathfrak{m}$ *has Cartan subalgebra* $\mathbb{R} \otimes_{\mathbb{Z}} II_{1,1} = \mathbb{R} \oplus \mathbb{R} =:$ $\mathfrak{m}_{(0,0)}$ *and simple roots $\alpha_{i,k}$ for each $i \in \{-1, 1, 2, 3, \ldots\}$ and $1 \le k \le a_i$. Only $\alpha_{-1,1}$ is real. The root-space decomposition of* $\mathfrak{m}$ *is* $\mathfrak{m} = \oplus_{i,j=-\infty}^\infty \mathfrak{m}_{(i,j)}$. *The Monster $\mathbb{M}$ acts on $\mathfrak{m}$ as Lie algebra automorphisms. Each root space $\mathfrak{m}_{(i,j)}$ (for $(i, j) \ne (0, 0)$) is an $\mathbb{M}$-module isomorphic to the homogeneous space $(V^\natural)_{ij+1}$, while the Cartan subalgebra*

$\mathfrak{m}_{(0,0)} \cong \rho_0 \oplus \rho_0$ *as an* $\mathfrak{m}$*-module. The denominator identity of* $\mathfrak{m}$ *is given in (3.4.7a). Finally,* $\mathfrak{m}$ *has a vector-space decomposition* $\mathfrak{u}^+ \oplus \mathfrak{gl}_2 \oplus \mathfrak{u}^-$ *into a sum of Lie subalgebras, where* $\mathfrak{u}^\pm$ *are free Lie algebras with countably many generators.*

The proof is given explicitly in section 6.2 of [**323**], and involves the No-Ghost Theorem (see the appendix of [**323**]) – a result first proved in string theory and special to VOAs with central charge $c = 24$. In particular, the No-Ghost Theorem establishes the $\mathbb{M}$-module isomorphisms in Theorem 7.2.2. $\mathfrak{m}$ has only one positive real root, so its Weyl group is order 2 and sends (i, j) to (j, i); it is responsible for the difference on the right side of (3.4.7a) (the j-function is the correction due to imaginary simple roots). The positive roots are $(-1, 1)$ and the α_{ij} of type (i, j), and this gives the product on the left.

Similarly, the fake Monster Lie algebra is associated in the same way with the near-VOA $\mathcal{V}(\Lambda) \otimes \mathcal{V}_{1,1}$. Though it is certainly an interesting example of a Borcherds–Kac–Moody algebra, it plays no role in the theory. Its name arose because it was initially suspected as playing a role in the Moonshine proof, but like $\mathcal{V}(\Lambda)$ doesn't carry a natural action of $\mathbb{M}$ so was discarded.

This construction of $\mathfrak{m}$ from $V^\natural$ may seem indirect. An alternate approach uses *Moonshine cohomology* [**386**] – a functor assigning to certain $c = 2$ near-VOAs a Lie algebra carrying an action of $\mathbb{M}$. To $\mathcal{V}_{1,1}$ this functor assigns $\mathfrak{m}$. This functor was anticipated in [**72**] and [**73**] and was inspired by BRST ('Becchi–Rouet–Stora–Tyutin') cohomology in string theory, or the semi-infinite cohomology of Lie theory. In particular, the standard method for obtaining the space of physical states in a string theory involves tensoring the original space $\mathcal{H}$ (a CFT with $c = 26$) with a space $\mathcal{H}_{ghosts}$ of ghosts (with $c = -26$); on $\mathcal{H} \otimes \mathcal{H}_{ghosts}$ is an operator Q obeying $Q^2 = 0$, and the space $\mathcal{H}_{phys}$ of physical states is the cohomology $H^* = \ker Q / \text{im } Q$. In particular, $\mathfrak{m}$ is the space H^1 for $\mathcal{H} = V^\natural \otimes \mathcal{V}_{1,1}$. The Baby Monster Lie algebra [**72**], which plays the same role for $\mathbb{B}$ as $\mathfrak{m}$ plays for $\mathbb{M}$, can be obtained in a similar way [**290**].

Because of a cohomological interpretation of denominator identities valid for any Borcherds–Kac–Moody algebra, (3.4.7a) can be 'twisted' by any $g \in \mathbb{M}$. This is how Borcherds derived (7.1.8b). These formulae are equivalent to the replication formulae (7.1.8a) conjectured in Section 7.1.4. However, these identities are obtained by more elementary means – requiring less of the theory of Borcherds–Kac–Moody algebras – in [**324**], [**331**], permitting a simplification of Borcherds' proof at this stage. In particular, in [**324**] the replication formulae (7.1.8a) appear quite naturally because $\mathfrak{u}^\pm$ are free Lie algebras.

7.2.3 The algebraic meaning of genus 0

Now, it turns out that if we verify for each conjugacy class K_g of $\mathbb{M}$ that the first, second, third, fourth and sixth coefficients of the McKay–Thompson series T_g and the corresponding Hauptmodul J_{Γ_g} agree, then $T_g = J_{\Gamma_g}$. That is precisely what Borcherds then did: he compared finitely many coefficients, and as they all equal what they should, this concluded the proof of Monstrous Moonshine!

However, this case-by-case verification occurred at the critical point where the McKay–Thompson series were being compared directly to the Hauptmoduls, and so provides little insight into why the T_g are genus 0. Recall that the main purpose for the proof of Conjecture 7.1.1 was not to establish its logical validity – the numerical evidence was already quite strong. Rather, the proof is supposed to help us understand how the Monster could be related to Hauptmoduls. This case-by-case verification became known as the conceptual gap. The basic problem is that $V^\natural$, $\mathfrak{m}$ and (7.1.8b) are *algebraic*, and the genus-0 property is *topological*. Fortunately, a more conceptual explanation of the equality $T_g = J_{\Gamma_g}$ – a conversion of the Hauptmodul property into an algebraic statement – has been found [**122**], replacing Borcherds' coefficient check with a general theorem.

Let p be prime. Exactly as in the argument of (7.1.6a), we find that the quantity

$$J(p\tau)^k + J\left(\frac{\tau}{p}\right)^k + J\left(\frac{\tau+1}{p}\right)^k + \cdots + J\left(\frac{\tau+p-1}{p}\right)^k \tag{7.2.2a}$$

is a degree-pk polynomial in $J(\tau)$. This uses the Hauptmodul property of J. Thus there is a polynomial $F_p(X, Y)$, of degree p in both X and Y, defined by

$$F_p(X, J(\tau)) = (X - J(p\tau)) \prod_{i=0}^{p-1} \left(X - J\left(\frac{\tau+i}{p}\right)\right). \tag{7.2.2b}$$

Indeed, the coefficients of $F_p(X, J(\tau))$ are symmetric polynomials in the roots $J(p\tau)$, $J\left(\frac{\tau+i}{p}\right)$, and so can be expressed polynomially using (7.2.2a). For example,

$$F_2(X, Y) = (X^2 - Y)(Y^2 - X) - 393768\,(X^2 + Y^2) - 42987520\,XY$$
$$- 40491318744\,(X + Y) + 120981708338256.$$

Definition 7.2.3 *Consider a formal series $f(\tau) = q^{-1} + \sum_{n=1}^\infty b_n q^n$ ('formal' means we don't worry about whether it converges). An* order-n modular equation *for f is a monic polynomial $F_n(x, y)$ in two variables, of degree $\psi(n) := n \prod_{\text{primes } p|n}(1 + 1/p)$, such that*

$$F_n\left(f(\tau), f\left(\frac{a\tau+b}{d}\right)\right) = 0$$

for all integers $a, b, d \geq 0$ such that $ad = n$, $\gcd(a, b, d) = 1$ and $0 \leq b < d$.

This definition looks a little obscure, but it is natural. The degree $\psi(n)$ is precisely the number of those triples (a, b, d). These triples come from the coset expansion

$$\Gamma_0(K) \begin{pmatrix} n & 0 \\ 0 & 1 \end{pmatrix} \Gamma_0(K) = \bigcup_{a,b,d} \begin{pmatrix} a & b \\ 0 & d \end{pmatrix} \Gamma_0(K),$$

for any K obeying $n \equiv 1 \pmod K$. Modular equations necessarily obey $F_n(x, y) = \pm F_n(y, x)$.

Thus $J(\tau)$ obeys a modular equation for all n. Note that this property depends crucially on it being a Hauptmodul. Conversely, does the existence of modular equations imply the Hauptmodul property? Unfortunately not: the exponential function $f(\tau) = q^{-1}$ also obeys one for every n. For example, for p prime, take $F_p(x, y) = (x^p - y)(x - y^p)$ (see also Question 7.2.5).

Beautiful and unexpected is that the only functions $f(\tau) = q^{-1} + b_1 q + \cdots$ to obey modular equations for all n are $J(\tau)$ and the 'modular fictions' q^{-1} and $q^{-1} \pm q$ (which are essentially exp, cos and sin) [360]. More generally, we have the following:

Theorem 7.2.4 [122]   *Let $f(\tau)$ be a formal series $q^{-1} + \sum_{n=1}^{\infty} b_n q^n$, $b_i \in \mathbb{C}$. Suppose f satisfies a modular equation of order n for all $n \equiv 1$ (mod N). Then:*
(a) *f converges to a holomorphic function on $\mathbb{H}$.*
(b) *If the symmetry group $\Gamma(f) := \{\alpha \in \mathrm{SL}_2(\mathbb{R}) \mid f(\alpha.\tau) = f(\tau)\}$ consists only of the translations $\pm \begin{pmatrix} 1 & t \\ 0 & 1 \end{pmatrix}$, then $f(\tau) = q^{-1} + \xi q$ for some coefficient $\xi \in \mathbb{C}$; if the coefficient ξ is an algebraic number, then $\xi = 0$ or $\xi^{\gcd(24,N)} = 1$.*
(c) *If the symmetry group $\Gamma(f)$ does not only contain translations, then $\Gamma(f)$ is genus 0 and f is a Hauptmodul for $\Gamma(f)$. Moreover, $\Gamma(f)$ contains some subgroup $\Gamma_0(K)$, for $K|N^\infty$.*

Conversely, if f is a Hauptmodul for some subgroup Γ of $\mathrm{SL}_2(\mathbb{R})$ containing $\Gamma_0(K)$, and all coefficients b_i lie in the cyclotomic field $\mathbb{Q}[\xi_K]$, then f obeys a modular equation for every $n \equiv 1$ (mod K). For the other n coprime to K, there is also a modular equation involving twisting by the Galois group, as in (2.3.14). See [122] for details. The condition $K|N^\infty$ means all primes dividing K also divide N.

The denominator identity argument tells us each T_g obeys a modular equation for each $n \equiv 1$ modulo the order $N = o(g)$ of g, so Theorem 7.2.4 concludes the proof of Monstrous Moonshine, and replaces Borcherds' coefficient check.

The proof of Theorem 7.2.4 is difficult. First, it is established that f is holomorphic on $\mathbb{H}$. This implies that whenever $f(\tau_1) = f(\tau_2)$, there is a diffeomorphism α defined locally about τ_1, such that $\alpha(\tau_1) = \tau_2$ and $f(\alpha(\tau)) = f(\tau)$. The hard part of the proof is to show α extends to all of $\mathbb{H}$. Once that is done, we know α is a Möbius transformation, and the rest of the argument is reasonably straightforward.

In [120] it is shown that if f obeys a modular equation for any n, all of whose prime divisors are congruent to 1 (mod N), then either $f = q + \xi q^{-1}$ for some ξ, or f is the Hauptmodul for a group containing some $\Gamma(N')$. However, computer calculations by [102] indicate that the hypothesis of these theorems can be considerably weakened:

Conjecture 7.2.5 [102], [120]   *Let $f(\tau) = q^{-1} + \sum_{n=1}^{\infty} b_n q^n$ be a formal series and p, p' any two distinct primes. If f satisfies modular equations for both p and p', then f converges in $\mathbb{H}$ to a holomorphic function, and either $f(\tau) = q^{-1} + \xi q$ for $\xi^{\gcd(p-1,p'-1)+1} = \xi$, or f is the Hauptmodul for a genus-0 group containing $\Gamma(N)$ for N coprime to pp'.*

This conjecture is completely out of reach at present.

Finding modular equations was a passion of Ramanujan, who filled his notebooks with them. See [82] for an application of Ramanujan's modular equations (namely, for the function $p(\tau) = \frac{d}{d\tau} \log \eta(\tau)$) to computing the first billion or so digits of π.

In Section 1.7.2 we show that although radicals can be used to solve (i.e. find closed expressions for the roots of) arbitrary polynomials of degree 4 or less, they are inadequate to solve all polynomials of degree 5 or higher. However, much as the relation $\cos(3\theta) = 4\cos(\theta)^3 - 3\cos(\theta)$ yields the solution to cubics, a modular equation relating τ and 5τ for $\sqrt{\theta_3/\eta}$ can be used to solve quintic polynomials (see e.g. chapter 7 of [464]).

Many of the applications of the j-function have to do with its modular equations. For instance, recall from Theorem 1.7.1 that each abelian extension of $\mathbb{Q}$ lies inside some cyclotomic field $\mathbb{Q}[\xi_n]$, in other words is generated by the values of the exponential function $\exp[2\pi i\alpha]$ when α is rational. Likewise, the abelian extensions of the imaginary quadratic fields $\mathbb{Q}[\sqrt{-d}]$ are generated by the values of $J(\tau)$ for special τ. See [117] for a review of this part of what is called class field theory. Modular equations are used to establish properties of those special values of $J(\tau)$ (see Question 7.2.2).

Generalising a little a definition of McKay (recall Conjecture 7.1.5), we get:

Definition 7.2.6 *By a* modular fiction *we mean any function of the form* $f(\tau) = q^{-1} + \xi q$, *where either* $\xi = 0$ *or* $\xi^{24} = 1$.

The point is that these behave like the modular functions T_g – more precisely [122], these are precisely the non-Hauptmoduls with cyclotomic integer coefficients, which obey (Galois-twisted) modular equations for each n (see [122] for more details). Perhaps, exceptional though they are, they shouldn't be ignored. This suggests the following:

Problem *What is the VOA-related question, for which '24' is the answer?*

More precisely, out of which VOA-like structure can we obtain the modular fictions, in a way analogous to how the T_g are obtained from $V^\natural$? That structure would complete Moonshine for the modular fictions. Incidentally, it is manifest in the proof of Theorem 7.2.4 that this 24 arises there through the usual exponent-2 property of Section 2.5.1.

7.2.4 Braided #7: speculations on a second proof

Monstrous Moonshine began with the challenge to understand how the Monster (the right side of (7.1.1)) could be related conceptually to modular functions (the left side of (7.1.1)). We have seen that VOAs constitute a bridge between the two sides: the Monster is the symmetry of a VOA $V^\natural$ whose graded dimension is the J-function.

That argument is still the only proof we have of Monstrous Moonshine. But does that put our finger on the essence of the mystery? The indirect argument sketched in the previous three subsections leaves the special role of the Monster unclear. As we'll see shortly, it also ignores what CFT has tried to teach us regarding modularity. It should also be remarked that a VOA is quite a complicated beast – do we really need all of its rich structure, if all we care about is Moonshine? Is there a simpler explanation that, by requiring less machinery, is both more general and more conceptual and that more directly connects $\mathbb{M}$ to a Hauptmodul property?

For these reasons, we should look for a second proof of Monstrous Moonshine. But what would it look like? To get a hint, let's recall the CFT explanation of modularity.

Two essentially equivalent formulations of quantum field theory are:

(i) The Hamiltonian formulation (canonical quantisation), which presents us with a state space V, carrying a representation of the symmetry algebra of the theory, and includes among other things a Hamiltonian (energy operator) H.

(ii) The Feynman formulation, which interprets the amplitudes using path integrals.

In RCFT, the Hamiltonian formulation describes concretely the space V, graded by H, on which we take the trace $\mathrm{tr}_V q^H$, and hence gives us the coefficients of our q-expansions. The Feynman path formalism, on the other hand, interprets these graded traces as functions over moduli spaces, and hence makes their modularity manifest. According to RCFT, *the modularity in Moonshine is the conjunction of these two formulations.*

On the Hamiltonian side of CFT, the space V is a module for the chiral algebra (VOA) $\mathcal{V}$. As such, it is a module of the Virasoro algebra (3.1.5) (giving us the Hamiltonian $H = L_0$), as well as possibly other algebras (e.g. Kac–Moody) and groups (e.g. $\mathbb{M}$). In our hypothetical second proof, we would like to avoid the full VOA structure, but probably the presence of $\mathfrak{Vir}$ is fundamental if we want to give meaning to the coefficients in the q-expansion, that is the grading of the modules. Thanks to the theory of VOAs, we understand fairly well the Virasoro side. The remainder of this subsection will be devoted to the more mysterious question: what is the key ingredient of the Feynman side?

In any treatment of RCFT (e.g. [436], [207], [131], [530], [32]), we read that $\mathcal{V}$-characters (5.3.13) are '1-point functions on the torus'. By this is meant that they are chiral blocks in $\mathfrak{B}_{\mathcal{V}}^{(1,1)}$ for the torus with one marked point, with that point labelled with the 'vacuum' module $\mathcal{V}$ itself (see e.g. Sections 4.3.3 and 5.3.4 for the physical description). Verlinde's formula tells us that space has dimension equal to the number of irreducible modules M of the chiral algebra $\mathcal{V}$, and indeed the characters χ_M form a natural basis for it. As explained in Section 2.1.4, its (enhanced) mapping class group $\widehat{\Gamma}_{1,1}$ is the braid group $\mathcal{B}_3$. Thus $\mathcal{B}_3$ will act on the characters of the RCFT. From this, using (1.1.10a), we obtain the action of the modular group $\mathrm{SL}_2(\mathbb{Z})$.

To see this $\mathcal{B}_3$ action explicitly, we have to undo a simplification we performed in Definition 5.3.6. The 1-point functions χ_M are actually functions of the triple (τ, v, z), where τ lies in the Teichmüller space $\mathbb{H}$ of the torus with 1 puncture, $v \in \mathcal{V}$ is the insertion state and $z \in \mathbb{C}$ is a local coordinate at the puncture. Explicitly, as explain in Section 5.3.4, for $v \in \mathcal{V}_{[k]}$ we get

$$\chi_M(\tau, v, z) := \mathrm{tr}_M Y(v, e^{2\pi i z}) q^{L_0 - c/24} = e^{-2k\pi i z}\, \mathrm{tr}_M o(v)\, q^{L_0 - c/24}, \qquad (7.2.3a)$$

using the notation of Section 5.3.3 (compare with (5.3.13)). The group $\widehat{\Gamma}_{1,1}$ is (like any mapping class group) generated by the Dehn twists, and as mentioned we obtain

$$\widehat{\Gamma}_{1,1} = \langle \sigma_1, \sigma_2 \mid \sigma_1\sigma_2\sigma_1 = \sigma_2\sigma_1\sigma_2 \rangle \cong \mathcal{B}_3, \qquad (7.2.3b)$$

where σ_i are the Dehn twists of Figure 2.8. The action of σ_i on the characters is then

$$\sigma_1 \cdot \chi_M(\tau, v, z) = e^{-2\pi i k/12}\, \chi_M(\tau + 1, v, z), \qquad (7.2.4a)$$

$$\sigma_2 \cdot \chi_M(\tau, v, z) = e^{-2\pi i k/12}\, \chi_M\left(\frac{\tau}{1-\tau}, \frac{v}{(1-\tau)^k}, z\right), \qquad (7.2.4b)$$

so in particular we get

$$(\sigma_1\sigma_2\sigma_1)^2.\chi_M(\tau, v, z) = e^{-2\pi i k/2} \chi_M(\tau, (-1)^k v, z), \tag{7.2.4c}$$

$$(\sigma_1\sigma_2\sigma_1)^4.\chi_M(\tau, v, z) = e^{-2\pi i k} \chi_M(\tau, v, z). \tag{7.2.4d}$$

The combination $(\sigma_1\sigma_2\sigma_1)^4$, which is trivial in the (unenhanced) mapping class group $\Gamma_{1,1} \cong \mathrm{SL}_2(\mathbb{Z})$, here equals the Dehn twist about the puncture, which for the logarithmic parameter z of course sends z to $z + 1$. The actions of σ_i on τ and v are determined from the homomorphism $\mathcal{B}_3 \to \mathrm{SL}_2(\mathbb{Z})$ given by (1.1.11b) at $w = -1$. This should be very reminiscent of Section 2.4.3.

Of course here the state v comes from the vacuum sector $\mathcal{V}$ so the conformal weight k is an integer. We are in the situation of Section 2.4.3, where our $\mathcal{B}_3$ action collapses to one of $\mathrm{PSL}_2(\mathbb{Z})$, since the centre (7.2.4c) acts trivially. This is why the z-dependence of χ_M could be safely ignored in Definition 5.3.6. As before, the more interesting case is when the weight k of the modular form is not integral. Here, that will happen when we insert states from other $\mathcal{V}$-modules, that is when we consider chiral blocks from the other $\mathfrak{B}_M^{(1,1)}$. In CFT these are equally fundamental. In this case, $v \in M$ will have rational conformal weight $k \in h_M + \mathbb{N}$, and here the Dehn twist about the puncture will typically not act trivially. As happened with the Dedekind eta function in (2.4.14), we will then see nontrivial $\mathcal{B}_3$ actions[2] (involving e.g. the $S_{(a)}$ of Figure 6.4).

It should be clear that in RCFT, modularity is a topological effect. Zhu's Theorem 5.3.8 generalises the appearance of $\mathrm{SL}_2(\mathbb{Z})$ in RCFT to any RVOA, but as we recall from Section 5.3.5, the proof follows closely the intuition of RCFT: modularity in VOAs arises through that $\mathrm{SL}_2(\mathbb{Z})$ action on the space of chiral blocks, which is inherited from the topological $\widehat{\Gamma}_{1,1}$-action mentioned above, once we drop (as Zhu did) the dependence on z.

A toy model of this idea is provided by the proof in Section 2.4.2 of the modularity of θ_3: we can interpret this action of $\mathrm{SL}_2(\mathbb{Z})$ as an action of $\mathcal{B}_3$. Note that this action of $\mathrm{SL}_2(\mathbb{Z})$ on the Heisenberg group H is really the action of $\mathcal{B}_3$ on the group $\mathbb{R}^2$ given in (2.4.15b); it factors through to $\mathrm{SL}_2(\mathbb{Z})$ because $\mathbb{R}^2$ is abelian.

The relation of the Hamiltonian ($\mathfrak{Vir}$) side to that of Feynman ($\mathcal{B}_3$) is that the Virasoro algebra acts naturally on the enhanced moduli space $\widehat{\mathcal{M}}_{1,1}$ (see Section 3.1.2), whose mapping class group is $\mathcal{B}_3$. This $\mathfrak{Vir}$-action leads to the KZ equations, which are partial differential equations obeyed by the chiral blocks in $\mathfrak{B}_\mathcal{V}^{(1,1)}$, that is by the VOA characters. The monodromy group of those equations is $\widehat{\Gamma}_{1,1} \cong \mathcal{B}_3$, and thus $\mathcal{B}_3$ acts on $\mathfrak{B}_\mathcal{V}^{(1,1)}$.

Of course the reason Borcherds chose a different route in [72] is that we need more than merely modularity: we need the genus-0 property. But as we will see in Section 7.3.3, Norton has proposed a possible relationship between the Monster and the genus-0 property, and his method also involves the $\mathcal{B}_3$ action given in (2.4.15b). Finally, we argue in Section 6.3.3 that the $\Gamma_\mathbb{Q}$-action associated with $\mathcal{B}_3$ underlies the Galois action in RCFT. In all of these examples, the modular group arose from an underlying appearance of the braid group $\mathcal{B}_3$. Is this the same $\mathcal{B}_3$? We suggest that this braid group action (together

---

[2] The thought that, for example, topological field theory really sees $\mathcal{B}_3$ and not $\mathrm{SL}_2(\mathbb{Z})$ is also made in [404].

with a compatible Virasoro action) somehow underlies Moonshine, and pursuing this thought would lead to a second, more conceptual proof of Monstrous Moonshine.

Question 7.2.1. Verify that any replicable function is uniquely determined by finitely many coefficients.

Question 7.2.2. (a) Verify that $J(\tau)$ obeys a modular equation for every $n = 2, 3, 4, \ldots$ (b) Suppose $\tau_0 = r + i\sqrt{s}$ for rational r, s, where $s > 0$. Use part (a) to prove that $J(\tau_0)$ is an algebraic number.

Question 7.2.3. Verify that any replicable function obeys a modular equation.

Question 7.2.4. Prove that $\mathcal{PV}_1/\mathrm{rad}(\star|\star)$ is a Lie subalgebra of $\mathcal{PV}_1/\mathcal{PV}_0$.

Question 7.2.5. For each n, find the modular equations obeyed by the modular fictions (a) $f(\tau) = q^{-1}$; (b) $f(\tau) = q^{-1} + q$; (c) $f(\tau) = q^{-1} - q$.

Question 7.2.6. Arguably, what makes two-dimensional quantum field theory so unique is the possibility of braid statistics. Could those braid groups directly be responsible for the $\mathcal{B}_3$ action of Section 7.2.4?

Question 7.2.7. Call any VOA $\mathcal{V}$ obeying the hypotheses of Corollary 6.2.5, 'nice'. Prove that a nice $\mathcal{V}$ is holomorphic iff its graded dimension $\chi_\mathcal{V}(\tau)$ is invariant under $\tau \mapsto -1/\tau$. Use this to show, for the class of nice VOAs, that Conjecture 7.2.1 is true iff $V^\natural$ is the unique nice VOA with graded dimension $J(\tau)$.

7.3 More Monstrous Moonshine

We give in this section a quick sketch of further developments and conjectures. As we know, Moonshine is an area where it is much easier to conjecture than to prove.

7.3.1 Mini-Moonshine

It is natural to ask about Moonshine for other groups. Of course any subgroup of $\mathbb{M}$ automatically inherits Moonshine by restriction, but this isn't at all interesting. A very accessible sporadic is M_{24} – see, for example, chapters 10 and 11 of [**113**]. Most constructions of the Leech lattice start with M_{24}, and most constructions of the Monster involve the Leech lattice. Thus we are led to the following natural hierarchy of (most) sporadics:

- M_{24} (from which we can get $M_{11}, M_{12}, M_{22}, M_{23}$); which leads to
- $Co_0 \cong 2.Co_1$ (from which we get $HJ, HS, McL, Suz, Co_3, Co_2$); which leads to
- $\mathbb{M}$ (from which we get $He, Fi_{22}, Fi_{23}, Fi'_{24}, HN, Th, \mathbb{B}$).

It can thus be argued that we could approach problems in Monstrous Moonshine by first addressing in order M_{24} and Co_1, which should be much simpler. Indeed, Moonshine for M_{24} has been completely established in [**153**].

Largely by trial and error, Queen [**466**] established Moonshine for the following groups (all essentially centralisers of elements of $\mathbb{M}$): Co_0, Th, $3.2.Suz$, $2.HJ$, HN, $2.\mathcal{A}_7$, He, M_{12}. In particular, to each element g of these groups, there corresponds a series $Q_g(\tau) = q^{-1} + \sum_{n=0}^{\infty} a_n(g)q^n$, which is a Hauptmodul for some modular group of Moonshine-type, and where each $g \mapsto a_n(g)$ is a virtual character. For Co_0, $3.2.Suz$, $2.HJ$ and $2..\mathcal{A}_7$, it is only a virtual character. Other differences with Monstrous Moonshine are that there can be a preferred nonzero value for the constant term a_0, and that although $\Gamma_0(N)$ will be a subgroup of the fixing group, it won't necessarily be normal.

For example, Queen's series Q_e for Co_0 is the Hauptmodul (2.2.17a) for the genus-0 group $\Gamma_0(2)$. Checking the tables in [**109**], we see that $276, 299, 1771, 2024$ and 8855 are dimensions of irreducible modules of the Conway group Co_1 (hence its $\mathbb{Z}_2$-extension Co_0), and 24 is the dimension of the Co_0 representation associated with the Leech lattice (it's only a projective representation of Co_1). We find $11\,202 = 8855 + 1771 + 299 + 276 + 1$, and the ambiguity $2048 = 1771 + 276 + 1 = 2024 + 24$ is resolved in favour of the latter by considering other character values and comparing to the list of Hauptmoduls. That a virtual character is needed for Co_0 is clear from the minus signs in (2.2.17a). This Hauptmodul is better known as the McKay–Thompson series T_{2B} (and the centraliser of 2B involves Co_0, which isn't a coincidence), but about half of Queen's Hauptmoduls Q_g for Co_0 do not arise as T_g for $\mathbb{M}$. Nevertheless, next subsection we see how to interpret them through the Moonshine for $\mathbb{M}$.

The Hauptmodul for $\Gamma_0(2)+$ looks like

$$q^{-1} + 4372q + 96256q^2 + 12\,40002q^3 + \cdots \qquad (7.3.1a)$$

and we find the relations

$$4372 = 4371 + 1, \quad 96\,256 = 96\,255 + 1, \quad 1\,240\,002 = 1\,139\,374 + 4371 + 2 \cdot 1, \qquad (7.3.1b)$$

where $1, 4371, 96\,255$ and $1\,139\,374$ are all dimensions of irreducible representations of the Baby Monster $\mathbb{B}$. Thus we may expect Moonshine for $\mathbb{B}$. This should actually fall into Queen's scheme because (7.3.1a) is the McKay–Thompson series associated with class 2A of $\mathbb{M}$, and the centraliser of an element in 2A is a double cover of $\mathbb{B}$.

However, there can't be a VOA $\mathcal{V} = \oplus_n \mathcal{V}_n$ with graded dimension (7.3.1a) and automorphism $\mathbb{B}$, because, for example, the $\mathbb{B}$-module $\mathcal{V}_3$ doesn't contain $\mathcal{V}_2$ as a submodule (recall Question 5.2.1). Nevertheless, Höhn deepened the analogy between $\mathbb{M}$ and $\mathbb{B}$ by constructing a vertex operator superalgebra $V\mathbb{B}^\natural$ of central charge $c = 23.5$, called the *shorter Moonshine module*, closely related to $V^\natural$ (see e.g. [**289**]). Like $V^\natural$ it is holomorphic (i.e. it has only one irreducible module), with automorphism group $\mathbb{Z}_2 \times \mathbb{B}$ and graded dimension

$$\chi_{V\mathbb{B}^\natural}(\tau) = q^{-47/48}\left(1 + 4371q^{3/2} + 96256q^2 + 1143745q^{5/2} + \cdots\right). \qquad (7.3.2a)$$

Of course the strange $-47/48$ is $-c/24$; the half-integer powers of q come from the odd (i.e. fermionic) part of $V\mathbb{B}^\natural$. Just as $\mathbb{M}$ is the automorphism group of the Griess algebra $V_2^\natural$, so is $\mathbb{B}$ the automorphism group of the algebra $(V\mathbb{B}^\natural)_2$. Just as $V^\natural$ is associated with the Leech lattice Λ, so is $V\mathbb{B}^{\natural}$ associated with the shorter Leech lattice O_{23}, the unique 23-dimensional positive-definite self-dual lattice with no vectors of length-squared 2 or 1 (see chapter 6 of [113]). The automorphism group of O_{23} is a central extension of Co_2 by $\mathbb{Z}_2$. The relation between (7.3.2a) and (7.3.1a) will be clearer next subsection.

Similarly, Duncan [163] constructs a vertex operator superalgebra $A^{f\natural}$ with $c = 12$ and automorphism group Co_1. Again it is holomorphic, and has graded superdimension

$$\chi_{A^{f\natural}}(\tau) = q^{-1/2}\left(1 + 276\,q - 2048\,q^{3/2} + 11202\,q^2 - 49152\,q^{5/2} + \cdots\right), \quad (7.3.2b)$$

i.e. is given by (2.2.17a) with $\tau \mapsto \tau/2$ and hence is fixed by a genus-0 subgroup of $\mathrm{SL}_2(\mathbb{R})$ (see Question 7.3.1). It is the unique 'nice' holomorphic vertex operator superalgebra with $c = 12$ and no elements with conformal weight $1/2$, in perfect analogy with the conjectured uniqueness of $V^{\natural}$ (Conjecture 7.2.1). The algebra $A^{f\natural}$ plays the same role for Co_1 that $V^{\natural}$ plays for $\mathbb{M}$. In particular, just as $V^{\natural}$ is obtained from a $\mathbb{Z}_2$-orbifold, so is $A^{f\natural}$, and this removes the constant term and enhances the symmetry. From this construction of $A^{f\natural}$, it is then straightforward (see Theorem 7.1 in [163]) to compute explicit finite expressions for the Thompson twists of (7.3.2b) by $g \in Co_1$, using Frame shapes as described in [111]. In this way, a genus-0 Moonshine for Co_1 is established (as expected, the arguments are far simpler than that for $\mathbb{M}$).

There has been no interesting Moonshine rumoured for the remaining six sporadics (the *pariahs* J_1, J_3, Ru, ON, Ly, J_4). There is some sort of weaker Moonshine for any group that is an automorphism group of a vertex operator algebra (so this means any finite group [152]!). Many finite groups of Lie type should arise as automorphism groups of VOAs associated with affine algebras except defined over finite fields. But apparently the known finite group examples of genus-0 Moonshine are limited to those involved with $\mathbb{M}$.

7.3.2 Twisted #7: Maxi-Moonshine

In an important announcement [450], on par with [111], Norton unified and generalised Queen's work. Unfortunately he called it 'Generalised Moonshine', but we won't (recall the diatribe in Section 3.3.1).

About a third of the McKay–Thompson series T_g will have some negative coefficients. In Section 7.3.5 we see that Borcherds interprets them as dimensions of superspaces (which automatically come with signs). Norton proposed that, although $T_g(-1/\tau)$ will not usually be another McKay–Thompson series, it will always have nonnegative integer q-coefficients, and these can be interpreted as ordinary dimensions. In the process, he extended the $g \mapsto T_g$ assignment to commuting pairs $(g, h) \in \mathbb{M} \times \mathbb{M}$.

Conjecture 7.3.1 (Norton [450])   *To each pair* $g, h \in \mathbb{M}$, $gh = hg$, *we have a function* $N_{(g,h)}(\tau)$ *such that*

$$N_{(g^a h^c, g^b h^d)}(\tau) = \alpha\, N_{(g,h)}\left(\frac{a\tau + b}{c\tau + d}\right), \qquad \forall \begin{pmatrix} a & b \\ c & d \end{pmatrix} \in \mathrm{SL}_2(\mathbb{Z}), \qquad (7.3.3)$$

for some root of unity α (of order dividing 24, and depending on g, h, a, b, c, d). $N_{(g,h)}(\tau)$ is either constant, or generates the modular functions for a genus-0 subgroup of $SL_2(\mathbb{R})$ containing some $\Gamma(M)$. Constants $N_{(g,h)}(\tau)$ arise when all elements of the form $g^a h^b$ (for $\gcd(a, b) = 1$) are 'non-Fricke' (defined below). Each $N_{(g,h)}(\tau)$ has a $q^{\frac{1}{M}}$-expansion for that M; the coefficients of this expansion are characters evaluated at h of some central extension of the centraliser $C_{\mathbb{M}}(g)$. Simultaneous conjugation of g, h leaves the function unchanged: $N_{(aga^{-1},aha^{-1})}(\tau) = N_{(g,h)}(\tau)$.

We call $N_{(g,h)}(\tau)$ the *Norton series*. An element $g \in \mathbb{M}$ is called *Fricke* if the group Γ_g fixing T_g contains an element sending 0 to i∞. In terms of the notation of Conjecture 7.1.1, $g \in \mathbb{M}$ is Fricke iff the invariance group Γ_g contains the *Fricke involution* $\tau \mapsto -1/(M\tau)$. The identity e is Fricke, as are 120 of the 171 Γ_g. For example, the classes pA, for p prime, are Fricke, while the classes pB are not.

The McKay–Thompson series are recovered by the $g = e$ specialisation: $N_{(e,h)}(\tau) = T_h(\tau)$. Unlike the McKay–Thompson series, the Norton series can have cyclotomic integer coefficients, and the groups fixing them may not contain $\Gamma_0(M)$. If g is Fricke, then clearly $N_{(g,e)}(\tau) = T_g(\tau/M)$. The action (7.3.3) of $SL_2(\mathbb{Z})$ is related to its natural action on the fundamental group $\mathbb{Z}^2$ of the torus, as we saw in Section 6.3.1, as well as a natural action of the braid group, as we'll see next subsection.

For example, when $\langle g, h \rangle \cong \mathbb{Z}_2 \times \mathbb{Z}_2$ and g, h, gh are all in class 2A, then

$$N_{(e,g)}(\tau) = N_{(e,h)}(\tau) = T_g(\tau) = q^{-1} + 4372q + 96256q^2 + \cdots, \qquad (7.3.4a)$$

$$N_{(g,e)}(\tau) = N_{(h,e)}(\tau) = T_g(\tau/2) = q^{-1/2} + 4372q^{1/2} + 96256q + \cdots, \qquad (7.3.4b)$$

$$N_{(g,h)}(\tau) = \sqrt{J(\tau) - 984} = q^{-1/2} - 492q^{1/2} - 22590q^{3/2} + \cdots, \qquad (7.3.4c)$$

$$N_{(g,g)}(\tau) = N_{(h,h)}(\tau) = q^{-1/2} + 4372q^{1/2} - 96256q + \cdots. \qquad (7.3.4d)$$

Hence $N_{(g,e)}(\tau + 1) = i N_{(g,g)}(\tau)$, giving us an example of a nontrivial α in (7.3.3).

The basic tool we have for approaching Moonshine conjectures is the theory of VOAs, so we need to understand Norton's suggestion from that point of view. This is done using twisted modules (Section 5.3.6). For each $g \in \mathbb{M}$, there is a unique g-twisted module of $V^{\natural}$ [150] – call this twisted module $V^{\natural}(g)$. This generalises the holomorphicity of $V^{\natural}$ mentioned in Section 7.2.1. Given any automorphism $h \in \text{Aut}(V^{\natural})$ commuting with g, we can perform Thompson's trick (5.3.23) and write

$$q^{-1}\text{tr}_{V^{\natural}(g)}h \, q^{L_0} =: \mathcal{Z}(g, h; \tau). \qquad (7.3.5)$$

Then $\mathcal{Z}(g, h) = N_{(g,h)}$.

[150] proves that, whenever the subgroup $\langle g, h \rangle$ generated by g and h is cyclic, then $N_{(g,h)}$ will be a Hauptmodul satisfying (7.3.3). This will happen, for instance, whenever the orders of g and h are coprime. [150] proves this by reducing it to Conjecture 7.1.1 (which is now a theorem). Extending [150] to all commuting pairs g, h is one of the most pressing tasks in Moonshine.

Höhn [290] verified Conjecture 7.3.1 for g in class 2A and $h \in C_{\mathbb{M}}(g) \cong 2.\mathbb{B}$. In particular, those 247 functions $N_{(g,h)}(2\tau)$ are Hauptmoduls for genus-0 groups of Moonshine-type (see Question 7.3.1). The proof mirrors that of [72] fairly closely. There is a simple relation between the twisted module $V^\natural(g)$ and the shorter Moonshine module $V\mathbb{B}^\natural$, and from this the 286 Thompson twists of (7.3.2a) can be obtained [**290**]. Verifying Conjecture 7.3.1 for g in class 2B should likewise be possible.

More satisfying though would be a uniform proof of Conjecture 7.3.1, for example, by considering the full orbifold $V^\natural/\mathbb{M}$. It appears that the 3-cocycle α corresponding to this orbifold (recall the cohomological twist of Section 5.3.6) will have to be nontrivial – in fact, its order in $H^3(\mathbb{M}, \mathbb{C}^\times)$ should be a multiple of 12 [**408**]. Suggestive is that the permutation orbifold $\mathbb{M}^{\otimes n}/\langle g \rangle$ gives a natural interpretation of the left-half of the definition (7.1.9) of a replicable function.

The orbifold theory for M_{24} is established in [**153**] (the relevant series $\mathcal{Z}(g, h)$ had already been constructed in [**407**]). Next up should be the orbifold theory for Conway's group Co_1, but that seems out of reach right now, in spite of [**163**].

As has been alluded to elsewhere in this book, the subfactor approach complements that of VOAs. In particular, orbifolds seem more accessible for them [**157**], [**332**].

7.3.3 Why the Monster?

That $\mathbb{M}$ is associated with *modular functions* can be explained mathematically by it being the automorphism group of the vertex operator algebra $V^\natural$. But what is so special about that group $\mathbb{M}$ that these modular functions T_g and $N_{(g,h)}$ should be Hauptmoduls? In fact, every group known to have rich genus-0 Moonshine properties is contained in the Monster. To what extent can we derive $\mathbb{M}$ from Monstrous Moonshine? Our understanding of this seemingly central role of $\mathbb{M}$ is still poor.

The most interesting approach to this important question is due to Norton, and was first (cryptically) stated in [**450**]: the Monster is probably the largest (in a sense) group with the 6-transposition property. A *k-transposition group* G is one generated by a conjugacy class K of involutions, where the product gh of any two elements of K has order $\leq k$. For example, take K to be the transpositions in the symmetric group S_n, that is, K is the set of all permutations (ij). Since $\pi \circ (ij) \circ \pi^{-1} = (\pi i, \pi j)$, K is a conjugacy class in S_n. An easy induction on n confirms that S_n is generated by K. Moreover, $(ij)(k\ell)$ has order 1, 2, 3, respectively iff the set $\{i, j\} \cup \{k, \ell\}$ has cardinality 2, 4, 3. Thus S_n is a 3-transposition group (this example is the source of the name 'k-transposition'). The Monster $\mathbb{M}$ is 6-transposition, for the choice of class $K = 2A$ (see Section 7.3.6 for more details). Transposition groups were used in the finite simple group classification by Fischer to great effect. The simplest relation known to this author, of the number '6' to genus 0, is given in Question 7.3.2.

The group $\Gamma = PSL_2(\mathbb{Z})$ is isomorphic to the free product $\mathbb{Z}_3 * \mathbb{Z}_2$ generated by an order 3 element $u = \begin{pmatrix} 0 & -1 \\ 1 & 1 \end{pmatrix}$ and an order 2 element $v = \begin{pmatrix} 0 & -1 \\ 1 & 0 \end{pmatrix}$. A transitive action of Γ on a finite set X with one distinguished point $x_0 \in X$ is equivalent to specifying a finite index subgroup Γ_0 of Γ. In particular, Γ_0 is the stabiliser $\{g \in \Gamma \mid g.x_0 = x_0\}$ of x_0, X can be identified with the cosets $\Gamma_0 \backslash \Gamma$ and x_0 with the coset Γ_0. (If we avoid specifying x_0, then Γ_0 will be identified only up to conjugation.) As an abstract group, Γ_0 will be a free product of a certain number of $\mathbb{Z}_2$'s, $\mathbb{Z}_3$'s and $\mathbb{Z}$'s (e.g. $\mathcal{F}_n = \mathbb{Z} * \mathbb{Z} * \cdots * \mathbb{Z}$ n times).

To such an action, we can associate a directed graph $\mathcal{G}$: its vertices are labelled by the set X, and we draw a solid edge directed from x to $u.x$, and a dotted undirected edge between x and $v.x$. Choose any spanning tree $\mathcal{T}$ of $\mathcal{G}$ (i.e. a connected subgraph of $\mathcal{G}$ containing all vertices of $\mathcal{G}$ and the minimum possible number ($\|X\| - 1$) of edges). Then the Reidemeister–Schreier method (see e.g. the appendix to [**292**] or section I.3 of [**103**]) gives a presentation for Γ_0, with one generator for every edge in $\mathcal{G}$ not in $\mathcal{T}$.

We are more interested though in a triangulation of the closed surface $\Gamma_0 \backslash \overline{\mathbb{H}}$, called a (modular) *quilt*, which we can canonically associate with the action of Γ in X. The definition, originally due to Norton and further developed by Parker, Conway and Hsu, is somewhat involved and will be avoided here (but see especially chapter 3 of [**292**]). It is so-named because there is a polygonal 'patch' covering every cusp of $\Gamma_0 \backslash \mathbb{H}$, and the closed surface is formed by sewing together the patches along their edges ('seams'). There are a total of $2n$ triangles and n seams in the triangulation, where n is the index $\|\Gamma_0 \backslash \Gamma\| = \|X\|$. The boundary of each patch has an even number of edges, namely the double of the corresponding cusp width. The formula (2.2.16) for the genus g of $\Gamma_0 \backslash \mathbb{H}$ in terms of the index n and the numbers n_i of Γ_0-orbits of fixed points of order i, can be interpreted in terms of the data of the quilt (see (6.2.3) of [**292**]), and we find in particular that if every patch of the quilt has at most six sides, then the genus will be 0 or 1, and genus 1 only exceptionally.

The quilt picture was specifically designed for one class of these Γ-actions (actually an $SL_2(\mathbb{Z})$-action, but this doesn't matter). Fix a finite group G (we're most interested in the choice $G = \mathbb{M}$). Recall from (2.4.15) the right action of $\mathcal{B}_3$ on triples $(g_1, g_2, g_3) \in G^3$, and the equivalent reduced action of $\mathcal{B}_3$ on G^2. We will be interested in this action on the subset of G^3 where all $g_i \in G$ are involutions. The modular group $SL_2(\mathbb{Z})$ is related to $\mathcal{B}_3$ by (1.1.10a). From this, we can get an action of $SL_2(\mathbb{Z})$ in two ways: either (i) by restricting to commuting pairs g, h; or (ii) by identifying each pair (g, h) with all conjugates (aga^{-1}, aha^{-1}). Norton's $SL_2(\mathbb{Z})$ action (7.3.3) arises from the $\mathcal{B}_3$ action of (2.4.15b), when we combine both (i) and (ii).

The number of sides in each patch of the corresponding quilt is determined by the orders of the g, h in these pairs. Taking G to be the Monster, and the involutions g_i from class 2A, then each patch will have ≤ 6 sides, and the corresponding genus will be 0 (usually) or 1 (exceptionally). In this way we can relate the Monster with a genus-0 property. This approach to genus 0 faces the same challenge of any other: how to incorporate the Atkin–Lehner involutions of Proposition 7.1.2(ii).

Based on the $\mathcal{B}_3$ actions (2.4.15), Norton hopes for some analogue of Moonshine valid for noncommuting pairs. Although the resulting series are always modular, they may not be Hauptmoduls, their fixing group may not contain some $\Gamma(N)$, and the coefficients won't always be cyclotomic integers. CFT considerations ('higher-genus orbifolds') alluded to in Section 6.3.1 suggest that this might be more natural to do using, for example, noncommuting quadruples $(g_1, g_2, h_1, h_2) \in \mathbb{M}^4$ obeying $g_1 h_1 g_1^{-1} h_1^{-1} = h_2 g_2 h_2^{-1} g_2^{-1}$; the role of $SL_2(\mathbb{Z})$ is then played by higher-genus mapping class groups.

An important question is, how much does Monstrous Moonshine determine the Monster? How much of $\mathbb{M}$'s structure can be deduced from, for example, McKay's $\widehat{E}_8$ Dynkin diagram observation (Section 7.3.6), and/or the (complete) replicability of the T_g, and/or Conjecture 7.3.1, and/or Modular Moonshine in Section 7.3.5 below? A small start towards this is taken in [**452**], where some control on the subgroups of $\mathbb{M}$ isomorphic to $\mathbb{Z}_p \times \mathbb{Z}_p$ (p prime) is obtained, using only the properties of the $N_{(g,h)}$. See also chapter 8 of [**292**].

7.3.4 Genus 0 revisited

Tuite [**532**] suggests a very intriguing reformulation of the genus-0 property, directly in terms of VOAs. Assume the uniqueness conjecture: $V^\natural$ is the only $c = 24$ VOA with graded dimension J (Section 7.2.1). He argues from this that, for each $g \in \mathbb{M}$, the McKay–Thompson series T_g will be a Hauptmodul iff the only orbifolds of $V^\natural$ are the Leech lattice VOA $\mathcal{V}(\Lambda)$ and $V^\natural$ itself. More precisely, orbifolding $V^\natural$ by $\langle g \rangle$ should be $V^\natural$ if g is Fricke, and $\mathcal{V}(\Lambda)$ if g is non-Fricke ('Fricke' is defined in Section 7.3.2).

In, for example, [**313**], this analysis is extended to the genus-0 property of some Norton series $N_{(g,h)}$, when the subgroup $\langle g, h \rangle$ is not cyclic (thus going beyond [**150**]), although again assuming the uniqueness conjecture. Tuite is thus suggesting that the genus-0 property of the Monstrous Moonshine functions T_g and $N_{(g,h)}$ seems to be equivalent to a single principle. These arguments emphasise the importance of establishing the uniqueness conjecture of $V^\natural$. Unfortunately, that still seems out of reach.

7.3.5 Modular Moonshine

Consider an element $g \in \mathbb{M}$. We know from [**466**], [**450**], [**150**] that there is a Moonshine for the centraliser $C_\mathbb{M}(g)$ of g in $\mathbb{M}$, governed by the g-twisted module $V^\natural(g)$. Unfortunately, $V^\natural(g)$ is not usually itself a VOA, so the analogy with $\mathbb{M}$ is not perfect. Ryba found it interesting that, for $g \in \mathbb{M}$ of prime order p, Norton's series $N_{(g,h)}$ can be transformed into a McKay–Thompson series (and has all the associated nice properties) whenever h is p-*regular* (i.e. h has order coprime to p) – as we know, in this case $\langle g, h \rangle$ is cyclic. This special behaviour of p-regular elements suggested to him to look at modular representations, for reasons we'll soon see.

Let's begin by reviewing the basics of modular representations and Brauer characters (see also [**446**], [**308**]). A *modular representation* ρ of a group G is a representation defined over a field of positive characteristic p dividing the order $\|G\|$ of G. This is precisely the class of finite-dimensional representations where the usual properties break down. Such representations possess many special (that is to say, unpleasant) features.

For one thing, they are no longer completely reducible, so Theorem 1.1.2 breaks down. For a simple example, let p be any prime and consider $G = \mathbb{Z}_p$; then over any field of characteristic p, the map

$$a \mapsto \begin{pmatrix} 1 & a \\ 0 & 1 \end{pmatrix} \tag{7.3.6}$$

defines a two-dimensional representation of G that is indecomposable but not irre-ducible. It's not irreducible because it maps the x-axis to itself, and so contains the one-dimensional identity representation as a subrepresentation. Before, given a repre-sentation we could simplify it enough merely by writing it as a direct sum of indecom-posables, but here there are far too many indecomposables. In other words, there are other more complicated ways to combine irreducibles than direct sum. The familiar role of irreducibles as direct summands is replaced here by their role as composition factors. It is completely analogous to, and simpler than, the role of simple groups in finite group theory (recall Section 1.1.2). Completely reducible representations (as in Theorem 1.1.2) are equivalent to a representation with blocks down the diagonal and zero-blocks above and below the diagonal; the diagonal blocks are its irreducible summands. On the other hand, a modular representation ρ is equivalent to a matrix with zero-blocks below the diagonal; the blocks along the diagonal (e.g. two copies of the trivial representation (1) for the representation in (7.3.6)) are the composition factors, and the blocks above the diagonal describe how these glue together.

Another complication is that the familiar character χ_ρ of (1.1.5) loses its usefulness. As we saw at the end of Section 1.1.3, very different modular representations can have identical characters. Instead, the more subtle *Brauer character* $\beta(\rho)$ is used. It can be defined as follows. Let m be the order $\|G\|$ of G, and write $m = p^a p'$ where p and p' are coprime. Let K be the cyclotomic field $\mathbb{Q}[\xi_m]$, and let $R = \mathbb{Z}[\xi_m]$ be the ring of cyclotomic integers. A finite field k of characteristic p can be obtained from R by choosing any prime ideal $\mathfrak{p}$ of R containing pR; then $k = R/\mathfrak{p}$. This construction of k defines a ring homomorphism $\phi_\mathfrak{p} : R \to k$. In particular, put $\xi := \xi_{p'} \in R$; then $\bar{\xi} = \phi_\mathfrak{p}(\xi)$ will be a primitive p'th root of unity in k.

Suppose ρ is some n-dimensional modular representation of G over k. Let $G_{p'}$ be the set of all p-regular elements in G. The field k defined above is big enough that the $n \times n$ matrix $\rho(g)$, for any $g \in G_{p'}$, is diagonalisable over k. More precisely, its n eigenvalues (counting multiplicities) are all p'th roots of unity in k, and so can be written as $\bar{\xi}^{\ell_i}$ for some integers ℓ_i, $1 \le i \le n$.

The Brauer character $\beta(\rho)$ of ρ is defined to be

$$\beta(\rho)(g) := \sum_{i=1}^{n} \xi^{\ell_i} \in R \subset \mathbb{C}, \qquad \forall g \in G_{p'}.$$

It is a well-defined class-function on $G_{p'}$, and in fact the Brauer characters form a basis for the space of class functions on $G_{p'}$. Two representations have the same Brauer character iff they have the same composition factors. Brauer characters were introduced by Brauer and his student Nesbitt in 1937. Apart from their role in modular representations, they also relate p-subgroups of G with properties of the usual character table. See Question 7.3.4 for an example.

Theorem 7.3.2 ([484], [79], [77]) *Let $g \in \mathbb{M}$ be any element of prime order p, for any p dividing $\|\mathbb{M}\|$. Then there is a vertex operator superalgebra ${}^g V = \oplus_{n \in \mathbb{Z}} {}^g V_n$ defined over the finite field $\mathbb{F}_p$ and carrying a (projective) representation of the centraliser $C_\mathbb{M}(g)$.*

If $h \in C_{\mathbb{M}}(g)$ is p-regular, then the graded Brauer character

$$R(g, h; \tau) := q^{-1} \sum_{n \in \mathbb{Z}} \beta({}^g\mathcal{V}_n)(h) \, q^n$$

equals the McKay–Thompson series $T_{gh}(\tau)$. Moreover, for g belonging to any conjugacy class in $\mathbb{M}$ except 2B, 3B, 5B, 7B or 13B, this is in fact an ordinary VOA (i.e. the 'odd' part vanishes), while in those remaining cases the graded Brauer characters of both the odd and even parts can be expressed separately using McKay–Thompson series.

We defined vertex operator superalgebras in Section 5.1.3. The centralisers $C_{\mathbb{M}}(g)$ in the theorem are quite nice: for example, for groups of type 2A, 2B, 3A, 3B, 3C, 5A, 5B, 7A, 11A these are extensions of the sporadic groups $\mathbb{B}$, Co_1, Fi'_{24}, Suz, Th, HN, HJ, He and M_{12}, respectively. The proof for $p = 2$ is not complete as it relies on a still-unproven hypothesis. The conjectures in [**484**] concerning modular analogues of the Griess algebra for several sporadic groups follow from Theorem 7.3.2.

Can these modular ${}^g\mathcal{V}$'s be interpreted as a reduction mod p of (super)algebras in characteristic 0? What can we say about elements g of composite order in $\mathbb{M}$?

Conjecture 7.3.3 (Borcherds [77]) *Choose any $g \in \mathbb{M}$ and let n denote its order. Then there is a $\frac{1}{n}\mathbb{Z}$-graded superspace ${}^g\widehat{\mathcal{V}} = \oplus_{i \in \frac{1}{n}\mathbb{Z}} {}^g\widehat{\mathcal{V}}_i$ over the ring of cyclotomic integers $\mathbb{Z}[e^{2\pi i/n}]$. It is often (but probably not always) a vertex operator superalgebra – in particular, ${}^1\widehat{\mathcal{V}}$ is an integral form of the Moonshine module $V^\natural$. Each ${}^g\widehat{\mathcal{V}}$ carries a representation of a central extension of $C_{\mathbb{M}}(g)$ by $\mathbb{Z}_n$. Define the graded trace*

$$B(g, h; \tau) = q^{-1} \sum_{i \in \frac{1}{n}\mathbb{Z}} \mathrm{ch}_{{}^g\widehat{\mathcal{V}}_i}(h) \, q^i.$$

*If $g, h \in \mathbb{M}$ commute and have coprime orders, then $B(g, h; \tau) = T_{gh}(\tau)$. If all q-coefficients of T_g are nonnegative, then the 'odd' part of ${}^g\widehat{\mathcal{V}}$ vanishes, so it is an ordinary space, and should equal the g-twisted module $V^\natural(g)$ of [**150**]. If g has prime order p, then the reduction mod p of ${}^g\widehat{\mathcal{V}}$ is the modular vertex operator superalgebra ${}^g\mathcal{V}$ of Theorem 7.3.2.*

More precisely, ${}^g\widehat{\mathcal{V}}$ is to be a free module over the ring $\mathbb{Z}[e^{2\pi i/n}]$, and each graded piece is finite-dimensional over that ring. When we say ${}^1\widehat{\mathcal{V}}$ is an integral form for $V^\natural$, we mean that ${}^1\widehat{\mathcal{V}}$ has the same structure as a VOA, with everything defined over $\mathbb{Z}$, and tensoring it with $\mathbb{C}$ gives $V^\natural$. Borcherds' conjecture, which beautifully tries to explain Theorem 7.3.2, is completely open. It provides the analogue for $V^\natural$ of the surprising Lie algebra Theorems 1.5.4 and 3.4.1.

7.3.6 McKay on Dynkin diagrams

McKay found other relationships with Lie theory [**411**], [**75**], [**247**], reminiscent of his A–D–E correspondence with finite subgroups of $SU_2(\mathbb{C})$ (see Section 2.5.2). As we see from Table 7.2, $\mathbb{M}$ has two conjugacy classes of involutions. Let K be the smaller one, called '2A' in [**109**] (the alternative, class '2B', has almost 100 million times more elements). The product of any two elements of K will lie in one of nine conjugacy classes: namely, 1A, 2A, 2B, 3A, 3C, 4A, 4B, 5A, 6A. These conjugacy classes are of elements of orders 1, 2, 2, 3, 3, 4, 4, 5, 6. It is remarkable that, for such a complicated group as $\mathbb{M}$, that list stops at only 6 – as we know from Section 7.3.3, we call $\mathbb{M}$ a 6-transposition group for this reason. The punchline: McKay noticed that those nine numbers are precisely the labels a_i of the affine E_8 diagram (see Figure 3.2). Thus we can attach a conjugacy class of $\mathbb{M}$ to each vertex of the $E_8{}^{(1)}$ diagram. A direct interpretation of the *edges* in the $E_8{}^{(1)}$ diagram, in terms of $\mathbb{M}$, is unfortunately not yet known, though [**247**], [**365**] establish how to unambiguously assign classes to the nodes.

We can't get the affine E_7 labels in a similar way, but McKay noticed that an order 2 folding of affine E_7 gives the affine F_4 diagram, and we can obtain its labels using the Baby Monster $\mathbb{B}$ (the second largest sporadic). In particular, let K now be the smallest conjugacy class of involutions in $\mathbb{B}$ (also labelled '2A' in [**109**]); the conjugacy classes in KK have orders 1, 2, 2, 3, 4 ($\mathbb{B}$ is a 4-transposition group) – these are the labels of $F_4{}^{(1)}$. Of course we'd prefer $E_7{}^{(1)}$ to $F_4{}^{(1)}$, but perhaps that *two*-folding has something to do with the fact that an order-2 central extension of $\mathbb{B}$ is the centraliser of an element $g \in \mathbb{M}$ of order 2.

Now, the *triple*-folding of affine E_6 is affine G_2. The Monster has three conjugacy classes of order 3. The smallest of these ('3A') has a centraliser that is a *triple* cover of the Fischer group $Fi'_{24}.2$. Taking the smallest conjugacy class of involutions in $Fi'_{24}.2$, and multiplying it by itself, gives conjugacy classes with orders 1, 2, 3 ($Fi'_{24}.2$ is a 3-transposition group) – and those not surprisingly are the labels of $G_2{}^{(1)}$!

McKay's $E_8{}^{(1)}$, $F_4{}^{(1)}$, $G_2{}^{(1)}$ observations still have no explanation. In [**247**] these patterns are extended, by relating various simple groups to the $E_8{}^{(1)}$ diagram with deleted nodes. More recently, [**365**] relate the $E_8{}^{(1)}$ observation to VOAs, by applying [**425**] to the lattice VOA $\mathcal{V}(\sqrt{2}E_8)$; the connection with $V^{\natural}$ is plausible but not yet completely established. As we know from Section 1.5.4, the folding of Coxeter–Dynkin diagrams arises when we restrict to the invariant subalgebras of automorphisms, so perhaps that provides a clue how to attack the $F_4{}^{(1)}$ and $G_2{}^{(1)}$ observations.

7.3.7 Hirzebruch's prize question

Algebra is the mathematics of structure, and so of course it has a profound relationship with every area of mathematics. Therefore the trick for finding possible fingerprints of Moonshine in, say, geometry is to look there for modular functions. And that search quickly leads to the elliptic genus.

We briefly discuss this in Section 5.4.2, where we mention several deep relationships between elliptic genera and the material covered elsewhere in this book. Let us simply mention here that the genus of a manifold will typically involve negative coefficients and be the graded dimension of a vertex operator superalgebra. This certainly doesn't preclude Moonshine-like behaviour – for example, Moonshine for Co_1 involves as we know the vertex operator superalgebra $A^{f\natural}$. However, the genera of even-dimensional projective spaces has nonnegative integer coefficients [**400**]; it would be interesting to study the representation-theoretic questions associated with them.

Hirzebruch's 'prize question' (page 86 of [**287**]) asks for the construction of a 24-dimensional manifold M with Witten- or $\widehat{A}$-genus J (after being normalised by η^{24}). We would like $\mathbb{M}$ to act on M by diffeomorphisms, and the twisted Witten genera to be the McKay–Thompson series T_g. See also [**151**]. It would also be nice to associate Norton's series $N_{(g,h)}$ with this Moonshine manifold. Constructing such a manifold would realise the geometry underlying Monstrous Moonshine, and as such is perhaps the remaining Holy Grail in the subject.

Hirzebruch's question was partially answered by Mahowald–Hopkins [**399**], who constructed a manifold with Witten genus J, but couldn't show it would support an effective action of $\mathbb{M}$. Related work is [**21**], who constructed several actions of $\mathbb{M}$ on, for example, 24-dimensional manifolds (but none of which could have genus J), and [**364**], who showed the graded dimensions of the subspaces $V_{\pm}^{\natural}$ of the Moonshine module are twisted $\widehat{A}$-genera of Milnor–Kervaire's manifold M_0^8 (the $\widehat{A}$-genus is the specialisation of elliptic genus to the cusp $\mathrm{i}\infty$).

Related to elliptic genus is elliptic cohomology, which is described beautifully in [**499**]. Mason's constructions [**407**] associated with Moonshine for the Mathieu group M_{24} have been interpreted as providing a geometric model ('elliptic system') for elliptic cohomology $\mathrm{Ell}^*(BM_{24})$ of the classifying space of M_{24} [**523**], [**154**].

7.3.8 Mirror Moonshine

There has been a second conjectured relationship between geometry and Monstrous Moonshine. Calabi–Yau manifolds (see e.g. [**299**]) are a class of complex manifolds with an unusually rich mathematical structure – for example, in dimensions 1 and 2 they are elliptic curves and K3 surfaces, respectively. Specifying a Calabi–Yau manifold X means choosing a complex structure, as well as a Kähler class $[\omega] \in H^2(X, \mathbb{C})$. In the case of an elliptic curve (i.e. a torus), this corresponds to choosing parameters $\tau, \sigma \in \mathbb{H}$. Mirror symmetry [**291**] says that most Calabi–Yau manifolds come in closely related pairs, where the roles of the complex structure and Kähler structure are switched. In the case of elliptic curves, it relates the pair (τ, σ) to the pair (σ, τ) and implies the modularity of certain generating functions for Gromov–Witten invariants – see [**132**] for a review. This unexpected modularity is, of course, reminiscent of Moonshine, and it is tempting to look for a concrete connection.

Consider a one-parameter family X_λ of Calabi–Yau manifolds, with mirror X^* given by the resolution of an orbifold X/G for G finite and abelian. Then the Hodge numbers $h^{1,1}(X)$ and $h^{2,1}(X^*)$ will be equal, and more precisely the moduli space of (complexified) Kähler structures on X will be locally isometric to the moduli space of complex structures on X^*. The 'mirror map' $\lambda(q)$, which can be defined using the Picard–Fuchs equation [**438**], is a canonical map between those moduli spaces. For example, $x_1^4 + x_2^4 + x_3^4 + x_4^4 + \lambda^{-1/4} x_1 x_2 x_3 x_4 = 0$ is such a family of K3 surfaces, where $G = \mathbb{Z}_4 \times \mathbb{Z}_4$. Its mirror map is given by

$$\lambda(q) = q - 104q^2 + 6444q^3 - 311\,744q^4 + 13\,018\,830q^5 - 493\,025\,760q^6 + \cdots .$$
$$(7.3.7)$$

Lian–Yau [**385**] noticed that the reciprocal $1/\lambda(q)$ of the mirror map in (7.3.7) equals the McKay–Thompson series $T_{2A}(\tau) + 104$. After looking at several other examples with similar conclusions, they proposed their *Mirror Moonshine Conjecture*: The reciprocal $1/\lambda$ of the mirror map of a one-parameter family of K3 surfaces with an orbifold mirror will be a McKay–Thompson series (up to an additive constant).

A counterexample (and more examples) are given in section 7 of [**544**]. In particular, although there are relations between mirror symmetry and modular functions (see e.g. [**266**] and [**275**]), there doesn't seem to be any special relation with $\mathbb{M}$. Doran [**158**] 'demystifies the Mirror Moonshine phenomenon' by finding necessary and sufficient conditions for $1/\lambda$ to be a modular function for a modular group commensurable with $SL_2(\mathbb{Z})$.

This focus on K3 surfaces is not significant. Calabi–Yau 3-folds are the real meat of mirror symmetry, but it is much harder to find explicit families. Some of the interesting number theory of Calabi–Yau manifolds and mirror symmetry is reviewed in [**571**].

7.3.9 Physics and Moonshine

The physical side of Moonshine (namely, perturbative string theory and conformal field theory) was noticed early on, and has profoundly influenced the development of Moonshine and VOAs. This effectiveness of physical interpretations isn't magic – it merely tells us that finite-dimensional objects are sometimes seen much more clearly when studied through infinite-dimensional structures (often by being 'looped'). Of course Monstrous Moonshine, which teaches us to study the finite group $\mathbb{M}$ via its infinite-dimensional module $V^\natural$, fits perfectly into this picture.

Throughout this book we've described various points-of-contact between mathematics and physics. Because $V^\natural$ is so mathematically special, it may be expected that it corresponds somehow to interesting physics. Although there have been some attempts to directly interpret Monstrous Moonshine in the context of physics, we still have no evidence Nature concurs.

There is a $c = 24$ RCFT whose anti-holomorphic chiral algebra is trivial, and whose holomorphic one, as well as the state space $\mathcal{H}$, are both $V^\natural$ (this is possible because $V^\natural$ is holomorphic). This RCFT is nicely described in [**142**]; its symmetry is the Monster. The Bimonster $\mathbb{M} \wr \mathbb{Z}_2 = (\mathbb{M} \times \mathbb{M}) \rtimes \mathbb{Z}_2$ (Section 7.1.1) is the symmetry of a $c = \bar{c} = 24$ RCFT with state space $\mathcal{H} = V^\natural \otimes \overline{V^\natural}$. The paper [**119**] finds the D-branes (boundary states) of lowest mass for this theory; they are in one-to-one correspondence $g \mapsto \|g\rangle\!\rangle$ with the elements of $\mathbb{M}$. The Bimonster permutes them: $(h, k).\|g\rangle\!\rangle = \|hgk^{-1}\rangle\!\rangle$, while the remaining involution sends $\|g\rangle\!\rangle$ to $\|g^{-1}\rangle\!\rangle$. Most interestingly, their 'overlaps' $\langle\!\langle g\|q^{\frac{1}{2}(L_0+\bar{L}_0-\frac{c}{24})}\|h\rangle\!\rangle$ equal the McKay–Thompson series $T_{g^{-1}h}$. We largely ignored D-branes (surfaces on which endpoints of open strings rest) in Chapter 4, but they are a natural ingredient in string theory. Much as every natural property of the Wess–Zumino–Witten string translates nicely into Lie theory, it would appear that the same holds with the string theory $\mathcal{H} = V^\natural \otimes \overline{V^\natural}$ and the Monster $\mathbb{M}$. Surely it would be interesting to continue that investigation. Other suggestions for the physics of Monstrous Moonshine are [**99**], [**274**], [**96**], [**260**], [**281**].

Question 7.3.1. Let $f(\tau)$ be a Hauptmodul for some genus-0 group Γ. For any $a > 0$, prove that $f(a\tau)$ is fixed by a genus-0 group (call it Γ_a), and any modular function for Γ_a will be a rational function in $f(a\tau)$.

Question 7.3.2. Let G be any group with exponent $k < 6$ (i.e. $g^k = e$ for all $g \in G$). Suppose there are a set of functions $N_{(g,h)}(\tau)$ associated with every commuting pair $g, h \in G$, with the property that equation (7.3.3) always holds with $\alpha = 1$. Prove that each of these functions is fixed by a genus-0 subgroup of $\mathrm{SL}_2(\mathbb{Z})$.

Question 7.3.3. Assume for simplicity that $g \in \mathbb{M}$ is such that $C_\mathbb{M}(g)$ acts linearly (i.e. nonprojectively) on the twisted module $V^\natural(g)$. Then for $h \in C_\mathbb{M}(g)$ of order n, the q-coefficients of $\mathcal{Z}(g, h)$ all lie in the field $\mathbb{Q}[\xi_n]$. Fix any Galois automorphism $\sigma \in \mathrm{Gal}(\mathbb{Q}[\xi_n]/\mathbb{Q})$, and let $\sigma \mathcal{Z}(g, h)$ denote the q-expansion obtained by formally applying σ term-by-term to $\mathcal{Z}(g, h)$: $\sigma(\sum_i a_i q^i) = \sum_i \sigma(a_i) q^i$. Show that $\sigma \mathcal{Z}(g, h)$ equals another series $\mathcal{Z}(g', h')$, for some $g' \in \mathbb{M}$, $h' \in C_\mathbb{M}(g')$.

Question 7.3.4. Consider the usual representation ρ of $G = \mathcal{S}_3$ by 3×3 permutation matrices, associating with $\pi \in \mathcal{S}_3$ the matrix $\rho(\pi)$ obtained from the identity matrix by applying π to the components of each column. For example,

$$\rho(123) = \begin{pmatrix} 0 & 0 & 1 \\ 1 & 0 & 0 \\ 0 & 1 & 0 \end{pmatrix}.$$

Show that ρ is completely reducible when considered as a modular representation over characteristic 2, but is not completely reducible when considered as a modular representation over characteristic 3. For both characteristic 2 and 3, compute its Brauer character using the definition given in Section 7.3.5.

Epilogue, or the squirrel who got away?

So, has Monstrous Moonshine been explained? According to most of the fathers of the subject, it hasn't. They consider VOAs in general, and $V^\natural$ in particular, to be too complicated to be God-given. The progress, though impressive, has broadened not lessened the fundamental mystery, they would argue.

For what it's worth, I don't completely agree. Explaining away a mystery is a little like grasping a bar of soap in a bathtub, or quenching a child's curiosity. Only extreme measures like pulling the plug, or growing up, ever really work. True progress means displacing the mystery, usually from the particular to the general. Why is the sky blue? Because of how light scatters in gases. Why are Hauptmoduls attached to each $g \in \mathbb{M}$? Because of $V^\natural$. Mystery exists wherever we can ask 'why' – like beauty, it's in the beholder's eye.

Understanding doesn't put an end to questions, it spices them. There's always a horizon, no matter how high you climb, beyond which everything is still hidden.

However, have we really isolated the key conjunction of properties needed for Moonshine to arise? Can we derive the Monster from Monstrous Moonshine? In Section 7.2.4 we make the case for a more direct, topological explanation for Moonshine involving compatible actions of the Virasoro algebra and the braid group $\mathcal{B}_3$. In any case, we need a second independent proof of Monstrous Moonshine.

Moonshine is now 'leaving the nest'. We are entering a consolidation phase, tidying up, generalising, simplifying, clarifying, working out more examples, climbing a few metres higher. Important and interesting discoveries will be made in the next few years, and yes, there still is mystery, but no longer does a Moonshiner feel like an illicit distiller: Moonshine is now a day-job!

Question and Answer in the Mountains[1]

They ask me why I live in the green mountains.
I smile and don't reply; my heart's at ease.
Peach blossoms flow downstream, leaving no trace –
And there are other earths and skies than these.

Li Bai 701 AD

---

[1] Translated by Vikram Seth, *Three Chinese Poets* (London, Faber and Faber, 1992).

Notation

Common notation	Meaning
$\mathrm{Re}\,z$, $\mathrm{Im}\,z$	real, imaginary parts of $z \in \mathbb{C}$
$\lfloor x \rfloor$, $\lceil x \rceil$	largest (smallest) integer $\leq x$ ($\geq x$)
$\langle \alpha, u \rangle$	Hermitian form
$(u\|v) = u \cdot v$	inner-product
M^t	transpose of matrix M
$M^\dagger$	matrix-adjoint $= \overline{M}^t$
$\|S\|$	cardinality of a set, order of a group
$\bar{z}$	complex conjugation
$\gcd(a, b)$	greatest common divisor
$\mathcal{B}_n$	the braid group on n strands (1.1.9)
$\mathbb{C}$	the complex numbers
$\mathbb{C}^\times$	the multiplicative group of nonzero $z \in \mathbb{C}$
CFT	conformal field theory
δ_{ij}	Kronecker delta: 1 if $i = j$, otherwise 0
$\delta(x)$	Dirac delta distribution
$\eta(\tau)$	Dedekind eta function (2.2.6b)
$\mathbb{H}$	the upper half-plane (0.1.1)
$\overline{\mathbb{H}}$	the upper half-plane with cusps (0.1.3)
I_n	$n \times n$ identity matrix
$j, J = j - 744$	Hauptmoduls for $\mathrm{SL}_2(\mathbb{Z})$ (0.1.8)
Λ	Leech lattice (Section 1.2.1)
$\mathbb{M}$	Monster finite simple group
$\mathbb{N}$	the nonnegative integers $\{0, 1, 2, \ldots\}$
$\mathbb{P}^1(\mathbb{C})$	Riemann sphere $\mathbb{C} \cup \{\infty\}$
q	$e^{2\pi i \tau}$, $\tau \in \mathbb{H}$
$\mathbb{Q}$	the rational numbers
$\mathbb{R}$	the real numbers
RCFT	rational conformal field theory
Σ	Riemann surface, complex curve
$\mathrm{SL}_n(R)$	$n \times n$ det $= 1$ matrices, entries in R
τ	point in $\mathbb{H}$
θ_3	Jacobi theta function (2.2.6a)
T_g	McKay–Thompson series (0.3.3)

$V^\natural$	Moonshine module (Section 7.2.1)
VOA	vertex operator algebra (Definition 5.1.3)
ξ_n	root of unity $\exp(2\pi i/n)$
$\mathbb{Z}$	the integers

Section 1.1

e	the identity in a group
$G \cong H$	groups G and H are isomorphic
$\mathbb{Z}_n$	ring and additive group $\mathbb{Z}/n\mathbb{Z}$
$\mathbb{F}_q$	finite field with q elements
$N \triangleleft G$	N normal subgroup of G
$H < G$	H subgroup of G
$\mathcal{F}_n$	the free group on n generators
$\langle g_1, \ldots, g_n \rangle$	the group generated by elements g_i
$\mathcal{D}_n$	dihedral group (1.1.1)
$N \times H$	direct product
$N \rtimes H$	semi-direct product
$\mathcal{S}_n$	symmetric group
$Z(G)$	centre of G
$\mathcal{A}_n$	alternating group
$M_{11}, \ldots, M_{24}$	Mathieu sporadics
$\mathrm{GL}_n(\mathbb{K})$	invertible matrices over field $\mathbb{K}$
K_g	conjugacy class $\{hgh^{-1}\}$
$\mathbb{C}G$	group algebra
$\mathcal{P}_n$	pure braid group
$\mathbb{C}[w, w^{-1}] = \mathbb{C}[w^\pm]$	Laurent polynomials
$[G, G]$	commutator subgroup $\langle ghg^{-1}h^{-1} \rangle$

Section 1.2

L	a lattice (Section 1.2.1)
L^*	the dual of a lattice (Section 1.2.1)
$II_{m,n}$	indefinite even self-dual lattice
$|L|$	determinant of lattice
$L_1 \oplus L_2$	orthogonal direct sum of lattices
S^n	the n-sphere
C^∞	smooth; all partials are continuous
$C^\infty(U)$	C^∞-functions $f : U \to \mathbb{R}$
$T_p(M)$	tangent space at $p \in M$
TM	tangent bundle
$\mathrm{Vect}(M)$	vector fields
$T_p^*(M)$	differential 1-forms
T^*M	cotangent bundle
$\pi_1(M, v) = \pi_1(M)$	fundamental group

$\mathbb{P}^n(\mathbb{R})$, $\mathbb{P}^n(\mathbb{C})$	projective n-space
$\mathfrak{C}_n$	configuration space (1.2.6)

Section 1.3

$\mathcal{H}$	complex separable Hilbert space
$\ell^2(\infty)$	Hilbert space of square-summable sequences
$\mathcal{C}_{cs}^{\infty}(\mathbb{R}^n)$	smooth functions with compact support
$\mathcal{S}(\mathbb{R}^n)$	Schwartz space
$L^2(X)$	Hilbert space of square-integrable functions
$\mathrm{d}\mu(x)$	Lebesgue measure
T^*	adjoint of operator T
$\int_X \mathcal{H}(x)\mathrm{d}\mu(x)$	direct integral of Hilbert spaces
$\mathcal{L}(\mathcal{H})$	bounded operators on $\mathcal{H}$
S'	commutant of set S
$M_n(\mathbb{C})$	$n \times n$ matrices over $\mathbb{C}$
type I_n, II_1, II_∞, III_λ	families of factors
$M \rtimes G$	crossed-product (1.3.4)

Section 1.4

$\mathfrak{g}$	a Lie algebra
$[xy]$	bracket (multiplication) in Lie algebra
$\mathfrak{Heis}$	Heisenberg algebra (1.4.3)
$\mathrm{SO}_{3,1}^+(\mathbb{R})$	Lorentz group
$\widetilde{G}$	universal cover of G
$\mathfrak{gl}_n$	Lie algebra of $n \times n$ matrices
$[\mathfrak{g}\mathfrak{h}]$	span $[xy]$, $x \in \mathfrak{g}$, $y \in \mathfrak{h}$
$Z(\mathfrak{g})$	centre of $\mathfrak{g}$
$\mathrm{ad}\,x$	adjoint operator $(\mathrm{ad}\,x)(y) = [xy]$
$\kappa(x\vert y)$	Killing form on $\mathfrak{g}$
$\mathfrak{g}(A)$	Lie algebra associated with Cartan matrix A
A_r, B_r, C_r, D_r	Lie algebras $\mathfrak{sl}_{r+1}(\mathbb{C})$, $\mathfrak{so}_{2r+1}(\mathbb{C})$, $\mathfrak{sp}_{2r}(\mathbb{C})$, $\mathfrak{so}_{2r}(\mathbb{C})$
E_6, E_7, E_8, F_4, G_2	exceptional simple Lie algebras
$\mathfrak{Witt}$	Witt algebra (1.4.9)
$\mathrm{Diff}(M)$, $\mathrm{Diff}^+(M)$	(orientation-preserving) diffeomorphism group

Section 1.5

$L(\lambda)$	irreducible module with highest weight λ
$P_+(\mathfrak{g})$	dominant integral weights
$M(\lambda)$	Verma module with highest weight λ
$\mathfrak{h}$	a Cartan subalgebra
$\alpha \in \Phi$	roots
$\mathfrak{g}_\alpha$	root-space (1.5.5b)
$(\alpha\vert\beta)$	Killing form on $\mathfrak{h}^*$
r_α	Weyl reflection (1.5.5c)
W	Weyl group

$\alpha_i \in \Delta$	simple roots in a base
ω_i	fundamental weights
$\beta, \mu \in \Omega(\rho)$	weights of representation ρ
V_β	weight-spaces (1.5.6b) in module V
$U(\mathfrak{g})$	universal enveloping algebra
$\mathrm{ch}_V(z)$	character (1.5.9a) of module V
ch_λ	character of $L(\lambda)$
ρ	representation; also, the Weyl vector $\sum_i \omega_i$
$\mathrm{ch}_\lambda^\gamma$	γ-twisted character
z, h	elements in $\mathfrak{h}$
$\gamma \in \mathrm{Aut}(\mathfrak{g})$	automorphisms of $\mathfrak{g}$
$\mathcal{B}(\mathcal{H})$	bounded operators with bounded inverse
$\widehat{G}$	unitary dual of Lie group G

Section 1.6

$\mathrm{Hom}(A, B)$	set of arrows = morphisms
Vect	category of vector spaces
Riem	category of Riemann surfaces
Braid	category of braids
a_{UVW}	associativity constraint
c_{UV}	commutativity constraint
Ribbon	category of ribbons
Ribbon$_S$	category of ribbons labelled from S

Section 1.7

$\mathbb{K}, \mathbb{L}$	fields
$\mathbb{K}[\alpha_1, \ldots, \alpha_n]$	field of polynomials in (algebraic) α_i
$[\mathbb{L} : \mathbb{K}]$	degree of field extension $\mathbb{K} \subset \mathbb{L}$
$\mathrm{Gal}(\mathbb{L}/\mathbb{K})$	Galois group
$\mathbb{Q}[\xi_n]$	cyclotomic field

Section 2.1

$\mathcal{K}(S)$	field of meromorphic functions
$\mathfrak{p}(z)$	Weierstrass function (2.1.6a)
$\theta_{r,s}(\tau, z)$	theta functions (2.1.7a)
$\mathcal{H}^k(S), \mathcal{M}^k(S)$	holomorphic/meromorphic k-differentials
$\mathfrak{M}_{g,n}$	moduli space for genus g, n punctures
$\Gamma_{g,n}$	mapping class group for genus g, n punctures
$\mathrm{Jac}(\Sigma)$	Jacobian variety
$\overline{\mathfrak{M}_{g,n}}$	Deligne–Mumford compactification
$\widehat{\mathfrak{M}}_{g,n}$	enhanced moduli space
$\widehat{\Gamma}_{g,n}$	enhanced mapping class group

Section 2.2

$\zeta(s)$	Riemann zeta function (2.2.3c)
$\Gamma(N)$	the principal congruence subgroup (2.2.4a)

$\Gamma_0(N)$	a congruence subgroup (2.2.4b)
Γ_θ	(2.2.5)
Θ_{t+L}	lattice theta function (2.2.11a), (2.3.7)

Section 2.3

$\Gamma(s)$	Gamma function
$\theta_1, \theta_2, \theta_4$	Jacobi theta functions

Section 2.4

$\widehat{\mathbb{Z}}_p$	p-adic integers
$\lim_\leftarrow$	projective limit
$\mathrm{Mp}_2(\mathbb{R})$	metaplectic group (2.4.9)

Section 2.5

A-D-E	the A_n, D_n, E_6, E_7, E_8 meta-pattern
PF2, PF2$^-$	condition on graphs

Section 3.1

$\mathfrak{Witt}$	Witt algebra (1.4.9)
ℓ_n	standard basis for $\mathfrak{Witt}$
$\mathfrak{Vir}$	Virasoro algebra (3.1.5)
L_n, C	standard basis of $\mathfrak{Vir}$
L_0	Hamilton operator, gives grading on $\mathfrak{Vir}$-modules
c, h	central charge, conformal weight
$M(c, h)$	Verma module
$V(c, h)$	irreducible module
$c_m, h_{m;rs}$	c, h for discrete series (3.1.6)
$\mathrm{ch}_{c,h}(\tau)$	the character of $V(c, h)$
$\mathrm{Diff}^+(S^1)$	diffeomorphism group of S^1

Section 3.2

$\overline{\mathfrak{g}}$	finite-dimensional semi-simple Lie algebra
$\mathcal{L}_{poly}\overline{\mathfrak{g}}$	polynomial loop algebra $S^1 \to \overline{\mathfrak{g}}$
$\overline{\mathfrak{g}}^{(1)} = X_r^{(1)}$	nontwisted affine algebra
$a_i, a_i^\vee$	labels, co-labels (Figure 3.2)
$\mathfrak{h}, \overline{\mathfrak{h}}$	Cartan subalgebras of $\mathfrak{g}$ and $\overline{\mathfrak{g}}$
$\overline{\mathfrak{g}}^{(N)}, N > 1$	twisted affine algebra
$M(\lambda) = M(\overline{\lambda}, k, u)$	Verma module with highest weight λ
$L(\lambda)$	irreducible module with highest weight λ
k	level
δ	imaginary root
P_+^k	integrable level k highest weights (3.2.8)
χ_λ	character of $L(\lambda)$
c_λ, h_λ	central charge, conformal weight (3.2.9)
$\lambda, \mu \in \Omega(V)$	weights

$h^\vee$	dual Coxeter number
KZ	Knizhnik–Zamolodchikov (Section 3.2.4)
$\mathcal{L}G$	loop group (Section 3.2.6)

Section 3.3

$\mathfrak{g}^e, \mathfrak{h}^e$	$\mathfrak{g}, \mathfrak{h}$ extended by derivations
$\mathfrak{g}_\tau$	toroidal algebra associated with 2-cocycle τ
$\mathfrak{g}_{\Sigma,P}$	Krichever–Novikov algebra

Section 3.4

χ_λ^α	α-twisted character
$\mathfrak{m}$	Monster Lie algebra (Sections 3.3.2, 7.2.2)

Section 4.1

L	Lagrangian
H	Hamiltonian
p^i	momentum components
c	speed of light
$\mathcal{L}$	Lagrangian density

Section 4.2

$\psi(\mathbf{x}, t)$	wave-function
$\hbar$	Planck's constant
$\mathcal{H}$	state-space (Hilbert space)
Ω	Fock space
$\widehat{a}, \widehat{a}^\dagger$	annihilation, creation operators
$\widehat{H}$	Hamiltonian operator
$|0\rangle$	vacuum
$|\star\rangle$	state
$|\mathrm{in}\rangle, |\mathrm{out}\rangle$	incoming, outgoing states
$\langle \mathrm{out}|\mathrm{in}\rangle$	transition amplitude
$\varphi(x), \phi(x), \psi(x)$	quantum fields
P^μ	energy–momentum operators
QED	quantum electrodynamics

Section 4.3

OPE	operator product expansion
$\mathfrak{B}^{(g,n)}$	space of chiral blocks
$\mathcal{V}$	chiral algebra (VOA)
$T(z)$	stress–energy tensor
WZW	Wess–Zumino–Witten model
$M \in \Phi(\mathcal{V})$	irreducible $\mathcal{V}$-module
$\chi_M(\tau)$	graded dimension (4.3.8a)
$\mathcal{Z}(\tau, \overline{\tau})$	1-loop partition function (4.3.8b)
$\mathcal{V}^h$	h-twisted sector

$\mathcal{V}^G$	fixed-point subalgebra for group G
$\mathcal{Z}_{(g,h)}(\tau)$	(4.3.14b)
$\chi_{(h,\rho)}(\tau)$	(4.3.15c)

Section 4.4

C_m	disjoint union of m circles
$\mathbf{C}_{g,k}$	connected component in Segal's $\text{Hom}(C_m, C_n)$
$\mathbf{Vect}_f$	category of finite-dimensional spaces

Section 5.1

$W[z^{\pm 1}]$	Laurent polynomials in z, coefficients in W
$W[[z^{\pm 1}]]$	formal power series in z
$\delta(z)$	multiplicative Dirac delta (5.1.2)
$\text{Res}_z(f)$	residue (5.1.3a)
$\mathcal{V}$	a VOA
$Y(a, z)$	vertex operator
$\mathcal{V}_n$	space of conformal weight n vectors
$u_{(n)}$	mode of $u \in \mathcal{V}$
ω	conformal vector
$\mathbf{1}$	vacuum vector in $\mathcal{V}$
$\chi_{\mathcal{V}}(\tau)$	graded dimension (5.1.10b)
$(\star|\star)$	invariant bilinear form (5.1.11)

Section 5.2

$o(u)$	the 'zero-mode' $u_{(n-1)}$, for $u \in \mathcal{V}_n$
$\mathcal{PV}_n$	conformal primaries (5.2.3)
$\text{Aut}(\mathcal{V})$	automorphism group of $\mathcal{V}$
$\mathcal{V}(\mathbb{C}^n)$	Heisenberg VOA
$\mathcal{V}(\mathfrak{g}, k)$	affine algebra VOA
$\mathcal{V}(L)$	lattice VOA
$\mathcal{V}^G$	G-fixed points in $\mathcal{V}$

Section 5.3

M	V-module
$Y_M(u, z)$	vertex operator for M
M_α	conformal weight α space
$M^\star$	contragredient ($=$ dual) module
$\Phi(\mathcal{V})$	irreducible $\mathcal{V}$-modules
$M \boxtimes N$	fusion
$\mathcal{N}_{MN}^P$	fusion multiplicities (5.3.3)
$h(M)$	conformal weight of M
$M_{n \times n}$	the algebra of $n \times n$ matrices
$A(\mathcal{V})$	Zhu's algebra
$\chi_M(\tau, v)$	character of M (5.3.13)
$\chi_M^J(\tau, u, v)$	Jacobi character (5.3.14)

$L[n]$	a second Virasoro action on $\mathcal{V}$
$\mathcal{V}[n]$	homogeneous spaces for $L[0]$
$\mathrm{Inn}(\mathcal{V})$	inner-automorphisms of $\mathcal{V}$
$\mathcal{Z}(M, h; \tau)$	h-twisted graded trace (5.3.23)

Section 5.4

X	smooth complex variety
$\mathcal{MSV}_X$	chiral de Rham complex
$\mathcal{MSV}_X(X)$	global sections
$\mathcal{T}(X)$	Tamanoi's invariant

Section 6.1

$\mathcal{N}_{ab}^c$	fusion multiplicities
$S = (S_{ab})$	modular matrix
$\mathcal{N}_{a_1,\ldots,a_n}^{(g,m+n)\ b_1,\ldots,b_m}$	Verlinde dimensions (6.1.2)
$T = (T_{ab})$	diagonal matrix in modular data
$\mathcal{Z} = (\mathcal{Z}_{ab})$	modular invariant
$\mathcal{Y}(w, z)$	intertwining operator
$\mathcal{V}\left(\begin{smallmatrix} c \\ a\ b \end{smallmatrix}\right)$	space of intertwining operators of type $\left(\begin{smallmatrix} c \\ a\ b \end{smallmatrix}\right)$

Section 6.2

$\mathrm{Gr}(m, n)$	Grassmannian
$L(x)$	Roger's dilogarithm
RVOA	rational vertex operator algebra
$U_q(\mathfrak{g})$	quantum group
$R(q)$	family of solutions to (6.2.8)
$C_G(g)$	centraliser of g in G
$N \subset M$	subfactor
$[M : N]$	Jones index
$_M X_N$	$M - N$ bimodule

Section 6.3

F_n	Fermat curve $x^n + y^n + z^n = 0$
$\mathrm{Jac}(F_n)$	Jacobian of Fermat curve
$\Gamma_{\mathbb{Q}}$	absolute Galois group of $\mathbb{Q}$
$\overline{\mathbb{Q}}$	algebraic closure of $\mathbb{Q}$
χ^{cyclo}	cyclotomic character
$\widehat{G}$	profinite completion of G

Section 7.1

$\mathbb{B}$	Baby Monster
Fi_{24}'	a Fischer group
$\mathbb{M} \wr \mathbb{Z}_2$	Bimonster
Γ_g	fixing group of T_g
$o(g)$	order of g

$\Gamma_0(p)+$	(7.1.5)
$k\|N^\infty$	any prime dividing k divides N
Section 7.3	
$A^{f\natural}$	vertex operator superalgebra for Co_1
$V\mathbb{B}^\natural$	Baby Monster Moonshine module
$N_{(g,h)}(\tau)$	Norton series
$\beta(g)$	Brauer character
$B(g,h;\tau)$	Modular Moonshine series
$\|\star\rangle\rangle$	boundary states in CFT

References

[1] T. Abe, G. Buhl and C. Dong, 'Rationality, regularity and C_2 co-finiteness', *Trans. Amer. Math. Soc.* **356** (2004) 3391–402.

[2] D. Adamović and A. Milas, 'Vertex operator algebras associated to modular invariant representations for $A_1^{(1)}$', *Math. Res. Lett.* **2** (1995) 563–75.

[3] I. Affleck, 'Universal term in the free energy at a critical point and the conformal anomaly', *Phys. Rev. Lett.* **56** (1986) 746–8.

[4] S. Agnihotri and C. Woodward, 'Eigenvalues of products of unitary matrices and quantum Schubert calculus', *Math. Res. Lett.* **5** (1998) 817–36.

[5] O. Aharony, S. Gubser, J. Maldacena, H. Ooguri and Y. Oz, 'Large N field theories, string theory and gravity', *Phys. Rep.* **323** (2000) 183–386.

[6] D. Alexander, C. Cummins, J. McKay and C. Simons, 'Completely replicable functions', *Groups, Combinatorics and Geometry (Durham, 1990)* (Cambridge, Cambridge University Press, 1992) 87–98.

[7] D. Altschuler, P. Ruelle and E. Thiran, 'On parity functions in conformal field theories', *J. Phys. A: Math. Gen.* **32** (1999) 3555–70.

[8] O. Alvarez, 'Theory of strings with boundaries: fluctuations, topology and quantum geometry', *Nucl. Phys.* **B216** (1983) 125–84.

[9] L. Alvarez-Gaumé, G. Moore and C. Vafa, 'Theta functions, modular invariance and strings', *Commun. Math. Phys.* **106** (1986) 1–40.

[10] H. H. Anderson and J. Paradowski, 'Fusion categories arising from semi-simple Lie algebras', *Commun. Math. Phys.* **169** (1995) 563–88.

[11] J. W. Anderson, *Hyperbolic Geometry* (London, Springer, 1999).

[12] E. Arbarello and C. DeConcini, 'On a set of equations characterising Riemann surfaces', *Ann. Math.* **120** (1984) 119–40.

[13] E. Arbarello, C. DeConcini, V. G. Kac and C. Procesi, 'Moduli spaces of curves and representation theory', *Commun. Math. Phys.* **117** (1988) 1–36.

[14] E. Ardonne, P. Bouwknegt and P. Dawson, 'K-matrices for 2D conformal field theories', *Nucl. Phys.* **B660** (2003) 473–531.

[15] V. I. Arnold, *Mathematical Methods of Classical Mechanics* (New York, Springer, 1978).

[16] V. I. Arnold, *Catastrophe Theory*, 2nd edn (Berlin, Springer, 1997).

[17] V. I. Arnold, 'From Hilbert's superposition problem to dynamical systems', *The Arnoldfest (Toronto, 1997)*, Fields Inst. Commun. **24** (Providence, American Mathematical Society, 1999) 1–18.

[18] V. I. Arnold, 'Symplectization, complexification and mathematical trinities', *The Arnoldfest (Toronto, 1997)*, Fields Inst. Commun. **24** (Providence, American Mathematical Society, 1999) 23–37.

[19] V. I. Arnold, V. V. Gorynuov, O. V. Lyashko and V. A. Vasil'ev, *Singularity Theory I* (Berlin, Springer, 1998).

[20] J. Arthur, 'The trace formula and Hecke operators', *Number Theory, Trace Formulas and Discrete Groups (Oslo, 1987)* (Boston, Academic Press, 1989) 11–27.

[21] M. G. Aschbacher, 'Finite groups acting on homology manifolds', *Olga Taussky-Todd: in Memoriam, Pacific J. Math. Special Issue* (Berkeley, Pacific Journal of Mathematics, 1997) 3–36.

[22] M. Aschbacher, 'The status of the classification of the finite simple groups', *Notices Amer. Math. Soc.* **51** (2004) 736–40.

[23] T. Asai, 'The reciprocity of Dedekind sums and the factor set for the universal covering group of $SL_2(\mathbb{R})$', *Nagoya Math. J.* **37** (1970) 67–80.

[24] M. Atiyah, 'The logarithm of the Dedekind η-function', *Math. Ann.* **278** (1987) 335–80.

[25] M. Atiyah, *The Geometry and Physics of Knots* (Cambridge, Cambridge University Press, 1990).

[26] H. Awata and Y. Yamada, 'Fusion rules for the fractional level $\widehat{sl(2)}$ algebra', *Mod. Phys. Lett.* **A7** (1992) 1185–95.

[27] J. A. de Azcárraga and J. M. Izquierdo, *Lie Groups, Lie Algebras, Cohomology and Some Applications in Physics* (Cambridge, Cambridge University Press, 1995).

[28] J. C. Baez, 'Higher-dimensional algebra and Planck-scale physics', *Physics Meets Philosophy at the Planck Scale* (Cambridge, Cambridge University Press, 2001) 177–95.

[29] J. C. Baez, 'The octonions', *Bull. Amer. Math. Soc.* **39** (2002) 145–205.

[30] J. C. Baez and J. Dolan, 'From finite sets to Feynman diagrams', *Mathematics Unlimited – 2001 and Beyond* (Berlin, Springer, 2001) 29–50.

[31] F. A. Bais, P. van Driel and M. de Wild Propitus, 'Anyons in discrete gauge theories with Chern–Simons terms', *Nucl. Phys.* **393** (1993) 547–70.

[32] B. Bakalov and A. Kirillov, Jr., *Lectures on Tensor Categories and Modular Functors* (Providence, American Mathematical Society, 2001).

[33] E. Bannai, 'Association schemes and fusion algebras (an introduction)', *J. Alg. Combin.* **2** (1993) 327–44.

[34] P. Bántay, 'Orbifolds, Hopf algebras and the Moonshine', *Lett. Math. Phys.* **22** (1991) 187–94.

[35] P. Bántay, 'Algebraic aspects of orbifold models', *Int. J. Mod. Phys.* **A9** (1994) 1443–56.

[36] P. Bántay, 'Higher genus moonshine', *Moonshine, the Monster, and Related Topics (South Hadley, 1994)*, Contemp. Math. **193** (Providence, American Mathematical Society, 1996) 1–8.

[37] P. Bántay, 'The kernel of the modular representation and the Galois action in RCFT', *Commun. Math. Phys.* **233** (2003) 423–38.

[38] D. Bar-Natan, 'On the Vassiliev knot invariants', *Topol.* **34** (1995) 423–72.

[39] D. Bar-Natan, 'On associators and the Grothendieck–Teichmüller group, I', *Selecta Math. (N.S.)* **4** (1998) 183–212.

[40] D. Bar-Natan, T. Q. T. Le and D. P. Thurston, 'Two applications of elementary knot theory to Lie algebras and Vassiliev invariants', *Geom. Topol.* **7** (2003) 1–31.

[41] H. Barcelo and A. Ram, 'Combinatorial representation theory', *New Perspectives in Algebraic Combinatorics (Berkeley, 1996–97)* (Cambridge, Cambridge University Press, 1999) 23–90.

[42] K. Bardacki and M. Halpern, 'New dual quark models', *Phys. Rev.* **D3** (1971) 2493–509.

[43] J. Barge and E. Ghys, 'Cocycles d'Euler et de Maslov', *Math. Ann.* **294** (1992) 235–65.

[44] V. Bargmann, 'Irreducible unitary representations of the Lorentz group', *Ann. Math.* **48** (1947) 568–640.

[45] I. G. Bashmakova, *Diophantus and Diophantine Equations* (Washington, Mathematical Association of America, 1997).

[46] M. Bauer, A. Coste, C. Itzykson and P. Ruelle, 'Comments on the links between SU(3) modular invariants, simple factors in the Jacobian of Fermat curves, and rational triangular billiards', *J. Geom. Phys.* **22** (1997) 134–89.

[47] R. E. Behrend, P. A. Pearce, V. B. Petkova and J.-B. Zuber, 'Boundary conditions in rational conformal field theories', *Nucl. Phys.* **B579** (2000) 707–73.

[48] A. A. Beilinson and V. Drinfel'd, *Chiral Algebras* (Providence, American Mathematical Society, 2004).

[49] A. A. Beilinson and V. V. Schechtman, 'Determinant bundles and Virasoro algebras', *Commun. Math. Phys.* **118** (1988) 651–701.

[50] A. A. Belavin, A. M. Polyakov and A. B. Zamolodchikov, 'Infinite conformal symmetry in two-dimensional quantum field theory', *Nucl. Phys.* **B241** (1984) 33–380.

[51] G. M. Bergman, 'Everybody knows what a Hopf algebra is', *Group Actions on Rings (Brunswick, 1984)*, Contemp. Math. **43** (Providence, American Mathematical Society, 1985) 25–48.

[52] N. Berline, E. Getzler and M. Vergne, *Heat Kernels and Dirac Operators* (Berlin, Springer, 1992).

[53] S. Berman and Y. Billig, 'Irreducible representations for toroidal Lie algebras', *J. Alg.* **221** (1999) 188–231.

[54] S. Berman, Y. Billig and J. Szmigielski, 'Vertex operator algebras and the representation theory of toroidal algebras', *Recent Developments in Infinite-Dimensional Lie Algebras and Conformal Field Theory (Charlottesville, 2000)*, Contemp. Math. **297** (Providence, American Mathematical Society, 2002) 1–26.

[55] S. Berman and K. H. Parshall, 'Victor Kac and Robert Moody: their paths to Kac–Moody Lie algebras', *Math. Intell.* **24** (2002) 50–60.

[56] L. Bers, 'Finite dimensional Teichmüller spaces and generalisations', *Bull. Amer. Math. Soc.* **5** (1981) 131–72.

[57] A. Bertram, 'Quantum Schubert calculus', *Adv. Math.* **128** (1997) 289–305.

[58] L. Birke, J. Fuchs and C. Schweigert, 'Symmetry breaking boundary conditions and WZW orbifolds', *Adv. Theor. Math. Phys.* **3** (1999) 671–726.

[59] J. S. Birman, *Braids, Links, and Mapping Class Groups* (Princeton, Princeton University Press, 1974).

[60] J. S. Birman, 'Mapping class groups of surfaces', *Braids (Santa Cruz, 1986)*, Contemp. Math. **78** (Providence, American Mathematical Society, 1988) 13–43.

[61] J. S. Birman, 'New points of view in knot theory', *Bull. Amer. Math. Soc.* **28** (1993) 253–87.

[62] D. Birmingham, M. Blau, M. Rakowski and G. Thompson, 'Topological field theory', *Phys. Rep.* **209** (1991) 129–340.

[63] S. Bloch, 'Zeta values and differential operators on the circle', *J. Alg.* **182** (1996) 476–500.

[64] D. M. Bloom, 'On the coefficients of the cyclotomic polynomials', *Amer. Math. Monthly* **75** (1968) 372–7.

[65] J. Böckenhauer and D. E. Evans, 'Modular invariants, graphs and α-induction for nets of subfactors III', *Commun. Math. Phys.* **205** (1999) 183–228.

[66] J. Böckenhauer and D. E. Evans, 'Subfactors and modular invariants', *Mathematical Physics in Mathematics and Physics (Sienna, 2000)* (Providence, American Mathematical Society, 2001) 11–37.

[67] N. N. Bogulubov, A. A. Logunov and I. T. Todorov, *Introduction to Axiomatic Quantum Field Theory* (Reading, W. A. Benjamin Inc., 1975).

[68] R. E. Borcherds, 'Vertex algebras, Kac–Moody algebras, and the Monster', *Proc. Natl. Acad. Sci. (USA)* **83** (1986) 3068–71.

[69] R. E. Borcherds, 'Generalized Kac–Moody algebras', *J. Algebra* **115** (1988) 501–12.

[70] R. E. Borcherds, 'The monster Lie algebra', *Adv. Math.* **83** (1990) 30–47.

[71] R. E. Borcherds, 'Central extensions of generalized Kac–Moody algebras', *J. Algebra* **140** (1991) 330–35.

[72] R. E. Borcherds, 'Monstrous moonshine and monstrous Lie superalgebras', *Invent. Math.* **109** (1992) 405–44.

[73] R. E. Borcherds, 'Sporadic groups and string theory', *First European Congress of Math. (Paris, 1992)*, Vol. I (Basel, Birkhäuser, 1994) 411–21.

[74] R. E. Borcherds, 'Automorphic forms on $O_{s+2,2}(\mathbb{R})^+$ and generalized Kac–Moody algebras', *Proceedings of the International Congress of Mathematicians (Zürich, 1994)* (Basel, Birkhäuser, 1995) 744–52.

[75] R. E. Borcherds, 'What is moonshine?', *Proceedings of the International Congress of Mathematicians (Berlin, 1998)* (Bielefeld, Documenta Mathematica, 1998) 607–15.

[76] R. E. Borcherds, 'Automorphic forms with singularities on Grassmannians', *Invent. Math.* **132** (1998) 491–562.

[77] R. E. Borcherds, 'Modular moonshine III', *Duke Math. J.* **93** (1998) 129–54.

[78] R. E. Borcherds, 'Problems in Moonshine', *First International Congress of Chinese Mathematics (Beijing, 1998)* (Providence, American Mathematical Society, 2001) 3–10.

[79] R. E. Borcherds and A. J. E. Ryba, 'Modular moonshine II', *Duke Math. J.* **83** (1996) 435–59.

[80] A. Borel *et al.*, *Algebraic D-Modules* (Orlando, Academic Press, 1987).

[81] L. A. Borisov and A. Libgober, 'Elliptic genera of toric varieties and applications to mirror symmetry', *Invent. Math.* **140** (2000) 453–85.

[82] J. M. Borwein, P. B. Borwein and D. H. Bailey, 'Ramanujan, modular equations, and approximations to pi or How to compute one billion digits of pi', *Amer. Math. Monthly* **96** (1989) 201–19.

[83] R. Bott, 'On induced representations', *The Mathematical Heritage of Hermann Weyl*, Proc. Symp. Pure Math. **48** (Providence, American Mathematical Society, 1988) 1–13.

[84] N. Bourbaki, *Lie Groups and Algebras,* chapters 4–6 (Berlin, Springer, 2002).

[85] M. J. Bowick and S. G. Rajeev, 'The holomorphic geometry of closed bosonic string theory and Diff S^1/S^1', *Nucl. Phys.* **293** (1987) 348–84.

[86] E. Brieskorn, 'Singular elements of semi-simple algebraic groups', *Actes Congrès International des Mathématiciens, Vol. 2 (Nice, 1970)* (Paris, Gauthier-Villars, 1971) 279–84.

[87] D. Brungs and W. Nahm, 'The associative algebras of conformal field theory', *Lett. Math. Phys.* **47** (1999) 379–83.

[88] A. S. Buch, A. Kresch and H. Tamvakis, 'Gromov–Witten invariants on Grassmannians', *J. Amer. Math. Soc.* **16** (2003) 901–15.

[89] D. Bump, *Automomorphic Forms and Representations* (Cambridge, Cambridge University Press, 1997).

[90] D. Bump, J. W. Cagwell, D. Gaitsgory, E. de Shalit, E. Kowalski and S. S. Kudla, *An Introduction to the Langlands Program* (Boston, Birkhäuser, 2003).

[91] A. Cappelli, C. Itzykson and J.-B. Zuber, 'The A-D-E classification of $A_1^{(1)}$ and minimal conformal field theories', *Commun. Math. Phys.* **113** (1987) 1–26.

[92] R. Carter, G. Segal and I. M. Macdonald, *Lectures on Lie Groups and Lie Algebras* (Cambridge, Cambridge University Press, 1995).

[93] P. Cartier, 'A mad day's work: from Grothendieck to Connes and Kontsevich. The evolution of concepts of space and symmetry', *Bull. Amer. Math. Soc.* **38** (2001) 389–408.

[94] A. Cayley, *An Elementary Treatise on Elliptic Functions*, 2nd edn (New York, Dover, 1961).

[95] S.-P. Chan, M.-L. Lang and C.-H. Lim, 'Some modular functions associated to Lie algebra E_8', *Math. Z.* **211** (1992) 223–46.

[96] G. Chapline, 'Unification of gravity and elementary particle interactions in 26 dimensions?', *Phys. Lett.* **B158** (1985) 393–6.

[97] V. Chari, 'Integrable representations of affine Lie algebras', *Invent. Math.* **85** (1986) 317–35.

[98] V. Chari and A. Pressley, *A Guide to Quantum Groups* (Cambridge, Cambridge University Press, 1994).

[99] S. Chaudhuri and D. A. Lowe, 'Monstrous string-string duality', *Nucl. Phys.* **B469** (1996) 21–36.

[100] G. Y. Chen, 'A new characterization of finite simple groups', *Chinese Sci. Bull.* **40** (1995) 446–50.

[101] I. Chen and N. Yui, 'Singular values of Thompson series', *Groups, Difference Sets, and the Monster (Columbus, 1993)* (Berlin, de Gruyter, 1996) 255–326.

[102] H. Cohn and J. McKay, 'Spontaneous generation of modular invariants', *Math. of Comput.* **65** (1996) 1295–309.

[103] D. J. Collins, R. I. Grigorchuk, P. F. Kurchanov and H. Zieschang, *Combinatorial Group Theory and Applications to Geometry* (Berlin, Springer, 1998).

[104] L. Conlon, *Differentiable Manifolds*, 2nd edn (Boston, Birkhäuser, 2001).

[105] A. Connes and D. Kreimer, 'From local perturbation theory to Hopf- and Lie-algebras of Feynman graphs', *Mathematical Physics in Mathematics and Physics (Siena, 2000)* (Providence, American Mathematical Society, 2001) 105–14.

[106] A. Connes and M. Marcolli, 'Renormalization and motivic Galois theory', *Int. Math. Res. Not.* **2004**, no. 76, 4073–91.

[107] J. H. Conway, 'Monsters and Moonshine', *Math. Intelligencer* **2** (1980) 165–71.

[108] J. H. Conway, *The Sensual (Quadratic) Form* (Washington, Mathematical Association of America, 1997).

[109] J. H. Conway, R. T. Curtis, S. P. Norton, R. A. Parker and R. A. Wilson, *An Atlas of Finite Groups* (Oxford, Oxford University Press, 1985).

[110] J. H. Conway, J. McKay and A. Sebbar, 'On the discrete groups of moonshine', *Proc. Amer. Math. Soc.* **132** (2004) 2233–40.

[111] J. H. Conway and S. P. Norton, 'Monstrous moonshine', *Bull. London Math. Soc.* **11** (1979) 308–39.

[112] J. H. Conway, S. P. Norton and L. H. Soicher, 'The Bimonster, the group Y_{555}, and the projective plane of order 3', *Computers in Algebra* (New York, Dekker, 1988) 27–50.

[113] J. H. Conway and N. J. A. Sloane, *Sphere Packings, Lattices and Groups*, 3rd edn (Berlin, Springer, 1999).

[114] A. Coste and T. Gannon, 'Remarks on Galois in rational conformal field theories', *Phys. Lett.* **B323** (1994) 316–21.

[115] A. Coste, T. Gannon and P. Ruelle, 'Finite group modular data', *Nucl. Phys.* **B581** (2000) 679–717.

[116] S. C. Coutinho, *A Primer of Algebraic D-Modules* (Cambridge, Cambridge University Press, 1995).

[117] D. Cox, *Primes of the Form $x^2 + ny^2$* (New York, Wiley, 1989).

[118] L. Crane and D. N. Yetter, 'Deformations of (bi)tensor categories', *Cah. Top. Géom. Diff. Catég.* **39** (1998) 163–80.

[119] B. Craps, M. R. Gaberdiel and J. A. Harvey, 'Monstrous branes', *Commun. Math. Phys.* **234** (2003) 229–51.

[120] C. J. Cummins, 'Modular equations and discrete, genus-zero subgroups of SL(2, $\mathbb{R}$) containing $\Gamma(N)$', *Canad. Math. Bull.* **45** (2002) 36–45.

[121] C. J. Cummins, 'Congruence subgroups of groups commensurable with $PSL(2, \mathbb{Z})$ of genus 0 and 1', *Experiment. Math.* **13** (2004) 361–82.

[122] C. J. Cummins and T. Gannon, 'Modular equations and the genus zero property', *Invent. Math.* **129** (1997) 413–43.

[123] C. J. Cummins and S. P. Norton, 'Rational Hauptmodul are replicable', *Canad. J. Math.* **47** (1995) 1201–18.

[124] D. G. Currie, T. F. Jordan and E. C. G. Sudarshan, 'Relativistic invariance and Hamiltonian theories of interacting particles', *Rev. Modern Phys.* **35** (1963) 350–75.

[125] C. W. Curtis and I. Reiner, *Methods of Representation Theory with Applications to Finite Groups and Orders*, Vol. I (New York, Wiley, 1981).

[126] P. Cvitanović, *Group Theory* (web-book available at http://www.nbi.dk/GroupTheory).

[127] T. Damour, M. Henneaux and H. Nicolai, 'E_{10} and a small tension expansion of M theory', *Phys. Rev. Lett.* **89** (2002) 221–601.

[128] P. Degiovanni, 'Equations de Moore et Seiberg, théories topologiques et théorie de Galois', *Helv. Phys. Acta* **67** (1994) 799–883.

[129] P. Deligne and B. H. Gross, 'La série exceptionelle des groupes de Lie', *C. R. Acad. Sci. Paris* **335** (2002) 877–81.

[130] E. D'Hoker and D. H. Phong, 'On determinants of Laplacians on Riemann surfaces', *Commun. Math. Phys.* **104** (1986) 537–45.

[131] P. Di Francesco, P. Mathieu and D. Sénéchal, *Conformal Field Theory* (New York, Springer, 1996).

[132] R. Dijkgraaf, 'Mirror symmetry and elliptic curves', *The Moduli Space of Curves (Texel Island, 1994)* (Boston, Birkhäuser, 1995) 149–63.

[133] R. Dijkgraaf, 'Fields, strings and duality', *Quantum Symmetries (Les Houches, 1995)* (Amsterdam, Elsevier, 1998) 3–147.

[134] R. Dijkgraaf, 'The mathematics of fivebranes', *Proceedings of the International Congress of Mathematicians (Berlin, 1998)*, Vol. III (Bielefeld, Documenta Mathematica, 1998) 133–42.

[135] R. Dijkgraaf, V. Pasquier and P. Roche, 'Quasi-Hopf groups, group cohomology and orbifold models', *Nucl. Phys. (Proc. Suppl.)* **B18** (1991) 60–72.

[136] R. Dijkgraaf, C. Vafa, E. Verlinde and H. Verlinde, 'The operator algebra of orbifold models', *Commun. Math. Phys.* **123** (1989) 485–526.

[137] R. Dijkgraaf, E. Verlinde and H. Verlinde, '$c = 1$ conformal field theories on Riemann surfaces', *Commun. Math. Phys.* **115** (1988) 649–90.

[138] R. Dijkgraaf and E. Witten, 'Topological gauge theories and group cohomology', *Commun. Math. Phys.* **129** (1990) 393–429.

[139] P. A. M. Dirac, 'The relation between mathematics and physics', *Proc. Roy. Soc. Edinburgh* **59** (1939) 122–9.

[140] P. A. M. Dirac, 'The large number hypothesis and the Einstein theory of gravitation', *Proc. Roy. Soc. London* **A365** (1979) 19–30.

[141] G. Dito and D. Sternheimer, 'Deformation quantization: genesis, developments and metamorphoses', *Deformation Quantization* (Berlin, Walter de Gruyter, 2002) 9–54.

[142] L. Dixon, P. Ginsparg and J. Harvey, 'Beauty and the beast: superconformal symmetry in a monster module', *Commun. Math. Phys.* **119** (1988) 221–41.

[143] L. Dixon, J. A. Harvey, C. Vafa and E. Witten, 'Strings on orbifolds', *Nucl. Phys.* **B261** (1985) 678–86.

[144] C. Dong, 'Vertex algebras associated with even lattices', *J. Alg.* **160** (1993) 245–65.

[145] C. Dong, 'Representations of the Moonshine module vertex operator algebra', *Mathematical Aspects of Conformal and Topological Field Theories and Quantum Groups (Mount Holyoke, 1992)*, Contemp. Math. **175** (Providence, American Mathematical Society, 1994) 27–36.

[146] C. Dong, R. L. Griess, Jr. and C. H. Lam, 'On the uniqueness of the Moonshine vertex operator algebra', Preprint (arXiv: math.QA/0506321).

[147] C. Dong, H. Li and G. Mason, 'Simple currents and extensions of vertex operator algebras', *Commun. Math. Phys.* **180** (1996) 671–707.

[148] C. Dong H. Li and G. Mason, 'Vertex operator algebras associated to admissible representations of $\widehat{sl_2}$', *Commun. Math. Phys.* **184** (1997) 65–93.

[149] C. Dong, H. Li and G. Mason, 'Vertex operator algebras and associative algebras', *J. Alg.* **206** (1998) 67–96.

[150] C. Dong, H. Li and G. Mason, 'Modular-invariance of trace functions in orbifold theory and generalised moonshine', *Commun. Math. Phys.* **214** (2000) 1–56.

[151] C. Dong, K. Liu and X. Ma, 'Elliptic genus and vertex operator algebras', *Pure Appl. Math. Q.* **1** (2005) 791–815.

[152] C. Dong and G. Mason, 'Nonabelian orbifolds and the boson–fermion correspondence', *Commun. Math. Phys.* **163** (1994) 523–59.

[153] C. Dong and G. Mason, 'An orbifold theory of genus zero associated to the sporadic group M_{24}', *Commun. Math. Phys.* **164** (1994) 87–104.

[154] C. Dong and G. Mason, 'Vertex operator algebras and moonshine: a survey', *Progress in Algebraic Combinatorics*, Adv. Stud. Pure Math. **24** (Tokyo, Mathematical Society of Japan, 1996) 101–36.

[155] C. Dong and G. Mason, 'Monstrous moonshine of higher weight', *Acta Math.* **185** (2000) 101–21.

[156] C. Dong and G. Mason, 'Rational vertex operator algebras and the effective central charge', *Int. Math. Res. Not.* **2004** 2989–3008.

[157] C. Dong and F. Xu, 'Conformal nets associated with lattices and their orbifolds', Preprint (arXiv: math.QA/0411499).

[158] C. F. Doran, 'Picard–Fuchs uniformization and modularity of the mirror map', *Commun. Math. Phys.* **212** (2000) 625–47.

[159] N. Dorey, T. J. Hollowood, V. V. Khoze and M. P. Mattis, 'The calculus of many instantons', *Phys. Rep.* **371** (2002) 231–459.

[160] V. G. Drinfel'd, 'Quantum groups', *Proceedings of the International Congress of Mathematicians (Berkeley, 1986)* (Providence, American Mathematical Society, 1987) 798–820.

[161] V. G. Drinfel'd, 'On quasitriangular quasi-Hopf algebras on a group that is closely related with Gal $(\overline{\mathbb{Q}}/\mathbb{Q})$', *Leningrad. Math. J.* **2** (1991) 829–60.

[162] D. S. Dummit and R. M. Foote, *Abstract Algebra*, 3rd edn (Chichester, Wiley, 2004).

[163] J. F. Duncan, 'Super-Moonshine for Conway's largest sporadic group', Preprint (arXiv: math.RT/0502267).

[164] J. L. Dupont and C.-H. Sah, 'Dilogarithm identities in conformal field theory and group homology', *Commun. Math. Phys.* **161** (1994) 265–282.

[165] P. DuVal, 'On isolated singularities which do not affect the conditions of adjunction', *Proc. Cambridge Phil. Soc.* **30** (1934) 453–59.

[166] F. Dyson, 'Missed opportunities', *Bull. Amer. Math. Soc.* **78** (1972) 635–52.

[167] F. J. Dyson, 'Unfashionable pursuits', *Math. Intelligencer* **5**, no. 3 (1983) 47–54.

[168] J. Eells, 'Automorphisms of the circle – and Teichmüller theory', *Collection of Papers on Geometry, Analysis and Mathematical Physics* (River Edge, World Scientific, 1997) 44–52.

[169] W. Eholzer and N.-P. Skoruppa, 'Conformal characters and theta series', *Lett. Math. Phys.* **35** (1995) 197–211.

[170] M. Eichler and D. Zagier, *The Theory of Jacobi Forms* (Boston, Birkhäuser, 1985).

[171] F. Englert, L. Houart, A. Taormina and P. West, 'The symmetry of M-theories', *J. High Energy Phys.* **9** (2003) 020.

[172] J. G. Esteve, 'Anomalies in conservation laws in the Hamiltonian formalism', *Phys. Rev.* **D34** (1986) 674–7.

[173] S. Eswara Rao and R. V. Moody, 'Vertex representations for n-toroidal Lie algebras and a generalisation of the Virasoro algebra', *Commun. Math. Phys.* **159** (1994) 239–64.

[174] P. Etingof, I. B. Frenkel and A. A. Kirillov, Jr., *Lectures on Representation Theory and Knizhnik–Zamolodchikov Equations* (Providence, American Mathematical Society, 1998).

[175] P. Etingof and S. Gelaki, 'Isocategorical groups', *Intern. Math. Res. Notices* **2001**, no. 2 (2001) 59–76.

[176] P. Etingof, D. Kazhdan and A. Polishchuk, 'When is the Fourier transform of an elementary function elementary?', *Selecta Math. (N.S.)* **8** (2002) 27–66.

[177] D. E. Evans and Y. Kawahigashi, *Quantum Symmetries on Operator Algebras* (Oxford, Oxford University Press, 1998).

[178] D. E. Evans and P. R. Pinto, 'Subfactor realization of modular invariants', *Commun. Math. Phys.* **237** (2003) 309–63.

[179] G. Faltings, 'A proof for the Verlinde formula', *I. Alg. Geom.* **3** (1994) 347–74.

[180] H. M. Farkas and I. Kra, *Riemann Surfaces* (New York, Springer, 1980).

[181] H. D. Fegan, 'The heat equation on a compact Lie group', *Trans. Amer. Math. Soc.* **246** (1978) 339–57.

[182] H. D. Fegan, 'The heat equation and modular forms', *J. Diff. Geom.* **13** (1978) 589–602.

[183] B. L. Feigin and D. B. Fuchs, *Cohomologies of Lie Groups and Lie Algebras*, Encycl. Math. Sci. **21** (Berlin, Springer, 2000) 125–215.

[184] B. Feigin and F. Malikov, 'Modular functor and representation theory of $\widehat{sl}(2)$ at a rational level', *Operads (Hartford/Luminy, 1995)*, Contemp. Math. **202** (Providence, American Mathematical Society, 1997) 357–405.

[185] A. J. Feingold, I. B. Frenkel and J. F. X. Ries, 'Spinor construction of vertex operator algebras, triality and $E_8^{(1)}$', *Contemp. Math.* **121** (Providence, American Mathematical Society, 1991).

[186] G. Felder, 'The KZB equations on Riemann surfaces', *Quantum Symmetries (Les Houches, 1995)* (Amsterdam, Elsevier, 1998) 687–725.

[187] G. Felder, J. Fröhlich and G. Keller, 'On the existence of unitary conformal field theory I. Existence of conformal blocks', *Commun. Math. Phys.* **124** (1989) 417–63.

[188] R. P. Feynman, R. B. Leighton and M. Sands, *The Feynman Lectures on Physics*, Vols I, II, III (Reading, Addison-Wesley, 1964).

[189] J. M. Figueroa-O'Farrill and S. Stanciu, 'On the structure of symmetric self-dual Lie algebras', *J. Math. Phys.* **37** (1996) 4121–34.

[190] M. Finkelberg, 'An equivalence of fusion categories', *Geom. Funct. Anal.* **6** (1996) 249–67.

[191] K. Fredenhagen, K. H. Rehren and B. Schroer, 'Superselection sectors with braid group statistics and exchange algebras. I. General theory', *Commun. Math. Phys.* **125** (1989) 201–26.

[192] D. S. Freed, 'On determinant line bundles', *Mathematical Aspects of String Theory* (Singapore, World Scientific, 1987) 189–238.

[193] D. S. Freed, M. J. Hopkins and C. Teleman, 'Twisted equivariant K-theory with complex coefficients', Preprint (arXiv: math.AT/0206257).

[194] D. S. Freed and F. Quinn, 'Chern–Simons theory with finite gauge group', *Commun. Math. Phys.* **156** (1993) 435–72.

[195] D. S. Freed and K. K. Uhlenbeck, *Instantons and Four-Manifolds* (New York, Springer, 1984).

[196] D. S. Freed and C. Vafa, 'Global anomalies on orbifolds', *Commun. Math. Phys.* **110** (1987) 349–89.

[197] E. Frenkel and D. Ben-Zvi, *Vertex Algebras and Algebraic Curves* (Providence, American Mathematical Society, 2001).

[198] I. B. Frenkel, 'Orbital theory for affine Lie algebras', *Invent. Math.* **77** (1984) 301–52.

[199] I. B. Frenkel, Y.-Z. Huang and J. Lepowsky, *On Axiomatic Approaches to Vertex Operator Algebras and Modules*, Mem. Amer. Math. Soc. **494** (Providence, American Mathematical Society, 1993).

[200] I. Frenkel, J. Lepowsky and A. Meurman, 'A natural representation of the Fischer–Griess monster with the modular function J as character', *Proc. Natl. Acad. Sci. USA* **81** (1984) 3256–60.

[201] I. Frenkel, J. Lepowsky and A. Meurman, *Vertex Operator Algebras and the Monster* (San Diego, Academic Press, 1988).

[202] I. B. Frenkel and Y. Zhu, 'Vertex operator algebras associated to representations of affine and Virasoro algebras', *Duke Math. J.* **66** (1992) 123–68.

[203] D. Friedan and S. Shenker, 'The analytic geometry of two-dimensional conformal field theory', *Nucl. Phys.* **B281** (1987) 509–45.

[204] J. Fröhlich, 'Statistics and monodromy in two- and three-dimensional quantum field theory', *Differential Geometric Methods in Theoretical Physics (Como, 1987)* (Dordrecht, Kluwer Academic Press, 1988) 173–86.

[205] J. Fröhlich and T. Kerler, *Quantum Groups, Quantum Categories, and Quantum Field Theory*, Lecture Notes in Math. **1542** (Berlin, Springer, 1993).

[206] J. Fuchs, 'More on the super WZW theory', *Nucl. Phys.* **B318** (1989) 631–54.

[207] J. Fuchs, *Affine Lie Algebras and Quantum Groups* (Cambridge, Cambridge University Press, 1992).

[208] J. Fuchs, 'Fusion rules in conformal field theory', *Fortsch. Phys.* **42** (1994) 1–48.

[209] J. Fuchs, 'Lectures on conformal field theory and Kac–Moody algebras', *Conformal Field Theories and Integrable Models (Budapest, 1996)* (Berlin, Springer, 1997) 1–54.

[210] J. Fuchs, U. Ray and C. Schweigert, 'Some automorphisms of generalized Kac–Moody algebras', *J. Alg.* **191** (1997) 511-40.

[211] J. Fuchs, I. Runkel and C. Schweigert, 'Boundaries, defects and Frobenius algebras', *Fortsch. Phys.* **51** (2003) 850–55.

[212] J. Fuchs, A. N. Schellekens and C. Schweigert, 'A matrix S for all simple current extensions', *Nucl. Phys.* **B473** (1996) 323–66.

[213] J. Fuchs, A. N. Schellekens and C. Schweigert, 'From Dynkin diagram symmetries to fixed point structures', *Commun. Math. Phys.* **180** (1996) 39–97.

[214] J. Fuchs and C. Schweigert, *Symmetries, Lie Algebras, and Representations* (Cambridge, Cambridge University Press, 1997).

[215] J. Fuchs and C. Schweigert, 'Category theory for conformal boundary conditions', *Vertex Operator Algebras in Mathematics and Physics (Toronto, 2000)* (Providence, American Mathematical Society, 2003) 25–70.

[216] J. Fuchs and C. Schweigert, 'The world-sheet revisited', *Vertex Operator Algebras in Mathematics and Physics (Toronto, 2000)* (Providence, American Mathematical Society, 2003) 241–9.

[217] W. Fulton, *Young Tableaux* (Cambridge, Cambridge University Press, 1997).

[218] W. Fulton, 'Eigenvalues, invariant factors, highest weights, and Schubert calculus', *Bull. Amer. Math. Soc.* **37** (2000) 209–49.

[219] W. Fulton and J. Harris, *Representation Theory: A First Course* (New York, Springer, 1996).

[220] L. Funar, 'On the TQFT representations of the mapping class groups', *Pacif. J. Math.* **188** (1999) 251–74.

[221] P. Furlan, A. Ganchev and V. B. Petkova, 'An extension of the character ring of sl(3) and its quantisation', *Commun. Math. Phys.* **202** (1999) 701–33.

[222] M. R. Gaberdiel, 'Fusion rules of chiral algebras', *Nucl. Phys.* **B417** (1994) 130–50.

[223] M. R. Gaberdiel, 'A general transformation formula for conformal fields', *Phys. Lett.* **B325** (1994) 366–70.

[224] M. R. Gaberdiel, 'Introduction to conformal field theory', *Rep. Prog. Phys.* **63** (2000) 607–67.

[225] M. R. Gaberdiel, 'Fusion rules and logarithmic representations of a WZW model at fractional level', *Nucl. Phys.* **B618** (2001) 407–36.

[226] M. R. Gaberdiel and T. Gannon, 'Boundary states for WZW models', *Nucl. Phys.* **B639** (2002) 471–501.

[227] M. R. Gaberdiel and P. Goddard, 'Axiomatic conformal field theory', *Commun. Math. Phys.* **209** (2000) 549–94.

[228] M. R. Gaberdiel and H. G. Kausch, 'A rational logarithmic conformal field theory', *Phys. Lett.* **B386** (1996) 131–7.

[229] M. R. Gaberdiel and A. Neitzke, 'Rationality, quasirationality and finite W-algebras', *Commun. Math. Phys.* **238** (2003) 305–31.

[230] D. Gaitsgory, 'Notes on two dimensional conformal field theory and string theory', *Quantum Fields and Strings: A Course for Mathematicians*, Vol. 2 (Providence, American Mathematical Society, 1999) 1017–89.

[230a] W. Gajda, 'On $K_*(\mathbb{Z})$ and classical conjectures in the arithmetic of cyclotomic fields,' *Homotopy Theory* (Northwestern, 2002) (Providence, American Mathematical Society, 2004).

[231] R. Gangolli, 'Asymptotic behaviour of spectra of compact quotients of certain symmetric spaces', *Acta Math.* **121** (1968) 151–92.

[232] T. Gannon, 'The classification of SU(3) modular invariants revisited', *Ann. Inst. H. Poincaré Phys. Théor.* **65** (1996) 15–55.

[233] T. Gannon, 'Monstrous moonshine and the classification of conformal field theories', *Conformal Field Theory* (Cambridge, MA, Perseus Publishing, 2000) 66 pages.

[234] T. Gannon, 'The Cappelli–Itzykson–Zuber A-D-E classification', *Rev. Math. Phys.* **12** (2000) 739–48.

[235] T. Gannon, 'Boundary conformal field theory and fusion ring representations', *Nucl. Phys.* **B627** (2002) 506–64.

[236] T. Gannon, 'Modular data: the algebraic combinatorics of conformal field theory', *J. Alg. Combin.* **22** (2005) 211–50.

[237] T. Gannon, 'Monstrous Moonshine: the first twenty-five years', *Bull. London Math. Soc.* **38** (2006) 1–33.

[238] T. Gannon and C. S. Lam, 'Gluing and shifting lattice constructions and rational equivalence', *Rev. Math. Phys.* **3** (1991) 331–69.

[239] K. Gawedzki, 'Conformal field theory', Séminaire Bourbaki, *Astérisque* **177–178** (1989) 95–126.

[240] K. Gawedzki, 'SU(2) WZW theory at higher genera', *Commun. Math. Phys.* **169** (1995) 329–71.

[241] K. Gawedzki, 'Lectures on conformal field theory', *Quantum Fields and Strings: A Course for Mathematicians*, Vol. 2 (Providence, American Mathematical Society, 1999) 727–805.

[242] R. W. Gebert, 'Introduction to vertex algebras, Borcherds algebras, and the monster Lie algebra', *Intern. J. Mod. Phys.* **A8** (1993) 5441–503.

[243] I. M. Gel'fand, M. I. Graev and I. I. Piatetski-Shapiro, *Representation Theory and Automorphic Functions* (Boston, Academic Press, 1990).

[244] I. M. Gel'fand and N. Ya. Vilenkin, *Generalized Functions*, Vol. 4 (New York, Academic Press, 1964).

[245] D. Gepner and E. Witten, 'Strings on group manifolds', *Nucl. Phys.* **B278** (1986) 493–549.

[246] P. Ginsparg, 'Applied conformal field theory', *Champs, cordes et phénomènes critiques (Les Houches, 1988)* (Amsterdam, North-Holland, 1990) 1–168.

[247] G. Glauberman and S. P. Norton, 'On McKay's connection between the affine E_8 diagram and the Monster', *Proceedings on Moonshine and Related Topics (Montréal, 1999)* (Providence, American Mathematical Society, 2001) 37–42.

[248] P. Goddard, 'Meromorphic conformal field theory', *Infinite Dimensional Lie Algebras and Groups (Luminy–Marseille, 1988)* (Teaneck, World Scientific, 1989) 556–87.

[249] P. Goddard, 'The work of Richard Ewen Borcherds', *Proceedings of the International Congress of Mathematicians (Berlin, 1998)*, Vol. I (Bielefeld, Documenta Mathematica, 1998) 99–108.

[250] P. Goddard, A. Kent and D. Olive, 'Unitary representations of Virasoro and super-Virasoro algebras', *Commun. Math. Phys.* **103** (1986) 105–19.

[251] P. Goddard and D. I. Olive, 'Kac–Moody and Virasoro algebras in relation to quantum physics', *Int. Journ. Mod. Phys.* **A1** (1986) 303–414.

[252] D. M. Goldschmidt and V. F. R. Jones, 'Metaplectic link invariants', *Geom. Ded.* **31** (1989) 165–91.

[253] C. Gómez, M. Ruiz-Altaba and G. Sierra, *Quantum Groups in Two-dimensional Physics* (Cambridge, Cambridge University Press, 1996).

[254] G. Gonzalez-Springberg and J. L. Verdier, 'Construction géométrique de la correspondance de McKay', *Ann. Scient. Ec. Norm. Sup.* **16** (1983) 409–49.

[255] F. M. Goodman and H. Wenzl, 'Littlewood–Richardson coefficients for Hecke algebras at roots of unity', *Adv. Math.* **82** (1990) 244–65.

[256] D. Gorenstein, *Finite Simple Groups: An Introduction to their Classification* (New York, Plenum, 1982).

[257] F. Q. Gouvéa, *p-adic Numbers: An Introduction*, 2nd edn (Berlin, Springer, 1997).

[258] I. S. Gradshteyn, I. M. Ryzhik and A. Jeffrey, *Table of Integrals, Series and Products*, 5th edn (London, Academic Press, 1994).

[259] J. J. Gray, *Linear Differential Equations and Group Theory from Riemann to Poincaré*, 2nd edn (Boston, Birkhäuser, 1999).

[260] M. B. Green and D. Kutasov, 'Monstrous heterotic quantum mechanics', *J. High Energy Phys.* **9801** (1998) 012.

[261] M. B. Green, J. H. Schwarz, and E. Witten, *Superstring Theory*, Vols 1, 2, 2nd edn (Cambridge, Cambridge University Press, 1988).

[262] B. Greene, *The Elegant Universe* (New York, W. W. Norton, 1999).

[263] R. L. Griess, Jr., 'The friendly giant', *Invent. Math.* **69** (1982) 1–102.

[264] V. A. Gritsenko and V. V. Nikulin, 'Automorphic forms and Loentzian Kac–Moody algebras, Part I', *Int. J. Math.* **9** (1998) 153–99.

[265] V. A. Gritsenko and V. V. Nikulin, 'Automorphic forms and Loentzian Kac–Moody algebras, Part II', *Int. J. Math.* **9** (1998) 201–75.

[266] V. A. Gritsenko and V. V. Nikulin, 'The arithmetic mirror symmetry and Calabi–Yau manifolds', *Commun. Math. Phys.* **210** (2000) 1–11.

[267] A. Grothendieck, 'Esquisse d'un programme', *Geometric Galois Actions*, Vol. 1 (Cambridge, Cambridge University Press, 1997) 5–48.

[268] S. Gukov and C. Vafa, 'Rational conformal field theories and complex multiplication', *Commun. Math. Phys.* **246** (2004) 181–210.

[269] R. Haag, *Local Quantum Physics*, 2nd edn (Berlin, Springer, 1996).

[270] R. Hain, 'Moduli of Riemann surfaces, transcendental aspects', *School on Algebraic Geometry (Trieste, 1999)* (Trieste, Abdus Salam Intern. Center Theor. Phys., 2000) 293–353.

[271] A. Hanany and Y.-H. He, 'Non-abelian finite gauge theories', *J. High Energy Phys.* 1999, no. 2, Paper 13, 31 pp.

[272] K. Harada, M. Miyamoto and H. Yamada, 'A generalization of Kac–Moody algebras', *Groups, Difference Sets, and the Monster (Columbus, 1993)* (Berlin, de Gruyter, 1996) 377–408.

[273] J. Harris, 'An introduction to the moduli space of curves', *Mathematical Aspects of String Theory* (Singapore, World Scientific, 1987) 285–312.

[274] J. A. Harvey, 'Twisting the heterotic string', *Workshop on Unified String Theories (Santa Barbara, 1985)* (Singapore, World Scientific, 1986) 704–20.

[275] J. A. Harvey and G. Moore, 'Algebras, BPS states, and strings', *Nucl. Phys.* **B463** (1996) 315–68.

[276] J. A. Harvey and G. Moore, 'On the algebras of BPS states', *Commun. Math. Phys.* **197** (1998) 489–519.

[277] A. Hatcher, *Algebraic Topology* (Cambridge, Cambridge University Press, 2002).

[278] A. Hatcher and W. Thurston, 'A presentation for the mapping class group of a closed orientable surface', *Topol.* **19** (1980) 221–37.

[279] M. Hazewinkel, 'The philosophy of deformations: introductory remarks and a guide to this volume', *Deformation Theory of Algebras and Structures and Applications (Il Ciocco, 1986)* (Dordrecht, Kluwer, 1988) 1–7.

[280] M. Hazewinkel, W. Hesselink, D. Siersma and F. D. Veldkamp, 'The ubiquity of Coxeter–Dynkin diagrams (an introduction to the A-D-E problem)', *Nieuw Arch. Wisk.* **25** (1977) 257–307.

[281] Y.-H. He and V. Jejjala, 'Modular matrix models', Preprint (arXiv: hep-th/0307293).

[282] E. Hecke, *Lectures on the Theory of Algebraic Numbers* (New York, Springer, 1981).

[283] G. Hemion, *The Classification of Knots and Three-dimensional Spaces* (New York, Oxford University Press, 1992).

[284] D. W. Henderson and D. Taimina, 'Crocheting the hyperbolic plane', *Math. Intell.* **23** (2001) 17–27.

[285] P. Henry-Labordere, B. Julia and L. Paulot, 'Real Borcherds superalgebras and M-theory', *J. High Energy Phys.* **0304** (2003) 060.

[286] M.-R. Herman, 'Simplicité du groupe des difféomorphismes de classe C^∞, isotopes à l'identité, du tore de dimension n', *C. R. Acad. Paris* **A273** (1971) 232–4.

[287] F. Hirzebruch, T. Berger and R. Jung, *Manifolds and Modular Forms*, 2nd edn (Aspects of Math, Braunschweig, Vieweg, 1994).

[288] N. Hitchin, 'Flat connections and geometric quantization', *Commun. Math. Phys.* **131** (1990) 347–80.

[289] G. Höhn, 'The group of symmetries of the shorter Moonshine module', Preprint (arXiv: math.QA/0210076).

[290] G. Höhn, 'Generalized moonshine for the Baby Monster', Preprint (2003).

[291] K. Hori, S. Katz, A. Klemm, R. Pandharipande, R. Thomas, C. Vafa, R. Vakil and E. Zaslow, *Mirror Symmetry* (Providence, American Mathematical Society, 2003).

[292] T. Hsu, *Quilts: Central Extensions, Braid Actions, and Finite Groups*, Lecture Notes in Math. **1731** (Berlin, Springer, 2000).

[293] Y.-Z. Huang, 'Applications of the geometric interpretation of vertex operator algebras', *Proceedings of the 20th International Conference on Differential Geometric Methods in Theoretical Physics (New York, 1991)* (Singapore, World Scientific, 1992) 333–43.

[294] Y.-Z. Huang, 'Binary trees and finite-dimensional Lie algebras', *Algebraic Groups and their Generalizations: Quantum and Infinite-Dimensional Methods*, Proc. Symp. Pure Math. **56**, pt. 2 (Providence, American Mathematical Society, 1994) 337–48.

[295] Y.-Z. Huang, *Two-Dimensional Conformal Geometry and Vertex Operator Algebras* (Boston, Birkhäuser, 1997).

[296] Y.-Z. Huang, 'Riemann surfaces with boundaries and the theory of vertex operator algebras', Preprint (arXiv: math.QA/0212308).

[297] Y.-Z. Huang, 'Vertex operator algebras, the Verlinde conjecture and modular tensor categories', *Proc. Natl. Acad. Sci. USA* **102** (2005) 5352–6.

[298] Y.-Z. Huang and J. Lepowsky, 'Tensor products of modules for a vertex operator algebra and tensor categories', *Lie Theory and Geometry in Honor of Bertram Kostant* (Boston, Birkhäuser, 1994) 349–83.

[299] T. Hübsch, *Calabi–Yau Manifolds* (River Edge, World Scientific, 1992).

[300] J. E. Humphreys, *Introduction to Lie Algebras and Representation Theory* (New York, Springer, 1994).

[301] J. E. Humphreys, *Reflection Groups and Coxeter Groups* (Cambridge, Cambridge University Press, 1990).

[302] J. E. Humphreys, 'Modular representations of simple Lie algebras', *Bull. Amer. Math. Soc.* **53** (1998) 105–22.

[303] T. I. Ibukiyama, 'Modular forms of rationalweights and modular varieties', *Abh. Math. Sem. Univ. Hamburg* **70** (2000) 315–39.

[304] Y. Ihara, 'Profinite braid groups, Galois representations and complex multiplication', *Ann. Math.* **123** (1986) 43–106.

[305] Y. Ihara, 'Braids, Galois groups, and some arithmetic functions', *Proceedings of the International Congress of Mathematicians (Kyoto, 1990)* (Tokyo, Japan Mathematical Society, 1991) 120–99.

[306] T. Inami, H. Kanno, T. Ueno and C.-S. Xiong, 'Two-toroidal Lie algebra as current algebra of the four-dimensional Kähler WZW model', *Phys. Lett.* **B399** (1997) 97–104.

[307] E. L. Ince, *Ordinary Differential Equations* (New York, Dover, 1956).

[308] I. M. Isaacs, *Character Theory of Finite Groups* (New York, Academic Press, 1976).

[309] D. Israël, A. Pakman and J. Troost, 'Extended SL(2,ℝ)/U(1) characters, or modular properties of a simple non-rational conformal field theory', *J. High Energy Phys.* **0404** (2004) 045.

[310] C. Itzykson and J.-B. Zuber, *Quantum Field Theory* (New York, McGraw-Hill, 1990).

[311] A. A. Ivanov, 'Geometric presentations of groups with an application to the Monster', *Proceedings of the International Congress of Mathematicians (Kyoto, 1990)*, Vol. II (Hong Kong, Springer, 1991) 1443–53.

[312] A. A. Ivanov, 'Y-groups via transitive extension', *J. Alg.* **218** (1999) 412–35.

[313] R. Ivanov and M. P. Tuite, 'Some irrational generalized Moonshine from orbifolds', *Nucl. Phys.* **B635** (2002) 473–91.

[314] N. Jacobson, *Lie Algebras* (New York, Dover, 1979).

[315] K. W. Johnson, 'The Dedekind–Frobenius group determinant: new life in an old problem', *Groups St. Andrews (Bath, 1997)*, Vol. II (Cambridge, Cambridge University Press, 1999) 417–28.

[316] V. F. R. Jones, 'Hecke algebra representations of braid groups and link polynomials', *Ann. Math.* **126** (1987) 335–88.

[317] V. F. R. Jones, *Subfactors and Knots* (Providence, American Mathematical Society, 1991).

[318] V. F. R. Jones, 'Three lectures on knots and von Neumann algebras', *Infinite-dimensional Geometry, Non-Commutative Geometry, Operator Algebras, and Fundamental Interactions* (Singapore, World Scientific, 1995) 96–113.

[319] V. Jones and V. S. Sunder, *Introduction to Subfactors* (Cambridge, Cambridge University Press, 1997).

[320] J. Jorgenson and S. Lang, 'The ubiquitous heat kernel', *Mathematics Unlimited – 2001 and Beyond* (Berlin, Springer, 2001) 655–83.

[321] A. Joyal and R. Street, 'Braided tensor categories', *Adv. Math.* **102** (1993) 20–78.

[322] E. Jurisich, 'An exposition of generalized Kac–Moody algebras', *Lie Algebras and their Representations (Seoul, 1995)*, Contemp. Math. **194** (Providence, American Mathematical Society, 1996) 121–59.

[323] E. Jurisich, 'Generalized Kac–Moody Lie algebras, free Lie algebras and the structure of the Monster Lie algebra', *J. Pure Appl. Alg.* **126** (1998) 233–66.

[324] E. Jurisich, J. Lepowsky and R. L. Wilson, 'Realizations of the Monster Lie algebra', *Selecta Math. (NS)* **1** (1995) 129–61.

[325] V. G. Kac, 'Simple irreducible graded Lie algebras of finite growth', *Math USSR-Izv.* **2** (1968) 1271–311.

[326] V. G. Kac, 'An elucidation of: infinite-dimensional algebras, Dedekind's η-function, classical Möbius function and the very strange formula. $E_8^{(1)}$ and the cube root of the modular invariant j', *Adv. Math* **35** (1980) 264–73.

[327] V. G. Kac, 'Simple Lie groups and the Legendre symbol', *Algebra*, Lecture Notes in Math. **848** (New York, Springer, 1981) 110–23.

[328] V. G. Kac, *Infinite Dimensional Lie Algebras*, 3rd edn (Cambridge, Cambridge University Press, 1990).

[329] V. G. Kac, 'The idea of locality', *Physical Applications and Mathematical Aspects of Geometry, Groups and Algebras* (Singapore, World Scientific, 1997) 16–32.

[330] V. G. Kac, *Vertex Algebras for Beginners*, 2nd edn (Providence, American Mathematical Society, 1998).

[331] V. G. Kac and S.-J. Kang, 'Trace formula for graded Lie algebras and Monstrous Moonshine', *Representations of Groups (Banff, 1994)* (Providence, American Mathematical Society, 1995) 141–54.

[332] V. G. Kac, R. Longo and F. Xu, 'Solitons in affine and permutation orbifolds', *Commun. Math. Phys.* **253** (2005) 732–64.

[333] V. G. Kac and D. H. Peterson, 'Infinite dimensional Lie algebras, theta functions, and modular forms', *Adv. Math.* **53** (1984) 125–264.

[334] V. G. Kac and A. K. Raina, *Highest Weight Representations of Infinite Dimensional Lie Algebras* (Singapore, World Scientific, 1987).

[335] V. G. Kac and M. Wakimoto, 'Classification of modular invariant representations of affine algebras', *Infinite-dimensional Lie Algebras and Groups (Luminy–Marseille, 1988)* (Teaneck, World Scientific, 1989) 138–77.

[336] V. G. Kac and M. Wakimoto, 'Integrable highest weight modules over affine superalgebras and number theory', *Lie Theory and Geometry in Honor of Bertram Kostant*, Progress in Math. **123** (Boston, Birkhäuser, 1994) 415–56.

[337] S. Kass, R. V. Moody, J. Patera and R. Slansky, *Affine Lie Algebras, Weight Multiplicities, and Branching Rules*, Vol. I (Berkeley, University of California Press, 1990).

[338] C. Kassel, *Quantum Groups* (New York, Springer, 1995).

[339] C. Kassel and V. Turaev, 'Chord diagram invariants of tangles and graphs', *Duke Math. J.* **92** (1998) 497–552.

[340] Y. Kawahigashi and R. Longo, 'Classification of local conformal nets. Case $c < 1$', *Ann. Math.* **160** (2004) 493–522.

[341] Y. Kawahigashi, R. Longo and M. Müger, 'Multi-internal subfactors and modularity of representations in conformal field theory', *Commun. Math. Phys.* **219** (2001) 631–69.

[342] T. Kawai, 'String duality and modular forms', *Phys. Lett.* **B397** (1997) 51–62.

[343] S. V. Ketov, *Quantum Non-Linear Sigma-Models* (Berlin, Springer, 2000).

[344] B. Khesin, 'A hierarchy of centrally extended algebras and the logarithm of the derivative operator', *Intern. Math. Res. Notices* **1992** 1–5.

[345] A. Khurana, 'Bosons condense and fermions exclude, but anyons...?', *Phys. Today* **Nov.** (1989) 17–21.

[346] A. A. Kirillov, *Lectures on the Orbit Method* (Providence, American Mathematical Society, 2004).

[347] A. N. Kirillov, 'Dilogarithm identities', *Quantum Field Theory, Integrable Models and Beyond (Kyoto, 1994)* (Progress of Theoretical Physics Supplement, Kyoto, 1995) 61–142.

[348] A. W. Knapp, *Lie Groups: Beyond an Introduction* (Boston, Birkhäuser, 1996).

[349] A. W. Knapp and P. E. Trapa, 'Representations of semisimple Lie groups', *Representation Theory of Lie Groups (Park City, 1998)* (Providence, American Mathematical Society, 2000) 7–87.

[350] M. Knopp and G. Mason, 'Generalized modular forms', *J. Number Theory* **99** (2003) 1–28.

[351] N. Koblitz and D. Rohrlich, 'Simple factors in the Jacobian of a Fermat curve', *Canad. J. Math.* **30** (1978) 1183–205.

[352] N. Koblitz, *Introduction to Elliptic Curves and Modular Forms*, 2nd edn (New York, Springer, 1993).

[353] J. Kock, *Frobenius Algebras and 2D Topological Quantum Field Theories* (Cambridge, Cambridge University Press, 2003).

[354] V. Kodiyalam and V. S. Sunder, *Topological Quantum Field Theories from Subfactors* (New York, Chapman & Hall, 2001).

[355] T. Kohno, *Conformal Field Theory and Topology* (Providence, American Mathematical Society, 2002).

[356] M. Koike, 'Moonshine for $PSL_2(\mathbb{F}_7)$', *Automorphic Forms and Number Theory*, Adv. Stud. Pure Math. **7** (Providence, American Mathematical Society, 1995) 103–11.

[357] M. L. Kontsevich, 'Virasoro algebra and Teichmüller spaces', *Funct. Anal. Appl.* **21** (1987) 156–7.

[358] M. Kontsevich, 'Product formulas for modular forms on $O(2, n)$ [after R. Borcherds],' *Séminaire Bourbaki 1996/97, no. 821, Astérisque* **245** (1997) 41–56.

[359] M. Kontsevich and Yu. Manin, 'Gromov–Witten classes, quantum cohomology, and enumerative geometry', *Mirror Symmetry II* (Providence, American Mathematical Society, 1997) 607–53.

[360] D. N. Kozlov, 'On completely replicable functions and extremal poset theory', MSc thesis, University of Lund, Sweden, 1994.

[361] D. Kreimer, *Knots and Feynman Diagrams* (Cambridge, Cambridge University Press, 2000).

[362] P. B. Kronheimer and H. Nakajima, 'Yang–Mills instantons on ALE gravitational instantons', *Math. Ann.* **288** (1990) 263–307.

[363] M. Kuga, *Galois' Dream: Group Theory and Differential Equations* (Boston, Birkhäuser, 1993).

[364] R. Kultze, 'Elliptic genera and the moonshine module', *Math. Z.* **223** (1996) 463–71.

[365] C. H. Lam, H. Yamada and H. Yamauchi, 'Vertex operator algebras, extended E_8 diagram, and McKay's observation on the Monster simple group', Preprint (arXiv: math.QA/0403010).

[366] M.-L. Lang, 'On a question raised by Conway–Norton', *J. Math. Soc. Japan* **41** (1989) 263–84.

[367] S. Lang, *Complex Multiplication* (New York, Springer, 1983).

[368] S. Lang, *Elliptic Functions*, 2nd edn (New York, Springer, 1997).

[369] H. B. Lawson, Jr. and M. L. Michelsohn, *Spin Geometry* (Princeton, Princeton University Press, 1989).

[370] F. W. Lawvere and S. H. Schanuel, *Conceptual Mathematics: A First Introduction to Categories* (Cambridge, Cambridge University Press, 1997).

[371] P. P. Lax and R. S. Phillips, *Scattering Theory for Automorphic Functions* (Princeton, Princeton University Press, 1976).

[372] F. Lemmermeyer, 'Conics – a poor man's elliptic curves', Preprint (arXiv: math.NT/0311306).

[373] J. Lepowsky, 'Euclidean Lie algebras and the modular function j', *The Santa Cruz Conference on Finite Groups (Santa Cruz, 1979)*, Proc. Sympos. Pure Math. **37** (Providence, American Mathematical Society, 1980) 567–70.

[374] J. Lepowsky, 'Application of the numerator formula to k-rowed plane partitions', *Adv. Math.* **35** (1980) 179–94.

[375] J. Lepowsky, 'Vertex operator algebras and the zeta function', *Recent Developments in Quantum Affine Algebras and Related Topics (Raleigh, 1999)*, Contemp. Math. **248** (Providence, American Mathematical Society, 1999) 327–40.

[376] J. Lepowsky and H. Li, *Introduction to Vertex Operator Algebras and their Representations* (Boston, Birkhäuser, 2004).

[377] J. Lepowsky and R. L. Wilson, 'Construction of the affine Lie algebra $A_1^{(1)}$', *Commun. Math. Phys.* **62** (1978) 43–53.

[378] F. Lesage, P. Mathieu, J. Rasmussen and H. Saleur, 'Logarithmic lift of the $SU(2)_{-1/2}$ model', *Nucl. Phys.* **B686** (2004) 313–46.

[379] J. B. Lewis and D. Zagier, 'Period functions for Maass wave forms I', *Ann. Math.* **153** (2001) 191–258.

[380] H. Li, 'Symmetric invariant bilinear forms on vertex operator algebras', *J. Pure Appl. Alg.* **96** (1994) 279–97.

[381] H. Li, 'Some finiteness properties of regular vertex operator algebras', *J. Alg.* **212** (1999) 495–514.

[382] H. Li, 'Determining fusion rules by $A(V)$-modules and bimodules', *J. Alg.* **212** (1999) 515–56.

[383] S.-P. Li, R. V. Moody, M. Nicolescu and J. Patera, 'Verma bases for representations of classical simple Lie algebras', *J. Math. Phys.* **27** (1986) 666–77.

[384] B. H. Lian, 'On classification of simple vertex operator algebras', *Commun. Math. Phys.* **163** (1994) 307–57.

[385] B. H. Lian and S.-T. Yau, 'Arithmetic properties of mirror map and quantum coupling', *Commun. Math. Phys.* **176** (1996) 163–91.

[386] B. H. Lian and G. J. Zuckerman, 'Moonshine cohomology', *Moonshine and Vertex Operator Algebras* (Surikaisekikenkyusho Kokyuroku no. 904, Kyoto, 1995) 87–115.

[387] G. Lion and M. Vergne, *The Weil Representation, Maslov Index and Theta Series* (Boston, Birkhäuser, 1980).

[388] K. Liu, 'On modular invariance and rigidity theorems', *J. Diff. Geom.* **41** (1995) 343–96.

[389] K. Liu, 'Heat kernels, symplectic geometry, moduli spaces and finite groups', *Surveys in Differential Geometry: Differential Geometry Inspired by String Theory* (Boston, International Press, 1999) 527–42.

[390] R. Longo and K.-H. Rehren, 'Nets of subfactors', *Rev. Math. Phys.* **7** (1995) 567–97.

[391] G. Lusztig, 'Leading coefficients of character values of Hecke algebras', *The Arcata Conference on Representations of Finite Groups (Arcata, 1986)*, Proc. Symp. Pure Math. **47** (Providence, American Mathematical Society, 1987) 235–62.

[392] G. Lusztig, *Introduction to Quantum Groups* (Boston, Birkhäuser, 1994).

[393] G. Lusztig, 'Exotic Fourier transform', *Duke Math. J.* **73** (1994) 227–41.

[394] Z.-Q. Ma, *Yang–Baxter Equation and Quantum Enveloping Algebras* (Singapore, World Scientific, 1993).

[395] H. Maass, *Siegel's Modular Forms and Dirichlet Series*, Lecture Notes in Math. **216** (Berlin, Springer, 1971).

[396] I. G. Macdonald, 'Affine root systems and Dedekind's η-function', *Invent. Math.* **15** (1972) 91–143.

[397] S. MacLane, *Categories for the Working Mathematician*, 2nd edn (New York, Springer, 1998).

[398] S. Majid, *Foundations of Quantum Group Theory* (Cambridge, Cambridge University Press, 1995).

[399] M. Mahowald and M. Hopkins, 'The structure of 24 dimensional manifolds having normal bundles which lift to $B O[8]$', *Recent Progress in Homotopy Theory*, Contemp. Math. **293** (Providence, American Mathematical Society, 2002) 89–110.

[400] F. Malikov and V. Schechtman, 'Deformations of vertex algebras, quantum cohomology of toric varieties, and elliptic genus', *Commun. Math. Phys.* **234** (2003) 77–100.

[401] F. Malikov, V. Schechtman and A. Vaintrob, 'Chiral de Rham complex', *Commun. Math. Phys.* **204** (1999) 439–73.

[402] Yu. I. Manin, 'Critical dimensions of the string theories and the dualizing sheaf on the moduli space of (super) curves', *Funct. Anal. Appl.* **20** (1986) 244–6.

[403] M. Mariño, 'Enumerative geometry and knot invariants', *Infinite Dimensional Groups and Manifolds*, 27–92 (Berlin, de Gruyter, 2004).

[404] G. Masbaum and J. D. Roberts, 'On central extensions of mapping class groups', *Math. Ann.* **302** (1995) 131–50.

[405] G. Mason, 'Finite groups and modular functions', *The Arcata Conference on Representations of Finite Groups (Arcata, 1986)*, Proc. Sympos. Pure Math. **47** (Providence, American Mathematical Society, 1987) 181–209.

[406] G. Mason, 'The quantum double of a finite group and its role in conformal field theory' *Groups '93 (Galway, 1993)* (Cambridge, Cambridge University Press, 1995) 405–17.

[407] G. Mason, 'On a system of elliptic modular forms attached to the large Mathieu group', *Nagoya Math. J.* **118** (1990) 177–93.

[408] G. Mason, talk, Erwin-Schrödinger-Institut (Vienna), June 2004.

[409] O. Mathieu, 'Classification of simple graded Lie algebras of finite growth', *Invent. Math.* **108** (1992) 455–519.

[410] A. Matsuo, 'On generalizations of Zhu's algebra and the zero-mode algebra' (talk, Edinburgh, July 2004).

[411] J. McKay, 'Graphs, singularities, and finite groups', *The Santa Cruz Conference on Finite Groups (Santa Cruz, 1979)*, Proc. Sympos. Pure Math. **37** (Providence, American Mathematical Society, 1980) 183–6.

[412] J. McKay, 'The essentials of Monstrous Moonshine', *Groups and Combinatorics – in Memory of M. Suzuki*, Adv. Studies Pure Math. **32** (Tokyo, Mathematical Society of Japan, 2001) 347–53.

[413] J. McKay and H. Strauss, 'The q-series of monstrous moonshine and the decomposition of the head characters', *Commun. Alg.* **18** (1990) 253–78.

[414] H. McKean and V. Moll, *Elliptic Curves: Function Theory, Geometry, Arithmetic* (Cambridge, Cambridge University Press, 1999).

[415] A. Medina and P. Revoy, 'Algèbres de Lie et produit scalaire invariant', *Ann. Scient. Éc. Norm. Sup.* **18** (1985) 553–61.

[416] J.-F. Mestre, R. Schoof, L. Washington and D. Zagier, 'Quotients Homophones des groupes libres/Homophonic quotients of free groups', *Experiment. Math.* **2** (1993) 153–5.

[417] D. Miličić, 'Algebraic $\mathcal{D}$-modules and representation theory of semisimple Lie groups', Contemp. Math. **154** (Providence, American Mathematical Society, 1993) 133–68.

[418] G. A. Miller, 'Determination of all the groups of order 64', *Amer. J. Math.* **52** (1930) 617–34.

[419] J. Milnor, *Introduction to Algebraic K-Theory*, Annals of Math. Studies **72** (Princeton, Princeton University Press, 1971).

[420] H. Minc, *Nonnegative Matrices* (New York, Wiley, 1988).

[421] R. Mirman, *Group Theory: An Intuitive Approach* (River Edge, World Scientific, 1995).

[422] C. W. Misner, K. S. Thorne and J. A. Wheeler, *Gravitation* (San Francisco, W. H. Freeman and Co., 1973).

[423] T. Miwa, M. Jimbo and E. Date, *Solitons, Differential Equations and Infinite-Dimensional Algebras* (Cambridge, Cambridge University Press, 1999).

[424] M. Miyamoto, '21 involutions acting on the Moonshine module', *J. Alg.* **175** (1995) 941–65.

[425] M. Miyamoto, 'Griess algebras and conformal vectors in vertex operator algebras', *J. Alg.* **179** (1996) 523–48.

[426] M. Miyamoto, 'A modular invariance on the theta functions defined on vertex operator algebras', *Duke Math. J.* **101** (2000) 221–36.

[427] M. Miyamoto, 'A new construction of the Moonshine vertex operator algebras over the real number field', *Ann. Math.* **159** (2004) 535–96.

[428] E. J. Mlawer, S. G. Naculich, H. A. Riggs and H. J. Schnitzer, 'Group-level duality of WZW fusion coefficients and Chern–Simons link observables', *Nucl. Phys.* **B352** (1991) 863.

[429] P. S. Montague, 'On representations of conformal field theories and the construction of orbifolds', *Lett. Math. Phys.* **38** (1996) 1–11.

[430] R. V. Moody, 'Lie algebras associated with generalized Cartan matrices', *Bull. Amer. Math. Soc.* **73** (1967) 217–21.

[431] R. V. Moody and J. Patera, 'Computation of character decompositions of class functions on compact semi-simple Lie groups', *Math. Comput.* **48** (1987) 799–827.

[432] R. V. Moody and A. Pianzola, *Lie Algebras with Triangular Decompositions* (New York, John Wiley & Sons, 1995).

[433] G. Moore, 'Atkin–Lehner symmetry', *Nucl. Phys.* **B293** (1987) 139–88.

[434] G. W. Moore, 'String duality, automorphic forms, and generalized Kac-Moody algebras', *Nucl. Phys. B: Proc. Suppl.* **67** (1998) 56–67.

[435] G. Moore, 'Les Houches lectures on strings and arithmetic', Preprint (arXiv: hep-th/0401049).

[436] G. Moore and N. Seiberg, 'Classical and quantum conformal field theory', *Commun. Math. Phys.* **123** (1989) 177–254.

[437] G. Moore and N. Seiberg, 'Lectures on RCFT', *Physics, Geometry and Topology* (New York, Plenum Press, 1990) 263–361.

[438] D. R. Morrison, 'Picard–Fuchs equations and mirror maps for hypersurfaces', *Essays on Mirror Manifolds* (Hong Kong, International Press, 1992) 241–64.

[439] D. Mumford, *Tata Lectures on Theta I* (Boston, Birkhäuser, 1983).

[440] D. Mumford, M. Nori and P. Norman, *Tata Lectures on Theta III* (Boston, Birkhäuser, 1991).

[441] W. Nahm, 'Lie group exponents and SU(2) current algebras', *Commun. Math. Phys.* **118** (1988) 171–6.

[442] W. Nahm, 'A proof of modular invariance', *Intern. J. Mod. Phys.* **A6** (1991) 2837–45.

[443] W. Nahm, 'Conformal field theory: a bridge over troubled waters', *Quantum Field Theory* (New Delhi, Hindustan Book Agency, 2000) 571–604.

[444] W. Nahm, 'Conformal field theory and torsion elements of the Bloch group', Preprint (arXiv: hep-th/0404120).

[445] H. Nakajima, 'Geometric construction of representations of affine algebras', *Proceedings of the International Congress of Mathematicians, Vol. I (Beijing, 2002)* (Beijing, Higher Education Press, 2002) 423–38.

[446] G. Navarro, *Characters and Blocks of Finite Groups* (Cambridge, Cambridge University Press, 1998).

[447] P. Nelson, 'Lectures on strings and moduli spaces', *Phys. Reports* **149** (1987).

[448] Yu. A. Neretin, *Representations of Virasoro and Affine Lie algebras*, Encycl. Math. Sci. **22** (New York, Springer, 1994) 157–234.

[449] S. P. Norton, 'More on Moonshine', *Computational Group Theory (Durham, 1982)* (New York, Academic Press, 1984) 185–93.

[450] S. P. Norton, 'Generalized moonshine', *The Arcata Conference on Representations of Finite Groups (Arcata, 1986)*, Proc. Symp. Pure Math. **47** (Providence, American Mathematical Society, 1987) 208–9.

[451] S. P. Norton, 'Constructing the Monster', *Groups, Combinatorics, and Geometry (Durham, 1990)* (Cambridge, Cambridge University Press, 1992) 63–76.

[452] S. P. Norton, 'From moonshine to the Monster', *Proceedings on Moonshine and Related Topics (Montréal, 1999)* (Providence, American Mathematical Society, 2001) 163–71.

[453] A. Ocneanu, 'Quantized groups, string algebras and Galois theory for algebras', *Operator Algebras and Applications*, Vol. 2 (Cambridge, Cambridge University Press, 1988) 119–72.

[454] A. Ocneanu, 'Paths on Coxeter diagrams: from Platonic solids and singularities to minimal models and subfactors', *Lectures on Operator Theory* (Providence, American Mathematical Society, 1999).

[455] A. Ocneanu, 'The classification of subgroups of quantum $SU(N)$', *Quantum Symmetries in Theoretical Physics and Mathematics* (Providence, American Mathematical Society, 2002).

[456] A. Ogg, *Modular Forms and Dirichlet Series* (New York, W. A. Benjamin, 1969).

[457] V. Ostrik, 'Module categories over the Drinfeld double of a finite group', *Int. Math. Res. Not.* (2003) 1507–20.

[458] V. Ostrik, 'Module categories, weak Hopf algebras and modular invariants', *Transform. Groups* **8** (2003) 177–206.

[459] J. Paris and L. Harrington, 'A mathematical incompleteness in Peano arithmetic', *Handbook of Mathematical Logic* (Amsterdam, North-Holland, 1977) 1133–42.

[460] O. Pekonen, 'Universal Teichmüller space in geometry and physics', *J. Geom. Phys.* **15** (1995) 227–51.

[461] V. B. Petkova and J.-B. Zuber, 'Conformal boundary conditions and what they teach us', *Nonperturbative Quantum Field Theory Methods and Their Applications* (Singapore, World Scientific, 2001) 1–35.

[462] B. Pioline and A. Waldron, 'Automorphic forms: a physicist's survey', Preprint (hep-th/9312068).

[463] J. Polchinski, *String Theory*, Vol. I and II (Cambridge, Cambridge University Press, 1998).

[464] V. Prasolov and Y. Solovyev, *Elliptic Functions and Elliptic Integrals* (Providence, American Mathematical Society, 1997).

[465] A. Pressley and G. Segal, *Loop Groups* (Oxford, Oxford University Press, 1988).

[466] L. Queen, 'Modular functions arising from some finite groups', *Math. of Comput.* **37** (1981) 547–80.

[467] F. Quinn, 'Lectures on axiomatic topological quantum field theory', *Geometry and Quantum Field Theory (Park City, 1991)* (Providence, American Mathematical Society, 1995) 323–453.

[468] H. Rademacher and E. Grosswald, *Dedekind Sums* (Washington, Mathematical Association of America, 1972).

[469] U. Ray, 'Generalized Kac–Moody algebras and some related topics', *Bull. Amer. Math. Soc.* **38** (2000) 1–42.

[470] K.-H. Rehren, 'Braid group statistics and their superselection rules', *The Algebraic Theory of Superselection Sectors* (Singapore, World Scientific, 1990) 333–55.

[471] M. Reid, 'La correspondance de McKay', *Séminaire Bourbaki, Astérisque* **276** (2002) 53–72.

[472] I. Reiten, 'Dynkin diagrams and the representation theory of algebras', *Notices Amer. Math. Soc.* **44** (1997) 546–56.

[473] N. Yu. Reshetikhin and V. G. Turaev, 'Ribbon graphs and their invariants derived from quantum groups', *Commun. Math. Phys.* **127** (1990) 1–26.

[474] N. Yu. Reshetikhin and V. G. Turaev, 'Invariants of 3-manifolds via link polynomials and quantum groups', *Invent. Math.* **103** (1991) 547–97.

[475] C. Reutenauer, *Free Lie Algebras* (New York, Oxford University Press, 1993).

[476] W. F. Reynolds, 'Thompson's characterization of characters and sets of primes', *J. Alg.* **156** (1993) 237–43.

[477] A. Rocha-Caridi, 'Vacuum vector representations of the Virasoro algebra', *Vertex Operators in Mathematics and Physics (Berkeley, 1983)* (New York, Springer, 1985) 451–73.

[478] D. Rolfsen, *Knots and Links* (Berkeley, Publish or Perish, 1976).

[479] P. Roman, *Introduction to Quantum Field Theory* (New York, John Wiley & Sons, 1969).

[480] M. Rosso, 'Quantum groups and braid groups', *Quantum Symmetries (Les Houches, 1995)* (Amsterdam, Elsevier, 1998) 757–85.

[481] W. Rudin, *Real and Complex Analysis*, 3rd edn (New York, McGraw–Hill, 1987).

[482] P. Ruelle, E. Thiran and J. Weyers, 'Implications of an arithmetic symmetry of the commutant for modular invariants', *Nucl. Phys.* **B402** (1993) 693–708.

[483] I. Runkel and G. M. T. Watts, 'A non-rational CFT with central charge 1', *Fortsch. Phys.* **50** (2002) 959–65.

[484] A. J. E. Ryba, 'Modular moonshine?', *Moonshine, the Monster, and Related Topics (South Hadley, 1994)* Contemp. Math. **193** (Providence, American Mathematical Society, 1996) 307–36.

[485] D. G. Saari and Z. Xia, 'Off to infinity in finite time', *Notices Amer. Math. Soc.* **42** (1995) 538–46.

[486] G. Sansone and J. Gerretsen, *Lectures on the Theory of Functions of a Complex Variable*, Vol. II (Groningen, Wolters-Noordhoff, 1969).

[487] S. Sawin, 'Links, quantum groups, and TQFTs', *Bull. Amer. Math. Soc.* **33** (1996) 413–45.

[488] A. N. Schellekens, 'Meromorphic $c = 24$ conformal field theories', *Commun. Math. Phys.* **153** (1993) 159–85.

[489] A. N. Schellekens and S. Yankielowicz, 'Extended chiral algebras and modular invariant partition functions', *Nucl. Phys.* **B327** (1989) 673–703.

[490] R. Schimmrigk and S. Underwood, 'The Shimura–Taniyama Conjecture and conformal field theory', *J. Geom. Phys.* **48** (2003) 169–89.

[491] M. Schlichenmaier and O. K. Sheinman, 'The Wess–Zumino–Witten–Novikov theory, Knizhnik–Zamolodchikov equations, and Krichever–Novikov algebras, I', *Russian Math. Surveys* **54** (1999) 213–49.

[492] M. Schlichenmaier and O. K. Sheinman, 'The Wess–Zumino–Witten–Novikov theory, Knizhnik–Zamolodchikov equations, and Krichever–Novikov algebras, II', Preprint (arXiv: math.MG/0410048).

[493] L. Schneps, 'The Grothendieck–Teichmüller group $\widehat{GT}$: a survey', *Geometric Galois Actions*, Vol. 1 (Cambridge, Cambridge University Press, 1997) 183–203.

[494] L. Schneps, 'Fundamental groupoids of genus zero moduli spaces and braided tensor categories', *Moduli Spaces of Curves, Mapping Class Groups, and Field Theory* (Providence, American Mathematical Society, 2003) 59–104.

[495] M. Schottenloher, *A Mathematical Introduction to Conformal Field Theory* (Berlin, Springer, 1997).

[496] C. Schweigert, J. Fuchs and J. Walcher, 'Conformal field theory, boundary conditions and applications to string theory', Preprint (arXiv: hep-th/0011109).

[497] P. Scott, 'The geometries of 3-manifolds', *Bull. London Math Soc.* **15** (1983) 401–87.

[498] G. Segal, 'The definition of conformal field theory', *Differential Geometric Methods in Theoretical Physics (Como, 1987)* (Boston, Academic Press, 1988) 165–71.

[499] G. Segal, 'Elliptic cohomology', *Séminaire Bourbaki 1987-88, no. 695, Astérisque* **161–162** (1988) 187–201.

[500] G. Segal, 'Two-dimensional conformal field theories and modular functors', *IXth Proceedings of the International Congress of Mathematical Physics (Swansea, 1988)* (Bristol, Hilger, 1989) 22–37.

[501] G. Segal, 'Geometric aspects of quantum field theories', *Proceedings of the International Congress of Mathematicians (Kyoto, 1990)* (Hong Kong, Springer, 1991) 1387–96.

[502] G. Segal, 'The definition of conformal field theory', *Topology, Geometry and Quantum Field Theory (Oxford, 2002)* (Cambridge, Cambridge University Press, 2004) 423–577.

[503] J.-P. Serre, *A Course in Arithmetic* (Berlin, Springer, 1973).

[504] I. R. Shafarevich, *Algebra I. Basic Notions of Algebra*, Encycl. Math. Sci. **11** (New York, Springer, 1990).

[505] G. Shimura, *Introduction to the Arithmetic Theory of Automorphic Functions* (Princeton, Princeton University Press, 1971).

[506] G. Shimura, *Abelian Varieties and Complex Multiplication* (Princeton, Princeton University Press, 1998).

[507] C. L. Siegel, 'A simple proof of $\eta(-1/\tau) = \eta(\tau)\sqrt{\tau/i}$', *Mathematika* **1** (1954) 4.

[508] S. Singh, *Fermat's Enigma* (London, Penguin Books, 1997).

[509] P. Slodowy, 'Platonic solids, Kleinian singularities, and Lie groups', *Algebraic Geometry (Ann Arbor, 1981)* Lecture Notes in Math. **1008** (Berlin, Springer, 1983) 102–38.

[510] G. W. Smith, 'Replicant powers for higher genera', *Moonshine, the Monster, and Related Topics (South Hadley, 1994)* Contemp. Math. **193** (Providence, American Mathematical Society, 1996) 337–52.

[511] S. D. Smith, 'On the head characters of the Monster simple group', *Finite Groups – Coming of Age (Montréal, 1982)*, Contemp. Math. **45** (Providence, American Mathematical Society, 1996).

[512] R. Solomon, 'A brief history of the classification of the finite simple groups', *Bull. Amer. Math. Soc.* **38** (2001) 315–52.

[513] J. Stasheff, 'Differential graded Lie algebras, quasi-Hopf algebras and higher homotopy algebras', *Quantum Groups (Leningrad, 1990)*, Lecture Notes in Math. **1510** (Berlin, Springer, 1992) 120–37.

[514] J. Stasheff, 'Homological (ghost) methods in mathematical physics', *Infinite-dimensional Geometry, Non-commutative Geometry, Operator Algebras, and Fundamental Interactions (Saint-Francois, 1993)* (Singapore, World Scientific, 1995) 242–64.

[515] I. N. Stewart and D. O. Tall, *Algebraic Number Theory* (London, Chapman & Hall, 1979).

[516] P. F. Stiller, 'Classical automorphic functions and hypergeometric functions', *J. Number Theory* **28** (1988) 219–32.

[517] S. Stolz and P. Teichner, 'What is an elliptic object?', *Topology, Geometry and Quantum Field Theory (Oxford, 2002)* (Cambridge, Cambridge University Press, 2004) 247–343.

[518] R. F. Streater and A. S. Wightman, *PCT, Spin and Statistics, and All That* (Princeton, Princeton University Press, 1989).

[519] T. Takayanagi, 'Modular invariance of strings on pp-waves with RR-flux', *J. High Energy Physics* **0212** (2002) 022.

[520] L. A. Takhtajan, 'Quantum field theory on an algebraic curve', *Lett. Math. Phys.* **52** (2000) 79–91.

[521] H. Tamanoi, *Elliptic Genera and Vertex Operator Superalgebras*, Lecture Notes in Math. **1704** (Berlin, Springer, 1999).

[522] A. Terras, *Fourier Analysis on Finite Groups and Applications* (Cambridge, Cambridge University Press, 1999).

[523] C. B. Thomas, *Elliptic Cohomology* (New York, Kluwer, 1999).

[524] J. G. Thompson, 'Finite groups and modular functions', *Bull. London Math. Soc.* **11** (1979) 347–51.

[525] J. G. Thompson, 'Some numerology between the Fischer–Griess Monster and the elliptic modular function', *Bull. Lond. Math. Soc.* **11** (1979) 352–3.

[526] J. G. Thompson, 'A finiteness theorem for subgroups of PSL(2,$\mathbb{R}$) which are commensurable with PSL(2, $\mathbb{Z}$)', *Santa Cruz Conference on Finite Groups (Santa Cruz, 1979)*, Proc. Symp. Pure Math. **37** (Providence, American Mathematical Society, 1980) 533–55.

[527] W. Thurston, *Three-dimensional Geometry and Topology*, Vol. 1 (Princeton, Princeton University Press, 1997).

[528] J. Tits, 'On R. Griess' "Friendly Giant"', *Invent. Math.* **78** (1984) 491–9.

[529] L. Toti Rigatelli, *Evariste Galois* (Basel, Birkhäuser, 1996).

[530] A. Tsuchiya, K. Ueno and Y. Yamada, 'Conformal field theory on universal family of stable curves with gauge symmetries', *Integrable Systems in Quantum Field Theory and Statistical Mechanics*, Adv. Stud. Pure Math. **19** (Boston, Academic Press, 1989) 459–566.

[531] I. Tuba and H. Wenzl, 'Representations of the braid group B_3 and of SL(2,$\mathbb{Z}$)', *Pac. J. Math.* **197** (2001) 491–510.

[532] M. Tuite, 'On the relationship between Monstrous moonshine and the uniqueness of the Moonshine module', *Commun. Math. Phys.* **166** (1995) 495–532.

[533] M. P. Tuite, 'Genus two meromorphic conformal field theory', *Proceedings on Moonshine and Related Topics (Montréal, 1999)* (Providence, American Mathematical Society, 2001) 231–51.

[534] V. G. Turaev, *Quantum Invariants of Knots and 3-manifolds* (Berlin, de Gruyter, 1994).

[535] V. G. Turaev and O. Y. Viro, 'State sum invariants of 3-manifolds and quantum $6j$-symbols', *Topology* **31** (1992) 865–902.

[536] K. Ueno, 'Introduction to conformal field theory with gauge symmetries', *Geometry and Physics (Aarhus, 1995)* (New York, Dekker, 1997) 603–745.

[537] K. Ueno, *Algebraic Geometry 1 and 2* (Providence, American Mathematical Society, 1997).

[538] C. Vafa, 'Modular invariance and discrete torsion on orbifolds', *Nucl. Phys.* **B273** (1986) 592–606.

[539] C. Vafa, 'Conformal theories and punctured surfaces', *Phys. Lett.* **B199** (1987) 195–202.

[540] C. Vafa and E. Witten, 'A strong coupling test of S-duality', *Nucl. Phys.* **B431** (1994) 3–77.

[541] A. Varchenko, *Multidimensional Hypergeometric Functions and Representation Theory of Lie Groups and Quantum Groups* (Singapore, World Scientific, 1995).

[542] E. Verlinde, 'Fusion rules and modular transformations in 2D conformal field theory', *Nucl. Phys.* **B300** (1988) 360–76.

[543] D.-N. Verma, 'The rôle of affine Weyl groups in the representation theory of algebraic Chevalley groups and their Lie groups', *Lie Groups and Their Representations (Budapest, 1971)* (New York, Halsted, 1975) 653–705.

[544] H. Verrill and N. Yui, 'Thompson series and the mirror maps of pencils of K3 surfaces', *The Arithmetic and Geometry of Algebraic Cycles (Banff, 1998)* (Providence, American Mathematical Society, 2000) 399–432.

[545] V. S. Vladimirov, I. V. Volovich and E. I. Zelenov, *p-adic Analysis and Mathematical Physics* (Singapore, World Scientific, 1994).

[546] S. G. Vladut, *Kronecker's Jugendtraum and Modular Forms* (Amsterdam, Gordon & Breach, 1991).

[547] P. Vogel, 'The universal Lie algebra' (Preprint, 1999).

[548] H. Völklein, *Groups as Galois Groups. AnIntroduction* (Cambridge, Cambridge University Press, 1996).

[549] H. Völklein, 'The braid group and linear rigidity', *Geom. Dedic.* **84** (2001) 135–50.

[550] B. Wajnryb, 'A simple presentation for the mapping class group of an orientable surface', *Israel J. Math.* **45** (1983) 157–74; errata: *Israel J. Math.* **88** (1994) 425–7.

[551] M. Wakimoto, *Infinite-Dimensional Lie Algebras* (Providence, American Mathematical Society, 2001).

[552] M. A. Walton, 'Algorithm for WZW fusion rules: a proof', *Phys. Lett.* **B241** (1990) 365–8.

[553] M. A. Walton, 'Affine Kac–Moody algebras and the Wess–Zumino–Witten model', *Conformal Field Theory* (Cambridge, MA, Perseus Publishing, 2000) 67 pages.

[554] A. J. Wassermann, 'Operator algebras and conformal field theory', *Proceedings of the International Congress of Mathematicians (Zürich, 1994)* (Basel, Birkhäuser, 1995) 966–79.

[554a] A. Weil, 'Sur certains groupes d'opérateurs unitaires,' *Acta Math* **111** (1964) 143–211.

[555] S. Weinberg, *The Quantum Theory of Fields*, Vols I, II (Cambridge, Cambridge University Press, 1995).

[556] J. A. Wheeler and W. H. Zurek (eds), *Quantum Theory and Measurement* (Princeton, Princeton University Press, 1983).

[557] F. Wilczek, *Fractional Statistics and Anyon Superconductivity* (Singapore, World Scientific, 1990).

[558] K. G. Wilson, 'Non-Lagrangian models of current algebras', *Phys. Rev.* **179** (1969) 1499–512.

[559] R. L. Wilson, 'Simple Lie algebras over fields of prime characteristic', *Proceedings of the International Congress of Mathematicians (Berkeley, 1986)* (Providence, American Mathematical Society, 1987) 407–16.

[560] E. Witten, 'Physics and geometry', *Proc. Intern. Congr. Math. (Berkeley, 1986)* (Providence, American Mathematical Society, 1987) 267–303.

[561] E. Witten, 'Elliptic genera and quantum field theory', *Commun. Math. Phys.* **109** (1987) 525–36.

[562] E. Witten, 'Quantum field theory, Grassmannians, and algebraic curves', *Commun. Math. Phys.* **113** (1988) 529–600.

[563] E. Witten, 'Coadjoint orbits of the Virasoro group', *Commun. Math. Phys.* **114** (1988) 1–53.

[564] E. Witten, 'Geometry and quantum field theory', *Mathematics into the Twenty-first Century*, Vol. II (Providence, American Mathematical Society, 1992) 479–91.

[565] E. Witten, 'The Verlinde formula and the cohomology of the Grassmannian', *Geometry, Topology and Physics (Harvard, 1993)* (Cambridge, International Press, 1995) 357–422.

[566] E. Witten, 'Magic, mystery, and matrix', *Notices Amer. Math. Soc.* **45** (1998) 1124–9.

[567] E. Witten, 'Perturbative quantum field theory', *Quantum Fields and Strings: A Course for Mathematicians*, Vol. 1 (Providence, American Mathematical Society, 1999) 419–73.

[568] F. Xu, 'Algebraic orbifold conformal field theories', *Mathematical Physics in Mathematics and Physics (Sienna, 2000)* (Providence, American Mathematical Society, 2001) 429–48.

[569] C. T. Yang, 'Hilbert's fifth problem and related problems on transformation groups', *Mathematical Developments Arising from Hilbert Problems*, Proc. Symp. Pure Math. **28** (Providence, American Mathematical Society, 1976) 142–6.

[570] D. N. Yetter, *Functorial Knot Theory* (Singapore, World Scientific, 2001).

[571] N. Yui, 'Update on the modularity of Calabi–Yau varieties', *Calabi–Yau Varieties and Mirror Symmetry (Toronto, 2001)* (Providence, American Mathematical Society, 2003) 307–62.

[572] D. Zagier, 'A one-sentence proof that every prime $p \equiv 1$ (mod 4) is a sum of two squares', *Amer. Math. Monthly* **97** (1990) 144.

[573] Y. Zhu, 'Global vertex operators on Riemann surfaces', *Commun. Math. Phys.* **165** (1994) 485–531.

[574] Y. Zhu, 'Modular invariance of characters of vertex operator algebras', *J. Amer. Math. Soc.* **9** (1996) 237–302.

[575] J.-B. Zuber, 'CFT, BCFT, ADE and all that', *Quantum Symmetries in Theoretical Physics and Mathematics (Bariloche, 2000)* (Providence, American Mathematical Society, 2002) 233–66.

Index